EJB Reviews 1996

European Journal of Biochemistry

EJB Reviews 1996

Contributing Authors:

F. G. Andersen, Gentofte
N. Blume, Gentofte
R. Bruzzone, Paris
A. C. Culhane, Keele
A. J. Hale, Keele
C. Hall, London
L. Hendriks, Antwerp
F. Hucho, Berlin
J. Jensen, Gentofte
P. B. Jensen, Gentofte
P. Jollès, Paris
C. Karlsen, Gentofte
S. M. Kingsman, Oxford
A. J. Kingsman, Oxford
H. Kleinkauf, Berlin
T. A. Kunkel, North Carolina
L.-I. Larsson, Copenhagen
T. Leung, Singapore
L. Lim, London, Singapore
V. L. Longthorne, Keele
K. Lund, Gentofte
J. Machold, Berlin
O. D. Madsen, Gentofte
E. Manser, Singapore
F. Maraboeuf, Orsay
B. Nordén, Gothenburg
J. H. A. Nugent, London
V. M. Pain, Brighton
D. L. Paul, Boston
H. V. Peterson, Gentofte
J. Piatigorsky, Bethesda
S. Reinbothe, Zürich
C. Reinbothe, Zürich
P. Serup, Gentofte
C. A. Smith, Keele
V. E. A. Stoneman, Keele
L. C. Sutherland, Keele
M. Takahashi, Orsay
S. Tomarev, Bethesda
V. I. Tsetlin, Moscow
A. Udvardy, Szeged
A. Umar, North Carolina
C. Van Broeckhoven, Antwerp
H. von Döhren, Berlin
G. H. Werner, Gif-sur Yvette
T. W. White, Genève
G. T. Williams, Staffordshire

Springer-Verlag Berlin Heidelberg GmbH

Professor Dr. P. Christen
Biochemisches Institut der Universität Zürich
Winterthurerstrasse 190
CH-8057 Zürich

Professor Dr. E. Hofmann
Institut für Biochemie der Universität Leipzig
Liebigstraße 16
D-04103 Leipzig

ISBN 978-3-540-62051-8 ISBN 978-3-642-60659-5 (eBook)
DOI 10.1007/978-3-642-60659-5

Originally published by Federation of European Biochemical Societies in 1997
Softcover reprint of the hardcover 1st edition 1997

Typesetting, printing and binding: Wiesbadener Graphische Betriebe GmbH, Wiesbaden
31/3130-5 4 3 2 1 0 – Printed on acid-free paper

Articles published in EJB Reviews 1989–1996

Protein chemistry and structure

Structure and biological activity of basement membrane proteins
R. Timpl (1989) 180, 487–502
NMR studies of mobility within protein structure
R. J. P. Williams (1989) 183, 479–497
Engineering of protein bound iron-sulfur clusters
H. Beinert and M. C. Kennedy (1990) 186, 5–15
Current approaches to macromolecular crystallization
A. McPherson (1990) 189, 1–23
Protein stability and molecular adaption to extreme conditions
R. Jaenicke (1991) 202, 715–728
Protein interaction with ice
C. L. Hew and D. S. C. Yang (1992) 203, 33–42
Natural protein proteinase inhibitors and their interaction with proteinases
W. Bode and R. Huber (1992) 204, 433–451
The protein kinase C family
A. Azzi, D. Boscoboinik and C. Hensey (1992) 208, 547–557
Advances in metallo-procarboxypeptidases – Emerging details on the inhibition mechanism and on the activation process
F. X. Avilés, J. Vendrell, A. Guasch, M. Coll and R. Huber (1993) 211, 381–389
The peripheral cholecystokinin receptors
Sandrine Silvente-Poirot, Marlène Dufresne, Nicole Vaysse and Daniel Fourmy (1993) 215, 513–529
X-ray crystal structures of cytosolic glutathione *S*-transferases – Implications for protein architecture, substrate recognition and catalytic function
H. Dirr, P. Reinemer and R. Huber (1994) 220, 645–661
Proteins under pressure – The influence of high hydrostatic pressure on structure, function and assembly of proteins and protein complexes
M. Gross and R. Jaenicke (1994) 221, 617–630
The functions and consensus motifs of nine types of peptide segments that form different types of nucleotide-binding sites
T. W. Traut (1994) 222, 9–19
Hemoglobin function under extreme life conditions
M. E. Clementi, S. G. Condò, M. Castagnola and B. Giardina (1994) 223, 309–317
Structure and modifications of the junior chaperone α-crystallin – From lens transparency to molecular pathology
P. J. T. A. Groenen, K. B. Merck, W. W. de Jong and H. Bloemendal (1994) 225, 1–19
The molecular biology of multidomain proteins – Selected examples
Alastair R. Hawkins and Heather K. Lamb (1995) 232, 7–18
Homologous nuclear-encoded mitochondrial and cytosolic isoproteins – A review of structure, biosynthesis and genes
Rolf Jaussi (1995) 228, 551–561

Nucleic acids, protein synthesis and molecular genetics

A chromosomal basis of lymphoid malignancy in man
T. Boehm and T. H. Rabbitts (1989) 185, 1–17

Enzymology

Molecular cell biology

Carbohydrates, lipids and other natural products

Physical and inorganic biochemistry

Membranes and bioenergetics

Metabolism and metabolic regulation

Signal transduction and molecular neurobiology

Developmental biochemistry and immunology

Contents

Eur. J. Biochem. 235, 449–465 (1996)

Review

Lens crystallins of invertebrates

Diversity and recruitment from detoxification enzymes and novel proteins

Stanislav I. TOMAREV and Joram PIATIGORSKY

Laboratory of Molecular and Developmental Biology, National Eye Institute, National Institutes of Health, Bethesda MD, USA

(Received 31 July 1995) – EJB 95 1264/0

The major proteins (crystallins) of the transparent, refractive eye lens of vertebrates are a surprisingly diverse group of multifunctional proteins. A number of lens crystallins display taxon-specificity. In general, vertebrate crystallins have been recruited from stress-protective proteins (i.e. the small heat-shock proteins) and a number of metabolic enzymes by a gene-sharing mechanism. Despite the existence of refractive lenses in the complex and compound eyes of many invertebrates, relatively little is known about their crystallins. Here we review for the first time the state of knowledge of invertebrate crystallins. The major cephalopod (squid, octopus, and cuttlefish) crystallins (S-crystallins) have, like vertebrate crystallins, been recruited from a stress protective metabolic enzyme, glutathione *S*-transferase. The presence of overlapping AP-1 and antioxidant responsive-like sequences that appear functional in transfected vertebrate cells suggests that the recruitment of glutathione *S*-transferase to S-crystallins involved response to oxidative stress. Cephalopods also have at least two taxon-specific crystallins: Ω-crystallin, related to aldehyde dehydrogenase, and O-crystallin, related to a superfamily of lipid-binding proteins. L-crystallin (probably identical to Ω-crystallin) is the major protein of the lens of the squid photophore, a specialized structure for emitting light. The use of L/Ω-crystallin in the ectodermal lens of the eye and the mesodermal lens of the photophore of the squid contrasts with the recruitment of different crystallins in the ectodermal lenses of the eye and photophore of fish. S- and Ω-crystallins appear to be lens-specific (some S-crystallins are also expressed in cornea) and, except for one S-crystallin polypeptide (SL11/Lops4; possibly a molecular fossil), lack enzymatic activity. The S-crystallins (except SL11/Lops4) contain a variable peptide that has been inserted by exon shuffling. The only other invertebrate crystallins that have been examined are in one marine gastropod (*Aplysia*, a sea hare), in jellyfish and in the compound eyes of some arthropods; all are different and novel proteins. Drosocrystallin is one of three calcium binding taxon-specific crystallins found selectively in the acellular corneal lens of *Drosophila*, while antigen 3G6 is a highly conserved protein present in the ommatidial crystallin cone and central nervous system of numerous arthropods. Cubomedusan jellyfish have three novel crystallin families (the J-crystallins); the J1-crystallins are encoded in three very similar intronless genes with markedly different 5′ flanking sequences despite their almost identical encoded proteins and high lens expression. The numerous refractive structures that have evolved in the eyes of invertebrates contrast markedly with the limited information on their protein composition, making this field as exciting as it is underdeveloped. The similar requirement of Pax-6 (and possibly other common transcription factors) for eye development as well as the diversity, taxon-specificity and recruitment of stress-protective enzymes as crystallins suggest that borrowing multifunctional proteins for refraction by a gene sharing strategy may have occurred in invertebrates as it did in vertebrates.

Keywords: glutathione *S*-transferase; crystallin; lens; invertebrates; aldehyde dehydrogenase.

Recent progress in molecular and developmental biology has brought renewed vigor into the fields of comparative biology and evolution. There are growing examples of species as distant as nematodes, insects, and mammals sharing basic molecular strategies for their development (see McGinnis and Krumlauf, 1992; Krumlauf, 1993, 1994; Kenyon, 1994; Zuckerkandl, 1994; Scott, 1994). The development of analogous organs in evolutionarily distant species may also be controlled by similar or overlapping sets of genes despite differences in the developmental course and final structure of the organs (see Scott, 1994, for further references).

In view of its biological importance and complexity, the eye has been studied intensively and considered a paradigm for evolution. The similarities in structure and function of vertebrate eyes have established that these are homologous organs; by contrast, invertebrates have acquired eyes with different structures, including simple photoreceptive pits lacking a lens or containing

Correspondence to J. Piatigorsky, Laboratory of Molecular and Developmental Biology, National Eye Institute, NIH, 6 Center Drive, MSC 2730, MD 20892-2730, Bethesda, USA

Abbreviations. ALDH, aldehyde dehydrogenase; GST, glutathione *S*-transferase; shsp, small heat-shock protein.

a pinhole for focussing, a complex lens-containing eye analogous to the camera-type eye of vertebrates, mirror-reflecting eyes, and multiple compound eyes composed of numerous ommatidia (Ali, 1984; Land, 1984, 1988; Cronin, 1986; Nilsson, 1989, 1990; Land and Fernald, 1992; Wolken, 1995). Indeed, even unicellular algae (Foster and Smyth, 1980) and dinoflagellates (Francis, 1967; Greuet, 1968) have developed sophisticated 'eyes' with refractive lenses (see Couillard, 1984; Wolken, 1995). It has been assumed, due to the structural diversity of invertebrate eyes and the uncertainty in the phylogenetic relationships among invertebrates (see Willmer, 1990; Morris, 1993), that the eyes of vertebrates and invertebrates have evolved convergently. The cephalopod (squid, octopus, cuttlefish) eye, especially, has such structural resemblances to the vertebrate eye that it has been considered a classical example of independent, convergent evolution (Packard, 1972). A detailed morphological analysis of eyes of many different species suggested that photoreceptors arose independently 40–65 times during evolution, reinforcing the idea that vertebrate and invertebrate eyes evolved convergently (Salvini-Plawen and Mayr, 1977). However, after classifying photoreceptors of many species on the basis of being either ciliary (present in most species, including vertebrates) or rhabdomeric (membranous; present in mollusks, annelids and arthropods) in design, Eakin (1979) challenged the idea that eye evolution was mostly convergent and suggested that photoreceptors share a common ancestor.

Recent molecular genetic studies have demonstrated a critical, possibly master switch, role for Pax-6 (a transcription factor containing paired and homeodomains) for eye development in vertebrates (Hill et al., 1991; Ton et al., 1991; Li et al., 1994; Grindley et al., 1995) and *Drosophila* (Quiring et al., 1994; Halder et al., 1995). In addition, genes involved in early eye morphogenesis acting downstream of Pax-6 may also be used similarly in different systematic groups (Zuker, 1994; Hanson and van Heyningen, 1995). These data showing similarity of gene regulation during eye development imply that the eyes of vertebrates and invertebrates may in fact share a common origin and that their differences may have evolved by parallel evolution after divergence from an early ancestral form, making their developmental processes and possibly some of their structural components partly homologous (see Zuker, 1994; Zuckerkandl, 1994; Dickinson, 1995). The idea that major anatomical differences result in evolution by changes in gene regulation has been advanced earlier (Prager and Wilson, 1975; King and Wilson, 1975; Wilson et al., 1977), and a recent computer-assisted model has suggested that the eye could evolve relatively quickly (Nilsson and Pelger, 1994; Dawkins, 1994).

Eyes in the animal kingdom employ a variety of imaging systems using transparent lenses. Although the optics of these ingenious refractive devices are beyond the scope of this review and are discussed elsewhere (see Land, 1981, 1988, 1991, 1995; Nilsson, 1990; Wolken, 1995; Ott and Schaeffel, 1995), some brief comments are useful in order to appreciate the biology of eye lenses. In terrestrial vertebrates about two-thirds of the refraction is done by an outer cornea and one third by the cellular lens, which is responsible for focussing or accommodation, while in aquatic vertebrates the cornea is optically neutral and the spherical lens accounts for all of the refraction. The cellular lenses of vertebrates are avascular, non-innervated transparent structures having a gradient of refractive index decreasing smoothly from the center to the periphery. This shortens the focal length/radius ratio to approximately 2.5 (Matthiessen's ratio), allows light to be refracted while passing through the lens as well as at the surfaces, and virtually eliminates spherical aberration. Lens inhomogeneity in vertebrates is created by the differential accumulation of soluble proteins (crystallins) in successive layers of fiber cells which are denucleated in the center of the lens (see Piatigorsky, 1981; Kuszak and Brown, 1994). The concentration of the abundant water-soluble crystallins in the center of the lens may reach as high as 70%, exceeding that often reached in the mother liquor of protein crystals (Jaenicke, 1994). The crystallins must be thermodynamically stable since they are trapped within an encapsulated lens whose central fiber cells are denucleated and thus cannot sustain protein turnover for the lifespan of the animal. Lens transparency is attained by minimizing the fluctuations in protein concentration and having the crystallins display short-range interactions (Benedek, 1971; Delaye and Tardieu, 1983; Clark, 1994).

The imaging systems used by invertebrates are more varied but have some similarities to those used by vertebrates. In particular, the eye lenses of cephalopods contain a refractive index gradient, as do lenses of vertebrates, which neutralizes spherical aberration (Sivak, 1982, 1991). An interesting case exists in the oceanic squid, *Histioteuthidae*, which has both a large and a small eye with differing Matthieson ratios (Denton and Warren, 1968). The mirror eyes of the scallop, *Pecten*, have lenses with very low refractive power that accounts for only minor adjustments in order to compensate for spherical aberration created by the reflective surface behind the retina (Land, 1965, 1978). Gradients of refractive index critical for focussing light are also used by the crystalline cones within ommatidia of compound eyes (see Land, 1988; Nilsson, 1990). Gradients of refractive index were demonstrated in the compound eyes of insects (McIntyre and Caveney, 1985), crustaceans (Land and Burton, 1979; Nilsson et al., 1983), the horseshoe crab, *Limulus* (Land, 1979), and fan worms (Nilsson, 1994). They are smaller than in vertebrate or cephalopod lenses and are usually formed as a protein gradient in each cell of the crystalline cone (see Nilsson, 1990). The mechanism establishing the gradient of refractive index in the crystalline cone must differ from that used in vertebrate lenses in view of their different cellular compositions (Fahrenbach, 1968). The eye of the copepod, *Pontella*, presents another invertebrate innovation where three transparent lenses with homogeneous refractive indices are placed in series, avoiding spherical aberration by the curvature of the lens surfaces (see Land, 1984, 1988).

Clearly there is a dazzling variety of optical tricks that are used by invertebrates for vision, raising many interesting questions with respect to evolutionary relationships and biophysics of transparency and refraction. While extensive molecular analyses of the vertebrate crystallins and their genes have been conducted (see Wistow and Piatigorsky, 1988; Bloemendal and de Jong, 1991; de Jong et al., 1994; Sax and Piatigorsky, 1994; Jaenicke, 1994; Groenen et al., 1994), this is for but one phylum, whereas studies of the much more diverse invertebrate phyla have only just begun. The most advanced molecular studies on invertebrate crystallins have been performed on cephalopods, because of their availability, the structural similarities between cephalopod and vertebrate eyes, and the large size of their lenses. Of particular importance has been the finding that cephalopods have used the same strategy as vertebrates (see Piatigorsky and Wistow, 1989; de Jong et al., 1989) for recruiting enzymes as crystallins (Wistow and Piatigorsky, 1987; Tomarev and Zinovieva, 1988). The purpose of this review is to summarize the literature on invertebrate crystallins.

Brief summary of crystallin diversity and gene sharing in vertebrates

Two unexpected findings about lens crystallins of vertebrates were their diversity and their additional non-refractive functions outside of the lens (Table 1; see Wistow and Piatigorsky, 1988;

Table 1. Vertebrate crystallins.

Distribution	Eye design	Crystallin	[Related] or identical proteins
All vertebrates	CORNEA, LENS, RETINA	α	small heat shock proteins (αB) molecular chaperones (αA, αB)
		β/γ	[epidermis differentiation specific protein] [*Myxococcos xanthus* protein S] [*Physarum polycephalum* spherulin 3a]
Birds and reptiles		δ	argininosuccinate lyase (δ2) [argininosuccinate lyase] (δ1)
Crocodiles, many birds		ε	lactate dehydrogenase B4
Guinea pig, degu, rock cavy, camel, llama		ζ	NADPH: quinone oxidoreductase
Elephant shrews		η	aldehyde dehydrogenase I
Rabbit, hare		λ	[hydroxyacyl CoA dehydrogenase]
Kangaroos, quoll		μ	[bacterial ornithine cyclodeaminase]
Diurnal geckos		π	glyceraldehyde 3-phosphate dehydrogenase
Frogs (*Rana*)		ϱ	[NADPH-dependent reductase]
Lamprey, turtle, some fish and birds		τ	α-enolase
Photophore of fish (*Porichthys notatus*)	REFLECTOR, CORNEA, LENS, PHOTOGENIC TISSUE	35 kDa 33 kDa	? ?

Piatigorsky and Wistow, 1989; de Jong et al., 1989; Piatigorsky, 1992; Wistow, 1993). While all vertebrates have α- and β/γ-families of crystallins in differing proportions in their lenses, selected species in different taxonomic groups accumulate in addition high concentrations of various other proteins for refraction. These taxon-specific crystallins of vertebrates are either related or identical to metabolic enzymes. The first hint that crystallins are borrowed proteins used in other tissues was the discovery that the two α-crystallins (αA and αB) have sequence similarity with the small heat-shock proteins (shsp) of *Drosophila* (Ingolia and Craig, 1982). The α-crystallins can also behave like molecular chaperones to protect proteins from various stresses (Horwitz, 1992), suggesting that they protect the lens and its proteins against age-related deteriorative changes and cataract. Among their numerous posttranslational modifications (Groenen et al., 1994), the α-crystallins are also substrates for cAMP-dependent phosphorylation (Spector et al., 1985; Voorter et al., 1986) and have autokinase activity (Kantorow and Piatigorsky, 1994; Kantorow et al., 1995), raising the possibility that they may participate in signal transduction events. αB-crystallin is a functional shsp which is expressed constitutively in many tissues, is induced by physiological stress and is overexpressed in numerous diseases, especially neurodegenerative diseases; by contrast, αA-crystallin is expressed principally, although not exclusively, in the lens and is not inducible by stress (see de Jong et al., 1993; Sax and Piatigorsky, 1994). The β- and γ-crystallins form a related superfamily which is highly specialized for lens expression (see Lubsen et al., 1988), although at least some members of these crystallins appear to be expressed to a limited extent outside of the lens (Head et al., 1991; Smolich et al., 1994). Structural analyses have indicated that the β/γ-crystallins, like the α-crystallins, are related to microbial stress proteins (Wistow et al., 1985; Wistow, 1990; Bagby et al., 1994). A relationship has also been found between the β/γ-crystallins and an amphibian epidermis differentiation protein (Wistow et al., 1995). No non-refractive function has been found yet for the β/γ-crystallins.

While the ubiquitous crystallins in vertebrates have undergone one (the α-crystallins) or multiple (the β/γ-crystallins) gene duplications, some of the taxon-specific enzyme-crystallins have acquired their new roles as structural, refractive proteins without gene duplication (see Piatigorsky and Wistow, 1991). Two examples of the recruitment of an active enzyme as crystallin without gene duplication include lactate dehydrogenase B_4/ε-crystallin (Wistow et al., 1987; Hendricks et al., 1988) and α-enolase/τ-crystallin (Wistow et al., 1988) in duck lenses. In some cases, gene duplication has occurred after recruitment as a lens crystallin with subsequent specialization of one of the daughter genes for its refractive role in the lens, such as in the shsp/α-crystallins (see Sax and Piatigorsky, 1994) and argininosuccinate lyase/δ-crystallins (Piatigorsky et al., 1988). The use of the same gene and its encoded protein for more than one role (i.e. shsp/

enzyme and refractive lens crystallin) has been called gene sharing (Piatigorsky et al., 1988; Piatigorsky and Wistow, 1989). One of the most interesting implications of gene sharing is that the innovation of a new function for a protein may be associated strictly and only with a change in the regulation of its gene (Piatigorsky and Wistow, 1991; Piatigorsky, 1992). Gene sharing may have consequences on the rate and nature of the structural changes of these multifunctional proteins during evolution since it places them under more than one selective pressure. Like the α- and possibly β/γ-crystallins, many of the enzymes used as crystallins have stress-protective roles in other tissues. It is not known yet to what extent these or other non-refractive functions of the enzyme-crystallins were important for their recruitment as refractive proteins or are used today for proper function of the lens (see Piatigorsky, 1992; Wistow, 1993; Wistow et al., 1994, for further discussion).

It is noteworthy that several hundred species of marine fish possess light organs or photophores. These photophores emit bioluminescence into the environment and are believed to be used in behavioral responses (Herring and Morin, 1978; see also Dove et al., 1993). Photophores resemble structurally the eyes of the fish: photophores contain photogenic tissue which is surrounded by supportive tissue reflecting the light, and have a lens and a corneal epidermis (Strum, 1969; Herring and Morin, 1978; see Table 1). The photophore lens of *Porichthys notatus*, like the eye lens, is developmentally derived from epithelial ectoderm and has high (29%) protein concentration. 60% of water-soluble proteins in the photophore lens are two related crystallins with molecular masses of 33 kDa and 35 kDa (Dove et al., 1993). These polypeptides are different from any known crystallins in eye lenses of vertebrates.

Cephalopods (octopus, squid and cuttlefish)

The eye lenses of cephalopods are composed of electrically uncoupled (Jacob and Duncan, 1981, 1984) anterior and posterior segments comprising cellular processes derived from the overlying layer of ectodermal cells (Arnold, 1967; Willekens et al., 1984; Meinertzhagen, 1990). The posterior lens primordium appears first during development and is derived from extensions of the group 2 lentigenic cells which eventually fuse into plate-like elements (West et al., 1995). The anterior lens primordium is derived from extensions of the adjacent group 1 lentigenic cells slightly later in development. Early tests demonstrated that the cephalopod crystallins are immunologically distinct from vertebrate crystallins (Wollman et al., 1938; Halbert and Manski, 1963; Manski and Halbert, 1965; Bon et al., 1967; Smith, 1969). Developmental studies using antisera derived from total lens extracts (Brahma, 1978) or purified squid crystallins (West et al., 1994) indicated that the cephalopod crystallins accumulate in both the lentigenic cells of the ectoderm and their processes comprising the lens. It also appears as if the adult cephalopod lens continues to grow and synthesize crystallins (West et al., 1994, 1995).

S-crystallins. Initial studies on cephalopod lens proteins gave conflicting results as to the number of major components and their possible antigenic relationships among themselves and with vertebrate crystallins (see Brahma and Lancieri, 1979). The immunological results of Brahma and Lancieri (1979) did suggest, however, that cephalopod lens crystallins were relatively homogeneous soluble proteins and that those of squid (*Loligo vulgaris*) and cuttlefish (*Sepia officinalis*) were more similar to each other than to those of octopus (*Octopus vulgaris*). Subsequent physicochemical characterizations of the water-soluble lens crystallins of the squids, *Nototodarus gouldi* (Siezen and Shaw, 1982) and *Loligo pealei* (Chiou, 1984), revealed the presence of only one major class of crystallins in the lens, which was called S(for squid)-crystallin (see Table 2). The major S-crystallins (S_{III}) were dimeric proteins around 60 kDa containing subunits with molecular masses ranging over 22−30 kDa. Less abundant higher-molecular-mass aggregates of S-crystallins (S_I and S_{II}) contained additional subunits of 35−40 kDa. The isoelectric points of the S-crystallin subunits ranged from 5 to 10, with the more acidic polypeptides contained in the higher aggregates. The N-termini of S-crystallins contain proline and, like γ-crystallins of vertebrates, are not blocked. The near- and far-ultraviolet circular-dichroic spectra of S_I, S_{II} and S_{III} indicated that they were structurally similar to each other and differed from the known vertebrate crystallins. Interestingly, S-crystallins have a methionine content as high as 12%, similar to the 10−15% value found in γ-crystallin of fish (Croft, 1973; Chiou et al., 1986, 1988; Pan et al., 1994). Siezen and Shaw (1982) speculated that high pressure, low temperature and high protein concentration may be essential for proper crystallin structures and lens transparency in deep sea squids. The extent of posttranslational modifications that might occur and change the physicochemical properties of S-crystallins, as occurs among the vertebrate crystallins (Zigler, 1994; Groenen et al., 1994), is not known.

Computer-assisted analyses of N-terminal (Wistow and Piatigorsky, 1987) and cDNA (Tomarev and Zinovieva, 1988) sequences indicated that S-crystallins are related to glutathione *S*-transferases (GST). GST is one of the most important family of multifunctional enzymes involved in the cellular detoxification of physiological electrophilic and xenobiotic substances (see Rushmore and Pickett, 1993; Armstrong, 1994; Dirr et al., 1994a; Wilce and Parker, 1994), making the strategy of using metabolic enzymes responsive to physiological stress as refractive crystallins similar to that of vertebrates. Further cDNA cloning experiments showed that the family of GST-related S-crystallins consists of at least two members in *O. vulgaris* (Lin and Chiou, 1992), 6 members in *Octopus dofleini* (Tomarev et al., 1991), 10−12 members in *Ommastrephes sloani pacificus* (Tomarev et al., 1992) and 24 members in *Loligo opalescens* (Tomarev et al., 1995a). The encoded protein sequences of the octopus and squid S-crystallins are, with few exceptions, considerably more similar within species than between them (Tomarev et al., 1991), suggesting that most members of these lens protein families diverged more recently than the 200−300 million year estimate (based on fossil evidence, Clarke and Trueman, 1988) for the divergence of these octapods and decapods (Tomarev et al., 1991). The deduced S-crystallin polypeptides of *L. opalescens* are 46−99% identical to each other, consistent with some members having appeared by gene duplication more recently than other members. Alternatively, the relative differences among the S-crystallins may be due to differences in gene conversion among the different family members, or even differences that have been maintained for long periods of time by selective pressures. These differences among the S-crystallins are not due to the presence of a central peptide (see below) which was not considered in the above calculations.

Almost all S-crystallins have a variable sequence ranging in size from 9 to 193 amino acids inserted into the central region of the polypeptide (Tomarev and Zinovieva, 1988; Tomarev et al., 1992, 1995a). No central peptide sequences of the S-crystallins matched any sequence in the database (Tomarev et al., 1992; Tomarev, unpublished). In each species examined, however, there is at least one S-crystallin that lacks this central insert. An alignment of the squid GST with two S-crystallins containing a central insert (Sl20-1 and Lops12) and two S-crystallins lacking a central insert (SL11 and Lops4) are shown in Fig. 1A.

Table 2. Invertebrate crystallins.

Phylum distribution	Eye design	Crystallin	[Related] or identical proteins
Mollusca cephalopods: Squid, Octopus, Cuttlefish	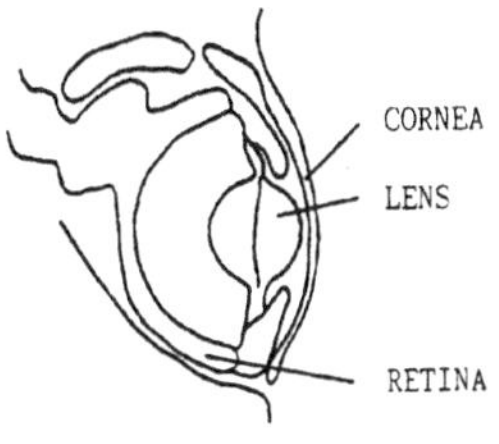 CORNEA LENS	S	glutathione *S*-transferase (only Sl11/Lops4) [glutathione *S*-transferase] (other family members)
Octopus	RETINA	Ω O	[aldehyde dehydrogenase] [lipid-binding protein]
Squid Photophore (Light Organ) (*Euprymna scolopes*)	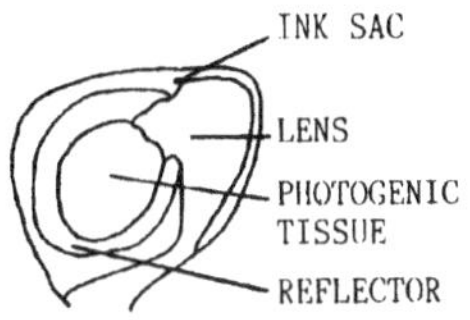 INK SAC LENS PHOTOGENIC TISSUE REFLECTOR	L	[aldehyde dehydrogenase] (probably identical to Ω)
Gastropod (Sea hare; *Aplysia california*)	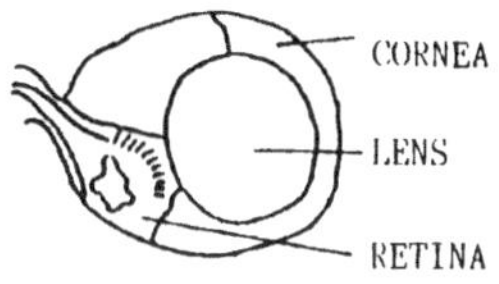 CORNEA LENS RETINA	80 kDa 63 kDa 28 kDa	? ? ?
Cnidaria Cubomedusa (Jellyfish; *Tripedalia cystophora*)	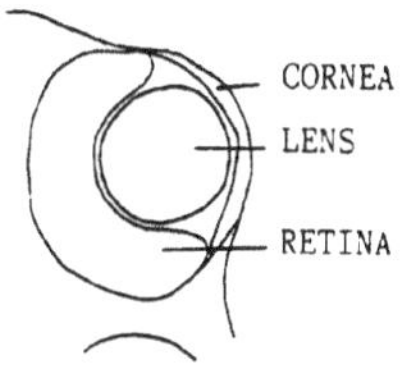 CORNEA LENS RETINA	J1A, B, C J2 J3	? ? ?
Hydromedusa (Jellyfish; *Cladonema radiatum*)	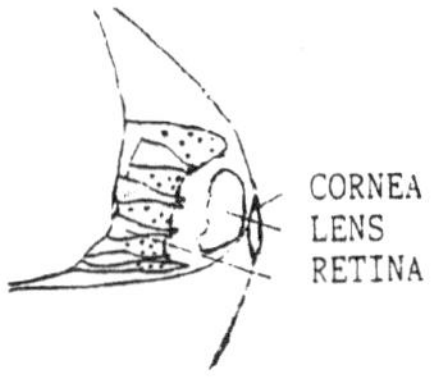 CORNEA LENS RETINA	70 kDa 60 kDa 40 kDa	? ? ?
Arthropoda Insecta (*Drosophila melanogaster*)	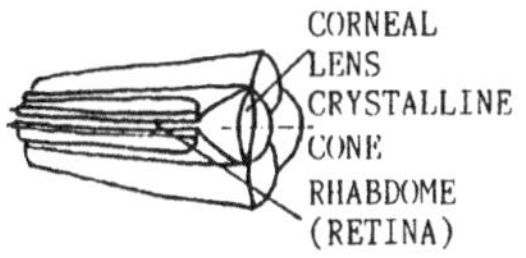 CORNEAL LENS CRYSTALLINE CONE RHABDOME (RETINA)	antigen 3G6 52 kDa (drosocrystallin) 47 kDa 45 kDa	? ? ? ?

A diagrammatic scheme of GST and S-crystallin evolution is presented in Fig. 2. It is interesting to note that GSTs share two sequence motifs with numerous other proteins and enzymes, some of which have stress-related functions (see Koonin et al., 1994; Billaut-Mulot et al., 1994). These include the γ subunit of the eukaryotic elongation factor 1 (EF1γ), bacterial reductive dehalogenases and β-etherases, plant stress-induced proteins, bacterial stringent starvation proteins, yeast nitrogen metabolism regulator URE2 and several other uncharacterized proteins. Koonin et al. (1994) have pointed out that EF1γ contains an N-terminal GST-related domain that may be necessary to keep the GST catalytic domain in proper conformation, and that GST activity by the homodimeric proteins may regulate the assembly of multisubunit complexes by shifting the oxidative forms of disulfides in the cell. Thus, they propose that GSTs (or GST-related proteins) may play a role in regulating the folding and associations of proteins as do molecular chaperones. This idea deserves further attention in view of the importance of the chaperone-like role of α-crystallin in the vertebrate lens (see above). Macrophage migration inhibitory factor (MIF) has been reported to have some enzymatic activity and N-terminal sequence similarity with GST (Blocki et al., 1992, 1993). Although a rigorous connection between the GSTs and MIF remains to be established, it is interesting to note that MIF is a prominent protein in mouse and human lenses (Wistow et al., 1993).

The reasons for the multiplicity and heterogeneity of S-crystallins are not known. They are considerably larger than any single family of crystallins in vertebrate lenses. Possibly individual S-crystallins have different structural or functional roles in the lens, such as the ability to bind water or reach certain critical concentrations to achieve the needed refractive index. The variability of S-crystallins may prevent crystallization at the concentrations and high hydrostatic pressures present in the squid lens (Gross and Jaenicke, 1994).

Structural and functional relationships between S-crystallins and active GSTs. The homodimeric GST has been isolated from the digestive gland of *L. vulgaris* (Harris et al., 1991),

A.

```
                *                              *   *
GST σ     PKYTLHYFPLMGRAELCRFVLAAHGEEFTDRVVEMADWPNLKATMYSNAM   50
SL20-1     N    Y NGR    I MLM  A VQY  KRF FNE  DKYRND P MCV   50
Lops12     N    Y NGR    I MLM VG VQY  KRF FNE  DKYRND P MCV   50
SL11       S    Y NGR    I MLF VASVQYQ KRI L E  TQF TK PCHML   50
Lops4      S    Y NGR    I MIF  AAIQYN K   LSE  TQF TK PCHML   50

                      **                               *
GST σ     PVLDIDG-TKMSQSMCIARHLAREFGLDGKTSLEKYRVDEITETLQDIFN   99
SL20-1           QN  PET A   Y    N YY  NNMDMF I Y CDCFYE LH  100
Lops12           QN  PET A   Y    H  Y  NNMDMF I Y CDCFYE MH  100
SL11       I E  TE QVP   A S Y      FY  NNMDMFK  CLCDS FEL    100
Lops4      I E  TD QVP   A S Y      FH  NNVDMFK  S CDS FEL    100

GST σ     DVVKIKFAPEAAK------------------EAVQQNYEKSCKRLAPFLE  131
SL20-1     YMRYFHTKNGRFMQGSGTDMSPDMDPTQMTSYI NR LDT R IL      150
Lops12     YMRYFHTKNGRFFENGKESEMNPATVI---PYM GRFMDT R VL      147
SL11       YMAVYNEKD   K-----------------TEL KRFQNT L VL YM   133
Lops4      YMTLFNEKD   K-----------------PDL KR MDT R VL Y    133

GST σ     GLLVSNGGGDGFFVGNSMTLADLHCYVALEVPLKHTPELLKDCPKIVALR  181
SL20-1    RT EMRN  KE  M DQ M C MM  CC  N MLEDQTTFNNF  LMS W  200
Lops12    KT ETKN  SQ  M DQ S C MM  CC  N MMENQSMFTSH  LM  W  197
SL11      KT EA K  A W I DQIL C MMTHA   N IQENAN   EY  LA    183
Lops4     KT EA K  S W I DQ LFC MM  AG  N VQEN NF    Y  LAG   183

GST σ     KRVAECPKIAAYLKKRPVRDF                               202
SL20-1        SH   TP     NNTNW                               221
Lops12        AH   TG     ANTNW                               218
SL11      T   AH      I   NNTA                                204
Lops4     N   AH    S I   NNTA                                204
```

B.

Protein	Relative GST Activity
GST σ	100
GST σ + SL20-1 central insert	3
SL20-1	0
Lops12	0
SL11	0.025
Lops4	0.02

Fig. 1. Amino acid sequences and glutathione *S*-transferase activities of various S-crystallins. (A) Amino acid sequence alignment of GST σ (Tomarev et al., 1993a) and SL20-1 and SL11 (Tomarev and Zinovieva, 1988) of *O. sloani pacificus* and Lops4 and Lops12 (Tomarev et al., 1995a) of *L. opalescens*. The GST σ sequence is shown in full; only the differing amino acids are shown for the S-crystallins. Dashes show the gaps which were introduced to maximize similarity. * marks the residues essential for GST binding (the identical residues are also boxed in GST σ and in S-crystallins). (B) Relative GST activity of the indicated proteins derived by expression of cDNAs in *E. coli* (Tomarev et al., 1995a). Enzyme activities were obtained by using 1-chloro-2,4-dinitrobenzene as a substrate.

O. vulgaris (Tang et al., 1994) and *O. sloani pacificus* (Tomarev et al., 1993a). As expected, S-crystallins are more closely related (42−44%) to GST σ from cephalopod digestive glands than to the α, μ or π classes of vertebrate GSTs (20−25%). Of particular interest, GST of *O. pacificus* lacks a central insert and has a methionine content of only 3.5% (Tomarev et al., 1993a). A 33% identity in amino acid sequence is present between some S-crystallins and GST of the filarial parasite, *Onchocerca volvulus* (Liebau et al., 1994). S-crystallins are also distantly related to the θ class of GST (Meyer et al., 1991), which is related to GSTs of insects, plants and bacteria (Pemble and Taylor, 1992).

On the basis of X-ray crystallographic studies, it has been proposed that the squid GST diverged from θ or another α/μ/π precursor and should be placed in novel class of GSTs called σ (Ji et al., 1995), following the earlier suggestion of Buetler and Eaton (1992). As shown in Fig. 3, the X-ray data of squid GST σ revealed that its structure is similar in overall topology to the vertebrate enzymes. GST π has a three-dimensional structure which is closer to GST σ than to any other vertebrate GST (Ji et al., 1995; Wilce et al., 1995; see Fig. 3). However, squid GST σ possesses several characteristic features. It has a unique dimer interface with electrostatic rather than 'lock and key' hydrophobic interactions characteristic of all other cytosolic GSTs (Ji et al., 1995). The active site of GST σ is unusually open and compatible with its high enzymatic activity (Tomarev et al., 1993a, 1995a; Ji et al., 1995). Knowledge of the X-ray structure of GST σ (Ji et al., 1995) allows speculation on the structure of S-crystallins. Alignment of S-crystallin and GST σ sequences has indicated that the variable central inserts form a surface loop between α-helical regions 4 and 5 of the GST (Fig. 3). Insertion of a surface loop is one of the most efficient ways to modify a protein without inducing instability in its framework (El Hawrani et al., 1994).

GST σ activity is 10−100 times higher than the activity of other GSTs when 1-chloro-2,4-dinitrobenzene is used as a substrate (Harris et al., 1991; Tomarev et al., 1993a, 1995a). Octopus GST shows, in addition, high activity when ethacrynic acid is used as a substrate (Tang et al., 1994). Most of the residues

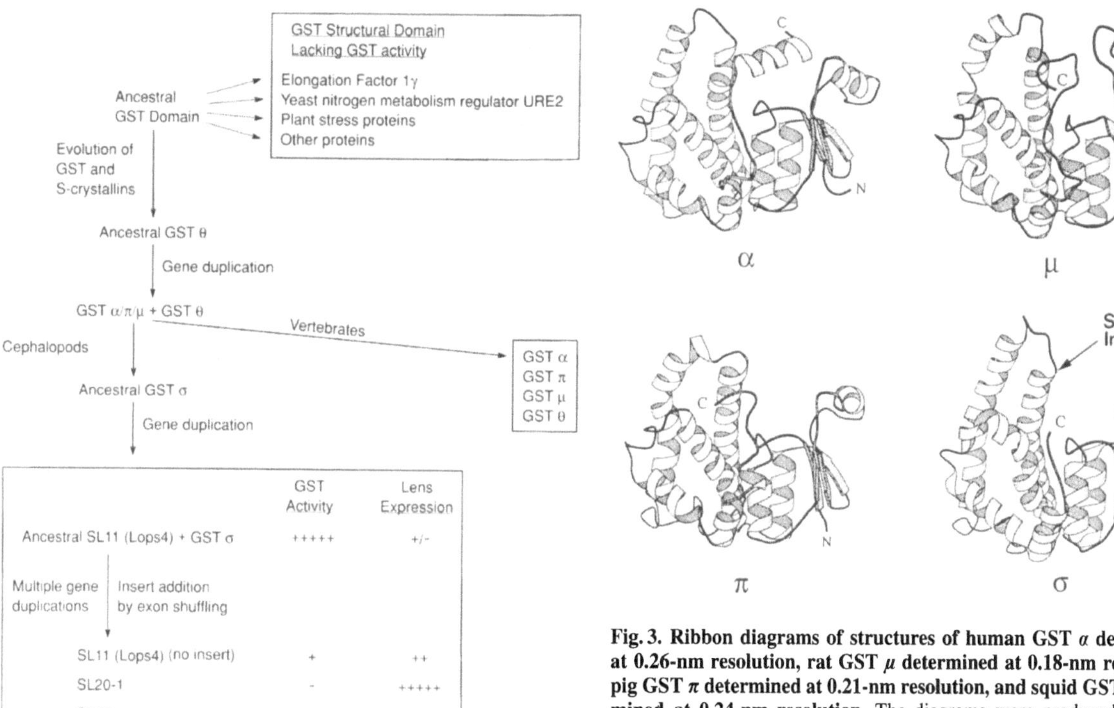

Fig. 2. Recruitment of lens S-crystallins in cephalopods. See text for explanations. Relative levels of GST activity and lens expression are indicated by + or −.

Fig. 3. Ribbon diagrams of structures of human GST *α* determined at 0.26-nm resolution, rat GST *μ* determined at 0.18-nm resolution, pig GST *π* determined at 0.21-nm resolution, and squid GST *σ* determined at 0.24-nm resolution. The diagrams were produced with the program MOLSCRIPT (Kraulis, 1991) using the Protein Database coordinate sets 1GUH for GST *α* (Sinning et al., 1993), 2GST for GST *μ* (Ji et al., 1994), 1GSR for GST *π* (Dirr et al., 1994b) and 1GSQ for GST *σ* (Ji et al., 1995). The arrow in GST *σ* indicates the position of the central inserts in the S-crystallins as deduced from their amino acid sequence alignments with GST *σ*; the X-ray structures of S-crystallins have not been obtained yet. (Diagrams were supplied by Drs Xinhua Ji and Gary L. Gilliland.)

which are essential for the activity or the three-dimensional structure of GST have been conserved in S-crystallins (see Fig. 1A). However, neither lens extracts (Tomarev et al., 1991, 1992; Lin and Chiou, 1992) nor expressed S-crystallin cDNAs (SL20-1 and Lops12) which contain central inserts (Tomarev et al., 1995a) showed GST activity when using chlorodinitrobenzene as a substrate. By contrast, recombinant SL11 and Lops4, the two S-crystallins which lack central inserts and have a lower (5%) methionine content (similar to GST), have low but detectable GST activity with the same substrate. Of the six GST residues (Y7, W38, K42, Q62, S63 and D96, see Fig. 1, asterisks) that have side chains directly involved in the binding of GSH (Ji et al., 1995), five have identical counterparts in the enzymatically active SL11 and Lops4 S-crystallins. The only exception is E97 of SL11/Lops4 which corresponds to D96 of GST. In the inactive SL20-1 and Lops12 S-crystallins there are three additional mutations in the residues necessary for GSH binding, consistent with their enzymatic inactivity and inability to bind to a hexylglutathione affinity column (Tomarev et al., 1995a). Tight binding of GSH by S-crystallins may be deleterious for lens cells, since sequestering of GSH could decrease its availability to protect against oxidative damage, one of the leading causes of cataract in vertebrates (Spector, 1991). Thus, site-specific changes of key residues for GST function have been associated with recruitment of S-crystallins from GST.

Introduction of the central inserts appears to have been another mechanism operative during evolution of S-crystallins that has contributed to their enzymatic inactivation. Addition of the insert peptide from SL20-1 into GST *σ* by *in vitro* mutagenesis reduced its specific activity by 30-fold (Tomarev et al., 1995a; see Fig. 1B). The reverse experiment, deletion of the central insert from SL20-1, did not promote GST activity in the recombinant protein, consistent with the notion that the loss of GST activity by S-crystallins resulted from both insertion of an extra peptide and by gradual drift in the sequence of the core protein. Further experiments are necessary to test the possibility that S-crystallins have enzymatic activity when different substrates are used.

Squid GST and S-crystallin genes. Southern blot hybridization studies have indicated that in *O. sloani pacificus* there is only one polymorphic or, at most, two genes for GST *σ* (Tomarev et al., 1993a) in contrast to the minimal number of 9 or 10 distinct genes for the S-crystallins (Tomarev et al., 1992); the genes for GST *σ* (Tomarev et al., 1993a) and several S-crystallins (Tomarev et al., 1992) have been cloned and characterized. Comparisons of the cDNA and gene sequences for S-crystallins have indicated, as expected for a wild population, considerable polymorphisms. Diagram of the exon-intron structures of the cloned genes together with the structure of rat GST *π* are shown in Fig. 4. The SL20-1 gene represents a member of what appears to be the most abundant subclass of S-crystallins, while the SL11 gene encodes an apparently much less abundant subfamily member of the S-crystallins. The structures of the GST *σ* and S-crystallin genes are very similar but not identical. In general, the exon-intron structure of the squid GST *σ* gene resembles the GST *π* gene more closely than any other vertebrate GST gene (Tomarev et al., 1993a). The introns in squid GST *σ* and vertebrate GST *π* are located between regions encoding *α*-helixes in the protein molecules. An especially important point is that the SL20-1 gene contains an extra exon (exon 4)

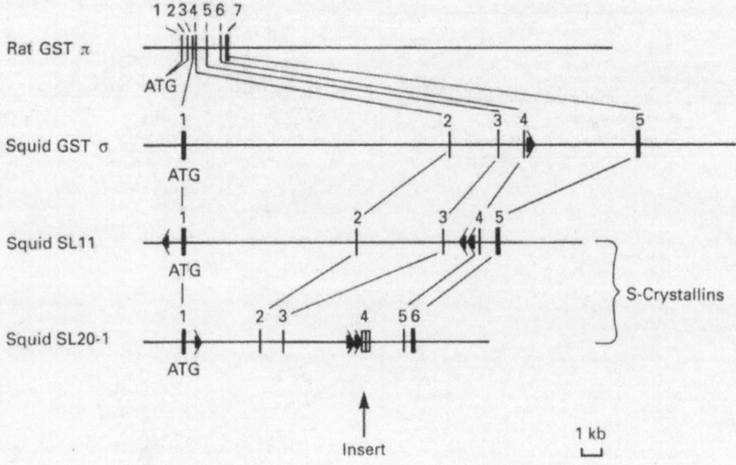

Fig. 4. Comparison of the exon-intron structure of the GST σ, SL20-1, and SL11 genes of squid and the GST π gene of rat. Thin lines between different genes connect homologous exons. Exon 4 in the SL20-1 gene encoding the central peptide is boxed. Large arrows mark the position and direction of repetitive sequences in the squid genes. The gene structure of rat GST π is taken from Okuda et al. (1987), squid GST σ from Tomarev et al. (1993), and squid SL11 and SL20-1 from Tomarev et al. (1992).

which is absent from the GST σ and SL11 genes and precisely encodes the central peptide of SL20-1. A shorter version of exon 4 encoding the central peptide was also present in another member of the SL20 subfamily (SL20-4) (Tomarev et al., 1993a). Thus, SL11 resembles GST σ more closely than SL20-1 (and other S-crystallins) in having some enzymatic activity, a low methionine content, no central peptide and a similar exon-intron gene structure, suggesting that it may represent an intermediate in the evolutionary process of recruiting GST family members as lens crystallins specialized for refraction.

Sequence analysis has suggested that unequal crossing over may have been the driving force leading to the insertion of exon 4, and consequently the central peptide, in the S-crystallin genes (Tomarev et al., 1993a, 1995a). The presence of tandem repetitive sequences in intron 3 of the SL20-1 gene preceding exon 4 and in intron 3 of the SL11 gene is consistent with an unequal crossing over mechanism for the generation of exon 4 (see Fig. 4). Tandem repetitive sequences in introns are believed to facilitate insertions or deletions by unequal crossing over (Stoppa-Lyonett et al., 1990; Olds et al., 1993). The variability of the length and sequence of the central peptides in different S-crystallins suggest that the insertion took place independently multiple times in the course of evolution.

Expression patterns of GST and S-crystallins. The tissue patterns for expression of GST σ and the S-crystallins are radically different. Northern blot hybridization experiments have shown that GST σ is expressed highly in the digestive gland but only very slightly in the mantle, testis and lens in the squid, *O. sloani pacificus* (Tomarev et al., 1993a). There was approximately 500 times less GST mRNA in the lens than in the digestive gland, clearly less than any other tissue examined. By contrast, SL20-1 and SL18 expression were very high in the lens and absent in brain, axon, ovary, heart and gills of *O. sloani pacificus* and *L. opalescens* (Tomarev et al., 1992). Northern blot hybridization of four S-crystallins of the octopus, *O. dofleini*, demonstrated expression specifically in the lens (Tomarev et al., 1991). Immunoblot analysis showed that the major S-crystallins of the squid, *Sepioteuthis lessioniana*, are the principal proteins of the cornea as well as the lens, and that these proteins are not present in the muscle of *O. sloani pacificus* (Cuthbertson et al., 1992). Thus, in contrast to the situation with enzyme-crystallins in vertebrates, where the enzyme is expressed at a high concentration in the lens when acting as a crystallin and at a low concentration in other tissues when required strictly for enzymatic functions (see above), GST σ/S-crystallins of cephalopods appear to have undergone nearly complete separation of function and expression patterns. A similar specialization for lens expression has occurred with an intermediate filament protein and with β-tubulin in cephalopods (Tomarev et al., 1993c).

Regulation of S-crystallin gene expression. The squid SL20-1, SL20-3 and SL11 crystallin promoters have markedly different sequences (Tomarev et al., 1992). The only striking features in common are TATA boxes and an overlapping AP-1/antioxidant responsive element (ARE) a few nucleotides further upstream. AREs and AP-1 sites are also present in the 5′ flanking sequence of the squid GST gene but these sites are located significantly further upstream than in the S-crystallin promoters (Tomarev et al., 1993a). ARE and AP-1 sites have been implicated for the expression of mammalian GST genes (see Rushmore and Pickett, 1993; Daniel, 1993; Nguyen et al., 1994; Morel et al., 1994).

Studies on S-crystallin gene expression have been hampered by the lack of a homologous cellular system for performing functional experiments. Some information concerning *cis*-control elements have been obtained in transfection experiments using the N/N1003A rabbit lens epithelial cell line (Tomarev et al., 1992) and primary cultures of embryonic chicken lens epithelial cells (Tomarev et al., 1993a). Minimal functional promoters were identified in the Sl20-1 (-84 to +37) and the SL11 (-111/+89) crystallin genes of *O. sloani pacificus* using the chloramphenicol acetyltransferase reporter gene. Longer 5′ flanking sequences gave less rather than more promoter activity, and truncation of the SL20-1 promoter fragment to -38 and of the SL11 promoter fragment to -60 abolished their activity in the transfected cells. The critical sequences for function in both promoters contain the overlapping AP-1/ARE sequence, and this was protected from DNase I digestion after incubation with a nuclear extract from embryonic chicken lenses (Tomarev et al., 1994). The AP-1/ARE sequence in the S-crystallin promoters resemble the PL-1 and PL-2 regulatory sequences in the lens-specific chicken βB1-crystallin promoter (Roth et al., 1991).

Site-specific mutations in the AP-1/ARE sequence eliminated activity of both the squid and chicken crystallin promoters in transfected lens cells (Tomarev et al., 1994). Moreover, the overlapping AP-1/ARE sequences of the S-crystallin and βB1-crystallin promoters generated at least one complex co-migrating in electrophoretic mobility with nuclear extracts from embryonic chicken lenses. The complexes formed with the related sequences cross-competed with each other and with consensus AP-1 and ARE sequences, suggesting that the S-crystallin promoter activities were dependent on AP-1 factors in the transfected cells. These data raise the intriguing possibility that the chicken βB1- and squid S-crystallin genes have evolved a similar *cis*-acting regulatory element (AP-1/ARE) for high expression in their respective vertebrate and invertebrate eye lenses. Although AP-1 has been implicated in the expression of many vertebrate crystallin genes (Piatigorsky and Zelenka, 1992; Cvekl et al., 1994), this is probably insufficient for explaining the lens-specific expression of the squid S-crystallin genes. However, the use of an AP-1/ARE *cis*-element for S-crystallin gene expression would link the recruitment of S-crystallins from GST in the squid to protection from oxidative stress, one of the major causes of cataract (Spector, 1991).

Current experiments are exploring the possibility that Pax-6 may be involved in expression of the squid S-crystallins, as it is in the expression of chicken (Cvekl et al., 1994) and mouse (Cvekl et al., 1995a) αA-, chicken δ1- (Cvekl et al., 1995b) and guinea pig ζ- (Richardson et al., 1995) crystallin genes. A squid Pax-6 gene has been cloned and shown by preliminary *in situ* hybridization experiments to be expressed in the developing squid eye (Tomarev, Zinovieva, Kos, Callaerts, Gehring and Piatigorsky, unpublished). It remains speculative at the time of writing whether or not the expression of S-crystallin genes involves Pax-6.

Speculations on the recruitment of S-crystallins from GST. We have proposed a model for the recruitment of S-crystallins from GST (Tomarev et al., 1995a; Fig. 2). The model depicts an ancestral GST domain being used in several proteins lacking GST activity as well as in ancestral GST θ having GST activity. Subsequent duplication of the ancestral GST θ gene gave rise to the GST $\alpha/\pi/\mu$; the squid GST σ was derived from GST $\alpha/\mu/\pi$ as was vertebrate α, π and μ classes of GST. The subsequent duplication of ancestral GST σ gave rise to the daughter gene (SL11/Lops4) still encoding an active GST, which became highly expressed in the lens, probably by acquisition or modification of gene regulatory elements. It is possible, of course, that both daughter genes were highly expressed in the lens and that one (GST σ) lost most of its lens expression and specialized for expression in the digestive gland later in evolution. It is likely that SL11/Lops4 was also expressed to some extent in other tissues. We imagine that the high lens expression of SL11/Lops4 was due to induction by physiological stress. Further duplications of the SL11/Lops4 gene and its progeny gave rise to the family of refractive S-crystallins whose expression became lens-specific and no longer dependent on induction by stress. Both sequence changes and insertions of a central peptide by exon shuffling led to specializations for crystallin function and loss of enzymatic activity. Thus, the residual GST activity and structural similarities to GST of SL11/Lops4 are consistent with it being considered a molecular fossil (albeit that it too has evolved) of the original GST/S-crystallin enzyme-crystallin of cephalopods.

Taxon-specific crystallins of cephalopods: Ω-, L- and O-crystallin. *Ω-crystallin.* As in vertebrates (see Tables 1 and 2), cephalopods display taxon-specificity in their lens crystallins. A homotetrameric protein (called Ω-crystallin) containing N-terminally blocked 59-kDa subunits comprises about 14% of the total soluble lens proteins of the octopus, *O. vulgaris*, but is not detectable in the lens of the squid, *Sepia esculenta* (Chiou, 1988). Subsequent studies identified small, non-crystallin amounts of Ω-crystallin in the lenses of the squids, *O. sloani pacificus* and *Euprymna scolopes* (Montgomery and McFall-Ngai, 1992; Zinovieva et al., 1993). Analysis of tryptic peptides (Tomarev et al., 1991) and cloned cDNAs (Zinovieva et al., 1993) showed that Ω-crystallin is 56–58% identical to both cytoplasmic aldehyde dehydrogenase (ALDH) 1 and mitochondrial ALDH 2 of vertebrates, but only about 25% identical to the liver tumor-inducible ALDH 3 or the microsomal ALDH 4 of vertebrates, making Ω-crystallin more highly conserved in evolution than the S-crystallins (the same is true for the active enzymes: ALDHs are more stringently conserved than GSTs). Enzymatic tests for ALDH in the lens using several different substrates have proved negative (Zinovieva et al., 1993).

In contrast to the GST/S-crystallin situation, ALDH has also been used as crystallin in the vertebrate lens. η-Crystallin, which is 89% identical in sequence to human cytosolic ALDH 1, comprises 10–25% (depending on the species) of the soluble protein of elephant shrew lenses (Wistow and Kim, 1991). However, η-crystallin, unlike Ω-crystallin, is enzymatically active (Wistow, personal communication). Interestingly, ALDH/η-crystallin was detected immunologically in modest amounts in the lenses of a number of different species of vertebrates that do not use this protein as an abundant crystallin, resembling the differences in abundance of ALDH/Ω-crystallin in lenses of octopus and squids. It is also of interest to note that enzymatically active ALDH 3 is present in crystallin concentrations (30–40% of cellular protein) in mammalian corneal epithelial cells, where it may protect against peroxidative damage as well as possibly provide refractive benefits (Evces and Lindahl, 1989; Abedinia et al., 1990; Cooper et al., 1991; Cuthbertson et al., 1992). Thus, ALDH is an example of an enzyme-crystallin that is used by both vertebrates and invertebrates.

Southern blot hybridization data suggest that Ω-crystallin is encoded in only one gene in cephalopods (Zinovieva et al., 1993). It is likely (but not established) that the Ω-crystallin gene differs from that for active ALDH. While Northern blot hybridization tests showed that Ω-crystallin is expressed highly in the lens and not in digestive gland (Zinovieva et al., 1993), ALDH activity was easily detected in the digestive gland (Montgomery and McFall-Ngai, 1992; Zinovieva et al., 1993). It remains possible however that ALDH activity of Ω-crystallin is suppressed in the lens by an inhibitor or posttranslational modifications.

L-crystallin. An ALDH-like protein, originally called L-crystallin, comprises 70% of the lens protein within the photophore or light organ in *E. scolopes* (Montgomery and McFall-Ngai, 1992; see Table 2). The light organ is present in the mantle cavity of the squid and emits bioluminescence produced by symbiotic bacteria as camouflage from predators while the squid feeds in the moonlight (McFall-Ngai and Ruby, 1991; Ruby and McFall-Ngai, 1992; McFall-Ngai, 1994). The lens refracts the emitted bioluminescent light from the ventral surface of the squid into the environment; the light is directed ventrally by a reflector behind the bacteria-containing tissue and the amount of light released from the squid is controlled by the surrounding ink sack. In contrast to the ectodermally derived eye lens, the transparent lens of the light organ is derived from mesodermal muscle tissue which thickens and loses organelles. Partial peptide (Montgomery and McFall-Ngai, 1992) and cDNA (Zinovieva et al., 1993) sequencing data are consistent with L-crystallin of the light organ lens being identical to Ω-crystallin in the eye lens of *E. scolopes*, although it remains possible that L-crystallin and Ω-crystallin are different proteins with almost identical sequences. In addition to sequence similarity, L- and Ω-crystallin share the property of lacking ALDH activity (Montgomery and McFall-Ngai, 1992; Zinovieva et al., 1993).

The expression of L-crystallin/Ω-crystallin has been examined within the light organ during development (Weis et al., 1993). In the adult, ALDH-like proteins were detected immunocytochemically in the lens and in the ciliated epithelial cells of the ducts connecting to the bacterial symbionts. Developmental studies indicated that the ALDH-like proteins begin to accumulate above background in the ciliated ducts and presumptive mesodermal lens within the first 10 days after hatching, during the time bacterial infection occurs.

Recruitment of Ω- and L-crystallin. Ω- and L-crystallin appear to be products of the same gene which encodes an enzyme-crystallin in the mesodermal lens of the light organ and the ectodermal lens of the eye. This differs from the situation in fish, where a bioluminescent photophore uses entirely different crystallins than in the eye (Dove et al., 1993; see Table 1). In contrast to the GST/S-crystallin evolutionary pathway employing multiple gene duplications (see Fig. 2), the ALDH/Ω-/L-crystallin pathway appears to have used few, possibly only one, gene duplication, as has occurred in the case of shsp/α-crystallin and argininosuccinate lyase/δ-crystallin in vertebrates (see Piatigorsky and Wistow, 1991; de Jong et al., 1993; Sax and Piatigorsky, 1994). Unlike the situation in vertebrates, and comparable to the situation with GST/S-crystallins in cephalopods, the gene encoding active ALDH does not appear to be used as a crystallin in the cephalopod eye or light organ. The use and major expression of L (or Ω) crystallin by two different tissues with different embryologic origins (mesoderm and ectoderm) could have arisen by at least two mechanisms. First, similar tissue-specific factors could have evolved in both the eye lens and the light organ to bind the same DNA sequences of the L-crystallin gene to enhance its expression. Alternatively, regulatory regions of the L-crystallin gene could have acquired the ability to respond to diferent tissue-specific factors already present in the eye lens and the light organ.

The recruitment of ALDH-like proteins as crystallins in the lenses of octopus (Chiou, 1988; Zinovieva et al., 1993) and elephant shrew (Wistow and Kim, 1991) eyes and the squid light organ (Montgomery and McFall-Ngai, 1992) is consistent with crystallins belonging to protein families involved in stress responses. Protection against oxidative stress may be especially important for cephalopod lenses since sea water contains high concentrations of photochemically generated oxygen radicals and redox-active transition metal ions (Heinecke and Shapiro, 1992). Moreover, the squid light organ contains a high concentration of peroxidase-like proteins (Tomarev et al., 1993b) that may generate potentially dangerous oxidants, such as superoxide anion, hydroxyl radicals and hypochlorous acid, threatening the lens. Thus the recruitment of Ω-/L-crystallin from ALDH may have initially involved induction of enzyme activity for stress protection and subsequent gene duplication and specialization for refraction in the eye and light organ.

O-crystallin. Another example of a taxon-specific crystallin reported here for the first time is a 22-kDa protein comprising several percent of the total protein of the lens of the octopus, *O. dofleini* (Tomarev, Zinovieva, Johnson, Piatigorsky, unpublished). We call this protein O-crystallin, since very little, if any, seems to be present in squid lenses. O-crystallin is 43% identical to a 21-kDa lipid-binding protein (Schoentgen et al., 1987) and a 23-kDa morphine-binding protein isolated from mammalian brain (Grandy et al., 1990). Octopus O-crystallin and its mammalian homologues are related to yeast TSF1 protein (Robinson and Tatchell, 1991; Schoentgen et al., 1993), which is a suppressor of cdc25 mutations in *Saccharomyces cerevisiae* and is implicated in ras/cdc25 signal transduction (Tripp et al., 1989; Robinson and Tatchell, 1991). Another related protein is an immunodominant antigen of the filarial worm, *O. volvulus* (Lobos et al., 1990). It is noteworthy that the GST of this parasite showed the highest similarity to cephalopod S-crystallins (see above). It was demonstrated recently that the bovine 21-kDa lipid binding protein forms complexes with ubiquitin and GST in the cell cytoplasm and is able to bind nucleotides, like many vertebrate crystallins (Zigler and Rao, 1991) and small GTP-binding proteins (Bucquoy et al., 1994). A 14-kDa cytosolic protein belonging to a lipid/retinoid binding protein superfamily is prominent in the bovine lens (Jaworski and Wistow, 1994), but O-crystallin is structurally different from this superfamily. The physiological roles of O-crystallin in the octopus lens, as well as the smaller lipid binding protein in the bovine lens, are not known.

Gastropods (snails)

A number of gastropod eyes contain lenses which may vary from soft with low refractive power to hard with enough refractive power to form an image (see Land, 1984; Messenger, 1991). The intertidal snail, *Littorina irrorata*, has an especially well developed eye with a spherical lens which corrects at least partially for spherical aberration and appears to contain a refractive index gradient from the core to the periphery (Hamilton et al., 1983). *L. irrorata* may indeed be able to discriminate between vertical and horizontal bars (Hamilton and Winter, 1982). *Aplysia brasiliana* is known to orient itself with respect to the position of the sun, a function which would not seem to require much visual resolving ability (Hamilton and Russell, 1982). The lens of the nudibranch, *Hermissenda crassicornis*, is a secreted spheroidal osmiophilic mass of fine granular substance (Eakin et al., 1967). The lens of the pulmonate snail, *Helix aspersa*, is secreted from retinal and corneal cells as fine droplets (Eakin and Brandenburger, 1967). The droplets are very small and densely packed at the surface of lens, which has a somewhat irregular surface; the droplets appear to fuse and are larger, coarser and less compact as they become more centrally located. It is unlikely that most snail lenses form images on the retina in view of the close apposition of these two structures (see Land, 1984; Messenger, 1991). Further discussion of the morphology of gastropod lenses can be found elsewhere (Eakin and Westfall, 1964; Eakin and Brandenburger, 1967).

To the best of our knowledge there is only one report on the lens crystallins of gastropods. Three major polypeptides with molecular masses of 80, 63 and 28 kDa were found in the acellular lens of the marine opisthobranch, *Aplysia californica* (sea hare), a nudibranch (Cox et al., 1991). Except for one ambiguity, the first 14 N-terminal amino acid sequences of the 80-kDa and 63-kDa polypeptides are identical to each other. Database computer-assisted searches using the first 24 N-terminal amino acid sequence of the 63-kDa polypeptide revealed a number of poor matches with different proteins, but no convincing homology. No sequence data were obtained for the 28-kDa polypeptide. A polyclonal rabbit antibody was made against a synthetic peptide comprising the 14 N-terminal amino acid sequence of the 80-kDa and 63-kDa proteins (Horwitz and Piatigorsky, unpublished). This antibody reacted with the 80-kDa and 63-kDa polypeptides but not with the 28-kDa polypeptide of the *Aplysia* lens or with bovine lens crystallins in Western immunoblots. Since no cloning experiments have been performed yet, it is not known whether the 80-kDa and 63-kDa *Aplysia* lens crystallins are products of different genes, of alternative RNA splicing or of posttranslational modifications.

Cnidarians (jellyfish)

One of the most unexpectedly complex eyes among invertebrates are the ocelli of cubomedusan jellyfish, whose gross structure has been known since the last century [see Piatigorsky et al. (1989) and Ali (1984) for further discussion and references]. The cubomedusan ocellus contains a pigmented retinal epithelium, a neuroretina, an encapsulated transparent cellular lens and an anterior epithelial layer that is analogous to an abbreviated cornea (Pearse and Pearse, 1978; Yamasu and Yoshida, 1976; Laska and Hundgen, 1982; Piatigorsky et al., 1989). The cubomedusae, *Tripedalia cystophora* and *Carybdea marsupialis*, have two ocelli of unequal sizes, situated at right angles to one another, within each of four equally spaced sensory structures called rhopalia located within small open depressions on the surface bell of the jellyfish. The ocelli are thus exposed directly to sea water. The cortical regions of the cellular lens of *T. cystophora* appear plate-like, while the central lens cells are more globular. As in vertebrates, there is a scarcity of organelles within the cytoplasm, which is generally homogeneous and granular when examined by transmission electron microscopy (Piatigorsky et al., 1989). It is likely that the lens is used more for regulating the amount of light entering the eye than to cast an image (Laska and Hundgen, 1982), since it is closely apposed to the retina. However, refractive studies have yet to be performed on jellyfish lenses, and it is not even known if there is a gradient of refractive index in jellyfish lenses.

Some hydromedusan jellyfish also have lens-like structures. One example, *Cladonema radiatum*, forms a transparent lens from cellular processes of the pigmented cells of the ocellus (Weber, 1981a). *Sarsia tubulosa* is another example (Singla and Weber, 1982); this hydomedusan contains an ocellus with a type of outer cornea derived from the swollen ciliary tips of microvilli of photoreceptor cells and a deeper amorphous homogeneous material resembling a lens.

A diffuse nerve net connects the jellyfish ocelli with the rest of the animal. Jellyfish ocelli respond to light of specific wavelengths, elicit electrical responses to reductions in light, and direct the swimming behavior by creating a system of dominance among the rhopalia (Singla, 1974; Satterlie, 1979).

J- and other jellyfish crystallins. The lens proteins have been studied in *C. radiatum* (Weber, 1981b), *T. cystophora* and *C. marsupialis* (Piatigorsky et al., 1989). A rabbit antiserum from the total soluble protein of the bovine lens was able to produce immunofluorescence specifically in the lens of the *C. radiatum* ocellus (Weber, 1981b). SDS/PAGE of ocellar proteins of *C. radiatum* showed three major polypeptides with apparent molecular masses ranging over 40–70 kDa, larger than the bovine crystallins from which the antiserum was raised, which range over 18–35 kDa. It is possible that the immunofluorescence was due to one or more common epitopes between jellyfish and vertebrate crystallins or to cross-reactivity between noncrystallin proteins.

Very different results were obtained when the lens proteins of *T. cystophora* and *C. marsupialis* were examined in a subsequent study (Piatigorsky et al., 1989). SDS/PAGE showed that the large ocellus of the former has three major protein bands with molecular masses near 35, 20 and 19 kDa (called J1-, J2- and J3-crystallin, respectively). The large ocellus of *C. marsupialis* had only the 35-kDa and 20-kDa protein bands, but it is possible that the lower-molecular-mass crystallins were unresolved in these gels. The lens from the smaller ocellus of *T. cystophora* had only J1-crystallin and lacked the smaller J2- and J3-crystallins. Gel filtration chromatography showed that the three classes of native *T. cystophora* crystallins are monomeric, and Western blotting gave no convincing cross-reactivity with antisera to squid S-, chicken δ- or mouse α-, β- or γ-crystallin. Conversely, a rabbit antiserum made against a synthetic peptide of J1-crystallin did not react with squid, frog and human crystallins or with J2- and J3-crystallins of the jellyfish. In contrast to the squid S-crystallins and fish γ-crystallins (see above), J1-crystallin has only a 2% methionine content. Interestingly, J1-crystallin has an unusually high leucine content of 13%, comparable to that of chicken δ-crystallin (Piatigorsky, 1984).

cDNA cloning experiments showed that J1-crystallin comprises three extremely similar proteins (J1A, J1B and J1C; Piatigorsky et al., 1993). The deduced amino acid sequences are 85–99% identical among J1A, J1B and J1C. Computer searches have not revealed any convincing matches of J1-crystallins with known proteins (Piatigorsky et al., 1993), although patches of similarity were noted with a group of ubiquitous chaperonins from eukaryotic and prokaryotic cells (Gupta et al., 1989) and a weak similarity was observed between J1-crystallin and the 60-kDa heat-shock protein family of molecular chaperones (Wistow, unpublished). Preliminary tests for chaperone activity of jellyfish crystallins have been conducted (Piatigorsky et al., 1993). J1-crystallin was stable when rhopalial extracts were heated at 50°C for 5 min but precipitated at 60°C. J2- and/or J3-crystallin appeared somewhat more thermostable than J1-crystallin but this needs further investigation. The addition of bovine α-crystallin to the extract protected all the proteins from thermal aggregation. Thus, if jellyfish crystallins have chaperone abilities, they are not as efficient or of the same type as α-crystallin.

A J3-crystallin cDNA has been obtained recently (Piatigorsky, Horwitz and Norman, unpublished). The deduced J3-crystallin amino acid sequence indicates a novel protein. J2-crystallin has not been cloned yet.

Structure and expression of J-crystallin genes. Unlike the coding sequences, the 5′ and 3′ untranslated sequences of the J1A-, J1B- and J1C-crystallin mRNAs are completely different (Piatigorsky et al., 1993), indicating that they were derived by gene duplication early within the 600 million years of jellyfish evolution (Robson, 1985; Field et al., 1988). Each J1-crystallin is encoded in a separate intronless gene (Piatigorsky et al., 1993), while J3-crystallin appears to be encoded in one gene with at least seven introns (Piatigorsky, Horwitz and Norman, unpublished). Although each J1-crystallin gene is expressed highly in the lens, their putative regulatory regions within their 5′ flanking sequences share only a TATA box. It is not known yet whether the jellyfish crystallins can be induced by physiological stresses or were derived from stress-induced proteins, as the cephalopod crystallins and many of the vertebrate crystallins (see above). Sequence analysis of 5 kbp of the J1B-crystallin gene revealed six potential AP-1 sites, consistent with the possibility of stress responsiveness.

Current studies are being performed to investigate the possibility that a Pax protein is involved in the expression of J-crystallin genes as it is in the expression of vertebrate crystallins (Cvekl et al., 1994, 1995a,b; Richardson et al., 1995). Approximately 2.5 kbp upstream of the J1B-crystallin gene there exists a consensus Pax-6 binding site which has not been examined yet functionally, and a genomic sequence encoding a Pax paired domain has been cloned from *T. cystophora* (Piatigorsky and Norman, unpublished). Preliminary experiments indicate that this Pax mRNA is present in the rhopalia. Another promising avenue of research for investigating the regulation of J-crystallin genes involves steroid hormone receptors. A member of this family most closely related to retinoid X receptor has been cloned from the *T. cystophora* genome (Kostrouch,

Kostrouchova and Rall, unpublished). A potential steroid hormone receptor binding sequence is present in the J1B-crystallin promoter. This is particularly interesting in view of the fact that steroid hormone receptors have been implicated in the regulation of the mouse γF-crystallin gene (Tini et al., 1993, 1994).

Compound eyes of arthropods

Ommatidial crystallins. In *Drosophila melanogaster* a group of 14 different proteins have been identified in the transparent cones of the ommatidia by two-dimensional gel electrophoresis (Fujita and Hotta, 1979; Fujita, 1980). These proteins differed from the group isolated from the photoreceptor cells. The relationships among the cone proteins are not known. In a more recent study, three calcium binding proteins with molecular masses of 52, 47 and 45 kDa were found specifically in the acellular transparent cornea of the ommatidia in *D. melanogaster* (Komori et al., 1992). There was approximately three times more of the 52-kDa protein than each of the smaller proteins in the fly cornea. Several properties distinguish these proteins from one another. First, only the 52-kDa protein is glycosylated; second, while all three proteins are high in leucine (as is chicken δ-crystallin) and glycine, only the 52-kDa and 45-kDa proteins are enriched for serine, and only the 47-kDa protein has a high content of arginine; finally, antibodies made to the denaturated 52-kDa protein or to its N-terminal end immunoreacted specifically with the 52-kDa protein, which was called drosocrystallin. Drosocrystallin is synthesized between 48–96 h of development of the pupa and remains confined to the cornea in the adult. Immunogold localization by electron microscopy showed further that drosocrystallin is enriched in the alternating bands of higher refractive index comprising the corneal structure and also indicated that drosocrystallin is secreted from vesicles of the primary pigment cells. While SDS/PAGE showed a prevalent protein band in the approximate 52-kDa range in numerous *Drosophila* species, only that from *D. melanogaster* reacted by Western immunoblotting with the N-terminal peptide antibody. Moreover, two species of dipteran flies, *Musca domestica* and *Calliphora erythrocephala*, did not even show an enriched 52-kDa protein in their eyes. These data suggest that drosocrystallin is one of several taxon-specific crystallins used among insects as refractive proteins in their corneas. Finally, it is of interest that drosocrystallin is a good substrate for both cholera- and pertussis-dependent ADP-ribosylation, possibly signifying that it was recruited from a protein with non-refractive functions such as a G protein (see Komori et al., 1992, for further discussion and references).

While drosocrystallin may represent a taxon-specific refractive protein in compound eyes, antigen 3G6 is a highly conserved protein in the crystalline cones of many arthropods (Edwards and Meyer, 1990). Antigen 3G6 was originally identified as a marker for neuropil glial cells in the nervous system of the house cricket, *Acheta domesticus* (Meyer et al., 1987). Crystalline cones are either cellular or secreted from cone cells, depending upon the species. In *A. domesticus* the cone comprises Semper cells which are ectodermal in origin. Screening with a monoclonal antibody (mAb 3G6) showed that antigen 3G6 is present in many structurally different refractive cones throughout insects, centipedes and various crustacea (i.e. hermit crabs, shore crabs and spiny lobsters) as well as in glial cells and in some cases retinal cells (Edwards and Meyer, 1990). The cone antigen 3G6 migrates as a single band of about 85 kDa. The 85-kDa band has a similar electrophoretic mobility to one of the doublet immunoreactive bands derived by SDS/PAGE of proteins from the central nervous system of *A. domesticus*. As drosocrystallin, the evidence suggests that antigen 3G6 is glycosylated. The fact that antigen 3G6 is present in the refractive crystalline cone as well as glial and retinal cells suggests that it may have more than one function, reminiscent of the multifunctional crystallins of vertebrates. The prevalence of antigen 3G6 throughout arthropods has been considered to support the idea that the arthropod compound eye has a monophyletic origin [see Edwards and Meyer (1990), Paulus (1979) and Meinertzhagen (1991) for further discussion].

Gene expression. Although no expression studies have been conducted on the crystallins of arthropods, there are several eye mutants in *Drosophila* that are of interest in that connection. First there is *eyeless* which has been shown to involve the putative master gene for eye development, Pax-6 (Quiring et al., 1994; Halder et al., 1995). Since Pax-6 appears directly involved in the expression of a number of vertebrate crystallins (Cvekl et al., 1994, 1995a,b; Richardson et al., 1995) it would seem to be a candidate for expression of crystallins in compound eyes. Another regulatory gene possibly involved with crystallin gene regulation in compound eyes is *prospero*. *Prospero* was first described in *Drosophila* (Doe et al., 1991; Vaessin et al., 1991; Matsuzaki et al., 1992) and subsequently shown to have a mouse homologue called Prox 1 (Oliver et al., 1993). *Prospero*/Prox 1 is a homeodomain protein which is used for development of the central nervous system in both flies and mice. It is also expressed in the cone cells of *Drosophila* and the lens of mice. We have cloned the homologue of Prox 1 from chicken and humans and shown that it is expressed in the lenses of these species as well (Tomarev et al., 1995b). It is not known whether Pax-6 or *prospero* is involved, directly or indirectly, in crystallin gene expression in *Drosophila* or other arthropods.

Final comments

After 100 years of detailed studies on lens crystallins of vertebrates, complementary investigations on crystallins of invertebrates have finally taken root. One of the first requirements for studying invertebrate crystallins is to characterize them in different species. Since there is a remarkable variety of eye structures and lifestyles among invertebrates, they are a rich source for the discovery of new proteins that nature has used for refraction. The most detailed studies so far have been on cephalopod crystallins, and these have shown that invertebrates, like vertebrates, have recruited crystallins from pre-existing metabolic enzymes (GST and ALDH) with stress-related functions. Unlike the vertebrate crystallins, however, there is no clear evidence that the cephalopod crystallins are still multifunctional proteins with non-lens roles. More studies, including X-ray diffraction, are needed in order to appreciate the structural properties of the cephalopod crystallins. Other invertebrates have chosen diverse proteins as novel lens crystallins. It remains to be shown whether these have also been recruited from stress proteins and whether they are multifunctional proteins. There are some initial suggestions that crystallins of compound eyes of arthropods are expressed outside of the lens. Thus, the field of invertebrate crystallins provides an unusual opportunity for descriptive and comparative studies. Of the many invertebrates with eye lenses, only cephalopods, one gastropod, three jellyfish and a few arthropods have been examined with respect to their crystallins, and only crystallins from squid and octopus and one species of jellyfish have been cloned. Even with this limited survey, it is clear that the crystallins of invertebrates, like those of vertebrates, are diverse and taxon-specific. The lens crystallins of dinoflagellates (Francis, 1967; Greuet, 1968), copepods (Wolken and Florida, 1969; Land, 1984, 1988), rotifers (Clement et al., 1983), scallops (Land, 1978), spiders (Blest and Land, 1977;

Uehara et al., 1994), annelid worms (Eakin and Westfall, 1964; Wald and Rayport, 1977), arthropod ocelli (Benzer, 1991) and many different apposition and superposition compound eyes (Land, 1988; Nilsson, 1990) have not been investigated. It would also be of great interest to examine the crystallins of larval ascidians, which have small lenses (Dilly, 1961; Eakin and Kuda, 1971; Barnes, 1974), since these invertebrates lie at the junction between invertebrates and vertebrates.

Finally, the diversity of lens crystallins used for refraction, the recruitment of pre-existing enzymes as lens crystallins, the multifunctional roles of lens crystallins, the involvement of stress induction in the recruitment of lens crystallins, and the evolutionary modifications of crystallins by gene duplications, exon shuffling and sequence modifications are among the important issues connected with crystallins of vertebrates and invertebrates. The variety of eye structures, or eye-like structures (i.e. the light organ) containing an array of refractive systems in invertebrates makes them ideally suited for approaching these key questions in biology and for investigating further the extent to which evolution has been driven by changes in control regions by a gene sharing strategy (Piatigorsky and Wistow, 1991; Piatigorsky, 1992).

We thank Drs Xinhua Ji and Gary L. Gilliland for providing Fig. 3, and Drs Joseph Horwitz, J. Edward Rall, Christina M. Sax, Jerome J. Wolken and J. Samuel Zigler Jr for critically reading the manuscript.

REFERENCES

Abedinia, M., Pain, T., Algar, E. M. & Holmes, R. S. (1990) Bovine corneal aldehyde dehydrogenase: the major soluble corneal protein with a possible dual protective role for the eye, *Exp. Eye Res. 51*, 419–426.

Ali, M. A. (ed.) (1984) Photoreception and vision in invertebrates, *Nato ASI Ser. A 74*, 1–858.

Armstrong, R. N. (1994) Glutathione *S*-transferases: structure and mechanism of an archetypical detoxification enzyme, *Adv. Enzymol. Relat. Areas Mol. Biol. 69*, 1–44.

Arnold, J. M. (1967) Fine structure of the development of the cephalopod lens, *J. Ultrastruct. Res. 17*, 527–543.

Bagby S., Harvey, T. S., Eagle, S. G., Inouye, S. & Ikura, M. (1994) Structural similarity of a developmentally regulated bacterial spore coat protein to β,γ-crystallins of the vertebrate eye lens, *Proc. Natl Acad. Sci. USA 91*, 4308–4312.

Barnes, S. N. (1974) Fine structure of the photoreceptor of the ascidian tadpole during development, *Cell Tiss. Res. 155*, 27–45.

Benedek, G. B. (1971) Theory of transparency of the eye, *Appl. Optics 10*, 459–473.

Benzer, S. (1991) Development of the visual system (Lam, D. M. & Shatz, C. J., eds) pp. 9–34, The MIT Press, Boston MA.

Billaut-Mulot, O., Schoneck, R., Fernandez-Gomez, R., Taibi, A., Capron, A., Pommier, V., Plumas-Marty, B., Loyens, M. & Ouaissi, A. (1994) Molecular and immunological characterization of a *Trypanosoma cruzi* protein homologous to mammalian elongation factor 1 γ, *Biol. Cell. 82*, 39–44.

Blest, A. D. & Land, M. F. (1977) The physiological optics of *Dinopis subrufus* L. Koch: a fish-lens in a spider, *Proc. R. Soc. Lond. 196B*, 197–222.

Blocki, F. A., Schlievert, P. M. & Wackett, L. P. (1992) Rat liver protein linking chemical and immunological detoxification systems, *Nature 360*, 269–270.

Blocki, F. A., Ellis, L. B. M. & Wackett, L. P. (1993) MIF proteins are θ-class glutathione *S*-transferase homologs, *Protein Sci. 2*, 2095–2102.

Bloemendal, H. & de Jong, W. W. (1991) Lens proteins and their genes, *Progr. Nucleic Acids Res. Mol. Biol. 41*, 259–281.

Bon, W. F., Dohrn, A. & Batnik, H. (1967) Lens proteins of a marine invertebrate *Octopus vulgaris*, *Biochim. Biophys. Acta 140*, 312–318.

Brahma, S. K. (1978) Ontogeny of lens crystallins in marine cephalopods, *J. Embryol. Exp. Morph. 46*, 111–118.

Brahma, S. K. & Lancieri, M. (1979) Isofocusing and immunological investigations on cephalopod lens proteins, *Exp. Eye Res. 29*, 671–678.

Bucquoy, S., Jolles, P. & Schoentgen, F. (1994) Relationships between molecular interactions (nucleotides, lipids and proteins) and structural features of the bovine brain 21-kDa protein, *Eur. J. Biochem. 225*, 1203–1210.

Buetler, T. M. & Eaton, D. L. (1992) Glutathione *S*-transferases: amino acid sequence comparison, classification and phylogenic relationship, *Environ. Carcinogen Ecotoxicol. Rev. C10*, 181–203.

Chiou, S.-H. (1984) Physicochemical characterization of a crystallin from the squid lens and its comparison with vertebrate lens crystallins, *J. Biochem. (Tokyo) 95*, 75–82.

Chiou, S.-H. (1988) A novel crystallin from octopus lens, *FEBS Lett. 241*, 261–264.

Chiou, S.-H., Chang, T., Chang, W.-C., Kuo, J. & Lo, T.-B. (1986) Characterization of lens crystallins and their mRNA from the carp lenses, *Biochim. Biophys. Acta 871*, 324–328.

Chiou, S.-H., Azari, P. & Himmel, M. E. (1988) Physicochemical characterization of γ-crystallins from bovine lens – hydrodynamic and biochemical properties, *J. Protein Chem. 7*, 67–80.

Clark, J. I. (1994) in *Principles and practice of ophthalmology* (Alberts, D. M. & Jakobiec, F. A., eds) pp. 114–123, W. B. Saunders Company, Philadelphia, London, Toronto, Montreal, Sydney, Tokyo.

Clarke, M. R. & Trueman, E. R. (eds) (1988) *The Mollusca*, vol. 12, 355 pp. Academic Press, San Diego.

Clement, P., Wurdak, E. & Amsellem, J. (1983) Behavior and ultrastructure of sensory organs in rotifers, *Hydrobiologia 104*, 89–130.

Cooper, D. L., Baptist, E. W., Enghild, J. J., Isola, N. R. & Klintworth, G. K. (1991) Bovine corneal protein 54 K (BCP54)is a homologue of the tumor- associated (class 3) rat aldehyde dehydrogenase (RAT-ALD), *Gene 98*, 201–207.

Couillard, P. (1984) in Photoreception and vision in invertebrates, *NATO ASI Ser. 74*, 115–130.

Cox, R. L., Glick, D. L. & Strumwasser, F. (1991) Isolation and protein sequence identification of *Aplysia californica* lens crystallins, *Biol. Bull. 181*, 333–335.

Croft, L. R. (1973) Amino acid carboxy terminal sequence of γ-crystallin from haddock lens, *Biochim. Biophys. Acta 295*, 174–177.

Cronin, T. W. (1986) Photoreception in marine invertebrates. *Am. Zool. 26*, 403–415.

Cuthbertson, R. A., Tomarev, S. I. & Piatigorsky, J. (1992) Taxon-specific recruitment of enzymes as major soluble proteins in the corneal epithelium of three mammals, chicken and squid, *Proc. Natl Acad. Sci. USA 89*, 4004–4008.

Cvekl, A., Sax, C. M., Bresnick, E. H. & Piatigorsky, J. (1994) A complex array of positive and negative elements regulates the chicken αA-crystallin gene: involvement of Pax-6, USF, CREB and/or CREM, and AP-1 proteins, *Mol. Cell. Biol. 14*, 7363–7376.

Cvekl, A., Kashanchi, F., Sax, C. M., Brady, J. N. & Piatigorsky, J. (1995a) Transcriptional regulation of the mouse αA-crystallin gene: activation dependent on cyclic AMP-responsive element (DE1/CRE) and Pax-6-binding site, *Mol. Cell. Biol. 15*, 653–660.

Cvekl, A., Sax, C. M., Li, X., McDermott, J. B. & Piatigorsky, J. (1995b) Pax-6 and lens-specific transcription of the chicken δ1-crystallin gene, *Proc. Natl Acad. Sci. USA 92*, 4681–4685.

Daniel, V. (1993) Glutathione *S*-transferases – gene structure and regulation of expression, *CRC Crit. Rev. Biochem. Mol. Biol. 28*, 173–208.

Dawkins, R. (1994) The eye in a twinkling, *Nature 368*, 690–691.

de Jong, W. W., Hendriks, W., Mulders, J. W. & Bloemendal, H. (1989) Evolution of eye lens crystallins: the stress connection, *Trends Biochem. Sci. 14*, 365–368.

de Jong, W. W., Leunissen, J. M. A. & Voorter, C. E. M. (1993) Evolution of the α-crystallin/small heat-shock family, *Mol. Biol. Evol. 10*, 103–126.

de Jong, W. W., Lubsen, N. H. & Kraft, H. (1994) Molecular evolution of the eye lens, *Progr. Retinal Eye Res. 13*, 391–442.

Delaye, M. & Tardieu, A. (1983) Short-range order of crystallin proteins accounts for eye lens transparency, *Nature 302*, 415–417.

Denton, E. J. & Warren, F. J. (1968) Eyes of the histioteuthidae, *Nature 219*, 400–401.

Dickinson, W. J. (1995) Molecules and morphology: where's the homology? *Trends Genetics 11*, 119–121.

Dilly, N. (1961) Electron microscope observations of the receptors in the sensory vesicle of the ascidian tadpole, *Nature 191*, 786–787.

Dirr, H., Reinemer, P. & Huber, R. (1994a) X-ray crystal structures of cytosolic glutathione *S*-transferases. Implications for protein architecture, substrate recognition and catalytic function, *Eur. J. Biochem. 220*, 645–661.

Dirr, H., Reinemer, P. & Huber, R. (1994b) Refined crystal structure of porcine class π glutathione *S*-transferase (pGST P1-1) at 2.1 Å resolution, *J. Mol. Biol. 243*, 72–92.

Doe, C. Q., Chu-Lagraff, Q., Wright, D. M. & Scott, M. P. (1991) The *prospero* gene specifies cell fates in the *Drosophila* central nervous system, *Cell 65*, 451–464.

Dove, S., Horwitz, J. & McFall-Ngai, M. (1993) A biochemical characterization of the photophore lenses of the midshipman fish, *Porichthys notatus* Girard, *J. Comp. Physiol. A 172*, 565–572.

Eakin, R. M. (1979) Evolutionary significance of photoreceptors: in retrospect, *Am. Zool. 19*, 647–653.

Eakin, R. M. & Westfall, J. A. (1964) Further observations of the fine structure of some invertebrate eyes, *Z. Zellforsch. 62*, 310–332.

Eakin, R. M. & Brandenburger, J. L. (1967) Differentiation in the eye of a pulmonate snail *Helix aspersa*, *J. Ultrastruct. Res. 18*, 391–421.

Eakin, R. M. & Kuda, A. (1971) Ultrastructure of sensory receptors in ascidian tadpoles, *Z. Zellforsch. 112*, 287–312.

Eakin, R. M., Westfall, J. A. & Dennis, M. J. (1967) Fine structure of the eye of a nudibranch mollusc, *Hermissenda crassicornis*, *J. Cell Sci. 2*, 349–358.

Edwards, J. S. & Meyer, M. R. (1990) Conservation of antigen 3G6: a crystalline cone constituent in the compound eye of arthropods, *J. Neurobiol. 21*, 441–452.

El Hawrani, A. S., Moreton, K. M., Sessions, R. B., Clarke, A. R. & Holbrook, J. J. (1994) Engineering surface loops of proteins – a preferred strategy for obtaining new enzyme function, *Trends Biotechnol. 12*, 207–211.

Evces, S. & Lindahl, R. (1989) Characterization of rat cornea aldehyde dehydrogenase, *Arch. Biochem. Biophys. 274*, 518–524.

Fahrenbach, W. H. (1968) The morphology of the eyes of *Limulus*. I. Cornea and epidermis of the compound eye, *Z. Zellforsch. 87*, 278–291.

Field, K. G., Olsen, G. J., Lane, D. J., Giovannoni, S. J., Ghiselin, M. T., Raff, E. C., Pace, N. R. & Raff, R. A. (1988) Molecular phylogeny of the animal kingdom, *Science 239*, 748–753.

Foster, K. W. & Smyth, R. D. (1980) Light antennas in phototactic algae, *Microbiol. Res. 44*, 572–630.

Francis, D. (1967) On the eyespot of the dinoflagellate, *Nematodinium*, *J. Exp. Biol. 47*, 495–501.

Fujita, S. C. (1980) Biochemical analysis of behavioral mutations of *Drosophila melanogaster*, *J. Metabol. 17, suppl. 1*, 109–114 (in Japanese).

Fujita, S. C. & Hotta, Y. (1979) Two-dimensional electrophoretic analysis of tissue-specific proteins of *Drosophila melanogaster*, *Proteins, Nucleic Acids, Enzymes 24*, 1336–1343 (in Japanese).

Grandy, D. K., Hanneman, E., Bunzow, J., Shih, M., Machida, C. A., Bidlack, J. M. & Civelli, O. (1990) Purification, cloning and tissue distribution of a 23-kDa rat protein isolated by morphine affinity chromatography, *Mol. Endocrinol. 4*, 1370–1376.

Greuet, C. (1968) *Leucopsis cylindrica*, Nov. Gen., Nov. Sp., reridinien *Warnowiidae* lindemann considerations phylogenetiques sur les *Warnowiidae*, *Protistologica 4*, 419–422.

Grindley, J. C., Davidson, D. R. & Hill, R. E. (1995) The role of *Pax-6* in eye and nasal development, *Development 121*, 1433–1442.

Groenen, P. J. T. A., Merck, K. B., de Jong, W. W. & Bloemendal, H. (1994) Structure and modifications of the junior chaperone α-crystallin. From lens transparency to molecular pathology, *Eur. J. Biochem. 225*, 1–19.

Gross, M. & Jaenicke, R. (1994) Proteins under pressure. The influence of high hydrostatic pressure on structure, function and assembly of proteins and protein complexes, *Eur. J. Biochem. 221*, 617–630.

Gupta, R. S., Picketts, D. J. & Ahmad, S. (1989) A novel ubiquitous protein 'chaperonin' supports the endosymbiotic origin of mitochondrion and plant chloroplast, *Biochem. Biophys. Res. Commun. 163*, 780–787.

Halbert, S. P. & Manski, P. (1963) Organ specificity with special reference to the lens, *Prog. Allergy 7*, 107–186.

Halder, G., Callaerts, P. & Gehring, W. J. (1995) Induction of ectopic eyes by targeted expression of the *eyeless* gene in *Drosophila*, *Science 267*, 1788–1792.

Hamilton, P. V. & Russell, B. J. (1982) Celestial orientation by surface-swimming *Aplysia brasiliana* Rang (Mollusca: Gastropoda), *J. Exp. Mar. Biol. Ecol. 56*, 145–152.

Hamilton, P. V. & Winter, M. A. (1982) Behavioral responses to visual stimuli by the snail *Littorina irrorata*, *Anim. Behav. 30*, 752–760.

Hamilton, P. V., Ardizzoni, S. C. & Penn, J. S. (1983) Eye structure and optics in the intertidal snail, *Littorina irrorata*, *J. Comp. Physiol. 152A*, 435–445.

Hanson, I. & van Heyningen, V. (1995) Pax6: more than meets the eye, *Trends Genet. 11*, 268–272.

Harris, J., Coles, B., Meyer, D. J. & Ketterer, B. (1991) The isolation and characterization of the major glutathione *S*-transferase from the squid *Loligo vulgaris*, *Comp. Biochem. Physiol. 98B*, 511–515.

Head, M. W., Peter, A. & Clayton, R. M. (1991) Evidence for the extralenticular expression of members of the β-crystallin gene family in the chick and a comparison with δ-crystallin during differentiation and transdifferentiation, *Differentiation 48*, 147–156.

Heinecke, J. W. & Shapiro, B. M. (1992) The respiratory burst oxidase of fertilization. A physiological target for regulation by protein kinase C, *J. Biol. Chem. 267*, 7959–7962.

Hendriks, W., Mulders, J. W. M., Bibby, M. A., Slinsby, C., Bloemendal, H. & De Jong, W. W. (1988) Duck lens ε-crystallin and lactate dehydrogenase B_4 are identical: a single-copy gene product with two distinct functions, *Proc. Natl Acad. Sci. USA 85*, 7114–7118.

Herring, P. J. & Morin, J. G. (1978) in *Bioluminescence in action* (Herring, P. J., ed.) pp. 273–329, Academic Press, London.

Hill, R. E., Favor, J., Hogan, B. L., Ton, C. C., Saunders, G. F., Hanson, I. M., Prosser, J., Jordan, T., Hastie, N. D. & van Heyningen, V. (1991) Mouse small eye results from mutations in a paired-like homeobox-containing gene, *Nature 354*, 522–525.

Horwitz, J. (1992) α-Crystallin can function as a molecular chaperone, *Proc. Natl Acad. Sci. USA 89*, 10449–10453.

Ingolia, T. D. & Craig, E. A. (1982) Four small *Drosophila* heat shock proteins are related to each other and to mammalian α-crystallin, *Proc. Natl Acad. Sci. USA 79*, 2360–2364.

Jacob, T. J. C. & Duncan, G. (1981) Electrical coupling between fibre cells in amphibian and cephalopod lenses, *Nature 290*, 704–706.

Jacob, T. J. C. & Duncan, G. (1984) A comparative study of the membrane permeability properties of amphibian and cephalopod mollusc lenses, *J. Comp. Physiol. 154B*, 333–341.

Jaenicke, R. (1994) Eye-lens proteins: structure, superstructure, stability, genetics, *Naturwissenschaften 81*, 423–429.

Jaworsky, C. J. & Wistow, G. J. (1994) LP2: A member of the lipid/retinoid binding protein superfamily in bovine lens, *Investigative Ophthalmol. 35*, 1706.

Ji, X., Johnson, W. W., Sesay, M., Dickert, L., Prasad, S. M., Ammon, H. L., Armstrong, R. N. & Gilliland, G. L. (1994) Structure and function of the xenobiotic substrate binding site of glutathione *S*-transferase as revealed by X-ray crystallographic analysis of product complexes with the diastereomers of 9-(*S*-glutathionyl)-10-hydroxy-9,10-dihydrophenanthrene, *Biochemistry 33*, 1043–1052.

Ji, X., von Rosenvinge, E. C., Johnson, W. W., Tomarev, S. I., Piatigorsky, J., Armstrong, R. N. & Gilliland, G. L. (1995) Three-dimensional structure, catalytic properties and evolution of a σ class glutathione transferase from squid, a progenitor of the lens S-crystallins of cephalopods, *Biochemistry 34*, 5317–5328.

Kantorow, M. & Piatigorsky, J. (1994) α-Crystallin/small heat-shock protein has autokinase activity, *Proc. Natl Acad. Sci. USA 91*, 3112–3116.

Kantorow, M., Horwitz, J., van Boekel, M. A. M., de Jong, W. W. & Piatigorsky, J. (1995) Conversion from oligomers to tetramers enhances autophosphorylation by lens αA-crystallin. Specificity between αA- and αB-crystallin subunits, *J. Biol. Chem. 270*, 17215–17220.

Kenyon, C. (1994) If birds can fly, why can't we? Homeotic genes and evolution, *Cell 78*, 175–180.

King, M. C. & Wilson, A. C. (1975) Evolution at two levels in humans and chimpanzees, *Science 188*, 107–116.

Komori, N., Usukura, J. & Matsumoto, H. (1992) Drosocrystallin, a major 52 kDa glycoprotein of the *Drosophila melanogaster* corneal lens. Purification, biochemical characterization, and subcellular localization, *J. Cell Sci. 102*, 191–201.

Koonin, E. V., Mushegian, A. R., Tatusov, R. L., Altschul, S. F., Bryant, S. H., Bork, P. & Valencia, A. (1994) Eukaryotic translation elongation factor 1 contains a glutathione transferase domain – study of a diverse, ancient protein superfamily using motif search and structural modeling, *Protein Sci. 3*, 2045–2054.

Kraulis, P. J. (1991) MOLSCRIPT: a program to produce both detailed and schematic plots of protein structures, *J. Appl. Crystallogr. 24*, 946–950.

Krumlauf, R. (1993) Hox genes and pattern formation in the branchial region of the vertebrate head, *Trends Genet. 9*, 106–112.

Krumlauf, R. (1994) Hox genes in vertebrate development, *Cell 78*, 191–201.

Kuszak, J. R. & Brown, H. G. (1994) in Principles and practice of ophthalmology (Albert, D. M. & Jakobiec, F. A., eds) pp. 82–96, W. B. Saunders Company, Philadelphia, London, Toronto, Montreal, Sydney, Tokyo.

Land, M. F. (1965) Image formation by a concave reflector in the eye of the scallop, *Pecten maximus*, *J. Physiol. (Lond.) 179*, 138–153.

Land, M. F. (1978) Animal eyes with mirror optics, *Sci. Am. 239*, 126–135?

Land, M. F. (1979) The optical mechanism of the eye of *Limulus*, *Nature 280*, 396–397.

Land, M. F. (1981) in *Handbook of sensory physiology* (Autrum, H.-J., ed.) vol. VII/6B, pp. 471–592, Springer-Verlag, Berlin.

Land, M. F. (1984) Photoreception and vision in invertebrates, *NATO ASI Ser. 74*, 401–438.

Land, M. F. (1988) The optics of animal eyes, *Contemp. Phys. 29*, 435–455.

Land, M. F. (1991) in *Evolution of the eye and visual system* (Crony-Dillon, J. R. & Gregory, R. L., eds) vol. 2, pp. 118–135, CRC Press, Boca Raton FL.

Land, M. F. (1995) Fast-focus telephoto eye, *Nature 373*, 658–659.

Land, M. F. & Burton, F. A. (1979) The refractive index gradient in the crystalline cones of the eyes of a euphausiid crustacean, *J. Exp. Biol. 82*, 395–398.

Land, M. F. & Fernald, R. D. (1992) The evolution of eyes, *Annu. Rev. Neurosci. 15*, 1–29.

Laska, V. G. & Hundgen, M. (1982) Morphologie und ultrastruktur der lichtsinnesorgane von *Tripedalia cystophora* Conant (Cnidaria, Cubozoa), *Zool. Jb. Anat. 108*, 107–123.

Li, H.-S., Yang, J.-M., Jacobson, R. D., Pasko, D. & Sundin, O. (1994) Pax-6 is first expressed in a region of ectoderm anterior to the early neural plate: implications for stepwise determination of the lens, *Dev. Biol. 162*, 181–194.

Liebau, E., Walter, R. D. & Henkle-Duhrsen, K. (1994) Isolation, sequence and expression of an *Onchocerca volvulus* glutathione *S*-transferase cDNA, *Mol. Biochem. Parasitol. 63*, 305–309.

Lin, C.-W. & Chiou, S.-H. (1992) Facile cloning and sequencing of S-crystallin genes from octopus lenses based on polymerase chain reaction, *Biochem. Int. 27*, 173–178.

Lobos, E., Altmann, M., Mengod, G., Weiss, N., Rudin, W. & Karam, M. (1990) Identification of an *Onchocerca volvulus* cDNA encoding a low-molecular-weight antigen uniquely recognized by onchocerciasis patient sera, *Mol. Biochem. Parasitol. 39*, 135–146.

Lubsen, N. H., Aarts, H. J. M. & Schoenmakers, J. G. G. (1988) The evolution of lenticular proteins: the β- and γ-crystallin super gene family, *Progr. Biophys. Mol. Biol. 51*, 47–76.

Manski, W. & Halbert, S. P. (1965) in *Protides of the biological fluids* (Peeters, H., ed.) pp. 117–134, Elsevier, Amsterdam.

Matsuzaki, F., Koizumi, K., Hama, C., Yoshioka, T. & Nabeshima, Y.-I. (1992) Cloning of the *Drosophila prospero* gene and its expression in ganglion mother cells, *Biochem. Biophys. Res. Commun. 182*, 1326–1332.

McFall-Ngai, M. J. (1994) Animal-bacterial interactions in the early life history of marine invertebrates: the *Euprymna scolopes/Vibrio fischeri* symbiosis, *Am. Zool. 34*, 554–561.

McFall-Ngai, M. J. & Ruby, E. G. (1991) Symbiont recognition and subsequent morphogenesis as early events in an animal-bacterial mutualism, *Science 254*, 1491–1494.

McGinnis, W. & Krumlauf, R. (1992) Homeobox genes and axial patterning, *Cell 68*, 283–302.

McIntyre, P. & Caveney, S. (1985) Graded-index optics are matched to optical geometry in the superposition eyes of scarab beetles, *Philos. Trans. R. Soc. Lond. 55B*, 85–90.

Meinertzhagen, I. A. (1990) in *Squid as experimental animals* (Gilbert, D. L., Adelman, W. J. & Arnold, J. M., eds) pp. 399–419, Plenum Press, New York and London.

Meinertzhagen, I. A. (1991) in *Evolution of the eye and visual system* (Crony-Dillon, J. R. & Gregory, R. L., eds) vol. 2, pp. 341–363, CRC Press, Boca Raton FL.

Messenger, J. B. (1991) in *Evolution of the eye and visual system* (Crony-Dillon, J. R. & Gregory, R. L., eds) vol. 2, pp. 364–397, CRC Press, Boca Raton FL.

Meyer, M. R., Reddy, G. R. & Edwards, J. S. (1987) Immunological probes reveal spatial and developmental diversity in insect neuroglia, *J. Neurosci. 7*, 512–521.

Meyer, D. J., Coles, B., Pemple, S. E., Gilmore, K. S., Fraser, G. M. & Ketterer, B. (1991) θ, a new class of glutathione transferases purified from rat and man, *Biochem. J. 274*, 409–414.

Montgomery, M. K. & McFall-Ngai, M. J. (1992) The muscle-derived lens of a squid bioluminescent organ is biochemically convergent with the ocular lens. Evidence for recruitment of aldehyde dehydrogenase as a predominant structural protein, *J. Biol. Chem. 267*, 20999–21003.

Morel, F., Schulz, W. A. & Sies, H. (1994) Gene structure and regulation of expression of human glutathione *S*-transferase α, *Biol. Chem. Hoppe-Seyler 375*, 641–649.

Morris, S. C. (1993) The fossil record and the early evolution of the Metazoa, *Nature 361*, 219–225.

Nguyen, T., Rushmore, T. H. & Pickett, C. B. (1994) Transcriptional regulation of a rat liver glutathione *S*-transferase Γa subunit gene. Analysis of the antioxidant response element and its activation by the phorbol ester 12-*O*-tetradecanoylphorbol-13-acetate, *J. Biol. Chem. 269*, 13656–13662.

Nilsson, D.-E. (1989) in *Facets of vision* (Stavenga, D. G. & Hardie, R. C., eds) pp. 30–73, Springer-Verlag, Berlin.

Nilsson, D.-E. (1990) From cornea to retinal image in invertebrate eyes, *Trends Neurosci. 13*, 55–64.

Nilsson, D.-E. (1994) Eyes as optical alarm systems in fan worms and ark clams *Phil. Trans. R. Soc. Lond. B346*, 195–212.

Nilsson, D.-E. & Pelger, S. (1994) A pessimistic estimate of the time required for an eye to evolve, *Proc. R. Soc. Lond. B256*, 53–58.

Nilsson, D.-E., Andersson, M., Hallberg, E. & McIntyre, P. (1983) A micro-interferometric method for analysis of rotation-symmetric refractive-index gradients in intact objects, *J. Microsc. 132*. 21–29.

Okuda, A., Sakai, M. & Muramatsu, M. (1987) The structure of the rat glutathione *S*-transferase P gene and related pseudogenes, *J. Biol. Chem. 262*, 3858–3863.

Olds, R. J., Lane, D. A., Chowdhury, V., De Stefano, V., Leone, G. & Thein, S. L. (1993) Complete nucleotide sequence of the antithrombin gene: evidence for homologous recombination causing thrombophilia, *Biochemistry 32*, 4216–4224.

Oliver, G., Sosa-Pineda, B., Geisendorf, S., Spana, E. P., Doe, C. Q. & Gruss, P. (1993) Prox 1, a prospero-related homeobox gene expressed during mouse development, *Mech. Dev. 44*, 3–16.

Ott, M. & Schaeffel, F. (1995) A negatively powered lens in the chameleon, *Nature 373*, 692–694.

Packard, A. (1972) Cephalopods and fish: the limits of convergence, *Biol. Rev. 47*, 241–307.

Pan, F.-M., Chang, W. C., Chao, Y.-K. & Chiou, S.-H. (1994) Characterization of γ-crystallins from a hybrid teleostean fish: Multiplicity of isoforms as revealed by cDNA sequence analysis, *Biochem. Biophys. Res. Commun. 202*, 527–534.

Paulus, H. F. (1979) Eye structure and the monphyly of the arthropoda, in *Arthropod phylogeny* (Gupta, A. P., ed.) pp. 299–383, Van Nostrand Reinhold, New York.

Pearse, J. S. & Pearse, V. B. (1978) Vision in cubomedusan jellyfishes, *Science 199*, 458.

Pemble, S. E. & Taylor, J. B. (1992) An evolutionary perspective on glutathione transferases inferred from class-Theta glutathione transferase cDNA sequences, *Biochem. J. 287*, 957–963.

Piatigorsky, J. (1981) Lens differentiation in vertebrates. A review of cellular and molecular features, *Differentiation 19*, 134–153.

Piatigorsky, J. (1984) δ-Crystallin and their nucleic acids, *Mol. Cell. Biochem. 59*, 33–56.

Piatigorsky, J. (1992) Lens crystallins. Innovation associated with changes in gene regulation, *J. Biol. Chem. 267*, 4277–4280.

Piatigorsky, J. & Wistow, G. (1989) Enzyme/crystallin: gene sharing as an evolutionary strategy, *Cell 57*, 197–199.

Piatigorsky, J. & Wistow, G. (1991) The recruitment of crystallins: new functions precede gene duplication, *Science 252*, 1078–1079.

Piatigorsky, J. & Zelenka, P. S. (1992) Transcriptional regulation of crystallin genes: cis elements, trans-factors, and signal transduction systems in the lens, *Adv. Dev. Biochem. 1*, 211–256.

Piatigorsky, J., O'Brien, W. E., Norman, B. L., Kalumuck, K., Wistow, G. J., Borras, T., Nickerson, J. M. & Wawrousek, E. F. (1988) Gene sharing by δ-crystallin and argininosuccinate lyase, *Proc. Natl Acad. Sci. USA 85*, 3479–3483.

Piatigorsky, J., Horwitz, J., Kuwabara, T. & Cutress, C. E. (1989) The cellular eye lens and crystallins of cubomedusan jellyfish, *J. Comp. Physiol. A 164*, 577–587.

Piatigorsky, J., Horwitz, J. & Norman, B. (1993) J1-crystallins of the cubomedusan jellyfish lens constitute a novel family encoded in at least three intronless gene, *J. Biol. Chem. 268*, 11894–11901.

Prager, E. M. & Wilson, A. C. (1975) Slow evolutionary loss of the potential for interspecific hybridization in birds: a manifestation of slow regulatory evolution, *Proc. Natl Acad. Sci. USA 72*, 200–204.

Quiring, R., Walldorf, U., Kloter, U. & Gehring, W. J. (1994) Homology of the *eyeless* gene of *Drosophila* to the *small eye* gene in mice and *aniridia* in humans, *Science 265*, 785–789.

Richardson, J., Cvekl, A. & Wistow, G. (1995) Pax-6 is essential for lens-specific expression of ζ-crystallin, *Proc. Natl Acad. Sci. USA 92*, 4676–4680.

Robinson, L. C. & Tatchell, K. (1991) TSF1: A suppressor of *cdc25* mutations in *Saccharomyces cerevisiae*, *Mol. Gen. Genet. 230*, 241–250.

Robson, E. A. (1985) in *The systematics association*, *The origin and relationships of lower invertebrates* (Morris, S. S., George, J. D., Gibson, R. & Platt, H. M., eds) special vol. 28, pp. 60–77, Clarendon Press, Oxford.

Roth, H. J., Das, G. C. & Piatigorsky, J. (1991) Chicken βB1-crystallin gene expression: presence of conserved functional polyomavirus enhancer-like and octamer binding-like promoter elements found in non-lens genes, *Mol. Cell. Biol. 11*, 1488–1499.

Ruby, E. G. & McFall-Ngai, M. J. (1992) A squid that glows in the night: development of an animal-bacterial mutualism, *J. Bacteriol. 174*, 4865–4870.

Rushmore, T. H. & Pickett, C. B. (1993) Glutathione *S*-transferases: structure, regulation, and therapeutic implications, *J. Biol. Chem. 268*, 11475–11478.

Salvini-Plawen, L. V. & Mayr, E. (1977) On the evolution of photoreceptors and eyes, *Evol. Biol. 10*, 207–263.

Satterlie, R. A. (1979) Central control of swimming in the cubomedusan jellyfish *Carybdea rastonii*, *J. Comp. Physiol. 133*, 357–367.

Sax, C. M. & Piatigorsky, J. (1994) Expression of the α-crystallin/small heat-shock protein/molecular chaperone genes in the lens and other tissues, *Adv. Enzymol. Related Areas Mol. Biol. 69*, 155–201.

Schoentgen, F., Saccoccio, F., Jolles, J., Bernier, I. & Jolles, P. (1987) Complete amino acid sequence of a basic 21-kDa protein from bovine brain cytosol, *Eur. J. Biochem. 166*, 333–338.

Schoentgen, F., Bucquoy, S., Seddiqi, N. & Jolles, P. (1993) Two cytosolic protein families implicated in lipid-binding: main structural and functional features, *Int. J. Biochem. 25*, 1699–1704.

Scott, M. P. (1994) Intimation of a creature, *Cell 79*, 1121–1124.

Siezen, R. J. & Shaw, D. C. (1982) Physicochemical characterization of lens proteins of the squid *Nototodarus gouldi* and comparison with vertebrate crystallins, *Biochim. Biophys. Acta 704*, 304–320.

Singla, C. L. (1974) Ocelli of hydromedusae, *Cell Tissue Res. 149*, 413–429.

Singla, C. L. & Weber, C. (1982) Fine structure of the ocellus of *Sarsia tubulosa* (Hydrozoa, Anthomedusae), *Zoomorphology 100*, 11–22.

Sinning, I., Kleywegt, G. J., Cowan, S. W., Reinemer, P., Dirr, H. W., Huber, R., Gilliland, G. L., Armstrong, R. N., Ji, X., Board, P., Olin, B., Mannervik, B. & Jones, T. A. (1993) Structure determination and refinement of human α class glutathione transferase A1-1, and a comparison with the μ and π class enzymes, *J. Mol. Biol. 232*, 192–212.

Sivak, J. G. (1982) Optical properties of the cephalopod eye (the short finned squid, *Illex illecebrosus*), *J. Comp. Physiol. A147*, 323–327.

Sivak, J. G. (1991) Shape and focal properties of the cephalopod ocular lens, *Can. J. Zool. 69*, 2501–2506.

Smith, A. C. (1969) An electrophoretic study of protein extracted in distilled water and in saline solution from the eye lens nucleus of the squid *Nototodarus hawaiiensis* (Berry), *Comp. Biochem. Physiol. 30*, 551–559.

Smolich, B. D., Tarkington, S. K., Saha, M. S. & Grainger, R. M. (1994) *Xenopus* γ-crystallin gene expression: evidence that the γ-crystallin gene family is transcribed in lens and non-lens tissues, *Mol. Cell. Biol. 14*, 1355–1363.

Spector, A. (1991) The lens and oxidative stress, in *Oxidative stress: oxidants and antioxidants* (Sies, H., ed.) pp. 529–558, Academic Press, New York.

Spector, A., Chiesa, R., Sredy, J. & Garner, W. (1985) cAMP-dependent phosphorylation of bovine lens α-crystallin, *Proc. Natl Acad. Sci. USA 82*, 4712–4716.

Stoppa-Lyonnet, D., Carter, P. E., Meo, T. & Tosi, M. (1990) Claster of intragenic Alu repeats predispose the human C1 inhibitor locus to deleterious rearrangement, *Proc. Natl Acad. Sci. USA 87*, 1551–1555.

Strum, J. M. (1969) Fine structure of the dermal luminescent organs, photophores, in the fish, *Porichthys notatus*, *Anat. Rec 164*, 433–462.

Tang, S.-S., Lin, C.-C. & Chang, G. G. (1994) Isolation and characterization of octopus hepatopancreatic glutathione *S*-transferase. Comparison of digestive gland enzyme with lens S-crystallin, *J. Protein Chem. 13*, 609–618.

Tini, M., Otulakowski, G., Breitman, M. L., Tsui, L.-C. & Giguere, V. (1993) An everted repeat mediates retinoic acid induction of the γF-crystallin gene: evidence of a direct role for retinoids in lens development, *Genes Dev. 7*, 295–307.

Tini, M., Tsui, L.-C. & Giguere, V. (1994) Heterodimeric interaction of the retinoic acid and thyroid hormone receptors in transcriptional regulation of the γF-crystallin everted retinoic response element, *Mol. Endocrinol. 8*, 1494–1506.

Tomarev, S. I. & Zinovieva, R. D. (1988) Squid major lens polypeptides are homologous to glutathione *S*-transferase subunits, *Nature 336*, 86–88.

Tomarev, S. I., Zinovieva, R. D. & Piatigorsky, J. (1991) Crystallins of the octopus lens. Recruitment from detoxification enzymes, *J. Biol. Chem. 266*, 24226–24231.

Tomarev, S. I., Zinovieva, R. D. & Piatigorsky, J. (1992) Characterization of squid crystallin genes. Comparison with mammalian glutathione *S*-transferase genes, *J. Biol. Chem. 267*, 8604–8612.

Tomarev, S. I., Zinovieva, R. D., Guo, K. & Piatigorsky, J. (1993a) Squid glutathione *S*-transferase. Relationships with other glutathione *S*-transferases and S-crystallins of cephalopods, *J. Biol. Chem. 268*, 4534–4542.

Tomarev, S. I., Zinovieva, R. D., Weis, V. M., Chepelinsky, A. B., Piatigorsky, J. & McFall-Ngai, M. J. (1993b) Abundant mRNAs in the squid light organ encode proteins with a high similarity to mammalian peroxidases, *Gene 132*, 219–226.

Tomarev, S. I., Zinovieva, R. D. & Piatigorsky, J. (1993c) Primary structure and lens-specific expression of genes for an intermediate filament protein and a β-tubulin in cephalopods, *Biochim. Biophys. Acta 1216*, 245–254.

Tomarev, S. I., Duncan, M. K., Roth, H. J., Cvekl, A. & Piatigorsky, J. (1994) Convergent evolution of crystallin gene regulation in squid and chicken: the AP-1/ARE connection, *J. Mol. Evol. 39*, 134–143.

Tomarev, S. I., Chung, S. & Piatigorsky, J. (1995a) Glutathione *S*-transferase and S-crystallins of cephalopods: Evolution from active enzyme to lens refractive protein, *J. Mol. Evol.*, in the press.

Tomarev, S. I., Zinovieva, R. D., Duncan, M. K., Banerjee-Basu, S., Johnson, T., Sundin, O., Yang, J.-M. & Piatigorsky, J. (1995b) Homeobox gene Prox 1 and lens development, *Invest. Ophthalm. Vis. Sci. 36*, S844.

Ton, C. C. T., Hirvonen, H., Miwa, H., Weil, M. M., Monaghan, P., Jordan, T., van Heyningen, V., Hastie, N. D., Meijers-Heijboer, H., Drechsler, M., Royer-Pokora, B., Collins, F., Swaroop, A., Strong, L. C. & Saunders, G. F. (1991) Positional cloning and characterization of a paired box- and homeobox-containing gene from the aniridia region, *Cell 67*, 1059–1074.

Tripp, M. L., Bouchard, R. A. & Pinon, R. (1989) Cloning and characterization of NSP1, a locus encoding a component of a CDC25-dependent nutrient-responsive pathway in *Saccharomyces cerevisiae*, *Mol. Microbiol. 3*, 1319–1327.

Vaessin, H., Grell, E., Wolff, E., Bier, E., Jan, L. Y. & Jan, Y. N. (1991) *prospero* is expressed in neuronal precursors and encodes a nuclear protein that is involved in the control of axonal outgrowth in *Drosophila*, *Cell 67*, 941–953.

Voorter, C. E. M., Mulders, J. W. M., Bloemendal, H. & de Jong, W. W. (1986) Some aspects of the phosphorylation of α-crystallin, *Eur. J. Biochem. 160*, 203–210.

Uehara, A., Uehara, K. & Ogawa, K. (1994) Fine structure of the anteromedial eye of the liphistiid spider, *Heptathela kimurai*, *Anat. Rec. 240*, 141–147.

Wald, G. & Rayport, S. (1977) Vision in annelid worms, *Science 196*, 1434–1439.

Weber, C. (1981a) Structure, biochemistry, ontogenetic development, and regeneration of the ocellus of *Cladonema radiatum* Dujardin (Cnidaria, Hydrozoa, Anthonedusae), *J. Morphol. 167*, 313–331.

Weber, C. (1981b) Lens of the hydromedusan *Cladonema* studied by SDS gel electrophoresis and immunofluorescent technique, *J. Exp. Zool. 217*, 15–21.

Weis, V. M., Montgomery, M. K. & McFall-Ngai, M. J. (1993) Enhanced production of ALDH-like protein in the bacterial light organ of the sepiolid squid *Euprymna scolopes*, *Biol. Bull. 184*, 309–321.

West, J. A., Sivak, J. G., Pasternak, J. & Piatigorsky, J. (1994) Immunolocalization of S-crystallins in the developing squid (*Loligo opalescens*) lens, *Dev. Dynam. 199*, 85–92.

West, J. A., Sivak, J. G. & Doughty, M. J. (1995) Microscopical Evaluation of the crystalline lens of the squid (*Loligo opalescens*) during embryonic development, *Exp. Eye Res. 60*, 19–35.

Wilce, M. C. J. & Parker, M. W. (1994) Structure and function of glutathione *S*-transferases, *Biochim. Biophys. Acta 1205*, 1–18.

Wilce, M. C. J., Board, P. G., Feil, S. C. & Parker, M. W. (1995) Crystal structure of a θ-class glutathione transferase, *EMBO J. 14*, 2133–2143.

Willekens, B., Vrensen, G., Jacob, T. & Duncan, G. (1984) The ultrastructure of the lens of the cephalopod *Sepiola*: a scanning electron microscopic study, *Tissue & Cell 16*, 941–950.

Willmer, P. (1990) *Invertebrate relationships. Pattern in animal evolution*, Cambridge University Press, Cambridge, New York, Port Chester, Melbourne, Sydney.

Wilson, A. C., Carlson, S. S. & White, T. J. (1977) Biochemical evolution, *Annu. Rev. Biochem. 46*, 573–693.

Wistow, G. J. (1990) Evolution of protein superfamily: relationships between vertebrate lens crystallins and micro-organism dormancy protein, *J. Mol. Evol. 30*, 140–145.

Wistow, G. J. (1993) Lens crystallins: gene recruitment and evolutionary dynamism, *Trends Biochem. Sci. 18*, 301–306.

Wistow, G. J. & Piatigorsky, J. (1987) Recruitment of enzymes as lens structural proteins, *Science 236*, 1554–1556.

Wistow, G. & Piatigorsky, J. (1988) Lens crystallins: the evolution and expression of proteins for a highly specialized tissue, *Annu. Rev. Biochem. 57*, 479–504.

Wistow, G. & Kim, H. (1991) Lens proteins expression in mammals: taxon-specificity and the recruitment of crystallins, *J. Mol. Evol. 32*, 262–269.

Wistow, G., Summers, L. & Blundell, T. (1985) *Myxococcus xanthus* spore coat protein S may have a similar structure to vertebrate lens βγ-crystallins, *Nature 315*, 771–773.

Wistow, G. J., Mulders, J. W. M. & de Jong W. W. (1987) The enzyme lactate dehydrogenase as a structural protein in avian and crocodilian lenses, *Nature 326*, 622–624.

Wistow, G. J., Lietman, T., Williams, L. A., Stapel, S. O., de Jong, W. W., Horwitz, J. & Piatigorsky, J. (1988) τ-Crystallin/α-enolale: one gene encodes both an enzyme and a lens structural protein, *J. Cell Biol. 107*, 2729–2736.

Wistow, G. J., Shaughnessy, M. P., Lee, D. G., Hodin, J. & Zelenka, P. S. (1993) A macrophage migration inhibitory factor is expressed in the differentiating cells of the eye lens, *Proc. Natl Acad. Sci. USA 90*, 1272–1275.

Wistow, G., Richardson, J., Jaworski, C., Graham, C., Sharon-Friling, R. & Segovia, L. (1994) Crystallins: The over-expression of functional enzymes and stress proteins in the eye lens, *Biotech. Genet. Eng. Rev. 12*, 1–38.

Wistow, G., Jaworsky, C. & Rao, P. V. (1995) A non-lens member of the β/γ-crystallin superfamily in the vertebrate *Cynops*, *Exp. Eye Res. 61*, 637–639.

Wolken, J. J. (1995) *Light detectors, photoreceptors, and imaging systems in nature*, Oxford University Press, New York, Oxford.

Wolken, J. J. & Florida, R. G. (1969) The eye structure and optical system of the crustacean copepod, *Copilia*, *J. Cell Biol. 40*, 279–285.

Wollman, E., Gonzales, P. & Ducrest, P. (1938) Recherches serologiques sur les milieux transparents de l'oeil. Proprietes specifiques du cristallin, *C. R. Soc. Biol. 127*, 668–670.

Yamasu, T. & Yoshida, M. (1976) Fine structure of complex ocelli of acubomedusan, *Tamoya bursaria* Haeckel, *Cell Tissue Res. 170*, 325–339.

Zigler, J. S. Jr (1994) in *Principles and practice of ophthalmology* (Albert, D. M. & Jakobiec, F. A., eds) pp. 97–113, W. B. Saunders Company, Philadelphia, London, Toronto, Montreal, Sydney, Tokyo.

Zigler, J. S. Jr & Rao, P. V. (1991) Nucleotide-binding enzyme-crystallins lead to extremely high levels of pyridine nucleotides in the ocular lens, *FASEB J. 5*, 223–225.

Zinovieva, R. D., Tomarev, S. I. & Piatigorsky, J. (1993) Aldehyde dehydrogenase-derived crystallin of squid and octopus. Specialization for lens expression, *J. Biol. Chem. 268*, 11449–11455.

Zuckerkandl, E. (1994) Molecular pathways to parallel evolution: I. Gene nexuses and their morphological correlates, *J. Mol. Evol. 39*, 661–678.

Zuker, C. S. (1994) On the evolution of eyes: would you like it simple or compound? *Science 265*, 742–743.

Eur. J. Biochem. 236, 1–26 (1996)

Review

Apoptosis: molecular regulation of cell death

Annette J. HALE, Christopher A. SMITH, Leslie C. SUTHERLAND, Victoria E. A. STONEMAN, Vanessa L. LONGTHORNE, Aedín C. CULHANE and Gwyn T. WILLIAMS

Biological Sciences Department, Keele University, Staffordshire, UK

(Received 10 August 1995) – EJB 95 1338/0

The field of apoptosis is unusual in several respects. Firstly, its general importance has been widely recognised only in the past few years and its surprising significance is still being evaluated in a number of areas of biology. Secondly, although apoptosis is now accepted as a critical element in the repertoire of potential cellular responses, the picture of the intra-cellular processes involved is probably still incomplete, not just in its details, but also in the basic outline of the process as a whole. It is therefore a very interesting and active area at present and is likely to progress rapidly in the next two or three years.

This review emphasises recent work on the molecular mechanisms of apoptosis and, in particular, on the intracellular interactions which control this process. This latter area is of crucial importance since dysfunction of the normal control machinery is likely to have serious pathological consequences, probably including oncogenesis, autoimmunity and degenerative disease. The genetic analysis of programmed cell death during the development of the nematode *Caenorhabditis elegans* has proved very useful in identifying important events in the cell death programme. Recently defined genetic connections between *C. elegans* cell death and mammalian apoptosis have emphasised the value of this system as a model for cell death in mammalian cells, which, inevitably, is more complex.

The signals inducing apoptosis are very varied and the same signals can induce differentiation and proliferation in other situations. However, some pathways appear to be of particular significance in the control of cell death; recent analysis of the apoptosis induced through the cell-surface Fas receptor has been especially important for immunology.

Two gene families are dealt with in particular detail because of their likely importance in apoptosis control. These are, first, the genes encoding the interleukin-1β-converting enzyme family of cysteine proteases and, second, those related to the proto-oncogene *bcl-2*. Both of these families are homologous to cell death genes in *C. elegans*. In mammalian cells the number of members of both families which have been identified is growing rapidly and considerable effort is being directed towards establishing the roles played by each member and the ways in which they interact to regulate apoptosis.

Other genes with established roles in the regulation of proliferation and differentiation are also important in controlling apoptosis. Several of these are known proto-oncogenes, e.g. *c-myc*, or tumour suppressors, e.g. p53, an observation which is consistent with the importance of defective apoptosis in the development of cancer. Viral manipulation of the apoptosis of host cells frequently involves interactions with these cellular proteins.

Finally, the biochemistry of the closely controlled cellular self-destruction which ensues when the apoptosis programme has been engaged is also very important. The biochemical changes involved in inducing phagocytosis of the apoptotic cell, for example, allow the process to be neatly integrated within the tissues, under physiological conditions. Molecular defects in this area too may have important pathological consequences.

Keywords: apoptosis; oncogenes; tumour suppressors; *bcl-2*; proteases.

Apoptosis (Kerr et al., 1972) is a crucial element in the behaviour of mammalian cells in many different situations. Indeed it is now clear that the complete analysis of any population of mammalian cells must include consideration of the possible importance of active cell death by apoptosis, together with the other obvious controlling processes such as cell differentiation and multiplication. This acceptance of the fundamental importance of apoptosis has come about only recently: less than 10 years ago apoptosis, when it was considered at all, was seen as an esoteric area of very limited importance.

The rather sudden collective recognition of the significance of apoptosis implies that a fundamentally important element of cell biology had previously been neglected. Viewed in a positive light this means that we now have an opportunity to investigate not just the details but the unexplored core mechanisms of a

---

Correspondence to G. T. Williams, Biological Sciences Department, Keele University, Keele, Staffordshire, UK ST5 5BG

Fax: +44 1782 630007.

Abbreviations. EBV, Epstein Barr virus; EGF, epidermal growth factor; GM-CSF, granulocyte macrophage colony stimulating factor; HPV, human papilloma virus; ICE, interleukin-1β-converting enzyme; IL, interleukin; PARP, poly(ADP-ribose) polymerase; PtdInsP_2, phosphatidylinositol 4,5-bisphosphate; PKC, protein kinase C; PTK, protein tyrosine kinase; SV40, simian virus 40; TNF, tumour necrosis factor; TosLysCH_2Cl, *N*-tosyl-L-lysine chloromethane; TosPheCH_2Cl, *N*-tosyl-L-phenylalanine chloromethane.

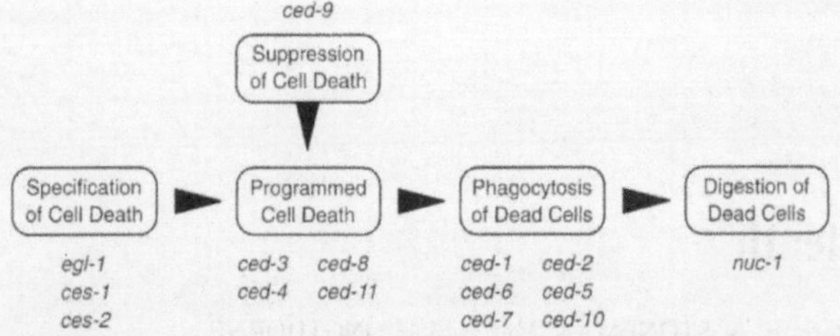

Fig. 1. Genes involved at different stages in developmental programmed cell death in *Caenorhabditis elegans*. The genes involved in specification of cell death determine whether or not particular classes of cells die. Expression of *nuc-1* is required in the phagocyte to degrade the DNA of the dying cell: in mammalian cells, the endonuclease is derived from the dying cell.

central element of cell biology. Almost all the work described in this review has therefore been performed in the past six years and the field is continuing to develop rapidly at the present time.

The integration of apoptosis into cell biology has changed, in some cases revolutionised, many areas (Tomei and Cope, 1991, 1994). Apoptosis is particularly important in the developing nervous system (Raff et al., 1993) and in both the development and the effective functioning of the immune system (reviewed by Golstein et al., 1991; Williams, 1994). Failure or suppression of apoptosis is likely to contribute both to the initial development of cancer and to the appearance of tumour cells resistant to cytotoxic therapy (reviewed by Williams, 1991). Failure of apoptosis may also be important in the development of auto-immune disease (reviewed by Williams, 1994) and, on the other hand, inappropriate induction of apoptosis may be involved in degenerative diseases. Apoptosis dysfunction may also play an important role in the HIV-induced pathology of AIDS (reviewed by Thompson, 1995). The molecular dissection of apoptosis is therefore one of our most important challenges from both theoretical and applied points of view.

Caenorhabditis elegans – a model system for active cell death

The nematode *C. elegans* is a very powerful model system for analysis of the cell and molecular biology of multicellular animals. The embryonic development of *C. elegans* is reproducible and has been precisely mapped. 1090 somatic cells are eventually formed in the adult hermaphrodite and of these 131 undergo programmed cell death (Ellis et al., 1991). The death of these cells is just as predictable as the division and differentiation occurring during development and, since cell death can be observed in the living animals by Nomarski microscopy, analysis of *C. elegans* programmed cell death has produced some very valuable information. This developmentally programmed cell death shows several areas of similarity with apoptosis in mammalian cells and the recently demonstrated genetic connections with apoptosis outlined below make this a particularly useful animal model.

Mutant nematodes have been identified with defects in different parts of the cell death process and this has allowed a genetic pathway of cell death to be produced (Fig. 1). The genes identified can be divided into several groups: genes involved in triggering cell death; those involved in the cell death process itself; those required for engulfment and finally those implicated in the disposal of the cell corpse. This analysis has provided a useful framework for investigation of the molecular mechanisms of apoptosis in mammalian cells, where inevitably the process is more complicated and involves a larger number of genes. Several of the nematode cell death genes have been cloned and sequenced (reviewed by Yuan, 1995). Ced-9, a suppressor of apoptosis, is homologous to the Bcl-2 family of proteins and Ced-3 is homologous to the interleukin-1β-converting enzyme (ICE) family of cysteine proteases. The *ced-4* gene has also been cloned and sequenced (Yuan and Horvitz, 1992). The deduced amino acid sequence of Ced-4 contains two potential calcium-binding domains but shows no significant sequence similarity to known mammalian proteins. It may be that Ced-4 and some of the other *C. elegans* proteins are related to unknown families of mammalian proteins. The *C. elegans* pathway is therefore particularly useful in highlighting the basic elements of the complex interacting pathways involved in apoptosis in mammalian cells. Predictably, however, it is not a perfect model; for example, no *C. elegans* equivalent of Reaper, a protein which is very important in programmed cell death in *Drosophila* (White et al., 1994b), has been identified.

Signalling in apoptosis

A cell will undergo apoptosis as a result of information received from its environment interpreted in the context of internal information, such as its cell type, state of maturity and developmental history (reviewed by Williams and Smith, 1993). The external information triggering apoptosis can take many forms, e.g. the appearance or disappearance of hormones or cytokines, or a change in direct intercellular interactions.

Since a particular external stimulus is only part of the information influencing the decision between self-destruction and survival, such stimuli are not, in general, exclusively involved in the control of apoptosis. Similarly, the intracellular signals involved in induction of apoptosis are often involved in promotion of proliferation or differentiation in other cellular contexts, although certain intracellular signalling pathways, e.g. the Fas-mediated pathway (Fas is also known as APO-1 or CD95), are of particular importance in controlling apoptosis in certain cell types.

The involvement of key molecules in the induction of both apoptosis and proliferation is a recurring theme (see, for example, sections below dealing with protein kinase C, ceramide, *c-myc* and p53), which reflects in part the adaptation of many signalling pathways to the control of different responses in different cell types and under different conditions.

Cell-surface receptor-mediated mechanisms which control apoptosis often act through a signal transduction system which involves the stimulation of the receptor, the activation of protein kinase/phosphatase cascades, and the release of second messengers to upregulate or suppress the transcription of specific genes. Evidence is accumulating that these signalling pathways can intersect, a process known as crosstalk, and therefore greatly alter a cell's response to a given stimulus.

Although the great variety of external signals which can control apoptosis means that many signalling systems can be involved, there is considerable evidence to suggest that the signalling pathways converge to one, or very few, common final pathways. The broad spectrum of apoptosis systems affected by the *bcl-2* family (see below), for example, is consistent with such convergence of signalling pathways at the point where the corresponding gene products act. There may also be some rather more surprising areas of convergence; Guenette and Tenniswood (1995) have suggested that apoptosis induced in the prostate by androgen withdrawal could be due to blocking of the insulin-like growth factor required for cell survival through the production of specific binding proteins for the cytokine.

Fas (APO-1, CD95). Particular interest has focused on the control of cell death and survival through engaging members of the Fas/TNF receptor 1 (TNFR1) family of cell-surface receptors (Nagata and Golstein, 1995), and these provide a useful model of apoptosis controlled by cell-surface receptors. Ligation of some of the members of this family promotes cell survival whereas ligation of other members, particularly Fas and TNFR1, is largely, but not exclusively, associated with the induction of apoptosis. Induction of apoptosis through aggregation of cell-surface Fas, e.g. by binding to Fas-ligand, is particularly important in regulation of the immune system (reviewed by Nagata and Golstein, 1995). It appears that apoptosis induced by engaging the T-cell receptor on T-lymphocyte cell lines (e.g. Brunner et al., 1995) is mediated through interaction between Fas-ligand and Fas. This system may therefore be important in down-regulating the immune response and/or eliminating self-reactive mature lymphocytes.

Recent studies on the primary structure of Fas and related proteins has highlighted some intriguing observations. TNFR1 and Fas both contain a cytoplasmic region of 60–70 amino acids which is important in inducing apoptosis and has been termed the 'death domain'; Golstein et al. (1995) noted that this region is homologous to the protein Reaper (White et al., 1994b) which is essential for most programmed cell death in *Drosophila*. This suggested that Fas, TNFR, and potentially other as yet undiscovered proteins, were produced in evolution after gene fusion of exons encoding an inducer of cell death related to Reaper with those encoding a surface receptor. The resulting hybrid protein would then allow induction of apoptosis by external stimuli by linking them to the pre-existing cell death pathway. The central role played by Reaper in programmed cell death in *Drosophila* suggests that dissection of this pathway may prove crucial to understanding apoptosis in mammalian cells.

Several proteins which may form part of the intracellular pathway induced by Fas have been identified recently. The genes encoding three proteins, MORT1/FADD (Boldin et al., 1995; Chinnaiyan et al., 1995), TRADD (Hsu et al., 1995) and RIP (Stanger et al., 1995) have been cloned using the yeast two-hybrid technique which allows identification of proteins which associate physically with a given 'bait' polypeptide (in this case the Fas cytoplasmic domain). All three predicted gene products contain the death domain (reviewed by Cleveland and Ihle, 1995) and it is likely that these domains associate with the death domains of Fas and/or TNFR1. Over-expression of any of the three genes in transient expression systems induces apoptosis. The deduced amino acid sequence of RIP suggests that it may have protein kinase activity but otherwise there are few clues to the mechanism of action of these proteins. This will undoubtedly be a focus of considerable attention in the immediate future. Other elements of the Fas-stimulated pathway, e.g. the role of cysteine proteases of the ICE family, are discussed below.

Protein tyrosine kinases. The protein tyrosine kinases (PTKs) play a critical role in the transmission of many signals from cell-surface receptors to the nucleus. These may be growth factor receptors which possess intrinsic PTK activity or receptors which lack intrinsic kinase activity but which can recruit and/or activate intracellular PTKs in response to ligand stimulation. There are many review articles concerning intracellular signalling via PTKs, including recent reviews by Erpel and Courtneidge (1995) and Ziemiecki et al. (1994).

Many cytokine receptors, e.g. receptors for the interferons and the majority of those for haematopoietic growth factors, do not have intrinsic tyrosine kinase activity. Oligomerization of the receptors follows ligand binding, and PTKs are recruited to the intracellular domain of the receptors to activate the signalling cascade of phosphorylation. PTKs involved in this cascade include the Src family of PTKs and the Janus kinase family (Jak); the latter having been associated with a number of cytokine signalling pathways controlling apoptosis e.g. interferons, interleukin-3 (IL-3), leukaemia inhibitory factor (reviewed by Ziemiecki et al., 1994) and granulocyte/macrophage-colony-stimulating factor (GM-CSF) (Yousefi et al., 1994). More recently, the investigations of Barber and D'Andrea (1994) have demonstrated that erythropoietin and IL-2, whose receptors are related, act through tyrosine phophorylation of two distinct Jak kinases, Jak2 and Jak1 respectively.

There is increasing experimental evidence implicating PTK in signalling pathways controlling apoptosis. For example: ionising radiation promoted PTK activation that was necessary for apoptosis in B cells (Uckun et al., 1992); and cells in which the active PTK Abl (a member of Src family) was overproduced were resistant both to apoptosis induced by growth-factor withdrawal and to apoptosis induced by chemotherapeutic agents (Evans et al., 1993). The ability of the chronic myelogenous leukaemia cell line K562 to undergo apoptosis was restored by antisense *abl* expression (McGahon et al., 1994). More recently, Owen-Lynch et al. (1995) studied a human leukaemia in which continuous functioning of v-Abl was required for IL-3 independence, and for phosphorylation of phospholipase C (PLC) and the Ras regulatory protein p120 GTPase-activating protein. Yao and Cooper (1995) have also demonstrated the requirement for phosphatidylinositol-3-kinase in the prevention of apoptosis by nerve growth factor and Somma et al. (1995) have shown that stimulation of apoptosis by the engagement of the T-cell receptor involves the PTK $p56^{lck}$.

Other kinases. As well as tyrosine kinases, serine/threonine kinases have been implicated in apoptosis. For instance, the serine/threonine kinase $p34^{cdc2}$ controls cell entry into mitosis by complexing with the cyclins A and B and initiates dissolution of the nuclear membrane and chromatin condensation, changes which are also observed in apoptosis. The protease Fragmentin-2, found in the granules of natural killer cells, prematurely activated (but did not cleave) $p34^{cdc2}$ kinase and dephosphorylation of the kinase was seen at the initiation of apoptosis. The DNA fragmentation characteristic of apoptosis could be prevented by blocking $p34^{cdc2}$ kinase activity (Shi et al., 1994).

The Ras signalling pathway. Ras is a member of a superfamily containing more than 50 small GTPases, functioning as GDP/GTP-regulated switches. They are implicated in the regulation of cellular proliferation, differentiation, intracellular transportation processes, and cytoskeletal organisation. The three Ras proteins H-, K- and N-Ras, have a well studied pivotal role in signal transduction, linking receptor and non-receptor tyrosine kinases to downstream serine/threonine kinases, including the mitogen-activated protein (MAP) kinases. Ras itself is immediately upstream of the protein kinase encoded by the oncogene *c-raf-1* (the regulation and function of Ras and Ras-related proteins have been summarised by Khosravi-Far and Der, 1994, and Khosravi-Far et al., 1995).

There are now indications that the Ras/Raf/MAP kinase cascade has a role in the inhibition of apoptosis. For instance, Cleveland et al. (1994) have demonstrated that the Raf-1 kinase promoted the proliferation of IL-3-dependent myeloid cells and suppressed apoptosis induced by IL-3 withdrawal; and Owen-Lynch et al. (1995) observed that IL-3 stimulated the Ras activation of MAP kinase via the Src homology-domain-containing (SHC) protein and the GTPase-activating protein through their phosphorylation.

Further studies have established roles for Ras-related proteins in the signalling processes relevant to the inhibition of apoptosis, often concentrating on the regulation of the Bcl-2 protein and the consequences of this regulation (Wang et al., 1994a, 1995a).

Recent studies by Kinoshita et al. (1995a,b) have implicated Rassignalling in the regulation of *bcl-2* in the inhibition of haematopoietic cell apoptosis following growth factor deprivation. The induction of *ras* mRNA (mutated at codon 12 to activate the *ras* oncogene) using an inducible expression vector restored the levels of *bcl-2* mRNA following IL-3 deprivation, and also blocked the degradation of the Bcl-2 protein that had previously been seen following IL-3 deprivation (Kinoshita et al., 1995b). Ras^{val12} could induce mRNA encoding the Bcl-2-related protein Bcl-x_L, though neither the induction of Ras^{val12} nor IL-3 deprivation affected the level of mRNA of another Bcl-2 related protein, Bax. As the induction of *ras* mRNA preceded that of the *bcl-2* and *bcl-x_L* mRNA, and was sustained at a level which correlated with that for Bcl-2 and Bcl-x_L proteins, Kinoshita et al. (1995b) concluded that this mechanism of regulation of Bcl-2/Bcl-x_L by Ras appeared to be fundamental to IL-3/GM-CSF inhibition of apoptosis. The central role of the Bcl-2 protein family is discussed in detail in a later section.

Protein kinase C and calcium. Surface receptors are coupled to the Ca^{2+} signalling pathway by G proteins and PTKs, which activate phospholipases C (PLC) and D (PLD) by the hydrolysis of phosphatidylinositol 4,5-bisphosphate (PtdInsP_2) producing inositol trisphosphate and diacylglycerol. Inositol trisphosphate promotes the release of Ca^{2+} from the endoplasmic reticulum and Ca^{2+} influx through the plasma membrane and diacylglycerol activates the serine/threonine kinase and protein kinase C (PKC). In thymocytes, chemicals that stimulate increases in PLC and intracellular calcium concentration $[Ca^{2+}]_i$ induce endonuclease activation and cell death (e.g. Smith et al., 1989). It is known that the PKC signalling pathway can interact with other signalling pathways. This is illustrated by observations such as those of Kyriakis et al. where PKC activation results in the rapid phosphorylation and activation of c-Raf kinase, which in turn activates the protein kinase cascade by phosphorylating and activating MAP kinases (Kyriakis et al., 1992); and the recent publication of Ojeda et al. (1995) providing evidence for the interaction of PKC and the protein kinase A signal transduction pathways. This crosstalk between signalling pathways could account for apparent contradictions concerning the roles of key molecules, such as PKC, in the induction of apoptosis and proliferation.

The above observations, plus the fact that the PtdInsP_2, phosphatidylcholine and sphingomyelin metabolic pathways (see below) all activate PKC, implicate PKC in the regulation of apoptosis in many cells. The experimental evidence which suggests that PKC is involved in apoptosis as well as in cell differentiation and proliferation is largely based on the use of phorbol esters as activators of PKC (reviewed in McConkey and Orrenius, 1994). Incubation of thymocytes with phorbol esters has been reported both to induce apoptosis and to block apoptosis. Similar apparent contradictions are seen with the frequently used inhibitors of PKC, such as staurosporine, which can induce apoptosis (Jacobson et al., 1993) and protect against phorbol-ester-induced apoptosis (Munn et al., 1995). Notably, apoptosis induced by ceramide (Hannun and Obeid, 1995) (see below) and nitric oxide (Messmer et al., 1995) is inhibited by PKC activators such as 12-tetradecanoyl-phorbol 13-acetate. These observations may be explained, at least in part, by the lack of specificity of the reagents used and by the complex interactions of several pathways regulating apoptosis.

In natural killer cells, stimulation of the IgG receptor FcγRIII-A on IL-2-activated natural killers, stimulated the PTK $p56^{lck}$, increased Ca^{2+} release from intracellular stores, and induced apoptosis. Further investigation demonstrated that this increase in $[Ca^{2+}]_i$ was necessary for other Fcγ-receptor-mediated signals, including transcription of cytokine genes (Azzoni et al., 1995).

However, there are several reports of failure to detect alterations in $[Ca^{2+}]_i$. Beaver and Waring (1994) concluded that there was no correlation between an early increase in cytosolic calcium ion concentration and apoptosis in thymocytes treated with gliotoxin (which may act through Ras) and thapsigargin (an inhibitor of the endoplasmic reticulum Ca^{2+} ATPase that did not induce apoptosis via a direct effect on $[Ca^{2+}]_i$). Incubation with dexamethasone (McConkey et al., 1989), showed no significant rise in $[Ca^{2+}]_i$ above the control. These studies do not rule out more subtle changes, such as redistribution of calcium within the cell, which could nevertheless have important consequences.

Ceramide. Ceramide can play a crucial role in the induction of proliferation and apoptosis in different situations (Kolesnick and Fuks, 1995) and its role as a signalling molecule has been reviewed recently by Hannun and Obeid (1995), Pushkareva et al. (1995) and by Wright and Kolesnick (1995).

All sphingolipids contain ceramide as a hydrophilic component. Ceramide is a second messenger, which is released by the hydrolysis of membrane sphingolipids (especially sphingomyelin) and which activates a protein kinase cascade. Ceramide inhibits growth and induces differentiation of leukaemia cells, modulates protein phosphorylation and regulates gene transcription. The IL-1, interferon-γ and tumor-necrosis factor-α (TNF-α) cytokine-stimulated pathways are likely to involve ceramide (reviewed by Hannun and Obeid, 1995); indeed, the cytotoxicity of ceramide mimics that of TNF-α (Jarvis et al., 1994). It is possible that ceramide has many roles in signalling, as discussed in the review of Hannun and Obeid (1995).

More recent work has detailed further aspects of ceramide signalling. For instance, the human promonocytic U937 cell line has been treated with C2-ceramide, which caused characteristic DNA fragmentation indistinguishable from that caused by TNF-α (Ji et al., 1995). This C2-ceramide-induced DNA fragmentation, cell shrinkage and condensation of the nucleus could be inhibited by Zn^{2+} and 12-tetradecanoyl-phorbol 13-acetate (suggesting the involvement of a Ca^{2+}-dependent endonuclease). Furthermore, insufficient production of ceramide has been implicated in resistance to certain types of apoptosis (Gottschalk et al., 1995). The study of Wong et al. (1995) indicated that ceramide inhibited superoxide release in neutrophils by activation of a type 2A protein phosphatase. The sphingolipid sphingosine has also been implicated in growth factor signalling by epidermal growth factor (EGF) in epidermal carcinoma, where this lipid enhanced the EGF-stimulated calcium influx in a dose-dependent manner over a range of concentrations of EGF (Hudson et al., 1994).

cAMP. cAMP acts mostly through cAMP-dependent protein kinase (Spaulding, 1993). Factors that elevate cAMP in T-lymphocytes stimulate DNA fragmentation via calcium-dependent protein kinase, though there are examples where cAMP can inhibit apoptosis (reviewed by McConkey and Orrenius, 1994); therefore elevations in cAMP, as with other components of signalling pathways, can have different effects depending on the type of cell.

An interesting study of apoptosis-defective mutants has indicated that apoptosis induced by cAMP, dexamethasone and ionising radiation merge into a common pathway prior to the point where Bcl-2 exerts its protective effect (Flomerfelt and Miesfeld, 1994). The existence of such a final common pathway, as discussed above, is an attractive possibility because of the many common features of apoptosis induced by very different stimuli.

Oxygen radicals and oxidative stress. Oxidative stress is important both in cytotoxicity and in cellular activation. The role of oxygen radicals and oxidation in apoptosis, including the observation that hydrogen peroxide can activate the transcription factor NF-κB, has been reviewed by McConkey and Orrenius (1994). It has been suggested that reactive oxygen species could be generally important in apoptosis (Hockenbery et al., 1993), but some recent observations are not consistent with this suggestion.

Greenlund et al. (1995) induced apoptosis in neurons by nerve growth factor (NGF) deprivation, where there was a transient increase (with a peak at 3 h after removal of the factor) in reactive oxygen species, including superoxide. Superoxide dismutase protein protected the cells against apoptosis. The importance of oxidative stress in induction of apoptosis has also been suggested by Slater et al. (1995), who reported that thymocyte apoptosis was inhibited by chemicals with antioxidant properties.

However, there is now direct evidence from studies performed in highly hypoxic conditions that oxygen radicals are not generally required for apoptosis (Jacobson and Raff, 1995; Muschel et al., 1995). The former investigators concluded that apoptosis induced by withdrawal of growth factor did not require reactive oxygen species, nor did the protection from apoptosis provided by Bcl-2 depend upon reactive oxygen species. The apparently contradictory observations in this area may be partly explained by other effects produced by the compounds used as anti-oxidants.

Nitric oxide is also important in inducing apoptosis in some situations (Mannick et al., 1994; Messmer et al., 1995): Genaro et al. (1995) summarised the effects of nitric oxide, including the possibility that it can amplify calcium-induced gene transcription in neural cells. They suggested the interesting possibility that there may be a pathway linking nitric oxide signalling with *bcl-2* expression, as nitric oxide release in B lymphocytes prevented the decrease in *bcl-2* expression which occurred on antigen-induced apoptosis, apparently by increasing cGMP concentrations.

Proteases in apoptosis

The structure and activity of interleukin-1β-converting enzyme. The observation that the *C. elegans* cell death protein Ced-3 has a significant sequence similarity (29% amino acid identity) to the human interleukin-1β converting enzyme (ICE) suggested that ICE or an ICE-like protease might play a role in the control of programmed cell death in vertebrates (Yuan et al., 1993). The most highly conserved regions, amino acids 246–360 of the Ced-3 protein and 166–287 of human ICE, share 43% sequence identity. ICE is a cysteine protease with Cys285 in the active site of the enzyme. It was the first cysteine protease identified with specificity for aspartate in the P_1 position (Fig. 2). The amino acid pentapeptide QACRG, containing Cys285, is the longest peptide conserved between murine and human ICE and Ced-3 proteins.

ICE was first discovered in the cytosol of monocytes and monocyte-like cell lines (Kostura et al., 1989) and is a protease that cleaves the precursor of interleukin-1β (proIL-1β) from its inactive 31-kDa precursor to its active 17.5-kDa form. ProIL-1β is cleaved at Asp116-Ala117 and at Asp27-Gly28, cleavage at each site being dependent upon Asp in the P_1 position. IL-1β is a cytokine involved in mediating inflammation, septic shock, wound healing, haematopoiesis and the growth of certain leukemias (Dinarello, 1991).

ICE exists as a tetramer or heterodimer consisting of equal numbers of 10-kDa and 20-kDa subunits (Wilson et al., 1994). The subunits are derived from the 45-kDa proenzyme by cleavage at four sites where Asp is in the P_1 position. The 45-kDa proenzyme is capable of autoprocessing to produce the ICE subunits (Thornberry et al., 1992). The 20-kDa subunit contains the catalytic Cys residue, but both the 10-kDa and 20-kDa subunits are required for catalytic activity because both subunits contribute to the selective recognition of aspartic acid. Alternative splicing of ICE mRNA may provide a means by which ICE activity is controlled and regulated in different tissues (Alnemri et al., 1995).

The gene encoding ICE has been mapped to chromosome 11 band q22.2-q22.3 (Cerretti et al., 1994). This site is frequently involved in rearrangement in human cancers, including a number of leukemias and lymphomas (Cerretti et al., 1992). Genetic defects producing ataxia telangiectasia have been mapped to this region (Savitsky et al., 1995). Ataxia telangiectasia is a progressive disease featuring neurodegeneration, immunodeficiency, chromosomal instability, radiation hypersensitivity and an increased predisposition to cancer particularly lymphomas and leukemias. Since it has been suggested that it may be caused by an inability to correctly regulate apoptosis (Taylor et al., 1994), the possibility that ICE dysfunction could be involved in this disease merits further investigation.

Substrates and inhibitors of ICE. ICE has a strong preference for cleavage at a peptide sequence with an Asp residue in the P_1 position (Fig. 2). The P_2 His residue may be replaced by other hydrophobic amino acids such as Ala (Thornberry et al., 1992). The enzyme does not require an amino acid to the right of the cleavage site, a methylamine group being sufficient for cleavage to occur. The best substrate yet identified for the enzyme is Ac-Tyr-Val-Ala-Asp-NH-CH_3. The most potent and selective inhibitors of ICE also contain the tetrapeptide sequence, Tyr-Val-Ala-Asp. Peptide aldehydes such as Ac-Tyr-Val-Ala-Asp-CHO, peptide nitriles and ketones are reversible competitive inhibitors of ICE. The tetrapeptide Ac-Tyr-Val-Ala-Asp-CHO is the most potent reversible inhibitor described for ICE. Peptide α-substituted ketones with the structure peptide-CO-CH_2-X where X is a halogen, diazonium ion, or carboxylate leaving group, such as Ac-Tyr-Val-Ala-Asp-CH_2-OCOPh, act as irreversible inhibitors of ICE (Thornberry and Molineaux, 1995). Affinity labeling of THP.1 monocytic cytosol with Ac-Tyr-Val-Ala-Asp-CH_2-OCOPh resulted in labeling of only ICE, demonstrating its specificity (Thornberry et al., 1994). Tripeptide inhibitors can act as potent inhibitors of human ICE although they may also inhibit other related proteases (Graybill et al., 1994).

ICE is inactivated by the protease inhibitors *N*-tosyl-L-phenylalanine chloromethane and 3,4-dichloroisocoumarin which are usually considered to be selective for serine proteases (Wilson et al., 1994). ICE is also inhibited by CrmA, a 38-kDa protein encoded by the cowpox virus (Ray et al., 1992). This is important since CrmA inhibits apoptosis in several situations, e.g. on treatment with anti-Fas antibodies or TNF. The amino acid sequence of CrmA is similar to those of members of the serpin superfamily which usually inhibit serine proteases.

The ICE family. There is an emerging family of proteins showing sequence similarity to ICE. The protein with the highest sequence similarity so far discovered is TX (Faucheu et al., 1995), identified using a probe corresponding to exon 6 of ICE.

The TX open reading frame encodes a 377-amino-acid protein, 27 amino acids shorter than the p45 ICE precursor. TX is similar to the ICE 20-kDa and 10-kDa subunits (59% and 64% amino acid identity, respectively). The QACRG sequence of ICE is conserved and its cataytlic activity is dependent on the Cys residue. TX does not cleave proIL-1β but it does have proteolytic activity against ICE and itself, and transfection of TX into COS cells induces apoptosis. TX mRNA was expressed in all of the adult human tissues tested, excluding brain. Interestingly ICE and TX can form hetero-oligomers when coexpressed in the same cells and the ICE-TX hybrid is incapable of cleaving proIL-1β, suggesting a possible role for TX in negative regulation of ICE activity (Gu et al., 1995). TX has recently been independently identified by two groups and named ICE_{rel}-II (Munday et al., 1995) and Ich-2 (Kamens et al., 1995). Munday et al. cloned two proteases, ICE_{rel}-II and ICE_{rel}-III, from human monocytic cells. ICE_{rel}-II is identical to TX whereas ICE_{rel}-III is a further member of the ICE family of proteases with a 74% sequence identity to ICE_{rel}-II. Transfection of the full-length clones of ICE_{rel}-II and ICE_{rel}-III into Sf9 insect cells did not result in cleavage into the active form of the proteases and did not result in apoptosis. However, transfection of a truncated form, with the pro-domain removed, into HeLa fibroblast cells did induce apoptosis. Ich-2 was isolated using a probe containing the entire human ICE coding sequence (Kamens et al., 1995). In contrast to the results of Munday et al., transfection with either the full-length *ich-2* coding sequence or a truncated form induced apoptosis in Sf9 cells.

Nedd2 was first identified as a gene which is highly expressed during the embryonic development of mouse brain and down-regulated in adult brain (Kumar et al., 1992). Overexpression of the *nedd2* gene results in apoptosis in mouse fibroblast and neuroblastoma cell lines (Kumar et al., 1994). The human homologue *Ich-1* encodes two proteins, Ich-1_L and Ich-1_S, which are produced by alternative splicing of the precursor RNA (Wang et al., 1994b). Ich-1_L is a 435-amino-acid protein with sequence similarity to both the 20-kDa and 10-kDa ICE subunits (27% identity) and the entire Ced-3 protein (28% identity). Translation of the second mRNA species results in a 312-amino-acid protein, Ich-1_S, that terminates 21 amino acids after the pentapeptide QACRG of Ich-1_L. Overexpression of Ich-1_L induces apoptosis in Rat-1 fibroblast cells whereas overexpression of Ich-1_S suppresses Rat-1 cell death induced by serum withdrawal. Thus, alternative splicing of Ich-1 results in the production of a positive or a negative regulator of apoptosis. Cell death induced by overexpression of Ich-1_L is only weakly inhibited by CrmA in comparison with the inhibition of ICE-induced cell death by CrmA.

The U1 small ribonucleoprotein particle is essential for the splicing of precursor mRNA. A 70-kDa U1-specific protein is specifically cleaved in apoptotic cells to a 40-kDa protein (Casciola-Rosen et al., 1994). The protease which cleaves it has inhibition characteristics similar to ICE with non-specific protease inhibitors. Four sites in the 70-kDa U1 protein satisfy the ICE consensus requirements, cleavage at one of which would result in the production of a 40-kDa product. However, a tetrapeptide ICE inhibitor did not inhibit the cleavage of the 70-kDa U1 protein and purified ICE did not cleave the protein, implicating a different ICE-like protease.

The apoptosis-inducing properties of the ICE-like proteins described above have only been demonstrated by their overexpression in cultured cells. Whether these proteins will induce apoptosis at their physiological levels has yet to be determined. The physiological substrates of TX and Ich-1 are unknown. The cleavage of proIL-1β by ICE to produce mature IL-1β may serve to alert the immune system that the cell is destroying itself in response to a viral threat. However, IL-1β itself does not appear to play a crucial role in induction of apoptosis. ICE mRNA has also been detected in cell types that do not produce IL-1β (Black et al., 1989), and overexpression of ICE induces apoptosis in rat fibroblast cells which do not express proIL-1β (Miura et al., 1993). Therefore ICE probably has an additional as yet unidentified substrate which functions in the control of apoptosis in some circumstances.

```
...- P4 - P3 - P2 - P1    - P1' - P2'- P3' - P4' -...
                      116      117
...- Tyr - Val -His - Asp  -Ala -Pro - Val - Arg -...  pro-IL-1β
                      216      217
...-Asp- Gly - Val - Asp   -Gln - Val - Ala - Leu -...  PARP
                          ↑
                     cleavage site
```

Fig. 2. The ICE cleavage site of pro-IL-1β and the CPP32 cleavage site of human PARP.

Recent work has shown that a protease which can cleave poly(ADP-ribose) polymerase (PARP) plays an important role in apoptosis. PARP is involved in DNA repair and in the supervision of genome structure and integrity in stressed cells (Nicholson et al., 1995). In the presence of DNA strand breaks, PARP transfers ADP-ribose from NAD^+ to nuclear proteins. A protease resembling ICE (prICE) which cleaves PARP was first identified in a cell-free system (Lazebnik et al., 1994). Isolated HeLa nuclei were added to cytoplasmic extracts derived from chicken cells committed to apoptosis. A protease present in the cytoplasmic extract cleaved the nuclear enzyme PARP and reproduced the nuclear and DNA fragmentation observed in apoptosis. The cleavage of PARP to an 85-kDa fragment is an early event in apoptosis and is observed in virtually every form of programmed cell death examined. The cleavage of PARP results in the separation of its DNA binding motifs from the poly-(ADP-ribos)ylating catalytic domains. The Ca^{2+}/Mg^{2+}-dependent endonuclease implicated in the internucleosomal DNA cleavage which occurs during apoptosis is negatively regulated by poly(ADP-ribos)ylation; therefore loss of PARP function could result in the activation of the nuclease in dying cells. The protease cleaves between Asp216 and Gly217 of PARP at a tetrapeptide sequence similar to one of two ICE sites in proIL-β (Fig. 2). The protease activity was inhibited by a specific tetrapeptide ICE inhibitor. However, purified ICE did not cleave PARP, and prICE cannot cleave proIL-1β; therefore another ICE-like enzyme is indicated. prICE could lie near the start of an apoptotic pathway since cleavage of PARP is an early event in apoptosis.

A protease which cleaves PARP has been independently identified by three different research groups. Fernandes-Alnemri et al. (1994) and Tewari et al. (1995a) searched the GenBank-expressed sequence tags for entries showing sequence similarity to Ced-3 and ICE. Fernandes-Alnemri et al. (1994) isolated several cDNA clones from a human Jurkat T lymphocyte cDNA library. Isoform-α of CPP32 contains an open reading frame of 831 bp which encodes a 32-kDa 277-amino-acid protein. Isoform-β contains a deletion of nucleotides in the noncoding region and two base pair substitutions in the coding region resulting in a single amino acid substitution. Tewari et al. (1995a) identified a protein identical to CPP32β which they named Yama after the Hindu god of death. CPP32 has a 35% amino acid identity and a 58% similarity to Ced-3, a 30% identity and 53% similarity to Nedd2 and a 30% identity and 53% similarity to ICE. The highest degree of sequence similarity is in the region of the highly conserved pentapeptide QACRG. The 32-kDa precursor of Yama is proteolytically cleaved to form the subunits of the active enzyme, a 17-kDa and a 12-kDa subunit. This cleavage can be achieved by ICE. The subunits form an active complex which cleaves PARP to the 85-kDa form. Purified ICE did not cleave PARP but the cleavage of PARP was inhibited by CrmA suggesting that CrmA may block apoptosis by the inhibition of either Yama or ICE. Overexpression of the two subunits of CPP32 results in apoptosis in Sf9 insect cells (Fernandes-Alnemri et al., 1994). CPP32 mRNA was detected in all of the human cell lines tested. It is highly expressed in cell lines of the haematopoietic lineage and in cell lines of brain and embryonic origin. CPP32 can be cleaved to its active form by granzyme B, a protease present in cytotoxic T lymphocyte granules which induces apoptosis (Darmon et al., 1995).

Nicholson et al. (1995) determined to isolate the protease responsible for the cleavage of PARP from the cytosol of THP-1 cells. A tetrapeptide aldehyde inhibitor containing the amino acid sequence of the PARP cleavage site was used as an affinity ligand to purify the protein. The purified PARP cleavage enzyme was composed of 12-kDa and 17-kDa polypeptides. Sequencing of the protein identified the precursor of the subunits as CPP32β. CPP32β is cleaved between Asp28-Ser29 and Asp175-Ser176 to yield the 12-kDa and 17-kDa subunits of the enzyme. The active form of CPP32, which was named apopain, was able to cleave PARP. However, CrmA was found to be a weak inhibitor in contrast to the findings of Tewari et al. (1995a). The cytosolic fraction of apoptotic osteosarcoma cells was able to induce apoptosis in healthy nuclei and this activity was inhibited by a specific tetrapeptide inhibitor of apopain but not by an ICE inhibitor. CPP32 alone was not able to induce apoptotic changes in healthy nuclei indicating that CPP32 is necessary but not sufficient for apoptosis, or that a critical substrate of CPP32 is present only in the cytosol. CPP32 may be part of a proteolytic cascade whereby apoptotic signals lead to the activation of a protease which cleaves CPP32 to its active form. The protease responsible could be ICE or another member of the ICE family. CPP32 then cleaves PARP, and probably other substrates whose proteolysis culminates in apoptosis. PARP-null mice develop normally and therefore the cleavage of PARP cannot be essential in the induction of apoptosis (Wang et al., 1995b). ICE is also not required during normal development in mice suggesting that there is some redundancy in the genes involved in the control of apoptosis by the ICE family (Kuida et al., 1995).

A further mammalian gene related to Ced-3, Mch-2, has been cloned from human Jurkat T lymphocytes using degenerate oligonucleotides in a PCR approach (Fernandes-Alnemri et al., 1995). The α transcript of *mch-2* encodes the full-length 34-kDa Mch-2 protein. The full-length Mch-2α protein shows the highest sequence similarity to CPP32 (38% amino acid identity, 56% similarity). It is more related to Ced-3 than to the other members of the ICE family (35% amino acid identity, 56% similarity). Mch-2 can also cleave PARP and overexpression of *mch-2α* induces apoptosis in Sf9 insect cells.

The involvement of ICE in Fas- and TNF-mediated apoptosis. ICE, or an ICE-like protease, appears to play a role in Fas-mediated and TNF-induced apoptosis. An antibody to the Fas receptor induces apoptosis in normal thymocytes which is inhibited by a tetrapeptide ICE inhibitor. In addition, anti-Fas antibody fails to induce apoptosis in ICE-deficient thymocytes obtained from ICE-knockout mice (Kuida et al., 1995).

The involvement of an ICE-like protease in Fas-mediated apoptosis has been demonstrated in a number of cell lines including mouse W4 and human Jurkat cell lines (Enari et al., 1995), L929-APO-1 and B-lymphoblastoid SKW 6.4 cell lines (Los et al., 1995) and the human MCF-7 breast carcinoma and the B-cell lymphoma cell line BJAB (Tewari and Dixit, 1995). Tetrapeptide ICE inhibitors prevented anti-Fas-induced apoptosis in the W4, Jurkat, L929-APO-1 and SKW 6.4 cells. Overexpression of ICE by microinjection of L929-APO-1 cells increased apoptosis induced by anti-Fas. Expression of CrmA inhibited anti-Fas-induced apoptosis in MCF-7, L929-APO-1 and BJAB cells, or TNF-induced apoptosis in MCF-7 cells. Transient expression of an antisense ICE construct also inhibited anti-Fas-induced apoptosis in L929-APO-1 cells, suggesting that ICE itself may be involved.

The mechanism required for anti-Fas-induced apoptosis is present in cells which do not normally express Fas. Rat-2 fibroblast cell lines transfected with human Fas were killed by anti-Fas or TNF in the presence of actinomycin D (Enari et al., 1995). Co-transfection and expression of the ICE inhibitor CrmA protected the cells from both anti-Fas and TNF-induced cell death.

The involvement of ICE family proteases in apoptosis induced by other mechanisms. A role for ICE in apoptosis has been suggested in neuronal cell death induced by nerve growth factor withdrawal (Gagliardini et al., 1994); in apoptosis induced by growth factor withdrawal in cultured mammary epithelial cells (Boudreau et al., 1995); in cytotoxic T-lymphocyte-induced macrophage apoptosis (Hogquist et al., 1991); and in etoposide-induced U937 apoptosis (Mashima et al., 1995). Microinjection of ICE-induced apoptosis of chicken dorsal root ganglion neurons (Gagliardini et al., 1994). Microinjection of the *crmA* gene prevented neuronal cell death induced by deprivation of nerve growth factor. Transfection of mammary epithelial cells with *crmA* or treatment with an ICE inhibitor reduced the incidence of apoptosis in cultured epithelial cells (Boudreau et al., 1995). When macrophages are induced to undergo apoptosis they release mature IL-1β, yet when they are killed by a non-apoptotic mechanism they release only pro-IL-1β (Hogquist et al., 1991), suggesting that ICE or an ICE-like protease is activated when apoptosis is induced in macrophages. A protease inhibitor which preferentially inhibits ICE and ICE-like proteases blocks apoptosis of human myeloid leukemia U937 cells induced by TNF and anti-Fas antibody as well as that induced by etoposide and camptothecin (Mashima et al., 1995). Overexpression of the murine ICE gene or the *C. elegans ced-3* gene causes Rat-1 fibroblast cells to undergo apoptosis (Miura et al., 1993). The cell death caused by murine ICE is suppressed by overexpression of the *crmA* gene.

The particular importance of ICE itself in programmed cell death was investigated by the production of ICE knockout mice (Kuida et al., 1995). The mice displayed an apparently normal phenotype for the first 16 weeks of life, indicating that ICE is not required during development, and that normal negative selection of autoreactive T cells appears to occur in the absence of ICE. However, other ICE-like proteases may be involved in the control of programmed cell death in the developing mouse. The lack of ICE in isolated ICE-deficient thymocytes, although conferring resistance to anti-Fas antibodies, has no effect on apoptosis induced by glucocorticoid, aging or ionising radiation (Kuida et al., 1995; Li et al., 1995), nor on ATP-induced macrophage apoptosis (Li et al., 1995) suggesting that ICE itself does not play an essential role in apoptosis in general.

Bcl-2 and ICE probably affect the same apoptosis pathway. Cell death of Rat-1 fibroblast cells caused by ICE overexpression can be suppressed by overexpression of *bcl-2*, suggesting either that ICE lies upstream of Bcl-2 or that they act at the same stage (Miura et al., 1993). Bcl-2 can also at least partially inhibit Nedd2-induced apoptosis in mammalian cells (Kumar et al., 1994).

Further involvement of proteases in apoptosis. Proteolytic activity has been reported in a number of different apoptotic systems. Proteases expressed during apoptosis include calpain, cathepsins B and D, collagenase, tissue-type and urokinase-type plasminogen activators (Guenette et al., 1994; Squier et al., 1994). Calpain is activated prior to the appearance of the morphological changes of apoptosis and DNA fragmentation in dexamethasone-treated thymocytes (Squier et al., 1994). Prior incubation with specific calpain inhibitors prevents dexamethasone- and irradiation-induced apoptosis in thymocytes. In isolated rat liver nuclei inhibitors of calpain decrease the formation of 50-kb DNA fragments but do not prevent subsequent internucleosomal DNA fragmentation (Zhivotovsky et al., 1994). *In vitro* substrates for calpain include cytoskeletal proteins, growth factor receptors, transcription factors and signal transducing enzymes. The expression of cathepsin B, a thiol protease, is increased in apoptotic epithelial cells in regressing rat prostate and mammary glands (Guenette et al., 1994). Cathepsin B is necessary for the destruction of the extracellular matrix. It degrades components of the basement membrane including collagen, fibronectin and proteoglycans. Urokinase-type plasminogen, a serine protease, may be involved in mediating apoptosis induced by TNF (Kumar and Baglioni, 1991). Plasminogen activator inhibitor type-2 (PAI-2), whose only known target is urokinase-type plasminogen activator, induces resistance to TNF-induced apoptosis.

Serine and cysteine proteases are implicated as mediators of apoptosis in a variety of cells including thymocytes, promyelocytic leukemic HL-60 cells and T lymphocytes. Apoptosis induced in immature rat thymocytes by a variety of inducing agents can be inhibited by *N*-tosyl-L-lysine chloromethane ($TosLysCH_2Cl$), $TosPheCH_2Cl$, phenylmethylsulfonyl fluoride ($PheMeSO_2F$) and dichloroisocoumarin (Weaver et al., 1993; Fearnhead et al., 1995). $TosLysCH_2Cl$ and $TosPheCH_2Cl$ inhibit many serine and cysteine proteases. $PheMeSO_2F$ inhibits serine proteases and a few cysteine proteases, and dichloroisocoumarin inhibits a large number of serine proteases. A $TosPheCH_2Cl$-sensitive protease is required for internucleosomal cleavage and DNA laddering in thymocytes although they still die by apoptosis in the presence of $TosPheCH_2Cl$. $TosLysCH_2Cl$ inhibits all of the characteristic changes of apoptosis including the formation of large chromatin fragments, internucleosomal cleavage, cell shrinkage and plasma membrane changes, indicating that the protease which is inhibited is an effector of early apoptotic changes. Thus more than one protease is required for the apoptotic cell death of thymocytes. Inhibitors of ICE were ineffective at inhibiting dexamethasone- or etoposide-induced apoptosis, confirming that ICE is not involved. $TosPheCH_2Cl$, $TosLysCH_2Cl$, dichloroisocoumarin and calpain inhibitor I inhibit DNA fragmentation in isolated rat liver nuclei (Zhivotovsky et al., 1994; Cain et al., 1995). Serine and cysteine proteases have also been implicated in HL-60 cell (Bruno et al., 1992) and T lymphocyte (Sarin et al., 1993) apoptosis.

A pre-existing 24-kDa protease is involved in TNF- and ultraviolet-induced DNA fragmentation in the U937 histiocytic lymphoma as well as ultraviolet-induced DNA fragmentation in BT20 breast carcinoma, HL-60 myelocytic leukemia and 3T3 fibroblasts (Wright et al., 1994). The protease was inhibited by $TosPheCH_2Cl$ but not by a variety of other serine protease inhibitors, including $TosLysCH_2Cl$ and *N*-*α*-*p*-tosyl-L-arginine methyl ester.

Proteolysis may be a common event in apoptotic cell death. Proteins shown to be cleaved during apoptosis include poly-(ADP-ribose) polymerase (Lazebnik et al., 1994), lamin B (Neamati et al., 1995), the 70-kDa protein (Casciola-Rosen et al., 1994), topoisomerases I and II, histone H1 (Kaufmann, 1989), protein kinase Cβ1, cPLa2 (Voelkel-Johnson et al., 1995), fodrin (Martin et al., 1995) and a 58-kDa protein kinase PITSLRE (Lahti et al., 1995).

Mammalian *ced-9*-related apoptosis-modulating genes

Several mammalian genes have been identified that encode proteins with significant amino acid sequence similarity to the *C. elegans ced-9* gene. The existence of this multigene family in mammals, corresponding to the single *ced-9* gene which regulates developmental cell death in the nematode *C. elegans*, probably reflects the greater complexity of vertebrates, with different members of the family functioning in different tissues or at different developmental stages. The involvement of members of the same gene family in the regulation of cell death in both nematodes and vertebrates suggests that related genes may regulate cell death in other metazoans of intermediate complexity, but, surprisingly, no members of this gene family have yet been reported from the fruit fly *Drosophila*.

The human genes encoding Ced-9-related proteins reported to date are: *bcl-2* (Tsujimoto and Croce, 1986), *bcl-x* (Boise et al., 1993), *bax* (Oltvai et al., 1993), *mcl-1* (Kozopas et al., 1993), *bak* (Chittenden et al., 1995; Farrow et al., 1995; Kiefer et al., 1995) and *bak-2* (closely related to *bak* and possibly a pseudogene; Kiefer et al., 1995). There is also a *bak*-related pseudogene, designated *bak-3* (Kiefer et al., 1995). The independent isolation of *bak* by three groups using two different methods suggests that only a few further human genes encoding Ced-9-related proteins await discovery. In addition to murine homologues of *bcl-2*, *bcl-x*, *bax*, *mcl-1*, and *bak*, two additional murine genes are known that encode Ced-9-related proteins for which human homologues have not yet been reported, *A1* (Lin et al., 1993) and *bad* (Yang et al., 1995a). Conservation between mammals and birds is shown by the presence of homologues at least of *bcl-2* and *bcl-x* in chickens, and an additional member of the family without a known mammalian homologue, designated *NR-13*, has recently been reported in the quail (Gillet et al., 1995).

Several mammalian viruses also carry genes encoding Ced-9/Bcl-2-related proteins: the Epstein-Barr virus BHRF1 gene (Pearson et al., 1987); the African swine fever virus (ASFV) LMW5-HL gene (Neilan et al., 1993) also designated A179L (Yanez et al., 1995), and the herpesvirus saimiri ORF16 gene (Smith, 1995). The BHRF1 protein is able to inhibit apoptosis (Henderson et al., 1993), but modulation of apoptosis by the other two viral proteins has yet to be demonstrated. The apoptosis-inhibiting adenovirus E1B 19-kDa protein also appears able to fulfill the same function, although it shows only minimal amino acid sequence similarity to the family of Ced-9-related proteins (Boyd et al., 1994; Farrow et al., 1995). It is likely that these viral proteins, in common with other viral proteins that are unrelated to Ced-9, prevent or delay the apoptosis of infected cells in order to prolong viral replication (Alnemri et al., 1992b).

Comparison of the amino acid sequences of the Ced-9-related proteins has defined regions with very similar sequences, designated BH1 and BH2 (Fig. 3), which are believed to be of functional importance (Oltvai et al., 1993; Williams and Smith, 1993; Yin et al., 1994). Mutational analysis of these regions shows that changing amino acids in either can abolish the ability of human Bcl-2 to prevent apoptosis following various treatments (Sato et al., 1994; Yin et al., 1994). Bcl-2α, Baxα, Bcl-x_L, Bcl-x_S, Mcl-1, Bak and A1 all have C-terminal hy-

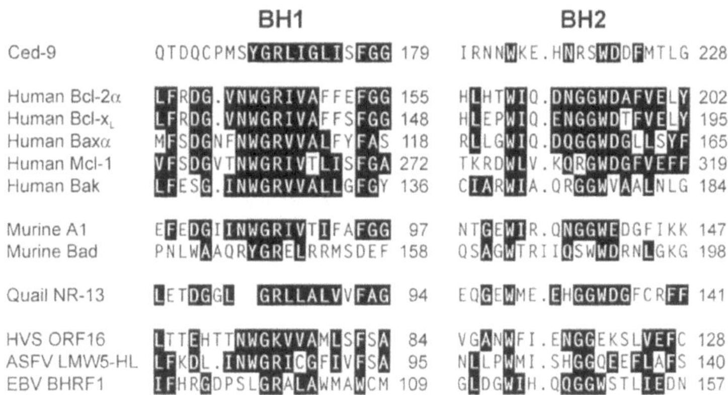

Fig. 3. The BH1 and BH2 regions of amino acid sequence similarity between the *C. elegans* apoptosis-regulator Ced-9 and mammalian, avian and viral proteins. The position of the last amino acid in each region is indicated and conserved or equivalent amino acids are blocked. The amino acids treated as equivalent for this figure are: D/E; W/Y/F; R/K; I/V/L; G/A; N/Q. The GenBank accession numbers of the nucleic acid sequences corresponding to the proteins are: *C. elegans* Ced-9, L26545; human Bcl-2α, M13994; human Bcl-x_L, Z23115; human Baxα, L22473; human Mcl-1, L08246; human Bak, X84213, U23765 or (designated Cdn-1) U16811; murine A1, S64373; murine Bad, L37296; quail NR-13, X84418; herpes virus saimiri ORF16, X64346; African swine fever virus LMW5-HL, L09548 or (designated A179L) U18466; Epstein-Barr virus BHRF1, V01555.

drophobic regions of about 20 amino acids terminated by one or more positively charged amino acids, although the actual sequences of these regions are not conserved. These regions are required for post-translational anchoring to membranes (Alnemri et al., 1992b; Hockenbery et al., 1993; Tanaka et al., 1993; Borner et al., 1994; Reynolds et al., 1994). There is some evidence that membrane association is usually necessary for the ability of Bcl-2 to inhibit apoptosis (Alnemri et al., 1992b; Tanaka et al., 1993; Nguyen et al., 1994). Determination of the tertiary structure of Bcl-2 is urgently needed to provide a framework for the interpretation of mutational studies and of the sequence homologies amongst Ced-9-related proteins.

Splicing variants. The human *bcl-2* gene encodes two gene products, Bcl-2α and Bcl-2β. Bcl-2β shares most of the amino acid sequence of Bcl-2α, incuding the BH1 and BH2 regions, but lacks the C-terminal hydrophobic membrane-anchor domain due to the failure to remove an intron (Tsujimoto and Croce, 1986; Tanaka et al., 1993).

The human *bcl-x* gene also produces two protein products, Bcl-x_L and Bcl-x_S, by alternative mRNA splicing using alternative 5′ splicing sites (Boise et al., 1993). Both Bcl-x_L and Bcl-x_S share the same C-terminal putative membrane-anchor domain, but Bcl-x_S lacks most of the BH1 and BH2 regions of amino acid sequence similarity. Two additional splicing variants, Bcl-$x_{\Delta TM}$, and Bcl-xβ have been reported in the mouse, which retain the BH1 and BH2 regions, but lack the putative membrane-anchor domain (Fang et al., 1994; Gonzalez-Garcia et al., 1994).

The main product of the *bax* gene is Baxα, but like the *bcl-2* and *bcl-x* genes, *bax* produces different products by alternative mRNA splicing. Baxβ shares most of the sequence of Baxα but lacks the putative C-terminal membrane anchor domain due to the failure to remove an intron from the mRNA. A third product of only 41 amino acids, Baxγ, is potentially encoded by mRNA species lacking the second exon (Oltvai et al., 1993).

Positive and negative regulation of apoptosis by Ced-9-related proteins. Although the *C. elegans ced-9* gene encodes a negative regulator of developmental cell death, the mammalian family of *ced-9*-related genes includes both inhibitors and promoters of apoptosis.

The *bcl-2* gene, first detected as a putative oncogene located near to the breakpoint of t(14;18)(q32;q21) translocations implicated in human follicular lymphoma (Tsujimoto and Croce, 1986), was the first gene shown to be involved in the regulation of apoptosis (Vaux et al., 1988). During early murine foetal development Bcl-2 is expressed widely in tissues derived from all three germ layers, but its expression becomes more restricted during development (Novack and Korsmeyer, 1994). In adults Bcl-2 is mainly expressed in tissues that are renewed from stem cells, have proliferative ability, or are long-lived (Krajewski et al., 1994c). Bcl-2, like Ced-9, is a negative regulator of cell death, able for example to prevent IL-3-dependent cells from undergoing apoptosis following deprivation of IL-3 (Vaux et al., 1988). The oncogene-like role of *bcl-2* results from its ability to prolong the survival of non-cycling cells and to prevent the death by apoptosis of cycling cells, and thus to cooperate with oncogenes such as c-*myc* which dysregulate the cell cycle (Vaux et al., 1988; Strasser et al., 1990), rather than from an independent ability to induce proliferation.

The *bcl-x* gene was isolated through its sequence similarity to the *bcl-2* gene (Boise et al., 1993). Bcl-x_L, the larger protein product of the human *bcl-x* gene, is also a negative regulator of apoptosis, able like Bcl-2 to prevent IL-3-dependent cells from undergoing apoptosis following deprivation of IL-3 (Boise et al., 1993). In contrast Bcl-x_S allows apoptosis to proceed in IL-3-deprived cells even in the presence of Bcl-2 (Boise et al., 1993). The pattern of expression of *bcl-x* differs from that of *bcl-2*, and different tissues show differential expression of Bcl-x_L and Bcl-x_S. Most tissues express Bcl-x_L, but tissues that show a high rate of apoptosis also express variable levels of Bcl-x_S (Boise et al., 1993; Krajewski et al., 1994b).

Human *mcl-1* was first identified as a gene which is expressed in a myeloid cell line during differentiation (Kozopas et al., 1993). Like Bcl-2 and Bcl-x_L, Mcl-1 is a negative regulator of apoptosis and can delay the death of cells overexpressing c-Myc (Reynolds et al., 1994). The Mcl-1 protein is larger than the other mammalian Ced-9-related proteins and has a non-homologous N-terminal region which contains sequences rich in the amino acids proline, glutamic acid, serine and threonine (PEST sequences) and pairs of arginine residues (Kozopas et al.,

1993). Such sequences are typical of proteins that have a rapid turnover, and the full size Mcl-1 protein has a half-life of about 1 h (Yang et al., 1995b) in contrast to the half-life of Bcl-2 protein of approximately 10 h in B cells (Merino et al., 1994). The pattern of Mcl-1 expression differs from those of Bcl-2 and Bcl-x_L; it is expressed in epithelial cells that are more differentiated than those expressing Bcl-2 and in various types of muscle and neuroendocrine cells as well as in lymph node germinal centre cells and differentiating myeloid cells (Krajewski et al., 1994, 1995). The short half-life of Mcl-1 may allow rapid modulation of the susceptibility of differentiating cells to apoptosis.

Human *bax* was isolated as the gene encoding a protein that binds to Bcl-2 (Oltvai et al., 1993). The principle product of the *bax* gene, Baxα, acts in opposition to Bcl-2 and overexpression of Baxα allows apoptosis to proceed after IL-3 withdrawal from IL-3-dependent cells even in the presence of dysregulated expression of Bcl-2 (Oltvai et al., 1993). Enforced expression of Baxα does not induce apoptosis in IL-3-dependent cells in the presence of IL-3, but it does increase the speed of apoptosis on withdrawal of IL-3. Bax is expressed more widely in the adult mouse than Bcl-2, particularly in cell types that show a high rate of apoptosis (Krajewski et al., 1994c).

The human *bak* gene was isolated both by amplifying sequences similar to coding sequences for both the BH1 and BH2 regions of Ced-9-related proteins using the polymerase chain reaction (Chittenden et al., 1995; Kiefer et al., 1995) and as the gene encoding a protein which binds to the adenovirus E1B 19-kDa protein (Farrow et al., 1995). Bak is expressed in a wide range of tissues (Farrow et al., 1995; Kiefer et al., 1995). Like Bax, Bak accelerates the apoptotic death of IL-3-deprived cells and partially blocks the inhibition of apoptosis by Bcl-2 (Chittenden et al., 1995). Bak also accelerates apoptosis in growth-factor-deprived neurons and blocks the protective effect of EIB 19-kDa protein (Farrow et al., 1995). However, Bak appears to be able to inhibit the apoptosis of serum-deprived Epstein-Barr-transformed cell lines (Kiefer et al., 1995).

Murine *bad* was isolated as a gene encoding a Bcl-2-binding protein (Yang et al., 1995a). Like Bax, Bak and Bcl-x_S, Bad promotes apoptosis. Bad allows apoptosis to proceed after IL-3 withdrawal from IL-3-dependent cells in the presence of Bcl-x_L but not in the presence of Bcl-2 (Yang et al., 1995a). The expression of Bad can therefore influence the relative ability of Bcl-x_L and Bcl-2 to prevent apoptosis. The amino acid sequence similarity between Bad and other mammalian members of the Ced-9-related family is limited to the BH1 and BH2 regions and Bad does not have a putative membrane anchor domain. The N-terminal region of Bad, like Mcl-1, contains arginine-bounded PEST sequences and it may therefore also have a short half-life.

The murine *A1* gene is induced by GM-CSF as an early response gene in several murine haematopoietic cell lineages, but whether it can modulate apoptosis has not been reported (Lin et al., 1993).

Transgenic mice. Homozygous *ced-9* loss-of-function mutant alleles are lethal in *C. elegans*, due to massive excess cell death during embryonic development (Yuan, 1995). In place of the single *C. elegans* Ced-9 protein which regulates cell death negatively, mammalian genomes encode a family of related proteins that modulate apoptosis, some negatively and others positively. Conceptually, developmental cell death or survival in mammals might be regulated either by a single member of this family (or by the ratio of a pair of members, one acting positively and the other negatively), or by different members (or pairs of members) in different tissues; or multiple members of the family might function in a redundant manner, so that disruption of any single gene would have little effect on development. The available evidence suggests that in mammals different members of the family of *ced-9*-related genes are responsible for controlling developmental apoptosis in different tissues. Transgenic mice deficient in either Bcl-2 or Bcl-x or Bax have been generated.

Homozygous *bcl-2*-null transgenic mice completely lacking both Bcl-2α and Bcl-2β are viable and develop nearly normally, although abnormalities become apparent after birth in various tissues due to defects in the survival of particular cell types (Nakayama et al., 1994; Kamada et al., 1995; Sorenson et al., 1995). Developmental abnormalities include polycystic kidneys due to excessive apoptosis during metanephric development and an abnormality of the small intestine, involving accelerated loss of epithelial cells and reduced numbers of mitotic precursors. Although lymphoid development in *bcl-2*-null mice is initially normal, lymphoid cells show increased sensitivity to the induction of apoptosis and lymphocytes gradually disappear, resulting in atrophy of the thymus and spleen. The coat of Bcl-2-null mice becomes hypopigmented after the first hair follicle cycle due to a defect in melanocyte cell survival.

In contrast to *bcl-2*-null mice, homozygous *bcl-x*-null transgenic mice die at about embryonic day 13 (Motoyama et al., 1995). Although *bcl-x*-null foetal mice show massive apoptosis both of haematopoietic cells in the foetal liver and of immature postmitotic neurons in the brain and spinal cord, they appear otherwise developmentally normal. Chimaeric mice formed from *rag-2*-null cells (blocked in lymphoid differentiation) and *bcl-x*-null cells are viable; their immature B cells and thymocytes show increased susceptibility to apoptosis, but they form reduced numbers of mature lymphocytes.

Like *bcl-2*-null mice, homozygous *bax*-null mice are viable and show nearly normal development (Knudson et al., 1995). The *bax*-null mice show thymocyte and B-cell hyperplasia, and females show excess granulosa cells in atretic ovarian follicles, consistent with excess cell survival and suggesting that Bax may not function solely as a positive regulator of apoptosis. However, males are infertile, lacking mature sperm and showing massive apoptotic death of precursors.

Thus neither *bcl-2*, *bcl-x* nor *bax* plays the same critical role in the control of developmental cell survival in mammals as *ced-9* does in the nematode *C. elegans*, but each is required for cell survival or deletion in particular tissues. The generation of gene-knockout mice defective in other *ced-9*-related genes should elucidate their roles in develpmental cell survival.

Several lines of transgenic mice have also been established which express *bcl-2* under the control of promotors or enhancers specific to particular tissues or cell types, generally leading to enhanced survival of the cell types in which it is expressed (Strasser et al., 1990; McDonnell and Korsmeyer, 1991; Sentman et al., 1991; Martinou et al., 1994).

Control of Bcl-2 and Bax by the tumour-suppressor p53. The tumour suppressor p53, implicated in induction of both growth arrest and apoptosis following DNA damage, is a positive regulator of *bax* expression and a negative regulator of *bcl-2* expression, and p53-null mice have higher levels of Bcl-2 and lower levels of Bax than normal mice in several tissues (Miyashita et al., 1994b; Selvakumaran et al., 1994). The *bax* promoter region contains four p53-binding motifs (Miyashita and Reed, 1995) and p53 causes immediate up-regulation of *bax* expression following DNA damage with ionizing radiation (Miyashita et al., 1994b; Selvakumaran et al., 1994). The *bcl-2* promoter alone coupled to a reporter gene does not show regulation by p53, but a 195-bp segment of the 5′ untranslated region of the *bcl-2* gene confers negative regulation by p53 (Miyashita et al., 1994a).

Homodimer and heterodimer formation by Ced-9-related proteins. The ability of members of the Ced-9 protein family to form heterodimers and homodimers has been investigated using yeast two-hybrid systems in which the ability of two proteins to interact is tested by fusing one to a DNA-binding domain and the other to a transcriptional-activating domain, allowing interaction to be detected phenotypically through the activation of a reporter gene (Sato et al., 1994; Farrow et al., 1995; Yang et al., 1995a; Zhang et al., 1995). Bcl-2 forms homodimers and heterodimerizes with Bcl-x_L, Bcl-x_S, Bax, Mcl-1 and Bad, while Bcl-x_L forms homodimers and heterodimerizes with Bcl-x_S, Bax, Mcl-1 and Bad as well as with Bcl-2. In some experiments, but not others, Bax and Bcl-x_S show homodimer and heterodimer formation (Oltvai et al., 1993; Zhang et al., 1995). Human Bak binds to E1B 19-kDa protein as well as to Bcl-x_L (Farrow et al., 1995) and may also interact more weakly with Bcl-2 (Chittenden et al., 1995). Bad heterodimerizes with Bcl-2 and Bcl-x_L but does not bind to Bax, Bcl-x_S, Mcl-1 or A1 and does not form homodimers. Bad binds more strongly to Bcl-x_L than to Bcl-2, and can displace Bax from Bcl-x_L (Yang et al., 1995a).

One way in which the internal state of a cell may control the apoptotic response is through the ratio of apoptosis-inhibiting to apoptosis-promoting proteins present. It has been proposed that the Bcl-2/Bax ratio in a cell acts as a rheostat which regulates susceptibility to apoptosis (Korsmeyer et al., 1993). This rheostat can be reset in favour of apoptosis by p53 following DNA damage, by repressing *bcl-2* and inducing *bax* expression (Miyashita et al., 1994a,b; Selvakumaran et al., 1994; Miyashita and Reed, 1995). The Bcl-x_L/Bcl-x_S ratio may be an alternative rheostat controlled by differential mRNA splicing (Boise et al., 1993). More generally, the ratio of the apoptosis-suppressing (Bcl-2, Bcl-x_L, Mcl-1) to apoptosis-promoting (Bax, Bcl-x_S, Bak, Bad) Ced-9-related proteins may control the susceptibility of mammalian cells to apoptosis. Thus differential expression of members of the Ced-9-related protein family, in some cases controlled at the level of mRNA splicing as well as transcription, may regulate susceptibility to apoptosis during development, during differentiation of renewing cell populations and in different tissues.

Scope of protection by Bcl-2. Although Bcl-2 does not appear to have a uniquely important position in the regulation of apoptosis in mammalian cells, it has been the subject of more study than the other mammalian Ced-9-related proteins because it was the first to be discovered. The ectopic or dysregulated expression of the *bcl-2* gene prevents or delays the induction of apoptosis by a wide variety of stimuli in a wide variety of cell types, but not in all cases.

In addition to protecting IL-3-dependent cell lines from apoptosis following IL-3-deprivation, the enforced expression of Bcl-2 allows many other, but not all, factor-dependent cells to survive in G_0 following factor withdrawal and to proliferate again when the growth factor is restored. Whether or not Bcl-2 is able to suppress apoptosis following growth-factor withdrawal may depend on which other members of the family of Ced-9-related proteins are being expressed in the cells. Bcl-2 can also prevent enforced expression of the c-*myc* oncogene from causing apoptosis (Wagner et al., 1993). Despite the ability of the tumor suppressor p53 to down-regulate the endogenous expression of the *bcl-2* gene, enforced over-expression of Bcl-2 can block or delay p53-dependent apoptosis (Chiou et al., 1994).

Enforced expression of Bcl-2 can also render cells resistant to killing by commonly used chemotherapeutic agents including methotrexate, cytosine arabinoside, etoposide, vincristine and cisplatin (Miyashita and Reed, 1993). Studies with human lymphoma cell lines which overexpress Bcl-2 due to a t(14;18) translocation, show that reducing Bcl-2 expression using either antisense oligonucleotides or plasmids with inducible antisense *bcl-2* transcription increases sensitivity to cytosine arabinoside or methotrexate (Kitada et al., 1994). Enforced expression of Bcl-2 also blocks the induction of apoptosis by glucocorticoids in immature thymocytes and pre-B cell lines (Sentman et al., 1991; Alnemri et al., 1992a).

Enforced Bcl-2 expression in target cells blocks killing either by cytotoxic cell granule extracts or by cytotoxic T lymphocytes under conditions where the Fas/APO-1/CD95 pathway is blocked, but does not protect against apoptotic cell death following treatment with anti-Fas antibodies or cytotoxic T lymphocytes competent to kill by the Fas pathway (Chiu et al., 1995). However, in some cases, expression of Bcl-2 does inhibit the induction of apoptosis by negative growth factors such as TNF-α and Fas/APO-1/CD95 ligands (Itoh et al., 1993). It is possible that levels of expression of the non-Ced-9-related Bcl-2-binding protein Bag-1 determine the ability of Bcl-2 to inhibit the induction of apoptosis by the Fas/APO-1/CD95 pathway (Takayama et al., 1995).

Cellular location of Ced-9-related proteins. Although some studies have suggested that Bcl-2α is associated with the inner mitochondrial membrane (Hockenbery et al., 1990; Tanaka et al., 1993), the bulk of recent studies using confocal microscopy or immuno-electronmicroscopy as well as fractionation have found that Bcl-2α is located at the cytoplasmic surfaces of the nuclear envelope, the outer mitochondrial membrane and the endoplasmic reticulum (Jacobson et al., 1993; Krajewski et al., 1993; Akao et al., 1994; Lithgow et al., 1994). *In vitro*, Bcl-2α inserts into the cytoplasmic surfaces of multiple intracellular membranes (Janiak et al., 1994). Bcl-2β, which lacks the C-terminal hydrophobic membrane-anchor domain of Bcl-2α, may be largely cytosolic (Tanaka et al., 1993). It has also been reported that Bcl-2 localizes strongly to the nucleus during mitosis and associates with the chromosomes during prophase, metaphase and anaphase before disappearing at telophase (Lu et al., 1994; Willingham and Bhalla, 1994).

There have been far fewer studies of the subcellular location of the other Ced-9-related proteins. One study found both Bcl-x_L and Bcl-x_S to have a predominantly mitochondrial location (Fang et al., 1994), while another found Bcl-x_L protein mainly located at the periphery of mitochondria and the nuclear envelope (Gonzalez-Garcia et al., 1994). Mcl-1 shows prominant mitochondrial localization, with less nuclear association than Bcl-2 but more association with the light membrane fraction (Krajewski et al., 1994c; Yang et al., 1995b).

Possible mechanisms of action of Bcl-2. The mechanism(s) by which Ced-9-related proteins modulate apoptosis are not known and several conflicting theories of the mode of action of Bcl-2 have been proposed. Bcl-2 is able to inhibit apoptosis caused by the dysregulated expression of the Ced-3-related proteases ICE (Miura et al., 1993) and Ich-1_L (Wang et al., 1994b), just as Ced-9 regulates Ced-3/Ced-4-dependent developmental cell death in *C. elegans*, but there is no published experimental evidence that it acts directly on these enzymes.

Since apoptosis is able to occur in cells lacking mitochondrial DNA and defective for most mitochondrial functions, and this apoptosis can be blocked by Bcl-2, theories based on the modulation of mitochondrial functions must be viewed with caution (Jacobson et al., 1993). It has been suggested that Bcl-2 controls levels of reactive oxygen species through an antioxidant pathway (Hockenbery et al., 1993; Korsmeyer et al., 1993). However, it is unlikely that modulation of an antioxidant pathway is the main mode of action of the Ced-9-related proteins, since apoptosis can occur, and can be prevented, by either Bcl-2 or Bcl-x_L, under nearly anaerobic conditions which minimise the formation of reactive oxygen species (Jacobson and Raff, 1995; Shimizu et al., 1995).

The reported assocation of Bcl-2 with condensed chromatin during mitosis suggests that it may act directly to prevent the chromatin cleavage (Lu et al., 1994; Willingham and Bhalla, 1994). However, the non-nuclear features of apoptosis can take place in enucleated cytoplasts and can be prevented by Bcl-2 (Jacobson et al., 1994; Schulze-Osthoff et al., 1994). This implies that the cytoplasmic features of apoptosis are controlled at the cytoplasmic level, without the obligate involvement of nuclear changes such as chromatin cleavage, and that Bcl-2 can act directly at this level.

Bcl-2α is not itself a GTP-binding protein (Monica et al., 1990), but coprecipitates with R-Ras, a member of the Ras family of small G-proteins, when either is immunoprecipitated (Fernandez-Sarabia and Bischoff, 1993). Bcl-2α also coprecipitates with Raf-1 (Wang et al., 1994a), a serine/threonine kinase that itself binds to and is activated by H-Ras in a GTP-dependent manner and in turn activates the microtubule-associated protein kinase cascade. The functional significance of these interactions is not clear. IL-3-dependent cell lines deprived of IL-3 undergo apoptosis more rapidly when transfected with an activated mutant R-*ras* and transfection with *bcl-2* prolongs the survival of R-*ras* transfectants (Wang et al., 1995a). However, Bcl-2 does not alter the ratio of GTP/GDP bound to R-Ras, does not affect the GTPase activity of R-Ras *in vitro* and does not interfere with the binding of R-Ras to Raf-1 or the activation of Raf-1 by R-Ras (Wang et al., 1995a). These results suggest that Bcl-2 must act downstream of Raf-1 to inhibit apoptosis.

Nip1, Nip2 and Nip3, unrelated human proteins that bind both to Bcl-2 and to the adenovirus apoptosis-inhibiting E1B 19-kDa protein, are likely to play some role in the control of apoptosis (Boyd et al., 1994). However, they show little amino acid sequence similarity to other known proteins and no functions have yet been assigned to them. The unrelated murine protein Bag-1 also binds to Bcl-2 and is certainly involved in blocking apoptosis, since enforced co-expression of Bag-1 and Bcl-2 provides stronger protection against the induction of apoptosis than the enforced expression of either alone, but its mode of action is unknown (Takayama et al., 1995).

It is also possible that Bcl-2 regulates cytosolic free Ca^{2+} concentration, possibly by inhibiting the release of Ca^{2+} from the endoplasmic reticulum to the cytosol (Baffy et al., 1993; Lam et al., 1994). Although preventing a rise in cytosolic free Ca^{2+} concentration can inhibit apoptosis (McConkey et al., 1990; Ray et al., 1993), this is not likely to be the sole mechanism by which Bcl-2 inhibits apoptosis, since Bcl-2 is able to block apoptosis despite a rise in cytosolic Ca^{2+} concentration in some cases (Zhong et al., 1993).

It has also been suggested that Bcl-2 modulates the transport of proteins through the pores of the nuclear envelope (Ryan et al., 1994). Both Bcl-2 and Bcl-x_L contain an amino acid sequence motif (FXFG) which is present in some nuclear pore proteins, and deletion of this motif reduces the anti-apoptotic activity of Bcl-2 (Borner et al., 1994).

Finally, Bcl-2 may not itself inhibit apoptosis, but rather, it may prevent Bax from inducing it. This mechanism was suggested by the finding that mutations which prevent Bcl-2 from forming heterodimers with Bax, but do not impair homodimerization, abolish its ability to prevent apoptosis following a variety of stimuli (Yin et al., 1994). Consistent with this mechanism, the expression of one Bax fusion protein is lethal to yeast unless Bcl-2, Bcl-x_L or Mcl-1 is co-expressed (Sato et al., 1994), although not all Bax fusion proteins are lethal (Zhang et al., 1995). However, overexpression of Bax does not by itself induce apoptosis in IL-3-dependent cells in the presence of IL-3, although it does increase the speed of apoptosis after withdrawal of IL-3 (Oltvai et al., 1993).

It is possible that Bcl-2 acts in more than one way either in order to prevent the induction of apoptosis by different stimuli or in order to control different aspects of the apoptotic effector pathway.

Other oncogenes modulating apoptosis

Other tumour-promoting genes which influence cell death include the genes encoding the GTP-binding protein Ras, the kinases Raf and Abl, the cellular transcription factors AP-1, Rel, Myc, Myb, E2F and Nur77, and the viral proteins latent membrane protein-1 (LMP-1), Tax, large T-antigen, E1A/B, E6 and E7 (see Table 1 for summary).

Signalling molecules. *Ras* genes encode a family of small guanine-nucleotide-binding proteins involved in signal transduction (for a review of the signal transduction mechanism see Izquierdo Pastor et al., 1995). The mutations which activate Ras proteins and give them oncogenic potential result in the protein being constitutively present in the GTP-bound, active, state (Barbacid, 1987; Saez et al., 1994). The same mutations also confer upon the proteins the ability to inhibit apoptosis. H-Ras has been shown to rescue cells from c-Myc-induced apoptosis in a serum-deprived environment (Wyllie et al., 1987), and overexpression of activated Ras inhibits apoptosis in the absence of IL-3, in the IL-3-dependent murine myeloid cell line H7 (Moore et al., 1993).

Transcription factors. Expression of the transcription factor Myc is important for cell proliferation, and is also involved in programmed cell death (reviewed by Green et al., 1994). Myc interacts with the retinoblastoma protein (pRB) and is able to override pRB-induced cell cycle arrest. Consequently, in the presence of certain growth-promoting cytokines such as IL-2, when Myc expression is induced cell proliferation occurs; overexpression results in uncontrolled cell proliferation. The continued presence of Myc under conditions of growth arrest, as in murine myeloid cells and Rat-1 fibroblasts deprived of growth factor, will induce apotosis which is p53-dependent (Askew et al., 1991; Evan et al., 1992; Hermeking and Eick, 1994), as c-Myc is a transcriptional activator of p53 (Reisman et al., 1993). Induction of apoptosis is dependent on the association of Myc with its heterodimeric partner Max (see Green et al., 1994) and is suppressed by Bcl-2 (Bissonnette et al., 1992; Fanidi et al., 1992).

Human T-cell hybridomas, HL-60 cells and rodent fibroblast cells which require Myc for activation-induced apoptosis, do not require Myc for apoptosis induced by glucocorticoid- and transforming growth factor-β (Warner et al., 1991; Shi et al., 1992). Myc is not, therefore, required for all forms of apoptosis, and cannot be considered an ultimate death gene. The cellular decision to proliferate or die in response to Myc overexpression appears to be influenced by other signals or other survival stimuli.

Recent work by Lee et al. (1995) has shown that Rel-family proteins are involved in the control of *c-myc* expression in the B-cell line WEHI 231. Cell-surface immunoglobulin-mediated apoptosis of WEHI 231 cells involves regulation of *c-myc* expression by both Rel and NF-κB, which have binding sites in the *myc* promoter. The Rel family of protooncogenes, which includes NF-κB (reviewed by Huguet et al., 1994), functions both to promote and to inhibit apoptosis. In primary avian fibroblasts, overexpression of c-*rel* leads to transformation, whereas in bone marrow cells, overexpression of c-*rel* leads to apoptosis. In contrast, the v-*rel* oncogene transforms bone marrow cells (see Huguet et al., 1994). White et al. (1995) report that v-*rel* blocks apoptosis in chicken spleen cells by preventing degradation of the Rel inhibitor protein IκB. Clearly, not only are the effects of *rel* overexpression cell-type-specific, but the overexpression of c-*rel* and the expression of v-*rel* have different effects in the same cell type.

Myb also transactivates Myc (Cogswell et al., 1993), Myb expression being highest in haematopoietic cells. Down-regulation of Myb leads to the inhibition of proliferation and, in neuroblastoma cells, inhibition of Myb expression can induce apoptosis (Piacentini et al., 1994b). In contrast, *c-myb* oncogene expression is increased after induction of apoptosis following IL-2 withdrawal from an IL-2-dependent cell line (see Cheng et al., 1995). As inhibition of *c-myb* expression induces apoptosis in neuroblastoma cells, and increased *c-myb* expression induces apoptosis in T cells, it is apparent that, as for Myc, Myb-induced apoptosis is, at the least, cell-type-specific, and dependent on the presence of other survival signals.

Expression of the proto-oncogenes encoding Fos and Jun, which together form the AP-1 transcription factor, is rapidly and transiently induced in IL-2- and IL-6-dependent mouse myeloma cell lines after induction of apoptosis by growth factor deprivation (Colotta et al., 1992). Treatment of these cells with antisense oligonucleotides directed against *fos* and *jun*, alone or in combination, protects the cells from apoptosis, suggesting an active role for *fos* and *jun* in the onset of apoptosis. A functional link between *fos* expression and apoptosis was established by examining the effect of serum-deprivation on *fos*-transformed cells; in rat fibroblasts deprived of serum, cell death was minimal (7%); however, in *fos*-transformed rat fibroblasts deprived of serum, cell death levels were dramatically increased (80%) (Smeyne et al., 1993). Activation of AP-1 is also associated with the death of many epithelial cells by apoptosis during involution of the mammary gland following lactation (Marti et al., 1994). This involution-related apoptosis can be a p53-dependent or independent mechanism. EGF and insulin have been found to act as survival factors for the p53-independent pathway, and it is conjectured that oncogene-mediated transformation of mammary cells may involve an oncogene-mediated effect on EGF and/or insulin levels (Merlo et al., 1995).

E2F is another class of oncogenic transcription factors known to modulate apoptosis (Zhu et al., 1995). It accomplishes this through interaction with pRB. pRB is a 105-kDa phosphoprotein which acts as a cell cycle regulator (De Caprio et al., 1989). During cell cycle progression, the phosphorylation status of pRB alters (Buchkovich et al., 1989; Mihara et al., 1989). Cells in G_0/G_1 predominantly express pRB in a hypophosphorylated form: in late G_1, pRB becomes hyperphosphorylated on serine and threonine residues by cyclin-dependent kinases. The hyperphosphorylated form remains throughout the S, G_2 and M phases. Importantly, hypophosphorylated pRB is believed to be active in growth suppression. In this form it can bind to a number of cellular proteins, including E2F which is required for transcription of cellular genes that participate in growth control and DNA synthesis, e.g. c-*myc*, c-*myb*, *cdc2*, thymidine kinase, dihydrofolate reductase and DNA polymerase-α (Nevins, 1992). Over-expression of E2F drives quiescent cells into S phase, prevents cycling cells from exiting the cell cycle, transforms rat embryo fibroblasts, and ultimately leads to apoptosis (Kowalik et al., 1995).

The nuclear steroid hormone receptor transcription factor Nur77 is induced during activation-induced apoptosis, is required for apoptosis, and may regulate the expression of downstream genes involved in apoptosis (Liu et al., 1994; Woronicz et al., 1994). It was recently shown by Yazdanbakhsh et al. (1995) that Nur77 must bind DNA to bring about the apoptotic response. Nur77 family members have recently been identified which bind the same DNA sequence. These include the human protein TINUR (which has also been shown to be induced during apoptosis; Okabe et al., 1995), a neuronal protein NOR-1 (Ohkura et al., 1994), and the human protein TR3. The TR3 promoter contains a novel enhancer element which may prove to be important in regulating apoptosis (Uemura et al., 1995).

The general pattern therefore emerging for oncogenic transcription factor modulation of apoptosis is one in which overexpression of factors induces the apoptotic response. The ultimate effects of these factors, whether oncogenic or apoptotic, are linked through the overlapping patterns of *trans*-activation produced. For instance, E2F regulates the expression of *myb* and *myc*, and Myb and AP-1 regulate the expression of *myc*. In addition, the Fas receptor gene, whose expression is upregulated following T cell activation, has been found to contain both a Myb consensus sequence (Cheng et al., 1995), and an AP-1 consensus sequence in its promoter (Emlen et al., 1994). It is therefore interesting to speculate whether the increase in Fas expression is a consequence of increased Myb and AP-1 levels and subsequent interaction with the Fas promoter. E2F and Myc interact with the cell-cycle regulator pRB, and the fact that overexpression of both Myb and Rel can either induce or suppress apoptosis depending on the cell type and even within the same cell type under different conditions, is evidence that the cellular response to their overexpression is-dependent on other signals which are cell-type-specific and perhaps cell-cycle-dependent.

Viral oncoproteins. DNA tumour viruses depend on host cell mechanisms to replicate the viral genome. As a consequence, these viruses encode early gene products capable of binding to specific cellular proteins that normally act to restrict cell growth, such as pRB and p53. Adenovirus, the human papilloma virus (HPV), Epstein Barr virus (EBV) and simian virus 40 (SV40) are all known to employ this mechanism. In addition to these cell-cycle-controlling viral/cellular protein interactions, some viruses have been found to express apoptosis-suppressor genes which inhibit the host cell response to infection.

The EBV latent membrane protein 1 (LMP-1), involved in the transformation of EBV infected B-cell lymphocytes and epithelium, also protects the cells from programmed cell death (Henderson et al., 1991). In B-cells, LMP-1 inhibits apoptosis by inducing expression of *bcl-2* (Henderson et al., 1991), although this is not true for epithelial cells (Dawson et al., 1995). In B-cells, LMP-1 also induces expression of the apoptosis suppressor gene *A20* (Tewari et al., 1995b). It is conceivable that particular apoptosis-suppressor genes are induced in order to suppress apoptosis which is induced through different pathways. For instance, A20 is known to be induced in B-cells following stimulation of the CD40 receptor (see Tewari et al., 1995b).

The viral oncoprotein Tax, of the human T cell leukaemia virus 1 (HTLV-1), appears to have both apoptosis-suppressing and -inducing capabilities. Expression of Tax leads to activation of NF-κB-like factors, can induce expression of *myc*, and upregulates the apoptosis-suppressing gene *A20* (Laherty et al., 1993). However, a recent paper by Chlichlia et al. (1995) shows that overexpression of Tax in Jurkat cells leads to apoptosis. The level of the apoptotic response is shown to be dependent on the duration of Tax stimulation: apoptosis resulting from a short-term exposure to Tax is dependent on prior T-cell activation,

Table 1. Some of the better-defined intracellular modulators of apoptosis.

Type	Protein	Apoptosis induction	Reference	Apoptosis suppression	Reference
Signalling molecules	RAS			+	Wyllie et al., 1987
	RAF			+	Camman et al., 1995
	ABL			+	McGahon et al., 1994
Transcription factors	MYC	+	Evan et al., 1992		
	REL	+	Abbadie et al., 1993	+	White et al., 1995
	MYB	+	Cheng et al., 1995	+	Piacentini et al., 1994b
	AP-1	+	Smeyne et al., 1993		
	E2F	+	Kowalik et al., 1995		
	NUR77	+	Liu et al., 1994		
Viral proteins	E1A	+	Rao et al., 1992		
	E1B			+	Rao et al., 1992
	LMP-1			+	Henderson et al., 1991
	TAX	+	Chlichlia et al., 1995	+	Laherty et al., 1993
	E7	+	White et al., 1994a		
	E6			+	Scheffner et al., 1990
	Large T antigen			+	Levine et al., 1990

whereas apoptosis resulting from a long-term exposure toTax is activation-independent (Chlichlia et al., 1995). Their findings support the proposal by Schwartz (1992) that Tax functions to suppress apoptosis by modulating the CD28 CD3/TCR co-stimulatory pathway.

Tumour suppressors and apoptosis

Properties of wild-type p53. The p53 phosphoprotein is the product of a tumour suppressor gene, whose inactivation represents the most common genetic lesion known in human cancer (Hollstein et al., 1991). The function of p53 as a tumour suppressor is exemplified by p53-null mice which develop normally but are prone to a variety of spontaneous tumours by 6 months of age (Donehower et al., 1992). Wild-type p53 possesses features of a transcription factor (Levine et al., 1991). When bound directly to DNA, it activates transcription; however, when tethered to the promoter via interactions with the TATA binding protein (TBP), p53 represses transcription (Horikoshi et al., 1995). The C-terminal region of p53 is required for sequence-specific and non-specific DNA binding, while the N-terminal region contains a potent transactivation domain (Zambetti and Levine, 1993). Activation of wild-type p53 has been shown to promote differentiation and apoptosis in certain cellular contexts (Yonish-Rouach et al., 1991). It also functions as a cell cycle regulator which can induce cell cycle arrest following DNA damage (Kastan et al., 1991).

Induction of apoptosis by p53 in response to DNA damage. p53 is an important participant in the cellular response to ionizing radiation and other DNA damaging agents. Exposure of mammalian cells to ionizing radiation can result in transient arrests of the cells at the G_1 and G_2 phases of the cell cycle (Kuerbitz et al., 1992) and in apoptotic cell death (Cohen et al., 1992). Following exposure, intracellular levels of the p53 protein increase resulting in cell cycle arrest at the G_1 phase (Kuerbitz et al., 1992). The role of p53 as a G_1 cell-cycle-control checkpoint was suggested by the observation that, whereas cells with wild-type p53 genes arrest in G_1 following ionizing radiation, cells with mutant p53 genes lack this checkpoint and arrest in G_2 (Kuerbitz et al., 1992). Furthermore, transfection of wild-type p53 genes into cells lacking endogenous p53 restores G_1 arrest after the radiation (Kuerbitz et al., 1992). This checkpoint presumably allows time for the cell to repair DNA damage or to commit to apoptosis if the damage is too great (Lane, 1992). This system would act to prevent propagation of extensive DNA damage to daughter cells, halting progression towards carcinogenesis. For this reason, p53 is now frequently referred to as the 'guardian of the genome' (Lane, 1992).

In normal cells, DNA strand breaks appear to be the necessary signal for induction of a p53-mediated response since other forms of DNA damage such as those produced by alkylating agents do not result in the accumulation of p53 (Lee and Bernstein, 1993). Protein synthesis inhibitors, such as cycloheximide, abrogate the accumulation of p53 following ionizing radiation (Kastan et al., 1991) and the DNA-damage-induced increase in the level of p53 appears to be due to a post-transcriptional stabilisation mechanism. Patients with ataxia telangiectasia, a genetic disorder predisposing to cancer, have a defect in the induction of p53 in response to DNA damage caused by ionizing radiation (Kastan et al., 1992). This implies that the optimal induction of p53 following such radiation requires gene products, missing or defective in ataxia telangiectasia, which function upstream of p53 in the signal transduction pathway.

The mechanism whereby p53 leads to G_1 arrest following DNA damage can be attributed to its role as a transcription factor. A radiation-induced increase in p53 levels leads to transcriptional upregulation of a number of proteins including $p21^{WAF1/CIP1}$, GADD45 and Mdm2 (reviewed by Canman and Kastan, 1995). $p21^{WAF1/CIP1}$ is a critical downstream effector in the p53-dependent G_1 checkpoint as it acts as a potent inhibitor of cyclin-dependent kinases, resulting in cell cycle arrest at the G_1/S border (el-Deiry et al., 1994). Induction of the GADD45 gene following ionizing radiation is believed to be important in suppression of cell cycle progression and is dependent on normal p53 function (Zhan et al., 1994). In contrast to $p21^{WAF1/CIP1}$ and GADD45, Mdm2 expression is not involved in cell cycle arrest but is thought to function in a regulatory negative feedback loop by binding to p53 and thereby suppressing its transcriptional activity (Chen et al., 1994). This would lead to down regulation of p53-dependent genes, such as those involved in cell cycle arrest, with the concomitant re-entry into the cell cycle presumably after the DNA damage has been repaired.

In addition to cell cycle arrest, apoptosis is a frequent consequence of the exposure of mammalian cells to DNA damage (Cohen et al., 1992) and often occurs through a p53-dependent pathway. The decision of a cell to undergo p53-mediated G_1 arrest or apoptosis in response to DNA damage is dependent on the cellular context. Immature thymocytes from p53-deficient mice, like those from control mice, undergo apoptosis in response to glucocorticoids or to activation with phorbol ester and ionomycin. However, in contrast to cells from control mice, immature thymocytes deficient in p53 are resistant to the induction of apoptosis by ionizing radiation (Lowe et al., 1993) or by the DNA-damaging topoisomerase II inhibitor etoposide (Clarke et al., 1993). Interestingly, thymocytes which contain only one wild-type p53 allele exhibit an intermediate sensitivity to apoptosis induction by ionizing radiation compared to normal thymocytes and those from p53-null mice. This suggests that the effect of p53 on this pathway is gene-dose-dependent. Furthermore, bone marrow cells (Lotem and Sachs, 1993) and intestinal epithelial cells (Clarke et al., 1994) from p53-deficient mice are also resistant to radiation-induced apoptosis as compared to wild-type p53 controls. These cells can undergo apoptosis in response to different stimuli, indicating that p53-dependent and -independent apoptotic pathways exist. Direct transcriptional regulation of members of the *bcl-2* family may also be important in the induction of apoptosis by p53.

The presence or absence of growth factors can influence p53-mediated growth arrest or apoptosis. For example, bone-marrow-derived haematopoietic cells are resistant to the induction of apoptosis by ionizing radiation when cultured in the presence of appropriate growth factors (Collins et al., 1992), but arrest in G_1. When cultured without these factors, apoptosis occurs through a p53-dependent pathway (Blandino et al., 1995) which is greatly accelerated upon exposure to DNA damage (Canman et al., 1995).

The inability to enter into apoptosis following DNA damage has important clinical implications. Loss of p53-mediated apoptosis may be one mechanism by which tumours with p53 mutations acquire resistance to radio-therapy and chemo-therapy (Lowe et al., 1993). Moreover, transgenic mice which express mutant p53 alleles have an increased susceptibility to radiation-induced tumour formation (Lee et al., 1994). Patients with Wilms' tumours containing mutant p53 have a poor prognosis and are more resistant to radio- and chemo-therapy compared to Wilms' tumours containing wild-type p53 (Bardeesy et al., 1995). Thus, if radiation is used to treat tumours in such individuals, an increase in radiation-induced mutation and consequent tumour progression may be an unfortunate side effect (Kastan et al., 1995).

Inhibition of apoptosis by pRB. pRB not only controls cell cycle progression but also plays a critical role in the development and differentiation of certain cell types. Mice homozygous for mutant pRB die between 10–14 days gestation and show increased levels of both cell division and cell death by apoptosis in the haematopoietic and nervous systems (Clarke et al., 1992). Inappropriate apoptosis has also been observed in the ocular lens of these pRB-deficient mice (Morgenbesser et al., 1994). These observations, together with the interactions between pRB and viral proteins, suggest that pRB not only inhibits proliferation and transformation but also apoptosis.

pRB has recently been shown to protect human SAOS-2 osteosarcoma cells from radiation-induced apoptosis by a p53-independent mechanism (Haas-Kogan et al., 1995). Transfection of pRB into SAOS-2 cells, which lack endogenous pRB and p53, results in a reduced level of radiation-induced apoptosis as compared to non-transfected controls. Moreover, SAOS-2 cells transfected with a mutant form of pRB which fails to bind the adenovirus E1A protein (see below), or the transcription factor E2F, are not protected from apoptosis. The mechanism by which pRB protects cells from radiation-induced apoptosis remains to be elucidated. One proposed model involves induction of a quiescent state in which cells are less sensitive to apoptosis (Clarke et al., 1992).

The product of the retinoblastoma gene therefore serves a dual function: control of cell cycle progression and suppression of apoptosis. Thus, pRB, and possibly other pRB-related proteins, are likely to play a major role in determining whether cells respond to a given stimulus by undergoing growth arrest or apoptosis.

Interactions of p53 and pRB with viral proteins. SV40, some adenoviruses and HPV type 16 and 18 encode oncoproteins that bind pRB and p53, indicating that impairment of both pRB and p53 function is essential for viral strategies. The SV40 large T antigen, the adenovirus E1A protein and the HPV E7 protein bind to and inactivate pRB (DeCaprio et al., 1988; Whyte et al., 1988; Dyson et al., 1989). Complex formation between pRB and the viral proteins requires sequences in the viral proteins that are also important for transformation. Interaction of E1A with pRB results in apoptosis (Rao et al., 1992) and the E1A sequences which are required for pRB binding and transformation are also the same ones responsible for E1A-mediated apoptosis (Haas-Kogan et al., 1995). When mutant forms of E1A, which retain the ability to inactivate pRB and the related proteins p107 and p130, are transfected into pluripotent P19 embryonal carcinoma cells, extensive apoptosis occurs upon induction of neural differentiation (Slack et al., 1995). E1A, SV40 large T antigen and E7 all interact with the hypophosphorylated form of pRB by binding to a pRB domain termed the A/B pocket through an amino acid motif, LXCXE (where X is any amino acid). Furthermore, the cellular transcription proteins E2F and BRG1 require the A/B pocket for pRB binding resulting in competition with viral proteins (Chellappan et al., 1992). It is likely that blocking pRB-mediated cell cycle control at the G_1 phase is required to push the infected cell into S phase and initiate viral DNA replication.

The adenovirus E1A oncoprotein also mediates transcription by competing with p53 for interaction with the same domain of the TATA binding protein (TBP). p53 inhibits transcription when bound to TBP. It is therefore conceivable that part of E1A's ability to induce cell proliferation and bring about apoptosis lies in its ability to dissociate p53 from TBP, and activate p53-repressed genes (Horikoshi et al., 1995). The adenovirus E1B 19-kDa protein, which does not bind to p53, has been shown to inhibit p53-dependent apoptosis mediated by E1A (Debbas and White, 1993; Sabbatini et al., 1995) allowing replication of the viral genome.

The SV40 large T antigen, the adenovirus E1B 55-kDa protein and the HPV E6 protein target p53 function by binding to p53, either directly or indirectly (Levine, 1990). Unlike the situation with viral oncoproteins and pRB, no consensus binding site for p53 has been identified. Moreover, each of the viral oncoproteins targets p53 by binding to different regions of the p53 protein (Prives and Manfredi, 1993). Subsequently, both the transcriptional and functional properties of p53 are inhibited by these oncoproteins. The binding of HPV E6 protein, in conjunction with the E6-associated protein (E6-AP) to p53 initiates its degradation by the ubiquitin pathway (Scheffner et al., 1990, 1993), whereas the adenovirus E1B 55-kDa protein and the SV40 large T antigen complex with p53 and inhibit its activity as a transcriptional activator, i.e. they sequester p53 in stable inactive complexes (Reich et al., 1983; Yew and Berk, 1992).

In contrast to the viruses discussed above, the Epstein-Barr virus (EBV) appears to have evolved a strategy which drives resting B cells into proliferation which is not dependent on the inactivation of p53 or pRB (Allday et al., 1995). In fact, EBV infection of resting B cells stimulates p53 synthesis. Furthermore, cells expressing EBV genes undergo p53-dependent apoptosis in response to DNA damage.

Cellular consequences of apoptosis

Gross cellular reorganisation occurs during apoptosis: the chromatin becomes fragmented and condensed, the organelles and the cell shrink, and the cell-surface blebs leading to budding off of membrane-bound apoptopic bodies. Although the mechanisms involved in inducing these processes are still largely unclear, it does appear likely that they proceed in parallel. DNA fragmentation, for example, is not required for the cytoplasmic changes of apoptosis (Jacobson et al., 1994).

Nuclear changes and DNA fragmentation. Chromatin condensation and nuclear envelope breakdown both occur during apoptosis, but the mechanisms by which these occur are unknown. During mitosis chromatin condensation and nuclear envelope breakdown occur via the solubilisation of nuclear lamin upon phosphorylation by $p34^{cdc2}$ kinases (Peter et al., 1990). Lamins are intermediate filaments that associate with the inner nuclear envelope. They organise the nucleoskeleton and provide a framework for the attachment of chromatin to the nuclear envelope. In contrast to mitosis, during apoptosis lamin disassembly occurs by proteolysis and is apparently irreversible (Kaufmann, 1989). In thymocytes, glucocorticoids promote the degradation of lamin B1 prior to the onset of DNA fragmentation (Neamati et al., 1995). The lamin protease responsible, LamP, is distinct from prICE but the inactivation of prICE does inactivate LamP (Earnshaw, 1995a). This suggests that either prICE activates LamP or that the two act in parallel. Lamin B binds to specific DNA sequence motifs called matrix attachment regions that mediate the interaction of chromatin with the nuclear matrix. Matrix attachment regions are regularly spaced in chromatin at 20–50-kb intervals. It has been suggested that the degradation of lamin may promote the formation of large fragments of DNA formation by allowing the release of matrix attachment regions to give access for endonucleases (Neamati et al., 1995). However, this is at present hypothetical.

The formation of distinct DNA fragments of oligonucleosomal size (180–200-bp lengths) is a biochemical hallmark of apoptosis in many cells. The DNA is cleaved between the nucleosomes resulting in a 'ladder' of DNA fragments of multiples of 180–200 bp. Recent observations also suggest that large DNA fragments (30–50 and 200–300 kb) and even single-strand-cleavage events occur during cell death. A novel nuclease which cleaves 100-bp fragments has also been reported (Solis-Recendez et al., 1995).

Large DNA fragments have been reported to occur both in the presence and absence of oligonucleosomal fragments (Oberhammer et al., 1993). The large fragments may not be precursors of the oligonucleosomal fragments as both types of fragments can be produced, apparently independently, under some conditions (Bortner et al., 1995). The nuclease responsible for large fragment formation is not inhibited by zinc (Earnshaw, 1995b). Topoisomerase II is thought to be important in the attachment of the 50-kb chromatin loop domain to the matrix attachment regions on the nuclear scaffold and it has been implicated as a nuclease involved in chromatin fragmentation (Arends and Wyllie, 1991; Sun and Cohen, 1994). However, topoisomerase II does not seem to play an active part in the formation of 50-kb fragments in HL-60 and MOLT-4 cells (Beere et al., 1995).

It is widely assumed that the DNA cleavage characteristic of apoptosis is the result of an endogenous neutral Ca^{2+}- and Mg^{2+}-dependent endonuclease activity capable of inducing double-strand breaks at internucleosomal sites (Arends et al., 1990). The endonuclease is inhibited by zinc and produces 3′-OH DNA breaks (Earnshaw, 1995b). Several candidate endonucleases of different cellular origins have been reported (see Table 2). For recent reviews on nucleases in apoptosis see Peitsch et al. (1994) and Bortner et al. (1995).

DNase I has been proposed as a candidate for the endonuclease activity during apoptosis (Peitsch et al., 1993a). DNase I is Ca^{2+}- and Mg^{2+}-dependent and has a pH optimum of 7.5 (range 5.5–9.0). Actin, a specific inhibitor of DNase I, and antibodies against DNase I reduce DNA laddering in thymocytes. The overexpression of DNase I induces the morphological and biochemical changes observed during apoptosis in transfected COS cells (Polzar et al., 1993). Single-stranded nicks spaced by 10 bp, characteristic of DNase I cleavage, are observed in apoptotic thymocyte DNA ladders (Peitsch et al., 1993b). The single-strand breaks occur in both the internucleosomal linker region and in the histone-associated DNA. However, cuts in the nucleosomal DNA are much less frequent. This results in fragments corresponding to multiples of 140–195 bp due to the close proximity of single-strand breaks in the linker region, with a low background of fragments with a spacing of approximately 10 bp. DNase I seems to require a cofactor for activity. Purified DNase I does not produce a clean ladder pattern, but if preincubated with nuclear extracts or normal serum it gains the ladder-forming activity (Peitsch et al., 1993a). Despite its proposed role in DNA fragmentation, DNase I is localised in the endoplasmic reticulum and the perinuclear region of the Golgi complex of the cell (Polzar et al., 1993), and the cDNA sequence of DNase I predicts the presence of a signal peptide suggesting its secretion from the cell. So how does it gain access to the nuclear DNA? Raised intracellular Ca^{2+} levels such as occur in the onset of apoptosis can cause the endoplasmic reticulum to disperse and dissociate. Also lamin solubilisation, which can lead to breakdown of the nuclear envelope, is an early event in apoptosis which could allow access of DNase I to the nucleus. However, whether DNase I can gain access to the nuclear DNA during apoptosis has yet to be proven.

DNase II is predominantly located in lysosomes, but is also found in the nucleus. It can function independently of Ca^{2+} and Mg^{2+} and requires an acid pH for activation (pH optimum 5.5; range 3.0–7.0). Purified DNase II can cause the apoptotic DNA ladder in isolated Chinese hamster ovary cell nuclei (Barry and Eastman, 1993). There is some experimental evidence for intracellular acidification sufficient to activate DNase II during apoptosis (Barry et al., 1993; Perez-Sala et al., 1995; Rebollo et al., 1995). However, although DNase II may function in some circumstances, intracellular acidification during apoptosis is not widely reported. It seems unlikely that DNase II would fit the role of an apoptotic nuclease because it generates 5′-OH ends on the DNA and is not inhibited by zinc which is known to inhibit DNA fragmentation (McGowan et al., 1994). However, zinc inhibits the intracellular acidification required for DNase II activity which may be an alternative means by which it inhibits DNA fragmentation (Morana et al., 1994).

Nuc18 was isolated from the nuclei of apoptotic thymocytes and is dependent on the glucocorticoid receptor (Caron-Leslie et al., 1991). It is a neutral endonuclease induced by dexamethasone and is dependent on Ca^{2+} and Mg^{2+} for activity. Nuc18 may exist in a complex with a repressor protein which separates from Nuc18 after glucocorticoid treatment. Its nuclear localisa-

Table 2. Candidate endonucleases for DNA cleavage in apoptosis. Abbreviations: NBC, *N*-bromosuccinamide; ATA, aurintricarboxylic acid; DEPC, diethylpyrocarbonate. This table has been adapted from Peitsch et al. (1994).

DNase	Ion requirements	Inhibitors	Cellular localisation	Tissue distribution	Inducing conditions
DNase I[a] 32–37 kDa	Ca^{2+} and Mg^{2+} (or Mn^{2+})	Zn^{2+}, EDTA, DEPC, actin	endoplasmic reticulum	widely distributed[d]	androgen withdrawal following castration in rats
DNase II[b] 29 kDa	none	NBC, ATA	lysosome, nucleus	widely distributed	
Nuc-18[c] 18 kDa	Ca^{2+} and Mg^{2+}	Zn^{2+}, EDTA, DEPC, ATA	nucleus	thymocytes	glucocorticoid, Ca^{2+} ionophore
Nuc-1		Ca^{2+}, Mg^{2+} and Mn^{2+}	not determined	identified in *C. elegans*	programmed cell death in the nematode

tion and cation dependence suggest that it plays a role in apoptosis. It shows sequence similarity to cyclophilin A and related proteins which are intracellular binding proteins for the immunosuppressive drug cyclosporin. Recombinant cyclophilins have also shown a Ca^{2+}/Mg^{2+}- dependent nuclease activity (Montague et al., 1994).

Several other candidate endonucleases have been proposed. Ribeiro and Carson (1993) described a 27-kDa nuclease purified from human spleen cell nuclei. The endonuclease had a pH optimum of 8.0, required Ca^{2+} and Mg^{2+} for activity and was inhibited by zinc. It was able to produce the mono- and oligonucleosomal fragmentation characteristic of apoptosis. Two neutral DNases designated nuc-40 and nuc-58 have been described in the extract of IL-2-deprived CTLL2 cells (Deng and Podack, 1995). Nuc-58 was dependent on Ca^{2+} and Mg^{2+} and found primarily in the cytoplasm. Nuc-40 showed a preferential nuclear localisation and was active in the presence of Mg^{2+} ions alone.

In the nematode *C. elegans* the enzyme Nuc-1 has been associated with DNA fragmentation but is not essential for programmed cell death. It is also expressed by the engulfing cell rather than by the dying cell itself (see Fig. 1).

Cytoplasmic changes. One of the most noticeable morphological features of apoptosis is the fragmentation of the cell into apoptotic bodies. A rearrangement of the microfilament network of the cell must occur during this process and may be responsible for some of the observed changes.

The microtubule-disrupting agents colchicine, vinblastine and nocodazole all induce apoptosis, suggesting that disruption of the microtubule network initiates events which lead to apoptosis (Martin and Cotter, 1990). Colchicine induces fragmentation of tubulins, increases in Ca^{2+} concentration, a change in the intracellular location of c-Fos and an increased expression of nitric oxide synthase in rat cerebellar granule cells (Bonfoco et al., 1995). Treatment with taxol to stabilise the microtubules prevented DNA laddering and apoptotic body formation induced by colchicine. Prior treatment of HL-60 cells with cytochalasin B, a microfilament-disrupting agent, followed by exposure to apoptotic stimuli results in nuclear fragmentation and DNA cleavage (Cotter et al., 1992). However, there is no production of apoptotic bodies, which suggests that actin microfilaments are essential in membrane breakdown.

Tissue-type or type II transglutaminase is a cytosolic protein of approximately 80 kDa which selectively accumulates in cells undergoing death by apoptosis. The role of transglutaminase in apoptosis has recently been reviewed by Piacentini et al. (1994a). Transglutaminases are a family of Ca^{2+}-dependent glutamine and glutamyl transferases. They catalyze the formation of N^{ε}-(γ-glutamyl) lysine isodipeptide cross links in proteins between the γ-amide of a donor glutamine and the ε-NH_2 of an acceptor lysine. They also catalyse the cross linking of glutamine residues through di- or polyamines. This post-translational modification forms biologically irreversible cross links which confer high resistance to mechanical breakage and chemical attack (Fesus et al., 1989). Five transglutaminases have been identified in humans (reviewed by Greenberg et al., 1991): tissue transglutaminase (TGase 2), blood coagulation factor XIII, TGase 4, keratinocyte transglutaminase (TGase 1) and epidermal transglutaminase (TGase 3). The enzymes appear to have very divergent functions, but all are important in protection of cell and tissue integrity.

At the onset of apoptosis there is a large increase in type II transglutaminase mRNA and an elevation in intracellular Ca^{2+} sufficient to activate the enzyme. The overexpression of transglutaminase in BALB/c-3T3 and neuroblastoma cells induces the cytoplasmic changes characteristic of cells undergoing apoptosis (Gentile et al., 1992; Melino et al., 1994). The cross linking of intracellular proteins will stabilise the cytoplasm of dying cells, preventing the leakage of harmful intracellular elements into the extracellular environment which could lead to an inflammatory response. Immunological characterisation of the cross-linked matrix indicates that actin, annexin II, vinculin, fibronectin and other uncharacterised proteins are components (Knight et al., 1993).

Cell membrane alterations in apoptosis. During apoptosis, there are extensive cell membrane alterations, the cell detaches from neighbouring cells, from the culture substrate *in vitro* or from the extracellular matrix *in vivo*, and the membrane loses specialised structures such as microvilli. Apoptotic cells display cell-surface markers that ensure their swift recognition and removal by phagocytic cells. This avoids the leakage of cell components which could cause inflammation and tissue damage. In mammalian cells the $\alpha_v\beta_3$ vitronectin receptor, CD36 and thrombospondin are important in phagocytic recognition (for reviews see Savill et al., 1993, and Savill, 1995). Phagocytic receptors also recognise changes in membrane sugars and lipid composition.

A number of different cells can act as eliminators of apoptotic bodies thereby ensuring their rapid clearance. Many healthy tissues, such as epithelia, do not have a large number of 'professional' macrophages with the capacity to phagocytose apoptotic cells, therefore other cells can also perform this function if required. These latter cells have been termed 'semiprofessional' or 'amateur' phagocytes, depending on the efficiency with which they remove the apoptotic cell (Savill, 1995).

The phagocytosis of apoptotic bodies can be inhibited by peptides and proteins containing an arginine-glycine-aspartic acid (RGD) sequence (Savill et al., 1990). This observation, together with the dependence on divalent cations, first implicated integrins in the recognition of apoptotic neutrophils by macrophages. The integrin family are transmembrane proteins which link the actin binding proteins in the cytoskeleton of the cell via linker molecules to the extracellular matrix of the cell. Almost all extracellular matrix linkers recognised and bound by integrin share an RGD sequence. Blocking by vitronectin receptor ($\alpha_v\beta_3$) specific monoclonal antibodies was used to show that the integrin vitronectin was involved in the phagocytosis. Acquisition of the vitronectin receptor gives monocyte-derived macrophages the ability to recognize apoptotic neutrophils, human lymphocytes and eosinophils (Savill, 1995). The ligand which mediates ingestion is an RGD-bearing adhesive glycoprotein secreted by macrophages, thrombospondin. It has been shown to bridge the vitronectin receptor on the macrophage and an as-yet-unidentified thrombospondin binding moiety on the apoptotic cell. Thrombospondin is encoded by at least four distinct genes which may be alternatively spliced in a tissue-specific manner to yield a range of different proteins. Macrophages also express CD36, an 88-kDa monomeric glycoprotein which co-operates with $\alpha_v\beta_3$ to bind thrombospondin. CD36 confers enhanced phagocytic capacity, since 'amateur' phagocytes transfected with CD36 display the phagocytic capacity of macrophages (Ren et al., 1995).

Phagocytic cells may also recognise apoptotic cells by a change in the lipid composition of their outer plasma membrane. Normally anionic phosphatidylserine is located in the inner plasma membrane, while the neutral phospholipids sphingomyelin and phosphatidylcholine are found on the outer membrane bilayer. During apoptosis phosphatidylserine is exposed on the outer membrane surface, which appears to be important in eliciting a phagocytic response (Fadok et al., 1992). It has recently been demonstrated that lipocortin 1 binds phophatidylserine. Lipocortin 1, a 35-kDa protein with anti-inflammatory and immune suppressive properties, is upregulated in apoptotic cells occurring in mammary regression (McKanna, 1995). It specifically binds Ca^{2+} and phosphatidylserine and may yet be implicated in phosphatidylserine-elicited phagocytosis. Apoptotic cells also acquire the ability to bind annexin V. Annexin V is a calcium-dependent phospholipid binding protein with high affinity for phosphatidylserine (Homburg et al., 1995).

Apoptotic cells have a reduced anionic charge which implies the loss of terminal sialic residues. The removal of these residues may unmask other sugar residues such as *N*-acetylglucosamine and *N*-acetylgalactosamine which could then interact with a phagocytic recognition lectin on the macrophage surface (Duvall et al., 1985). Lectins are proteins that specificallly bind sugar residues such as those of the glycan side chains of glycoproteins. A lectin appears to be important in the phagocytosis of apoptotic bodies by liver endothelial cells and, interestingly, IL-1β enhances their scavenger action (Dini et al., 1995). Fibroblasts can recognize apoptotic neutrophils via the vitronectin receptor and via a mannose/fucose-specific lectin pathway (Hall et al., 1994). This pathway, however, plays no part in phagocytosis of apoptotic neutrophils by macrophages, indicating that cell-specific mechanisms exist.

A role for other receptors in phagocytic recognition of apoptotic bodies has been suggested. Flora and Gregory (1994) have reported that the monocyte-derived macrophage surface antigen defined by the 61D3 monoclonal antibody mediates apoptotic cell recognition by a mechanism which is distinct from the $\alpha_v\beta_3$ receptor pathway. Also, neutrophils lose surface expression of FcγRIII (CD16), probably by proteolytic cleavage of part of the FcγRIII upon *in vitro* activation and the resulting FcγRIII-negative population undergoes apoptosis (Dransfield et al., 1994).

In the nematode *C. elegans* seven genes downstream of the *ced-3* and *ced-4* control point are involved in the process of engulfment of dying cells. Mutant studies on the *C. elegans* genes have suggested that two parallel and distinct modes of recognition of apoptotic cells might be employed (Ellis et al., 1991). The existence of seven different genes in the nematode suggests that several different mechanisms may be employed to phagocytose apoptotic bodies in mammalian tissues. For example, both the vitronectin and lectin mechanisms can be employed by fibroblasts (Hall et al., 1994). The recognition mechanisms described above may act in parallel or be specific to the phagocytosis of distinct cell types.

Outlook

The substantial body of literature on apoptosis, only a fraction of which is discussed here, has almost all been generated in the past few years, and the rate at which data is accumulating in this area is still increasing. It is now recognised that the molecular mechanisms which control apoptosis involve a large number of components, and that many more must remain to be analysed. It is safe to predict, for example, that the next two years will see further members of the *bcl-2* and/or *ice* families cloned. It is much more difficult to predict how rapidly progress will be made in other areas, e.g. in understanding the evasive molecular mechanisms of control by the Bcl-2 family, and in integrating the many different elements which influence apoptosis. The current intensity of research effort provides hope that important advances will be made in the near future. The growing understanding of this fundamental but long neglected area should then prove very valuable both in basic cell biological and biomedical research and, in the longer term, in the diagnosis and therapy of disease.

We are grateful for financial support from the Grand Charity of the National Grand Order of Freemasons, the Wellcome Trust, the Leukaemia Research Fund, the Royal Society and the Medical Research Council (UK).

REFERENCES

Abbadie, C., Kabrun, N., Bouali, F., Smardova, J., Stehelin. D., Vandenbunder, B. & Enrietto, P. J. (1993) High levels of c-*rel* expression are associated with programmed cell death in the developing avian embryo and in bone marrow cells *in vitro*, *Cell 75*, 899–912.

Akao, Y., Otsuki, Y., Kataoka, S., Ito, Y. & Tsujimoto, Y. (1994) Multiple subcellular localization of Bcl-2: detection in nuclear outer-membrane, endoplasmic reticulum membrane, and mitochondrial membranes, *Cancer Res. 54*, 2468–2471.

Allday, M. J., Sinclair, A., Parker, G., Crawford, D. H. & Farrell, P. J. (1995) Epstein-Barr virus efficently immortalizes human B cells without neutralising the function of p53, *EMBO J. 14*, 1382–1391.

Alnemri, E. S., Fernandes, T. F., Haldar, S., Croce, C. M. & Litwack, G. (1992a) Involvement of BCL-2 in glucocorticoid-induced apoptosis of human pre-B-leukemias, *Cancer Res. 52*, 491–495.

Alnemri, E. S., Robertson, N. M., Fernandes, T. F., Croce, C. M. & Litwack, G. (1992b) Overexpressed full-length human BCL2 extends the survival of baculovirus-infected Sf9 insect cells, *Proc. Natl Acad. Sci. USA 89*, 7295–7299.

Alnemri, E. S., Fernandes-Alnemri, T. & Litwack, G. (1995) Cloning and expression of four novel isoforms of human interleukin-1β converting enzyme with different apoptotic activities, *J. Biol. Chem. 270*, 4312–4317.

Arends, M. J., Morris, R. G. & Wyllie, A. H. (1990) Apoptosis. The role of the endonuclease, *Am. J. Pathol. 136*, 593–608.

Arends, M. J. & Wyllie, A. H. (1991) Apoptosis: mechanisms and roles in pathology, *Int. Rev. Exp. Pathol. 32*, 223–254.

Askew, D. S., Ashmun, R. A., Simmons, B. C. & Cleveland, J. L. (1991) Constitutive c-*myc* expression in an IL-3-dependent myeloid cell line suppresses cell cycle arrest and accelerates apoptosis, *Oncogene 6*, 1915–1922.

Azzoni, L., Anegon, I., Calabretta, B. & Perussia, B. (1995) Ligand binding to FcγR induces c-*myc*-dependent apoptosis in IL-2 stimulated NK cells, *J. Immunol. 154*, 491–499.

Baffy, G., Miyashita, T., Williamson, J. R. & Reed, J. C. (1993) Apoptosis induced by withdrawal of interleukin-3 (IL-3) from an IL-3-dependent hematopoietic cell line is associated with repartitioning of intracellular calcium and is blocked by enforced Bcl-2 oncoprotein production, *J. Biol. Chem. 268*, 6511–6519.

Barbacid, M. (1987) *ras* genes, *Annu. Rev. Biochem. 56*, 779–827.

Barber, D. L. & D'Andrea, A. D. (1994) Erythropoietin and interleukin-2 activate distinct JAK kinase family members, *Mol. Cell Biol. 14*, 6506–6514.

Bardeesy, N., Beckwith, J. B. & Pelletier, J. (1995) Clonal expansion and attenuated apoptosis in Wilms' tumors are assaciated with p53 gene mutations, *Cancer Res. 55*, 215–219.

Barry, M. A. & Eastman, A. (1993) Identification of deoxyribonuclease II as an endonuclease involved in apoptosis, *Arch. Biochem. Biophys. 300*, 440–450.

Barry, M. A., Reynolds, J. E. & Eastman, A. (1993) Etoposide-induced apoptosis in human HL-60 cells is associated with intracellular acidification, *Cancer Res. 53*, 2349–2357.

Beaver, J. P. & Waring, P. (1994) Lack of correlation between early intracellular calcium ion rises and the onset of apoptosis in thymocytes, *Immunol. Cell Biol. 72*, 489–499.

Beere, H. M., Chresta, C. M., Alejo-Herberg, A., Skladanowski, A., Dive, C., Larsen, A. K. & Hickman, J. A. (1995) Investigation of the mechanism of higher order chromatin fragmentation observed in drug-induced apoptosis, *Mol. Pharmacol. 45*, 986–996.

Bissonnette, R. P., Echeverri, F., Mahboubi, A. & Green, D. R. (1992) Apoptotic cell death induced by c-*myc* is inhibited by *bcl-2*, *Nature 359*, 552–554.

Black, R. A., Kronheim, S. R. & Sleath, P. R. (1989) Activation of interleukin-1β by a co-induced protease, *FEBS Lett. 247*, 386–390.

Blandino, G., Scardigli, R., Rizzo, M. G., Crescenzi, M., Soddu, S. & Sacchi, A. (1995) Wild-type p53 modulates apoptosis of normal, IL-3 deprived, hematopoietic cells, *Oncogene 10*, 731–737.

Boise, L. H., Gonzalez-Garcia, M., Postema, C. E., Ding, L., Lindsten, T., Turka, L. A., Mao, X., Nuñez, G. & Thompson, C. B. (1993) *bcl-x*, a *bcl-2*-related gene that functions as a dominant regulator of apoptotic cell death, *Cell 74*, 597–608.

Boldin, M. P., Varfolomeev, E. E., Pancer, Z., Mett, I. L., Camonis, J. H. & Wallach, D. (1995) A novel protein that interacts with the death domain of Fas/APO-1 contains a sequence motif related to the death domain, *J. Biol. Chem. 270*, 7795–7798.

Bonfoco, E., Ceccatelli, S., Manzo, L. & Nicotera, P. (1995) Colchicine induces apoptosis in cerebellar granule cells, *Exp. Cell Res. 218*, 189–200.

Borner, C., Martinou, I., Mattmann, C., Irmler, M., Schaerer, E., Martinou, J. C. & Tschopp, J. (1994) The protein Bcl-2α does not require membrane attachment, but two conserved domains to suppress apoptosis, *J. Cell Biol. 126*, 1059–1068.

Bortner, C. D., Oldenburg, N. B. E. & Cidlowski, J. A. (1995) The role of DNA fragmentation in apoptosis, *Trends Cell Biol. 5*, 21–26.

Boudreau, N., Sympson, C. J., Werb, Z. & Bissell, M. J. (1995) Suppression of ICE and apoptosis in mammary epithelial cells by extracellular matrix, *Science 267*, 891–893.

Boyd, J. M., Malstrom, S., Subramanian, T., Venkatesh, L. K., Schaeper, U., Elangovan, B., D'Sa-Eipper, C. & Chinnadurai, G. (1994) Adenovirus E1B 19 kDa and Bcl-2 proteins interact with a common set of cellular proteins, *Cell 79*, 341–351.

Boyd, J. M., Gallo, G. J., Elangovan, B., Houghton, A. B., Malstrom, S., Avery, B. J., Ebb, R. G., Subramanian, T., Chittenden, T., Lutz, R. J. & Chinnadurai, G. (1995) Bik, a novel death-inducing protein shares a distinct sequence motif with Bcl-2 family proteins and interacts with viral and cellular survival-promoting proteins, *Oncogene 11*, 1921–1928

Brunner, T., Mogil, R. J., LaFace, D., Yoo, N. J., Mahboubi, A., Echeverri, F., Martin, S. J., Force, W. R., Lynch, D. H., Ware, C. F. & Green, D. R. (1995) Cell-autonomous Fas (CD95)/Fas-ligand interaction mediates activation-induced apoptosis in T-cell hybridomas, *Nature 373*, 441–444.

Bruno, S., Del Bino, G., Lassota, P., Giaretti, W. & Darzynkiewicz, Z. (1992) Inhibitors of proteases prevent endonucleolysis accompanying apoptotic cell death of HL-60 leukemic cells and normal thymocytes, *Leukemia 6*, 1113–1120.

Buchkovich, K., Duffy, L. A. & Harlow, E. (1989) The retinoblastoma protein is phosphorylated during specific phases of the cell cycle, *Cell 58*, 1097–1105.

Cain, K., Inayat-Hussain, S. H., Kokileva, L. & Cohen, G. M. (1995) Multi-step DNA cleavage in rat liver nuclei is inhibited by thiol reactive agents, *FEBS Lett. 358*, 255–261.

Canman, C. E. & Kastan, M. B. (1995) Induction of apoptosis by tumor suppressor genes and oncogenes, *Semin. Cancer Biol. 6*. 17–25.

Canman, C. E., Gilmer, T. M., Coutts, S. B. & Kastan, M. B. (1995) Growth-factor modulation of p53-mediated growth arrest versus apoptosis, *Genes & Dev. 9*, 600–611.

Caron-Leslie, L. M., Schwartzman, R. A., Gaido, M. L., Compton, M. M. & Cidlowski, J. A. (1991) Identification and characterization of glucocorticoid-regulated nuclease(s) in lymphoid cells undergoing apoptosis, *J. Steroid Biochem. Mol. Biol. 40*, 661–671.

Casciola-Rosen, L. A., Miller, D. K., Anhalt, G. J. & Rosen, A. (1994) Specific cleavage of the 70-kDa protein component of the U1 small nuclear ribonucleoprotein is a characteristic biochemical feature of apoptotic cell death, *J. Biol. Chem. 269*, 30757–30760.

Cerretti, D. P., Kozlosky, C. J., Mosley, B., Nelson, N., Van Ness, K., Greenstreet, T. A., March, C. J., Kronheim, S. R., Druck, T., Cannizzaro, L. A., Huebner, K. & Black, R. A. (1992) Molecular cloning of the interleukin-1β converting enzyme, *Science 256*, 97–100.

Cerretti, D. P., Hollingsworth, L. T., Kozlosky, C. J., Valentine, M. B., Shapiro, D. N., Morris, S. W. & Nelson, N. (1994) Molecular characterisation of the gene for human interleukin-1β converting enzyme, *Genomics 20*, 468–473.

Chellappan, S., Kraus, V. B., Kroger, B., Munger, K., Howley, P. M., Phelps, W. C. & Nevins, J. R. (1992) Adenovirus E1A, simian virus 40 tumour antigen, and human papillomavirus E7 protein share the capacity to disrupt the interaction between transcription factor E2F and the retinoblastoma gene product, *Proc. Natl Acad. Sci. USA 89*, 4549–4553.

Chen, C. Y., Oliner, J. D., Zhan, Q., Fornace, A. J. Jr, Vogelstein, B. & Kastan, M. B. (1994) Interactions between p53 and MDM2 in a mammalian cell cycle checkpoint pathway, *Proc. Natl Acad. Sci. USA 91*, 2684–2688.

Cheng, J. H., Liu, C. D., Koopman, W. J., Mountz, J. D. (1995) Characterization of human Fas gene. Exon-intron organization and promoter region, *J. Immunol. 154*, 1239–1245.

Chinnaiyan, A. M., O'Rourke, K., Tewari, M. & Dixit, V. M. (1995) FADD, a novel death domain-containing protein, interacts with the death domain of Fas and initiates apoptosis, *Cell 81*, 505–512.

Chiou, S. K., Rao, L. & White, E. (1994) BCL-2 blocks p53-dependent apoptosis, *Mol. Cell Biol. 14*, 2556–2563.

Chittenden, T., Harrington, E. A., O'Connor, R., Flemington, C., Lutz, R. J., Evan, G. I. & Guild, B. C. (1995) Induction of apoptosis by the Bcl-2 homologue Bak, *Nature 374*, 733–736.

Chiu, V. K., Walsh, C. M., Liu, C. C., Reed, J. C. & Clark, W. R. (1995) Bcl-2 blocks degranulation but not Fas-based cell mediated cytotoxicity, *J. Immunol. 154*, 2023–2032.

Chlichlia, K., Moldenhauer, G., Daniel, P. T., Busslinger, M., Gazzolo, L., Schirrmacher, V. & Khazaie, K. (1995) Immediate effects of reversible HTLV-1 tax function: T-cell activation and apoptosis, *Oncogene 10*, 269–277.

Choi, S. S., Park, I. C., Yun, J. W., Sung, Y. C., Hong, S. I. & Shin, H. S. (1995) A novel Bcl-2 related gene, Bfl-1, is overexpressed in stomach cancer and preferentially expressed in bone marrow, *Oncogene 11*, 1693–1698.

Clarke, A. R., Maandag, E. R., van Roon, M., van der Lugt, N. M. T., van der Valk, M., Hooper, M. L., Berns, A. & te Riele, H. (1992) Requirement for a functional Rb-1 gene in murine development, *Nature 359*, 328–330.

Clarke, A. R., Purdie, C. A., Harrison, D. J., Morris, R. G., Bird, C. C., Hooper, M. L. & Wyllie, A. H. (1993) Thymocyte apoptosis induced by p53-dependent and independent pathways, *Nature 362*, 849–852.

Clarke, A. R., Gledhill, S., Hooper, M. L., Bird, C. C. & Wyllie, A. H. (1994) p53 dependence of early apoptotic and proliferative responses within the mouse intestinal epithelium following γ-irradiation, *Oncogene 9*, 1767–1773.

Cleveland, J. L., Troppmair, J., Packham, G., Askew, D. S., Lloyd, P., Gonzalez-Garcia, M., Nuñez, G., Ihle, J. N. & Rapp, U. R. (1994) v-*raf* suppresses apoptosis and promotes growth of interleukin-3-dependent myeloid cells, *Oncogene 9*, 2217–2226.

Cleveland, J. L. & Ihle, J. N. (1995) Contenders in FasL/TNF death signaling, *Cell 81*, 479–482.

Cogswell, J. P., Cogswell, P. C., Kuehl, W. M., Cuddihy, A. M., Bender, T. M., Engelke, U., Marcu, K. B. & Ting, J. P.-Y. (1993) Mechanism of c-*myc* regulation by c-*myb* in different cell lineages, *Mol. Cell. Biol. 13*, 2858–2869.

Cohen, J. J., Duke, R. C., Fadok, V. A. & Sellins, K. S. (1992) Apoptosis and programmed cell death in immunity, *Annu. Rev. Immunol. 10*, 267–293.

Collins, M. K. L., Marvel, J., Malde, P. & Lopez-Rivas, A. (1992) Interleukin 3 protects murine bone marrow cells from apoptosis induced by DNA damaging agents, *J. Exp. Med. 176*, 1043–1051.

Colotta, F., Polentarutti, N., Sironi, M. & Mantovani, A. (1992) Expression and involvement of c-*fos* and c-*jun* protooncogenes in programmed cell death induced by growth factor deprivation in lymphoid cell lines, *J. Biol. Chem. 267*, 18278–18283.

Cotter, T. G., Lennon, S. V., Glynn, J. M. & Green, D. R. (1992) Microfilament-disrupting agents prevent the formation of apoptotic bodies in tumor cells undergoing apoptosis, *Cancer Res.* 52, 997–1005.

Darmon, A. J., Nicholson, D. W. & Bleackley, R. C. (1995) Activation of the apoptotic protease CPP32 by cytotoxic T-cell-derived granzyme B, *Nature 377*, 446–448.

Dawson, C. W., Eliopoulos, A. G., Dawson, J. & Young, L. S. (1995) BHRF1, a viral homologue of the Bcl-2 oncogene, disturbs epithelial cell differentiation, *Oncogene 9*, 69–77.

Debbas, M. & White, E. (1993) Wild-type p53 mediates apoptosis by E1A which is inhibited by E1B, *Genes & Dev. 7*, 546–554.

DeCaprio, J. A., Ludlow, J. W., Figge, J., Shew, J.-Y., Huang, C.-M. Lee, W.-H., Marsilio, E., Paucha, E. & Livingston, D. M. (1988) SV40 large tumour antigen forms a specific complex with the product of the retinoblastoma susceptibility gene, *Cell 54*, 275–283.

DeCaprio, J. A., Ludlow, J. W., Lynch, D., Furukawa, Y., Griffin, L., Piwnica-Worms, H., Huang, C.-M. & Livingston, D. M. (1989) The product of the retinoblastoma susceptibility gene has properties of a cell cycle regulatory element, *Cell 58*, 1085–1095.

Deng, G. E. & Podack, E. R. (1995) Deoxyribonuclease induction in apoptotic cytotoxic T-lymphocytes, *FASEB J. 9*, 665–669.

Dinarello, C. A. (1991) Interleukin-1 and interleukin-1 antagonism, *Blood 77*, 1627–1652.

Dini, L., Lentini, A., Diez, G. D., Rocha, M., Falasca, Serafino, L. & Vidalvanaclocha, F. (1995) Phagocytosis of apoptopic bodies by liver endothelial cells, *J. Cell Sci. 108*, 967–973.

Donehower, L. A., Harvey, M., Slagle, B. L., McArthur, M. J., Montgomery, C. A. Jr, Butel, J. S. & Bradley, A. (1992) Mice deficient for p53 are developmentally normal but susceptible to spontaneous tumours, *Nature 356*, 215–221.

Dransfield, I., Buckle, A. M., Savill, J. S., McDowall, A., Haslett, C. & Hogg, N. (1994) Neutrophil apoptosis is associated with a reduction in CD16 (FcγRIII) expression, *J. Immunol. 153*, 1254–1263.

Duvall, E., Wyllie, A. H. & Morris, R. G. (1985) Macrophage recognition of cells undergoing programmed cell death (apoptosis), *Immunology 56*, 351–358.

Dyson, N., Howley, P., Munger, K. & Harlow, E. (1989) The human papilloma virus 16 E7 oncoprotein binds to the retinoblastoma gene product, *Science 242*, 934–937.

Earnshaw, W. C. (1995 a) Nuclear changes in apoptosis, *Curr. Opin. Cell Biol. 7*, 337–343.

Earnshaw, W. C. (1995 b) Apoptosis: Lessons from *in vitro* systems, *Trends Cell Biol. 5*, 217–220.

Eastman, A. E., Barry, M. A., Demarcq, C., Li, J. & Reynolds J. E. (1994) Endonuclease associated with apoptosis, in *Apoptosis* (Mihich, E. & Schimke, R. T., eds) pp. 249–259. Plenum Press, New York.

el-Deiry, W. S., Harper, J. W., O'Conner, P. M., Velculescu, V. E., Canman, C. E., Jackman, J., Pietenpol, J. A., Burrell, M., Hill, D. E., Wang, Y. S., Wiman, K. G., Mercer, W. E., Kastan, M. B., Kohn, K. W., Elledge, S. J., Kinzler, K. W. & Vogelstein, B. (1994) WAF1/CIP1 is induced in p53-mediated G1 arrest and apoptosis, *Cancer Res. 54*, 1169–1174.

Ellis, R. E., Jacobson, D. M. & Horvitz, H. R. (1991) Genes required for the engulfment of cell corpses during programmed cell death in *Caenorhabditis elegans*, *Genetics 129*, 79–94.

Emlen, W., Niebur, J. & Kadera, R. (1994) Accelerated *in vitro* apoptosis of lymphocytes from patients with systemic lupus erythematosus, *J. Immunol. 152*, 3685–3692.

Enari, M., Hug, H. & Nagata, S. (1995) Involvement of an ICE-like protease in Fas-mediated apoptosis, *Nature 375*, 78–83.

Erpel, T. & Courtneidge, S. A. (1995) Src family tyrosine kinases and cellular signal transduction pathways, *Curr. Opin. Cell Biol. 7*, 176–182

Evan, G. I., Wyllie, A. H., Gilbert, C. S., Littlewood, T. D., Land, H., Brooks, M., Waters, C. M., Penn, L. Z. & Hancock, D. C. (1992) Induction of apoptosis in fibroblasts by c-*myc* protein, *Cell 69*, 119–128.

Evans, C. A., Owen-Lynch, J. P., Whetton, A. D. & Dive, C. (1993) Activation of the Abelson tyrosine kinase-activity is associated with suppression of apoptosis in hemopoietic, *Cancer Res. 53*, 1735–1738.

Fadok, V. A., Voelker, D. R., Campbell, P. A., Cohen, J. J., Bratton, D. L. & Henson, P. M. (1992) Exposure of phosphatidylserine on the surface of apoptopic lymphocytes triggers specific recognition and removal by macrophages, *J. Immunol. 148*, 2207–2216

Fang, W., Rivard, J. J., Mueller, D. L. & Behrens, T. W. (1994) Cloning and molecular characterization of mouse bcl-X in B lymphocytes and T lymphocytes, *J. Immunol. 153*, 4388–4398.

Fanidi, A., Harrington, E. A. & Evan, G. I. (1992) Cooperative interaction between c-*myc* and *bcl-2* proto-oncogenes, *Nature 359*, 554–556.

Farrow, S. N., White, J. H. M., Martinou, I., Raven, T., Pun, K. T., Grinham, C. J., Martinou, J. C. & Brown, R. (1995) Cloning of a *bcl-2* homolog by interaction with adenovirus E1B 19 K, *Nature 374*, 731–733.

Faucheu, C., Diu, A., Chan, A. W. E., Blanchet, A.-M., Miossec, C., Herve, F., Collard-Dutilleul, V., Gu, Y., Aldape, R. A., Lippke, J. A., Rocher, C., Su, M. S.-S., Livingston, D. J., Hercend, T. & Lalanne, J.-L. (1995) A novel human protease similar to the interleukin-1β converting enzyme induces apoptosis in transfected cells, *EMBO J. 14*, 1914–1922.

Fearnhead, H. O., Rivett, A. J., Dinsdale, D. & Cohen, G. M. (1995) A pre-existing protease is a common effector of thymocyte apoptosis mediated by diverse stimuli, *FEBS Lett. 357*, 242–246.

Fernandes-Alnemri, T., Litwack, G. & Alnemri, E. S. (1994) CPP32, a novel human apoptotic protein with homology to *Caenorhabditis elegans* cell death protein ced-3 and mammalian interleukin-1β-converting enzyme, *J. Biol. Chem. 269*, 30761–30764.

Fernandes-Alnemri, T., Litwack, G. & Alnemri, E. S. (1995) Mch2, a new member of the apoptotic Ced-3/Ice cysteine protease gene family, *Cancer Res. 55*, 2737–2742.

Fernandez-Sarabia, M. J. & Bischoff, J. R. (1993) Bcl-2 associates with the *ras*-related protein R-*ras* p23, *Nature 366*, 274–275.

Fesus, L., Thomazy, V., Autuori, F., Cerù, M. P., Tarcsa, E. & Piacentini, M. (1989) Apoptotic hepatocytes become insoluble in detergents and chaotropic agents as a result of transglutaminase action, *FEBS Lett. 245*, 150–154.

Flomerfelt, F. A. & Miesfeld, R. L. (1994) Recessive mutations in a common pathway block thymocyte apoptosis induced by multiple signals, *J. Cell Biol. 127*, 1729–1742.

Flora, P. K. & Gregory, C. D. (1994) Recognition of apoptotic cells by human macrophages: inhibition by a monocyte/macrophage- specific monoclonal-antibody, *Eur. J. Immunol. 24*, 2625–2632.

Gagliardini, V., Fernandez, P.-A., Lee, R. K. K., Drexler, H. C. A., Rotello, R. J., Fishman, M. C. & Yuan, J. (1994) Prevention of vertebrate neuronal death by the *crmA* gene, *Science 263*, 826–828.

Genaro, A. M., Hortelano, S., Alvarez, A., Martinez-A, C. & Bosca, L. (1995) Splenic B lymphocyte programmed cell death is prevetned by nitric oxide release through mechanisms involving sustained Bcl-2 levels, *J. Clin. Invest. 95*, 1884–1890.

Gentile, V., Thomazy, V., Piacentini, M., Fesus, L. & Davies, P. J. A. (1992) Expression of tissue transglutaminase in Balb-C 3T3 fibroblasts: effects on cellular morphology and adhesion, *J. Cell Biol. 119*, 463–474.

Gillet, G., Guerin, M., Trembleau, A. & Brun, G. (1995) A Bcl-2-related gene is activated in avian cells transformed by the Rous sarcoma virus, *EMBO J. 14*, 1372–1382.

Golstein, P., Ojcius, D. M. & Young, J. D. (1991) Cell death mechanisms and the immune system, *Immunol. Rev. 121*, 29–65.

Golstein, P., Marguet, D. & Depraetere, V. (1995) Homology between reaper and the cell-death domains of Fas and TNFR1, *Cell 81*, 185–186.

Gonzalez-Garcia, M., Perez-Ballestero, R., Ding, L. Y., Duan, L., Boise, L. H., Thompson, C. B. & Nuñez, G. (1994) Bcl-X_L is the major *bcl-x* mRNA form expressed during murine development and its product localizes to mitochondria, *Development 120*, 3033–3042.

Gottschalk, A. R., McShan, C. L., Kilkus, J., Dawson, G. & Quintans, J. (1995) Resistance to anti-IgM-induced apoptosis in a WEH1-231 subline is due to insufficient production of ceramide, *Eur. J. Immunol. 25*, 1032–1038.

Graybill, T. L., Dolle, R. E., Helaszek, C. T., Miller, R. E. & Ator, M. A. (1994) Preparation and evaluation of peptidic aspartyl hemiacetals as reversible inhibitors of interleukin-1β converting enzyme (ICE), *Int. J. Peptide Protein Res. 44*, 173–182

Green, D. R., Mahboubi, A., Nishioka, W., Oja, S., Echeverri, F., Shi, Y., Glynn, J., Yang, Y., Ashwell, J. & Bissonnette, R. (1994) Promotion and inhibition of activation-induced apoptosis in T-cell hybridomas by oncogenes and related signals, *Immunol. Rev. 142*, 321–342.

Greenberg, C. S., Birckbichler, P. J., Rice, R. H. (1991) Transglutaninases: Multi-functional cross-linking enzymes that stabilize tissues, *FASEB J. 5*, 3071–3077.

Greenlund, L. J. S., Deckwerth, T. L. & Johnson, E. M. Jr (1995) Superoxide dismutase delays neuronal apoptosis: a role for reactive oxygen species in programmed neuronal death, *Neuron 14*, 303–315.

Gu, Y., Wu, J., Faucheu, C., Lalanne, J.-L., Diu, A., Livingston, D. J. & Su, M. S.-S. (1995) Interleukin-1β converting enzyme requires oligomerization for activity of processed forms *in vivo*, *EMBO J. 14*, 1923–1931.

Guenette, R. S. & Tenniswood, M. (1995) The role of growth factors in the suppression of active cell death in the prostate: an hypothesis, *Biochem. Cell Biol. 72*, 553–559.

Guenette, R. S., Mooibroek, M., Wong, K., Wong, P. & Tenniswood, M. (1994) Cathepsin B, a cysteine protease implicated in metastatic progression, is also expressed during regression of the rat prostate and mammary glands, *Eur. J. Biochem. 226*, 311–321.

Haas-Kogan, D. A., Kogan, S. C., Levi, D., Dazin, P., T'Ang, A., Fung, Y.-K. T. & Israel, M. A. (1995) Inhibition of apoptosis by the retinoblastoma gene product, *EMBO J. 14*, 461–472.

Hall, S. E., Savill, J. S., Henson, P. M., Haslett, C. (1994) Apoptopic neutrophils are phagocytosed by fibroblasts with particpation of the fibroblast vitronectin receptor and involvement of a mannose/fucose-specific lectin, *J. Immunol. 153*, 3218–3227

Hannun, Y. A. & Obeid, L. M. (1995): Ceramide: an intracellular signal for apoptosis, *Trends Biochem. Sci. 20*, 73–77.

Henderson, S., Rowe, M., Gregory, C., Croom-Carter, D., Wang, F., Longnecker, R., Keiff, E. & Rickinson, A. (1991) Induction of *bcl-2* expression by Epstein-Barr virus latent membrane protein 1 protects infected B cells from programmed cell death, *Cell 65*, 1107–1115.

Henderson, S., Huen, D., Rowe, M., Dawson, C., Johnson, G. & Rickinson, A. (1993) Epstein-Barr virus-coded BHRF1 protein, a viral homologue of Bcl-2, protects human B cells from programmed cell death, *Proc. Natl Acad. Sci. USA 90*, 8479–8483.

Hermeking, H. & Eick, D. (1994) Mediation of c-Myc-induced apoptosis by p53, *Science 265*, 2091–2093.

Hockenbery, D., Nuñez, G., Milliman, C., Schreiber, R. D. & Korsmeyer, S. J. (1990) Bcl-2 is an inner mitochondrial membrane protein that blocks programmed cell death, *Nature 348*, 334–336.

Hockenbery, D. M., Oltvai, Z. N., Yin, X. M., Milliman, C. L. & Korsmeyer, S. J. (1993) Bcl-2 functions in an antioxidant pathway to prevent apoptosis, *Cell 75*, 241–251.

Hogquist, K. A., Nett, M. A., Unanue, E. R. & Chaplin, D. D. (1991) Interleukin 1 is processed and released during apoptosis, *Proc. Natl Acad. Sci. USA 88*, 8485–8489.

Hollstein, M., Sidransky, D., Vogelstein, B. & Harris, C. C. (1991) p53 mutations in human cancers, *Science 253*, 49–53.

Homburg, C. H. E., de Haas, M., von dem Borne, A. E. G. K., Verhoeven, A. J., Reutelingsperger, C. P. M. & Roos, D. (1995) Human neutrophils lose their surface FcγRIII and accquire Annexin V binding sites during apoptosis *in vitro*, *Blood 85*, 532–540

Horikoshi, N., Usheva, A., Chen, J., Levine, A. J., Weinmann, R. & Shenk, T. (1995) Two domains of p53 interact with the TATA-binding protein, and the adenovirus 13S E1A protein disrupts the association, relieving p53-mediated transcriptional repression, *Mol. Cell. Biol. 15*, 227–234.

Hsu, H. L., Xiong, J. & Goeddel, D. V. (1995) The TNF receptor 1-associated protein TRADD signals cell death and NFκB activation, *Cell 81*, 495–504.

Hudson, P. L., Pedersen, W. A., Saltsman, W. S., Liscovitch, M., MacLaughlin, D. T., Donahoe, P. K. & Blusztajn, J. K. (1994) Modulation by sphinoglipids of calcium signals evoked by epidermal growth-factor, *J. Biol. Chem. 269*, 21885–21890.

Huguet, C., Enrietto, P., Vandenbunder, B. & Abbadie, C. (1994) C-rel: a multifunctional transcription factor? *Cell Death Diff. 1*, 71–76.

Itoh, N., Tsujimoto, Y. & Nagata, S. (1993) Effect of *bcl-2* on Fas antigen-mediated cell death, *J. Immunol. 151*, 621–627.

Izquierdo Pastor, M., Reif, K. & Cantrell, D. (1995) The regulation and function of p21ras during T-cell activation and growth, *Immunol. Today 16*, 159–164.

Jacobson, M. D., Burne, J. F., King, M. P., Miyashita, T., Reed, J. C. & Raff, M. C. (1993) Bcl-2 blocks apoptosis in cells lacking mitochondrial DNA, *Nature 361*, 365–369.

Jacobson, M. D., Burne, J. F. & Raff, M. C. (1994) Programmed cell-death and Bcl-2 protection in the absence of a nucleus, *EMBO J. 13*, 1899–1910.

Jacobson, M. D. & Raff, M. C. (1995) Programmed cell death and Bcl-2 protection in very low oxygen, *Nature 374*, 814–816.

Janiak, F., Leber, B. & Andrews, D. W. (1994) Assembly of Bcl-2 into microsomal and outer mitochondrial membranes, *J. Biol. Chem. 269*, 9842–9849.

Jarvis, W. D., Kolesnick, R. N., Fornari, F. A., Traylor, R. S., Gewirtz, D. A. & Grant, S. (1994) Induction of apoptotic DNA damage and cell death by activation of the sphingomyelin pathway, *Proc. Natl Acad. Sci. USA 91*, 73–77.

Ji, L., Zhang, G., Uematsu, S., Akahori, Y. & Hirabayashi, Y. (1995) Induction of apoptotic DNA fragmentation and cell death by natural ceramide, *FEBS Lett. 358*, 211–214.

Kamada, S., Shimono, A., Shinto, Y., Tsujimura, T., Takahashi, T., Noda, T., Kitamura, Y., Kondoh, H. & Tsujimoto, Y. (1995) *bcl-2* deficiency in mice leads to pleiotropic abnormalities: accelerated lymphoid cell death in thymus and spleen, polycystic kidney, hair hypopigmentation, and distorted small intestine, *Cancer Res. 55*, 354–359.

Kamens, J., Paskind, M., Hugunin, M., Talanian, R. V., Allen, H., Banach, D., Bump, N., Hackett, M., Johnston, C. G., Li, P., Mankovich, J. A., Terranova, M. & Ghayur, T. (1995) Identification and characterization of ICH-2, a novel member of the interleukin-1β-converting enzyme family of cysteine proteases, *J. Biol. Chem. 279*, 15250–15256.

Kastan, M. B., Onyekwere, O., Sidransky, D., Vogelstein, B. & Craig, R. W. (1991) Participation of p53 protein in the cellular response to DNA damage, *Cancer Res. 51*, 6304–6311.

Kastan, M. B., Zhan, Q., El-Deiry, W. S., Carrier, F., Jacks, T., Walsh, W. V., Plunkett, B. S., Vogelstein, B. & Fornace, A. J. Jr (1992) A mammalian cell cycle checkpoint pathway utilizing p53 and GADD45 is defective in ataxia telangiectasia, *Cell 71*, 587–597.

Kastan, M. B., Canman, C. E. & Leonard, C. J. (1995) p53, cell-cycle control and apoptosis: implications for cancer, *Cancer Metast. Rev. 14*, 3–15.

Kaufmann, S. H. (1989) Induction of endonucleolytic DNA cleavage in human acute myelogenous leukemia cells by etoposide, camptothecin, and other cytotoxic anticancer drugs, *Cancer Res. 49*, 5870–5878.

Kerr, J. F. R., Wyllie, A. H. & Currie, A. R. (1972) Apoptosis: a basic biological phenomenon with wide-ranging implications in tissue kinetics, *Br. J. Cancer 26*, 239–257.

Khosravi-Far, R. & Der, C. J. (1994) The *Ras* signal transduction pathway, *Cancer Metastasis Rev. 13*, 67–89.

Khosravi-Far, R., Chrzanowska-Wodnicka, M., Solski, P. A., Eva, A., Burridge, K. & Der, C. J. (1995) Dbl and Vav mediate transformation via mitogen-activated protein kinase pathways that are distinct from those activated by oncogenic Ras, *Mol. Cell Biol. 14*, 6848–6857.

Kiefer, M. C., Brauer, M. J., Powers, V. C., Wu, J. J., Umansky, S. R., Tomei, L. D. & Barr, P. J. (1995) Modulation of apoptosis by the widely distributed Bcl-2 homologue Bak, *Nature 374*, 736–739.

Kinoshita, T., Yokota, T., Arai, K. & Miyajima, A. (1995a) Suppression of apoptotic death in hematopoietic-cells by signalling through the IL3/GM-CSF receptors, *EMBO J. 14*, 266–275.

Kinoshita, T., Yokata, T., Arai, K. & Miyajima, A. (1995b) Regulation of Bcl-2 expression by oncogenic Ras protein in hematopoietic cells, *Oncogene 10*, 2207–2212.

Kitada, S., Takayama, S., De Riel, K., Tanaka, S. & Reed, J. C. (1994) Reversal of chemoresistance of lymphoma cells by antisense-mediated reduction of *bcl-2* gene expression, *Antisense Res. Dev. 4*, 71–79.

Knight, R. L., Hand, D., Piacentini, M. & Griffin, M. (1993) Characterization of the transglutaminase-mediated large molecular mass polymer from rat liver; its relationship to apoptosis, *Eur. J. Cell Biol. 60*, 210–216.

Knudson, C. M., Tung, K. S. K., Tourtellotte, W. G., Brown, G. A. J. & Korsmeyer, S. J (1995) Bax-deficient mice with lymphoid hyperplasia and germ cell death, *Science 270*, 96–99.

Kolesnick, R. & Fuks, Z. (1995) Ceramide: a signal for apoptosis or mitogenesis? *J. Exp. Med. 181*, 1949–1952.

Korsmeyer, S. J., Shutter, J. R., Veis, D. J., Merry, D. E. & Oltvai, Z. N. (1993) Bcl-2/Bax: a rheostat that regulates an anti-oxidant pathway and cell death, *Semin. Cancer Biol. 4*, 327–332.

Kostura, M. J., Tocci, M. J., Limjuco, G., Chin, J., Cameron, P., Hillman, A. G., Chartrain, N. A. & Schmidt, J. A. (1989) Identification of a monocyte specific pre-interleukin 1β convertase enzyme, *Proc. Natl Acad. Sci. USA 86*, 5227–5231.

Kowalik, T. F., DeGregori, J., Schwarz, J. K. & Nevins, J. R. (1995) E2F1 overexpression in quiescent fibroblasts leads to induction of cellular DNA synthesis and apoptosis, *J. Virol. 69*, 2491–2500.

Kozopas, K. M., Yang, T., Buchan, H. L., Zhou, P. & Craig, R. W. (1993) MCL1, a gene expressed in programmed myeloid cell differentiation, has sequence similarity to BCL2, *Proc. Natl Acad. Sci. USA 90*, 3516–3520.

Krajewski, S., Tanaka, S., Takayama, S., Schibler, M. J., Fenton, W. & Reed, J. C. (1993) Investigation of the subcellular distribution of the Bcl-2 oncoprotein: residence in the nuclear envelope, endoplasmic reticulum and outer mitochondrial membranes, *Cancer Res. 53*, 4701–4714.

Krajewski, S., Bodrug, S., Gascoyne, R., Berean, K., Krajewska, M. & Reed, J. C. (1994a) Immunohistochemical analysis of Mcl-1 and Bcl-2 proteins in normal and neoplastic lymph nodes, *Am. J. Pathol. 145*, 515–525.

Krajewski, S., Krajewska, M., Shabaik, A., Wang, H. G., Irie, S., Fong, L. & Reed, J. C. (1994b) Immunohistochemical analysis of *in vivo* patterns of Bcl-x expression, *Cancer Res. 54*, 5501–5507.

Krajewski, S., Krejewska, M., Shabaik, A., Miyashita, T., Wang, H. G. & Reed, J. C. (1994c) Immunohistochemical determination of *in vivo* distribution of Bax, a dominant inhibitor of Bcl-2, *Am. J. Pathol. 145*, 1323–1336.

Krajewski, S., Bodrug, S., Krajewska, M., Shabaik, A., Gascoyne, R., Berean, K. & Reed, J. C. (1995) Immunohistochemical analysis of Mcl-1 protein in human tissues-differential regulation of Mcl-1 and Bcl-2 protein-production suggests a unique role for Mcl-1 in control of programmed cell-death *in vivo*, *Am. J. Pathol. 146*, 1309–1319.

Kuerbitz, S. J., Plunkett, B. S., Walsh, W. V. & Kastan, M. B. (1992) Wild-type p53 is a cell cycle checkpoint determinant following irradiation, *Proc. Natl Acad. Sci. USA 89*, 7491–7495.

Kuida, K., Lippke, J. A., Ku, G., Harding, M. W., Livingston, D. J., Su, M. S.-S. & Flavell, R. A. (1995) Altered cytokine export and apoptosis in mice deficient in interleukin-1β converting enzyme, *Science 267*, 2000–2003.

Kumar, S. & Baglioni, C. (1991) Protection from tumor necrosis factor-mediated cytolysis by overexpression of plasminogen activator inhibitor type-2, *J. Biol. Chem. 31*, 20960–20964.

Kumar, S., Tomooka, Y. & Noda, M. (1992) Identifiaction of a set of genes with developmentally down-regulated expression in the mouse brain, *Biochem. Biophys. Res. Commun. 185*, 1155–1161.

Kumar, S., Kinoshita, M., Noda, M., Copeland, N. G. & Jenkins, N. A. (1994) Induction of apoptosis by the mouse *Nedd2* gene, which encodes a protein similar to the product of the *Caenorhabditis elegans* cell death gene *ced-3* and the mammalian IL-1β-converting enzyme, *Genes & Dev. 8*, 1613–1626.

Kyriakis, J. M., App, H., Zhang, X. F., Banerjee, P., Brautigan, D. L., Rapp, U. R. & Avruch, J. (1992) *Raf-*1 activates MAP kinase-kinase, *Nature 358*, 417–421.

Laherty, C. D., Perkins, N. D. & Dixit, V. M. (1993) Human T cell leukemia virus 1 Tax and phorbol 12-myristate 13-acetate induce expression of the A20 zinc finger protein by distinct mechanisms involving nuclear factor-κB, *J. Biol. Chem. 268*, 5032–5039.

Lahti, J. M., Xiang, J., Heath, L. S., Campana, D. & Kidd, V. J. (1995) PITSLRE protein kinase activity is associated with apoptosis, *Mol. Cell Biol. 15*, 1–11.

Lam, M., Dubyak, G., Chen, L., Nuñez, G., Miesfeld, R. L. & Distelhorst, C. W. (1994) Evidence that BCL-2 represses apoptosis by regulating endoplasmic reticulum-associated Ca^{2+} fluxes, *Proc. Natl Acad. Sci. USA 91*, 6569–6573.

Lane, D. P. (1992) p53, guardian of the genome, *Nature 358*, 15–16.

Lazebnik, Y. A., Kaufmann, S. H., Desnoyers, S., Poirier, G. G. & Earnshaw, W. C. (1994) Cleavage of poly(ADP-ribose) polymerase by a proteinase with properties like ICE, *Nature 371*, 346–347.

Lee, J. M. & Bernstein, A. (1993) p53 mutations increase resistance to ionizing radiation, *Proc. Natl Acad. Sci. USA 90*, 5742–5746.

Lee, J. M., Abrahamson, J. L. A., Kandel, R., Donehower, L. A. & Bernstein, A. (1994) Susceptibility to radiation-carcinogenesis and accumulation of chromosomal breakage in p53 deficient mice, *Oncogene 9*, 3731–3736.

Lee, H., Arsura, M., Wu, M., Duyao, M., Buckler, A. J. & Sonenshein, G. E. (1995) Role of Rel-related factors in control of *c-myc* gene transcription in receptor-mediated apoptosis of the murine B cell WEHI 231 line, *J. Exp. Med. 181*, 1169–1177.

Levine, A. J. (1990) The p53 protein and its interactions with the oncogene products of the small DNA tumour viruses, *Virology 177*, 419–426.

Levine, A. J., Momand, J. & Finlay, C. A. (1991) The p53 tumour suppressor gene, *Nature 351*, 453–456.

Li, P., Allen, H., Banerjee, S., Franklin, S., Herzog, L., Johnston, C., McDowell, J., Paskind, M., Rodman, L., Salfeld, J., Towne, E., Tracey, D., Wardwell, S., Wei, F.-Y., Wong, W., Kamen, R. & Seshadri, T. (1995) Mice deficient in IL-1β-converting enzyme are defective in production of mature IL-1β and resistant to endotoxic shock, *Cell 80*, 401–411.

Lin, E. Y., Orlofsky, A., Berger, M. S. & Prystowsky, M. B. (1993) Characterization of *A1*, a novel hemopoietic-specific early-response gene with sequence similarity to *bcl-2*, *J. Immunol. 151*, 1979–1988.

Lithgow, T., van Driel, R., Bertram, J. F. & Strasser, A. (1994) The protein product of the oncogene *bcl-2* is a component of the nuclear envelope, the endoplasmic reticulum, and the outer mitochondrial membrane, *Cell Growth Differ. 5*, 411–417.

Liu, Z.-G., Smith, S. W., McLaughlin, K. A., Schwartz, L. M. & Osborne, B. A. (1994) Apoptotic signals delivered through the T-cell receptor of a T-cell hybrid require the immediate-early gene *nur77*, *Nature 367*, 281–284.

Los, M., Van de Craen, M., Penning, L. C., Schenk, H., Westendorp, M., Baeuerle, P. A., Droge, W., Krammer, P. H., Fiers, W. & Schulze-Osthoff, K. (1995) Requirement of an ICE/CED-3 protease for Fas/APO-1-mediated apoptosis, *Nature 375*, 81–83.

Lotem, J. & Sachs, L. (1993) Hematopoietic cells from mice deficient in wild-type p53 are more resistant to induction of apoptosis by some agents, *Blood 82*, 1092–1096.

Lowe, S. W., Schmitt, E. M., Smith, S. W., Osborne, B. A. & Jacks, T. (1993) p53 is required for radiation-induced apoptosis in mouse thymocytes, *Nature 362*, 847–849.

Lu, Q. L., Hanby, A. M., Nasser Hajibagheri, M. A., Gschmeissner, S. E., Lu, P. J., Taylor-Papadimitiou, J., Krajewski, S., Reed, J. C. & Wright, N. A. (1994) Bcl-2 protein localizes to the chromosomes of mitotic nuclei and is correlated with the cell cycle in cultured epithelial cell lines, *J. Cell Sci. 107*, 363–371.

Mannick, J. B., Asano, K., Izumi, K., Kieff, E. & Stamler, J. S. (1994) Nitric-oxide produced by human B-lymphocytes inhibits apoptosis and Epstein-Barr virus reactivation, *Cell 79*, 1137–1146

Marti, A., Jehn, B., Costello, E., Keon, N., Ke, G., Martin, F. & Jaggi, R. (1994) Protein kinase A and AP-1 (c-Fos/JunD) are induced during apoptosis of mammary epithelial cells, *Oncogene 9*, 1213–1223.

Martin, S. J. & Cotter, T. G. (1990) Disruption of microtubules induces an endogenous suicide pathway in human leukaemia HL-60 cells, *Cell Tissue Kinet. 23*, 545–559.

Martin, S. J., O'Brien, G. A., Nishioka, W. K., McGahon, A. J., Mahboubi, A., Saido, T. C. & Green, D. R. (1995) Proteolysis of fodrin (non-erythroid spectrin) during apoptosis, *J. Biol. Chem. 270*, 6425–6428.

Martinou, J. C., Dubois-Dauphin, M., Staple, J. K., Rodriguez, I., Frankowski, H., Missotten, M., Albertini, P., Talabot, D., Catsicas, S., Pietra, C. & Huarte, J. (1994) Overexpression of BCL-2 in transgenic mice protects neurons from naturally occurring cell death and experimental ischemia, *Neuron 13*, 1017–1030.

Mashima, T., Naito, M., Kataoka, S., Kawai, H. & Tsuruo, T. (1995) Aspartate-based inhibitor of interleukin-1-β-converting enzyme prevents antitumor agent-induced apoptosis in human myeloid leukemia U937 cells, *Biochem. Biophys. Res. Commun. 209*, 907–915.

McConkey, D. J., Nicotera, P., Hartzell, P., Bellomo, G., Wyllie, A. H. & Orrenius, S. (1989) Glucocorticoids activate a suicide process in thymocytes through elevation of cytolsolic Ca^{2+} concentration, *Arch. Biochem. Biophys. 269*, 365–370.

McConkey, D. J., Chow, S. C., Orrenius, S. & Jondal, M. (1990) NK cell-induced cytotoxicity is dependent on a Ca^{2+} increase in the target, *FASEB J. 4*, 2661–2664.

McConkey, D. J., Nicotera, P. & Orrenius, S. (1994) Signaling and chromatin fragmentation in thymocyte apoptosis, *Immunol. Rev. 142*, 343–363.

McDonnell, T. J. & Korsmeyer, S. J. (1991) Progression from lymphoid hyperplasia to high-grade malignant lymphoma in mice transgenic for the t(14; 18), *Nature 349*, 254–256.

McGahon, A., Bissonnette, R., Schmitt, M., Cotter, K. M., Green, D. R. & Cotter, T. G. (1994) BCR-ABL maintains resistence of chronic myelogenous leukemia cells to apoptotic cell death, *Blood 83*, 1179–1187.

McGowan, A. J., Fernandes, R. S., Verhaegen, S. & Cotter, T. G. (1994) Zinc inhibits UV radiation-induced apoptosis but fails to prevent subsequent cell death, *Int. J. Radiat. Biol. 66*, 343–349.

McKanna, J. A. (1995) Lipocortin-1 in apotosis: mammary regression, *Anat. Rec. 242*, 1–10.

Melino, G., Annicchiarico-Petruzzelli, M., Piredda, L., Candi, E., Gentile, V., Davies, P. J. A., Piacentini, M. (1994) Tissue transglutamiase and apoptosis: Sense and anti-sense transfection studies with human neuroblastona cells, *Mol. Cell Biol. 14*, 6584–6596.

Merino, R., Ding, L. Y., Veis, D. J., Korsmeyer, S. J. & Nuñez, G. (1994) Developmental regulation of the Bcl-2 protein and susceptibility to cell death in B lymphocytes, *EMBO J. 13*, 683–691.

Merlo, G. R., Basolo, F., Fiore, L., Duboc, L. & Hynes, N. E. (1995) p53-dependent and p53-independent activation of apoptosis in mammary epithial cells reveals a survival function of EGF and insulin, *J. Cell. Biol. 128*, 1185–1196.

Messmer, U. K., Lapetina, E. G. & Brune, B. (1995) Nitric oxide-induced apoptosis in RAW-264. 7 macrophages is antagonized by protein-kinase-C- and protein kinase-A-activating compounds, *Mol. Pharm. 47*, 757–765.

Mihara, K., Cao, X.-R., Yen, A., Chandler, S., Driscoll, B., Murphree, A. L., T'Ang, A. & Fung, Y.-K. (1989) Cell cycle-dependent regulation of phosphorylation of the human retinoblastoma gene product, *Science 246*, 1300–1303.

Miura, M., Zhu, H., Rotello, R., Hartwieg, E. A. & Yuan, J. (1993) Induction of apoptosis in fibroblasts by IL-1β-converting enzyme, a mammalian homolog of the *C. elegans* cell death gene *ced-3*, *Cell 75*, 653–660.

Miyashita, T. & Reed, J. C. (1993) Bcl-2 oncoprotein blocks chemotherapy-induced apoptosis in a human leukemia cell line, *Blood 81*, 151–157.

Miyashita, T., Harigai, M., Hanada, M. & Reed, J. C. (1994a) Identification of a p53-dependent negative response element in the *bcl-2* gene, *Cancer Res. 54*, 3131–3135.

Miyashita, T., Krajewski, S., Krajewska, M., Wang, H. G., Lin, H. K., Liebermann, D. A., Hoffman, B. & Reed, J. C. (1994b) Tumor-suppressor p53 is a regulator of *bcl-2* and *bax* gene-expression *in vitro* and *in vivo*, *Oncogene 9*, 1799–1805.

Miyashita, T. & Reed, J. C. (1995) Tumor-suppressor p53 is a direct transcriptional activator of the human *bax* gene, *Cell 80*, 293–299.

Monica, K., Chen-Levy, Z. & Cleary, M. L. (1990) Small G proteins are expressed ubiquitously in lymphoid cells and do not correspond to Bcl-2, *Nature 346*, 189–191.

Montague, J. W., Gaido, M. L., Frye, C. & Cidlowski, J. A. (1994) A calcium-dependent nuclease from apoptotic rat thymocytes is homologous with cyclophilin. Recombinant cyclophilins A, B and C have nuclease activity, *J. Biol. Chem. 269*, 18877–18880.

Moore, J., Boswell, S., Hoffman, R., Burgess, G. & Hromas, R. (1993) Mutant H-ras over-expression inhibits a random apoptotic nuclease in myeloid leukemia cells, *Leuk. Res. 17*, 703–709.

Morana, S., Li, J. F., Springer, E. W. & Eastman, A. (1994) The inhibition of etoposide induced apoptosis by zinc is associated with modulation of intracellular pH, *Int. J. Oncol. 5*, 153–158.

Morgenbesser, S. D., Williams, B. O., Jacks, T. & DePinho, R. A. (1994) p53-dependent apoptosis produced by RB-deficiency in the developing mouse lens, *Nature 371*, 72–74.

Motoyama, N., Wang, F. P., Roth, K. A., Sawa, H., Nakayama, K., Nakayama, K., Negishi, I., Senju, S., Zhang, Q., Fujii, S. & Loh, D. Y. (1995) Massive cell death of immature hematopoietic cells and neurons in Bcl-x-deficient mice, *Science 267*, 1506–1510.

Munday, N. A., Vaillancourt, J. P., Ali, A., Casano, F. J., Miller, D. K., Molineaux, S. M., Yamin, T.-T., Yu, V. L. & Nicholson, D W. (1995) Molecular cloning and pro-apoptotic activity of $ICE_{rel}II$ and $ICE_{rel}III$, members of the ICE/CED-3 family of cysteine proteases, *J. Biol. Chem. 270*, 15870–15876.

Munn, D. H., Beall, A. C., Song, D., Wrenn, R. W. & Throckmorton, D. C. (1995) Activation-induced apoptosis in human macrophages: developmental regulation of a novel cell-death pathway by macrophage-colony-stimulating factor and interferon-gamma, *J. Exp. Med. 181*, 127–136.

Muschel, R. J., Bernhard, E. J., Garza, L., McKenna, W. G. & Koch, C. J. (1995) Induction of apoptosis at different oxygen-tensions: evidence that oxygen radicals do not mediate apoptotic signaling, *Cancer Res. 55*, 995–998.

Nagata, S. & Golstein, P. (1995) The Fas death factor, *Science 267*, 1449–1456.

Nakayama, K., Nakayama, K., Negishi, I., Kuida, K., Sawa, H. & Loh, D. Y. (1994) Targeted disruption of Bcl-2-α-β in mice: occurrence of gray hair, polycystic kidney disease, and lymphocytopenia, *Proc. Natl Acad. Sci. USA 91*, 3700–3704.

Neamati, N., Fernandez, A., Wright, S., Kiefer, J. & McConkey, D. J. (1995) Degradation of lamin B_1 precedes oligonucleosomal DNA fragmentation in apoptotic thymocytes and isolated thymocyte nuclei, *J. Immunol. 154*, 3788–3795.

Neilan, J. G., Lu, Z., Afonso, C. L., Kutish, G. F., Sussman, M. D. & Rock, D. L. (1993) An African swine fever virus gene with similarity to the proto-oncogene *bcl-2* and the Epstein-Barr virus gene *BHRF1*, *J. Virol. 67*, 4391–4394.

Nevins, J. R. (1992) E2F: a link between the Rb tumour suppressor protein and viral oncoproteins, *Science 258*, 424–429.

Nguyen, M., Branton, P. E., Walton, P. A., Oltvai, Z. N., Korsmeyer, S. J. & Shore, G. C. (1994) Role of membrane anchor domain of Bcl-2 in suppression of apoptosis caused by E1B-deffective adenovirus, *J. Biol. Chem. 269*, 16521–16524.

Nicholson, D. W., Ali, A., Thornberry, N. A., Vaillancourt, J. P., Ding, C. K., Gallant, M., Gareau, Y., Griffin, P. R., Labelle, M., Lazebnik, Y. A., Munday, N. A., Raju, S. M., Smulson, M. E., Yamin, N. A., Yu, V. L. & Miller, D. K. (1995) Identification and inhibition of the ICE/CED-3 protease necessary for mammalian apoptosis, *Nature 376*, 37–43.

Novack, D. V. & Korsmeyer, S. J. (1994) Bcl-2 protein expression during murine development, *Am. J. Pathol. 145*, 61–73.

Oberhammer, F., Wilson, J. W., Dive, C., Morris, I. D., Hickman, J. A., Wakeling, A. E., Walker, P. R. & Sikorska, M. (1993) Apoptotic death in epithelial cells: Cleavage of DNA to 300 and/or 50 kb fragments prior to or in the absence of internucleosomal fragmentation, *EMBO J. 12*, 3679–3684.

Ohkura, N., Hijikuro, M., Yamamoto, A. & Miki, K. (1994) Molecular-cloning of a novel thyroid/steroid-receptor superfamily gene from cultured rat neuronal cells, *Biochem. Biophys. Res. Commun. 205*, 1959–1965.

Ojeda, F., Folch, H., Guarda, M. I., Jastorff, B. & Diehl, H. A. (1995) Induction of apoptosis in thymocytes: new evidence for an interaction of PKC and PKA pathways, *Biol. Chem. Hoppe-Seyler 376*, 389–393.

Okabe, T., Takayanagi, R., Imasaki, K., Haji, M., Nawata, H. & Watanabe, T. (1995) cDNA cloning of a NGFI-B/nur77-related transcription factor from an apoptotic human T-cell line, *J. Immunol. 154*, 3871–3879.

Oltvai, Z. N., Milliman, C. L. & Korsmeyer, S. J. (1993) Bcl-2 heterodimerizes *in vivo* with a conserved homolog, Bax, that accelerates programed cell death, *Cell 74*, 609–619.

Owen-Lynch, P. J., Wong, A. K.-Y. & Whetton, A. D. (1995) v-Abl-mediated apoptotic suppression is associated with SHC phosphorylation without concomitant mitogen-activated protein kinase activation, *J. Biol. Chem 270*, 5956–5962.

Pearson, G. R., Luka, J., Petti, L., Sample, J., Birkenbach, M., Braun, D. & Kieff, E. (1987) Identification of an Epstein-Barr virus early gene encoding a second component of the restricted early antigen complex, *Virology 160*, 151–161.

Peitsch, M. C., Polzar, B., Stephan, H., Crompton, T., MacDonald, H. R., Mannhertz, H. G. & Tschopp, J. (1993a) Characterisation of the endogeneous deoxyriboneclease involved in nuclear DNA degradation during apoptosis (programmed cell death), *EMBO J. 12*, 371–377.

Peitsch, M. C., Muller, C. & Tschopp, J. (1993b) DNA fragmentation during apoptosis is caused by frequent single-strand cuts, *Nucleic Acids Res. 21*, 4206–4209.

Peitsch, M. C., Mannherz, H. G. & Tschopp, J. (1994) The apoptosis endonucleases: cleaning up after cell death? *Trends Cell Biol. 4*, 37–41.

Peitsch, M. C., Irmler, M., French, L. E. & Tschopp, J. (1995) Genomic organization and expression of mouse deoxyribonuclease I, *Biochem. Biophys. Res. Commun. 207*, 62–68.

Perez-Sala, D., Cellado-Escobar, D. & Mollinedo, F. (1995) Intracellular alkalinization supresses lovastatin induced apoptosis in HL-60 cells through the inactivation of a pH-dependent endonuclease, *J. Biol. Chem. 270*, 6235–6242.

Peter, M., Nakagawa, J., Doreé, M., Labbé, J. C. & Nigg, E. A. (1990) *In vitro* disassembly of the nuclear lamina and M phase-specific phosphorylation of lamins by *cdc2* kinase, *Cell 61*, 591–602.

Piacentini, M., Davies, P. J. A. & Fesus, L. (1994a) Tissue transglutamase in cells undergoing apoptosis, in *Apoptosis II: The molecular basis of apoptosis in disease* (Tomei, D. & Cope, F. O., eds) pp. 143–163, Cold Spring Harbor Laboratory Press, NY.

Piacentini, M., Raschella, G., Calabretta, B. & Melino, G. (1994b) c-Myb down regulation is associated with apoptosis in human retinoblastoma cells, *Cell Death Diff. 1*, 85–92.

Polzar, B., Peitsch, M. C., Loos, R., Tschopp, J. & Mannherz, H. G. (1993) Overexpression of deoxyribonuclease I (DNase I) transfected into COS cells: its distribution during apoptotic cell-death, *Eur. J. Cell Biol. 62*, 397–405.

Polzar, B., Zanotti, S., Stephan, H., Rauch, F., Peitsch, M. C., Irmler, M., Tschopp, J. & Mannherz, H. G. (1994) Distribution of deoxyribonuclease I in rat tissues and its correlation to cellular-turnover and apoptosis (programmed cell death), *Eur. J. Cell Biol. 64*, 200–210.

Prives, C. & Manfredi, J. J. (1993) The p53 tumour suppressor protein: meeting review, *Genes & Dev. 7*, 529–534.

Pushkareva, M., Obeid, L. M. & Hannun, Y. A. (1995) Ceramide: an endogenous regulator of apoptosis and growth suppression, *Immunol. Today 16*, 294–297.

Raff, M. C., Barres, B. A., Burne, J. F., Coles, H. S., Ishizaki, Y. & Jacobson, M. D. (1993) Programmed cell death and the control of cell survival, *Science 262*, 695–700.

Rao, L., Debbas, M., Sabbatini, P., Hockenberg, D., Korsmeyer, S. & White, E. (1992) The adenovirus E1A proteins induce apoptosis which is inhibited by the E1B 19-kDa and Bcl-2 proteins, *Proc. Natl Acad. Sci. USA 89*, 7742–7746.

Ray, C. A., Black, R. A., Kronheim, S. R., Greenstreet, T. A., Sleath, P. R., Salvesen, G. S. & Pickup, D. J. (1992) Viral inhibition of inflammation: cowpox virus encodes an inhibitor of the interleukin-1β converting enzyme, *Cell 69*, 597–604.

Ray, S. D., Kamendulis, L. M., Gurule, M. W., Yorkin, R. D. & Corcoran, G. B. (1993) Ca^{2+} antagonists inhibit DNA fragmentation and toxic cell death induced by acetaminophen, *FASEB J. 7*, 453–463.

Rebollo, A., Gomez, J., De-Aragon, A. M., Lastres, P., Silva, A. & Perez-Sala, D. (1995) Apoptosis induced by IL-2 withdrawal is associated with an intracellular acidification, *Exp. Cell Res. 218*, 581–585.

Reich, N. C., Oren, M. & Levine, A. J. (1983) Two distinct mechanisms regulate the levels of a cellular tumor antigen, *Mol. Cell. Biol. 3*, 2143–2150.

Reisman, D., Elkind, N. B., Roy, B., Beamon, J. & Rotter, V. (1993) c-Myc trans-activates the p53 promoter through a required downstream CACGTG motif, *Cell Growth Differ. 4*, 57–65.

Ren, Y., Silverstein, R. L., Allen, J. & Savill, J. (1995) CD36 gene-transfer confers capacity for phagocytosis of cells undergoing apoptosis, *J. Exp. Med. 181*, 1857–1862.

Reynolds, J. E., Yang, T., Qian, L. P., Jenkinson, J. D., Zhou, P., Eastman, A. & Craig, R. W. (1994) Mcl-1, a member of the Bcl- 2 family, delays apoptosis induced by c-Myc overexpression in Chinese hamster ovary cells, *Cancer Res. 54*, 6348–6352.

Ribeiro, J. M. & Carson, D. A. (1993) Ca^{2+}/Mg^{2+}-dependent endonuclease from human spleen: purification, properties, and role in apoptosis, *Biochemistry 32*, 9129–9136.

Ryan, J. J., Prochownik, E., Gottlieb, C. A., Apel, I. J., Merino, R., Nuñez, G. & Clarke, M. F. (1994) c-*myc* and *bcl-2* modulate p53 function by altering p53 subcellular trafficking during the cell cycle, *Proc. Natl Acad. Sci. USA 91*, 5878–5882.

Sabbatini, P., Chiou, S. K., Rao, L. & White, E. (1995) Modulation of p53-mediated transcriptional repression and apoptosis by the adenovirus E1B 19 K protein, *Mol. Cell Biol. 15*, 1060–1070.

Saez, R., Chan, A. M.-L., Miki, T. & Aaronson, S. A. (1994) Oncogenic activaton of human R-ras by point mutations analogous to those of prototype H-ras oncogenes, *Oncogene 9*, 2977–2982.

Sarin, A., Adams, D. H. & Henkart, P. A. (1993) Protease inhibitors selectively block T cell receptor-triggered programmed cell death in a murine T cell hybridoma and activated peripheral T cells, *J. Exp. Med. 178*, 1693–1700.

Sato, T., Hanada, M., Bodrug, S., Irie, S. J., Iwana, N., Boise, L. H., Thompson, C. B., Golemis, E., Fong, L., Wang, H. G. & Reed, J. C. (1994) Interactions among members of the Bcl-2 protein family analyzed with a yeast two-hybrid system, *Proc. Natl Acad. Sci. USA 91*, 9238–9242.

Savill, J. S., Dransfield, I., Hogg, N. & Haslett, C. (1990) Vitronectin receptor-mediated phagocytosis of cells undergoing apoptosis, *Nature 343*, 170–173.

Savill, J., Fadok, V., Henson, P. & Haslett, C. (1993) Phayocyte recognition of cells undergoing apoptosis, *Immunol. Today 14*, 131–136.

Savill, J. (1995) The innate immune system: recognition of apoptotic cells, in *Apoptosis and the immune response* (Gregory, C. D., ed.) pp. 341–369, Wiley-Liss Inc., New York.

Savitsky, K., Bar-Shira, A., Gilad, S., Rotman, G., Ziv, Y., Vanagaite, L., Tagle, D. A., Smith, S., Uziel, T., Sfez, S., Ashenazi, M., Pecker, I., Frydman, M., Harnik, R., Sankhavaram, R. P., Simmons, A., Clines, G. A., Sartiel, A., Gatti, R. A., Chessa, L., Sanal, O., Lavin, M. F., Jaspers, N. G. J., Taylor, A. M. R., Arlett, C. F., Miki, T., Weissman, S. M., Lovett, M., Collins, F. S. & Shiloh, Y. (1995) A single ataxia telangiectasia gene with a product similar to PI-3 kinase, *Science 268*, 1749–1753.

Scheffner, M., Werness, B. A., Huibregtse, J. M., Levine, A. J. & Howley, P. M. (1990) The E6 oncoprotein encoded by human papillomavirus types 16 and 18 promotes the degradation of p53, *Cell 63*, 1129–1136.

Scheffner, M., Huibregtse, J. M., Vierstra, R. D. & Howley, P. M. (1993) The HPV-16 E6 and E6-AP complex functions as a ubiquitin-protein ligase in the ubiquitination of p53, *Cell 75*, 495–505.

Schulze-Osthoff, K., Walczak, H., Droge, W. & Krammer, P. H. (1994) Cell nucleus and DNA fragmentation are not required for apoptosis, *J. Cell Biol. 127*, 15–20.

Schwartz, R. H. (1992) Costimulation of T lymphocytes: the role of CD28, CTLA-4, and B7/BB1 in interleukin-2 production and immunotherapy, *Cell 71*, 1065–1068.

Selvakumaran, M., Lin, H. K., Miyashita, T., Wang, H. G., Krajewski, S., Reed, J. C., Hoffman, B. & Liebermann, D. (1994) Immediate early up-regulation of *bax* expression by p53 but not $TGF\beta 1$: a paradigm for distinct apoptotic pathways, *Oncogene 9*, 1791–1798.

Sentman, C. L., Shutter, J. R., Hockenbery, D., Kanagawa, O. & Korsmeyer, S. J. (1991) Bcl-2 inhibits multiple forms of apoptosis but not negative selection in thymocytes, *Cell 67*, 879–888.

Shi, Y., Glynn, J. M., Guilbert, L. J., Cotter, T. G., Green, D. R. & Bissonnette, R. P. (1992) Role for c-*myc* in activation-induced apoptotic cell death in T-cell hybridomas, *Science 257*, 212–214.

Shi, L., Nishioka, W. K., Th'ng, J., Bradbury, E. M., Litchfield, D. W. & Greenberg, A. H. (1994) Premature $p34^{cdc2}$ activation required for apoptosis, *Science 263*, 1143–1145.

Shimizu, S., Eguchi, Y., Kosaka, H., Kamiike, W., Matsuda, H. & Tsujimoto, Y. (1995) Prevention of hypoxia-induced cell death by Bcl-2 and Bcl-x_L, *Nature 374*, 811–813.

Slack, R. S., Skerjanc, I. S., Lach, B., Craig, J., Jardine, K. & McBurney, M. W. (1995) Cells differentiating into neuroectoderm undergo apoptosis in the absence of functional retinoblastoma family proteins, *Blood 129*, 779–788.

Slater, A. F. G., Nobel, C. S. I., Maellaro, E., Bustamante, J., Kimland, M. & Orrenius, S. (1995) Nitrogen spin traps and a nitroxide antioxidant inhibit a common pathway of thymocyte apoptosis, *Biochem. J. 306*, 771–778.

Smeyne, R. J., Vendrell, M., Hayward, M., Baker, S. J., Miao, G. G., Schilling, K., Robertson, L. M., Curran, T. & Morgan, J. I. (1993) Continuous c-*fos* expression precedes programmed cell death *in vivo*, *Nature 363*, 166–169.

Smith, C. A. (1995) A Novel Viral Homologue of Bcl-2 and Ced-9, *Trends Cell Biol. 5*, 344.

Smith, C. A., Williams, G. T., Kingston, R., Jenkinson, E. J. & Owen, J. J. T. (1989) Antibodies to CD3/T-cell receptor complex induce death by apoptosis in immature T cells in thymic cultures, *Nature 337*, 181–184.

Solis-Recendez, M. G., Perani, A., D'Habit, B., Stacey, G. N. & Maugras, M. (1995) Hybridoma cell cultures continously undergo apoptosis and reveal a novel 100 bp DNA fragment, *J. Biotechnol. 38*, 117–127.

Somma, M. M. D., Nuti, S., Telford, J. L. & Baldari, C. T. (1995) $p56^{lck}$ plays a key role in transducing apoptotic signals in T cells, *FEBS Lett. 363*, 101–104.

Sorenson, C. M., Rogers, S. A., Korsmeyer, S. J. & Hammerman, M. R. (1995) Fulminant metanephric apoptosis and abnormal kidney development in *bcl-2*-deficient mice, *Am. J. Physiol. 268*, F73–F81.

Spaulding, S. W. (1993) The ways in which hormones changes cyclic adenosine 3′,5′-monophosphate-dependent protein kinase subunits, and how such changes affect cell behaviour, *Endocr. Rev. 14*, 632–650.

Squier, M. K. T., Miller, A. C. K., Malkinson, A. M. & Cohen, J. J. (1994) Calpain activation in apoptosis, *J. Cell Physiol. 159*, 229–237.

Stanger, B. Z., Leder, P., Lee, T. H., Kim, E. & Seed, B. (1995) RIP: a novel protein containing a death domain that interacts with Fas/APO-1 (CD95) in yeast and causes cell death, *Cell 81*, 513–523.

Strasser, A., Harris, A. W., Vaux, D. L., Webb, E., Bath, M. L., Adams, J. M. & Cory, S. (1990) Abnormalities of the immune system induced by dysregulated *bcl-2* expression in transgenic mice, *Curr. Top. Microbiol. Immunol. 166*, 175–181.

Sun, X. M. & Cohen, G. M. (1994) Mg^{2+}-dependent cleavage of DNA into kilobase pair fragments is responsible for the initial degradation of DNA in apoptosis, *J. Biol. Chem. 269*, 14857–14860.

Takayama, S., Sato, T., Krajewski, S., Kochel, K., Irie, S., Millan, J. A. & Reed, J. C. (1995) Cloning and functional analysis of BAG-1: a novel Bcl-2-binding protein with anti-cell death activity, *Cell 80*, 279–284.

Tanaka, S., Saito, K. & Reed, J. C. (1993) Structure-function analysis of the Bcl-2 oncoprotein. Addition of a heterologous transmembrane domain to portions of the Bcl-2β protein restores function as a regulator of cell survival, *J. Biol. Chem. 268*, 10920–10926.

Taylor, A. M. R., Byrd, P. J., McConville, C. M. & Thacker, S. (1994) Genetic and cellular features of ataxia telangiectasia, *Int. J. Radiat. Biol. 65*, 65–70.

Tewari, M. & Dixit, V. M. (1995) Fas- and tumor necrosis factor-induced apoptosis is inhibited by the poxvirus *crmA* gene product, *J. Biol. Chem. 270*, 3255–3260.

Tewari, M., Quan, L. T., O'Rourke, K., Desnoyers, S., Zeng, Z., Beidler, D. R., Poirier, G. G., Salvesen, G. S. & Dixit, V. M. (1995a) Yama/CPP32β, a mammalian homolog of CED-3, is a CrmA-inhibitable protease that cleaves the death substrate poly(ADP-ribose) polymerase, *Cell 81*, 801–809.

Tewari, M., Wolf, F. W., Seldin, M. F., O'Shea, K. S., Dixit, V. M. & Turka, L. A. (1995b) Lymphoid expression and regulation of A20, an inhibitor of programmed cell death, *J. Immunol. 154*, 1699–1706.

Thompson, C. B. (1995) Apoptosis in the pathogenesis and treatment of disease, *Science 267*, 1456–1462.

Thornberry, N. A. & Molineaux, S. M. (1995) Interleukin-1β converting enzyme: a novel cysteine protease required for IL-1β production and implicated in programmed cell death, *Protein Sci. 4*, 3–12.

Thornberry, N. A., Bull, H. G., Calaycay, J. R., Chapman, K. T., Howard, A. D., Kostura, M. J., Miller, D. K., Molineaux, S. M., Weidner, J. R., Aunins, J., Elliston, K. O., Ayala, J. M., Casano, F. J., Chin, J., Ding, G. J.-F., Egger, L. A., Gaffney, E. P., Limjuco, G., Palyha, O. C., Raju, S. M., Rolando, A. M., Salley, J. P., Yamin, T.-T., Lee, T. D., Shively, J. E., MacCross, M., Mumford, R. A., Schmidt, J. A. & Tocci, M. J. (1992) A novel heterodimeric cysteine protease is required for interleukin-1β processing in monocytes, *Nature 356*, 768–774.

Thornberry, N. A., Peterson, E. P., Zhao, J. J., Howard, A. D., Griffin, P. R. & Chapman, K. T. (1994) Inactivation of interleukin-1β converting enzyme by peptide (acyloxy)methyl ketones, *Biochemistry 33*, 3934–3940.

Tomei, L. D. & Cope, F. O. (1991) *Apoptosis: the molecular basis of cell death*, Cold Spring Harbor Laboratory Press, Cold Spring Harbor NY.

Tomei, L. D. & Cope, F. O. (1994) *Apoptosis II: the molecular basis of apoptosis in disease*, Cold Spring Harbor Laboratory Press, Cold Spring Harbor NY.

Tsujimoto, Y. & Croce, C. M. (1986) Analysis of the structure, transcripts, and protein products of *bcl-2*, the gene involved in human follicular lymphoma, *Proc. Natl Acad. Sci. USA 83*, 5214–5218.

Uckun, F. M., Tuel-Ahlgren, L., Song, C. W., Waddick, K., Myers, D. E., Kirihara, J., Ledbetter, J. A. & Schieven, G. L. (1992) Ionizing radiation stimulates unidentified tyrosine-specific protein-kinases in human B-lymphocyte precursors, triggering apoptosis and clonogenic cell death, *Proc. Natl Acad. Sci. USA 89*, 9005–9009.

Uemura, H., Mizokami, A. & Chang, C. (1995) Identification of a new enhancer in the promoter region of human TR3 orphan receptor gene, a member of the steroid receptor superfamily, *J. Biol. Chem. 270*, 5427–5433.

Vaux, D. L., Cory, S. & Adams, J. M. (1988) Bcl-2 gene promotes haemopoietic cell survival and cooperates with c-*myc* to immortalize pre-B cells, *Nature 335*, 440–442.

Voelkel-Johnson, C., Entingh, A. J., Wold, W. S. M., Gooding, L. R. & Laster, S. M. (1995) Activation of intracellular proteases is an early event in TNF-induced apoptosis, *J. Immunol. 154*, 1707–1716.

Wagner, A. J., Small, M. B. & Hay, N. (1993) Myc-mediated apoptosis is blocked by ectopic expression of Bcl-2, *Mol. Cell Biol. 13*, 2432–2440.

Wang, H. G., Miyashita, T., Takayama, S., Sato, T., Torigoe, T., Krajewski, S., Tanaka, S., Hovey, L., Troppmair, J., Rapp, U. R. & Reed, J. C. (1994a) Apoptosis regulation by interaction of Bcl- 2 protein and Raf-1 kinase, *Oncogene 9*, 2751–2756.

Wang, L., Miura, M., Bergeron, L., Zhu, H. & Yuan, J. (1994b) *Ich-1*, an *Ice/ced-3*-related gene, encodes both positive and negative regulators of programmed cell death, *Cell 78*, 739–750.

Wang, H. G., Millan, J. A., Cox, A. D., Der, C. J., Rapp, U. R., Beck, T., Zha, H. B. & Reed, J. C. (1995a) R-Ras promotes apoptosis caused by growth factor deprivation via a Bcl-2 suppressible mechanism, *J. Cell Biol. 129*, 1103–1114.

Wang, Z. Q., Auer, B., Stingl, L., Berghammer, H., Haidacher, D., Schweiger, M. & Wagner, E. F. (1995b) Mice lacking ADPRT and poly(ADP-ribosyl)ation develop normally but are susceptible to skin disease, *Genes & Dev. 9*, 509–520.

Warner, G. L., Ludlow, J. W., Nelson, D. A., Gaur, A. & Scott, D. W. (1991) Anti-immunoglobulin treatment of murine B-cell lymphomas induces active transforming growth factor β but pRB hypophosphorylation is transforming growth factor β independent, *Cell Growth Differ. 3*, 175–181.

Weaver, V. M., Lach, B., Walker, P. R. & Sikorska, M. (1993) Role of proteolysis in apoptosis-involvement of serine proteases in internucleosomal DNA fragmentation in immature thymocytes, *Biochem. Cell Biol. 71*, 488–500.

White, A. E., Livanos, E. M. & Tlsty, T. D. (1994a) Differential disruptions of genomic integrity and cell cycle regulation in normal human fibroblasts by the HPV oncoproteins, *Genes & Dev. 8*, 666–677.

White, K., Grether, M. E., Abrams, J. M., Young, L., Farrell, K. & Steller, H. (1994b) Genetic control of programmed cell death in *Drosophila, Science 264*, 677–683.

White, D. W., Roy, A. & Gilmore, T. D. (1995) The v-Rel oncoprotein blocks apoptosis and proteolysis of IκBα in transformed chicken spleen-cells, *Oncogene 10*, 857–868.

Whyte, P., Buchkovich, K. J., Horowitz, J. M., Friend, S. H., Raybuck, M., Weinberg, R. A. & Harlow, E. (1988) Association between an oncogene and an anti-oncogene: the adenovirus E1A protein binds to the retinoblastoma gene product, *Nature 334*, 124–129.

Williams, G. T. & Smith, C. A. (1993) Molecular regulation of apoptosis: genetic controls on cell death, *Cell 74*, 777–779.

Williams, G. T. (1991) Programmed cell death: apoptosis and oncogenesis, *Cell 65*, 1097–1098.

Williams, G. T. (1994) Apoptosis in the immune system, *J. Pathol. 173*, 1–4.

Willingham, M. C. & Bhalla, K. (1994) Transient mitotic phase localization of *bcl-2* oncoprotein in human carcinoma cells and its possible role in prevention of apoptosis, *J. Histochem. Cytochem. 42*, 441–450.

Wilson, K. P., Black, J. A., Thomson, J. A., Kim, E. E., Griffith, J. P., Navia, M. A., Murcko, M. A., Chambers, S. P., Aldape, R. A., Raybuck, S. A. & Livingston, D. J. (1994) Structure and mechanism of interleukin-1β converting enzyme, *Nature 370*, 270–275.

Wong, K., Li, X. B. & Hunchuk, N. (1995) *N*-Acetylsphingosine (C2-ceramide) inhibited neutrophil superoxide formation and calcium influx, *J. Biol. Chem. 270*, 3056–3062.

Woronicz, J. D., Calnan, B., Ngo, V. & Winoto, A. (1994) Requirement for the orphan steroid receptor Nur77 in apoptosis of T-cell hybridomas, *Nature 367*, 277–281.

Wright, S. C., Wei, Q. S., Zhong, J., Zheng, H., Kinder, D. H. & Larrick, J. W. (1994) Purification of a 24-kD protease from apoptotic tumor cells that activates DNA fragmentation, *J. Exp. Med. 180*, 2113–2123.

Wright, S. D. & Kolesnick, R. N. (1995) Does endotoxin stimulate cells by mimicking ceramide? *Immunol. Today 16*, 297–302.

Wyllie, A. H., Rose, K. A., Morris, R. G., Steel, C. M., Forster, E. & Spandidos, D. A. (1987) Rodent fibroblast tumors expressing human *myc* and *ras* genes: growth, metastasis and endogenous oncogene expression, *Br. J. Cancer 56*, 251–259.

Yanez, R. J., Rodriguez, J. M., Nogal, M. L., Yuste, L., Enriquez, C., Rodriguez, J. F. & Vinuela, E. (1995) Analysis of the complete nucleotide sequence of African swine fever virus, *Virology 208*, 249–278.

Yang, E., Zha, J. P., Jockel, J., Boise, L. H., Thompson, C. B. & Korsmeyer, S. J. (1995a) Bad, a heterodimeric partner for $Bcl\text{-}X_L$ and Bcl-2, displaces Bax and promotes death, *Cell 80*, 285–291.

Yang, T., Kozopas, K. M. & Craig, R. W. (1995b) The intracellular distribution and pattern of expression of Mcl-1 overlap with, but are not identical to, those of Bcl-2, *J. Cell Biol. 128*, 1173–1184.

Yao, K. & Cooper, G. M. (1995) Requirement for phosphatidylinositol-3 kinase in the prevention of apoptosis by nerve growth factor, *Science 267*, 2003–2006.

Yazdanbakhsh, K., Choi, J.-W., Li, Y., Lau, L. F. & Choi, Y. (1995) Cyclosporin A blocks apoptosis by inhibiting the DNA-binding activity of the transcription factor Nur77, *Proc. Natl Acad. Sci. USA 92*, 437–441.

Yew, P. R. & Berk, A. J. (1992) Inhibition of p53 transactivation required for transformation by adenovirus early 1B protein, *Nature 357*, 82–85.

Yin, X. M., Oltvai, Z. N. & Korsmeyer, S. J. (1994) BH1 and BH2 domains of Bcl-2 are required for inhibition of apoptosis and heterodimerization with Bax, *Nature 369*, 321–323.

Yonish-Rouach, E., Resnitzky, D., Lotem, J., Sachs, L., Kimchi, A. & Oren, M. (1991) Wild-type p53 induces apoptosis of myeloid leukaemic cells that is inhibited by interleukin-6, *Nature 352*, 345–347.

Yousefi, S., Green, D. R., Blaser, K. & Simon, H. U. (1994) Protein-tyrosine phosphorylation regulates apoptosis in human eosinophils and neutrophils, *Proc. Natl Acad. Sci. USA 91*, 10868–10872.

Yuan, J. & Horvitz, H. R. (1992) The *Caenrhabditis elegans* gene *ced-4* encodes a novel protein and is expressed during the period of extensive programmed cell death, *Development 116*, 309–320.

Yuan, J., Shaham, S., Ledoux, S., Ellis, H. M. & Horvitz, H. R. (1993) The *C. elegans* cell death gene *ced-3* encodes a protein similar to mammalian interleukin-1β-converting enzyme, *Cell 75*, 641–652.

Yuan, J. (1995) Molecular control of life and death, *Curr. Opin. Cell Biol. 7*, 211–214.

Zambetti, G. P. & Levine, A. J. (1993) A comparison of the biolgical activities of wild-type and mutant p53, *FASEB J. 7*, 855–865.

Zhan, Q., Bae, I., Kastan, M. B. & Fornace, A. J. Jr (1994) The p53 dependent γ-ray response of GADD45, *Cancer Res. 54*, 2755–2760.

Zhang, H. C., Saeed, B. & Ng, S. C. (1995) Combinatorial interaction of human Bcl-2 related proteins – Mapping of regions important for $Bcl\text{-}2/Bcl\text{-}x_S$ interaction, *Biochem. Biophys. Res. Commun. 208*, 950–956.

Zhivotovsky, B., Wade, D., Gahm, A., Orrenius, S. & Nicotera, P. (1994) Formation of 50 kbp chromatin fragments in isolated liver nuclei is mediated by protease and endonuclease activation, *FEBS Lett. 351*, 150–154.

Zhong, L. T., Sarafian, T., Kane, D. J., Charles, A. C., Mah, S. P., Edwards, R. H. & Bredesen, D. E. (1993) *bcl-2* inhibits death of central neural cells induced by multiple agents, *Proc. Natl Acad. Sci. USA 90*, 4533–4537.

Zhu, L., Enders, G., Lees, J. A., Beijersbergen, R. L., Bernards, R. & Harlow, E. (1995) The pRB-related protein p107 contains two growth suppression domains:independent interactions with E2F and cyclin/cdk complexes, *EMBO J. 14*, 1904–1913.

Ziemiecki, A., Harpur, A. G. & Wilks, A. F. (1994) JAK protein tyrosine kinases: their role in cytokine signalling, *Trends Cell Biol. 4*, 207–213.

Note added in proof. Recently, Choi et al. (1995) have described a new *bcl-2*-related gene *bfl-1*, which may be the human homologue of the mouse *A1* gene. Boyd et al. (1995) have described a death-inducing protein, Bik, which shares BH3, a 9-amino-acid doman, with Bax and Bak.

Eur. J. Biochem. *236*, 335–351 (1996)

Review

A nonribosomal system of peptide biosynthesis

Horst KLEINKAUF and Hans VON DÖHREN

Institute of Biochemistry and Molecular Biology, Technical University Berlin, Germany

(Received 21 June/18 September 1995) – EJB 95 1003/0

This review covers peptide structures originating from the concerted action of enzyme systems without the direct participation of nucleic acids. Biosynthesis proceeds by formation of linear peptidyl intermediates which may be enzymatically modified as well as transformed into specific cyclic structures. The respective enzyme systems are constructed of biosynthetic modules integrated into multienzyme structures. Genetic and DNA-sequence analysis of biosynthetic gene clusters have revealed extensive similarities between prokaryotic and eukaryotic systems, conserved principles of organisation, and a unique mechanism of transport of intermediates during elongation and modification steps involving 4′-phosphopantetheine. These similarities permit the identification of peptide synthetases and related aminoacyl-ligases and acyl-ligases from sequence data. Similarities to other biosynthetic systems involved in the assembly of polyketide metabolites are discussed.

Keywords: biosynthesis; multienzyme systems; nonribosomal peptide synthesis; peptide synthetases; peptides.

In our preceding review on the synthesis of peptide antibiotics, the thiotemplate multienzymic mechanism was summarized from an enzymological approach and it was suggested in conclusion that similarly operating multienzymes should be considered as a nonribosomal system [1]. Since that review, the missing structural data to justify this terminology have been derived from molecular genetic studies of a variety of peptide biosynthetic systems. Analysis of sequence data has provided an understanding of the modular construction of peptide and depsipeptide synthetases, and their relationship to other biosynthetic systems that catalyse the sequential condensation of acyl precursors such as fatty acids and other polyketide metabolites. Some of these studies have been reviewed previously at a more restricted level [2–11]. Here, we discuss the structural data in relation to the classical and more recent enzyme studies.

Ribosomal and nonribosomal systems

Several hundred peptides of different structural types are known, which by virtue of their composition cannot originate from the ribosomal system, the latter being restricted to the 20 protein amino acids and selenocysteine and their possible modification products. More than three hundred direct precursors including hydroxy acids are known and modifications such as acylation, *N*-methylation, epimerisation, and hydroxylation are frequently observed. Nevertheless, peptides arising from the different biosynthetic machineries are formed by head growth [1, 12] as linear precursors. Until recently, the size of enzymatically formed peptides seemed to be limited to 20 residues as no longer peptides of the nonribosomal type had been described [13]. Recently, the structure of a 48-residue peptide of marine origin from the sponge *Theonella swinhoei* has been determined [14], which also should originate from an as yet unspecified enzymic system (Fig. 1).

Several unusual peptides originating from ribosomal precursors are known including lantibiotics (containing lanthionine and methyl-lanthionine [15–20]), cyclic structures [21–23], and toxins with positionally defined D-amino acids [24–26] (Fig. 1). All residues arise from all-L-precursors by enzymatic modification.

In a classical essay, Lipmann outlined a *process evolution* from fatty acid biosynthesis and the thiotemplate peptide-forming polyenzymes to the ribosomal system [27]. These ideas, which address many basic questions in biochemistry, were based on functional rather than structural data. In the near future, sufficient sequence and structural information will be available to evaluate such evolutionary concepts in more detail.

General scheme of multienzymatic peptide biosynthesis

Molecular genetic analyses of peptide synthetase genes have led to a revision of the general scheme of enzymatic peptide bond formation [2–11, 28] which has been substantiated partially by protein chemical and protein engineering data [29–33] (Scheme 1). The sequence of reactions is (a) carboxyl activation by adenylate formation, (b) acylation of enzyme-attached pantothenoyl-thiols, and (c) directed transfer to the next acyl intermediate with condensation. In peptide synthesis initiation events, such transfers could also be catalysed by a specific transferase from acyl-CoA derivatives or enzyme-stabilized adenylates. Epimerisation or modifications of acyl intermediates are observed both before and after the directed transfer (see below). The completed peptide is covalently attached to a specific cysteamine group, and has to be removed by cyclisation, amidation or hydrolysis. Thioesters of peptidyl intermediates are thus released

Correspondence to H. von Döhren, Institute of Biochemistry and Molecular Biology, Technical University Berlin, Franklinstrasse 29, D-10587 Berlin, Germany

Abbreviation. Aad-Cys-D-Val, 4-(L-2-aminoadipyl)-L-cysteinyl-D-valine.

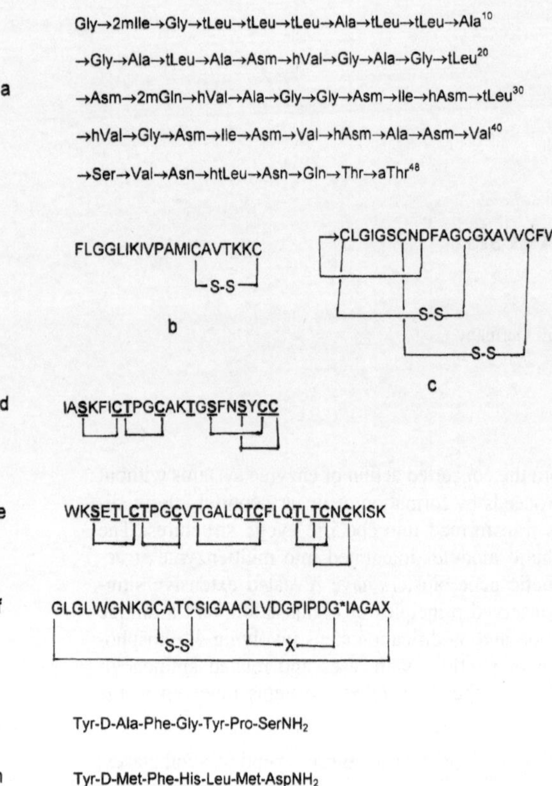

Fig. 1. Unusual structures of nonribosomal (a) and ribosomal origin (b–h). (a) Polytheonamide, the longest peptide reported of presumably enzymatic origin; (b) ranalexin, an antimicrobial peptide from the skin of *Rana catesbeiana* mimicking the antibiotic polymyxin of bacterial origin; (c) aborycin or RP71955, a tricyclic antibacterial peptide active against HIV-1, produced by *Streptomyces griseoflavus*; (d) and (e) the lantibiotics epidermin and subtilin, forming unique cyclic structures by dehydration of serine and threonine residues (S→**S**, T→**T**), followed by thioether formation with cysteines (**C**); (f) subtilosin, a cyclic structure found in *Bacillus subtilis* 168, with nonidentified bridging chemistry; (g) and (h) dermorphin and dermenkephalin, two D-amino-acid-containing opioid peptides from the skin of the frog *Phyllomedusa sauvagii*, in which epimerization is thought to proceed enzymatically in the processed peptide. For references see text. Unusual amino acids are indicated as follows: Asm, *N*-methyl-asparagine; aThr, allothreonine; hAsm, *threo*-2-hydroxy-*N*-methyl-asparagine; hVal, hydroxymethylvaline; htLeu, 3-hydroxy-t-leucine; mGln, methylglutamine; mIle, 2-methylisoleucine; tLeu, t-leucine.

by cyclisation with terminal or internal peptide or ester bond formation, or alternatively by modification or hydrolysis of the activated C-terminus. In contrast to the ribosomal elongation scheme, no extra energy is required for the directed transfer: detailed measurements of tripeptide formation by 4-(L-2-aminoadipyl)-L-cysteinyl-D-valine (Aad-Cys-D-Val) synthetase have confirmed earlier estimates on crude extracts of *Bacillus* systems (gramicidin S [34, 35]) that 1 mol ATP is consumed per peptide bond formed [36].

Amino acid activation in the ribosomal and nonribosomal systems. Activation of the carboxyl groups of amino, hydroxy or carboxylic acids proceeds by the formation of mixed anhydrides with phosphoric acid derivatives. These derivatives originate from adenosine or guanosine triphosphates, and are either adenylate or phosphate intermediates. The intermediates are stabilized as enzyme complexes and transferred as stable esters to acceptor hydroxyls (tRNA) or thiols (CoA or enzyme-attached pantetheine). An overview shown in Table 1 groups the respective enzymes according to their functional and structural properties. The main conclusions are as follows: although amino acid activation via aminoacyl adenylate and subsequent ester formation is formally identical in the ribosomal and nonribosomal systems, the enzymes involved have no obvious structural similarities [37]; a specific group of acyl-CoA synthetases and peptide synthetases share a structurally similar adenylate-forming domain [38, 39] (this is discussed in more detail below); similar enzyme structures activate amino, hydroxy, or carboxylic acids; peptides are generally activated as phosphates.

It remains to be shown, however, whether or not structural similarities exist between activating enzymes despite the lack of sequence similarities.

The modular construction of enzyme systems. The use of the terms domain, module, multienzyme, multienzyme complex, synthase and synthetase has not been uniform. In a key paper on polyketide synthase construction [40], Donadio et al. introduced the term module for a DNA segment encoding all the functions required to complete a single elongation step. These functions are assembled on a protein segment called a synthase unit. Each unit contains the necessary catalytic domains, which in case of polyketide systems are acyltransferase, acyl carrier protein, β-ketoacyl-ACP synthase, and, optionally, β-keto reductase, dehydratase, enoylreductase, and thioesterase. If this system was applied to peptide synthetases, a module would constitute the DNA segment encoding the information for the addition of one amino acid (or hydroxy acid, respectively), together with its possible modification. The respective protein segment would be termed a synthetase unit and the catalytic domains would comprise the adenylate domain, the acyl transfer domain, the condensation domain, the epimerase domain, and the *N*-methyltransferase domain. Should this terminology be applied? Preceding the polyketide module definition, the term module has occasionally been used in connection with protein structures [41]. Modules have been introduced as particular kinds of domains associated with exon shuffling and duplication events. The mechanisms of module exchange or amplification are clearly related to the respective gene structures and thus the intended meaning should address the DNA segment. In peptide synthetase genes, no introns have been found by computer analysis, despite their considerable sizes. Acyl-CoA synthetases and insect luciferases show mostly conserved intron/exon structures [42–45] though their relationships to the functions of the adenylate catalytic domain are not yet understood [39]. Recent consideration of the exon shuffling hypothesis concluded that there is no obvious relationship between exon-derived protein structure and function [46]. It thus seems appropriate to use the term module for the respective gene pieces. The term domain is unfortunately seldom used in its structural implication, i.e. distinct globular intrachain region [47] but mostly in the sense of a subregion that is autonomous in the sense that it possesses all characteristics of a complete globular protein [48]. Often, protein fragments obtained by limited proteolysis are referred to as domains, indicating their structural integrity and perhaps functional autonomy. Such domains may well be composed of several distinct structural subdomains. Hence the use of domain to refer to functionally defined regions of synthetase units reflects well the current conventions [49]. In the following discussion, the term module will be restricted to the DNA level, while the term domain is used in connection with protein structure.

The adenylate domain

Enzymatic properties. Sequence analysis of peptide synthetases has revealed the presence of a region highly similar to

Scheme 1. Reactions catalysed by peptide synthetases.

(a) simple elongation (no modification).
Numbering of enzymes indicates either domains or individual multienzymes.
M stands for functionally suitable divalent metal ions, usually Mg^{2+}, Mn^{2+}, Ca^{2+}.

acyladenylate — aminoacylation

$$E^1 + A^1 + MATP^{2-} \rightleftarrows E^1\{A^1AMP)\} + MPP_i^{2-} \rightarrow E^1\text{-}S^1A^1 + AMP$$

$$E^2 + A^2 + MATP^{2-} \rightleftarrows E^2\{A^2AMP\} + MPP_i^{2-} \rightarrow E^2\text{-}S^2A^2 + AMP^-$$

condensation

$$E^1\text{-}S^1A^1 + E^2\text{-}S^2A^2 \rightarrow E^2\text{-}S^2A^2A^1 + E^1\text{-}SH$$

General elongation scheme (nth step)

$$E^{n-1} + A^{n-1} + MATP^{2-} \rightleftarrows E^{n-1}\{A^{n-1}\,AMP\} + MPP_i^{2-} + E^{n-1}\text{-}S^{n-1}A^{n-1} + AMP$$

$$E^n + A^n + MATP^{2-} \rightleftarrows E^n\{A^nAMP\} + MPP_i^2 \rightarrow E^n\text{-}S^nA^n + AMP \rightarrow E^n\text{-}S^nA^nA^{n-1}\ldots A^1$$

$$E^{n-1}\text{-}S^{n-1}A^{n-1} + E^n\text{-}S^nA^n \rightarrow E^n\text{-}S^nA^nA^{n-1}\ldots A^1 + E^{n-1}\text{-}SH$$

(b) elongation preceded by epimerization

$$E^1 + A^1 + MATP^{2-} \rightleftarrows E^1\{A^1AMP\} + MPP_i^2 \rightarrow E^1\text{-}S^1A^1 + AMP \rightarrow E^1\text{-}S^1\text{D,L-}A^1$$

$$E^2 + A^2 + MATP^{2-} \rightleftarrows E^2\{A^2AMP\} + MPP_i^{2-} \rightarrow E^2\text{-}S^2A^2 + AMP^- \rightarrow E^2\text{-}S^2A^2\text{D-}A^1$$

$$E^1\text{-}S^1\text{D,L-}A^1 + E^2\text{-}S^2A^2 \rightarrow E^2\text{-}S^2A^2\text{D-}A^1 + E^1\text{-}SH$$

(c) epimerization during elongation

$$E^n + A^n + MATP^{2-} \rightleftarrows E^n\{A^nAMP\} + MPP_i^{2-} \rightarrow E^n\text{-}S^nA^n + AMP$$

$$E^n\text{-}S^nA^n + E^{n-1}\text{-}S^{n-1}A^{n-1}\ldots A^1 \rightarrow E^n\text{-}S^nA^nA^{n-1}\ldots A^1 + E^{n-1}\text{-}SH$$

$$E^n\text{-}S^nA^nA^{n-1}\ldots A^1 \rightarrow E^n\text{-}S^n\text{D,L-}AnAn^{-1}\ldots A^1 \rightarrow E^{n+1}\text{-}S^{n+1}A^{n+1}\text{D-}AnAn^{-1}\ldots A^1$$

(d) modification (N-methylation)

$$E^1 + A^1\ MATP^{2-} \rightleftarrows E^1\{A^1AMP\} + MPP_i^{2-} \rightarrow E^1\text{-}S^1A + AMP \rightarrow E^1\text{-}S^1\text{NMe-}A^1$$

$$E^1\text{-}S^1A^1 + SAM \rightarrow E^1\text{-}S^1\text{NMe-}A^1 + SAhC$$

acyl-CoA synthetases (Fig. 2). Such regions are characterized by a collection of motifs that are conserved within the groups [2, 6–11, 39, 51–56]. Most prominent is the AMP-binding motif, readily identified by the signature or core sequence SGT/STGXPKG [50]. Enzymes containing this sequence have been shown to catalyse reactions (1) and (2):

$$E + RCOOH + MATP^{2-} \rightarrow E\{RCO\text{-}AMP\} + MPP_i^{2-}, \quad (1)$$

$$E\{RCO\text{-}AMP\} + CoASH \rightarrow E + AMP + RCO\text{-}SCoA. \quad (2)$$

Although these enzymes, which are referred to as the tyrocidine synthetase/insect luciferase family, share equidistantly spaced conserved sequences, there are minor differences in their catalytic activities. Hence there might be stable adenylated forms, as can be concluded from ATP/PP_i exchange in the absence of substrates in some acetyl-CoA synthetases [see reaction (3)]. Reversal of acyladenylate formation by adenosine phosphates, other nucleotide phosphates and polyphosphates can be restricted to substrates with a certain chain length [reactions (4–6)] [57–59]. Gramicidin S synthetases have been shown to catalyse ADP and ATP formation from aminoacyl intermediates [60]. Reactions (3–6) are as follows:

$$E + MATP^{2-} \rightarrow E\{AMP\} + MPP_i^{2-}, \quad (3)$$

$$E\{RCO\text{-}AMP\} + P_n \rightarrow E + P_{n+1} + RCOOH, \quad (4)$$

$$E + MATP^{2-} + P_n \rightarrow E + AP_{n+1} + MPP_i^{2-}, \quad (5)$$

$$E\{RCO\text{-}AMP\} + NTP \rightarrow E + Ap_4N + RCOOH. \quad (6)$$

The size of the adenylate domain has only been roughly estimated. Its C-terminal boundary, which is followed by the 4′-phosphopantetheine attachment motif J, has been inferred from alignments with acyl-CoA synthetase sequences, proteolytic studies, and the expression of stable fragments. Alignments of tyrocidine synthetase 1 with acyl-CoA synthetases and acyl-carrier proteins reveal a stretch of nonconserved amino acids located between motifs I and J [39]. The C-terminus of a 115-kDa proline-activating fragment of gramicidin S synthetase 2 [61] can also be traced into this region using the available sequence data from the multienzyme polypeptide [63, 65]. Saito et al.

Table 1. Types of carboxyl activation reactions. For type I (AMP), carboxyl activation occurs by cleavage of ATP to AMP and PP_i. For type II (NDP), carboxyl activation occurs by cleavage of NTP to NDP and P_i (ATP if not indicated differently). Unusual amino acids are indicated as follows: Aad, L-2-aminoadipic acid; Aib, aminoisobutyric acid; aIle, alloisoleucine; Bmt, (4*R*)-4[(*E*)-2-butenyl]-4-methyl-L-threonine; Dab, 2,4-diaminobutyric acid; Dap, 2,6-diaminopimelic acid; Dbu, diaminobutyric acid; Dha, 2,6-diamino-7-hydroxyazelaic acid; Dhb, dihydroxybutyric acid; D-Hiv, D-hydroxyisovaleric acid; Dpm, diaminopimelic acid; Hse, homoserine; Ise, isoserine; Kyn, kynurenine; Pip, pipecolic acid.

Enzyme	Class	Activated substrates	
		type I (AMP)	type II (NDP)
Aminoacyl-tRNA synthetases	class I	Arg, Cys, Glu, Gln, Ile, Leu, Met, Trp, Tyr, Val	
	class II	Ala, Asn, Asp, Gly, His, Lys, Phe, Pro, Ser, Thr	
Acyl-CoA synthetases		acetate, medium chain C4−C12), intermediate chains (C3−C7), long chains (C12−C18), benzoate, cholate, luciferin, lysergate, phenylacetate, oxalate, biotin	succinate, succinate (GDP), medium chains (C4−C12) (GDP), glutarate
Peptide synthetases	1-amino acids α-carboxyls	Aib, Ala, D-Ala, Asn, Asp, Bmt, Cys, Dab, Dap, Glu, Gln, Gly, His, Ile, aIle, Kyn, Leu, D-Leu, Lys, Orn, Phe, D-Phe, Pip, Pro, Hyp, Ser, Thr, Trp, Tyr, Val	Pro, D-Ala
2-amino acids other carboxyls	Aad, Dha, Ise, βTyr	βAla, D-Asp(β), Glu(γ), D-Glu(γ)	
3-peptides α-carboxyls	AcLeu-Leu	Glu(γ)-Cys, UDP-Mur-NAcGly(Ala), UDP-Mur-NAc-A1-D-Glu-A3 (A3 = Lys, Dpm, Dbu, Orn, Hse)	
4-other carboxyls		UDP-MurNAc-A1-D-Glu(γ)	
5-carboxylic acids	Dhb, D-Hiv, pantoate		

have subcloned and sequenced this fragment and confirmed its properties [63]. Subtilisin digestion of Aad-Cys-D-Val synthetase from *Aspergillus nidulans* to determine internal sequences also provided cleavage sites between motifs I and J, as well as between J and K [64]. Following these observations, Dieckmann et al. obtained a 60.8-kDa fragment of tyrocidine synthetase 1 by C-terminal deletion [39]. As shown by kinetic analysis, the fragment displayed similar adenylate-forming properties to the apo form of the multienzyme (produced by overexpression in *Escherichia coli* [56]). This size, resembling acyl-CoA synthetases, may be reduced further by terminal reductions to arrive at fragments catalysing adenylate formation but lacking the aminoacylation function.

Site-directed mutagenesis studies. Several site-directed mutagenesis and affinity labelling studies have been carried out to investigate the functions of the various motifs. Gocht and Marahiel [32] have modified single amino acids within the C-, E- and F-motifs. In the signature sequence SGTTGKPKG, the mutations S(G→A)TT(G→A)K(P→A)K*(G→A) had little effect on the rates of the amino-acid-dependent ATP/PP_i exchange [32]. However, introduction of a negative charge by mutagenesis of the first G to D in the valine-activating domain of gramicidin S synthetase 2 led to complete loss of adenylate formation [67]. Dramatic rate decreases were found upon exchange of K* to R (90%) or T (99.5%), which suggests an essential role for the lysine residue. However, in the valine-activating fourth module of surfactin synthetase, a reduction of rate of only 61% was observed upon K→Q exchange [68]. The signature sequence has similarities with the phosphate-binding P-loop found in ATP- and GTP-binding proteins [69].

Charge-introducing conversions of the conserved G(→D) between the C- and D-motifs and the conserved G in the D-motif [YG(→E)XTE], lead to inactivation of the adenylation reaction [67]. The highly conserved TGD sequence within the F-motif has also been subject to mutagenesis. Due to its occurrence in various ATPase sequences, it has sometimes been termed the ATPase motif [7]. While a D→N change retained 78% activity, D→S showed an 88% reduction. The multiple alteration TGDLA→TCSLS led to a 97% reduction [32]. A gramicidin S synthetase 2 mutation resulting in loss of proline-dependent ATP/PP_i exchange reaction has been traced by sequencing to a G870→E substitution [70]. This glycine is part of the highly conserved G-motif EXXGRXGXQVKXR**G**XRIEXGEIE. Labelling studies with fluorescein isothiocyanate and 2-azido-ATP [55, 71], identified specific residues within the G, H and I-boxes. It has not been established if all of these sites are strictly related to binding of $MATP^{2-}$ as substrate in carboxyl activation, since domain insertions are observed between the G- and H-motifs indicating perhaps an additional nucleotide-binding site (see below). However, mutagenesis of a highly conserved lysine to glutamine within the H-motif (LTXNG**K**XXXXXLP) in the valine-activating module of surfactin synthetase led to a reduction of the reverse rate of adenylate formation by 94% [68]. A double mutant with L→Q alterations in the C- and H-boxes was inactive.

Although the Walker nucleotide-binding motifs A and B [72, 73] are not obviously present, extensive replacements by

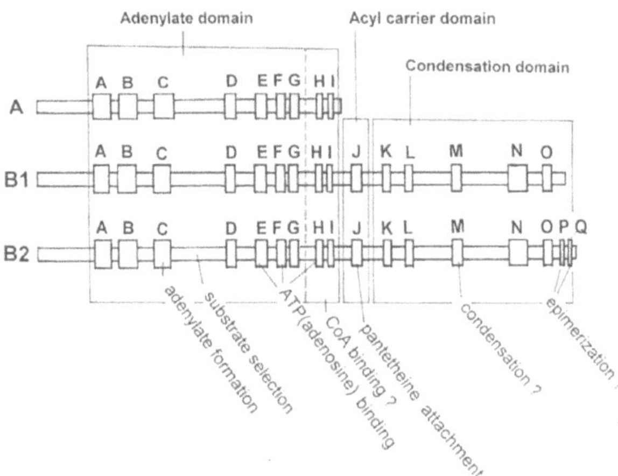

Fig. 2. Similarities of acyl-CoA synthetases (A) and peptide synthetase domains (B1) as well as peptide synthetase domains with epimerizing function (B2). The motifs A–I are shared within the adenylate-forming domains, while peptide synthetases contain at least two additional domains with the motifs J (acyl-carrier domain) and K–Q (involved in condensation and epimerization reactions). The current state of functional identification is indicated. The respective consensus sequences are as follows: A, LTXXELXXXAXXLXR; B, AVXXAXAXYVXIDXXYPXER; C, YTSGTTGXPKG; D, IIXXYGXT; E, GELXIXGXXVAR; F, RLYRTGDL; G, IEYLGRXDXQVKIRXXRIELGEIE; H, LXXYMVP; I, LTXXGKLXRKAL; J, LGGXSIXAI; K(B1), YPSVXXQXRMYIL; K(B2), LXPIQXWF; L(B1), LIXRHEXL; L(B2), LXXXHD; M(B1), DMHHIIXDGXSXXI; M(B2), HHXXVDXVSWXIL; N(B1), LSKXGQXDIIXGTPXAGR; N(B2), VXXEGHGRE; O(B1), IXGMFVNTXLALR; O(B2), TVGWFTXXXPXXL; P, PXXGXGX; Q, VXFNYLG.

alanine-([168]DEGKRGGKLI→DEAARGAALI and [268]KYVFDIHDGDRY→KYVAAIHAAARY) have been conducted in related motifs in acetyl-CoA synthetase from *Neurospora crassa* and the resultant synthetases were inactive in acylation of CoA [74].

Specificity of acyl-group substrates: the nonribosomal code. Structures of substrates vary and wide differences in substrate specificity are observed. Varying degrees of selection between L- and D-forms of amino acids can be found. Some enzymes such as tyrocidine synthetase 1 accept both isomers, the L-form being epimerized and channelled to initiate peptide synthesis [1]. An L-leucine-activating site of gramicidin S synthetase 2 has been shown to accept D-leucine but is then blocked for continued elongation [75]. Most activation sites show strict stereospecificity, and no adenylate formation is found with the alternative isomer. Although generally regarded as a system of low specificity, activating sites of the nonribosomal system may display unexpected selectivity. Hence, in echinocandin biosynthesis it is not possible to generate proline analogues of the hydroxyproline-containing peptide [76]; the multienzyme system discriminates both analogues (Truglowski, B. and von Döhren, H., unpublished results). In contrast, a proline-containing analogue of viridogrisein can be manufactured by culture feeding [77]. Surprisingly, selection in the ribosomal system does not provide greater discrimination. However, the subsequent steps of aminoacylation and translation permit proof reading [78, 79].

Conserved amino acid sequences are found outside the motifs A–I when domains of comparable substrate specificity are aligned. A major determinant of the substrate-binding site has been tentatively assigned to the region between the C- and D-motifs [80]. Tree alignments of adenylate domains of various synthetases group these according to enzyme system and substrate [53, 54, 65]. Several sequences are available from acetyl-, long-chain fatty acyl, 4-coumaroyl-CoA synthetases, insect luciferases, and peptide synthetase domains that activate leucine, valine, cysteine, and aminoadipate. Cosmina et al. [80] have suggested that from alignments of activating domains from *Bacillus* systems it is possible to predict specificity. It is also obvious from alignments of the five Aad-Cys-D-Val synthetases sequenced to date, that each domain can be identified with certainty in any Aad-Cys-D-Val synthetase, regardless of its prokaryotic or eukaryotic origin [2, 8, 9, 81]. In long-chain acyl-CoA synthetases and luciferases, substrate-binding sites have been predicted in regions of high hydrophobicity also located between C- and D-motifs. Alignments of adenylate domains together with other sequence data will thus help to correlate gene structures with enzyme functions, especially in cases where product structures are already available (Fig. 3).

Additional functions of adenylate domains. In peptide synthetases, aminoacyl adenylates are formed and transferred to the cysteamine moiety of 4′-phosphopantetheine, which is attached to the adjacent acyl carrier domain (Fig. 2). Thus, binding of CoA and catalysis of the transfer reaction has to be located in this region, spanning 500–600 amino acid residues. Kinetic studies with tyrocidine 1 synthetase have shown non-competitive inhibition of CoA on the amino-acid-dependent ATP/PP_i exchange reaction [37]. Labelling studies with fluorescein isothiocyanate and 2-azido-ATP [55, 71] have identified residues within the G, H and I-boxes. Some of these sites might not be involved in adenylate formation but instead involved in CoA binding during acylation of CoA or post-translational modification of peptide synthetases.

The acyl or aminoacyl transfer to the cysteamine moiety of CoA or pantetheine must be catalysed by the formation of a thiolate anion. This can be concluded from the pH dependence of the aminoacylation of tyrocidine 1 synthetase [56]. So far, no conserved groups for this deprotonation have been proposed. To investigate the possible participation of a cysteine residue in aminoacylation and epimerization, as had been proposed by Kanda et al. [82], Gocht and Marahiel have changed the E-motif Cys364 of tyrocidine synthetase 1 to serine [32]. This significantly increased aminoacylation levels but had no effect on the epimerization of enzyme-bound phenylalanine. Aminoacylation-deficient and epimerization-deficient mutants of gramicidin S synthetase 1 have been sequenced by Hori et al. [83] and shown to have unaltered primary structures. These data confirm the post-translational origin of the aminoacylated thiol and the requirement of post-translational modification for epimerization.

Peptide bond formation

Transport of activated amino acids in the enzymatic and ribosomal systems. The transfer of aminoacyl groups of enzyme-bound adenylates to the cysteamine moiety of CoA or pantetheine may indicate that CoA in the nonribosomal system functions equivalently to tRNA in the ribosomal system. This had already been proposed by Reanney [84]. However, the known high reactivity of aminoacyl thioesters probably prohibits their function as intermediates in cellular metabolism [85].

The general occurrence of protein-stabilized thioesters may be implied from the recent work of Neuhaus and Heaton on the biosynthesis of D-alanyl-lipoteichoic acid [86]. D-Alanine is activated by an enzyme containing the adenylate domain. In-

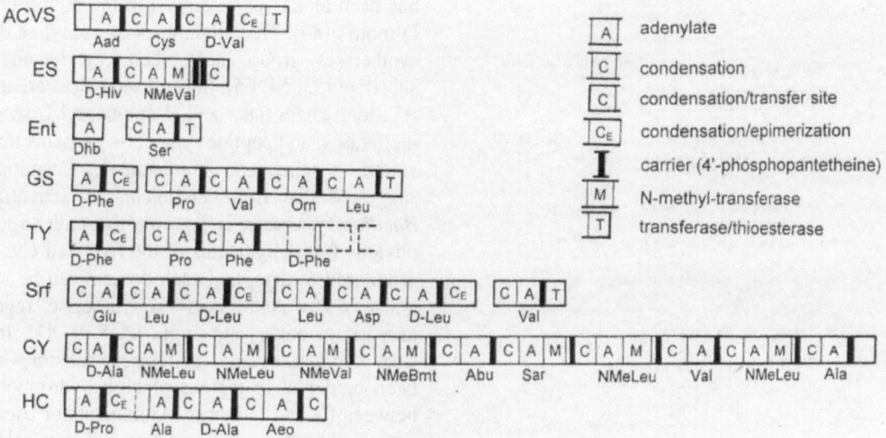

Fig. 3. Organisation of peptide synthetases derived from DNA sequencing and various functional studies. ACVS, Aad-Cys-D-Val synthetase; ES, enniatin synthetase; Ent, enterobactin biosynthetic enzymes EntE and EntF; GS, gramicidin S synthetases 1 and 2; TY, tyrocidine synthetases 1 and 2; Srf, surfactin synthetases 1, 2 and 3; CY, cyclosporin synthetase; HC, HC-toxin synthetase. The specificity of the domains is indicated. Note that three different types of condensation domains are found, slightly deviating in their consensus sequences. Unusual amino acids are indicated as follows: Aad, L-2-aminoadipic acid; Abu, 2-aminobutyric acid; Aeo, 2-amino-9,10-epoxi-8-oxodecanoic acid; Bmt, (4*R*)-4[(*E*)-2-butenyl]-4-methyl-L-threonine; Dhb, dihydroxybutyric acid; D-Hiv, D-hydroxyisovaleric acid; MeLeu, 2-methylisoleucine; MeVal, methylvaline; Sar, sarcosine.

stead of CoA, a protein similar to the acyl-carrier protein of polyketide biosynthesis acts as the acceptor with its pantetheine cofactor. Surprisingly, alternatives to this aminoacyl carrier protein were acyl-carrier proteins from *E. coli* and *Saccharopolyspora erythrea*. Thus, we encounter the situation observed in polyketide formation in the type II system of nonintegrated modules, where a carrier protein has to interact with a set of enzymes catalysing condensation and modification reactions. This is contrasted by the type I synthase systems with partially or fully integrated modules, where the carrier protein is flanked by adjacent domains [87]. So far, type II systems have been rarely found in peptide biosynthesis.

By comparison, the ribosomal system uses aminoacyl-tRNA, in which aminoacyl groups are stabilized as esters of the 2′- or 3′-hydroxyl groups of ribose. Whether additional stabilization of the charged tRNA by elongation factors is essential has not been analysed in detail.

Aminoacylation and the multiple carrier concept. The thiotemplate principle, forwarded by analogy with the early model of multienzymic fatty acid formation, proposed peripheral cysteine residues as aminoacyl acceptors. Elongation was proposed to proceed by successive transpeptidation-transthiolation steps [1, 88, 89]. However, DNA sequencing of peptide synthetase genes revealed the absence of conserved cysteines. Thioesters were thus connected to the J-motif-resembling pantetheine attachment sites of polyketide synthetases [5–11, 29, 32, 33, 39, 81]. Two types of putative cofactor attachment sites were postulated but one site did not show strict conservation of serine [6] nor similarity to carrier protein domains outside the attachment motif [39]. Affinity labelling of aminoacylated residues, hampered by the instability of aminoacyl-thioesters, was accomplished by Vater et al. [28, 29, 90, 91] employing *N*-methylmaleimide after selective blocking steps. Mass spectrometric analysis, however, led to the detection of dehydroalanine residues which could have resulted from seryl- or cysteinyl-residues. Gene sequencing proved that these residues were serines [28]. The more advanced methodology using electrospray mass spectrometry permitted direct analysis of the cofactor attachment site in gramicidin S synthetases 1 and 2 [29, 91]. This direct proof of modification was even successful in gramicidin S synthetase 1, which like tyrocidine synthetase 1 has been repeatedly analysed for the presence of pantothenate by the *Lactobacillus* growth assay [32, 91–93]. In addition, labelling of peptide synthetases has been achieved by feeding β-alanine to cultures of a panD mutant of *E. coli* defective in β-alanine biosynthesis [32, 94, 95]. However, only in the case of EntF (the serine-activating component of the enterobactin biosynthetic complex) has the structural identity of the label been confirmed [95]. The procedure has also been used recently to demonstrate the presence of pantetheine in polyhydroxybutyrate synthase [96].

Thus, the thiotemplate model has been extended to a multiple carrier mechanism [29]. The linear arrangement of domains should restrict the spatial interaction of acyl intermediates with adjacent cofactors. The control of the direction of transport could be exercised simply by the arrangement of functional groups catalyzing the condensation reaction (Fig. 4). This model also convincingly explains the accumulation of peptidyl-intermediates observed upon cycle interruption [88]. While in the single cofactor model each peptidyl-transfer would have to be considered a specific reversible reaction [1], the multiple carrier model assumes only an irreversible directed transfer [8].

Apoprotein and holoprotein forms of peptide synthetases. Experiments demonstrating indirectly the requirement of post-translational modification for aminoacylation and peptide formation originated mainly from the expression of tyrocidine and gramicidin S synthetase 1 genes in *E. coli.* These multienzymes initiate peptide biosynthesis by introducing the starter D-amino acid into the pathway. Gramicidin S synthetase 1 had been classified as an ATP-dependent phenylalanine racemase [97] (EC 5.1.1.11) although phenylalanine is only an intermediate of the reaction cycle. The sequence of reactions has been elucidated by Kurahashi et al. [97, 98] and evaluated in more detail by Lee and Lipmann [100] and Saito et al. [82, 83]. After adenylate formation [reaction (7)], phenylalanine is epimerized at the thioester stage implying transfer of the intermediate to the respective catalytic site [reactions (8, 9)]. Only the D-isomer is transferred to the subsequent acyl intermediate on the adjacent multienzyme site on tyrocidine synthetase 2 [reaction (10)]:

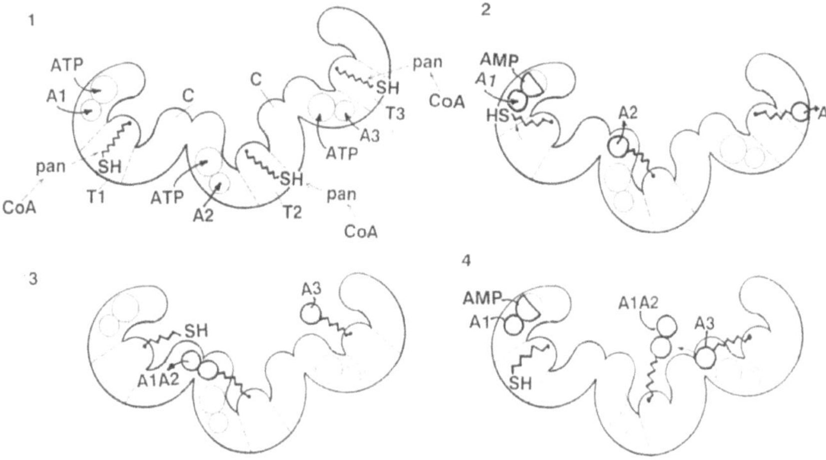

Fig. 4. Scheme of the multiple carrier mechanism using a tripeptide synthetase (e.g. Aad-Cys-D-Val synthetase). (1) View of three connected peptide synthetase domains activating the amino acids A1–A3. Each domain contains a central carrier subdomain, which has to be charged with pantetheine (pan) employing CoA. (2) The sites can be charged independently with aminoacyl residues; in activation site A1, an adenylate has been formed (A1AMP), which will be transferred to the first thiol group; the sites A2 and A3 are already aminoacylated. (3) Peptide bond formation occurs at the contact site of the activating domains. (4) Formation of the second peptide bond at the contact site of the second and third domain, and recharging of the first domain.

$$E^1 + \text{L-Phe} + \text{MATP}^{2-} \rightleftarrows E^1\{\text{L-Phe-AMP}\} + \text{MPP}_i^{2-}, \quad (7)$$

$$E^1\{\text{L-Phe-AMP}\} \rightarrow E^1\text{-S-L-Phe} + \text{AMP}, \quad (8)$$

$$E^1\text{-S-L-Phe} \rightleftarrows E^1\text{-S-D,L-Phe}, \quad (9)$$

$$E^1\text{-S-D,L-Phe} + E^2\text{-S}^2\text{Pro} \rightarrow E^2\text{-S}^2\text{-Pro-D-Phe} + E^1. \quad (10)$$

Hori et al. found a 90% reduction of epimerase activity in bacterially expressed gramicidin S synthetase 1 [101]. Pfeifer et al. have isolated native and overexpressed tyrocidine synthetase 1, and compared both multienzymes [102]. Only about 1% of the overexpressed synthetase is post-translationally modified and levels of aminoacylation are thus reduced. In addition, the rate of the amino-acid-dependent ATP/PP_i exchange is reduced. It has been proposed that this reverse rate is accelerated in the aminoacylated multienzyme reaction (11)]:

$$E^1\text{-S-D,L-Phe} + \text{L-Phe} + \text{MATP}^{2-}\ E^1\text{-S-D,L-Phe}\{\text{L-Phe-AMP}\} + \text{MPP}_i^{2-}. \quad (11)$$

The principle emerging is that acyl-carrier proteins mediate the modification of acyl intermediates by interaction with the respective modifying enzymes. Such enzymes could be intrinsic catalytic domains, such as epimerases, *N*-methyltransferases, or hydroxylases. Similarly, acyl-CoA compounds are formed as intermediates in many metabolic transformations, e.g. in the dehalogenation of aromatic acids [103–105]. In the case of chlorobenzoate, the acyl derivative is formed, interacts with the dehalogenase, and the hydroxybenzoyl-CoA formed is released by the action of a thioesterase [105].

Modification of amino acid residues: epimerization. Epimerization of residues during multienzymatic synthesis has long been thought to occur at the stage of the acyl precursor, the aminoacyl thioester. This was concluded from the work on gramicidin S synthetase and tyrocidine synthetases. In both systems, the initiating phenylalanine is epimerized before dipeptide formation occurs. At the aminoacylated stage, the two isomers have been detected in a ratio of 2:1 (D:L). Condensation occurs exclusively with the D-form.

As long ago as 1978, Loy [106] presented evidence for an epimerization at the peptidyl stage in a nonapeptide intermediate of bacitracin biosynthesis. More recently, Stindl and Keller [107–109] studied the formation of the acyl-dipeptide in actinomycin formation and have demonstrated the presence of both LL- and LD-peptide intermediates; they concluded that epimerizations in the chain are generally catalysed at the peptidyl stage. This is in agreement with the failure to detect D-valine as an intermediate in Aad-Cys-D-Val biosynthesis [81, 110, 111]. In this latter case, it has been assumed that epimerization takes place at the tripeptide stage. However, the rapid hydrolytic removal of the LLD-tripeptide did not permit the configurational analysis of the tripeptide intermediate. Recent reviews have adapted this new view on the timing of epimerization [10]. Alignments of the amino acid sequences of five [80, 112] and ten [56, 102] synthetase units which introduce D-amino acid residues indicate two sets of conserved motifs: (a) five positionally conserved motifs in D- and L-isomer adding steps, with different amino acid sequences (K, L, M, N, O), and (b) two additional motifs in a region extending from the domain adjacent to the acyl carrier domain (P, Q). So far, these motifs have been tentatively assigned to condensation and epimerization reactions based on the knowledge of the chemistry involved (Fig. 5). Condensation requires successive proton transfers for the nucleophilic attack by the amino nitrogen at the thioester carboxyl, followed by restoration of the thiol group. Epimerization involves hydrogen abstraction and readdition from a spatially different site. The dependency of peptide biosynthesis on pH with highest rates at pH 7–8, indicate the likely involvement of histidine side chains. The motifs L, M, and N contain conserved histidine residues. The spacer motif M identified by Zuber et al. [112] may well resemble the actual condensing motif of the peptide synthetases [83]. These respective M-motifs differ slightly not only in L- and D-isomer adding domains, but also in acyl-, aminoacyl- and peptidyl group accepting condensing domains (Fig. 3).

Thus, initiating D-amino acids may enter the biosynthetic cycle either as D-isomers (e.g. D-Ala in cyclosporin, D-Hyp in viridogrisein) or as L-isomers with epimerization at the activated thioester stage (Phe in gramicidin S and tyrocidine). D-Amino acids in chain positions seem to be epimerized in the activated peptidyl stage and not at the aminoacyl stage (Scheme 1).

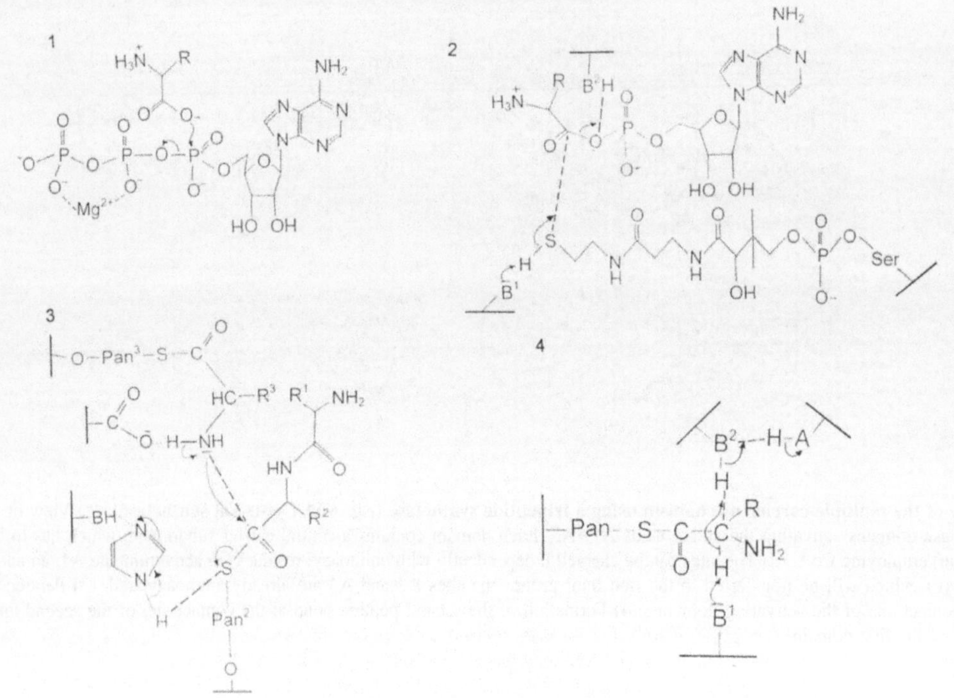

Fig. 5. Schematic view of reactions involved in peptide biosynthesis. (1) Adenylate formation involving nucleophilic attack of the carboxyl group at the a-phosphate of the $MgATP^{2-}$ complex with release of $MgPP_i^{2-}$; (2) aminoacylation of the pantetheine (Pan) cofactor by formation of the thiolate anion, attack of the mixed anhydride and release of AMP; (3) tentative view of the peptide bond formation by nucleophilic attack of the aminoacyl nitrogen at the preceding thioester carboxyl, involving deprotonation-protonation; (4) epimerization of an aminoacyl thioester, a reaction differing from those catalysed by the well-characterized amino acid racemases.

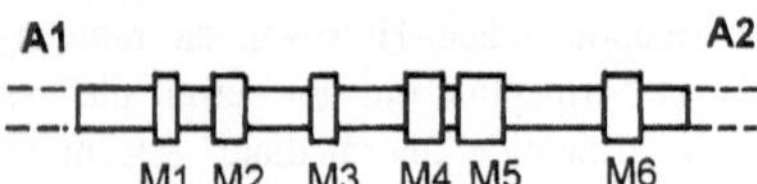

Fig. 6. Motifs shared by *N*-methyltransferase domains from enniatin and cyclosporin synthetases. Consensus sequences are as follows: M1, GXDFXXWTSMYDG; M2, LEIGTGTGMVLFNLXXXXGL; M3, VXNSVAQYFP; M4, EXXEDEXLXXPAFF; M5, HVEXXPKXMX-XXNELSXYRYXAV; M6, GXXVEXSXARQXGXLD.

The mechanism of epimerization remains to be solved and does not involve pyridoxal phosphate, which is the common mechanism in amino acid racemases. The hydrogens in the α-position will partially exchange with enzymic protons and, likewise, tritiated water will be exchanged [100, 101, 107].

Inserted domains: *N*-methylation. *N*-methylation of cyclopeptides has been studied in detail by Zocher et al. [113−115] and the enzymology has been reviewed [116]. More recently, genetic studies on enniatin [117] and cyclosporin biosynthetic systems [118] have revealed the presence of an *N*-methyltransferase domain in the peptide synthetases. This domain is inserted between the G- and H-motifs and the eight available sequences of the transferase inserts contain six highly conserved motifs, M1−M6, with M2 showing similarities to other *N*-methyltransferases [117] (Fig. 6). Ultraviolet-induced affinity labelling with *S*-adenosylmethionine of bacterially expressed fragments of enniatin synthetase was used to locate unambiguously this region [119].

N-Methylation occurs at the aminoacyl thioester state, preceding peptide bond formation. In the absence of *S*-adenosylmethionine or in the presence of a competitive inhibitor like sinefungin or *S*-adenosylhomocysteine as a noncompetitive inhibitor, unmethylated or undermethylated enniatins have been produced [115].

Intramolecular and intermolecular acyl and peptidyl transfer. While prokaryotic peptide synthetase systems with the exception of Aad-Cys-D-Val synthetase display an as yet unpredictable state of integration, which requires the interaction of the various synthetases, the known eukaryotic peptide synthetases are fully integrated. Intermolecular acyl- and peptidyl transfers are restricted to bacterial systems and several lines of evidence indicate specific associations of multienzymes, perhaps with the involvement of additional proteins:

(a) although isofunctional with respect to L- and D-phenylalanine activation, epimerization and transfer to a prolyl residue on the following multienzyme polypeptide, tyrocidine synthetase 1 cannot replace gramicidin S synthetase 1 in gramicidin S biosynthesis. The major differences between the two multienzyme polypeptides are their N-terminal sequences [101, 120]; (b) during the simultaneous production of at least two different peptides (e.g. tyrocidine and linear gramicidin, surfactin and iturin [10, 121]), no mixed products are detected *in vivo*. The same holds for various polyketide forming systems. However, *in vitro* studies are still lacking; (c) copurification of peptide synthetases and chaperones has been observed in the cases of gramicidin S, tyrocidine (von Döhren, H. and Pfeifer, E., unpublished results) and enniatin synthetases (Zocher, R. and Pieper, R., unpublished results). Separation of gramicidin S synthetase 2

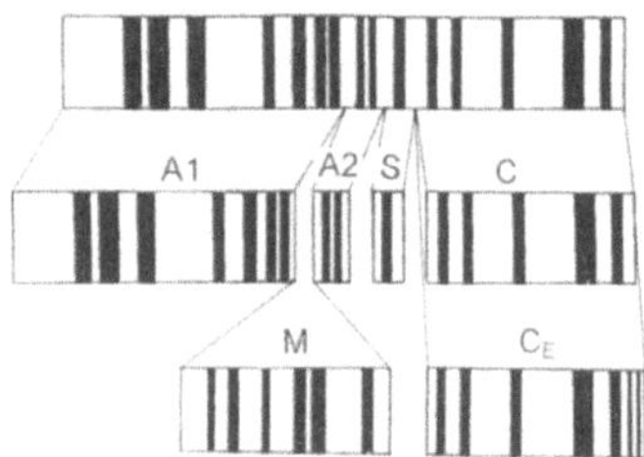

Fig. 7. Peptide synthetase modules. A peptide synthetase module (top) can be dissected into submodules catalysing adenylate formation (A1 and A2), acyl-carrier module (S), and condensing module (C); condensing modules with epimerizing function have two additional motifs (C_E). *N*-Methyltransferase modules (M) are inserted between A1 and A2.

from endogenous GroEL can be achieved by gradient centrifugation [122]. Complexes between gramicidin S synthetase 2 and GroEL have been detected in electron micrographs (von Döhren, H. and Wecke, J., unpublished results). Gradient centrifugation of crude gramicidin S biosynthetic systems yields complexes containing gramicidin S synthetases 1 and 2 and GroEL.

The interaction of peptide synthetases during intermolecular transfer of aminoacyl or peptidyl intermediates seems to require at least the interaction of two domains containing the M-motif. Also, in the case of initiation of peptide synthesis by acyl transfer, for example by surfactin and polymyxin, acyl transfer seems to require the interaction of a respective acyltransferase with a peptide synthetase domain. This can be concluded from the juxtaposition of modules in the systems forming gramicidin S, tyrocidine, and surfactin considering intermolecular condensation steps (Fig. 4). However, the positioning of these modules is not in all cases understood. Whilst in Aad-Cys-D-Val synthetases the first condensation domain is inserted between the first and second activation domains, a condensation domain precedes the first activation domain in the cyclosporin synthetase gene [118]. Since in cyclosporin synthetase a total of eleven condensation domains are present, this first domain could also function in the terminating cyclization reaction. The actual spatial position and time of operation of the expressed module may not therefore correspond to the module position within the gene. Alignment of amino acid sequences of condensation modules containing the M-motif shows at least three different types of domains: (a) N-terminal domains, (b) L-amino-acid-adding domains; (c) D-amino-acid-adding domains (Fig. 7).

Survey of biosynthetic systems studied

The peptide-forming and related systems including aminoacyl addition steps currently studied are compiled in Table 2. It seems surprising that although acyl-CoA synthetases are common to all species, so far no peptide synthetases of archaebacteria, plant or animal systems have been studied. Pioneering work in this context is in progress in the authors' laboratories on bouvardin (Zocher, R. and Weckwerth, W., unpublished results) from the medicinal plant *Bouvardia terniflora*, and the toxin lophyrotomin, a benzoylated octapeptide from the sawfly larvae *Lophyrotoma interrupta* [243] and *Arge pullata* [244]. It cannot be expected, however, that every organism contains such enzyme systems. In the recently released genome sequence of *Haemophilus influenzae*, no respective genes have been detected. Concerning the work summarized here, we point to a few selected aspects. For more details the reader is referred to a forthcoming more extensive review (von Döhren, H., Pfeifer, E. and van Liempt, H., in preparation). The grouping of peptides is sometimes difficult if several properties are combined. For a biochemical approach, we prefer here to consider structural features instead of discussing available examples in detail.

Linear peptides. Major advances have been made in the analysis of the penicillin tripeptide precursor-forming Aad-Cys-D-Val synthetase [2]. Multienzymes purified from *Streptomyces clavuligerus* [2, 81, 110, 111, 147–152] and *Acremonium chrysogenum* [144–146] have been studied. Contrary to earlier assumptions, valine is apparently epimerized at the tripeptide stage since no D-valine intermediate has been detected [81, 111]. The terminating thioesterase function, presumably located in the C-terminal region of Aad-Cys-D-Val synthetases, has thus been proposed to specifically release the LLD-tripeptide. Several tripeptide analogues have been synthesized [146], and the stoichiometry of 3 mol ATP consumed for each tripeptide formed under optimal conditions has been confirmed [36]. A surprising result recently obtained was the formation of *O*-methyl-seryl-D,L-valine by Aad-Cys-D-Val synthetase upon replacement of cysteine by *O*-methylserine [245]. The result seems to indicate the initial formation of the second peptide bond. The assignment of activating functions to the domains has been achieved by fragment expression in bacteria [81, 110] (Uhlmann, V., van Liempt, H. and von Döhren, H., unpublished results) and limited proteolysis [246]. However, bacterially expressed and isolated fragments containing domains required renaturation and activities measured were poor. Current work is performed with reference to domain boundaries [39].

In the biosynthetic system forming the tripeptide bialaphos or phosphinothricyl-alanylalanine, a peptide synthetase gene has been cloned [158], but this gene still has to be functionally characterized. Contrary to Aad-Cys-D-Val biosynthesis, alanylation of the phosphinothricine moiety is accomplished by the products of multiple genes [159]. This gene cluster also contains two thioesterase-like genes, possibly involved in the release of the peptide [159, 160].

A similar organization of modules with a terminating thioesterase function can be expected for the vancomycin-type of glycosylated heptapeptides, which also have a free C-terminus. In these systems, phenylglycine is a direct precursor, which is supplied by a polyketide-type pathway. Potential polyketide biosynthetic genes have been identified in the ardacin producer *Kibdelosporangium aridum* [166]. Other linear peptides have modified terminal regions. Biosynthesis of linear gramicidins is initiated by formylation of the valine-thioester intermediate and termination occurs by aminolysis of the pentadecapeptide with phosphatidylethanolamine [169]. Alamethicin biosynthesis is initiated by acetyl transfer to thioester-activated aminoisobutyrate, and terminated by the addition of phenylalaninol [170]. Termination by aminoalcohol binding is a common reaction in peptaibol formation. In the recently described tolaasin, a toxin produced by the mushroom pathogen *Pseudomonas tolaasii*, N-terminal acylation and lactonization of the C-terminus with threonyl-hydroxyl is observed, forming a five-membered-ring structure. Tolaasin is unique because an analogue has been reported with a four-amino-acid deletion within the chain [231]. Mutant analysis has provided evidence for the involvement of at least three multienzymes of 465, 440, and 435 kDa in its biosynthesis [232].

Cyclic peptides and peptidolactones. The complete peptide synthetase DNA sequences of gramicidin S [65, 66, 83], surfactin [80, 112], HC-toxin [174], and cyclosporin [118] have been determined. The sizes of their modules corresponded to those found for the single-module tyrocidine synthetase 1 and gramicidin S synthetase 1, as well as for the three modules of Aad-Cys-D-Val synthetase. The sequencing efforts have provided reliable

Table 2. Current state of research on peptide synthetases. For additional references see [1]. The abbreviations used are as follows: P, peptide; C, cyclopeptide; L, lactone; D, depsipeptide; E, ester; R, acyl; M, modified. The structural types are defined by the number of amino, imino or hydroxy acids in the precursor chain. The ring sizes of cyclic structures are indicated by the number following C, L, D, or E, defining the type of ring closure.

Peptide	Organism	Structural type	Gene cloned	Enzymology	References
Linear					
Carnosine	rat	P-2	−	+	123−125
Bacilysin	*Bacillus subtilis*	P-2-M	(+)	(+)	126
Aad-Cys-D-Val		P-3	+	+	2, 6, 127−133
	Aspergillus nidulans		+	+	64, 134−137
	Penicillium chrysogenum		+	(+)	138−143
	Acremonium chrysogenum		+	+	144−146
	Streptomyces clavuligerus		+	−	81, 110, 111, 147−153
	Nocardia lactamgenus		+	−	154
	Lysobacter sp.		+	−	155
Ergopeptides	*Claviceps purpurea*	R-P-3-M	−	+	156
Bialaphos	*Streptomyces hygroscopicus*	P-3	+	−	157−160
Edeine	*Bacillus brevis* Vm4	P-5-M	−	+	1
Anguibactin	*Vibrio anguillarum*	R-P-2-M	+	−	161−163
Yersiniabactin	*Yersinia enterocolitica*	R-P-3-M	+	−	164
Phaselotoxin	*Pseudomonas syringae* pv. *ph.*	P-4-M	(+)	−	165
Ardacin	*Kibdelosporangium aridum*	P-7-M	(+)	−	166
Pyoverdin	*Pseudomonas aeruginosa*	R-P-8-M	+	−	167, 168
Linear gramicidin	*Bacillus brevis*	P-15-M	−	+	169
Alamethicin	*Trichoderma viride*	P-19-M	−	+	170
Cyclopeptides					
Cyclopeptin	*Penicillium cyclopium*	C-2	−	+	1
Enterochelin	*Escherichia coli*	P-C-E-3	+	+	171−173
HC-toxin	*Cochliobolus carbonum*	C-4	+	+	174−178
Tentoxin	*Alternaria alternata*	C-4	−	+	179−181
Ferrichrome	*Aspergillus quadricinctus*	C-6	−	+	182
Echinocandin	*Aspergillus nidulans*	R-C-6	−	(+)	76, 183
Microcystin	*Microcystis aeruginosa*	C-7	(+)	−	184
Iturin	*Bacillus subtilis*	C-8	(+)	−	185
Gramicidin S	*Bacillus brevis* ATCC 9999	$C-(P-5)_2$	+	+	28, 29, 61−63, 65, 67, 82, 83, 91, 101, 186, 187
Tyrocidine	*Bacillus brevis* ATCC 8185	C-10	+	+	32, 39, 55, 56, 71, 188
Cyclosporin	*Tolypocladium nivea*	C-11	+	+	118, 189
Mycobacillin	*Bacillus subtilis*	C-13	−	+	1
Lactones					
Actinomycin	*Streptomyces clavuligerus*	$R-(L-5)_2$	−	+	107−109, 200, 101
Destruxin	*Metarhizium anisopliae*	L-6	+	(+)	202
Etamycin	*Streptomyces griseus*	R-L-7	−	(+)	203
Surfactin	*Bacillus subtilis*	L-8	+	+	30, 31, 33, 204−207, 211
Quinomycin	*Streptomyces echinatus*	$(R-P-4)_2$	−	+	212
R106	*Aureobasidium pullulans*	L-9	−	(+)	183
Syringomycin and syringostatin	*Pseudomonas syringae*	R-L-9	(+)	−	213−216
SDZ90-215	*Septoria* sp.	L-10	−	+	217
SDZ214-103	*Cylindrotrichum oligosporum*	L-11	(+)	+	199, 218
Branched polypeptides					
Polymyxin	*Bacillus polymyxa*	R-P-10-C-7	−	+	219
Bacitracin	*Bacillus licheniformis*	R-P-12-C-7	(+)	+	1, 220−223
Nosiheptide	*Streptomyces actuosus*	R-P-13-C-10-M	−	+	224, 225
Thiostrepton	*Streptomyces laurentii*	R-P-17-C-10-M	−	+	224, 225
Branched peptidolactones					
Lysobactin	*Lysobacter* sp.	P-11-L-9	(+)	+	226−228
A21798C	*Streptomyces roseosporus*	R-P-13-L-10	(+)	+	224, 225
A54145	*Streptomyces fradiae*	R-P-13-L-10	−	+	225
Tolaasin	*Pseudomonas tolaasii*	R-P-18-L-5	(+)	−	226, 227
Depsipeptides					
Enniatin	*Fuarsium oxysporum, F. scirpi*	D-6	+	+	109−113, 115, 228, 229
Beauvericin	*Beauveria bassiana*	D-6	−	+	230, 231
Related systems					
D-Ala addition to lipoteichoic acid	*Lactobacillus casei*	ester	+	+	82, 232
Coronatin	*Pseudomonas syringae*	acyl amino acid	+	(+)	233−235
FK506 (immunomycin)	*Streptomyces tsukubaensis*	macrolide	+	+	236, 237

data for molecular size estimations and sizes of all the peptide synthetases have been revised. Cyclosporin synthetase is encoded by a 45.8-kb intron-free reading frame and thus represents the largest known enzyme with a size of 1689243 Da (before post-translational modification). It catalyses a total of 40 reactions and all peptidyl intermediates remain covalently attached during the cycle [190]. Colinearity of modules and peptide structure is evident from the positioning of the seven *N*-methyltransferase modules in the gene.

The HC-toxin synthetase gene *hts-1* (15.7 kb), with an estimated gene product of 570 kDa [174, 175], resides in two functional copies on the same chromosome in the corn pathogen *Cochliobolus carbonum* [175]. Initial work on multienzyme characterization has led to the production of two partly functional fragments, which have been used for precursor identification, sequence information, and antibody production [176–178]. The sequence data suggest that the first domain activates proline since the peptide contains D-proline, L-proline is activated, and epimerizing motifs can be identified. Lack of additional epimerization motifs indicates that D-alanine is a direct precursor and this agrees with the observed activation of D-alanine. In addition to *hts-1*, genes have been identified for export and thus resistance to the toxin (*TOXA*) and biosynthesis of Aoe (*TOXC*) [174]. In addition to HC-toxin (TOX cluster) in *Cochliobolus carbonum*, related strains produce T-toxin (TOX1, *C. heterostrophus*), and the cyclopeptide victorin (TOX3, *C. victoriae*). What cannot be deduced from the sequence data, is how cyclization is directed. Like Aad-Cys-D-Val synthetases, gramicidin S synthetase 2 and surfactin synthetase 3 contain a putative thioesterase/transferase motif GXSXG in the C-terminal region. In addition, thioesterase-like genes are components of their clusters [66, 120], and some evidence suggests the participation of the thioesterase GrsT in the catalytic cycle of gramicidin S formation [7]. This motif, however, is missing in the HC-toxin and cyclosporin sequences, suggesting the involvement of as yet uncharacterized proteins assisting peptide cyclization. The gramicidin S synthetase 2 gene (13.4 kb) has been sequenced from both the original Russian strain (ATCC 9999) and the corresponding Japanese isolate (Nagano strain) [64, 66]. Only minor differences in sequence were found (comparing the latest revisions of sequences). The structures revealed four adenylate-forming domains, four carrier protein domains, and four condensing domains. An unexpected feature was a C-terminal transferase motif, while the postulated waiting site for the pentapeptide intermediate was missing. The mode of dimerization of the two pentapeptides is still not resolved. Colinearity of the peptide structure and activating modules, which had already been concluded from limited proteolysis and analysis of early termination mutants [247], has now been verified by fragment expression [63, 66].

The surfactin system is the first example of an acyl-peptidolactone biosynthetic system the DNA sequence of which was determined, but the acyl-transferase gene remains to be identified. Sequencing of the operon and structural studies have been stimulated by its presumed involvement in competence development [68, 80, 205, 207]. Fine structure mapping and the generation of loss-of-function mutations within the synthetase region finally revealed that a small gene, *comS*, is responsible for an as yet uncharacterized signalling process [68]. This gene, encoding a 46-residue peptide, is superimposed on the second synthetase gene *srfAB*. The physiological function(s) of surfactin, which is not required for competence, remain(s) to be established. A second 55-residue peptide competence pheromone (ComX) has been identified and this has been proposed to function in the ComP-ComA regulatory system in the induction of the *srfA* operon [248]. Recently, it has been shown that *srfP* controls both iturin and surfactin biosynthesis in *B. subtilis RB14* [185]. Hence, the production of both peptides, which may act synergistically, may be significant in this strain. However, not all strains of *B. subtilis* seem to produce these two peptides; many other peptides have been described [10].

The surfactin system has several remarkable features: the first activating domain is preceded by a condensation domain, presumably involved in acyl transfer. An acyltransferase has been partially purified and is functionally comparable to an acyltransferase operating in the polymyxin biosynthetic system [219]. These transferases employ acyl-CoA, while several aromatic acyl groups of peptides seem to be introduced as enzyme-bound adenylates [200]. Such systems, including enterobactin (EntE) [172], actinomycin (ACMSI) [201], triostin (TRSI) [212], mikamycin (MKMS) [203], nosiheptide and thiostreptone [224, 225] are involved in the initiation of cyclic chromopeptides by acylation of the first amino acid, thioester bound to the respective peptide synthetase. This reaction has been studied by Keller and Stindl and requires the presence of the cofactor phosphopantetheine [200].

In several systems, multienzyme polypeptides have been partially characterized or their presence has been demonstrated. In others, peptide synthetase genes have been detected but their respective products remain to be identified.

Mixed polyketide synthase/peptide synthetase systems. Several systems condensing amino acids with polyketides are known. The addition of amino acids to proteins or polymers such as lipoteichoic acid may also involve enzymes containing the well-known adenylate module. Recent work concerns the phytotoxin coronatin [238–240], the immunomodulator immunomycin (FK506) [241, 242], and D-alanine addition to cell wall polymers [86, 237]

Pathway prediction and cloning strategies

The work carried out so far permits the prediction of the sequence of biosynthetic reactions in peptide formation. The nature of the direct precursors, the timing of their modification, and the extent of integration of modules can be tentatively suggested. For certain, the even spacing of adenylate-forming modules points to a peptide-forming system. Additional motif information permits the secure prediction of the size of peptide, an occasionally good guess at the amino acids in defined positions, and their respective modifications. However, the direct incorporation of D-amino acids may escape this analysis. Despite this limitation, the gene sequence provides a wealth of information.

The cloning of peptide biosynthetic clusters has been successfully addressed using core motifs as boundaries for the PCR production of specific gene probes [221]. Earlier approaches using the expression of multigene fragments with immunodetection [249] are still valuable, since the structural similarities of synthetases permits the identification of a wide variety of multienzymes. A useful approach has also been the employment of peptide-specific antibodies, e.g. directed against the key motif of the adenylate domain, SGTTGXPKG (Etchegaray, A. and Turner, G., unpublished data).

Alteration and construction of biosynthetic systems

The availability of advanced genetic tools in the *B. subtilis* system has helped Stachelhaus et al. [250] and Grandi et al. [251] to demonstrate the manipulation of peptide synthetase programming using the surfactin system. Disruption of the gene encoding the addition of the terminal leucine led to formation of the linear hexapeptide precursor. Introduction of alternative amino-acid-activating modules by recombinational integration of modules from the gramicidin S system (*Bacillus brevis*) or the penicillin system (*Penicillium chrysogenum*) restored the formation of cyclopeptides, now with respectively exchanged residues in position 7 [250]. Likewise, the exchange of domains within the sequence has been carried out employing surfactin modules (Grandi, G., personal communication [251]).

With these experiments, protein templates have been successfully altered for the production of new peptides and cyclopeptides not found in nature. It can thus be predicted that the construction of protein templates for any given peptide will be accomplished in the not too distant future.

Regulatory aspects, resistance, and transport functions

Functions of peptides. The function of secondary metabolite production in general is barely understood. Speculations have gradually shifted from the 'purpose-free game of nature' to the functionally specialized compound that coevolved with other cell components with a defined function [252]. This shift can be understood from the impressive complexity of biosynthetic gene clusters and also from the equally impressive chemical complexity of natural products. However, with the exception of siderophores only minor advances have been made in the assignment of defined function. The most fascinating experimental work of the past years has been the putative involvement of surfactin in the development of competence in *B. subtilis* [68, 112, 205–207].

Induction of peptide biosynthesis. More details are accumulating on the regulation of events related to secondary metabolism by diffusible factors. A-factor analogues have been found to be involved in antibiotic or pigment production and sporulation especially in Actinomycetes [253]. Formation of the virginiamycins and presumably a variety of homologous compounds from various strains of Streptomycetes depends on the presence of virginia butanolides [254]. Likewise, homoserine lactones control various activities in different prokaroytes, including the synthesis of carbapenems in *Erwinia carotovora* and *E. herbicola*, *Pseudomonas aeruginosa*, and *Serratia marcescens* [255]. A specific A-factor receptor resembling the known factor from the streptomycin producer *Streptomyces griseus* has been recently identified in the virginiamycin producer [256], indicating the involvement of signalling systems based on known protein kinase systems. The 35-kDa receptor VbrA may constitute a part of an essential gene cluster encoding components of the transcriptional and translational machineries.

The synthesis of toxins and their secretion are considered to be essential properties for plant-invading microbes. In strains of *Pseudomonas syringae*, several classes of toxins have been identified, including peptides (phaseolotoxin, tabtoxin, syringomycin, syringostatin, syringotoxin, and syringopeptin) and amino acid conjugates (coronatin) [238]. Genes associated with syringomycin production have been shown to be conserved as single copies in strains of the pathovar *syringae* and not to be present in a set of different pathovars. Structural differences within the respective DNA regions indicate the relatedness of genes associated with the production of peptide analogues (syringotoxin, syringostatin). Production of the phytotoxin is induced and enhanced in the presence of plant signal molecules, such as arbutin, flavonoid glycosides, and D-fructose [257, 258]. Internal metabolites may also be involved in the regulation of peptide expression. When indole 3-acetic acid biosynthesis in *Pseudomonas syringae pv. syringae* is inactivated by Tn5 insertion, syringomycin biosynthesis is diminished [257]. This strain also markedly differs in its growth properties *in planta*.

Several peptides have also been identified in phytopathogenic fungi [259]. The identification of a resistance gene in maize, which inactivates HC-toxin by reduction of the 8-keto group, is particularly remarkable [260, 261].

Global regulatory signals. Similar reading frames of about 0.7 kb have been detected in the operons involved in the formation of surfactin and iturin (sfp, *Bacillus subtilis*) [185, 262], gramicidin S (gsp, *Bacillus brevis*) [263], enterobactin (entD, *E. coli*) [262], and a secondary metabolite possibly involved in heterocyst formation in *Anabaena* sp. (hetI) yet to be identified [264]. Disruption of *sfp* prevents formation of surfactin [263], or surfactin and iturin [185] and biosynthesis can be restored by *gsp* replacement [263]. It has been suggested that the gene product is involved in the export of the peptides formed. In the case of gramicidin S, however, the peptide is accumulated inside the cell. Release can be triggered by shedding of the S-layer protein by high salt concentration, inducing the formation of L-forms [265].

Peptide-induced gene expression. Recently, Thompson et al. [266] reported the expression of several unknown genes in *Streptomyces lividans* upon treatment with thiostrepton or related peptides. The promoter of one of these genes (*tipA*) has been used to screen for new thiopeptides employing kanamycin resistance as a reporter [267, 268]

Evolutionary considerations

Several discussions on groupings and tree constructs of peptide synthetases, acyl-CoA synthetases, and luciferases restrict themselves to the adenylate domain, which besides catalyzing the formation of acyladenylates and transferring these to CoA may also catalyze the reformation of the phosphoester bond to synthesize ATP, adenosine polyphosphates or diadenosine polyphosphates. In peptide synthetases, aminoacyl adenylates are formed and transferred to the cysteamine moiety of 4′-phosphopantetheine, which is attached to the adjacent acyl carrier domain (Fig. 2). Thus, the binding of CoA and catalysis of the transfer reaction has to be located in this region spanning 500–600 amino acid residues.

Acyl-CoA synthetase and peptide synthetases share at least the acyladenylate forming domain and have been grouped as AMP-binding proteins [50, 54]. Trees constructed with 25 [65] and 42 sequences [54] show three and four major branches, respectively. The gross similarity however, does not permit a reliable grouping of some of the proteins. In the peptide synthetase groups, subgroups are observed reflecting both organism and specificity of the integrated systems. Thus all domains of Bacilli form a subgroup and sub-subgroups reflect substrate specificity [80].

Phylogenetic approaches from the polyketide biosynthetic field have included acyl-carrier proteins either as autonomous (type II) or integrated structures (type I), depending on the state of organization of the respective synthase system. Hopwood and colleagues have pointed out similarities between peptide synthetases and acyl-carrier proteins and constructed alignments [Hopwood, D. (1992) lecture at the American Society of Microbiology Meeting, Bloomington, unpublished results). Dieckmann has compared various acyl-carrier protein structures, and found similar spatial structure predictions, regardless of the origin of the protein (Dieckmann, R. and von Döhren, H., unpublished results).

Table 3. Peptide synthetase structural elements and their occurrence in related enzymes and enzyme systems.

Element	Enzymes	Functions
Adenylate formation (A)	acyl-CoA ligases	providing acyl-CoA for a variety of biosynthetic reactions
	insect luciferases [adenylate forming, like EntE (enterobactin)], peptide synthetases, polyketide synthases	luminescence, siderophore biosynthesis, carboxyl activation
4′-Phosphopantetheine-binding site (carrier protein)	peptide synthetases, polyketide synthases, polyhydroxybutyrate synthase	covalent transport of acyl intermediates
Epimerase	peptide synthetases, polyketide synthases?	epimerization of acyl intermediates
Thioesterase (T)	thioesterases, polyketide synthases	providing free carboxylates for a variety of biosynthetic routes
	peptide synthetases	possible transferase function (cyclization), stereospecific peptide release?
N-Methyltransferase	methyltransferases	various methyl transfers
	peptide synthetases	

In the case of Aad-Cys-D-Val synthetase, sequence data from both prokaryotic and eukaryotic origins form a subgroup, with the positionally conserved domains grouped in the respective sub-branches. This can been taken as evidence for horizontal gene transfer from prokaryotes to eukaryotes in the β-lactam biosynthetic cluster [65]. The transfer hypothesis has been put forward from analysis of the peptide-transforming enzymes isopenicillin *N*-synthase (IPNS) and deacetoxycephalosporin synthase (DAOCS) [269–272]. A scheme of evolution and gene transfers has been proposed based on the relatedness of bacterial and fungal enzymes involved in penicillin and cephalosporin biosynthesis [270–272]. In addition, while peptide synthetase genes known so far are intron free despite their size, the acyltransferase gene involved in penicillin formation contains three introns and has been designated a eukaryotic gene or a fused prokaryotic and eukaryotic gene linked to the prokaryotic biosynthetic genes of Aad-Cys-D-Val synthetase and isopenicillin *N*-synthase. This hypothesis has been questioned by Smith et al. [273], who root isopenicillin *N*-synthase and deacetoxycephalosporin synthase data with each other and claimed to obtain a conventional tree. No definite conclusion can be reached, since the rate of evolution remains an uncertain factor. The occurrence of structurally related peptides within diverse microorganisms implies horizontal transfer events of biosynthetic clusters.

Regardless of possible transfer events, biosynthetic enzymes are composed of common structural elements (Table 3). The timing of the assembly of these elements into defined genes and clusters will eventually be traced. The defined positions of catalytic modules within multienzymes and their respective domains indicate, however, an ancient fusion process (Graur, D., personal communication). No evidence for the frequent exchange of modules and a flexible operating system has been obtained.

Conclusions

The nonribosomal formation of peptides is directed by a set of DNA modules encoding multienzyme systems. These multienzymes are protein templates and catalyse the sequential condensation of amino acids, including their activation and modification. The sequence of events is determined by the spatial arrangement of catalytic domains but is also reflected by the order of modules at the gene level. Similar organizational principles have been found in polyketide-forming systems. Thus, the biosynthesis of metabolites formed by sequential condensations may be altered by defined genetic changes. Such alterations, however, require the detailed study of multienzyme systems some of which exceed 10^4 amino acid residues and the manipulation of the corresponding genes. The modular construction of the enzymatic systems and their regulatory signals will accelerate research on a variety of fundamental processes. Such processes include not only the synthesis of natural products and their exploitation in drug development but also the roles of many metabolites in invasive and defensive processes.

We thank our colleagues for suggestions and helpful discussions, and for providing access to unpublished data. Special thanks to Eva Pfeifer and Andrew MacCabe for their invaluable efforts in improving the manuscript. The work from the authors' laboratory was supported by grants from the *Deutsche Forschungsgemeinschaft*, *Bundesministerium für Forschung und Technologie*, European Community Biotech program, the German-Israeli Research Foundation, fellowships from EMBO and DAAD, and research cooperations with Gist-Brocades (Delft), Eli Lilly Research Laboratories (Indianapolis), Sandoz (Basel), Biochemie (Kundl), InViTek GmbH (Berlin), and Molnar GmbH (Berlin).

REFERENCES

1. Kleinkauf, H. & von Döhren, H. (1990) *Eur. J. Biochem. 192*, 1–15.
2. Aharonowitz, Y., Bergmeyer, J., Cantoral, J. M., Cohen, G., Demain, A. L., Fink, U., Kinghorn, J., Kleinkauf, H., MacCabe, A., Palissa, H., Pfeifer, E., Schwecke, T., van Liempt, H., von Döhren, H., Wolfe, S. & Zhang, J. (1993) *Bio/Technology 11*, 807–810.
3. Zuber, P. (1991) *Curr. Opin. Cell. Biol. 3*, 1046–1050.
4. Kleinkauf, H., van Liempt, H., Palissa, H. & von Döhren, H. (1992) *Naturwissenschaften 79*, 153–162.
5. Marahiel, M. A. (1992) *Naturwissenschaften 79*, 202–212.
6. van Liempt, H., Palissa, H., Schwecke, T., Aharonowitz, Y., Pfeifer, E., von Döhren, H. & Kleinkauf, H. (1993) in *50 years of penicillin application – history and trends* (Kleinkauf, H. & von Döhren, H., eds) pp. 136–144, Publica, Prague.
7. Marahiel, M. A. (1992) *FEBS Lett. 307*, 40–43.
8. von Döhren, H., Pfeifer, E., van Liempt, H., Lee, Y.-O., Pavela-Vrancic, M. & Schwecke, T. (1993) in *Industrial microorganisms: basic and applied molecular genetics* (Baltz, R. H.,

Hegeman, G. D. & Skatrud, P. L., eds) pp. 159–167, American Society of Microbiology, Washington DC.
9. von Döhren, H. (1993) *Biochem. Soc. Trans. 21*, 214–217.
10. von Döhren, H. (1995) in *Genetics and biochemistry of antibiotic production* (Vining, L. C. & Stuttard, C., eds) pp. 129–172, Butterworth-Heinemann, Boston.
11. Stachelhaus, T. & Marahiel, M. A. (1995) *FEMS Microbiol. Lett. 125*, 3–14.
12. Lipmann, F. (1968) *Essays Biochem. 4*, 1–23.
13. von Döhren, H. (1990) in *Biochemistry of peptide antibiotics* (Kleinkauf, H. & von Döhren, H., eds) pp. 411–507, de Gruyter, Berlin.
14. Hamada, T., Sugawaka, T., Masunagi, S. & Fusetani, N. (1994) *Tetrahedron Lett. 35*, 719–720.
15. Hansen, J. N. (1993) *Annu. Rev. Microbiol. 47*, 535–564.
16. Jung, G. & Sahl, H.-G. (1991) *Nisin and novel lantibiotics*, Leiden, the Netherlands, Escom.
17. Entian, K. D. & Klein, C. (1993) *Naturwissenschaften 80*, 454–460.
18. Gasson, M. J. (1995) in *Genetics and biochemistry of antibiotic production* (Vining, L. C. & Stuttard, C., eds) pp. 283–306, Butterworth-Heinemann, Boston.
19. Sahl, H.-G., Jack, R. W. & Bierbaum, G. (1995) *Eur. J. Biochem. 230*, 827–853.
20. Jack, R. W. & Sahl, H.-G. (1995) *Trends Biotechnol. 13*, 269–278.
21. Martínez-Bueno, M., Maqueda, M., Gálvez, A., Samyn, B., van Beeumen, J., Coyette, J. & Valdivia, E. (1994) *J. Bacteriol. 176*, 6334–6339.
22. Potterat, O., Stephan, H., Metzger, J. W., Gnau, V., Zähner, H. & Jung, G. (1994) *Liebigs Ann. Chem.*, 741–743.
23. Fréchet, D., Guitton, J. D., Herman, F., Faucher, D., Helynck, G., Monegier de Sorbier, B., Ridoux, J. P., James-Surcouf, E. & Vuilhorgne, M. (1994) *Biochemistry 33*, 42–50.
24. Richter, K., Egger, R. & Kreil, G. (1987) *Science 248*, 200–202.
25. Kreil, G. (1994) *Science 266*, 996–997.
26. Heck, S. D., Siok, C. J., Krapcho, K. J., Kelbaugh, P. R., Thadeio, P. F., Welch, M. J., Williams, R. D., Ganong, A. H., Kelly, M. E., Lanzetti, A. J., Gray, W. R., Phillips, D., Parks, T. N., Jackson, H., Ahlijanian, M. K., Saccomano, N. A. & Volkmann, R. A. (1994) *Science 266*, 1065–1068.
27. Lipmann, F. (1973) *Science 173*, 875–884.
28. Schlumbohm, W., Stein, T., Ullrich, C., Vater, J., Krause, M., Marahiel, M. A., Kruft, V. & Wittmann-Liebold, B. (1991), *J. Biol. Chem. 266*, 23135–23141.
29. Stein, T., Vater, J., Kruft, V., Wittmann-Liebold, B., Franke, P., Panico, M., McDowell, R. & Morris, H. R. (1994) *FEBS Lett. 340*, 39–44.
30. Vollenbroich, D., Kluge, B., D'Souza, C., Zuber, P. & Vater, J. (1993) *FEBS Lett. 325*, 220–224.
31. Vollenbroich, D., Mehta, N., Zuber, P., Vater, J. & Kamp, R. M. (1994) *J. Bacteriol. 176*, 395–400.
32. Gocht, M. & Marahiel, M. A. (1994) *J. Bacteriol. 176*, 2654–2662.
33. D'Souza, C., Nakano, M. M., Corbell, N. & Zuber, P. (1993) *J. Bacteriol. 175*, 3502–3510.
34. Fujikawa, K., Suzuki, T. & Kurahashi, K. (1968) *Biochim. Biophys. Acta 161*, 232–246.
35. Wang, D. I. C., Stramondo, J. & Fleischaker, R. (1976) in *Biotechnological applications of proteins and enzymes* (Sharon, N. & Bohak, Z., eds) pp. 183–201, Academic Press, New York.
36. Kallow, W., von Döhren, H., Kennedy, J. & Turner, G. (1994) *Abstr. 7th IUMS Congr.*, Prague.
37. Pavela-Vrancic, M., Pfeifer, E., Freist, W. & von Döhren, H. (1994) *Eur. J. Biochem. 220*, 535–542.
38. Toh, H. (1990) *Protein Sequences & Data Anal. 3*, 517–521.
39. Dieckmann, R., Lee, Y.-O., van Liempt, H., von Döhren, H. & Kleinkauf, H. (1995) *FEBS Lett. 357*, 212–216.
40. Donadio, S., Staver, M. J., McAlpine, J. B., Swanson, S. J. & Katz, L. (1991) *Science 252*, 675–679.
41. Baron, M., Norman, D. G. & Campbell, I. D. (1991) *Trends Biochem. Sci. 16*, 13–17.
42. de Wet, J. R., Wood, K. V., DeLuca, M., Helinski, D. R. & Subramani, S. (1987) *Mol. Cell. Biol. 7*, 725–737.
43. Martinez-Blanco, H., Orejas, M., Reglero, A., Luengo, J. M. & Peñalva, M. A. (1993) *Gene (Amst.) 130*, 265–270.
44. Connerton, I. F., Fincham, J. R. S., Sandeman, R. A. & Hynes, M. J. (1990) *Mol. Microbiol. 4*, 451–460.
45. Garre, V., Murillo, F. J. & Torres-Martinez, S. (1994) *Mol. & Gen. Genet. 244*, 278–286.
46. Stoltzfus, A., Spencer, D. F., Zuker, M., Logsdon, J. M. Jr & Doolittle, W. F. (1994) *Science 265*, 202–207.
47. Wetlaufer, D. B. (1973) *Proc. Natl Acad. Sci. USA 70*, 697–701.
48. Schultz, G. E. & Schirmer, R. H. (1979) *Principles of protein structure*, Springer-Verlag, Berlin.
49. Bork, P. (1991) *FEBS Lett. 286*, 47–54.
50. Bairoch, A. (1991) Procite release no. 7.
51. Eggen, R. I. L., Geerling, A. C. M., Boshoven, A. B. P. & deVos, W. M. (1991) *J. Bacteriol. 173*, 6383–6389.
52. Babitt, P. C., Kenyon, G. L., Martin, B. M., Charest, H., Sylvestre, M., Scholten, J. D., Chang, K.-H., Liang, P. H. & Dunaway-Mariano, D. (1992) *Biochemistry 31*, 5594–5604.
53. Guilvout, I., Mercereau-Puijalon, O., Bonnefoy, S., Pugsley, A. P. & Carniel, E. (1993) *J. Bacteriol. 175*, 5488–5504.
54. Fulda, M., Heinz, E. & Wolter, F. P. (1994) *Mol. & Gen. Genet. 242*, 241–249.
55. Pavela-Vrancic, M., Pfeifer, E., van Liempt, H., Schäfer, H.-J., von Döhren, H. & Kleinkauf, H. (1994) *Biochemistry 33*, 6276–6283.
56. Pfeifer, E. (1994) Ph.D. Thesis, Technical University Berlin.
57. Sillero, M. A. G., Guranowski, A. & Sillero, A. (1991) *Eur. J. Biochem. 202*, 507–513.
58. Ortiz, B., Sillero, A. & Sillero, M. A. G. (1993) *Eur. J. Biochem. 212*, 263–270.
59. Guranowski, A., Günther-Sillero, M. A. & Sillero, A. (1994) *J. Bacteriol. 176*, 2986–2990.
60. Rapaport, E., Remy, P., Kleinkauf, H., Vater, J. & Zamecnik, P. C. (1987) *Proc. Natl Acad. Sci. USA 84*, 7891–7895.
61. Skarpeid, H.-J., Zimmer, T.-L., Shen, B. & von Döhren, H. (1990) *Eur. J. Biochem. 187*, 627–633.
62. Skarpeid, H.-J., Zimmer, T.-L. & von Döhren, H. (1990) *Eur. J. Biochem. 189*, 517–522.
63. Hori, K., Yamamoto, Y., Tokita, K., Saito, F., Kurotsu, T., Kanda, M., Okamura, K., Furuyama, J. & Saito, Y. (1991) *J. Biochem. (Tokyo) 110*, 111–119.
64. MacCabe, A., van Liempt, H., Palissa, H., Unkles, S. E., Riach, M. B. R., Pfeifer, E., von Döhren, H. & Kinghorn. J. (1991) *J. Biol. Chem. 266*, 12646–12654.
65. Turgay, K., Krause, M. & Marahiel, M. A. (1992) *Mol. Microbiol. 6*, 529–546.
66. Saito, F., Hori, K., Kanda, M., Kurotsu, T. & Saito, Y. (1994) *J. Biochem. (Tokyo) 116*, 357–367.
67. Saito, M., Hori, K., Kurotsu, T., Kanda, M. & Saito, Y. (1995) *J. Biochem. (Tokyo) 117*, 276–282.
68. Hamoen, L. W., Eshuis, H., Jongbloed, J., Venema, G. & Van Sinderen, D. (1995) *Mol. Microbiol. 15*, 55–63.
69. Traut, T. W. (1994) *Eur. J. Biochem. 222*, 9–19.
70. Tokita, K., Hori, K., Kurotsu, T., Kanda, M. & Saito, Y. (1994) *J. Biochem. (Tokyo) 114*, 522–527.
71. Pavela-Vrancic, M., Pfeifer, E., Schröder, W., von Döhren, H. & Kleinkauf, H. (1994) *J. Biol. Chem. 269*, 14962–14966.
72. Walker, J. E., Saraste, M., Runswick, M. J. & Gay, N. J. (1982) *EMBO J. 1*, 945–951.
73. Yoshida, M. & Amano, T. (1995) *FEBS Lett. 359*, 1–5.
74. Chaure, P. T. & Connerton, I. F. (1994) *Abstr. 2nd ECFG Meeting*, Lunteren, Abstract A17.
75. Saxholm, H., Zimmer, T.-L. & Laland, S. G. (1972) *Eur. J. Biochem. 20*, 138–144.
76. Adefarati, A. A., Giacobbe, R. A., Hensens, O. D. & Tkacz, J. S. (1991) *J. Am. Chem. Soc. 113*, 3542–3545.
77. Okumura, Y. (1990) in *Biochemistry of peptide antibiotics* (Kleinkauf, H. & von Döhren, H., eds) pp. 365–378, de Gruyter Berlin.
78. Jakubowski, H. & Goldman, E. (1992) *Microbiol. Rev. 56*, 412–429.
79. Giegé, R., Puglisi, J. D. & Florentz, C. (1993) *Prog. Nucleic Acids Res. Mol. Biol. 45*, 129–206.

80. Cosmina, P., Rodriguez, F., de Ferra, F., Grandi, G., Perego, M., Venema, G. & van Sinderen, D. (1993) *Mol. Microbiol. 8*, 821–831.
81. Schwecke, T. (1993) Ph.D. Thesis, Technical University Berlin.
82. Kanda, M., Hori, K., Kurotsu, T., Miura, S. & Saito, Y. (1989) *J. Biochem. (Tokyo) 105*, 653–659.
83. Hori, K., Saito, F., Tokita, K., Kurotsu, T., Kanda, M. & Saito, Y. (1994) *J. Biochem. (Tokyo) 116*, 1202–1204.
84. Reanney, D. C. (1977) *J. Theor. Biol. 65*, 555–569.
85. Wieland, T. (1988) in *The roots of modern biochemistry* (Kleinkauf, H., von Döhren, H. & Jaenicke, L., eds) pp. 213–221, de Gruyter Berlin.
86. Heaton, M. P. & Neuhaus, F. C. (1994) *J. Bacteriol. 176*, 681–690.
87. Hopwood, D. & Sherman, D. H. (1990) *Annu. Rev. Genet. 24*, 37–66.
88. Lipmann, F., Gevers, W., Kleinkauf, H. & Roskoski, R. Jr (1971) *Adv. Enzymol. 35*, 1–34.
89. Laland, S. G. & Zimmer, T.-L. (1973) *Essays Biochem. 9*, 31–57.
90. Schlumbohm, W. (1987) Ph. D. Thesis, Technical University Berlin.
91. Stein, T., Kluge, B., Vater, J., Franke, P., Otto, A. & Wittmann-Liebold, B. (1995) *Biochemistry 34*, 4633–4642.
92. Gilhuus-Moe, C. C., Kristensen, T., Bredesen, J. E., Zimmer, T.-L. & Laland, S. G. (1970) *FEBS Lett. 7*, 287–290.
93. Kleinkauf, H., Gevers, W., Roskoski, R. Jr & Lipmann, F. (1970) *Biochem. Biophys. Res. Commun. 41*, 1218–1221.
94. Rock, C. O. (1982) *J. Bacteriol. 152*, 1298–1300.
95. Rusnack, F., Sakaitani, M., Drueckhammer, D., Reichert, J. & Walsh, C. (1991) *Biochemistry 30*, 2916–2927.
96. Gerngross, T. U., Snell, K. D. & Peoples, O. P. (1994) *Biochemistry 33*, 9311–9320.
97. Yamada, M. & Kurahashi, K. (1968) *J. Biochem. (Tokyo) 63*, 59–69.
98. Yamada, M. & Kurahashi, K. (1969) *J. Biochem. (Tokyo) 66*, 529–540.
99. Kanda, M., Hori, K., Kurotsu, T., Miura, S., Yamada, Y. & Saito, Y. (1981) *J. Biochem. (Tokyo) 90*, 765–771.
100. Lee, S. G. & Lipmann, F. (1976) *Abstr. 10th Int. Congr. Biochem.*, Hamburg, Abstract 4-3-352.
101. Hori, K., Yamamoto, Y., Minetoki, T., Kurotsu, T., Kanda, M., Miura, S., Okamura, K., Furuyama, J. & Saito, Y. (1989) *J. Biochem. (Tokyo) 106*, 639–645.
102. Pfeifer, E., Pavela-Vrancic, M., von Döhren, H. & Kleinkauf, H. (1995) *Biochemistry 34*, 7450–7459.
103. Babbitt, P. C., Kenyon, G. L., Martin, B. M., Charest, H., Sylvestre, M., Scholten, J. D., Chang, K.-H., Liang, P.-H. & Dunaway-Mariano, D. (1992) *Biochemistry 31*, 5594–5604.
104. Chang, K.-H., Liang, P.-H., Beck, W., Scholten, J. D. & Dunaway-Mariano, D. (1992) *Biochemistry 31*, 5605–5610.
105. Scholten, J. D., Chang, K.-H., Babbitt, P. C., Charest, H., Sylvestre, M. & Dunaway-Mariano, D. (1991) *Science 253*, 182–185.
106. Loy, E. (1978) Diplomarbeit, Technical University Berlin.
107. Stindl, A. & Keller, U. (1994) *Biochemistry 33*, 9358–9364.
108. Stindl, A. & Keller, U. (1993) *J. Biol. Chem. 268*, 10612–10620.
109. Stindl, A. (1994) Ph.D. Thesis, Technical University Berlin.
110. Schwecke, T., Tobin, M. B., Kovacevic, S., Miller, J. R., Skatrud, P. L. & Jensen, S. E. (1992) in *Industrial microorganisms: basic and applied molecular genetics* (Baltz, R. H., Hegeman, G. D. & Skatrud, P. L., eds) p. 291, American Society of Microbiology, Washington DC.
111. Schwecke, T., Aharonowitz, Y., Palissa, H., von Döhren, H., Kleinkauf, H. & van Liempt, H. (1992) *Eur. J. Biochem. 205*, 687–694.
112. Fuma, S., Fujishima, Y., Corbell, N., D'Souza, C., Nakano, M., Zuber, P. & Yamane, K. (1993) *Nucleic Acids Res. 21*, 93–97.
113. Zocher, R., Keller, U. & Kleinkauf, H. (1982) *Biochemistry 21*, 43–48.
114. Zocher, R., Keller, U. & Kleinkauf, H. (1983) *Biochem. Biophys. Res. Commun. 110*, 292–299.
115. Billich, A. & Zocher, R. (1987) *Biochemistry 26*, 8417–8423.
116. Billich, A. & Zocher, R. (1990) in *Biochemistry of peptide antibiotics* (Kleinkauf, H. & von Döhren, H., eds) pp. 57–80, de Gruyter, Berlin.
117. Haese, A., Schubert, M., Herrmann, M. & Zocher, R. (1993) *Mol. Microbiol. 7*, 905–914.
118. Weber, G., Schmorgendörfer, K., Schneider-Scherzer, E. & Leitner, E. (1994) *Curr. Genet. 26*, 120–125.
119. Haese, A., Pieper, R., von Ostrowski, T. & Zocher, R. (1994) *J. Mol. Biol. 243*, 116–122.
120. Krätzschmar, J., Krause, M. & Marahiel, M. A. (1989) *J. Bacteriol. 171*, 5422–5429.
121. von Döhren, H. & Marahiel, M. A. (1990) in *Proc. Gen. Ind. Microbiol. Meet.*, vol. 2, pp. 691–702, Strasbourg.
122. Koischwitz, H. & Kleinkauf, H. (1976) *Biochim. Biophys. Acta 429*, 1041–1051.
123. Jungblut, P. (1981) Ph.D. Thesis, Technical University Berlin.
124. Bauer, K. & Schulz, M. (1994) *Eur. J. Biochem. 219*, 42–47.
125. Crowe, M. J. & Pixley, S. K. (1991) *Brain Res. 538*, 147–151.
126. Sakajoh, M., Solomon, N. A. & Demain, A. L. (1987) *J. Ind. Microbiol. 2*, 201–208.
127. Zhang, J. Y. & Demain, A. L. (1990) *Chin. J. Biotechnol. 6*, 1–9.
128. Baldwin, J. E., Bird, J. W., Field, R. A., O'Callaghan, N. M. & Schofield, C. J. (1990) *J. Antibiot. 43*, 1055–1057.
129. Queener, S. W. (1990) *Antimicrob. Agents Chemother. 34*, 943–948.
130. Martin, J. F. (1991) *Ann. N. Y. Acad. Sci. 646*, 193–201.
131. Zhang, J. & Demain, A. L. (1992) *Crit. Rev. Biotechnol. 12*, 245–260.
132. White, R. L., DeMarco, A. C., Shapiro, S., Vining, L. C. & Wolfe, S. (1989) *Anal. Biochem. 178*, 399–403.
133. Smith, D. J., Burnham, M. K., Bull, J. H., Hodgson, J. E., Ward, J. M., Browne, P., Brown, J., Barton, B., Earl, A. J. & Turner, G. (1990) *EMBO J. 9*, 2492–2499.
134. van Liempt, H., von Döhren, H. & Kleinkauf, H. (1989) *J. Biol. Chem. 264*, 3680–3684.
135. MacCabe, A. P., Riach, M. B. & Kinghorn, J. R. (1990) *EMBO J. 9*, 279–287.
136. Brakhage, A. A., Browne, P. & Turner, G. (1992) *J. Bacteriol. 174*, 3789–3799.
137. Brakhage, A. A., Browne, P. & Turner, G. (1994) *Mol. & Gen. Genet. 242*, 57–64.
138. Diez, B., Gutierrez, S., Barredo, J. L., van Solingen. P., van der Voort, L. H. & Martin, J. F. (1990) *J. Biol. Chem. 265*, 16358–16365.
139. Smith, D. J., Earl, A. J. & Turner, G. (1990) *EMBO J. 9*, 2743–2750.
140. Müller, W. H., van der Krift, T. P., Krouwer, A. J., Wosten, H. A., van der Voort, L. H., Smaal, E. B. & Verkleij, A. J. (1991) *EMBO J. 10*, 489–495.
141. Montenegro, E., Fierro, F., Fernandez, F. J., Gutierrez, S. & Martin, J. F. (1992) *J. Bacteriol. 174*, 7063–7067.
142. Cantoral, J. M., Gutierrez, S., Fierro, F., Gil-Espinosa, S., van Liempt, H. & Martin, J. F. (1993) *J. Biol. Chem. 268*, 737–744.
143. Lendenfeld, T., Ghali, D., Wolschek, M., Kubicek-Pranz, E. M. & Kubicek, C. P. (1993) *J. Biol. Chem. 268*, 665–671.
144. Hoskins, J. A., O'Callaghan, N., Queener, S. W., Cantwell, C. A., Wood, J. S., Chen, V. J. & Skatrud, P. L. (1990) *Curr. Genet. 18*, 523–530.
145. Gutierrez, S., Diez, B., Montenegro, E. & Martin, J. F. (1991) *J. Bacteriol. 173*, 2354–2365.
146. Baldwin, J. E., Shiau, C. Y., Byford, M. F. & Schofield, C. J. (1994) *Biochem. J. 301*, 367–372.
147. Jensen, S. E., Wong, A., Rollins, M. J. & Westlake, D. W. (1990) *J. Bacteriol. 172*, 7269–7271.
148. Zhang, J. Y. & Demain, A. L. (1990) *Biochem. Biophys. Res. Commun. 169*, 1145–1152.
149. Zhang, J., Wolfe, S. & Demain, A. L. (1989) *FEMS Microbiol. Lett. 48*, 145–150.
150. Baldwin, J. E., Bird, J. W., Field, R. A., O'Callaghan, N. M., Schofield, C. J. & Willis, A. C. (1991) *J. Antibiot. 44*, 241–248.
151. Tobin, M. B., Kovacevic, S., Madduri, K., Haskins, J. A., Skatrud, P. L., Vining, L. C., Stuttard, C. & Miller, J. R. (1991) *J. Bacteriol. 173*, 6223–6229.
152. Zhang, J., Wolfe, S. & Demain, A. L. (1992) *Biochem. J. 283*, 691–698.

153. Petrich, A. K., Leskiw, B. K., Paradkar, A. S. & Jensen, S. E. (1994) *Gene (Amst.) 142*, 41–48.
154. Coque, J. J., Martin, J. F., Calzada, J. G. & Liras, P. (1991) *Mol. Microbiol. 5*, 1125–1133.
155. Kimura, H. (1989) Japanese patent no. 2-291274.
156. Keller, U., Zocher, R., Krengel, U. & Kleinkauf, H. (1984) *Biochem. J. 218*, 587–862.
157. Thompson, C. J. & Seto, H. (1995) in *Genetics and biochemistry of antibiotic production* (Vining, L. C. & Stuttard, C., eds) pp. 197–222, Butterworth-Heinemann, Boston.
158. Wohlleben, W., Alijah, R., Dorendorf, J., Hillemann, D., Nußbaumer, B. & Pelzer, S. (1992) *Gene (Amst.) 115*, 127–132.
159. Hara, O., Anzai, H., Imai, S., Kumada, Y., Murakami, T., Itoh, R., Takano, E., Satoh, A. & Nagaoka, K. (1988) *J. Antibiot. 41*, 538–547.
160. Raibaud, A., Zalacain, M., Holt, T. G., Tizard, R. & Thompson, C. J. (1991) *J. Bacteriol. 173*, 4454–4463.
161. Crosa, J. H. (1989) *Microbiol. Rev. 53*, 517–530.
162. Farrell, D. H., Mikesell, P., Actis, L. A. & Crosa, J. H. (1990) *Gene (Amst.) 86*, 45–51.
163. Tolmasky, M. E., Actis, L. A. & Crosa, J. H. (1995) *Mol. Microbiol. 15*, 87–95.
164. Guilvout, I., Mercereau-Puijalon, O., Bonnefoy, S., Pugsley, A. P. & Carniel, E. (1993) *J. Bacteriol. 175*, 5488–5504.
165. Turgay, K. & Marahiel, M. A. (1994) *Peptide Res. 7*, 238–240.
166. Piecq, M., Dehottay, P., Biot, A. & Dusart, J. (1994) *J. DNA Sequencing and Mapping 4*, 219–229.
167. Merriman, T. R., Merriman, M. E. & Lamont, I. L. (1995) *J. Bacteriol. 177*, 252–258.
168. Adams, C., Dowling, D. N., O'Sullivan, D. J. & O'Gara, F. (1994) *Mol. & Gen. Genet. 243*, 515–524.
169. Kubota, K. (1987) *Biochem. Biophys. Res. Commun. 144*, 203–209.
170. Mohr, H. & Kleinkauf, H. (1978) *Biochim. Biophys. Acta 526*, 375–386.
171. Rusnack, F., Sakaitani, M., Drueckhammer, D., Reichert, J. & Walsh, C. T. (1991) *Biochemistry 30*, 2916–2927.
172. Rusnack, F., Faraci, W. S. & Walsh, C. T. (1989) *Biochemistry 28*, 6827–6835.
173. Reichert, J., Sakaitani, M. & Walsh, C. T. (1992) *Protein Sci. 1*, 549–556.
174. Scott-Craig, J. S., Panaccione, D. G., Pocard, J.-A. & Walton, J. D. (1992) *J. Biol. Chem. 267*, 26044–26049.
175. Panaccione, D. G., Scott-Craig, J. S., Pocard, J.-A. & Walton, J. D. (1993) *Proc. Natl Acad. Sci. USA 89*, 6590–6594.
176. Walton, J. D., Akimitsu, K., Ahn, J. H. & Pitkin, J. W. (1994) in *Host-specific toxin: biosynthesis, receptor and molecular biology* (Kohmoto, K. & Yoder, O. C., eds) pp. 227–237, Tottori University, Tottori.
177. Walton, J. D. (1987) *Proc. Natl Acad. Sci. USA 84*, 8444–8447.
178. Walton, J. D. & Holden, F. R. (1988) *Mol. Plant-Microbe Interact. 1*, 128–134.
179. Ramm, K., Ramm, M., Liebermann, B. & Reuter, G. (1994) *Microbiology 140*, 3257–3266.
180. Ramm, K., Ramm, M., Liebermann, B. & Reuter, G. (1994) *Appl. Biochem. Biotechnol. 49*, 35–43.
181. Ramm, K., Brückner, B. & Liebermann, B. (1994) *Appl. Biochem. Biotechnol. 49*, 45–50.
182. Siegmund, K. D., Plattner, H. J. & Diekmann, H. (1991) *Biochim. Biophys. Acta 1076*, 123–129.
183. Reinstädler, D., von Döhren, H., van Liempt, H., Pfeifer, E., Truglowski, B., Schwecke, T., Xuei, X., Peery, R. & Skatrud, P. (1994) *Abstr. 7th IUMS Conf.*, Prague, p. 127.
184. Rouhiainen, L., Buikema, W., Paulin, L., Fonstein, M., Sivonen, K. & Haselkorn, R. (1994) *Abstr. 8th Int. Symp. Phototrophic Prokaryotes*, Urbino.
185. Huang, C.-C., Ano, T. & Shoda, M. (1993) *J. Ferment. Bioeng. 76*, 445–450.
186. Vater, J., Schlumbohm, W., Salnikow, J., Irrgang, K. D., Miklus, M., Choli, T. & Kleinkauf, H. (1989) *Biol. Chem. Hoppe-Seyler 370*, 1013–1018.
187. Kurotsu, T., Hori, K., Kanda, M. & Saito, Y. (1991) *J. Biochem. (Tokyo) 109*, 763–769.
188. Mittenhuber, G., Weckermann, R. & Marahiel, M. A. (1989) *J. Bacteriol. 171*, 4881–4887.
189. Dittmann, J., Lawen, A., Zocher, R. & Kleinkauf, H. (1990) *Biol. Chem. Hoppe-Seyler 371*, 829–834.
190. Lawen, R. & Zocher, R. (1990) *J. Biol. Chem. 265*, 11355–11360.
191. Kleinkauf, H., Dittmann, J. & Lawen, A. (1991) *Biomed. Biochim. Acta 50*, 219–224.
192. Lawen A., Traber, R. & Geyl, D. (1991) *J. Biol. Chem 266*, 15567–15570.
193. Dittmann, J., Wenger, R. M., Kleinkauf, H. & Lawen, A. (1994) *J. Biol. Chem. 269*, 2841–2846.
194. Offenzeller, M., Su, Z., Santer, G., Moser, H., Traber, R., Memmert, K. & Schneider-Scherzer, E. (1993) *J. Biol. Chem. 268*, 26127–26134.
195. Hoffmann, K., Schneider-Scherzer, E., Kleinkauf, H. & Zocher, R. (1994) *J. Biol. Chem. 269*, 12710–12714.
196. Schmidt, B., Riesner, D., Lawen, A. & Kleinkauf, H. (1992) *FEBS Lett. 307*, 355–360.
197. Lawen, A., Dittmann, J., Schmidt, B., Riesner, D. & Kleinkauf, H. (1992) *Biochimie (Paris) 74*, 511–516.
198. Lawen, A. & Traber, R. (1993) *J. Biol. Chem. 268*, 20452–20465.
199. Lawen, A., Traber, R., Reuille, R. & Ponelle, M. (1994) *Biochem. J. 300*, 395–399.
200. Keller, U. (1995) in *Genetics and biochemistry of antibiotic production* (Vining, L. C. & Stuttard, C., eds) pp. 173–196, Butterworth-Heinemann, Boston.
201. Keller, U. & Schlumbohm, W. (1992) *J. Biol. Chem. 267*, 11745–11752.
202. Rabie, M. (1990) Ph.D. Thesis, Technical University Berlin.
203. Schlumbohm, W. & Keller, U. (1990) *J. Biol. Chem. 265*, 2156–2161.
204. Ullrich, C., Kluge, B., Palacz, Z. & Vater, J. (1991) *Biochemistry 30*, 6503–6508.
205. van Sinderen, D., Galli, G., Cosmina, P., de Ferra, F., Withoff, S., Venema, G. & Grandi, G. (1993) *Mol. Microbiol. 8*, 833–841.
206. Menkhaus, M., Ullrich, C., Kluge, B., Vater, J., Vollenbroich, D. & Kamp, R. M. (1993) *J. Biol. Chem. 268*, 7678–7684.
207. van Sinderen, D. & Venema, G. (1994) *J. Bacteriol. 176*, 5762–5770.
208. Besson, F. (1994) *Biotechnol. Lett. 16*, 1269–1274.
209. Galli, G., Rodriguez, F., Cosmina, P., Pratesi, C., Nogarotto, R., de Ferra, F. & Grandi, G. (1994) *Biochim. Biophys. Acta 1205*, 19–28.
210. D'Souza, C., Nakano, M. M., Corbell, N. & Zuber, P. (1993) *J. Bacteriol. 175*, 3502–3510.
211. Menkhaus, M. (1994) Ph.D. Thesis, Technical University Berlin.
212. Glund, K., Schlumbohm, W., Bapat, M. & Keller, U. (1990) *Biochemistry 29*, 3522–3527.
213. Quigley, N. B. & Gross, D. C. (1994) *Mol. Plant-Microbe Interact. 7*, 78–90.
214. Mo, Y.-Y. & Gross, D. C. (1991) *J. Bacteriol. 173*, 5784–5792.
215. Morgan, M. K. & Chatterjee, A. K. (1988) *J. Bacteriol. 170*, 5689–5697.
216. Xu, G.-W. & Gross, D. C. (1988) *J. Bacteriol. 170*, 5680–5688.
217. Lee, C. & Lawen, A. (1993) *Biochem. Mol. Biol. Int. 31*, 797–805.
218. Lawen A., Traber, R. & Geyl, D. (1991) *Biomed. Biochim. Acta 50*, 260–263.
219. Komura, S. & Kurahashi, K. (1985) *J. Biochem. (Tokyo) 97*, 1409–1417.
220. Ishihara, H., Hara, N. & Iwabuchi, T. (1989) *J. Bacteriol. 171*, 1705–1711.
221. Borchert, S., Patil, S. S. & Marahiel, M. A. (1992) *FEMS Microbiol. Lett. 71*, 175–180.
222. Herzog-Velikonja, B., Podlesek, Z. & Grabnar, M. (1994) *FEMS Microbiol. Lett. 121*, 147–152.
223. Prágai, Z., Trân, S. L. P., Nagy, T., Fülöp, L., Holczinger, A. & Sík, T. (1994) *Microbiology 140*, 3091–3097.
224. Smith, T. D., Priestley, N. D., Knaggs, A. R., Nguyen, T. & Floss, H. G. (1993) *Chem. Commun.*, 1612–1614.
225. Strohl, W. R. & Floss, H. G. (1995) in *Genetics and biochemistry of antibiotic production* (Vining, L. C. & Stuttard, C., eds) pp. 223–238, Butterworth-Heinemann, Boston.

226. Heinze, K. (1993) Ph.D. Thesis, Technical University Berlin.
227. Kopiez, H. (1994) Ph.D. Thesis, Technical University Berlin.
228. Lück, B. (1994) Ph. D. Thesis, Technical University Berlin.
229. McHenney, M. A., Boll, H. A., Godfrey, O. W. & Baltz, R. H. (1992) in *Industrial microorganisms: basic and applied molecular genetics* (Baltz, R. H., Hegeman, G. D. & Skatrud, P. L., eds) p. 286, American Society of Microbiology, Washington DC.
230. Weßels, P., Pfeifer, E., von Döhren, H., McHenney, M. A. & Baltz, R. H. (1994) *Abstr. 7th IUMS Conf.*, Prague, p. 127.
231. Nutkins, J. C., Mortishire-Smith, R. J., Packman, L. C., Brodey, C. L., Rainey, P. B., Johnstone, K. & Williams, D. H. (1991) *J. Am. Chem. Soc. 113*, 2621–2627.
232. Rainey, P. B., Brodey, C. L. & Johnstone, K. (1993) *Mol. Microbiol. 8*, 643–652.
233. Pieper, R., Kleinkauf, H. & Zocher, R. (1992) *J. Antibiot. 45*, 1273–1277.
234. Lee, C., Görisch, H., Kleinkauf, H. & Zocher, R. (1992) *J. Biol. Chem. 267*, 11741–11744.
235. Peeters, H., Zocher, R. & Kleinkauf, H. (1988) *J. Antibiot. 41*, 352–359.
236. Peeters, H. (1988) Ph.D. Thesis, Technical University Berlin.
237. Heaton, M. P. & Neuhaus, F. C. (1992) *J. Bacteriol. 174*, 4707–4717.
238. Ullrich, M. & Bender, C. A. (1994) *J. Bacteriol. 176*, 7574–7586.
239. Ullrich, M., Guenzi, A. C., Mitchell, R. E. & Bender, C. A. (1994) *Appl. Environ. Microbiol. 60*, 2890–2897.
240. Liyanage, H., Penfold, C., Turner, J. & Bender, C. L. (1995) *Gene (Amst.) 153*, 17–23.
241. Nielsen, J. B., Hsu, M. J., Byrne, K. M. & Kaplan, L. (1991) *Biochemistry 30*, 5789–5796.
242. Motamedi, H., Cai, S.-J., Streicher, S. S. & Shafiee, A. (1994) *7th Int. Symp. Genet. Indust. Microorg.*, Montreal, Abstract 219.
243. Oelrichs, P. B., Vallely, P. J., MacLeod, J. K., Cable, J., Kiely, D. E. & Summons, R. E. (1977) *Lloydia 40*, 209–214.
244. Thamsborg, S. M., Jorgensen, R. J. & Brummerstedt, E. (1987) *Proc. Sheep Vet. Soc. 12*, 1.
245. Shiau, C.-Y., Baldwin, J. E., Byford, M. F., Sobey, W. J. & Schofield, C. J. (1995) *FEBS Lett. 358*, 97–100.
246. Tavanlar, M. A., van Liempt, H., von Döhren, H. & Raymundo, A. K. (1994) *Abstr. Proc. IUMS Congr.*, Prague, p. 125.
247. Koischwitz, H. (1979) *Hoppe-Seyler's Z. Physiol. Chem. 360*, 307.
248. Magnuson, R., Solomon, J. & Grossman, A. D. (1994) *Cell 77*, 207–216.
249. Krause, M., Marahiel, M., von Döhren, H. & Kleinkauf, H. (1985) *J. Bacteriol. 162*, 1120–1125.
250. Stachelhaus, T., Schneider, A. & Marahiel, M. A. (1995) *Science 269*, 69–72.
251. Grandi, G., Cosmina, P., Rodriuez, F., de Ferra, F., Pratesi, C. & Galli, G. (1993) *Abstr. 6th Int. Conf. on Bacillus*, Paris, Abstract L22.
252. Kleinkauf, H. & von Döhren, H. (1994) *Antonie Leeuwenhoek 67*, 229–242.
253. Horinouchi, S. & Beppu, T. (1992) *Annu. Rev. Microbiol. 46*, 377–398.
254. Fucqua, W. C., Winans, S. C. & Greenberg, E. P. (1994) *J. Bacteriol. 176*, 269–275.
255. Beppu, T. (1995) *Trends Biotechnol. 13*, 264–269.
256. Okamoto, S., Nihira, T., Kataoka, H., Suzuki, A. & Yamada, Y. (1994) *J. Biol. Chem. 267*, 1093–1098.
257. Mo, Y. Y., Geibel, M., Bonsall, R. F. & Gross, D. C. (1995) *Plant Physiol. (Bethesda) 107*, 603–612.
258. Mazzola, M. & White, F. F. (1994) *J. Bacteriol. 176*, 1374–1382.
259. Walton, J. D. (1990) in *Biochemistry of peptide antibiotics* (Kleinkauf, H. & von Döhren, H., eds) pp. 179–204, Walter de Gruyter, Berlin.
260. Meeley, R. B. & Walton, J. D. (1991) *Plant Physiol. (Bethesda) 97*, 1080–1086.
261. Meeley, R. B., Johal, G. S., Briggs, S. P. & Walton, J. D. (1992) *Plant Cell 4*, 71–77.
262. Grossman, T. H., Tuckman, M., Ellestad, S. & Osburne, M. S. (1993) *J. Bacteriol. 175*, 6203–6211.
263. Borchert, S., Stachelhaus, T. & Marahiel, M. A. (1994) *J. Bacteriol. 176*, 2458–2462.
264. Black, T. A. & Wolk, C. P. (1994) *J. Bacteriol. 176*, 2282–2292.
265. Lee, Y.-O. (1994). Ph.D. Thesis, Technical University Berlin.
266. Holmes, D. J., Caso, J. L. & Thompson, C. J. (1993) *EMBO J. 8*, 3183–3191.
267. Yun, B.-S., Hidaka, T., Furihata, K. & Seto, H. (1994) *J. Antibiot. 47*, 510–514.
268. Yun, B.-S., Hidaka, T., Furihata, K. & Seto, H. (1994) *J. Antibiot. 47*, 969–975.
269. Carr, L. G., Skatrud, P. L., Scheetz, M. E., Queener, S. W. & Ingolia, T. (1986) *Gene (Amst.) 48*, 257–266.
270. Peñalva, M. A., Moya, A., Dopazo, J. & Ramon, D. (1990) *Proc. R. Soc. London Ser. B Biol. Sci. 241*, 164–169.
271. Landan, G., Cohen, G., Aharonowitz, Y., Shuali, Y. & Graur, D. (1991) *Mol. Biol. Evol. 7*, 399–406.
272. Yeh, W.-K., Dotzlaf, J. E. & Huffman, G. W. (1993) in *50 years of penicillin application – history and trends* (Kleinkauf, H. & von Döhren, H., eds) pp. 208–223, Publica, Prague.
273. Smith, M. W., Feng, D.-F. & Doolittle, R. F. (1992) *Trends Biochem. Sci. 17*, 489–491.

Eur. J. Biochem. 236, 747–771 (1996)

Review

Initiation of protein synthesis in eukaryotic cells

Virginia M. PAIN

School of Biological Sciences, University of Sussex, Brighton, UK

(Received 9 October 1995/5 December 1995) – EJB 95 1641/0

It is becoming increasingly apparent that translational control plays an important role in the regulation of gene expression in eukaryotic cells. Most of the known physiological effects on translation are exerted at the level of polypeptide chain initiation. Research on initiation of translation over the past five years has yielded much new information, which can be divided into three main areas: (a) structure and function of initiation factors (including identification by sequencing studies of consensus domains and motifs) and investigation of protein–protein and protein–RNA interactions during initiation; (b) physiological regulation of initiation factor activities and (c) identification of features in the 5′ and 3′ untranslated regions of messenger RNA molecules that regulate the selection of these mRNAs for translation. This review aims to assess recent progress in these three areas and to explore their interrelationships.

Keywords: translation; initiation; mRNA; review, regulation.

In 1986 I published a review describing the mechanism and regulation of initiation of translation in mammalian cells [2]. By that time most of the polypeptide initiation factors catalysing this process had been extensively purified and their individual activities studied in various *in vitro* systems. Several of them had been shown to be phosphoproteins and, in one case, eukaryotic initiation factor-2 (eIF-2), the effects of phosphorylation had been elucidated and two physiological kinases had been identified. There seemed to be a feeling in some circles that the most interesting problems in protein synthesis had been solved, and that only a few rather boring nuts and bolts awaited discovery.

Over the intervening ten years there has been an explosion of research activity in this area, largely fuelled by information yielded by molecular biology and genetic techniques. Cloning of cDNAs encoding initiation factors has revealed domain structures indicative of function and potential regulatory mechanisms. Experiments exploiting the ability to elucidate and manipulate mRNA sequences have demonstrated that translational control contributes to changes in patterns of gene expression during growth, differentiation and development to an extent that would have seemed inconceivable in 1985. Such experiments have, in particular, revealed important roles for structural features in the 5′ and 3′ untranslated regions (UTRs) of mRNA molecules in the regulation of mRNA utilization. Experiments employing techniques of gene transfection have also implicated translation in the control of cell proliferation. Moreover, a completely novel mechanism of initiation has been defined for the translation of a class of viral RNAs, and there are now indications that such a mechanism may also be utilized by some cellular mRNAs.

In this ten-years-after review, I have first attempted to summarize recent major developments in our understanding of translational mechanisms. In a longer section on regulation I have then tried to draw together two related aspects that often tend to be considered separately, i.e. the control of initiation factor activity and the influence of structural features of mRNA molecules. Clearly in such a wide area I have had to make subjective selections in the material covered and, particularly, in literature citation. I apologise especially to younger workers whose original contributions have been consolidated into citation of reviews by their laboratory head! Readers are directed to two recent general reviews on initiation of translation [3, 4], to a number of others which, though less all-embracing, discuss multiple topics in this area [1, 5–16] and to a new review which, for the first time, covers comprehensively the initiation of protein synthesis in plants [16a].

MECHANISM OF INITIATION

Initiation of protein synthesis involves the sequential binding of first the 40S and then the 60S ribosomal subunit to a messenger RNA molecule. The process in eukaryotes can be divided into three stages (Fig. 1): (1) association of initiator tRNA (Met-$tRNA_f$) and several initiation factors with the 40S ribosomal subunit to form the 43S preinitiation complex; (2) the binding of this complex to mRNA, followed by its migration to the correct AUG initiation codon and (3) the addition of the 60S ribosomal subunit to assemble an 80S ribosome at the initiation codon, ready to commence translation of the coding sequence. This last step requires the prior release of the initiation factors bound

Correspondence to V. M. Pain, School of Biological Sciences, University of Sussex, Falmer, Brighton, BN1 9QG, United Kingdom

Fax: +44 1273 678433.

Abbreviations. eIF, eukaryotic initiation factor, with suffix denoting individual factors as recently revised by IUB working party (see [1]); UTR, untranslated region (in mRNA); RRM, RNA recognition motif; ORF, open reading frame; EMCV, encephalomyocarditis virus; FMDV, foot-and-mouth disease virus; IRES, internal ribosome entry segment; PTB, polypyrimidine tract binding protein; HCR, heme-controlled repressor; PKR, protein kinase activated by RNA; dsRNA, double-stranded RNA; $p70^{S6K}$ and $p90^{rsk}$, 70 kDa and 90–92 kDa ribosomal protein S6 kinases; GSK, glycogen synthase kinase; IRE, iron-regulatory element; IRP, iron regulatory protein; PABP, poly(A) binding protein; AMV, alfalfa mosaic virus; CPE, cytoplasmic polyadenylation element (also known as ACE, adenylation control element); MAP, mitogen-activated protein.

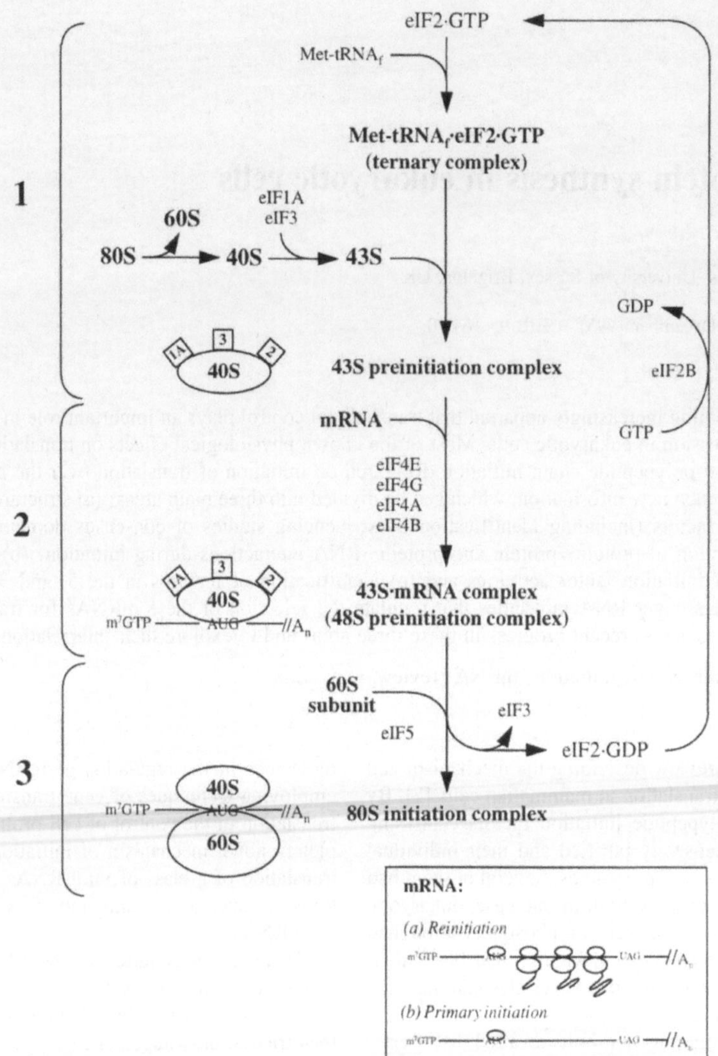

Fig. 1. Mechanism of initiation of protein synthesis. Stage 1, 80S ribosomes dissociate, and 40S subunits are captured for initiation by binding eIF1A and eIF3; the size of the latter causes the particle to sediment at 43S. Initiator tRNA (Met-tRNA$_f$) binds, in the form of a ternary complex with eIF2 and GTP, to give the 43S preinitiation complex. Stage 2, the 43S preinitiation complex binds to mRNA at the 5′ terminal m^{7}GTP cap structure, and then migrates along the mRNA towards the AUG initiation codon. The initial binding involves the factors eIF4E, eIF4G and eIF4A, which assemble at the 5′-end of mRNA, creating conditions that allow the melting of intramolecular secondary structures within the mRNA that would otherwise prevent the binding of the 43S preinitiation complex. In the vast majority of cases the most 5′ AUG codon is utilized for initiation. The term 48S preinitiation complex is frequently used, and refers to the 43S · globin-mRNA complex formed in the reticulocyte lysate (but see inset, below). Stage 3, when the 43S preinitiation complex stops at the initiation codon, the GTP molecule introduced as part of the eIF2 complex is hydrolysed to GDP, and this powers the ejection of the initiation factors bound to the 40S ribosomal subunit. The initiation factor eIF5 is involved in this process. The release of these factors permits the association of a native 60S ribosomal subunit, to reconstitute an 80S ribosome at the initiation codon poised to commence the elongation stage of translation. The continuity of initiation events requires the recycling of initiation factor molecules. eIF2 is released as a binary complex with GDP and requires a guanine nucleotide exchange factor, eIF2B, to catalyse the regeneration of the eIF2. GTP complex required to recruit the next Met-tRNA$_f$ molecule. Inset: (a) Most initiation events *in vivo* involve binding of the 43S preinitiation complex to mRNAs engaged in pre-existing polysomes (sometimes called reinitiation; note, however, that this term is also used to define a separate mechanism involving two initiation events on the same mRNA, as described later in this review). (b) primary initiation on to an mRNA molecule vacant of ribosomes is relatively rare *in vivo*, but is required for utilization of newly synthesized transcripts or for recruitment of mRNAs from an untranslated pool in response to a growth or differentiation signal.

to the 40S ribosomal subunit during the earlier stages; these factors are then recycled to catalyse further initiation events.

Formation of the 43S preinitiation complex

eIF2. This factor is a complex of three polypeptide chains, α, β and γ, which appear to remain associated throughout the initiation cycle. For all three polypeptides, cDNAs have been cloned and sequenced from both mammalian cells and *Saccharomyces cerevisiae* (Table 1); the α and β subunits have also been cloned from *Drosophila.* The most well-defined function of eIF2 is to recruit the initiator tRNA and conduct it as a Met-tRNA$_f$ · eIF2 · GTP ternary complex to the 43S ribosomal subunit (reviewed previously [2] and recently [4, 17]). Met-tRNA$_f$

Table 1. Cloning of initiation factor cDNAs. This table is updated from [8]. Accession numbers refer to GenBank and SwissProt databases. DM = *Drosophila melanogaster.*

Factor	Vertebrates				*S. cerevisiae*			Other species	
	former name(s)	species	mass	accession no.	name	mass	accession no.	species	accession no.
			kDa			kDa			
eIF1A	eIF-4C	human	16.5	L18960	TIF11	17.4	U11585	wheat	L08060
eIF2α		human	36.2	JO2645	SUI2	34.7	M25552	DM	L19196
		rat		36.1	JO2646				
eIF2β		human	38.4	M29536	SUI3	31.6	M21813	DM	L19197
		rabbit	38.3	X73836					
eIF2γ		human	51.8	L19161	GCD11	57.9	L04268		
eIF2Bα		rat	33.7	U05821	GCN3	34.0	M23356		
β		rabbit	39.0	Z48222	GCD7	42.6	L07116		
		rat		U31880					
γ		–			GCD1	65.7	X07846		
δ		rabbit	57.6	X75451	GCD2	70.9	X15658		
		rat	57.8	Z48225					
ε		rabbit	80.2	U23037	GCD6	81.2	L07115		
		rat	80.2	U19151					
eIF3					SUI1	12.3	M77514		
							S31245		
					GCD10	54.4	X83511		
					PRT1	88.1	J02674		
eIF4AI	eIF-4β	mouse	46.3	X14421	TIF1	45.0	X12813	DM	X69045
eIF4AII		human		D30655	TIF2	44.6	X12814	wheat	Z21510
		mouse	46.4	X12507				rice	D12627
								maize	U17979
eIF4B		human	69.2	S12566	TIF3	48.5	X71996		
eIF4E	CBP1, eIF-4α	human	25.1	M15353	CDC33	24.3	M29251	wheat	
								p26	Z12616
		mouse	25.0	A34295				p28	M95818
		rabbit	25.1	X61939				DM	L37034
eIF4G	p220	human	154	D12686	TIF4631	107	L16923	wheat	
					TIF4632	104	L16924	p82	M95747
	eIF-4γ	rabbit	154	L22090					
eIF5		rat	49.0	L11651	TIF5	45.2	L10840		

can only bind to a binary complex of eIF2 and GTP. Cross-linking studies originally suggested interaction of GTP with both β and γ subunits [4, 17], but both the human [18] and yeast [19] γ subunits are now known to contain all three consensus GTP-binding elements with the correct spacing and to show similarity to other GTP-binding proteins. The human β subunit also possesses two of these elements, but none of them occurs in this polypeptide from yeast [19, 20] or *Drosophila* [21]. The binding site for Met-tRNA$_f$ is still not fully identified [4]. Cross-linking studies again show close proximity of this ligand to the N-terminus of the γ subunit and the C-terminus of the β subunit [18]. Sequences in the C-terminal domain of the β subunit show similarity between the human, yeast and *Drosophila* proteins [21], and include a single zinc-finger-like motif [20–22], although active eIF2 is reported not to contain zinc [4]. Much attention has been given to the possibility that binding of eIF2 to mRNA, readily demonstrated *in vitro*, may have functional relevance (reviewed [17, 23]). This binding activity has been localized to the C-terminal domain of the β subunit [24], but appears to be relatively non-specific with respect to RNA. Potential interactions involving eIF2 could be with Met-tRNA$_f$ or with rRNA within the 40S ribosomal subunit as well as with mRNA. However, in yeast, mutations in each of the three subunits of eIF2 have been found to influence the fidelity of the interaction of the Met-tRNA$_f$ anticodon with codons serving as initiation codons [20, 25, 26]. The primary sequence of the α subunit of eIF2 reveals no consensus motifs connected with ligand binding; the most significant feature of this polypeptide is a conserved phosphorylation site (Ser51 in mammalian cells), which is the target for a family of protein kinases important in the regulation of protein synthesis (see below).

eIF3 and eIF1A. Under intracellular conditions the equilibrium between free, or native, ribosomal subunits and 80S couples is strongly weighted towards the latter. The factors eIF3 and eIF1A are thought to bind to newly dissociated 40S ribosomal subunits and to delay reassociation with 60S subunits for long enough to permit their recruitment for initiation. Sequence analysis of eIF1A from human, rabbit and wheat [27] and from *S. cerevisiae* [27a] identified no obvious ligand-binding motifs but revealed a dipolar molecule, with a basic N-terminus and an acidic C-terminus. This may be conducive to a role for the factor as a bridge between other protein factors or between an initiation factor and the ribosome [27]. In subsequent work Northwestern blotting analysis demonstrated strong interaction between recombinant eIF1A and RNA [28] but the specificity of this has yet to be examined.

eIF3 is a multimeric complex of total molecular mass 500–750 kDa, consisting of at least eight polypeptide chains in mammalian cells and ten polypeptide chains in wheat germ (reviewed [1, 3, 4]). Two recent reports describe purification of this factor from *S. cerevisiae*, identifying five [29] and eight [30] polypeptide chains, respectively. Recent analysis of complete or partial cDNA clones [31] (and Asano, K., Naranda, T. and Hershey, J. W. B., personal communication) has led to the identification of the yeast polypeptides p90, p62, p39 and p16 as PRT1, GCD10, p36 and SUI1, respectively; the yeast p39 subunit is homologous to the p36 subunit of human eIF3. p16/SUI1 has been identified as an initiation factor previously listed as eIF1 [32]. The best characterized polypeptide of eIF3 is yeast PRT1 [29, 30]. The N-terminal domain contains an RNA recognition motif and is probably involved in ribosome binding [33]. eIF3 has long been known to stabilize 43S preinitiation complexes *in vitro*, and also to be essential for the binding of these complexes to mRNA. The molecular basis of its role in mRNA binding is still not clear, but potentially important interactions have been observed between eIF3 and other initiation factors involved in this step [3, 34]. In addition the yeast eIF3 complex binds directly to RNA [29, 30], probably via the 62-kDa subunit [30]; surprisingly, this is not a homologue of the 66-kDa polypeptide in mammalian eIF3, which has recently been shown to bind to an mRNA transcript in a Northwestern blot assay [28] (Asano, K., Naranda, T. and Hershey, J. W. B., personal communication). Chemical cross-linking experiments on mammalian 48S preinitiation complexes trapped by blocking the 60S joining stage of initiation showed the 66-kDa eIF3 polypeptide to cross-link to globin mRNA within these particles and also to 18S ribosomal RNA of the 40S subunit [35]. Together with immunoelectron microscopy, these studies gave rise to models depicting the alignment of eIF3 and eIF2 on the 40S subunit, and identified relationships between these factors, 18S rRNA and surface domains of specific ribosomal structural proteins [35].

Guanine nucleotide exchange on eIF2: the recycling factor eIF2B. eIF2B, a complex of five polypeptide chains in both mammalian cells and yeast, catalyses the guanine nucleotide exchange reaction required to recycle the eIF2 released from initiation complexes as an eIF-2 · GDP complex to the eIF2 · GTP form capable of recruiting a new molecule of initiator tRNA (Fig. 1). The mechanism and regulation of this reaction are surveyed in an excellent review by Price and Proud [36], and it is clear that there is still a lot to be learned. Two opposing mechanisms have been proposed, one a substituted enzyme, or ping-pong, mechanism similar to that utilized by the elongation factors Tu/Ts and the other involving an intermediate quaternary complex GTP · eIF2B · eIF2 · GDP (see [36]). Dholakia et al. [37] presented evidence in support of the latter mechanism by demonstrating labelling of one of the polypeptide subunits of eIF2B with 8-azido analogues of GTP. However, with complete sequence information now available on all five polypeptide subunits in yeast (see Table 1 and references in [36]) and on four of the mammalian subunits [38-41], no conserved GTP binding elements have been identified. The complexity of both mammalian and yeast eIF2B relative to other guanine nucleotide exchange proteins may reflect the multiple mechanisms involved in the regulation of this step in protein synthesis (see below). The mammalian ε subunit may be responsible for the guanine nucleotide exchange activity, which can be blocked by monoclonal antibodies recognizing this polypeptide [42].

mRNA binding to ribosomes

The binding of the 40S subunit to mRNA involves several initiation factors and has potential for controlling both the overall rate of translation and the relative rates of utilization of different mRNA molecules in response to physiological signals. For most eukaryotic mRNAs, the initiating 40S ribosomal subunit binds at the 5′ end of the message and then migrates in a 5′-3′ direction towards the initiation codon. Sequence analysis of mRNA molecules indicates considerable potential for the formation of hairpin loops and other intramolecular secondary structures. Regions of stable secondary structure within the 5′ untranslated region (5′ UTR) of mRNA impede initiation of protein synthesis, particularly if located near to the 5′ end, where the initial binding of the 43S preinitiation complex takes place [43, 44]. The initiation factors catalysing this mRNA binding step are believed to act in concert to (a) locate the 5′ end of the mRNA, (b) unwind any secondary structure that would impede ribosome binding, (c) direct the binding of the 43S preinitiation complex and (d) melt any further secondary structure that might inhibit migration to the initiation codon. Step (a) is achieved by the specific interaction of eIF4E with the mRNA cap and step (b) by eIF4A and eIF4B. The binding and placement of the 43S preinitiation complex (step c) is most likely achieved by the ability of the factor eIF4G to form a bridge between the cap-bound eIF4E and the incoming 43S complex; this factor may also facilitate step (b) by linking eIF4A into the complex. Subsequent melting of downstream secondary structure (step d) may be achieved by the recruitment of additional eIF4A and eIF4B molecules. I will first outline the properties of the individual initiation factors and some of the evidence for this general mechanism, and then present some further speculation on possible models. Early work on this stage in initiation was reviewed previously [2] and useful reviews of more recent work are [1, 3, 4, 6, 8, 45–49].

eIF4E. The amino acid sequences are now known for this protein in mammalian cells, *S. cerevisiae*, wheat and *Drosophila* (Table 1, see [50]). Its most characteristic function is recognition of the 5′-terminal m^7GTP cap on mRNA; analysis of yeast mutants suggests that at least some of the eight highly conserved tryptophan residues are important in this interaction [51, 52], consistent with biophysical studies on the human protein (see [4]). The eIF4E gene is essential for viability in yeast; a cDNA encoding the mammalian protein, though only 35% conserved at the amino acid level, can substitute for the yeast gene *in vivo* [53]. The yeast cell cycle division mutation cdc33, which induces a G1 block at the non-permissive temperature, has been localized to a single amino acid substitution in eIF4E, close to one of the conserved tryptophan residues; the mutated protein shows reduced cap recognition activity [51]. The phosphorylation of eIF4E in mammalian cells has been a major focus of investigation and is reviewed below.

eIF-4A. Biochemical characterization of this factor showed it to bind ATP and to exhibit RNA-dependent ATPase and ATP-dependent RNA duplex unwinding activity (reviewed previously [2, 4]), leading to the conclusion that its function is to melt secondary structures in mRNA. This idea was strengthened by data from cDNA sequencing, which revealed a series of motifs conserved not just between eIF4A molecules from different organisms but between eIF4A and an extended family of around 70 RNA and (mainly) DNA helicases [45, 49]. Work in the laboratories of Sonenberg, Linder and Trachsel (reviewed in [45]) examined the functions of some of these conserved sequences by introducing mutations into recombinant yeast and mammalian eIF4A proteins, and led to a model in which the binding and hydrolysis of ATP permitted the binding of eIF4A to RNA and hence the unwinding activity. The ATPase A motif (AXXXXGKT), near the N-terminus, is required for ATP binding, and mutations abrogating this function also abolished the RNA binding and helicase activities of eIF4A. The ATPase B, or D-E-A-D, motif, appears to be involved in ATP hydrolysis. A C-terminal, arginine-rich, motif (HRIGRXXR) is important in RNA binding and is required for helicase activity. A fourth motif (SAT) is needed for helicase activity, but mutations in this area do not inactivate ATP binding, ATPase activity or RNA binding.

eIF-4B. *In vitro* assays of the RNA-dependent ATPase and ATP-dependent RNA unwinding functions of eIF4A show strong dependence of these activities on the factor eIF4B. The human cDNA encodes a 68-kDa protein containing an RNA recognition motif (RRM) near the N terminus and a central region rich in aspartate, arginine, tyrosine and glycine (the DRYG domain) [54, 55]. The polar C-terminal region bears several potential phosphorylation sites, but is not essential for RNA binding [55] and has no equivalent in the yeast homologue [56, 57]. The eIF4B gene is not essential for viability in yeast, but its disruption results in slow growth, and extracts derived from the defective cells show poor ability to translate mRNAs with structured 5′ UTRs [57]. Studies on mutant forms of the human protein indicate that the N-terminal RRM alone is not sufficient for RNA binding, and identify an arginine-rich sequence close to the DRYG region as important for both the RNA binding activity [55] and for the ability of eIF4B to stimulate the helicase activity of eIF4A [58]. The N-terminal RRM may function in an interaction of eIF4B with ribosomes [58, 58a].

eIF4G (eIF-4γ, p220) and the eIF4F complex. Part of the eIF4E in mammalian cells [4], yeast [59], plants [60] and *Drosophila* [61] is found in the form of high-molecular-mass complexes. In mammalian cells these complexes were originally characterized as having three components, eIF4E, eIF4A and eIF4G, and were referred to as eIF4F or eIF4 (reviewed [2, 4, 46]). However, some workers have isolated eIF4E · eIF4G complexes [4, 62], and the absence of an eIF4A polypeptide is also common to cap-binding complexes from yeast [59], plants [60] and *Drosophila* [61]. Emerging evidence now indicates that eIF4G functions to bring together, in the correct orientation and in close proximity to the cap, the components necessary to unwind secondary structure in the mRNA and place a 40S ribosomal subunit at the 5′ end. The N-terminal half of the molecule contains a binding site for eIF4E [34, 63], while the C-terminal half binds to ribosomes, possibly via interaction with ribosome-associated eIF3, and also has affinity for eIF4A [34]. Sequencing of a cDNA encoding human eIF4G revealed a 154-kDa polypeptide (Table 1). In *S. cerevisiae* two genes (*TIF4631* and *TIF4632*) were identified as encoding the largest polypeptide of an eIF4F complex that included eIF4E and a 20-kDa protein of unknown function [59]. The TIF4631 and 4632 proteins were 53% identical overall, but with 80% identity at the C-terminus. Both contained sequences resembling the RRM in other RNA binding proteins, but these were less clearly apparent in the sequence of the human homologue [64]. In plant cells the situation appears to be more complicated. Two distinct cap-binding complexes have been isolated; one form, known as eIF-4F contains polypeptides of 26 kDa and 220 kDa; the other, known as eIF-*iso*4F, contains 28-kDa and 82-kDa proteins [60]. The 26-kDa and 28-kDa proteins have cap-binding activity and show some sequence similarity to human and yeast eIF4E while, of the larger polypeptides, only the 82-kDa protein of eIF-*iso*4F has been sequenced and shows a small degree of similarity to the human and yeast forms of eIF4G [60].

Alternative models for the mRNA binding step. Detailed consideration of the mechanism of this step reveals uncertainty on the exact order of binding and dissociation events involved. Some possible models are shown in Fig. 2. A key question is whether the factors eIF4E, eIF4G and eIF4A associate to form the eIF4F complex before interacting with mRNA [model (a) in Fig. 2] or whether eIF4E first binds alone to the mRNA cap, with the subsequent addition of the other factors to assemble the eIF4F complex *in situ* at the 5′ end of the mRNA [model (b)] (for various views on this see [3, 4, 8, 46, 47]). A further variant of the latter model envisages eIF4G as part of the 43S preinitiation complex [47, 65] [model (c)]. The interaction of this molecule with the mRNA-bound eIF4E would then become a major determining event in the placing of the 43S preinitiation complex on the mRNA molecule. Evidence for this model has been presented by Rhoads' laboratory, who added radiolabelled eIF4E and eIF4G to reticulocyte lysate translation systems and found that the latter, but not the former, was incorporated into 43S preinitiation complexes [47, 65]. In contrast, others favour models involving the prior formation of an eIF4F complex, since *in vitro* assays with purified factors and mRNA indicate that the cap-binding activity of eIF4E and the RNA helicase activity of eIF4A are greater when the factors are in an eIF4F complex than when they are assayed as individual proteins [66, 67]. However, these assays tend to be performed in the absence of other components, such as ribosomal subunits and eIF3, which may themselves play influential roles *in vivo*. The function of eIF4B is still not clear. Even more uncertain is the timing of subsequent dissociation events (Fig. 2). Some models depict the release of eIF4E and eIF4G from the mRNA immediately after 43S binding [68], and a recent extension of this scheme proposes that these factors rapidly reassociate with cytosolic eIF4A (to form new eIF4F complexes) and then cycle back on to the mRNA in order to present additional eIF4A molecules required during scanning [46, 69]. However, work with radiolabelled factors indicates that both eIF4E and eIF4G are present on the 48S complexes that accumulate in the reticulocyte lysate in the presence of inhibitors of 60S subunit joining when, presumably, the 40S subunits have reached the initiation codon [65]. We have no evidence at present, however, on whether these factors are associated with the 40S subunit itself, or whether they remain bound to the mRNA cap (see Fig. 2).

Migration of the 43S ribosomal complex to the initiation codon

For over 95% of vertebrate and yeast mRNAs analysed, the most 5′ AUG codon is utilized as the initiation codon [6, 70, 71]. The migration or scanning process itself has so far proved difficult to examine. Scanning can be impeded by very stable secondary structures, but is less sensitive to inhibition by moderately stable hairpin loops than is the initial association between the 40S ribosomal subunit and mRNA [44]. This may indicate that the 40S subunit, once bound to mRNA, contributes to the unwinding of downstream regions of secondary structure [3, 4]. Most current models suggest that scanning involves further utilization of the ATP-dependent helicase activity of eIF4A, probably stimulated by eIF4B [1, 3, 4, 46, 57], although some data indicate that most of the ATP needed during initiation is utilized by the unwinding associated with the initial, cap-dependent mRNA binding stage [72].

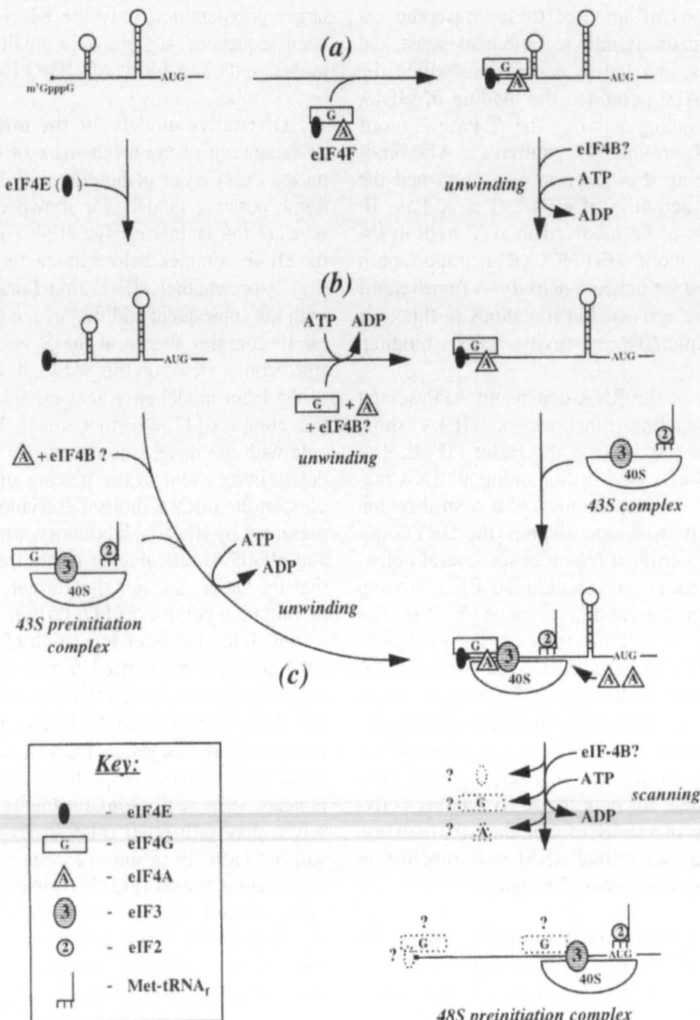

Fig. 2. Alternative pathways for binding the 43S preinitiation complex to mRNA. (a) An eIF4F complex between eIF4E, eIF4G and eIF4A is preformed in the cytosol and then binds to mRNA through the cap recognition function of eIF4E. Nearby secondary structure in the mRNA is unwound by the helicase activity of eIF4A, possibly stimulated by eIF4B, permitting the binding of the 40S ribosomal subunit and its subsequent migration to the AUG initiation codon. (b) eIF4E binds alone to the 5′ cap structure, followed by eIF4G and eIF4A to assemble the eIF4F complex on the mRNA itself. Subsequent steps are then as described for (a). (c) eIF4E binds alone to the 5′ cap, while eIF4G binds to the 43S preinitiation complex, probably by association with eIF3 already bound. After preliminary unwinding of the mRNA by eIF4A, the 43S preinitiation complex is directed to bind by virtue of protein−protein interaction between eIF4E and eIF4G. Symbols depicting initiation factors are listed in the key. Symbols with dotted outlines indicate uncertainty concerning the binding relationships between factors in the 48S preinitiation complex and the timing of factor dissociation from the initiation complex. Note that the 40S subunit covers about 30 nucleotides of mRNA.

Initiation at the 5′ proximal AUG codon. The next problem to consider is the mechanism that arrests the scanning 40S subunit at the AUG initiation codon. An obvious means of identifying an AUG codon is by codon-anticodon recognition involving the initiator tRNA, but this is clearly not the whole story. Kozak investigated the role of surrounding nucleotides in influencing the selection of AUG codons for initiation (reviewed in [43, 70]), using two complementary approaches: (a) analysis of sequences around the initiation codons of 699 vertebrate mRNAs and (b) examining the effects of sequence manipulation in the 5′ UTR on expression of a transfected gene. The sequence analysis revealed a consensus around the initiation codon: GCCGCCA/GCC<u>AUG</u>G, while the expression experiments showed that mutation to pyrimidines of the purine at −3 (relative to the first nucleotide of the initiation codon) or the G at +4 greatly decreased the chance of an AUG being recognized. Codons lacking these conserved, neighbouring nucleotides were described as being in a poor or weak context. Mutations in nucleotides immediately upstream, though less influential, were able to modulate the utilization of AUG codons in a weak context. Further experiments demonstrated the ability of an additional AUG codon in good context, inserted upstream, to intercept scanning 40S subunits. This provided a further assay for context effects, since changes in the sequence surrounding the inserted AUG codon could be assessed for their influence on its ability to intercept. Yeast mRNAs also tend to have a purine nucleotide, usually A, in the −3 position relative to the initiation codon, but they show weaker consensus at other positions [6, 73]. In plant mRNAs a G residue at the +4 position appears to be important [9].

CUG, ACG or GUG are occasionally used as initiation codons (see references in [74]); their recognition is again dependent on context. Studies with two viral RNAs identified the nucleotide in the +5 position (C or, preferably, A) to be as important as that in the +4 position in facilitating selection of non-AUG initiation codons [74, 75]. This position, not examined previously, may also be important for recognition of AUG codons in cellular mRNAs under some conditions, since reexamination of the sequences of the 699 mRNAs analysed by Kozak indicated a high frequency of A or C at position +5 [74]. Initiation at AUG codons in poor context, or at non-AUG codons, can be enhanced by base-paired structures around 14 nucleotides downstream [43]. Since the leading edge of a 40S ribosomal subunit poised over the AUG initiation codon would protrude about 12−15 nucleotides into the coding sequence, a hairpin structure at this position could serve to arrest or slow down the scanning particle at the appropriate position to allow more time for codon recognition [43, 71]. Under normal conditions, a strong surrounding context may in some way function similarly to slow down scanning in the region of the initiation codon. Finally, it should be noted that, in addition to codon−anticodon interaction between Met-tRNA$_f$ and mRNA, the factor eIF2 may also play a role in codon recognition (see above).

Initiation at downstream AUG codons. Although most initiation events take place at the first AUG codon encountered by a scanning 40S subunit, there are some notable exceptions, many of which have turned out to be of particular interest. These fall into three groups; two of these can be explained in terms of the scanning mechanism, but the third is incompatible with many of its features.

Leaky scanning. This occurs when the first AUG codon is in poor context and is consequently by-passed by many 40S subunits, which then initiate at an AUG codon in stronger context further downstream. This accounts for a number of cases where an RNA (often viral) appears to encode two products [76]. If the first and second AUG codons are in the same open reading frame, with no termination codon in between, the result is two translation products with the longer having an N-terminal extension. If the two AUGs are in different reading frames, two distinct products are translated.

Reinitiation. Some mRNAs have one or more AUG codons upstream of the authentic initiation codon. An upstream AUG in strong context will severely inhibit translation by intercepting scanning 40S subunits. Upstream AUG codons are frequently followed closely by an in-frame termination codon, such that a short peptide is encoded. After translation of some such minicistrons, some of the 40S subunits remain associated with the mRNA, continue scanning and eventually commence initiation at the authentic initiation codon (see [43, 77, 78] for reviews). The criteria that determine whether or not this will occur are not yet clear, but some clues are emerging. (a) Nucleotides around the termination codon of the minicistron or the actual peptide sequence encoded may determine whether or not 40S subunits are released or resume scanning [77, 79, 80]; in addition, some property of the downstream cistron may influence its effectiveness at recruiting reinitiating 40S subunits [81]. (b) The chances of initiation at the downstream AUG codon increase with the distance of this from the upstream open reading frame (ORF). This is thought to reflect the time needed for the 40S subunit to reacquire a Met-tRNA$_f$ · eIF2 · GTP ternary complex, which is needed for recognition of the authentic initiation codon [10, 81]. (c) Successful reinitiation appears to require the upstream minicistron to be relatively short. This may be due to retention of some essential factor for a short time after completion of the primary initiation event [10, 70, 82]. The reinitiation mechanism underlies the well-known translational control system for the yeast transcription factor GCN4 [83], which is discussed below.

Internal initiation. Among mRNAs translated in eukaryotic cells, those of picornaviruses are exceptional; they are uncapped and possess 5′ UTRs that are several hundred nucleotides long and include regions of secondary structure sufficiently stable to prevent 40S subunits scanning from the 5′ end. In each case their translation product is a single polyprotein that undergoes posttranslational cleavage to yield a distinct set of functional viral polypeptides. The 5′ UTRs of these mRNAs also contain AUG codons in good context that would be difficult to by-pass if actually encountered by scanning 40S subunits. In the late 1980s an alternative, cap-independent, initiation mechanism was proposed for picornavirus RNAs. This involves the binding of the 43S preinitiation complex directly to an internal site either very close to the authentic AUG initiation codon or up to 160 nucleotides upstream of it, depending on the species of virus concerned. At first this mechanism was vigorously challenged [10, 11], but the mass of detailed information on the involvement of RNA structure now provides compelling evidence in favour of internal initiation, at least in the case of picornavirus RNAs. Further evidence comes from the demonstration that elements from picornavirus 5′ UTRs can direct ribosomes to translate artificial circular mRNAs [84]. The mechanism of internal initiation has been reviewed in detail [85−91] and will be discussed relatively briefly here.

The basic test for the ability of a 5′ UTR to direct internal initiation is that, when placed between the two cistrons in a bicistronic construct, it can promote active translation of the downstream cistron, even under conditions that prevent cap-dependent translation of the upstream cistron (reviewed [85−87, 90]). Careful consideration of potential pitfalls is required when applying this test to new mRNAs [87]. Such experiments were extended, particularly in the case of poliovirus and encephalomyocarditis virus (EMCV) RNAs, to examine the effects of progressive deletions in the 5′ UTR and thus delineate the minimum region required to direct internal initiation of the downstream cistron. These essential regions, approximately 450 nucleotides long [85], and at first given the evocative name of ribosome landing pads [92], are now, rather more soberly, referred to as internal ribosome entry segments or IRESes. Most studies so far have focussed on two major classes of picornavirus that infect mammalian cells, the cardiovirus/aphthovirus group (including EMCV, Theiler's murine encephalomyelitis virus and foot-and-mouth-disease virus [FMDV]) and the enterovirus/rhinovirus group (including poliovirus, human rhinovirus and Coxsackie virus). The distinguishing features of these classes, and the structures of the 5′ UTRs of their RNAs, are thoroughly described by Jackson and colleagues [85, 86]. For both classes the entry site on the RNA for the incoming 43S preinitiation complex is an AUG codon near the 3′ end of the IRES. For cardiovirus RNAs the AUG at the entry site is utilized for initiation by most 40S subunits, and little or no scanning is required. Initiation on the RNA of the aphthovirus FMDV utilizes two closely placed AUG codons [93, 94]. For enterovirus RNAs the AUG at the entry site is not utilized for initiation, and the normal scanning mechanism is thought to be involved in the passage of the 40S subunit to the next AUG downstream. However, the mechanism may be more complicated [95]. In addition to the main classes of picornavirus mentioned above, internal initiation appears to be responsible for translation of the RNA of the third class, hepatitis A [91], and those of the non-picornaviruses hepatitis C [88] and pestiviruses [96]. However, there is some doubt in the case of plant coronaviruses and potyviruses [87].

In all the picornavirus RNAs examined the IRES regions in the 5′ UTR comprise a complex pattern of stem-loops organized into a series of highly structured domains [85, 86, 97–99]. The one widely conserved region of sequence conservation between the 5′ UTRs of picornavirus RNAs is an oligopyrimidine tract at the 3′ end of the IRES, followed by a spacer region of about 25 nucleotides, immediately upstream of the 40S subunit entry site. However, the role of this is still uncertain [85, 100]. Secondary structure is much more highly conserved, and mutational analysis demonstrates its importance in the function of IRES elements. A role for base-pairing interactions between picornavirus 5′ UTRs and the 3′ end of 18S ribosomal RNA has been suggested (reviewed [101]). Secondary and tertiary structure could facilitate this by providing the correct orientation, but to date this proposal awaits supporting evidence. While the secondary structure of the RNA may itself function sterically to form a ribosome binding site, reports are rapidly accumulating of general and tissue-specific polypeptides that interact with different elements of picornavirus IRESes [85, 89, 91, 97–99, 102, 103]. This binding mainly involves recognition of secondary and tertiary structural motifs, and it seems likely that protein complexes are assembled by virtue of protein–protein, as well as protein–RNA, interactions [91, 97–99]. However, caution may be required in the interpretation of cross-linking experiments involving IRES elements [87]. A recent, intriguing, report suggests that a small RNA, so far isolated only from yeast, may specifically inhibit poliovirus RNA translation by competing for essential IRES-binding proteins [104]. Surprisingly, two proteins found to interact with several IRESes, the 57-kDa polypyrimidine tract binding protein (PTB) and the autoantigen La [91, 102, 103, 105, 106], already have known functions in nuclear events such as RNA transcription and processing. La has been reported to increase in concentration in the cytosol during poliovirus infection, to have RNA unwinding activity *in vitro* [107] and to improve the fidelity of initiation at the correct AUG codon of poliovirus RNA [105, 106], although unphysiologically high concentrations of La are required for the last effect. However, for neither of these proteins has a precise role in internal initiation yet been established. Of equal interest is the potential role of *trans*-acting cellular proteins that may be specific to, or unequally distributed between, different cell types [88, 89]. Translation of poliovirus RNA is notoriously sluggish in the reticulocyte lysate translation system but is stimulated by the addition of crude HeLa cell extract; this may partly reflect differences in concentrations of La or PTB and associated proteins [85, 102], but it seems likely that other factors are involved [88]. The distribution of *trans*-acting proteins may be an important factor in determining relative degrees of virulence of picornaviruses in different cell types [89, 91, 98]. Since it is unlikely that cells and organisms would evolve proteins whose primary function was to facilitate the replication of invading viruses, an important question is whether the phenomenon of internal initiation extends beyond picornavirus RNA translation. Indeed, several cellular mRNAs have now been reported to utilize this mechanism (see below).

Joining of the 60S ribosomal subunit to the preinitiation complex

It would seem reasonable that the reassociation of a 60S subunit with the 40S particle bound at the initiation codon would require the prior release of the factors that reinforced dissociation of the ribosomal subunits earlier in the pathway. This was indeed found in early work using translation systems reconstituted from purified initiation factors (reviewed previously [2, 4]). The release of initiation factors is dependent on the hydrolysis of the GTP molecule bound to the eIF2 within the preinitiation complex, and this is catalysed by the factor eIF5. The amino acid sequences of both mammalian [108] and yeast [109] eIF5 are now known to include motifs typical of the GTPase superfamily. However, the factor only promotes GTP hydrolysis in the presence of 40S subunits. Both the mammalian and yeast factors are monomers (49 and 45 kDa respectively [108, 109]), but in earlier studies eIF5 activity from ribosome salt-wash preparations always behaved as a complex of around 150 kDa [110, 111]. Recent work suggests that this is due to a highly specific association of eIF5 with eIF2 [112]. Thus it seems likely that these two factors interact on the surface of the 40S subunit when the initiation complex is aligned over the initiation codon, resulting in activation of GTPase, hydrolysis of the eIF2-bound GTP molecule and ejection of eIF2 · GDP, eIF3 and probably other factors from the initiation complex. Presumably there is a mechanism to prevent this occurring prematurely on the 43S preinitiation complex or during mRNA binding or scanning. Very little attention has been given to the regulation of the 60S subunit joining step or its potential as a control point in translation. However, there are a few provocative observations. Firstly, eIF5 is a phosphoprotein, with at least two sites that can become metabolically labelled with [^{32}P]phosphate in mammalian cells [110]. Secondly, at least in the reticulocyte lysate, the joining of the 60S subunit seems to be a slow step relative to mRNA binding to the 40S subunit [113]. Finally, again in the reticulocyte lysate, this is the step at which the possession of a poly(A) tail seems to confer kinetic advantage on recruitment of mRNAs into polysomes [114] (see below).

REGULATION OF TRANSLATIONAL INITIATION

Translation is now recognized as an important site of regulation of gene expression, with the initiation stage as the most commonly observed target for physiological control. Modulation of initiation can influence both the overall, global, rate of protein synthesis (quantitative regulation) and the relative rates of synthesis of different proteins (qualitative regulation); frequently, controls at these two levels are superimposed. Control of the overall rate of protein synthesis is potentially important in achieving cell growth during the G1 phase of the cell cycle, while the concentrations of an increasing number of specific proteins involved in the control of cell proliferation or differentiation are now thought to be modulated at least in part at the translational level. Two particular steps of the initiation pathway appear to be hot spots for physiological regulation, the binding of Met-tRNA$_f$ to the 40S ribosomal subunit, mediated by eIF2, and the initial binding of the 43S preinitiation complex to the 5′ end of mRNA, mediated by eIF4E and associated factors. The first of these, which precedes mRNA involvement, is mainly, but not exclusively, relevant to quantitative regulation, whereas the mRNA binding step can, in addition, exert preferential effects on the translation of different mRNAs. Two recurrent themes repeatedly surface during investigation of translational regulation, namely phosphorylation of initiation factors and the influence of structural features in the 5′ and 3′ untranslated regions of mRNA molecules. Links between these two themes may be forged where features of an mRNA molecule may render its translation particularly sensitive to modulation of the activity of particular initiation factors. Several of the initiation factors are phosphoproteins, but the clearest links between phosphorylation and the regulation of translation concern the factors eIF2 and eIF4E. Regulatory features in mRNA molecules include structures that may act directly (for example by impeding 40S subunit binding or scanning) or indirectly, by providing a binding site for a *trans*-acting protein.

Table 2. Phosphorylation of initiation factors.

Factor	Site	Protein kinases	Physiological conditions	Effect on activity
eIF2α	Ser51	HCR	iron (heme) deficiency (reticulocytes)	increased affinity for eIF2B, which becomes sequestered in eIF2B · eIF2 · GDP complexes
		PKR	virus infection/interferon treatment; depletion of calcium stores	
		GCN2	amino acid starvation (*S. cerevisiae*)	
		?	heat shock	
		?	amino acid or serum starvation (mammalian cells)	
eIF4E	Ser209	not known (Protein kinase C or protamine kinase *in vitro*)	phosphorylation increased by: mitogenic stimulation of quiescent mammalian cells; overexpression of *src* or *ras* oncogenes; meiotic maturation of *Xenopus* oocytes phosphorylation decreased by: heat shock; entry into mitosis; infection by adenovirus or influenza virus.	? increased affinity for mRNA cap ? increased association with eIF4G

Regulation of eIF2 activity

Many physiological conditions that inhibit initiation of protein synthesis have been shown to decrease the activity of eIF2, and, consequently, to impair the formation of 43S initiation complexes (reviewed [2, 4, 7, 8, 17, 23, 115–121]). To date, the site of control of eIF2 activity has always been identified as the recycling step involving the guanine nucleotide exchange factor, eIF2B. Two different physiological mechanisms appear to regulate this step. The first involves phosphorylation of the α subunit of eIF2 at a single serine residue, Ser51 (Table 2); this results in increased affinity of eIF2 for eIF2B, but the complex formed fails to carry out guanine nucleotide exchange. The effect of this is to decrease the concentration of eIF2B available to recycle even the remaining non-phosphorylated eIF2 · GDP. This mechanism is rendered particularly effective by the low molar ratio of eIF2B/eIF2 in all mammalian tissues examined [42]. The second mechanism, discovered more recently, involves direct regulation of eIF2B activity, independent of changes in eIF2 phosphorylation.

Regulation of eIF2 phosphorylation. Three protein kinases that specifically phosphorylate eIF2α at Ser51 have been cloned and sequenced (Table 2; see reviews [116, 122–125] for comparisons of domains). These each possess 11 catalytic domains widely conserved between protein kinases, but show little sequence similarity elsewhere. Another common feature is the presence of an insertion sequence separating groups of kinase domains, though only a small portion of the insert shows significant similarity between the three kinases. Regions within each of the individual kinases have been suggested to have regulatory functions associated with their responses to different physiological signals.

The heme-controlled repressor (HCR; Table 2) [122, 123, 125–128] is a 70 kDa protein which behaves as a 90-kDa polypeptide on SDS/PAGE. Its physiological function is to prevent protein synthesis in erythroid cells (over 90% of which is devoted to the production of globin chains) in the absence of the heme prosthetic group. Its ability to phosphorylate eIF2α closely correlates with autophosphorylation at multiple sites. Interaction with heme results in loss of both autophosphorylation and kinase activity towards eIF2α, and is probably associated with inability to bind ATP. Heme-mediated inactivation is thought to involve the formation of disulfide links between HCR subunits (reviewed [122, 123, 125, 126]). However, interactions between HCR and heat shock proteins, particularly hsp90 and hsp70, have also been demonstrated [129–131] (reviewed [123, 125]). In addition a 67-kDa protein in the reticulocyte lysate that can inhibit the phosphorylation of eIF2α by activated HCR has been proposed to play a physiological role [132]. Chen and coworkers find that, in keeping with its proposed physiological function, expression of HCR at both the protein and the mRNA level is restricted to erythroid cells [123, 126, 133]. However, Mellor et al. [134] have succeeded in cloning a rat brain cDNA encoding an eIF2α kinase with over 80% similarity at the amino acid level with rabbit reticulocyte HCR. They also detected low levels of mRNA recognized by this cDNA in a number of other tissues. This raises the possibility that HCR-like kinases may exist in non-erythroid cells to mediate effects of one or more of the other physiological signals that result in increased phosphorylation of eIF2α, such as heat shock or nutrient starvation [23, 117, 118, 121] (Table 2).

The second eIF2α kinase, PKR (Protein Kinase activated by double-stranded RNA), is important in the defence of mammalian cell populations against viral invasion. It is markedly induced by transcriptional activation in response to interferons α or β released by neighbouring cells. Upon subsequent viral infection, the kinase is activated and severely inhibits translation by increasing eIF2α phosphorylation and blocking the recycling activity of eIF2B. This, while clearly deleterious to the individual cell, prevents the utilization of its translational apparatus for the production of viral proteins and hence restricts viral replication within the cell population as a whole. The structural features of PKR are described in recent reviews [116, 124, 125, 128, 135]. In addition to the catalytic domains common to protein kinases, it possesses two N-terminal domains involved in binding its activator, double-stranded RNA, of which the first is more critical for activation [116, 124, 125, 128, 135–140]. Activation is closely associated with autophosphorylation at multiple (but unidentified) sites and appears to involve dimerization. A popular model attributes the autophosphorylation to mutual phosphorylation between two PKR molecules brought into close proximity by binding to a single molecule of dsRNA. This model appears to be consistent with many observations [128, 135, 137, 138, 140a], but it seems likely that protein–protein interactions are also involved; recent data suggest that alternative mechanisms bringing about dimerization without the mediation of dsRNA may also exist [136, 139]. The nature and origin of the dsRNA molecules responsible for the physiological activation of PKR are not clear in all cases, although such molecules are produced as part of the replication cycle of many viruses.

Many viruses have evolved strategies to subvert the host defence mechanism mediated by PKR (see Table 3 and reviews [10, 128, 135, 141–146a]). These involve a variety of virally encoded molecules, including small RNAs with extensive sec-

Table 3. Strategies employed by viruses to ensure efficient translation of their own products during viral infection.

Strategy	Virus	Mechanism
Production or activation of agents that challenge host defences mediated by interferon-induced PKR	adenovirus	production of virus-associated RNA (VA-1) which interacts with PKR to prevent activation by double-stranded RNA
	influenza virus	activation of cellular protein (p58) which binds PKR and prevents activation
	poliovirus	degradation of PKR protein
	vaccinia virus	encodes two proteins: (i) E3L, which binds double-stranded RNA and prevents it activating PKR; (ii) K3L, which resembles part of eIF2 and probably acts as a decoy
	herpes simplex virus	encodes protein that inhibits PKR by unknown mechanism
Production of viral RNAs with high translational efficiency	alfalfa mosaic virus (AMV); adenovirus	viral mRNAs have unstructured 5′ UTRs, thought to exhibit lower initiation factor requirements than host mRNAs
	influenza virus	5′ UTRs of viral RNAs confer high translational efficiency in infected cells (unknown mechanism)
Inactivation of mRNA-binding initiation factors, conferring advantage on viral RNAs which exhibit low factor requirements	adenovirus influenza virus	decreased phosphorylation of eIF4E in infected cells; may have relatively low impact on translation of viral RNAs with unstructured 5′ UTR
	poliovirus human rhinovirus Coxsackie virus foot-and-mouth disease virus	Cellular eIF4G cleaved in infected cells, probably by virally encoded proteases; translation of uncapped picornavirus RNAs remains operative when eIF4G is cleaved.

ondary structures that bind PKR, proteins that bind and sequester dsRNA, a protease that degrades PKR in poliovirus-infected cells and a protein that resembles a truncated version of the eIF2α substrate of PKR. Infection by influenza virus activates a cellular inhibitor of PKR, p58, by somehow effecting its release from a complex with another cellular protein, I-p58, which normally holds it in an inactive form [146, 147].

While the importance of PKR in the battle between host cells and invading viruses is clearly evident, much excitement has been generated recently by observations implicating this protein kinase in regulation in *uninfected* cells. This would require activation either by cellular RNA molecules or by alternative mechanisms. Two reports implicate PKR in the increased eIF2α phosphorylation in mammalian cells treated with calcium-mobilizing agents [148, 149]. A role for this kinase in growth regulation was suggested by observations of increased PKR activity in 3T3 fibroblasts as they approached stationary phase [150]; the cloning of a cDNA encoding human PKR subsequently permitted more direct approaches. Overexpression of PKR in yeast cells was found to result in severe inhibition of growth, concomitant with increased eIF2α phosphorylation [151]. This experiment has not proved feasible for higher eukaryotic cells, probably because high levels of PKR are so inhibitory that one cannot obtain enough cells to examine. However, it has been possible to do the converse experiment, i.e. to suppress endogenous PKR activity, because several different mutations in PKR result in molecules that have *trans*-dominant inhibitory properties both *in vivo* and *in vitro*. These include a point mutation [152, 153] and a deletion [154] within the catalytic (kinase) domains and the deletion of the more critical of the two dsRNA binding domains [139, 140]. Mechanisms by which these mutant molecules might down-regulate endogenous PKR activity are reviewed in [135]. Expression of each of these mutant forms in NIH3T3 cells has been shown to result in malignant transformation [139, 152, 154]; similar results were obtained in cells where PKR was down-regulated by over-expression of the inhibitory protein, p58 [155] (see Table 3). The simplest conclusion from these experiments is that wild-type PKR in normal cells may be involved in restraint of cell growth. An important question is whether this potential tumour-suppressor role involves regulation of phosphorylation of eIF2α. Most of these *trans*-dominant mutants reduced the phosphorylation of endogenous eIF2α [139, 156], and this was also the case in cells overexpressing p58 [155]. If eIF2α phosphorylation does mediate a role for PKR in growth control, one might expect that overexpression of the non-phosphorylatable mutant of this factor, Ser51→Ala, would itself be tumourigenic. However, conflicting results have been reported [149, 156]. In any case, it now seems likely that PKR has other, possibly unrelated, roles in cellular regulation; recent data link this kinase to signal transduction pathways involving a number of growth factors and cytokines [136, 157, 158] and it can phosphorylate another target, IκB, which is involved in transcriptional control [136, 159, 160]. Moreover, a further challenge to a simple role for PKR as a tumour suppressor comes from the recent report that mice with the PKR gene inactivated failed to develop spontaneous tumours [161].

The third well-characterized eIF2α kinase, GCN2, is a *Saccharomyces cerevisiae* protein of 182 kDa, which is activated during chronic amino acid starvation. Expression of a constitutively active mutant of this enzyme in yeast results in inhibition of overall protein synthesis in a manner similar to that resulting from activation of HCR or PKR in mammalian cells. However, the physiological control mediated by GCN2 is more subtle: overall protein synthesis is scarcely inhibited when yeast is starved for amino acids, yet the perturbation of eIF2 function is able to switch on the translation of the mRNA encoding a specific protein, GCN4, which is not produced at all in fed cells. The physiological significance of this is that GCN4 is a transcription factor that promotes expression of genes encoding a number of enzymes involved in *de novo* amino acid synthesis; thus the ability to switch on its synthesis provides a mechanism for the yeast cells to compensate for their nutritional deficiencies.

The mechanism by which eIF2α phosphorylation, normally associated with inhibition of overall protein synthesis, can actually enhance the translation of GCN4 mRNA is summarized in several recent short reviews [83, 117, 124, 128] and, in more detail, with recent experimental evidence in [162]. The studies leading to the elucidation of this mechanism, in the laboratory of Hinnebusch and colleagues, illustrate particularly well the power of applying complementary genetic and biochemical approaches to translational regulation; Wek [124] provides a clear introduction to this aspect. At the heart of the mechanism is the location of the GCN4 coding sequence downstream of four short open reading frames (ORF 1–4), such that its translation is dependent on a reinitiation mechanism as described above. The most 5′ of these minicistrons has properties that permit about 50% of 40S ribosomal subunits to remain on the mRNA and resume scanning. Under fed conditions virtually all these reacquire a new Met-tRNA$_f$ · eIF2 · GTP ternary complex in time to translate the third or fourth ORF, both of which have properties leading to complete release of the ribosomes that translate them. Thus access of ribosomes to the GCN4 coding sequence is completely precluded. During amino acid starvation, the activation of GCN2 leads to an increase in eIF2α phosphorylation which, by impairing eIF2B activity, reduces the availability of ternary complexes. The reinitiating ribosomes therefore take longer to recapture a ternary complex, and some of them become competent to translate only after scanning past the inhibitory ORFs. These are able to reach and recognize the initiation codon of the GCN4 coding sequence, which is then translated.

An important consequence of the study of this system was the characterization of yeast eIF2γ and all the polypeptide subunits of yeast eIF2B, which were originally identified as products of genes involved in the regulation of GCN4 translation [36, 83, 117, 124, 128, 162]. This greatly assisted the characterization of mammalian eIF2B, and is beginning to shed light on the interactions of eIF2B with phosphorylated eIF2, which is also important for the understanding of regulation by HCR or PKR. The yeast homologue of the smallest subunit of eIF2B, eIF2Bα, is the product of the *GCN3* gene; genetic analysis predicts this to be non-essential for basic eIF2B function but important in the response to activation of GCN2. Recent work now indicates that two further subunits of eIF2B, GCD2 and GCD7 (corresponding to mammalian eIF2Bβ and δ), while essential for guanine nucleotide exchange activity, also play a similar regulatory role in recognizing the phosphorylated form of eIF2 [162]. Interestingly, regions of these three subunits exhibit some sequence similarity with each other in both yeast and mammalian eIF2B [36].

Two important questions concerning GCN4 regulation are still not fully answered. Firstly, the phosphorylation of eIF2α during amino acid starvation seems to delay the capture of ternary complexes by reinitiating 40S subunits under conditions where overall rates of initiation, including primary initiation at ORF1 of GCN4 mRNA, are hardly affected. Thus the main role of ORF1 seems to be to ensure that initiation at ORF3 or ORF4, which precludes GCN4 translation, is extra sensitive to a subtle change in eIF2 phosphorylation. But why is reinitiation more sensitive? One possibility, suggested by Hinnebusch [162], is that ternary complex binding is rate-limiting for reinitiation but not necessarily for primary initiation, which may instead be controlled at the mRNA binding step. The second question concerns the complex area of how GCN2 activation is triggered by amino acid starvation. Current ideas on this result from structural and mutational analysis of two domains in the C-terminal half of the kinase molecule that are essential for activation [124, 162]. One of these is a region of 530 amino acids that shows 22% identity and 45% similarity with yeast histidyl-tRNA synthetase, and evidence is beginning to appear for recognition by this domain of uncharged tRNA, thought to accumulate during amino acid starvation [163]. The other is a region of 120 amino acids at the C-terminus, required for association of GCN2 with ribosomes, which appears to be necessary for activation. Such association may bring GCN2 into closer contact with its activator (uncharged tRNA bound to the ribosomal A-site) and/or allow the kinase activity to be targeted specifically at ribosome-bound eIF2. Additional genes, *GCN1* and *GCN20*, recently identified as involved in the activation, are currently under investigation [162, 164]. A type-I protein phosphatase, GLC7, appears to reverse the effect of GCN2 by dephosphorylating eIF2α [165].

Direct regulation of eIF2B activity. Over recent years several examples have emerged where the guanine nucleotide exchange activity of eIF2B appears to be regulated in the absence of a change in eIF2α phosphorylation (reviewed [36, 117, 119, 120, 125]. A particularly well characterized example concerns regulation by insulin (reviewed [36, 120, 166, 167]). Treatment of fibroblasts, or Chinese hamster ovary cells overexpressing the insulin receptor, with the hormone increases the eIF2B activity detectable in crude cell extracts, apparently by down-regulating an inhibitory activity [168, 169]. Studies on the purification behaviour of this inhibitor led to the suggestion that it may be glycogen synthase kinase-3 (GSK-3), an insulin-regulated enzyme known to phosphorylate a number of other targets. Indeed, purified GSK-3 does phosphorylate the largest (ε) subunit of mammalian eIF2B, and also inhibits guanine nucleotide exchange activity when added to crude cell extracts [36, 169]. This evidence, together with other data on insulin regulation of GSK-3 activity, has led to the proposal of the pathway shown in Fig. 3 for the regulation of eIF2B activity by insulin. The ε subunit of eIF2B also contains consensus phosphorylation sites for casein kinase II, although reports differ on whether or not phosphorylation by this enzyme enhances eIF2B activity (reviewed [17]). The physiological importance of allosteric regulation of eIF2B by NADP/NADPH, polyamines and glucose 6-phosphate is also unclear at present [17].

Regulation of initiation factors involved in binding the 43S initiation complex to mRNA

Regulation of eIF4E activity. eIF4E has long been regarded as important in the regulation of initiation. Apart from the reasoning that a regulatory role would be expected for a factor that performs the initial step in mRNA recruitment, this belief is based on two lines of evidence. First, this factor undergoes regulated phosphorylation in response to a very wide range of physiological stimuli. Second, early work indicated that, in HeLa cells [170] and reticulocytes [171], eIF4E was present at very low molar concentrations relative to ribosomes and other initiation factors. However, even after much investigation in several laboratories of the potential regulatory roles of eIF4E phosphorylation and availability, many important questions remain unanswered, and it is worth subjecting each of these possibilities to critical scrutiny.

eIF4E phosphorylation. Increases in eIF4E phosphorylation are frequently observed when quiescent or dormant cells are stimulated by appropriate hormones, growth factors or mitogens; conversely, decreased phosphorylation is seen in some states where translation is inhibited (see Table 2 and more extensive lists in [8, 48, 67, 120, 172]). It is widely assumed that such changes in phosphorylation mediate changes in both the overall rate of protein synthesis and the pattern of recruitment of individual mRNAs, which are thought to differ in their dependence on the activity state of eIF4E by virtue of features such as 5′ secondary structure (see below). However, changes in overall translation rate are often relatively modest in cells undergoing growth stimulation, and in some of the better characterized conditions, e.g. insulin treatment [120, 166] and heat shock [118, 121, 173], regulation at the level of eIF2/2B activity also makes a major contribution. Most evidence indicates that the major site of phosphorylation of eIF4E is a single serine

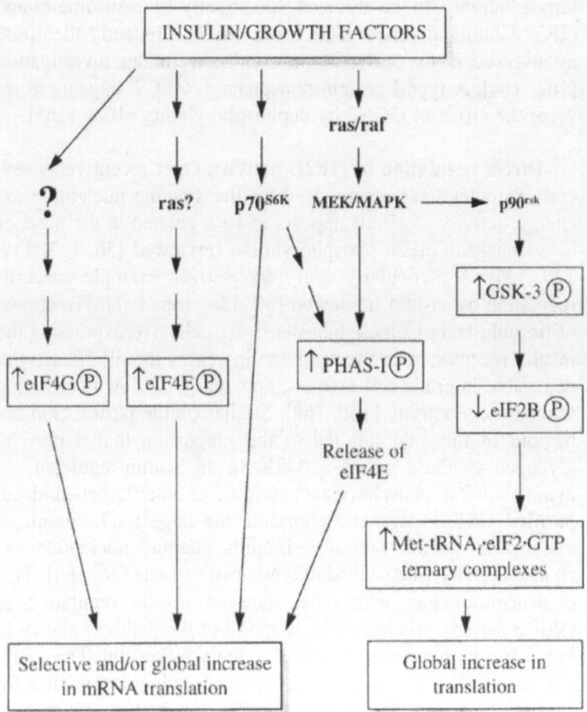

Fig. 3. Potential signalling pathways involved in the regulation of translation in mammalian cells by insulin and growth factors. Adapted from [48, 166]. Hormones and growth factors stimulate phosphorylation of eIF4E [8, 67, 120], eIF4G [176, 184] and ribosomal protein S6 [361] in a variety of cell types. In at least some cases, stimulation of eIF4E phosphorylation has been linked to activation of *ras* [67, 362]. Phosphorylation of PHAS-I is believed to result in the release of eIF4E sequestered in eIF4E. PHAS-I complexes. Both the MAP kinase [208] and p70^{S6K} [209] pathways have been implicated in PHAS-I phosphorylation following insulin treatment of cells. Stimulation by insulin of eIF2B activity has been attributed to down-regulation of an inhibitory kinase, with some evidence identifying this kinase as GSK-3 [36, 169]. GSK-3 can be inactivated by phosphorylation via the MAP kinase pathway, as shown here and discussed in [167], but others have suggested that p70^{S6K} can mediate regulation of this enzyme by insulin [363].

residue, but the search for physiological kinase(s) was hampered for several years by mis-identification of this site as Ser53; recently the phosphorylation site has been identified as Ser209 by labelling studies in the reticulocyte lysate [174], and this has been confirmed by analysis of eIF4E labelled in cultured cells stimulated with serum [175]. The nature of the signalling pathways regulating eIF4E phosphorylation is still unclear. There is evidence both for and against the involvement of protein kinase C [48, 172]. One might expect the mitogen-activated protein (MAP) kinase pathway to be involved in the elevation of eIF4E phosphorylation in cells overexpressing the *src* or *ras* oncoproteins [8, 67, 120, 172], and in *Xenopus* oocytes the time of onset of eIF4E phosphorylation coincides closely with that of MAP kinase activation during hormone-induced meiotic maturation [176]. However, attempts to phosphorylate the factor *in vitro* with MAP kinase, other cell-cycle-related kinases and, indeed, protein kinase C, have generally given poor results [172], suggesting that the physiological kinase has yet to be identified. An insulin-stimulated protamine kinase has recently been shown to phosphorylate eIF4E at Ser209 [177], but further evidence is needed on the general physiological relevance of this.

While the physiological correlations strongly suggest a regulatory role for eIF4E phosphorylation, it is by no means clear how this modification affects the activity of the factor in molecular terms. The factor participates in two key recognition events: the binding to the mRNA cap and the interaction with eIF4G and associated factors. In the case of cap recognition, there is clearly no all-or-nothing effect, since both phosphorylated and unphosphorylated eIF4E bind to m^{7}GTP-Sepharose. However, more subtle effects on the affinity of the factor for capped oligonucleotides were observed by Minich et al. [178], who separated the phosphorylated and unphosphorylated forms of the factor by chromatography on RNA-cellulose. In the case of complex formation with eIF4G and other factors, there have been several observations in intact cells of parallel changes in the extent of phosphorylation of eIF4E and its degree of engagement in high-molecular-mass complexes that include eIF4G [62, 176, 179–181]. Isolation of the fraction of eIF4E in mammalian cells present in high-molecular-mass eIF4F complexes, or associated with ribosomes, reveals a higher degree of phosphorylation than is seen for the free factor [62, 182, 182a], suggesting that phosphorylation may regulate complex formation. However, physiological stimulation of eIF4E phosphorylation tends to occur in parallel with increased phosphorylation of other factors, such as eIF4B, p120 of eIF3 and eIF4G [176, 180, 183–185]. This may also enhance the mRNA binding step in initiation; in the case of eIF4G there is direct *in vitro* evidence for this [186].

Direct comparison of the distribution of phosphorylated and unphosphorylated eIF4E within the reticulocyte lysate reveals that ribosome-bound eIF4E includes a substantial amount of unphosphorylated factor, and, conversely, at least half the phosphorylated eIF4E exists in low-molecular-mass form [182a]. Hence, while eIF4E phosphorylation may enhance its function, it is neither necessary nor sufficient for involvement of the factor in initiation complexes. A more fundamental challenge to an essential role for eIF4E phosphorylation, at least at Ser209, comes from studies on eIF4E from non-vertebrate sources. No phosphorylation site equivalent to Ser209 is seen in eIF4E from wheat or *S. cerevisiae* (see [50]). Labelling studies do suggest phosphorylation of the yeast factor, but the sites are at the N-terminus and do not appear to play a regulatory role [186a]. *Drosophila* eIF4E does have a serine residue at an equivalent location [50], although with differing surrounding sequence, but labelling studies suggest very low levels of phosphorylation of the factor *in vivo* [187].

It is difficult to assess the significance of this in terms of evaluating the role of eIF4E phosphorylation in the regulation of overall protein synthesis, since few studies of physiological regulation have been made in these invertebrate systems. However, mechanisms mediating the global shut-off of translation during moderate and severe heat shock have been extensively studied in both mammalian and *Drosophila* cells (reviewed [118, 121, 173]). Although eIF2α phosphorylation clearly plays an important role in this effect, it has been shown in both the *Drosophila* [188] and mammalian [62] systems that defective translation in extracts from heat-shocked cells is most effectively rescued by addition of eIF4E in the form of complexes with eIF4G. In both cell types, there is evidence for a fall in the proportion of eIF4E in eIF4E · eIF4G complexes during heat shock [170, 179, 188]; although this effect may be less pronounced in *Drosophila* cells [187], the observations suggest that such effects are not specific to cells clearly exhibiting regulation of eIF4E phosphorylation. However, the absence of the key phosphorylation site in at least some invertebrates seems much less of a challenge to the concept that eIF4E phosphorylation is important in the qualitative regulation of translation. Data available so far from sequence analysis suggest that 5′ UTRs of vertebrate mRNAs tend to have a higher G+C content than those of *S. cerevisiae*, *Drosophila* and plants [9, 73, 189], indicating that 5′-secondary structure is more likely to impede the mechanisms mediating mRNA binding in vertebrate cells. Thus there may be more potential for selective regulation at the level of mRNA unwinding in vertebrate cells than in *S. cerevisiae*, insects and plants, and an additional level of regulation involving eIF4E phosphorylation may be an advantage in providing a means of linking mRNA selection to cell signalling events.

Availability of eIF4E. As stated above, cellular concentrations of eIF4E have been estimated to be low relative to those of other components of the translational machinery, and the factor is widely regarded as potentially rate-limiting in amount [4, 67, 120, 170]. Recently, however, we have detected in reticulocyte lysates substantially higher concentrations of this factor than previously reported [182a], and, moreover, a large proportion of the eIF4E can be removed by treatment with m^7GTP affinity resin with little detriment to translational activity. However, some of the most exciting studies on translational factors in recent years have shown that gross manipulation of cellular eIF4E levels can have dramatic effects on growth control and protein synthesis. Initially, work in the laboratories of Sonenberg [190] and Rhoads [191] demonstrated that overexpression of this factor in several types of cultured cells variously led to changes in cell morphology, abrogation of growth control and acquisition of the ability to induce tumour formation in nude mice. A study using cells in primary culture indicated co-operativity between eIF4E overexpression and *v-myc* or E1A in the induction of tumourigenesis [192]. The effects of eIF4E appeared to be exerted via a *ras*-mediated pathway, since they were blocked by overexpression of the negative regulator of *ras*, GAP [193]. The generally favoured mechanism for this effect of eIF4E overexpression is based on the observation that mRNAs encoding proteins involved in growth or growth regulation often possess stable secondary structure in the 5′ UTR to an extent that would lead to a prediction of extremely low translational efficiency [194]. Translation of such mRNAs might be expected to be particularly dependent on the unwinding activity of initiation factors, and thus overexpression of eIF4E may lead to inappropriately high rates of translation [67]. Although evidence in favour of this model was provided by experiments with reporter constructs bearing highly structured 5′ UTRs [195], there are still relatively few convincing demonstrations of natural mRNAs that behave according to this model (see below). The best probably concerns the mRNA encoding ornithine decarboxylase [196], the rate-limiting enzyme in polyamine biosynthesis, which itself can induce cell transformation when overexpressed [197]. However, it is important to note that part of the eIF4E in both mammalian [198] and yeast [199] cells is located in the nucleus, indicating a possible role in other processes that may contribute to growth control.

In contrast to the effects of eIF4E overexpression, Rhoads' group have engineered severe depletion of the factor in HeLa cells by introducing anti-sense RNA [200–202]. This reduced protein synthesis to very low levels, with a pattern of products very similar to that in severely heat-shocked cells. This is easier to relate to the expectation that the dependence of mRNA on eIF4E for translation reflects its degree of 5′-secondary structure, since mRNAs encoding heat-shock proteins have relatively unstructured 5′ UTRs (see below). However, a surprising, and perhaps important, aspect of these anti-sense experiments is that the cellular level of eIF4G was also profoundly depleted [200].

The studies discussed above concern the effects of artificial manipulation of eIF4E levels, but it is now clear that effective concentrations of this factor can be physiologically modulated. First, the expression of mRNA encoding eIF4E (and also that encoding eIF2α) is enhanced in cells transformed with c-*myc* [203] and during T cell activation [204]. Elevated levels of eIF4E mRNA and protein have also been observed in a variety of transformed cell lines and tumours [205, 206], although it is not clear whether this effect is specific to eIF4E or part of a general increase in the translational apparatus. More recently, however, a mechanism for regulating the availability of eIF4E on an acute basis has been revealed, and it seems likely that this can play an important role in translational regulation. Sonenberg and collaborators identified cDNAs encoding two novel eIF4E-binding proteins, which shared a degree of sequence similarity. One of these, which they called eIF4E-BP1, was found to share extensive sequence similarity with a protein called PHAS-I, which had previously been identified on the basis of its rapid phosphorylation in adipose cells following insulin treatment [207, 208]. Evidence rapidly accumulated that insulin could regulate the binding between PHAS-I and eIF4E in responsive cells, and a model emerged in which association with PHAS-I serves to restrain eIF4E from participating in protein synthesis [63, 207, 208] (reviewed [48, 166, 167]). Sonenberg and co-workers have now demonstrated competition between PHAS-I and eIF4G for eIF4E binding [208a], and identified sequence similarity between a region of PHAS-I and the eIF4E binding site on eIF4G [63]. Insulin treatment appears to remove this block on eIF4E by promoting the phosphorylation of PHAS-I, which triggers the release of the associated eIF4E (Fig. 3). Initially this phosphorylation was proposed to be mediated by a signalling pathway involving MAP kinase, which directly phosphorylates PHAS-I on a serine residue, Ser64, very close to the eIF4E binding site [208]. However, recent studies with 3T3-L1 adipocytes treated with the inhibitors rapamycin and wortmannin suggest that multiple signal transduction pathways are involved, including the $p70^{S6K}$ pathway [209].

Regulation of initiation factors involved in mRNA binding during viral infection. As in the case of PKR, some viruses have evolved cunning ways to modify the eIF4 initiation factors in infected cells so as to favour translation of their own products. Cells infected by either adenovirus [210] or influenza virus (see Table 3) [211] show reduced phosphorylation of eIF4E. In each case, translation of the virally encoded mRNAs is likely to compete well with host cell mRNAs under these conditions. Late adenovirus messages bear a particularly unstructured 5′ UTR (see below) and show low dependence on eIF4E for translation [212, 213]. The 5′ UTR of influenza virus mRNAs also confers high efficiency of translation in infected cells; the structural basis of this is not yet clear [146, 214], but interaction with virally encoded proteins may be involved [215].

The other major example of viral interference with cap-dependent initiation mechanisms is the proteolytic degradation of eIF4G during infection with picornaviruses (reviewed previously [2, 141]). This process was functionally linked to the virus-encoded 2A protease, which is primarily responsible for the cleavage of the viral polyprotein translation product into multiple polypeptides [89, 216]. Later the same phenomenon was observed, and again linked to the 2A protease, in cells infected by other members of the enterovirus group, but it does not occur in cells infected by cardioviruses such as EMCV [217]. Cleavage of eIF4G is also seen in cells infected by the aphthovirus, FMDV, but in this case the leader (L) protease is responsible [218]. Original work with the poliovirus system suggested that the 2A proteinase did not act directly on eIF4G, but instead initiated an activation cascade of cellular proteases [89, 216, 219]. However, it is now clear that purified, recombinant 2A and L proteases can cleave eIF4G directly, each initially at a single site (between Arg486 and Gly487 in the case of 2A protease and Gly479 and Arg480 in the case of L protease) [220, 221].

These observations suggested a viral strategy whereby the cleavage of eIF4G would shut off the translation of capped cellular mRNAs, but permit the continued translation of picornavirus RNAs, which are uncapped and utilize the internal initiation mechanism (reviewed [89]). Moreover, the viral RNA translation would benefit from the removal of the competition of host protein synthesis for common initiation factors. However, this rather neat model may be rather simplistic, since there have been reports that at early stages in poliovirus infection, or in the presence of some inhibitors of viral replication, it is possible to observe comprehensive proteolysis of eIF4G without commensurate inhibition of host cell mRNA translation [222, 223]. Effects of 2A protease on other processes in infected cells [224, 225] have also been reported. Furthermore it seems increasingly likely that proteolysis of eIF4G may benefit picornavirus RNA translation in a more positive way than by merely removing the competition from host cell mRNAs, since evidence is accumulating for an active role of the cleavage products [89, 226, 227]. This is reflected in *in vitro* translation assays, where addition of the 2A or FMDV L proteases, which inhibit translation of capped transcripts, has little effect on that driven by IRES elements from cardioviruses or FMDV [228–231], and actually enhances translation driven by IRES elements from enteroviruses [231], or of uncapped transcripts encoding cellular proteins [230]. These data appear at variance with earlier reports of a role for the eIF4F complex even in internal initiation [232–234]. However, recent results from our laboratory indicate that the eIF4F requirement for IRES-driven internal initiation can be fulfilled by the C-terminal L proteinase cleavage product of eIF4G in the absence of either eIF4E or intact eIF4G [234a]. This C-terminal product is also responsible for the stimulatory activity towards the translation of uncapped cellular mRNA transcripts mentioned above [234a]; it lacks the eIF4E binding site, but remains bound to ribosomes in the reticulocyte lysate, and is thought to include the domains in eIF4G responsible for interactions with eIF3 and eIF4A [34].

STRUCTURAL FEATURES IN THE UNTRANSLATED REGIONS OF mRNAs THAT INFLUENCE TRANSLATIONAL EFFICIENCY

Structures in the 5′ untranslated region (5′ UTR) associated with inefficient translation

Secondary structure. Experimental studies with both *in vivo* and *in vitro* systems clearly demonstrate that mRNAs with a high potential to form stable secondary structure in the 5′ UTR tend to be translated inefficiently [43, 76]. This effect is position-dependent, suggesting that the initial binding of initiation factors or 40S ribosomal subunits to mRNA is more sensitive to impairment than subsequent scanning [14, 43, 44, 76, 235]. *In vitro* evidence has linked secondary structure close to the cap with poor cap-binding activity of eIF4E and eIF4B [236]. The group of natural mRNAs possessing highly structured 5′ UTRs includes a disproportionately high number of examples encoding proteins that take part in or regulate processes involved in cell proliferation; their translational inefficiency may play a crucial role in the maintenance of correct restraints on cell growth [76]. The significance of this may be linked with the frequent expression by tumour cells of alternative transcripts lacking the inhibitory regions [237, 238]. Mechanisms permitting the transient release of this translational repression during growth stimulation are under active investigation. An attractive possibility is the activation of general initiation factors involved in mRNA binding (e.g. by phosphorylation [8, 48, 67, 120, 172]). Alternatively, activation of the MAP kinase or $p70^{S6K}$ signalling pathways could lead to phosphorylation of the eIF4E binding protein, PHAS-I, resulting in a transient increase in the availability of eIF4E for initiation [48, 166, 167, 208, 209] (see above). If eIF4E (or eIF4F complex) were rate-limiting for translation, these effects could be expected to confer a selective advantage on mRNAs that normally competed poorly [68]. Such mechanisms seem likely in the case of ornithine decarboxylase mRNA; increased translation of this message has been correlated with phosphorylation of eIF4E and eIF4B in insulin-treated cells [183] and, moreover, is observed in eIF4E-overexpressing cells [196]. There is also some evidence for enhanced translation of ornithine aminotransferase [239], the growth-related protein P23 [240] and the cell-cycle-regulating protein cyclin D1 [241] in eIF4E-overexpressing cells, though in the last case the effect may not be primarily exerted at the level of translation [242]. In contrast, eIF4E overexpression failed to promote translation of mRNA constructs with structured 5′ UTRs in *S. cerevisiae* [199] or that of the highly structured *c-sis*/PDGF2 mRNA in NIH3T3 cells [243]. However, the most inhibitory element in *c-sis* mRNA was localized to a position a long way downstream of the cap [243], making a mechanism involving regulation of eIF4E activity less likely. An interesting report recently demonstrated that microinjection of eIF4E into early embryos of *Xenopus laevis* promoted mesoderm induction and specifically increased the translation of microinjected activin mRNA [244]. However, this seems difficult to relate to 5′ secondary structure in this mRNA, since, in the transcript employed, the activin coding sequence was inserted downstream of the 5′UTR of *Xenopus* *β*-globin mRNA, which is associated with high translational efficiency. Thus the jury is still out on the general importance of regulation of the eIF4 group of general initiation factors in the selective regulation of translation of growth- or differentiation-related mRNAs with structured 5′ UTRs, and further studies with different mRNAs and systems are needed.

It is increasingly clear, however, that translation of a number of mRNAs with structured 5′ UTRs is highly responsive to the cellular environment. Translation of coding sequences downstream of the 5′ UTR of *c-myc* is relatively efficient following transfection into immortalized cell lines, or in extracts prepared from HeLa cells, but elements within this 5′ UTR strongly inhibit translation in wheat germ extracts and reticulocyte lysates [245]. The inhibitory effect of the *c-myc* mRNA 5′ UTR seen in unstimulated *Xenopus* oocytes is relieved in mature or fertilized eggs [246], and, in another study, fertilization of *Xenopus* eggs increased translation of structured mRNA constructs [247]. Some of these observations would be consistent with negative regulation of PHAS-I by activated MAP kinase or $p70^{S6K}$, but no evidence is available as yet and other helix-unwinding activities could well play a role in such regulation [14, 45, 248]. In a different system the La autoantigen, which exhibits RNA helicase activity *in vitro* [107], has been proposed to contribute to alleviating translational repression by the highly structured 5′ TAR element in HIV-1 mRNAs [249]. Finally, several observations now suggest that, in addition to general initiation or unwinding factors, more specific interactions between proteins and 5′ UTRs may play a role in regulating translation of growth factor mRNAs [250–252].

Upstream AUG codons. In many cases these are followed by a termination codon to produce short open reading frames (see above). They are often encountered in mRNAs encoding proteins involved in growth control of mammalian cells [76], and occur at a strikingly high frequency in *Drosophila* mRNAs (see [78]), where their significance has been little studied as yet. Again, tissue specificity may play a regulatory role (see reviews

[77, 78]). For example, one group reported that an upstream AUG near the 5′ end of the mRNA encoding *S*-adenosylmethionine decarboxylase is ignored in all mammalian cell types except T-lymphocytes, where it exerts translational repression [79, 253]. However, other workers identified the same upstream open reading frame as a major contributor to poor translational performance of this mRNA in non-lymphoid cells and suggested that it may be important in regulation of synthesis of the enzyme by polyamines [254].

Elements recognized by specific binding proteins. Structural motifs in mRNA molecules can provide sites for the binding of specific proteins. In the case of the 5′ UTR, such binding can impede initiation by stabilizing an element of secondary structure not itself strong enough to be inhibitory. The potential for protein binding to the 5′ UTR to act as a negative regulator of translation is illustrated by work from Hentze's laboratory, in which various specific protein-binding sites, not normally involved in translational control, were inserted into mRNA constructs [255, 256]. The archetypal physiological example of this mechanism concerns translation of the mRNA encoding the iron-binding protein, ferritin. At the 5′ end of this mRNA is a small hairpin loop termed the iron-responsive element (IRE), which is recognized by specific proteins termed iron regulatory proteins (IRPs; formerly known as iron regulatory factors or IRE-binding proteins). The binding of the IRPs to the IRE in ferritin mRNA is responsible for a regulatory mechanism that ensures that ferritin is only synthesized in cells adequately supplied with iron (reviewed [257]). The IRE is only effective in regulating translation when placed at or near its natural position near the 5′ cap, consistent with the observation that association of IRP1 blocks the initial placement of the 43S preinitiation complex on the mRNA [255, 257]. The iron status of the cell regulates the association of the IRPs with the IRE in different ways. In the case of IRP1, iron status regulates its interconversion between high- and low-affinity forms [257], while the more recently identified IRP2 [258, 259] contains a sequence that confers rapid, proteasome-mediated degradation in iron-replete cells [259a]. The mRNA encoding another protein involved in iron metabolism, 5′-aminolevulinate synthase, also contains an IRE and is subject to the same regulation. Further interesting developments on the physiological regulation of this system seem likely in the near future, following reports that the IRE-binding activity of IRP1 can be regulated by the signalling molecule, nitric oxide, in an iron-independent manner [260–262]. Phosphorylation of IRP1, possibly mediated by protein kinase C, has also been reported [263].

Other studies are beginning to reveal that binding of proteins to mRNA 5′ UTRs can mediate autoregulation of translation by the protein product. In the case of thymidylate synthetase mRNA, one of two elements recognizing the cellular enzyme includes the AUG initiation codon, the other being within the coding sequence [264]. As with the IRE/IRP interaction, binding may be regulated by a redox mechanism involving protein -SH groups [265]. The poly(A)-binding protein (PABP) represses its own translation by binding to an oligo(A) sequence found in the 5′ UTR of all PABP mRNAs so far examined [266]. This repression can be relieved *in vitro* by addition of exogenous poly(A), which competes the binding protein off the 5′ UTR. It is thought that physiologically the sequence in the 5′ UTR competes unfavourably for the PABP with the poly(A) tails of cellular mRNAs, such that its synthesis is only repressed when PABP accumulates.

Oligopyrimidine tracts. Both prokaryotic and eukaryotic cells have evolved multiple mechanisms to ensure balanced production of the protein and RNA components of ribosomes. In vertebrate cells translational control contributes to the co-ordinate regulation of synthesis of ribosomal proteins. All known mRNAs encoding these proteins, and some encoding other polypeptides involved in translation, possess at the extreme 5′ end a short (5–14 nucleotides) oligopyrimidine tract that is predicted to assume a hairpin structure (reviewed [267]). Possession of this element confers a distinctive pattern of translational utilization: (a) relative underutilization of the mRNA, with a high proportion in the untranslated mRNP pool even in growing cells; (b) bimodal distribution across the polyribosome profile in which mRNA is either untranslated or utilized in full-size polyribosomes, suggesting specific translational control; (c) a selective shift into polysomes during growth stimulation [268]. Translation of these mRNAs may be cell-cycle-regulated [267]. The regulation of translation of ribosomal protein mRNAs has been linked with eIF4F function [269] or eIF4E phosphorylation [270], but overexpression of eIF4E did not prevent the translational repression of ribosomal protein mRNAs in NIH3T3 cells undergoing growth arrest in response to inhibitors of DNA synthesis [271]. An exciting, recent, observation is that the selective enhancement of translation of several of these mRNAs in growth-stimulated cells is impaired by the immunosuppressive drug, rapamycin [272, 273]. This drug blocks the signalling pathway involving the $p70^{S6K}$, which is responsible for the phosphorylation of ribosomal protein S6 in response to a variety of hormones and growth factors [274]. Thus the translation of mRNAs bearing 5′-oligopyrimidine tracts may be regulated by this pathway, perhaps by a phosphorylation/dephosphorylation mechanism involving a specific binding protein. Alternatively, phosphorylation of S6 itself, a protein located in the mRNA binding region of the 40S ribosomal subunit [35], may regulate the recruitment of this class of mRNA.

Potential internal ribosome entry segments (IRESes) in the 5′ UTR

The acceptance and increasing understanding of the cap-independent, internal initiation mechanism for the translation of picornavirus RNAs (see above) has been accompanied by growing support for the more radical proposal that such a mechanism may be utilized by a wider range of cellular and viral mRNAs. As with other examples of translational regulation by elements in the 5′ UTR, the candidate cellular mRNAs for an internal initiation mechanism encode an interesting selection of proteins involved in cell regulation. mRNAs reported as possessing IRESes that pass the test of conferring cap-independent translation behaviour on the downstream cistron of a bicistronic construct (see above) include those encoding the immunoglobulin heavy chain binding protein (BiP) [275], the growth factor FGF2 [276] (Vagner, S., Touriol, C., Gensac, M.-C., Amalric, F., Bayard, F., Prats, H. and Prats, A.-C., unpublished results) the *gag* precursor proteins of murine leukemia viruses [277, 278], the product of the *Drosophila* homeotic gene, *Antennapedia* [279], and two yeast transcription factors, TFIID and HAP4 [280]. However, the 5′ UTRs of these mRNAs do not resemble those of picornaviruses. In some of these cases the regulated utilization of the IRES-mediated mechanism is suggested to promote usage of alternative initiation codons and thus modulate the balance of synthesis of products or isoforms with different biological activities [276, 278] (Vagner et al., unpublished results). Transacting protein factors, which may be regulated in response to physiological conditions, have been implicated, but have yet to be fully characterized [276, 278] (Vagner et al., unpublished results).

5′ UTRs conferring high translational efficiency

Some mRNAs are translated very efficiently and show strong resistance to inhibition of translation by cellular stresses that impair overall protein synthesis. One might expect such mRNAs to possess 5′ UTRs relatively free of secondary structure, upstream ORFs and other impeding features. Indeed, Kozak [281] improved the translational efficiency of mRNA constructs simply by inserting repeats that extended the 5′ UTR without introducing secondary structure. The improved performance was accompanied by decreased cap dependence *in vitro*, which presumably indicated a lower requirement for eIF4E or eIF4F complex, although these may not necessarily be synonymous [9, 68]. In a later study, Kozak concluded that greater length, together with a lower potential to form secondary structure near the cap, formed the basis of the translational advantage of β-globin mRNA over α-globin mRNA [282].

The possession of a 5′ UTR conferring high translational efficiency can be of particular advantage to invading viruses, whose mRNAs need to commandeer the host cell translational machinery. Examples are the 5′ leaders of the plant viral mRNA encoding alfalfa mosaic virus coat protein 4 (AMV 4) [9] and the first segment of the 5′ UTR of late adenovirus mRNAs (the so-called adenovirus tripartite leader [213]). The efficiency of mRNAs encoding heat shock proteins is also conferred by their 5′ UTRs. These mRNAs can be translated actively under conditions where overall protein synthesis in severely down-regulated by cellular effects on both eIF2 and eIF4F, and earlier reports that heat shock, adenovirus and AMV 4 mRNAs can be translated in poliovirus-infected cells, or in extracts derived from them [283−285], suggested that they may utilize a cap-independent mechanism. However, more recent studies have challenged this view [87]. Expression of the FMDV L protease, which cleaves eIF4G, inhibited *in vitro* translation of mRNAs bearing the adenovirus tripartite leader [228] and substantially impaired that of *Drosophila* hsp70 mRNA [286]. In addition, mRNA electroporation experiments indicated that possession of a methylated cap was necessary for efficient translation of hsp70 mRNA in either normal or heat-shocked *Drosophila* cells [286]. Re-evaluation of the translation of AMV4 mRNA in extracts from poliovirus-infected cells also indicated that efficient translation was cap-dependent [287].

In the case of heat-shock-protein mRNAs, as with late adenovirus mRNAs, it is an attractive hypothesis to link the efficiency conferred by their relatively unstructured 5′ UTRs to their selective translation under conditions where eIF4E is underphosphorylated. However some holes are beginning to appear in this argument. Although selective survival of heat-shock mRNA translation was observed in mammalian cells in which eIF4E and eIF4G had been drastically down-regulated by antisense technology [200−202], preferential translation under heat shock conditions is much more pronounced in *Drosophila* cells [121], in which eIF4E phosphorylation seems less likely to play a major regulatory role [187]. Studies on the 5′ UTRs of *Drosophila* heat shock mRNAs suggest that a low degree of secondary structure is necessary, but not sufficient, to permit translation under heat shock conditions [288]. Specific regions in the 5′ UTR, including one near the cap, seem to be important, but attempts to identify a role for precise primary sequences or for hsp-mRNA-specific binding proteins have so far been unsuccessful [118, 121, 289]. A recent survey, however, has challenged the idea that the 5′ UTRs of heat shock mRNAs are significantly less burdened with secondary structure [189]; although many more samples of normal mRNAs are needed to substantiate this, it is clear that, for all the examples of insect mRNAs quoted, the proportion of G+C bases in the 5′ UTRs is very low by vertebrate standards. It may therefore be of doubtful relevance to extrapolate data from mammalian cells on the regulatory role of 5′ secondary structure to organisms like *Drosophila* and *S. cerevisiae*.

Regulation by elements in the 3′ UTR

It is relatively easy to envisage effects of structural features in the 5′ UTR of an mRNA on translational efficiency, since the 40S subunit comes into close contact with this region during initiation. However, in the last few years reports of regulatory elements in the 3′ UTR have constituted an even faster growth area. A few of these concern studies involving mRNAs in somatic cells. For example, cytokine mRNAs frequently contain U+A-rich sequences (often repeated) in the 3′ UTR. Although these tend to be associated with instability, they can also confer inefficient, and probably regulated, translation [290, 291]. Regulatory elements in the 3′, as well as 5′, UTRs of plant viral mRNAs have also been defined [9]. However, 3′-regulatory elements really come into their own in exerting translational control during development, as discussed in the following sections. Finally, the role of the poly(A) tail as a regulator of initiation is emerging as an important topic for all eukaryotic systems.

Translational repression and activation during development

Developmental stages up to and including early embryogenesis are characterized by pronounced changes in the pattern of gene expression in the absence of transcription. Oocytes, eggs and early embryos of a number of organisms have thus provided rich pickings for investigators of translational control. Oocytes and unfertilized eggs are packed with ribosomes and maternal mRNAs that are largely withheld from translation; during meiotic maturation of *Xenopus* oocytes, for example, the proportion of ribosomes engaged in polysomes rises from 1−2% to about 4% [292]. The mechanisms underlying this suppression are multiple and complex. Activation of protein synthesis during early development is a highly co-ordinated process, involving effects on general initiation factors (mainly eIF2/eIF2B and eIF4F [176, 293−295]) together with global mobilization of mRNA from the untranslated pool. A 56−60-kDa mRNP protein thought to be responsible for global repression of mRNA recruitment in *Xenopus* oocytes has been identified as belonging to the Y box group of transcription factors [291, 296−300]. Newly transcribed mRNAs interact with this and related proteins before export from the nucleus, but escape if they possess introns or are microinjected directly into the cytoplasm [301, 302]. The repressor proteins dissociate from the mRNA during progesterone-induced maturation, and their activity is probably regulated by phosphorylation/dephosphorylation [301, 303]. So far no specificity has been identified for this mechanism, either with respect to species of mRNA involved or binding sites within indvidual mRNA molecules [291, 300]. Interestingly, translational repression can be induced in somatic cells by overexpression of the Y box protein FRGY2 [304], and one of the major proteins of mRNP particles in reticulocytes has been identified as a member of this family [305].

Sequence-specific regulation of mRNA translation during development

This topic has been the subject of several excellent, specialized, reviews in recent years [15, 306−315]. In most organisms there are profound changes in translational priorities during meiotic maturation and/or fertilization in favour of a rather specific group of mRNAs encoding proteins required to bring about meiotic maturation and to support the rapid cycles of cell division that characterize early embryogenesis. Each mRNA appears to be programmed for translational activation at an appropriate time and, in many cases, for withdrawal from translation at a later stage in development. Conversely, mRNAs encoding housekeeping proteins, such as components of the cytoskeleton and ribosomal proteins, are translated in oocytes, but their production is less essential during maturation and early embryogenesis and their translation is then down-regulated. Signals conferring these translational patterns are almost invariably located in the 3′ UTR and appear to be of two main types. First, a wide range of sequence elements exerts negative regulation in *cis*; possession of one or more of these elements prevents translation of the mRNA until a critical developmental stage is reached. Second, a more homogeneous set of sequences confers stage-specific elongation of the poly(A) tail, which shows a strong correlation with translational activation.

Sequence elements conferring negative regulation. Elements in the 3′ UTR are involved in extremely complex patterns of translational control (Table 4). Evidence for interaction with specific proteins has often been obtained by genetic analysis, and detailed knowledge of the regulatory mechanism is confined to very few examples. In addition to repressing translation prior to a developmental signal (temporal control), elements in the 3′ UTR can confer spatial regulation by undergoing interactions that prevent the mRNA being translated until it reaches a specific location. This type of translational regulation is of immense importance in controlling pattern formation in the early embryo (reviewed [311, 312, 314]), and studies on some of the mRNAs listed in Table 4 are beginning to unravel amazing linked networks of regulated expression, particularly in the case of *Drosophila*. Other examples, such as the regulation of lipoxygenase mRNA translation in reticulocytes, indicate that this type of control is not limited to early development.

Sequences regulating poly(A) tail length during early development. A phenomenon first observed in general terms, but now understood in considerable molecular detail, is the strong correlation of translational activation of an mRNA during maturation or early embryogenesis with the elongation of its poly(A) tail from less than 100 residues to 100–200 residues (reviewed [15, 306, 308, 313–317]). Conversely, mRNAs whose translation is switched off during early development undergo parallel deadenyation or poly(A) shortening. Developmentally regulated elongation of the poly(A) tail is conferred by the possession of two sequence elements in the 3′ UTR, the consensus signal required for nuclear polyadenylation, AAUAAA, and, upstream from it, a more variable U-rich element, initially termed a cytoplasmic polyadenylation element, or CPE [315]. *Xenopus* mRNAs with these features include a group (G10, B4, D7) selected on the basis that they moved into polysomes during meiotic maturation [318–320], together with cyclins [321], cdk2 [322] and c-*mos*, whose translational activation is an important enabling step in maturation [321, 323]. Examples in mouse oocytes include the mRNAs encoding tissue-type plasminogen activator (tPA) [313] and, again, c-*mos* [324]. In all these cases deletion of the consensus elements prevents both poly(A) elongation and translational recruitment. Interesting insight into the function of these elements came from studies of the tPA mRNA in mouse oocytes. This mRNA receives a long (>300 nucleotides) poly(A) tail in the nucleus, which is immediately cut to <60 nucleotides upon export into the cytoplasm. The deadenylation, as well as the subsequent readenylation during meiotic maturation, is dependent on the U-rich CPE in the 3′ UTR [325]. This dual function has led to some groups re-naming this sequence as the adenylation control element, or ACE.

Within the general association between the possession of CPE/ACE elements and the ability to undergo poly(A) tail elongation during early development, the programming conferred by these sequences on an individual level is extremely subtle. Comparison of four such mRNAs in *Xenopus* oocytes during maturation showed each to have a different pattern of behaviour, both in the timing and the extent of polyadenylation during maturation [321]. Other mRNAs only undergo polyadenylation and translational activation during embryogenesis. For two of these, C11 and C12 mRNAs in *Xenopus*, manipulation of the distance between the CPE/ACE and the downstream AAUAAA elements had profound effects on both the timing and extent of polyadenylation [326, 327]. In the same mRNAs a much larger element, encompassing the CPE/ACE, was also found necessary to prevent premature polyadenylation during maturation [308, 326, 327]. A strong correlation between translational activation and poly(A) tail elongation during embryogenesis is also seen in the case of three *Drosophila* mRNAs, *bicoid*, *torso* and *toll*, although specific CPE/ACEs have yet to be delineated [328]. In *Xenopus*, some mRNAs activated during maturation subsequently undergo programmed shortening of their poly(A) tails and concomitant withdrawal from translation. This appears to require additional 3′ UTR sequences that do not overlap with the CPE [322, 329]. Proteins that interact with the CPE/ACE and AAUAAA elements clearly play a major role. The general mechanics of cytoplasmic poly(A) elongation require a poly(A) polymerase and a CPE/ACE binding protein, which are very closely related to enzymes that catalyse polyadenylation in the nucleus [314, 315, 330–332]; Richter [315] has proposed a model for their interaction on the mRNA. In addition, proteins binding CPE elements of specific mRNAs have been identified in *Xenopus* oocyte and egg extracts, and mechanisms regulating their interaction are beginning to be elucidated [314, 315, 333–335].

In contrast to mRNAs possessing CPE/ACE elements, mRNAs translated efficiently in oocytes but down-regulated in mature eggs and early embryos undergo poly(A) tail shortening after germinal vesicle breakdown, which releases a deadenylase enzyme, previously sequestered in the nucleus, and allows it access to cytoplasmic mRNAs (reviewed [308]). Deadenylation appears to be relatively non-specific and to act in default, affecting all mRNAs except those possessing a CPE in the 3′ UTR [314, 336, 337]. This would down-regulate translation of mRNAs whose products are not required during early embryogenesis.

Correlation between poly(A) tail length and translation. The strong correlation between poly(A) tail elongation and translation activation, right across the range of mRNAs and organisms investigated so far, provides powerful, but not conclusive, evidence for a causal relationship [15, 308, 314, 315]. There are, however, differences in detail between some of these examples. In several cases the increased efficiency of translation seems to depend solely on the possession of a longer poly(A) tail; thus if an mRNA transcript with an extended poly(A) tail is microinjected its translation is prematurely activated. This experiment works for a number of CPE/ACE-regulated mRNAs, e.g. B4 in *Xenopus* [338], tPA in mouse [325] and, partially, *bicoid* in *Drosophila* [328]. However, other mRNAs, e.g. G10 [333] and C12 [326] in *Xenopus*, seem to have to undergo the polyadenylation process itself to permit translational activation. The mechanisms underlying this distinction are not clear, but an effect similar to that observed for G10 and C12 mRNAs has now been observed for cellular mRNAs undergoing reactivation in somatic *Drosophila* cells following heat shock [339]. An important question still under investigation concerns the relationship between poly(A)-linked regulation and other processes required for translational activation. Translational activation of several mRNAs *in vivo* involves both liberation from sequence-specific repression and poly(A) tail elongation [328, 340], but in some cases these processes can be dissociated *in vitro* [15]. In other cases, changes in poly(A) tail length do not appear to

Table 4. Messenger RNAs with translational control elements in the 3′ UTR.

Organism/mRNA	3′ UTR element	Binding factor(s)	Function/target	References
Drosophila				
Oskar	not identified	Staufen	during oogenesis, directs movement of *oskar* mRNA to posterior pole, mediated by interaction with microtubules; may play role in preventing premature translational activation	[311, 312, 364]
	Bruno response elements (multiple)	Bruno	represses translation of *oskar* mRNA prior to localization	[311, 365]
Nanos	not identified	unidentified repressor throughout embryo; oskar and vasa, localized to posterior pole	localization of *nanos* mRNA to posterior pole during oogenesis; translational activation at fertilization, restricted to *nanos* mRNA localized to posterior pole, where it is protected from a repressor that prevents its translation elsewhere in the embryo; vasa is a homologue of eIF4A and has RNA helicase activity, but its function is unknown	[311, 314, 342]
Hunchback	Nanos response elements (2) (NREs)	pumilio; unidentified 55-kDa protein	represses translation of mRNA at posterior pole, leading to preferential expression at anterior pole; binding of pumilio and the 55-kDa protein to NREs thought to provide a landing pad for nanos	[311, 314, 366]
Bicoid	three regions predicted to form stem loops; BLE1 localization signal	Staufen	Staufen binds to *bicoid* mRNA to form particles that undergo microtubule-dependent localization to anterior pole after eggs laid; anchors mRNA at anterior pole; two copies of BLE1 element required for localization but not involved in anchoring; poly(A) elongation closely associated with translational activation (see text)	[364]
Cyclin B	two closely located segments of 94 and 87 nucleotides	not identified	required for localization to posterior pole during late oogenesis	[367]
	39-nucleotide region containing an NRE-like sequence	not identified	repression of translation until posterior pole cells commence proliferation during embryogenesis	[367]
C. elegans				
tra-2	two direct repeat elements (DREs)	DRE-binding factor	*tra-2* gene directs feminine development; repression of *tra-2* mRNA translation promotes spermatogenesis at one end of gonad	[314, 368]
fem-3	regulatory element with five critical nucleotides	unidentified binding activity in *C. elegans* extracts	represses masculine development; hence promotes oogenesis in gonad cells	[314]
glp-1	temporal control region within 3′ terminal 125 nucleotides	not identified	translation repressed until 2−4 cell stage	[369]
	61-nucleotide spatial control region containing NRE-like elements	not identified	translation confined to anterior blastomeres	[369]
lin-14	seven conserved sequences	two small, untranslated RNAs, 22 and 61 nucleotides (products of *lin-4*)	RNA products of *lin-4* have anti-sense complementarity to the seven sequences in *lin-14* mRNA; binding represses translation at later larval stages either directly or by creating site that binds (unknown) protein	[314, 370, 371]
Surf clam				
ribonucleotide reductase; cyclin A	masking boxes (≈ 130 nucleotides)	82-kDa binding protein	prevents translation until maturation; probably regulated by phosphorylation	[307, 372]
Mammalian erythroid cells				
lipoxygenase (LOX)	19-nucleotide sequence (10 repeats in rabbit, 4 in mouse and human)	48-kDa protein	represses translation until reticulocyte stage. LOX protein initiates breakdown of mitochondria during terminal differentiation to erythrocytes	[341]

be involved at all [341, 342]. On the other hand, as mentioned above, elongation of poly(A) tracts on some mRNAs is alone sufficient to increase translational activity in *Xenopus* and mouse oocytes. This complex question is considered in more detail in recent reviews [15, 306, 314].

General role of poly(A) and PABP in translation

Even before the studies on developmental regulation discussed above there were many reports in the literature that the poly(A) tail contributed to the translational efficiency of mRNA molecules (reviewed [114, 316]). The effects are often small in *in vitro* translation systems, but more pronounced differences have been seen in experiments where mRNAs were microinjected into *Xenopus* oocytes [343] or electroporated into cells [344] (see, however, [345] which describes rather different behaviour in yeast). In the reticulocyte lysate, a rather small increment in efficiency of recruitment of poly(A)-bearing mRNA was localized to the joining of the 60S ribosomal subunit to the 43S · mRNA complex [346], consistent with the observation that the effects of mutating the yeast poly(A)-binding protein gene were suppressed by a mutation in a structural protein of the 60S subunit [347]. Addition of exogenous poly(A) inhibits the activity of cell-free translation systems, probably by sequestering the PABP away from the poly(A) tails of the mRNAs [306, 316, 348]. Increasing attention is now being given to the idea that there may be interaction between poly(A) tails and mRNA caps or 5′ UTRs. This stemmed from observations of synergistic effects of caps and poly(A) tails on translational efficiency in mRNA-electroporated cells [344] and, later, from studies on mRNA degradation mechanisms in yeast which clearly implicated communication between these two features [310]. Such interactions are likely to be indirect, and could be mediated by PABP, by initiation factors binding mRNA 5′ UTRs or by unknown proteins. One report suggests association between poly(A) and initiation factors of the eIF4 group [349]. Various forms of a closed loop model, bringing the poly(A) tail close to the 5′ end of the mRNA, are considered in a recent review by Jacobson [316].

CONCLUDING REMARKS

I will finish this article with a summary of topics where I feel that interesting developments are imminent or where there are outstanding questions requiring elucidation.

Functions and molecular interactions involving initiation factors. There is still much to be learned here, even concerning relatively well-characterized factors like eIF2. The recent cloning and sequencing of cDNAs encoding the five subunits of eIF2B should facilitate elucidation of the individual roles of these polypeptides. Even more encouraging is the news of progress on the molecular characterization of the multiple subunits of eIF3, since lack of information on the function of this factor, clearly at the heart of the initiation process, has constituted a large hole in our understanding for so many years. For the better-known factors, a remaining outstanding question concerns the functions of the various RNA-binding domains identified in sequencing studies. How does each of these relate to the interactions of the factors with mRNA, rRNA and tRNA during initiation? In many cases RNA-binding studies have revealed little or no sequence specificity. The function of the RNA helicase activity of eIF4A is still not fully clear. This activity has always been regarded as responsible for unwinding secondary structure in the mRNA 5′ UTR. However, extracts prepared from *S. cerevisiae* strains with disrupted eIF4A genes show total dependence on exogenous eIF4A for the translation of any mRNA, whether or not the 5′ untranslated region is highly structured [350]. The enhanced unwinding activity of this factor *in vitro* when presented as part of an eIF4F complex [66] is probably explained by the ability of eIF4G to bring it into proximity to the RNA, either by direct eIF4G · RNA association or by binding to cap-associated eIF4E [34]. The significance of the 3′−5′ RNA duplex unwinding activity of eIF4F, demonstrated *in vitro* with uncapped RNA substrates [66], needs to be elucidated: possible functions could be in internal initiation or in transient melting of structures in exposed regions of rRNA involved in ribosome−mRNA interactions during initiation. For eIF4B, long regarded on the basis of *in vitro* assays as a mere facilitator of eIF4A-catalysed RNA unwinding, a whole new area of interest is opened up by a recent paper demonstrating duplex annealing activity of this factor [351]. It should be remembered, however, that a great deal of the experimental evidence currently available on the function of the individual initiation factors, particularly those of the eIF4 group, is derived from *in vitro* assays performed in the absence of other components, such as ribosomal subunits and eIF3, with which they would be associated in the intact cell; caution is therefore needed in extrapolating these data to the situation *in vivo*. Finally, as illustrated in Fig. 2, there is still much to be learned about the exact sequence of association and dissociation events involving initiation complexes.

Role of initiation factor phosphorylation in regulation. Many factors exhibit regulated phosphorylation, but for only one (eIF2α) has a clear role for this been established. For eIF4E, apart from one study [178], evidence indicating a functional role for phosphorylation remains entirely correlative. Work is needed to assess the relative importance of phosphorylation and PHAS-I-mediated sequestration in the regulation of this factor in response to physiological signals. The absence of an appropriate C-terminal phosphorylation site in *S. cerevisiae* eIF4E raises interesting questions, and information on this factor from other invertebrate species is needed. Does the importance of eIF4E phosphorylation in different organisms relate to the incidence of stable secondary structure in mRNA 5′ UTRs? It should be noted that regulation of eIF4E phosphorylation during early development of marine invertebrates shows strong resemblance to that seen in *Xenopus* [176, 295], though sequence information on phosphorylation sites is not yet available. Returning to yeast, it is perhaps of interest that *S. cerevisiae* homologues of ribosomal protein S6, and possibly eIF4B, also lack potential phosphorylation sites present at the C-termini of the mammalian proteins.

Translational control and growth regulation. Further work is needed to establish whether the apparent roles of eIF4E and PKR in the regulation of tumourigenesis are related to their functions in translation or whether the presence of these proteins in the cell nucleus indicates dual function. In the case of eIF4E one might ask whether one role for the PHAS-I-mediated sequestration mechanism is to regulate the amount of the protein available to enter the nucleus. For PKR, work on the effects of expression of dominant negative mutants could be criticized for having been limited so far to NIH-3T3 cells, and it is now essential to investigate the significance of the recent report that PKR-knockout mice do not exhibit a tumourigenic phenotype [161]. Another important area in which research will continue is the role of both general initiation factors and specific mRNA-binding proteins in controlling the translation of mRNAs encoding growth-regulatory proteins. Recent hints of the involvement of internal initiation in the translation of such mRNAs [276, 280] may also lead to interesting developments.

Cross-talk between the 5′ and 3′ ends of mRNA and spatial relationships within cells. An exciting aspect of recent work on the translational control of individual mRNAs has been the recognition of the role of regulatory elements at the 3′ end, including the poly(A) tail. Together with related work on the control of mRNA degradation [352], this leads to the inescapable conclusion that the two ends of mRNA interact, either directly or indirectly. Further, studies on mRNA localization during early development clearly implicate interactions of mRNAs and associated regulatory proteins with cytoskeletal components [314, 353, 354]. It is clear that future work on translational control both in embryos and in somatic cells will have to focus more and more on how attachment to cytoskeletal networks [355−359] and involvement in other macromolecular complexes [360] can orientate the interactions between mRNAs, ribosomes, tRNAs and initiation factors within cells.

Thanks are particularly due to Mike Clemens and to my lab colleagues Simon Morley, Mike Rau and Theo Ohlmann for many stimulating discussions, for keeping me aware of important publications, for allowing me to ransack their reprint collections and for thorough reading of the manuscript. Mike Rau produced the figures, and he and Simon Morley made major contributions to their design. I wish to thank numerous people who sent me articles and told me of data prior to publication, and who responded to various queries. Work from our laboratory was funded by a grant from the Wellcome Trust (034710/Z/91/Z/1.5).

REFERENCES

1. Merrick, W. C. & Hershey, J. W. B. (1996) in *Translational control* (Hershey, J. W. B., Mathews, M. B. & Sonenberg, N., eds) pp. 31−69, Cold Spring Harbor Laboratory Press, Cold Spring Harbor NY.
2. Pain, V. M. (1986) *Biochem. J. 235*, 625−637.
3. Hershey, J. W. B. (1991) *Annu. Rev. Biochem. 60*, 717−755.
4. Merrick, W. C. (1992) *Microbiol. Rev. 56*, 291−315.
5. Altmann, M. & Trachsel, H. (1993) *Trends Biochem. Sci. 18*, 429−432.
6. Altmann, M. & Trachsel, H. (1994) *Biochimie 76*, 853−861.
7. Hershey, J. W. B. (1989) *J. Biol. Chem. 264*, 20823−20826.
8. Rhoads, R. E. (1993) *J. Biol. Chem. 268*, 3017−3020.
9. Gallie, D. R. (1993) *Annu. Rev. Plant Physiol. Plant Mol. Biol. 44*, 77−105.
10. Kozak, M. (1992) *Annu. Rev. Cell Biol. 8*, 197−225.
11. Kozak, M. (1992) *Crit. Rev. Biochem. Mol. Biol. 27*, 385−402.
12. Sonenberg, N. (1994) *Curr. Opin. Genet. Dev. 4*, 310−315.
13. Hentze, M. W. (1995) *Curr. Opin. Cell Biol. 7*, 393−398.
14. Gray, N. K. & Hentze, M. W. (1994) *Mol. Biol. Rep. 19*, 195−200.
15. Standart, N. & Jackson, R. J. (1994) *Biochimie 76*, 867−879.
16. McCarthy, J. E. G. & Kollmus, H. (1995) *Trends Biochem. Sci. 20*, 191−197.

16a. Browning, K. S. (1996) *Plant Mol. Biol.*, in the press.

17. Trachsel, H. (1996) in *Translational control* (Hershey, J. W. B., Mathews, M. B. & Sonenberg, N., eds) pp. 113−138, Cold Spring Harbor Laboratory Press, Cold Spring Harbor NY.
18. Gaspar, N. J., Kinzy, T. G., Scherer, B. J., Hümbelin, M., Hershey, J. W. B. & Merrick, W. C. (1994) *J. Biol. Chem. 269*, 3415−3422.
19. Hannig, E. M., Cigan, A. M., Freeman, B. A. & Kinzy, T. G. (1993) *Mol. Cell. Biol. 13*, 506−520.
20. Donahue, T. F., Cigan, A. M., Pabich, E. K. & Valavicius, B. C. (1988) *Cell 54*, 621−632.
21. Ye, X. & Cavener, D. R. (1994) *Gene 142*, 271−274.
22. Pathak, V. K., Nielsen, P. J., Trachsel, H. & Hershey, J. W. B. (1988) *Cell 54*, 633−639.
23. Proud, C. G. (1992) *Curr. Top. Cell Regul. 32*, 243−368.
24. Flynn, A., Shatsky, I. N., Proud, C. G. & Kaminski, A. (1994) *Biochim. Biophys. Acta Gene Struct. Expression 1219*, 293−301.
25. Dorris, D. R., Erickson, F. L. & Hannig, E. M. (1995) *EMBO J. 14*, 2239−2249.
26. Cigan, A. M., Pabich, E. K., Feng, L. & Donahue, T. F. (1989) *Proc. Natl Acad. Sci. USA 86*, 2784−2788.
27. Dever, T. E., Wei, C.-L., Benkowski, L. A., Browning, K., Merrick, W. C. & Hershey, J. W. B. (1994) *J. Biol. Chem. 269*, 3212−3218.

27a. Wei, C.-L., Kainuma, M. & Hershey, J. W. B. (1995) *J. Biol. Chem. 270*, 22788−22794.

28. Wei, C.-L., MacMillan, S. E. & Hershey, J. W. B. (1995) *J. Biol. Chem. 270*, 5764−5771.
29. Danaie, P., Wittmer, B., Altmann, M. & Trachsel, H. (1995) *J. Biol. Chem. 270*, 4288−4292.
30. Naranda, T., MacMillan, S. E. & Hershey, J. W. B. (1994) *J. Biol. Chem. 269*, 32286−32292.
31. Garcia-Barrio, M. T., Naranda, T., Vazquez de Aldana, C. R., Cuesta, R., Hinnebusch, A. G., Hershey, J. W. B. & Tamame, M. (1995) *Genes & Dev. 9*, 1781−1796.
32. Kasperaitis, M. A. M., Voorma, H. O. & Thomas, A. A. M. (1995) *FEBS Lett. 365*, 47−50.
33. Evans, D. R. H., Rasmussen, C., Hanic-Joyce, P. J., Johnston, G. C., Singer, R. A. & Barnes, C. A. (1995) *Mol. Cell. Biol. 15*, 4525−4535.
34. Lamphear, B. J., Kirchweger, R., Skern, T. & Rhoads, R. E. (1995) *J. Biol. Chem. 270*, 21975−21983.
35. Bommer, U. A., Lutsch, G., Stahl, J. & Bielka, H. (1991) *Biochimie 73*, 1007−1019.
36. Price, N. & Proud, C. (1994) *Biochimie 76*, 748−760.
37. Dholakia, J. N., Francis, B. R., Haley, B. E. & Wahba, A. J. (1989) *J. Biol. Chem. 264*, 20638−20642.
38. Price, N. T., Francia, G., Hall, L. & Proud, C. G. (1994) *Biochim. Biophys. Acta Gene Struct. Expression 1217*, 207−210.
39. Flowers, K. M., Kimball, S. R., Feldhoff, R. C., Hinnebusch, A. G. & Jefferson, L. S. (1995) *Proc. Natl Acad. Sci. USA 92*, 4274−4278.
40. Bushman, J. L., Asuru, A. I., Matts, R. L. & Hinnebusch, A. G. (1993) *Mol. Cell. Biol. 13*, 1920−1932.
41. Craddock, B. L., Price, N. T. & Proud, C. G. (1995) *Biochem. J. 309*, 1009−1014.
42. Oldfield, S., Jones, B. L., Tanton, D. & Proud, C. G. (1994) *Eur. J. Biochem. 221*, 399−410.
43. Kozak, M. (1991) *J. Biol. Chem. 266*, 19867−19870.
44. Kozak, M. (1989) *Mol. Cell. Biol. 9*, 5134−5142.
45. Pause, A. & Sonenberg, N. (1993) *Curr. Opin. Struct. Biol. 3*, 953−959.
46. Merrick, W. C. (1994) *Biochimie 76*, 822−830.
47. Rhoads, R. E., Joshi, B. & Minich, W. B. (1994) *Biochimie 76*, 831−838.
48. Morley, S. J. (1996) in *Protein phosphorylation in cell growth regulation* (Clemens, M. J., ed.) Harwood Academic Publishers, Amsterdam, in the press.
49. Linder, P. (1992) *Antonie Van Leeuwenhoek 62*, 47−62.
50. Hernández, G. & Sierra, J. M. (1995) *Biochim. Biophys. Acta Gene Struct. Expression 1261*, 427−431.
51. Altmann, M. & Trachsel, H. (1989) *Nucleic Acids Res. 17*, 5923−5931.
52. Altmann, M., Edery, I., Trachsel, H. & Sonenberg, N. (1988) *J. Biol. Chem. 263*, 17229−17232.
53. Altmann, M., Muller, P. P., Pelletier, J., Sonenberg, N. & Trachsel, H. (1989) *J. Biol. Chem. 264*, 12145−12147.
54. Milburn, S. C., Hershey, J. W. B., Davies, M. V., Kelleher, K. & Kaufman, R. J. (1990) *EMBO J. 9*, 2783−2790.
55. Naranda, T., Strong, W. B., Menaya, J., Fabbri, B. J. & Hershey, J. W. B. (1994) *J. Biol. Chem. 269*, 14465−14472.
56. Coppolecchia, R., Buser, P., Stotz, A. & Linder, P. (1993) *EMBO J. 12*, 4005−4011.
57. Altmann, M., Müller, P. P., Wittmer, B., Ruchti, F., Lanker, S. & Trachsel, H. (1993) *EMBO J. 12*, 3997−4003.
58. Méthot, N., Pause, A., Hershey, J. W. B. & Sonenberg, N. (1994) *Mol. Cell. Biol. 14*, 2307−2316.

58a. Méthot, N., Pickett, G., Keene, J. D. & Sonenberg, N. (1996) *RNA*, in the press.
59. Lanker, S., Müller, P. P., Altmann, M., Goyer, C., Sonenberg, N. & Trachsel, H. (1992) *J. Biol. Chem. 267*, 21167–21171.
60. Allen, M. L., Metz, A. M., Timmer, R. T., Rhoads, R. E. & Browning, K. S. (1992) *J. Biol. Chem. 267*, 23232–23236.
61. Zapata, J. M., Martínez, M. A. & Sierra, J. M. (1994) *J. Biol. Chem. 269*, 18047–18052.
62. Lamphear, B. J. & Panniers, R. (1990) *J. Biol. Chem. 265*, 5333–5336.
63. Mader, S., Lee, H., Pause, A. & Sonenberg, N. (1995) *Mol. Cell. Biol. 15*, 4990–4997.
64. Yan, R., Rychlik, W., Etchison, D. & Rhoads, R. E. (1992) *J. Biol. Chem. 267*, 23226–23231.
65. Joshi, B., Yan, R. & Rhoads, R. E. (1994) *J. Biol. Chem. 269*, 2048–2055.
66. Rozen, F., Edery, I., Meerovitch, K., Dever, T. E., Merrick, W. C. & Sonenberg, N. (1990) *Mol. Cell. Biol. 10*, 1134–1144.
67. Sonenberg, N. (1994) *Biochimie 76*, 839–846.
68. Thach, R. E. (1992) *Cell 68*, 177–180.
69. Pause, A., Méthot, N., Svitkin, Y., Merrick, W. C. & Sonenberg, N. (1994) *EMBO J. 13*, 1205–1215.
70. Kozak, M. (1989) *J. Cell Biol. 108*, 229–241.
71. Kozak, M. (1994) *Biochimie 76*, 815–821.
72. Jackson, R. J. (1991) *Eur. J. Biochem. 200*, 285–294.
73. Cigan, A. M. & Donahue, T. F. (1987) *Gene 59*, 1–18.
74. Grünert, S. & Jackson, R. J. (1994) *EMBO J. 13*, 3618–3630.
75. Boeck, R. & Kolakofsky, D. (1994) *EMBO J. 13*, 3608–3617.
76. Kozak, M. (1991) *J. Cell Biol. 115*, 887–903.
77. Geballe, A. P. & Morris, D. R. (1994) *Trends Biochem. Sci. 19*, 159–164.
78. Geballe, A. P. (1996) in *Translational control* (Hershey, J. W. B., Mathews, M. B. & Sonenberg, N., eds) Cold Spring Harbor Laboratory Press, Cold Spring Harbor NY, in the press.
79. Hill, J. R. & Morris, D. R. (1993) *J. Biol. Chem. 268*, 726–731.
80. Grant, C. M. & Hinnebusch, A. G. (1994) *Mol. Cell. Biol. 14*, 606–618.
81. Grant, C. M., Miller, P. F. & Hinnebusch, A. G. (1994) *Mol. Cell. Biol. 14*, 2616–2628.
82. Luukkonen, B. G. M., Tan, W. & Schwartz, S. (1995) *J. Virol. 69*, 4086–4094.
83. Hinnebusch, A. G. (1993) *Mol. Microbiol. 10*, 215–223.
84. Chen, C. & Sarnow, P. (1995) *Science 268*, 415–417.
85. Jackson, R. J., Hunt, S. L., Gibbs, C. L. & Kaminski, A. (1994) *Mol. Biol. Rep. 19*, 147–159.
86. Jackson, R. J., Howell, M. T. & Kaminski, A. (1990) *Trends Biochem. Sci. 15*, 477–483.
87. Jackson, R. J., Hunt, S. L., Reynolds, J. E. & Kaminski, A. (1995) *Curr. Top. Microbiol. Immunol. 203*, 1–29.
88. Jackson, R. J. (1996) in *Translational control* (Hershey, J. W. B., Mathews, M. B. & Sonenberg, N., eds) 71–112, Cold Spring Harbor Laboratory Press, Cold Spring Harbor NY.
89. Ehrenfeld, E. (1996) in *Translational control* (Hershey, J. W. B., Mathews, M. B. & Sonenberg, N., eds) pp. 549–573, Cold Spring Harbor Laboratory Press, Cold Spring Harbor NY.
90. OH, S.-K. & Sarnow, P. (1993) *Curr. Opin. Genet. Dev. 3*, 295–300.
91. McBratney, S., Chen, C. Y. & Sarnow, P. (1993) *Curr. Opin. Cell Biol. 5*, 961–965.
92. Pelletier, J. & Sonenberg, N. (1988) *Nature 334*, 320–325.
93. Belsham, G. J. (1992) *EMBO J. 11*, 1105–1110.
94. Cao, X., Bergmann, I. E., Füllkrug, R. & Beck, E. (1995) *J. Virol. 69*, 560–563.
95. Hellen, C. U. T., Pestova, T. V. & Wimmer, E. (1994) *J. Virol. 68*, 6312–6322.
96. Poole, T. L., Wang, C., Popp, R. A., Potgieter, L. N. D., Siddiqui, A. & Collett, M. S. (1995) *Virology 206*, 750–754.
97. Witherell, G. W. & Wimmer, E. (1994) *J. Virol. 68*, 3183–3192.
98. Haller, A. A. & Semler, B. L. (1995) *Virology 206*, 923–934.
99. Blyn, L. B., Chen, R., Semler, B. L. & Ehrenfeld, E. E. (1995) *J. Virol. 69*, 4381–4389.
100. Kaminski, A., Belsham, G. J. & Jackson, R. J. (1994) *EMBO J. 13*, 1673–1681.
101. Scheper, G. C., Voorma, H. O. & Thomas, A. A. M. (1994) *FEBS Lett. 352*, 271–275.
102. Borman, A., Howell, M. T., Patton, J. G. & Jackson, R. J. (1993) *J. Gen. Virol. 74*, 1775–1788.
103. Hellen, C. U. T., Pestova, T. V., Litterst, M. & Wimmer, E. (1994) *J. Virol. 68*, 941–950.
104. Das, S., Coward, P. & Dasgupta, A. (1994) *J. Virol. 68*, 7200–7211.
105. Meerovitch, K., Svitkin, Y. V., Lee, H. S., Lejbkowicz, F., Kenan, D. J., Chan, E. K. L., Agol, V. I., Keene, J. D. & Sonenberg, N. (1993) *J. Virol. 67*, 3798–3807.
106. Svitkin, Y. V., Meerovitch, K., Lee, H. S., Dholakia, J. N., Kenan, D. J., Agol, V. I. & Sonenberg, N. (1994) *J. Virol. 68*, 1544–1550.
107. Xiao, Q., Sharp, T. V., Jeffrey, I. W., James, M. C., Pruijn, G. J. M., Van Venrooij, W. J. & Clemens, M. J. (1994) *Nucleic Acids Res. 22*, 2512–2518.
108. Das, K., Chevesich, J. & Maitra, U. (1993) *Proc. Natl Acad. Sci. USA 90*, 3058–3062.
109. Chakravarti, D. & Maitra, U. (1993) *J. Biol. Chem. 268*, 10524–10533.
110. Chevesich, J., Chaudhuri, J. & Maitra, U. (1993) *J. Biol. Chem. 268*, 20659–20667.
111. Chakravarti, D., Maiti, T. & Maitra, U. (1993) *J. Biol. Chem. 268*, 5754–5762.
112. Chaudhuri, J., Das, K. & Maitra, U. (1994) *Biochemistry 33*, 4794–4799.
113. Anthony, D. D. & Merrick, W. C. (1992) *J. Biol. Chem. 267*, 1554–1562.
114. Munroe, D. & Jacobson, A. (1990) *Gene 91*, 151–158.
115. Pain, V. M. & Clemens, M. J. (1991) in *Translation in eukaryotes* (Trachsel, H., ed.) pp. 293–324, CRC Press, Boca Raton FL.
116. Samuel, C. E. (1993) *J. Biol. Chem. 268*, 7603–7606.
117. Pain, V. M. (1994) *Biochimie 76*, 718–728.
118. Panniers, R. (1994) *Biochimie 76*, 737–747.
119. Kimball, S. R. & Jefferson, L. S. (1994) *Biochimie 76*, 729–736.
120. Redpath, N. T. & Proud, C. G. (1994) *Biochim. Biophys. Acta Mol. Cell Res. 1220*, 147–162.
121. Duncan, R. F. (1996) in *Translational control* (Hershey, J. W. B., Mathews, M. B. & Sonenberg, N., eds) pp. 271–293, Cold Spring Harbor Laboratory Press, Cold Spring Harbor NY.
122. Chen, J.-J. (1993) in *Translational regulation of gene expression 2* (Ilan, J., ed.) pp. 349–372, Plenum Press, New York.
123. Chen, J.-J. & London, I. M. (1995) *Trends Biochem. Sci. 20*, 105–108.
124. Wek, R. C. (1994) *Trends Biochem. Sci. 19*, 491–496.
125. Clemens, M. J. (1996) in *Translational control* (Hershey, J. W. B., Mathews, M. B. & Sonenberg, N., eds) pp. 139–172, Cold Spring Harbor Laboratory Press, Cold Spring Harbor NY.
126. Chen, J. J., Crosby, J. S. & London, I. M. (1994) *Biochimie 76*, 761–769.
127. Jackson, R. J. (1991) in *Translation in eukaryotes* (Trachsel, H., eds) pp. 193–229, CRC Press, Boca Raton FL.
128. Clemens, M. J. (1994) *Mol. Biol. Rep. 19*, 201–210.
129. Matts, R. L., Xu, Z., Pal, J. K. & Chen, J.-J. (1992) *J. Biol. Chem. 267*, 18160–18167.
130. Méndez, R. & De Haro, C. (1994) *J. Biol. Chem. 269*, 6170–6176.
131. Gross, M., Olin, A., Hessefort, S. & Bender, S. (1994) *J. Biol. Chem. 269*, 22738–22748.
132. Chakraborty, A., Saha, D., Bose, A., Chatterjee, M. & Gupta, N. K. (1994) *Biochemistry 33*, 6700–6706.
133. Crosby, J. S., Lee, K., London, I. M. & Chen, J.-J. (1994) *Mol. Cell. Biol. 14*, 3906–3914.
134. Mellor, H., Flowers, K. M., Kimball, S. R. & Jefferson, L. S. (1994) *J. Biol. Chem. 269*, 10201–10204.
135. Proud, C. G. (1995) *Trends Biochem. Sci. 20*, 241–246.
136. Williams, B. R. G. (1995) *Semin. Virol. 6*, 191–202.
137. Green, S. R., Manche, L. & Mathews, M. B. (1995) *Mol. Cell. Biol. 15*, 358–364.
138. Romano, P. R., Green, S. R., Barber, G. N., Mathews, M. B. & Hinnebusch, A. G. (1995) *Mol. Cell. Biol. 15*, 365–378.
139. Barber, G. N., Wambach, M., Thompson, S., Jagus, R. & Katze, M. G. (1995) *Mol. Cell. Biol. 15*, 3138–3146.

140. Barber, G. N., Jagus, R., Meurs, E. F., Hovanessian, A. G. & Katze, M. G. (1995) *J. Biol. Chem. 270*, 17423–17428.
140a. Cosentino, G. P., Venkatesan, S., Serluca, F. C., Green, S. R., Mathews, M. B. & Sonenberg, N. (1995) *Proc. Natl Acad. Sci. USA 92*, 9445–9449.
141. Sonenberg, N. (1990) *New Biologist 2*, 402–409.
142. Katze, M. G. (1992) *J. Interferon Res. 12*, 241–248.
143. Mathews, M. B. & Shenk, T. (1991) *J. Virol. 65*, 5657–5662.
144. Clemens, M. J., Laing, K. G., Jeffrey, I. W., Schofield, A., Sharp, T. V., Elia, A., Matys, V., James, M. C. & Tilleray, V. J. (1994) *Biochimie 76*, 770–778.
145. Jagus, R. & Gray, M. M. (1994) *Biochimie 76*, 779–791.
146. Katze, M. G. (1996) in *Translational control* (Hershey, J. W. B., Mathews, M. B. & Sonenberg, N., eds) pp. 607–630, Cold Spring Harbor Laboratory Press, Cold Spring Harbor NY.
146a. Chou, J., Chen, J.-J., Gross, M. & Roizman, B. (1995) *Proc. Natl Acad. Sci. USA 92*, 10516–10520.
147. Katze, M. G. (1995) *Trends Microbiol. 3*, 75–78.
148. Prostko, C. R., Dholakia, J. N., Brostrom, M. A. & Brostrom, C. O. (1995) *J. Biol. Chem. 270*, 6211–6215.
149. Srivastava, S. P., Davies, M. V. & Kaufman, R. J. (1995) *J. Biol. Chem. 270*, 16619–16624.
150. Judware, R. & Petryshyn, R. (1992) *J. Biol. Chem. 267*, 21685–21690.
151. Chong, K. L., Feng, L., Schappert, K., Meurs, E., Donahue, T. F., Friesen, J. D., Hovanessian, A. G. & Williams, B. R. G. (1992) *EMBO J. 11*, 1553–1562.
152. Meurs, E. F., Galabru, J., Barber, G., Katze, M. G. & Hovanessian, A. G. (1993) *Proc. Natl Acad. Sci. USA 90*, 232–236.
153. Sharp, T. V., Xiao, Q., Jeffrey, I., Gewert, D. R. & Clemens, M. J. (1993) *Eur. J. Biochem. 214*, 945–948.
154. Koromilas, A. E., Roy, S., Barber, G. N., Katze, M. G. & Sonenberg, N. (1992) *Science 257*, 1685–1689.
155. Barber, G. N., Thompson, S., Lee, T. G., Strom, T., Jagus, R., Darveau, A. & Katze, M. G. (1994) *Proc. Natl Acad. Sci. USA 91*, 4278–4282.
156. Donze, O., Jagus, R., Koromilas, A. E., Hershey, J. W. B. & Sonenberg, N. (1995) *EMBO J. 14*, 3828–3834.
157. Ito, T., Jagus, R. & May, W. S. (1994) *Proc. Natl Acad. Sci. USA 91*, 7455–7459.
158. Mundschau, L. J. & Faller, D. V. (1995) *J. Biol. Chem. 270*, 3100–3106.
159. Kumar, A., Haque, J., Lacoste, J., Hiscott, J. & Williams, B. R. G. (1994) *Proc. Natl Acad. Sci. USA 91*, 6288–6292.
160. Maran, A., Maitra, R. K., Kumar, A., Dong, B., Xiao, W., Li, G., Williams, B. R. G., Torrence, P. F. & Silverman, R. H. (1994) *Science 265*, 789–792.
161. Yang, Y.-L., Reis, L. F. L., Pavlovic, J., Aguzzi, A., Schafer, R., Kumar, A., Williams, B. R. G., Aguet, M. & Weissman, C. (1996) *EMBO J.*, in the press.
162. Hinnebusch, A. G. (1996) in *Translational control* (Hershey, J. W. B., Mathews, M. B. & Sonenberg, N., eds) pp. 199–244, Cold Spring Harbor Laboratory Press, Cold Spring Harbor NY.
163. Wek, S. A., Zhu, S. & Wek, R. C. (1995) *Mol. Cell. Biol. 15*, 4497–4506.
164. Vazquez de Aldana, C. R., Marton, M. J. & Hinnebusch, A. G. (1995) *EMBO J. 14*, 3184–3199.
165. Wek, R. C., Cannon, J. F., Dever, T. E. & Hinnebusch, A. G. (1992) *Mol. Cell. Biol. 12*, 5700–5710.
166. Proud, C. G. (1994) *Nature 371*, 747–748.
167. Denton, R. M. & Tavaré, J. M. (1995) *Eur. J. Biochem. 227*, 597–611.
168. Welsh, G. I. & Proud, C. G. (1992) *Biochem. J. 284*, 19–23.
169. Welsh, G. I. & Proud, C. G. (1993) *Biochem. J. 294*, 625–629.
170. Duncan, R. F., Milburn, S. C. & Hershey, J. W. B. (1987) *J. Biol. Chem. 262*, 380–388.
171. Hiremath, L., Webb, N. R. & Rhoads, R. E. (1985) *J. Biol. Chem. 260*, 7843–7849.
172. Morley, S. J. (1994) *Mol. Biol. Rep. 19*, 221–231.
173. Sierra, J. M. & Zapata, J. M. (1994) *Mol. Biol. Rep. 19*, 211–220.
174. Joshi, B., Cai, A.-L., Keiper, B. D., Minich, W. B., Mendez, R., Beach, C. M., Stepinski, J., Stolarski, R., Darzynkiewicz, E. & Rhoads, R. E. (1995) *J. Biol. Chem. 270*, 14597–14603.
175. Flynn, A. & Proud, C. G. (1995) *J. Biol. Chem. 270*, 21684–21688.
176. Morley, S. J. & Pain, V. M. (1995) *J. Cell Sci. 108*, 1751–1760.
177. Makkinje, A., Xiong, H., Li, M. & Damuni, Z. (1995) *J. Biol. Chem. 270*, 14824–14828.
178. Minich, W. B., Balasta, M. L., Goss, D. J. & Rhoads, R. E. (1994) *Proc. Natl Acad. Sci. USA 91*, 7668–7672.
179. Lamphear, B. J. & Panniers, R. (1991) *J. Biol. Chem. 266*, 2789–2794.
180. Bu, X., Haas, D. W. & Hagedorn, C. H. (1993) *J. Biol. Chem. 268*, 4975–4978.
181. Morley, S. J., Rau, M., Kay, J. E. & Pain, V. M. (1993) *Eur. J. Biochem. 218*, 39–48.
182. Joshi-Barve, S., Rychlik, W. & Rhoads, R. E. (1990) *J. Biol. Chem. 265*, 2979–2983.
182a. Rau, M., Ohlmann, T., Morley, S. J. & Pain, V. M. (1996) *J. Biol. Chem.*, in the press.
183. Manzella, J. M., Rychlik, W., Rhoads, R. E., Hershey, J. W. B. & Blackshear, P. J. (1991) *J. Biol. Chem. 266*, 2383–2389.
184. Morley, S. J. & Traugh, J. A. (1990) *J. Biol. Chem. 265*, 10611–10616.
185. Morley, S. J. & Pain, V. M. (1996) *Biochem. J.*, in the press.
186. Morley, S. J., Dever, T. E., Etchison, D. & Traugh, J. A. (1991) *J. Biol. Chem. 266*, 4669–4672.
186a. Zanchin, N. I. T. & McCarthy, J. E. G. (1995) *J. Biol. Chem. 270*, 26505–26510.
187. Duncan, R. F., Cavener, D. R. & Qu, S. (1995) *Biochemistry 34*, 2985–2997.
188. Zapata, J. M., Maroto, F. G. & Sierra, J. M. (1991) *J. Biol. Chem. 266*, 16007–16014.
189. Joshi, C. P. & Nguyen, H. T. (1995) *Nucleic Acids Res. 23*, 541–549.
190. Lazaris-Karatzas, A., Montine, K. S. & Sonenberg, N. (1990) *Nature 345*, 544–547.
191. De Benedetti, A. & Rhoads, R. E. (1990) *Proc. Natl Acad. Sci. USA 87*, 8212–8216.
192. Lazaris-Karatzas, A. & Sonenberg, N. (1992) *Mol. Cell. Biol. 12*, 1234–1238.
193. Lazaris-Karatzas, A., Smith, M. R., Frederickson, R. M., Jaramillo, M. L., Liu, Y., Kung, H. & Sonenberg, N. (1992) *Genes & Dev. 6*, 1631–1642.
194. Kozak, M. (1991) *J. Cell Biol. 115*, 887–903.
195. Koromilas, A. E., Lazaris-Karatzas, A. & Sonenberg, N. (1992) *EMBO J. 11*, 4153–4158.
196. Shantz, L. M. & Pegg, A. E. (1994) *Cancer Res. 54*, 2313–2316.
197. Auvinen, M., Paasinen, A., Andersson, C. & Holtta, E. (1992) *Nature 360*, 355–358.
198. Lejbkowicz, F., Goyer, C., Darveau, A., Neron, S., Lemieux, R. & Sonenberg, N. (1992) *Proc. Natl Acad. Sci. USA 89*, 9612–9616.
199. Lang, V., Zanchin, N. I. T., Lünsdorf, H., Tuite, M. & McCarthy, J. E. G. (1994) *J. Biol. Chem. 269*, 6117–6123.
200. De Benedetti, A., Joshi-Barve, S., Rinker-Schaeffer, C. & Rhoads, R. E. (1991) *Mol. Cell. Biol. 11*, 5435–5445.
201. Joshi-Barve, S., De Benedetti, A. & Rhoads, R. E. (1992) *J. Biol. Chem. 267*, 21038–21043.
202. Rhoads, R. E. & Lamphear, B. J. (1995) *Curr. Top. Microbiol. Immunol. 203*, 131–153.
203. Rosenwald, I. B., Rhoads, D. B., Callanan, L. D., Isselbacher, K. J. & Schmidt, E. V. (1993) *Proc. Natl Acad. Sci. USA 90*, 6175–6178.
204. Mao, X., Green, J. M., Safer, B., Lindsten, T., Frederickson, R. M., Miyamoto, S., Sonenberg, N. & Thompson, C. B. (1992) *J. Biol. Chem. 267*, 20444–20450.
205. Kerekatte, V., Smiley, K., Hu, B., Smith, A., Gelder, F. & De Benedetti, A. (1995) *Int. J. Cancer 64*, 27–31.
206. Miyagi, Y., Sugiyama, A., Asai, A., Okazaki, T., Kuchino, Y. & Kerr, S. J. (1995) *Cancer Lett. 91*, 247–252.
207. Pause, A., Belsham, G. J., Gingras, A.-C., Donzé, O., Lin, T.-A., Lawrence, J. C. Jr & Sonenberg, N. (1994) *Nature 371*, 762–767.
208. Lin, T.-A., Kong, X., Haystead, T. A. J., Pause, A., Belsham, G., Sonenberg, N. & Lawrence, J. C. Jr (1994) *Science 266*, 653–656.

208a. Haghighat, A., Mader, S., Pause, A. & Sonenberg, N. (1995) *EMBO J. 14*, 5701–5709.
209. Lin, T.-A., Kong, X., Saltiel, A. R., Blackshear, P. J. & Lawrence, J. C. (1995) *J. Biol. Chem. 270*, 18531–18538.
210. Zhang, Y., Feigenblum, D. & Schneider, R. J. (1994) *J. Virol. 68*, 7040–7050.
211. Feigenblum, D. & Schneider, R. J. (1993) *J. Virol. 67*, 3027–3035.
212. Dolph, P. J., Huang, J. & Schneider, R. J. (1990) *J. Virol. 64*, 2669–2677.
213. Zhang, Y., Dolph, P. J. & Schneider, R. J. (1989) *J. Biol. Chem. 264*, 10679–10684.
214. Garfinkel, M. S. & Katze, M. G. (1993) *J. Biol. Chem. 268*, 22223–22226.
215. De la Luna, S., Fortes, P., Beloso, A. & Ortín, J. (1995) *J. Virol. 69*, 2427–2433.
216. Krausslich, H. G., Micklin, M. J., Toyoda, D., Etchison, D. & Wimmer, E. (1987) *J. Virol. 61*, 2711–2718.
217. Lloyd, R. E., Grubman, M. & Ehrenfeld, E. (1988) *J. Virol. 62*, 4216–4223.
218. Devaney, M. A., Vakharia, V. N., Lloyd, R. E., Ehrenfeld, E. & Grubman, M. J. (1988) *J. Virol. 62*, 4407–4409.
219. Wyckoff, E. E., Lloyd, R. E. & Ehrenfeld, E. (1992) *J. Virol. 66*, 2943–2951.
220. Lamphear, B. J., Yan, R., Yang, F., Waters, D., Liebig, H.-D., Klump, H., Kuechler, E., Skern, T. & Rhoads, R. E. (1993) *J. Biol. Chem. 268*, 19200–19203.
221. Kirchweger, R., Ziegler, E., Lamphear, B. J., Waters, D., Liebig, H.-D., Sommergruber, W., Sobrino, F., Hohenadl, C., Blaas, D., Rhoads, R. E. & Skern, T. (1994) *J. Virol. 68*, 5677–5684.
222. Bonneau, A.-M. & Sonenberg, N. (1987) *J. Virol. 61*, 986–991.
223. Pérez, L. & Carrasco, L. (1992) *Virology 189*, 178–186.
224. Davies, M. V., Pelletier, J., Meerovitch, K., Sonenberg, N. & Kaufman, R. J. (1991) *J. Biol. Chem. 266*, 14714–14720.
225. Yu, S., Benton, P., Bovee, M., Sessions, J. & Lloyd, R. E. (1995) *J. Virol. 69*, 247–252.
226. Buckley, B. & Ehrenfeld, E. (1987) *J. Biol. Chem. 262*, 13599–13606.
227. Hambidge, S. J. & Sarnow, P. (1992) *Proc. Natl Acad. Sci. USA 89*, 10272–10276.
228. Thomas, A. A. M., Scheper, G. C., Kleijn, M., De Boer, M. & Voorma, H. O. (1992) *Eur. J. Biochem. 207*, 471–477.
229. Liebig, H.-D., Ziegler, E., Yan, R., Hartmuth, K., Klump, H., Kowalski, H., Blaas, D., Sommergruber, W., Frasel, L., Lamphear, B., Rhoads, R., Kuechler, E. & Skern, T. (1993) *Biochemistry 32*, 7581–7588.
230. Ohlmann, T., Rau, M., Morley, S. J. & Pain, V. M. (1995) *Nucleic Acids Res. 23*, 334–340.
231. Ziegler, E., Borman, A. M., Kirchweger, R., Skern, T. & Kean, K. M. (1995) *J. Virol. 69*, 3465–3474.
232. Anthony, D. D. & Merrick, W. C. (1991) *J. Biol. Chem. 266*, 10218–10226.
233. Thomas, A. A. M., Ter Haar, E., Wellink, J. & Voorma, H. O. (1991) *J. Virol. 65*, 2953–2959.
234. Scheper, G. C., Voorma, H. O. & Thomas, A. A. M. (1992) *J. Biol. Chem. 267*, 7269–7274.
234a. Ohlmann, T., Rau, M., Pain, V. M. & Morley, S. J. (1996) *EMBO J.*, in the press.
235. Laso, M. R. V., Zhu, D., Sagliocco, F., Brown, A. J. P., Tuite, M. F. & McCarthy, J. E. G. (1993) *J. Biol. Chem. 268*, 6453–6462.
236. Berben-Bloemheuvel, G., Kasperaitis, M. A. M., Van Heugten, H., Thomas, A. A. M., Van Steeg, H. & Voorma, H. O. (1992) *Eur. J. Biochem. 208*, 581–587.
237. Darveau, A., Pelletier, J. & Sonenberg, N. (1985) *Proc. Natl Acad. Sci. USA 82*, 2315–2319.
238. Arrick, B. A., Grendell, R. L. & Griffin, L. A. (1994) *Mol. Cell. Biol. 14*, 619–628.
239. Fagan, R. J., Lazaris-Karatzas, A., Sonenberg, N. & Rozen, R. (1991) *J. Biol. Chem. 266*, 16518–16523.
240. Bommer, U.-A., Lazaris-Karatzas, A., De Benedetti, A., Nurnberg, P., Benndorf, R., Bielka, H. & Sonenberg, N. (1994) *Cell. Mol. Biol. Res. 40*, 633–641.
241. Rosenwald, I. B., Lazaris-Karatzas, A., Sonenberg, N. & Schmidt, E. V. (1993) *Mol. Cell. Biol. 13*, 7358–7363.
242. Rosenwald, I. B., Kaspar, R., Rousseau, D., Gehrke, L., Leboulch, P., Chen, J.-J., Schmidt, E. V., Sonenberg, N. & London, I. M. (1995) *J. Biol. Chem. 270*, 21176–21180.
243. Bernstein, J., Shefler, I. & Elroy-Stein, O. (1995) *J. Biol. Chem. 270*, 10559–10565.
244. Klein, P. S. & Melton, D. A. (1994) *Science 265*, 803–806.
245. Parkin, N., Darveau, A., Nicholson, R. & Sonenberg, N. (1988) *Mol. Cell. Biol. 8*, 2875–2883.
246. Lazarus, P., Parkin, N. & Sonenberg, N. (1988) *Oncogene 3*, 517–521.
247. Fu, L., Ye, R., Browder, L. W. & Johnston, R. N. (1991) *Science 251*, 807–810.
248. Bass, B. L. & Weintraub, H. (1987) *Cell 48*, 607–613.
249. Svitkin, Y. V., Pause, A. & Sonenberg, N. (1994) *J. Virol. 68*, 7001–7007.
250. Romeo, D. S., Park, K., Roberts, A. B., Sporn, M. B. & Kim, S.-J. (1993) *Mol. Endocrinol. 7*, 759–766.
251. Teerink, H., Kasperaitis, M. A. M., De Moor, C. H., Voorma, H. O. & Thomas, A. A. M. (1994) *Biochem. J. 303*, 547–553.
252. De Moor, C. H., Jansen, M., Bonte, E. J., Thomas, A. A. M., Sussenbach, J. S. & Van den Brande, J. L. (1995) *Biochem. J. 307*, 225–231.
253. Ruan, H., Hill, J. R., Fatemie-Nainie, S. & Morris, D. R. (1994) *J. Biol. Chem. 269*, 17905–17910.
254. Shantz, L. M., Viswanath, R. & Pegg, A. E. (1994) *Biochem. J. 302*, 765–772.
255. Gray, N. K. & Hentze, M. W. (1994) *EMBO J. 13*, 3882–3891.
256. Stripecke, R., Oliveira, C. C., McCarthy, J. E. G. & Hentze, M. W. (1994) *Mol. Cell. Biol. 14*, 5898–5909.
257. Melefors, O. & Hentze, M. W. (1993) *BioEssays 15*, 85–90.
258. Guo, B., Yu, Y. & Leibold, E. A. (1994) *J. Biol. Chem. 269*, 24252–24260.
259. Kim, H.-Y., Klausner, R. D. & Rouault, T. A. (1995) *J. Biol. Chem. 270*, 4983–4986.
259a Iwai, K., Klausner, R. D. & Rouault, T. A. (1995) *EMBO J. 14*, 5350–5357.
260. Drapier, J.-C., Hirling, H., Wietzerbin, J., Kaldy, P. & Kühn, L. C. (1993) *EMBO J. 12*, 3643–3649.
261. Jaffrey, S. R., Cohen, N. A., Rouault, T. A., Klausner, R. D. & Snyder, S. H. (1994) *Proc. Natl Acad. Sci. USA 91*, 12994–12998.
262. Pantopoulos, K. & Hentze, M. W. (1995) *Proc. Natl Acad. Sci. USA 92*, 1267–1271.
263. Eisenstein, R. S., Tuazon, P. T., Schalinske, K. L., Anderson, S. A. & Traugh, J. A. (1993) *J. Biol. Chem. 268*, 27363–27370.
264. Chu, E., Voeller, D., Koeller, D. M., Drake, J. C., Takimoto, C. H., Maley, G. F., Maley, F. & Allegra, C. J. (1993) *Proc. Natl Acad. Sci. USA 90*, 517–521.
265. Chu, E., Voeller, D. M., Morrison, P. F., Jones, K. L., Takechi, T., Maley, G. F., Maley, F. & Allegra, C. J. (1994) *J. Biol. Chem. 269*, 20289–20293.
266. De Melo Neto, O. P., Standart, N. & Martins de Sa, C. (1995) *Nucleic Acids Res. 23*, 2198–2205.
267. Meyuhas, O., Avni, D. & Shama, S. (1995) in *Translational control* (Hershey, J. W. B., Mathews, M. B. & Sonenberg, N., eds) pp. 363–388, Cold Spring Harbor Laboratory Press, Cold Spring Harbor NY.
268. Jefferies, H. B. J. & Thomas, G. (1994) *J. Biol. Chem. 269*, 4367–4372.
269. Hammond, M. L., Merrick, W. & Bowman, L. H. (1991) *Genes & Dev. 5*, 1723–1736.
270. Kaspar, R. L., Rychlik, W., White, M. W., Rhoads, R. E. & Morris, D. R. (1990) *J. Biol. Chem. 265*, 3619–3622.
271. Shama, S., Avni, D., Frederickson, R. M., Sonenberg, N. & Meyuhas, O. (1995) *Gene Expression 4*, 241–252.
272. Jefferies, H. B. J., Reinhard, C., Kozma, S. C. & Thomas, G. (1994) *Proc. Natl Acad. Sci. USA 91*, 4441–4445.
273. Terada, N., Patel, H. R., Takase, K., Kohno, K., Nairn, A. C. & Gelfand, E. W. (1994) *Proc. Natl Acad. Sci. USA 91*, 11477–11481.
274. Stewart, M. J. & Thomas, G. (1994) *BioEssays 16*, 809–815.
275. Macejak, D. G. & Sarnow, P. (1991) *Nature 353*, 90–94.
276. Vagner, S., Gensac, M.-C., Maret, A., Bayard, F., Amalric, F., Prats, H. & Prats, A.-C. (1995) *Mol. Cell. Biol. 15*, 35–44.

277. Berlioz, C. & Darlix, J.-L. (1995) *J. Virol. 69*, 2214–2222.
278. Vagner, S., Waysbort, A., Marends, M., Gensac, M.-C., Amalric, F. & Prats, A.-C. (1995) *J. Biol. Chem. 270*, 20376–20383.
279. OH, S.-K., Scott, M. P. & Sarnow, P. (1992) *Genes & Dev. 6*, 1643–1653.
280. Iizuka, N., Najita, L., Franzusoff, A. & Sarnow, P. (1994) *Mol. Cell. Biol. 14*, 7322–7330.
281. Kozak, M. (1991) *Gene Expression 1*, 117–125.
282. Kozak, M. (1994) *J. Mol. Biol. 235*, 95–110.
283. Munoz, A., Alonso, M. A. & Carrasco, L. (1984) *Virology 137*, 150–159.
284. Dolph, P. J., Racaniello, V., Villamarin, A., Palladino, F. & Schneider, R. (1988) *J. Virol. 62*, 2059–2066.
285. Sonenberg, N., Guertin, D. & Lee, K. A. W. (1982) *Mol. Cell. Biol. 2*, 1633–1638.
286. Song, H.-J., Gallie, D. R. & Duncan, R. F. (1996) *Eur. J. Biochem.*, in the press.
287. Hann, L. E. & Gehrke, L. (1995) *J. Virol. 69*, 4986–4993.
288. Lindquist, S. & Petersen, R. (1990) *Enzyme 44*, 147–166.
289. Hess, M. A. & Duncan, R. F. (1994) *J. Biol. Chem. 269*, 10913–10922.
290. Kruys, V. & Huez, G. (1994) *Biochimie 76*, 862–866.
291. Spirin, A. S. (1996) in *Translational control* (Hershey, J. W. B., Mathews, M. B. & Sonenberg, N., eds) pp. 319–334, Cold Spring Harbor Laboratory Press, Cold Spring Harbor NY.
292. Woodland, H. R. (1974) *Dev. Biol. 40*, 90–101.
293. Hille, M. B., Dholakia, J. N., Wahba, A., Fanning, E., Stimler, L., Xu, Z. & Yablonka-Reuveni, Z. (1990) *J. Reprod. Fert. Suppl. 42*, 235–248.
294. Jagus, R., Huang, W.-I., Hansen, L. J. & Wilson, M. A. (1992) *J. Biol. Chem. 267*, 15530–15536.
295. Xu, Z., Dholakia, J. N. & Hille, M. B. (1993) *Dev. Genet. 14*, 424–439.
296. Murray, M. T., Schiller, D. L. & Franke, W. W. (1992) *Proc. Natl Acad. Sci. USA 89*, 11–15.
297. Deschamps, S., Viel, A., Garrigos, M., Denis, H. & Le Maire, M. (1992) *J. Biol. Chem. 267*, 13799–13802.
298. Standart, N. & Jackson, R. (1994) *Curr. Biol. 4*, 939–941.
299. Standart, N. & Hunt, T. (1990) *Enzyme 44*, 106–119.
300. Tafuri, S. R. & Wolffe, A. P. (1993) *J. Biol. Chem. 268*, 24255–24261.
301. Braddock, M., Muckenthaler, M., White, M. R. H., Thorburn, A. M., Sommerville, J., Kingsman, A. J. & Kingsman, S. M. (1994) *Nucleic Acids Res. 22*, 5255–5264.
302. Bouvet, P. & Wolffe, A. P. (1994) *Cell 77*, 931–942.
303. Kick, D., Barrett, P., Cummings, A. & Sommerville, J. (1987) *Nucleic Acids Res. 15*, 4099–4109.
304. Ranjan, M., Tafuri, S. R. & Wolffe, A. P. (1993) *Genes & Dev. 7*, 1725–1736.
305. Evdokimova, V. M., Wei, C.-L., Sitikov, A. S., Simonenko, P. N., Lazarev, O. A., Vasilenko, K. S., Ustinov, V. A., Hershey, J. W. B. & Ovchinnikov, L. P. (1995) *J. Biol. Chem. 270*, 3186–3192.
306. Jackson, R. J. & Standart, N. (1990) *Cell 62*, 15–24.
307. Standart, N. (1992) *Semin. Dev. Biol. 3*, 367–379.
308. Wormington, M. (1993) *Curr. Opin. Cell Biol. 5*, 950–954.
309. Jackson, R. J. (1993) *Cell 74*, 9–14.
310. Decker, C. J. & Parker, R. (1995) *Curr. Opin. Cell Biol. 7*, 386–392.
311. Curtis, D., Lehmann, R. & Zamore, P. D. (1995) *Cell 81*, 171–178.
312. St Johnston, D. (1995) *Cell 81*, 161–170.
313. Vassalli, J.-D. & Stutz, A. (1995) *Curr. Biol. 5*, 476–479.
314. Wickens, M., Kimble, J. & Strickland, S. (1996) in *Translational control* (Hershey, J. W. B., Mathews, M. B. & Sonenberg, N., eds) pp. 411–450, Cold Spring Harbor Laboratory Press, Cold Spring Harbor NY.
315. Richter, J. D. (1996) in *Translational control* (Hershey, J. W. B., Mathews, M. B. & Sonenberg, N., eds) pp. 481–503, Cold Spring Harbor Laboratory Press, Cold Spring Harbor NY.
316. Jacobson, A. (1995) in *Translational control* (Hershey, J. W. B., Mathews, M. B. & Sonenberg, N., eds) pp. 451–480, Cold Spring Harbor Laboratory Press, Cold Spring Harbor NY.
317. Bachvarova, R. F. (1992) *Cell 69*, 895–897.
318. Dworkin, M. B., Shrutkowski, A. & Dworkin-Rastl, E. (1985) *Proc. Natl Acad. Sci. USA 82*, 7636–7640.
319. McGrew, L., Dworkin-Rastl, E., Dworkin, M. & Richter, J. D. (1989) *Genes & Dev. 3*, 803–815.
320. Fox, C., Sheets, M. & Wickens, M. (1989) *Genes & Dev. 3*, 2151–2162.
321. Sheets, M. D., Fox, C. A., Hunt, T., Vande Woude, G. & Wickens, M. (1994) *Genes & Dev. 8*, 926–938.
322. Stebbins-Boaz, B. & Richter, J. D. (1994) *Mol. Cell. Biol. 14*, 5870–5880.
323. Sheets, M. D., Wu, M. & Wickens, M. (1995) *Nature 374*, 511–516.
324. Gebauer, F., Xu, W., Cooper, G. M. & Richter, J. D. (1994) *EMBO J. 13*, 5712–5720.
325. Huarte, J., Stutz, A., O'Connell, M. L., Gubler, P., Belin, D., Darrow, A. L., Strickland, S. & Vassalli, J.-D. (1992) *Cell 69*, 1021–1030.
326. Simon, R., Tassan, J.-P. & Richter, J. D. (1992) *Genes & Dev. 6*, 2580–2591.
327. Simon, R. & Richter, J. D. (1994) *Mol. Cell. Biol. 14*, 7867–7875.
328. Sallés, F. J., Lieberfarb, M. E., Wreden, C., Gergen, J. P. & Strickland, S. (1994) *Science 266*, 1996–1999.
329. Legagneux, V., Bouvet, P., Omilli, F., Chevalier, S. & Osborne, H. B. (1992) *Development 116*, 1193–1202.
330. Bilger, A., Fox, C. A., Wahle, E. & Wickens, M. (1994) *Genes & Dev. 8*, 1106–1116.
331. Ballantyne, S., Bilger, A., Astrom, J., Virtanen, A. & Wickens, M. (1995) *RNA 1*, 64–78.
332. Gebauer, F. & Richter, J. D. (1995) *Mol. Cell. Biol. 15*, 1422–1430.
333. McGrew, L. L. & Richter, J. D. (1990) *EMBO J. 9*, 3743–3751.
334. Paris, J., Swenson, K., Piwnica-Worms, H. & Richter, J. D. (1991) *Genes & Dev. 5*, 1697–1708.
335. Hake, L. E. & Richter, J. D. (1994) *Cell 79*, 617–627.
336. Varnum, S. M. & Wormington, W. M. (1990) *Genes & Dev. 4*, 2278–2286.
337. Fox, C. A. & Wickens, M. (1990) *Genes & Dev. 4*, 2287–2298.
338. Paris, J. & Richter, J. D. (1990) *Mol. Cell. Biol. 10*, 5634–5645.
339. Duncan, R. F. (1996) *Eur. J. Biochem.*, in the press.
340. Standart, N. & Dale, M. (1993) *Dev. Genet. 14*, 492–499.
341. Ostareck-Lederer, A., Ostareck, D. H., Standart, N. & Thiele, B. J. (1994) *EMBO J. 13*, 1476–1481.
342. Gavis, E. R. & Lehmann, R. (1994) *Nature 369*, 315–318.
343. Drummond, D. R., Armstrong, J. & Colman, A. (1985) *Nucleic Acids Res. 13*, 7375–7394.
344. Gallie, D. R. (1991) *Genes & Dev. 5*, 2108–2116.
345. Proweller, A. & Butler, S. (1994) *Genes & Dev. 8*, 2629–2640.
346. Munroe, D. & Jacobson, A. (1990) *Mol. Cell. Biol. 10*, 3441–3455.
347. Sachs, A. B. & Davis, R. W. (1989) *Cell 58*, 857–867.
348. Grossi de Sa, M., Standart, N., Martins de Sa, C., Akhayat, O., Huesca, M. & Scherrer, K. (1988) *Eur. J. Biochem. 176*, 521–526.
349. Gallie, D. R. & Tanguay, R. (1994) *J. Biol. Chem. 269*, 17166–17173.
350. Altmann, M., Blum, S., Pelletier, J., Sonenberg, N., Wilson, T. M. A. & Trachsel, H. (1990) *Biochim. Biophys. Acta Gene Struct. Expression 1050*, 155–159.
351. Altmann, M., Wittmer, B., Methot, N., Sonenberg, N. & Trachsel, H. (1995) *EMBO J. 14*, 3820–3827.
352. Decker, C. J. & Parker, R. (1994) *Trends Biochem. Sci. 19*, 336–340.
353. Singer, R. H. (1992) *Curr. Opin. Cell Biol. 4*, 15–19.
354. Wilhelm, J. E. & Vale, R. D. (1993) *J. Cell Biol. 123*, 269–274.
355. Nielsen, P., Goelz, S. & Trachsel, H. (1983) *Cell Biol. Int. Rep. 7*, 245–254.
356. Hesketh, J. E. & Pryme, I. F. (1991) *Biochem. J. 277*, 1–10.
357. Hesketh, J. E. (1994) *Mol. Biol. Rep. 19*, 233–243.
358. Luby-Phelps, K. (1993) *J. Cell. Biochem. 52*, 140–147.
359. Hamill, D., Davis, J., Drawbridge, J. & Suprenant, K. A. (1994) *J. Cell Biol. 127*, 973–984.
360. Filonenko, V. V. & Deutscher, M. P. (1994) *J. Biol. Chem. 269*, 17375–17378.

361. Stewart, M. J. & Thomas, G. (1994) *BioEssays 16*, 809–815.
362. Sonenberg, N. (1993) *Curr. Opin. Cell Biol. 5*, 955–960.
363. Shepherd, P. R., Nave, B. T. & Siddle, K. (1995) *Biochem. J. 305*, 25–28.
364. Ferrandon, D., Elphick, L., Nüsslein-Volhard, C. & St Johnston, D. (1994) *Cell 79*, 1221–1232.
365. Kim-Ha, J., Kerr, K. & Macdonald, P. M. (1995) *Cell 81*, 403–412.
366. Murata, Y. & Wharton, R. P. (1995) *Cell 80*, 747–756.
367. Dalby, B. & Glover, D. M. (1993) *EMBO J. 12*, 1219–1227.
368. Goodwin, E. B., Okkema, P. G., Evans, T. C. & Kimble, J. (1993) *Cell 75*, 329–339.
369. Evans, T. C., Crittenden, S. L., Kodoyianni, V. & Kimble, J. (1994) *Cell 77*, 183–194.
370. Lee, R., Feinbaum, R. & Ambros, V. (1993) *Cell 75*, 843–854.
371. Wightman, B., Ha, I. & Ruvkun, G. (1993) *Cell 75*, 855–862.
372. Standart, N., Dale, M., Stewart, E. & Hunt, T. (1990) *Genes & Dev. 4*, 2157–2168.

Eur. J. Biochem. *237*, 6–15 (1996)

Review

The βA4 amyloid precursor protein gene and Alzheimer's disease

Lydia HENDRIKS and Christine VAN BROECKHOVEN

Laboratory of Neurogenetics, Born-Bunge Foundation, University of Antwerp, Flemish Institute for Biotechnology, Department of Biochemistry, Belgium

(Received 11 August/25 October 1995) – EJB 95 1348/0

Alzheimer's disease is a senile dementia caused by progressive neurodegeneration of the central nervous system. One of the most prominent pathological characteristics is βA4 amyloid deposition in senile plaques in the brain parenchyma and in cerebral blood vessels. βA4 amyloid is processed from a larger integral membrane protein, the βA4 amyloid precursor protein. Different pathogenic mutations in this protein have been detected in a small number of Alzheimer's disease families. Here functional implications of these mutations on the processing of the precursor protein and the βA4 amyloid deposition will be discussed with respect to the pathogenesis of Alzheimer's disease and related disorders.

Keywords: Alzheimer's; proteolysis; βA4 amyloid precursor protein; βA4 amyloid; mutation.

Alzheimer's disease, Down syndrome and congophylic amyloid angiopathy

Alzheimer's disease was first described by Aloïs Alzheimer in 1907 in a 51-year-old female patient with memory deficits and orientation disturbances in place and time, perception difficulties and hallucinations (Alzheimer, 1907). Although the first patient was only middle-aged, today, Alzheimer's disease is the major form of progressive senile dementia in the elderly. It afflicts 10.8% of persons in the age group 80–89 years (Rocca et al., 1991). The first symptoms of the disease are short-term memory disturbances and disorientation. As the disease progresses, complete memory judgment and reasoning impairment follows. Clinically, the disease can be diagnosed on the basis of neurological examination, neuropsychological testing and brain imaging using techniques such as computed tomography, positron emission tomography, single-photon emission computed tomography and magnetic resonance imaging. However, a definite diagnosis of the disease can only be obtained by post-mortem examination of the brain. The pathology is characterized by extensive neuronal cell loss and by particular brain lesions in the neo-cortex, hippocampus and amygdala, neurofibrillary tangles (NFT) and senile plaques (SP) (Fig. 1). The NFT are intraneuronal inclusions of paired helical filaments composed of abnormally phosphorylated τ, a microtubuli-associated protein. The SP are mainly composed of βA4 amyloid deposited in the brain parenchyma. βA4 amyloid is a 4-kDa proteolysis product of a larger precursor protein, the βA4 amyloid precursor protein (APP). The SP or neuritic plaques comprise a dense βA4 amyloid core surrounded by dystrophic neurites. Besides the SP, diffuse plaques, lacking the dense βA4 amyloid core and surrounding altered neurites, are also observed (Yamaguchi et al., 1989; Tagliavini et al., 1988). Diffuse plaques are not only detected in the brain of Alzheimer's disease patients but also in the brain of normal aged persons. In Alzheimer's disease brain, congophylic amyloid angiopathy is also present resulting from the accumulation of βA4 amyloid in the leptomeningeal and cortical blood vessel walls.

Aged (30–40 years) Down syndrome patients (trisomy 21) tend to develop dementia and neuropathological features similar to those seen in Alzheimer's disease patients (Olson and Shaw, 1969; Glenner and Wong, 1984a; Wisniewski et al., 1985).

In congophylic amyloid angiopathy, βA4 amyloid is deposited in blood vessel walls where it provokes intracerebral haemorrhages. Most congophylic amyloid angiopathy cases occur sporadically but a particular form of congophylic amyloid angiopathy, hereditary cerebral haemorrhage with amyloidosis of the Dutch type (HCHWA-D), is inherited as an autosomal dominant trait in a small number of families from Katwijk and Scheveningen in The Netherlands. HCHWA-D is typified by recurrent haemorrhagic strokes occurring between the age of 45–65 years due to extensive βA4 amyloid deposition in the cerebral and leptomeningeal blood vessel walls. In addition diffuse plaques are also present in the brain of HCHWA-D patients. Further, no NFT have been observed in brains of HCHWA-D patients. Hence the neuronal cell population seems not to be involved (van Duinen et al., 1987; Luyendijk et al., 1988; Haan et al., 1989, 1991; Timmers et al., 1990). Approximately 50% of the HCHWA-D patients die at the first haemorrhagic stroke. Those patients that survive develop a progressive dementia of the multi-infarct type (Haan et al., 1990a,b). HCHWA-D has been referred to as the vascular form of Alzheimer's disease because of its predominant vascular involvement (van Duinen et al., 1987).

Structure of βA4 amyloid and APP

βA4 amyloid was first isolated from cerebral blood vessels from Alzheimer's disease and Down syndrome patients and par-

Correspondence to C. Van Broeckhoven, Neurogenetics Laboratory, Born-Bunge Foundation, University of Antwerp (UIA), Flemish Institute for Biotechnology, Department of Biochemistry, Universiteitsplein 1, B-2610 Antwerpen, Belgium

Abbreviations. APP, βA4 amyloid precursor protein; HCHWA-D, hereditary cerebral haemorrhage with amyloidosis of the Dutch type; NFT, neurofibrillary tangles; SP, senile plaques.

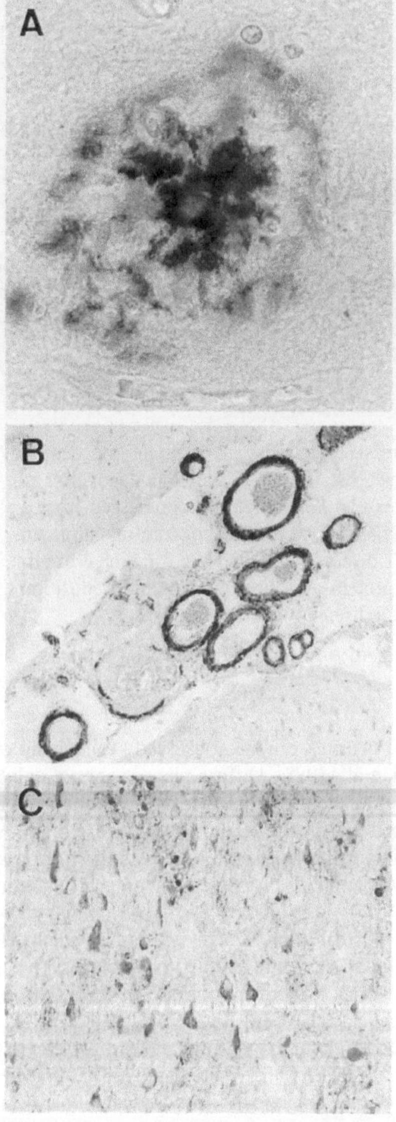

Fig. 1. Typical lesions in Alzheimer's disease brain. (A) Senile plaque with a dense core from the hippocampus (magnification ×675); (B) congiophylic angiopathy (magnification ×67), both stained with antiserum against βA4 amyloid; (C) NFT and SP with dystrophic neurites from brain area CA4, stained with mAb AT8 against abnormally phosphorylated τ (magnification ×135).

tially sequenced (Glenner and Wong, 1984a,b). The peptide is 40–43 amino acids long. The vascular βA4 amyloid from Down syndrome and Alzheimer's disease patients were identical except for βA4 amyloid codon 11 where Down syndrome βA4 amyloid has an Glu and Alzheimer's disease βA4 amyloid a Gln. Also the protein of the SP from the brains of Down syndrome and Alzheimer's disease patients was isolated and sequenced. These SP 4-kDa peptides are nearly identical to the vascular peptide but in the SP the peptides have ragged NH_2-termini (Masters et al., 1985). The 4-kDa peptide was designated Aβ-amyloid, β-amyloid, A4-amyloid, βA4 amyloid or other variations thereof. In this text we will use the abbreviation βA4 amyloid. More recent investigations on vascular βA4 amyloid also revealed COOH-terminal heterogeneity. The 1–42-residue peptide is the major component in congophylic amyloid angiopathy compared to the more soluble 1–40-residue form (Roher et al., 1993). The latter observation is in conflict with a previous study which claimed that the 1–40 βA4 amyloid is more abundant in congophylic amyloid angiopathy (Mori et al., 1992). These conflicting results are probably due to the use of different βA4 amyloid purification techniques. The SP βA4 amyloid is much more racemized and isomerized at aspartyl residues than the vascular βA4 amyloid, indicating that the vascular βA4 amyloid is 'younger' (Roher et al., 1993).

βA4 amyloid is derived by proteolytic processing of a larger transmembrane glycoprotein, APP. The APP gene is localized on chromosome 21 at 21q21.2 (Kang et al., 1987; Tanzi et al., 1987) and is encoded by 18 exons of which exons 16 and 17 encode in part for the βA4 amyloid peptide (Lemaire et al., 1989; Yoshikai et al., 1990). Three major splicing variants have been identified containing the βA4 amyloid sequence, i.e. the APP695, the APP751 and the APP770 isoform of which APP695 is the major isoform found in brain. The two larger isoforms include exon 7 which encodes a domain similar to a Kunitz protease inhibitor (Kitaguchi et al., 1988; Ponte et al., 1988; Tanzi et al., 1988). The APP770 isoform also includes exon 8 which is similar in sequence to the MRC OX-2 antigen (Clark et al., 1985; Ponte et al., 1988). Other, less abundant, isoforms of APP (752, 733, 714, 696, 677 amino acids) resulting from alternative splicing of exons 7, 8 and 15, also exist. APP belongs to a multi-gene family of APP-like proteins, but APP is the only one which comprises the βA4 amyloid region (Sandbrink et al., 1994).

The APP gene is a housekeeping gene since it is expressed in all cell types and tissues, especially in brain, heart, spleen, kidney and muscle (reviewed by Tanzi et al., 1989). The APP promotor lacks a TATA box and displays a high C+G content, two features typical for a housekeeping gene promotor (Salbaum et al., 1988).

The amino acid sequence of APP predicts that it is an integral membrane protein with a single transmembrane domain near the COOH-terminal side. The protein is N- and O-glycosylated and comprises an acidic domain (Fig. 2).

Processing and function of APP

APP is processed through different proteolytic pathways (reviewed by Haass and Selkoe, 1993; Fig. 3). The major pathway is the constitutive secretory pathway which cleaves APP between Lys687 and Leu688 (numbering refers to the APP770 isoform through the whole text) within the extracellular portion of the βA4 amyloid sequence (between residues 16 and 17) (Esch et al., 1990) producing soluble APP (APP_s) and a membrane-bound 10-kDa (P10) fragment containing only part of βA4 amyloid. Since this pathway does not produce intact βA4 amyloid, it is therefore non-amyloidogenic and cannot lead to Alzheimer's disease pathology.

For APP_s different functions have been proposed. It may bind the components of the extracellular matrix such as heparan sulfate proteoglycans, laminin and collagen(IV). In cell culture it has been demonstrated that APP_s possesses growth-promoting activities and has a positive effect on neuritic outgrowth which might be mediated through a heparin-binding domain near the APP NH_2-terminus (residues 96–110; reviewed by Small et al., 1994). APP_s may also have a protective effect on neuronal cells by lowering the intraneuronal Ca^{2+} levels (Mattson et al., 1993). APP_s comprising the Kunitz protease inhibitor domain or protease nexin 2 (Van Nostrand et al., 1989; Oltersdorf et al., 1989) is identical to the coagulation factor XIa inhibitor (Smith et al., 1990). APP is localized in the α-granules of blood platelets and APP_s is secreted upon activation of the platelets. Since APP_s

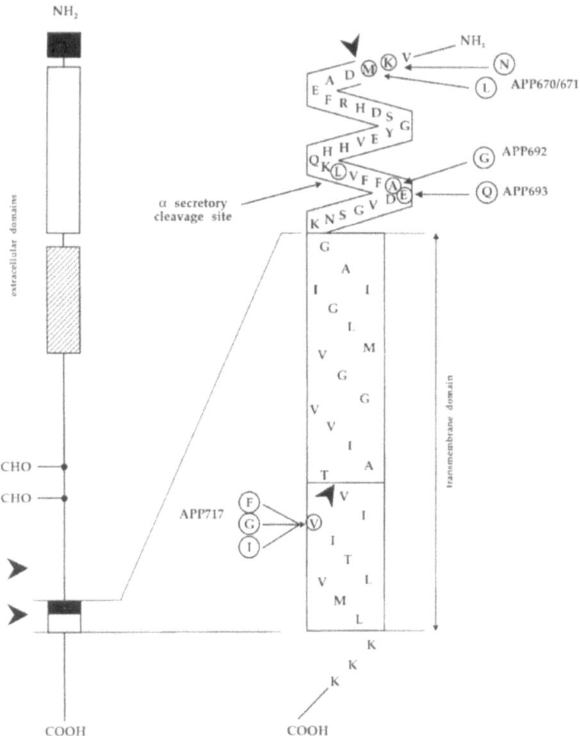

Fig. 2. Structure of APP. Proposed domain structure of APP as a cell-surface glycoprotein (Kang et al., 1987): black box, signal sequence; open box, cysteine-rich domain; hatched box, highly negatively charged domain (45% Asp and Glu residues); filled circles, N-glycosylation sites. The maximum length βA4 amyloid (1–43 residues) is between the arrow heads with its amino acid sequence. The amino acid substitutions involving the different APP mutations are shown.

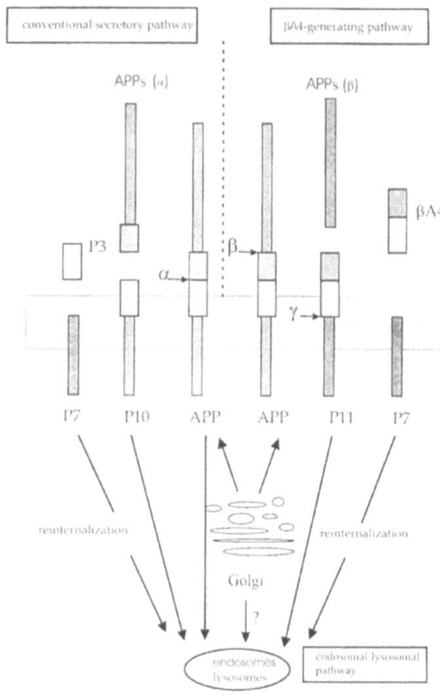

Fig. 3. APP processing pathways. The conventional secretory pathway (non-amyloidogenic) and βA4-amyloid-generating pathways are schematically represented here. The horizontal shaded box represents the cell membrane. The arrow heads designated α, β and γ represent the respective unknown secretases.

might also have a growth-stimulating function, it is likely that this protein is involved in wound repair (Van Nostrand et al., 1990).

In contrast, toxic effects have also been described for APP derivatives. Transfection studies with APP cDNA showed that differentiated neurons are vulnerable to overexpressed APP which generates potentially amyloidogenic fragments (Yoshikawa et al., 1992; Yanker et al., 1989). Studies in which synthetic βA4 amyloid or part thereof are administered to cell culture or injected into animal brains have led to contradicting results with respect to neurotoxicity (reviewed by Marx, 1992). The consensus about toxicity of βA4 amyloid now is that it partially depends on the aggregation state of βA4 amyloid. Aggregated peptides tend to be more neurotoxic than the soluble ones (reviewed by Selkoe, 1994).

Several potential amyloidogenic pathways have been described. An endosomal/lysosomal pathway for APP was proposed on the basis of pulse/chase experiments in mammalian cells (Chinese hamster ovary) transfected with APP cDNA and with fibroblasts defective in the lysosomal pathway. These experiments showed that APP is degraded by lysosomes producing βA4 amyloid containing 11–15-kDa COOH-terminal fragments of APP (Cole et al., 1992; Estus et al., 1992; Golde et al., 1992). In this pathway the membrane-bound APP may be reinternalized from the cell surface and is then degraded into potentially amyloidogenic fragments accumulating in the lysosomes (Haass et al., 1992a; Koo and Squazzo, 1994). APP has a NPXY consensus sequence in its cytoplasmic domain which can mediate the internalization of the protein. Besides this internalizing pathway, it is possible that APP and βA4 amyloid containing fragments from the Golgi or trans-Golgi network are conducted to lysosomes and endosomes for further degradation.

Previously it was believed that βA4 amyloid generation was an abnormal event, resulting from a pathological condition and that βA4 amyloid could only be released after cell injury. Now it has been shown that intact βA4 amyloid is released in the medium of normal neuronal and non-neuronal cultured cells and in cerebrospinal fluid of non-demented control persons (Haass et al., 1992b; Seubert et al., 1992, 1993; Shoji et al., 1992; Busciglio et al., 1993). These findings might explain the observation that in brain of aged non-demented persons, diffuse plaques are commonly seen.

βA4 amyloid is processed from APP by β- and γ-secretase. β-Secretase cleaves APP between Met671 and Asp672, at the NH$_2$-terminus of βA4 amyloid, releasing a truncated APP$_s$ (β-cleaved APP$_s$) in the medium. An 11-kDa, potentially amyloidogenic fragment (P11) is retained in the membrane (Seubert et al., 1993). This fragment could be further processed by γ-secretase into βA4 amyloid.

The proteolytic enzymes involved in the different APP metabolic pathways have not yet been identified. The α-, β- and γ-secretases may each represent a collection of enzymes. The β-secretase, cutting APP between Met771 and Asp772, is highly sequence-specific while the α-secretase is not (Citron et al., 1995; Selkoe, 1995). Recent data show that several proteases can cleave βA4 amyloid at or nearby the NH$_2$-terminus, resulting in βA4 amyloid peptides which are heterogeneous at their NH$_2$-terminal side (Haass et al., 1995). Also not much is known yet with respect to where in the cell the cleavages occur. Some investigators claim that the α-secretase cleavage occurs in the *trans*-Golgi network or post-Golgi vesicles (Sambamurti et al., 1992; De Strooper et al., 1993), others state that APP is 'α-cleaved' on the cell surface (Sisodia, 1992; Haass et al., 1992a). The most obscure enzyme is the γ-secretase, clipping βA4 amyloid at its NH_2-terminus which is embedded in the membrane. How and when this part is released from the membrane is not known but γ-secretase cleavage is preceded by β-secretase cleavage.

In polarized cells such as neurons and endothelial cells, APP is sorted. Also in polarized Mandin Darby canine kidney (MDCK) cells, eventually transfected with APP cDNA, 80–90% of APP_s, βA4 amyloid and P3 fragment are secreted basolaterally (Haass et al., 1994b; De Strooper et al., 1995). Deletions of cytoplasmic and transmembrane domain of APP did not disturb the sorting indicating that the basolateral sorting signal resides in the extracellular domain of APP (De Strooper et al., 1995). Since different secretases are probably located in different cell compartments, the routing of APP and amyloidogenic fragments is important in view of βA4 amyloid production.

Genetics of Alzheimer's disease

The etiology of Alzheimer's disease is complex and the primary biochemical causes of the disease have not yet been resolved. Although most Alzheimer's disease patients are sporadic cases, it has been recognized that genetic factors play an important role in the pathogenesis of Alzheimer's disease. Population survey studies revealed that 25–40% of the Alzheimer's disease cases have at least one other demented family member. In most families the disease is inherited in an autosomal dominant manner with age-dependent penetrance (Heston et al., 1981; Fitch et al., 1988; Hofman et al., 1989).

A simple autosomal dominant transmission pattern is found in families with an early-onset of the disease, before the age of 65 years (presenile or early-onset Alzheimer's disease). In most families with late onset of Alzheimer's disease, i.e. onset after the age of 65 years, the transmission pattern is not well defined because additional genetic and environmental factors may influence expression (penetrance) of the disease. The best approximation of the inheritance pattern in late-onset Alzheimer's disease families is autosomal dominant with a multifactorial component (Farrer et al., 1990). From the genetic point of view, Alzheimer's disease is a heterogeneous disorder. Hitherto, using molecular genetic approaches (reviewed by Van Broeckhoven, 1995a), two separate genetic loci have been identified in early onset Alzheimer's disease families by positional cloning: one on chromosome 21, i.e. the APP gene localized at 21q21.2, and a second on chromosome 14 localized at 14q24.3. More recently, a third locus for early-onset Alzheimer's disease has been localized at 1q31-42 (Levy-Lahad et al., 1995a).

The Alzheimer's disease gene (S182) on chromosome 14 has been identified and termed presenilin-1 (Sherrington et al., 1995). Shortly thereafter, using sequence similarity search strategy and linkage analysis with chromosome 1 markers, a gene homologous (STM-2 or E5-1) with presenilin-1 was identified on chomosome 1 and termed presenilin-2 (Levy-Lahad et al., 1995b; Rogaev et al., 1995). So far, 22 different pathogenic missense mutations have been identified in the presenilin-1 (for review see Van Broeckhoven, 1995b) and two distinct mutations in the presenilin-2 gene (Levy-Lahad et al., 1995b; Rogaev et al. 1995).

The apolipoprotein E4 allele has been identified as a genetic risk factor for Alzheimer's disease. ApoE could play a role in Alzheimer's disease pathogenesis: ApoE is present in Alzheimer's disease brain lesions and ApoE binds with high avidity to βA4 amyloid (Strittmatter et al., 1993a,b). Conceivably, the binding of apoE to βA4 amyloid makes the latter less soluble and thus more prone to deposition (Wisniewski and Frangione, 1992). The apolipoprotein E gene is located on chromosome 19 at 19q13.2. Three major alleles occur, i.e. E2, E3 and E4, giving rise to three isoforms. The E2 and E4 isoforms differ from the most common E3 isoform by single amino acid substitutions at positions 112 and 158: E3 (Cys112, Arg158), E2 (Arg158→Cys) and E3 (Cys112→Arg).

The E4 allele of the apoE gene (apolipoprotein E*4), significantly increases the risk for both early-onset (van Duijn et al., 1994) and late-onset Alzheimer's disease (Saunders et al., 1993; Strittmatter et al., 1993a) and risk is highest for apolipoprotein E*4 homozygotes. Also, in familial late-onset Alzheimer's disease, the onset age of the disease decreases as the number of apolipoprotein E*4 alleles increases (Corder et al., 1993; van Duijn et al., 1994). The E2 allele of the apolipoprotein E gene (apolipoprotein E*2) seems to exhibit a protective effect as shown by a decreased risk for Alzheimer's disease (Chartier-Harlin et al., 1994; Corder et al., 1994). However, the apolipoprotein E*2 protective effect is still controversial since other investigators have reported an increased risk for Alzheimer's disease also for the apolipoprotein E*2 allele (van Duijn et al., 1995).

APP mutations

Chromosome 21 was a candidate chromosome for carrying an Alzheimer's disease gene because of the pathological link between Down syndrome and Alzheimer's disease patients. Linkage analysis showed cosegregation of Alzheimer's disease with markers on the long arm of chromosome 21 in a fraction of early-onset Alzheimer's disease families (St George-Hyslop et al., 1987, 1990). Also a mutation coding Glu693→Gln (Dutch mutation) in the APP gene, at 21q21.2 located close to the Alzheimer's disease linked markers, was described in HCHWA-D patients (Levy et al., 1990; Van Broeckhoven et al., 1990). This led to the identification of point mutations (single base changes) in the APP gene in early-onset Alzheimer's disease families (Goate et al., 1991).

Up till now six different mutations have been found in exons 16 and 17 of the APP gene. The mutations in the APP gene change the coding for amino acids of APP i.e. Val717→Ile (Goate et al., 1991), Val717→Phe (Murrell et al., 1991) and Val717→Gly (Chartier-Harlin et al., 1991), and Lys670→Asn, Met671→Leu, Val717→Ile (Swedish mutation; Mullan et al., 1992). Only the Val717→Ile mutation has been observed in distinct early-onset Alzheimer's disease families of different ethnic background and therefore is the most common APP mutation in Alzheimer's disease. The prevalence of this mutation is 5.5% among familial early-onset Alzheimer's disease (Van Broeckhoven, 1995a). In early-onset Alzheimer's disease families in which mutations in the APP gene segregate, the mean age at onset of the disease occurs between 50–60 years.

A mutation in βA4 amyloid was also identified in patients with cerebral haemorrhages due to congophylic amyloid angiopathy and in Alzheimer's disease patients belonging to the same Dutch family. Here an Ala692→Gly mutation was detected in APP (Flemish mutation). The mean age at onset for all patients was 45.7 ± 7.3 years. Histopathological examination of brain biopsy material of a patient suffering from a cerebral haemorrhage showed extensive deposition of βA4 amyloid in cerebral blood vessels and in the brain parenchyma. Dystrophic neurites were also present but no NFT could be observed (Hendriks et al., 1992). Histopathological studies on brain autopsy material of a demented patient confirmed Alzheimer's disease. Numerous NFT and extensive congophylic amyloid angiopathy were observed. Compared to classical Alzheimer's disease, this form has

Table 1. Effect of APP mutations on APP processing.

APP mutation	Effect on αA4 amyloid production compared to wild type
APP670/671 (Swedish mutant)	3−8-fold more βA4 amyloid produced
APP717 (London mutant)	1.5−1.9-fold more βA4 amyloid (1−42) compared to βA4 amyloid (1−40)
APP692 (Flemish mutant)	relatively more βA4 amyloid compared to P3 by partial inhibition of α-secretase

a severe vascular component with more βA4 amyloid deposited in the SP and blood vessels (Cras et al., 1995).

Effect of APP mutations on APP processing

The early-onset Alzheimer's disease mutations, APP717 and APP670/671, are located outside the βA4 amyloid sequence, with APP717 close to the COOH-terminal γ-secretase cleavage site, within the transmembrane domain and APP670/671 near the NH_2-terminal β-secretase cleavage site within the extracellular part of APP (Figs 2 and 3). In contrast, the APP692 and APP693 mutations are located inside the βA4 amyloid sequence, next to the α-secretase cleavage site. These localizations led to the hypothesis that these mutations might influence the activity of the respective secretases, resulting in an aberrant processing of APP. *In vitro* studies have provided evidence for this hypothesis. The effects of the different mutations on βA4 amyloid generation are listed in Table 1.

APP670/671 (Swedish) mutation. Transfections with wild-type and Swedish mutant APP670/671 cDNA in neuronal and non-neuronal cells showed a 3−8-fold overproduction of βA4 amyloid and a decrease of the 3-kDa (P3) fragment, the COOH-terminal subpeptide of βA4 amyloid resulting from both α- and γ-secretase activity (Fig. 3) in the medium (Citron et al., 1992; Cai et al., 1993; Hendriks et al., 1995; Felsenstein et al., 1994). The Met671→Leu substitution is responsible for this increase (Citron et al., 1992). With the use of specific antibodies a qualitative change in the APP secreted products was shown, most likely resulting from a shift in the APP secretase activities favouring β-secretase activity in the APP670/671 mutant (Felsenstein et al., 1994). Amino acid sequence analysis of the P11 fragments and βA4 amyloid from both wild-type APP and the APP670/671 mutant, showed that the latter is cleaved at the same location as the wild type (Cheung et al., 1994; Dovey et al., 1993). Thus the increased production of P11 and βA4 amyloid indicated an accelerated β-secretase activity. It is likely that P11 is further processed into βA4 amyloid by γ-secretase. Mass spectrometric analysis and sequencing of the COOH-terminus of βA4 amyloid produced by APP-cDNA-transfected human cells, showed that the predominant form of βA4 amyloid is 1−40 amino acids long. However, a significant amount of βA4 amyloid of 1−42 amino acids was also present (Dovey et al., 1993). The overproduction of βA4 amyloid in transfected cells with the Swedish mutant was confirmed in primary skin fibroblasts from patients with the Swedish mutation (Citron et al., 1994; Johnston et al., 1994). The fact that the increase in βA4 amyloid production is detected in Alzheimer's disease patients, as well as in presymptomatic persons carrying the Swedish mutation, points to a causal effect of the βA4 amyloid overproduction rather than to a secondary one (Citron et al., 1994).

In 293 kidney cells transfected with Swedish mutant APP cDNA, the βA4 amyloid secretion level is lowered after treatment with the antibiotic bafilomycin A1, an agent which inhibits lysosomal protein degradation. In cells transfected with wild-type APP, however, bafilomycin provokes an increase in βA4 amyloid production. After sequencing, it became clear that in cells transfected with the wild-type an alternative secretase is used which cleaves βA4 amyloid at alternative positions. In cells expressing the Swedish mutant, no such alternative βA4 amyloid peptides could be formed. This differential effect of bafilomycin A1 is also seen in fibroblast from patients bearing the Swedish mutation compared to fibroblast of normal control persons (Haass et al., 1995).

When the Swedish mutant cDNA is transfected into MDCK cells the sorting polarity is affected: 20% of the β-cleaved APP is secreted on the apical side instead of at the basolateral side (De Strooper et al., 1995).

APP692 (Flemish) and APP693 (Dutch) mutations. *In vitro* aggregation studies with synthetic βA4 amyloid containing the APP693 mutation showed accelerated amyloid fibril formation (Wisniewski et al., 1991; Clements et al., 1993) as well as enhanced amyloid fibril stability (Fraser et al., 1992). For the APP692 mutation, however, no significant change in the rate of *in vitro* amyloid aggregation was detected. In fact, the aggregation of βA4 amyloid fibrils bearing the APP692 mutation was slightly slower than that of the wild type (Clements et al., 1993).

Recently, transfection experiments in human kidney 293 cells with the Flemish mutant APP cDNA demonstrated a relative increase in the secretion of βA4 amyloid and a decrease in P3 production. These data suggested a lowered activity for the α-secretase due to the presence of the APP692 mutation, most likely caused by a destabilization of the α-helical structure in this region. Substitution Ala692→Pro has a similar effect as the natural occurring Ala→Gly mutation, but the structurally conservative Ala692→Val substitution has no such effect (Haass et al., 1994a). In cells trasfected with mutant APP692 or APP693 cDNA, no increase in the βA4 amyloid production was noticed (Maruyama et al., 1991; Felsenstein and Lewis-Higgins, 1993; Hendriks et al., 1995).

APP717 mutation. In cells transfected with mutant APP717 cDNA (Val→Ile, Phe or Gly), the ratio of βA4 amyloid (1−42,43)/βA4 amyloid (1−40) is increased compared to cells transfected with wild-type APP (Suzuki et al., 1994; Tamaoka et al., 1994a). The amount of secreted βA4 amyloid peptides however remained the same (Cai et al., 1993; Felsenstein et al., 1994). *In vitro*, longer βA4 amyloid peptides tend to aggregate faster because the nucleation step is significantly accelerated (Jarrett et al., 1993).

βA4 amyloid was also purified from brain of two unrelated patients with the Val717→Ile mutation and it was shown that the ratio of βA4 amyloid (1−42,43)/βA4 amyloid (1−40) was markedly increased compared to sporadic Alzheimer's disease patients (Tamaoka et al., 1994b).

Apolipoprotein E and APP mutations

The effect of the apolipoprotein E genotype on the disease penetrance in families segregating an APP mutation was also examined. In the Swedish Alzheimer's disease family segregating the APP670/671 mutation, patients with the apolipoprotein E*4 allele had the earliest and patients with the apolipoprotein E*2 allele the latest onset age (Hardy et al., 1993). The APP670/671 mutation is a sufficient cause for Alzheimer's disease in this family but the age at onset is influenced by the apolipoprotein E genotype. In families with APP717 mutations, the apolipoprotein E allele frequencies did not differ significantly from population frequencies (Hardy et al., 1993; Saunders et al., 1993). In the HCHWA-D families and the family with Alzheimer's disease and congophylic amyloid angiopathy patients with the APP693 (Dutch) and APP692 (Flemish) mutations respectively, a high apolipoprotein E*4 allele frequency was observed. However, no correlation was found between number of apolipoprotein E*4 alleles and age at onset, age at death, occurrence of dementia or haemorrhage and number of strokes (Haan et al., 1994).

The Dutch and Flemish mutations are localized within the βA4 amyloid sequence, in the portion with the highest avidity for apolipoprotein E, i.e. amino acids 12–28 (Strittmatter et al., 1993a). It was hypothesized that the mutation weakens or even prevents the binding of apolipoprotein E to βA4 amyloid. Consequently, apolipoprotein E cannot interfere with βA4 amyloid deposition in these families. In contrast, the APP717 and APP670/671 mutations are located outside the βA4 amyloid sequence and hence no interference with the apolipoprotein E binding is expected.

Animal models

To elucidate the mechanisms by which Alzheimer's disease originates and to test therapeutic agents aiming at prevention or treatment of the disease, animal models are indispensable. Natural occurring models can be found. The brains of aged non-human primates (*Macaca mulatta*) contain SP similar to those found in Alzheimer's disease patient brains (e.g. Martin et al., 1991). Cognitive and memory dysfunction seem to occur in these animals (reviewed by Price and Sisodia, 1994). Several attempts have been undertaken to create a transgenic mouse model for Alzheimer's disease. The first models were based on overexpression of human wild-type APP cDNA (Quon et al., 1991) or COOH-terminal parts of it (C-100) (reviewed by Price and Sisodia, 1994). APP displays synaptotrophic effect in the brain of transgenic mouse which overexpress the wild-type cDNA (Mucke et al., 1994). Recently it was reported that mice overexpressing the APP695 and APP751 isoforms displayed reduced scores in cognitive testings such as the Morris water maze test. This, however was due to motor impairment and hypoactivity of these mice, rather than to memory impairment (Perry et al., 1995).

Only the mouse model described by Quon et al. (1991) showed extracellular βA4 amyloid deposition in the brain, which resembled plaques in early Down syndrome patients. There was staining with antibody Alz50, which recognizes abnormally phosphorylated τ, but no NFT were observed (Higgens et al., 1994).

The entire human wild-type APP genomic sequence, including 5′ and 3′ flanking sequences, cloned in a yeast artificial chromosome vector, was introduced into mouse using the embryonic stem cell transfection technique. Although overexpression of the exogenous gene was seen, no pathological features resembling those of Alzheimer's disease were detected (Lamb et al., 1993; Pearson and Choi, 1993). The fact that most mouse models do not exhibit Alzheimer's-disease-like pathology is probably due to low levels of APP expression (Price and Sisodia, 1994).

In order to study the neurotoxic effect of βA4 amyloid, LaFerla and coworkers (1995) have created transgenic mice expressing murine βA4 amyloid intracellularly. These mice live only half the lifespan of their control litter mates and show seizure activity. Pathologically, degenerative neurons were seen in the cerebral cortex and hippocampus, followed by apoptotic cell death and gliosis. No βA4 amyloid deposits or NFT were observed.

More recently researchers have established transgenic mice which express mutated APP cDNAs (e.g. Games et al., 1995). The mouse model created by Games and coworkers (1995) looks promising with respect to SP formation in the mouse brain. They used the APP cDNA with the Val717→Phe mutation and placed it after a strong promotor for platelet-derived growth factor (PDGF) β. Furthermore introns 6, 7 and 8 were introduced, enabling alternative splicing of exons 7 and 8. The pathological picture shows βA4 amyloid deposits, SP, synapses loss, astrocytosis and microgliosis. The observed pathology is progressive: 6–9-months-old mice have deposits in the hippocampus, corpus callosum and cerebral cortex; as they age the number of plaques increases drastically. No NFT, however, were observed. The Val717→Ile (Duff et al., 1994), Glu693→Gln and Ala692→Gly mutations have now been applied in the APP gene cloned in a yeast artificial chromosome vector (unpublished results). Embryonic stem cell transfections are on their way.

Further the APP gene has been knocked-out in mice by deleting the promotor sequence and first exon. Homozygous APP-null mice are fertile. Compared to their normal litter mates they have a reduced body mass (15–20%), reduced locomotor activity and forelimb grip strength. Some of the homozygous APP null mice brains showed gliosis, indicating an impaired neuronal function (Zheng et al., 1995).

DISCUSSION

It is widely accepted that Alzheimer's disease is a complex disorder in which many genetic and environmental factors influence the pathogenesis. There is substantial evidence from the literature indicating that βA4 amyloid deposition plays a central role in Alzheimer's disease pathogenesis. The reasons why the βA4 amyloid peptide has such a pivotal function is that SP or plaque-like structures, mainly composed out of βA4 amyloid, appear in familial and sporadic Alzheimer's disease, Down syndrome and congophylic amyloid angiopathy cases. Mutations in the APP gene are, in a limited number of families, responsible for Alzheimer's disease pathogenesis. In other amyloidoses, the precursor molecule of the amyloidogenic peptide has been identified as the primary cause of the disease. No other agents have been found to cause the disease (reviewed by Selkoe, 1994).

The cascade hypothesis states that APP or, more likely, amyloidogenic derivatives of it, are neurotoxic and trigger the neuronal response, i.e. formation of NFT and neuronal cell loss. The mechanisms by which this happens are mostly unknown (Hardy and Higgins, 1992). The fact that the severity of dementia positively correlates with the amount of NFT and not with the amount of SP can hereby be explained.

Since βA4 amyloid is produced during normal metabolism it seems that, in Alzheimer's disease patients, the APP is misprocessed and that the deposition of βA4 amyloid is accelerated. Little by little, the mechanisms behind this misprocessing become clear. Point mutations in the APP gene can mediate the preferential use of an amyloidogenic pathway compared to the non-amyloidogenic one. This leads to the production of longer and or more βA4 amyloid peptides which aggregate faster. Other factors can act in the pathway leading to βA4 amyloid deposition. One such factor is apolipoprotein E4 which might intensify the aggregation of βA4 amyloid peptides. Recently the early-onset Alzheimer's disease genes on chromosome 14 (presenilin-1) and on chromosome 1 (presenilin-2) have been identified. The presenilin-1 and presenilin-2 genes encode proteins of 467 and 448 amino acids, respectively, and show an amino acid similarity of 65%. So far, 22 different pathogenic missense mutations have been identified in the presenilin-1 (for review see Van

Broeckhoven, 1995b) and two distinct mutations in the presenilin-2 gene in familial early-onset Alzheimer's disease patients (Levy-Lahad et al., 1995b; Rogaev et al., 1995). The protein sequence predicts seven transmembrane regions, resembling the topology of a receptor-, a channel or a structural-membrane protein. Studies aimed at unravelling the function of this gene will render better insight into βA4 amyloid formation and deposition. In fibroblasts and plasma of Alzheimer's disease patients with mutations in the presenilin-1 gene, an increase in the amount of the longer βA4 amyloid (1−42,43) compared to the shorter (1−40) form has been detected (Scheuner et al., 1995; Song et al., 1995). However, these results have to be confirmed and therefore it is still possible that presenilins do not interfere with APP processing, imply that the amyloid cascade hypothesis and central role of βA4 amyloid have to be reconsidered.

The Alzheimer's disease research work in the laboratory of Neurogenetics (University of Antwerp) is funded by the Flemish Biotechnology Programme of the Ministry of Economy, the National Fund of Scientific Research, a concerted action of the Ministry of Education, Belgium, Biomed (Gene-CT93-0015) and Biotech (BIO2-CT94-2065) programmes of the European Community and by the Alzheimer's Disease Research Programme of the American Health Assistance Foundation.

REFERENCES

Alzheimer, A. (1907) Über eine eigenartige Erkrankung der Hirnrinde, *Centralblatt für Nervenheilkunde und Psychiatrie 30*, 177−179.

Busciglio, J., Gabuzda, D., Matsudeira, P. & Yanker, B. (1993) Generation of β-amyloid in the secretory pathway in neuronal and nonneuronal cells, *Proc. Natl Acad. Sci. USA 90*, 2092−2096.

Cai, X., Golde, T. E. & Younkin, S. G. (1993) Release of excess amyloid β protein from a mutant amyloid β protein precursor, *Science 259*, 514−516.

Chartier-Harlin, M., Crawford, F., Houlden, H., Warren, A., Hughes, D., Fidani, L., Goate, A., Rossor, M., Roques, P., Hardy, J. & Mullan, M. (1991) Early-onset Alzheimer's disease caused by mutations at codon 717 of the β-amyloid precursor protein gene, *Nature 353*, 844−846.

Chartier-Harlin, M., Parfitt, M., Legrain, S., Pérerz-Tur, J., Brousseau, T., Evans, A., Berr, C., Vidal, O., Roques, P., Gourlet, V., Fruchart, J., Delacourte, A., Rossor, M. & Amouyel, P. (1994) Apolipoprotein E, ε4 allele as a major risk factor for sporadic early and late-onset forms of Alzheimer's disease: analysis of the 19q13.2 chromosomal region, *Hum. Mol. Genet. 3*, 569−574.

Cheung, T. T., Ghiso, J., Shoji, M., Cai, X. D., Golde, T., Gandy, S., Frangione, B. & Younkin, S. (1994) Characterization by radiosequencing of the carboxyl-terminal derivatives produced from normal and mutant amyloid β-protein precursors, amyloid, *Int. J. Exp. Clin. Invest. 1*, 30−38.

Citron, M., Oltersdorf, T., Haass, C., McConlogue, L., Hung, A. Y., Seubert, P., Vigo-Pelfrey, C., Lieberburg, I. & Selkoe, D. J. (1992) Mutation of the β-amyloid precursor protein in familial Alzheimer's disease increases β-protein production, *Nature 360*, 672−674.

Citron, M., Vigo-Pelfrey, C., Teplow, D. B., Miller, C., Schenk, D., Johnston, J., Windblad, B., Venizelos, N., Lannfelt, L. & Selkoe, D. J. (1994) Excessive production of amyloid β-protein by peripheral cells of symptomatic and presymptomatic patients carrying the Swedish familial Alzheimer disease mutation, *Proc. Natl Acad. Sci. USA 91*, 11993−11997.

Citron, M., Teplow, D. B. & Selkoe, D. J. (1995) Generation of amyloid β protein from its precursor is sequence specific, *Neuron 14*, 661−670.

Clark, M. J., Gagnon, J., Williams, A. F. & Barcley, A. N. (1985) MRC OX-2 antigen: A lymphoid/neuronal membrane glycoprotein with a structure like a single immunoglobulin light chain, *EMBO J. 4*, 113−118.

Clements, A., Walsh, D. M., Williams, C. H. & Allsop, D. (1993) Effects of the mutations Glu^{22} to Gln and Ala^{21} to Gly on the aggregation of a synthetic fragment of the Alzheimer's amyloid β/A4 peptide, *Neurosci. Lett. 161*, 17−20.

Cole, G., Bell, L., Truong, Q. B. & Saithoh, T. (1992) An endosomal-lysosomal pathway for degradation of amyloid precursor protein, *Ann. N. Y. Acad. Sci. 674*, 103−117.

Corder, E. H., Saunders, A. M., Strittmatter, W. J., Schmechel, D. E., Gaskell, P. C., Small, G. W., Roses, A. D., Haines, J. L. & Pericak-Vance, M. A. (1993) Gene dose of apolipoprotein E type 4 allele and the risk of Alzheimer's disease in late onset families, *Science 261*, 921−923.

Corder, E. H., Saunders, A. M., Risch, N. J., Strittmatter, W. J., Schmechel, D. E., Gaskell, P. C., Rimmler, J. B., Locke, P. A., Conneally, P. M., Schmader, K. E., Small, G. W., Roses, A. D., Haines, J. L. & Pericak-Vance, M. A. (1994) Protective effect of apolipoprotein E type 2 allele for late onset Alzheimer disease, *Nature Genet. 7*, 180−184.

Cras, P., van Harskamp, F., Stefanko, S. Z., Hofman, A., van Duijn, C. M., Van Broeckhoven, C., Kros, J. M., Ceuterick, C. & Martin, J.-J. (1995) Amyloid angiopathy is associated with a large amyloid core type of senile plaques in the amyloid precursor protein 692Ala→Gly mutation, *J. Neuropathol. Exp. Neurol. 54*, 431.

De Strooper, B., Umans, L., Van Leuven, F. & Van Den Berghe, H. (1993) Study of the synthesis and secretion of normal and artificial mutants of murine amyloid precursor protein (APP): Cleavage of APP occurs in a late compartment of the default secretion pathway, *J. Cell Biol. 121*, 295−304.

De Strooper, B., Craessaerts, K., Dewachter, I., Moechars, D., Greenberg, B., Van Leuven, F. & Van Den Berghe, H. (1995) Basolateral secretion of amyloid precursor protein in Madin-Darby canine kidney cells is disturbed by alterations of intracellular pH and by introducing a mutation associated with familial Alzheimer's disease, *J. Biol. Chem. 270*, 4058−4065.

Dovey, H. F., Suomensaari-Chrysler, S., Lieberburg, I., Sinha, S. & Keim, P. S. (1993) Cells with a familial Alzheimer's disease mutation produce authentic β-peptide, *Neurochemistry 4*, 1039−1042.

Duff, K., McGuigan, A., Huxley, C., Schulz, F. & Hardy, J. (1994) Insertion of a pathogenic mutation into a yeast artificial chromosome containing the human amyloid precursor protein gene, *Gene Therapy 1*, 70−75.

Esch, F. S., Keim, P. S., Beattie, E. C., Blacher, R. W., Culwell, A. R., Oltersdorf, T., McClure, D. & Ward, P. J. (1990) Cleavage of amyloid β peptide during constitutive processing of its precursor, *Science 248*, 1122−1124.

Estus, S., Golde, T. E., Kunishita, T., Blades, D., Lowery, D., Eisen, M., Usiak, M., Qu, X., Tabira, T., Greenberg, B. D. & Younkin, S. G. (1992) Potentially amyloidogenic, carboxyl-terminal derivatives of the amyloid protein precursor, *Science 255*, 726−728.

Farrer, L. A., Myers, R. H., Cupples, L. A., St George-Hyslop, P. H., Bird, T. D., Rossor, M. N., Mullan, M. J., Polinsky, R., Nee, L., Heston, L., Van Broeckhoven, C., Martin, J., Crapper-McLachlan, D. & Growdon, J. H. (1990) Transmission and age-at-onset patterns in familial Alzheimer's disease: Evidence for heterogeneity, *Neurology 40*, 395−403.

Felsenstein, K. M., Hunihan, L. W. & Roberts, S. B. (1994) Altered cleavage and secretion of a recombinant β-APP bearing the swedish familial Alzheimer's disease mutation, *Nature Genet. 6*, 251−256.

Felsenstein, K. M. & Lewis-Higgins, L. L. (1993) Processing of the β-amyloid precursor protein carrying the familial, Dutch-type, and a novel recombinant C-terminal mutation, *Neurosci. Lett. 152*, 185−189.

Fitch, N., Becker, R. & Heller, A. (1988) The inheritance of Alzheimer's disease: A new interpretation, *Ann. Neurol. 23*, 14−19.

Fraser, P. E., Nguyen, J. T., Inouye, H., Surewicz, W. K., Selkoe, D. J., Podlisny, M. B. & Kirschner, D. A. (1992) Fibril formation by primate, rodent and Dutch-hemorrhagic analogues of Alzheimer amyloid β-protein, *Biochemistry 31*, 10716−10723.

Games, D., Adams, D., Alessandrini, R., Barbour, R., Berthelette, P., Blackwell, C., Carr, T., Clemens, J., Donalson, T., Gillespi, F., Guido, T., Hagoplan, S., Johnson-Wood, K., Khan, K., Lee, M., Leibowitz, P., Lieberburg, I., Little, S., Masliah, E., McConlogue, L., Montoya-Zavala, M., Mucke, L., Paganini, L., Penniman, E., Power, M., Schenk, D., Seubert, P., Snyder, B., Soriano, F., Tan, H., Vitale, J., Wadsworth, S., Wolozin, B. & Zhao, J. (1995) Alzheimer-type neuropathology in transgenic mice overexpressing V717F β-amyloid precorsor protein, *Nature 373*, 523−527.

Glenner, G. G. & Wong, C. W. (1984a) Alzheimer's disease: Initial report of the purification and characterization of a novel cerebrovascular amyloid protein, *Biochem. Biophys. Res. Commun. 122*, 885–890.

Glenner, G. G. & Wong, C. W. (1984b) Alzheimer's disease and Down's syndrome: Sharing of a unique cerebrovascular amyloid fibril protein, *Biochem. Biophys. Res. Commun. 122*, 1131–1135.

Goate, A., Chartier-Harlin, M., Mullan, M., Brown, J., Crawford, F., Fidani, L., Giuffra, L., Haynes, A., Irving, N., James, L., Mant, R., Newton, P., Rooke, K., Roques, P., Talbot, C., Pericak-Vance, M., Roses, A., Williamson, R., Rossor, M., Owen, M. & Hardy, J. (1991) Segregation of a missense mutation in the amyloid precursor protein gene with familial Alzheimer's disease, *Nature 349*, 704–706.

Golde, T. E., Estus, S., Younkin, L. H., Selkoe, D. J. & Younkin, S. G. (1992) Processing of the amyloid protein precursor to potentially amyloidogenic derivatives, *Science 255*, 728–730.

Haan, J., Roos, R. A. C., Briet, P. E., Herpers, M. J. H. M., Luyendijk, W. & Bots, G. T. A. M. (1989) Hereditary cerebral haemorrhage with amyloidosis of the Dutch type, *Clin. Neurol. Neurosurg. 81*, 285–290.

Haan, J., Algra, P. R. & Roos, R. A. C. (1990a) Hereditary cerebral haemorrhage with amyloidosis-Dutch type: Clinical and CT-analysis of 24 cases, *Arch. Neurol. 47*, 649–653.

Haan, J., Lanser, J. B. K., Zijderveld, I., Van Der Does, I. G. F. & Roos, R. A. C. (1990b) Dementia in hereditary cerebral haemorrhage with amyloidosis-Dutch type, *Arch. Neurol. 47*, 956–968.

Haan, J., Hardy, J. A. & Roos, R. A. C. (1991) Hereditary cerebral hemorrhage with amyloidosis-Dutch type: its importance for Alzheimer research, *Trends Nuerosci. 14*, 231–234.

Haan, J., Van Broeckhoven, C., van Duijn, C. M., Voorhoeve, E., van Harskamp, F., van Swieten, J. C., Maat-Schieman, M. L. C., Roos, R. A. C. & Bakker, E. (1994) The apolipoprotein E ε4 allele does not influence the clinical expression of the amyloid precursor protein-gene codon 693 or 692 mutations, *Ann. Neurol. 36*, 434–437.

Haass, C., Koo, E. H., Mellon, A., Hung, A. Y. & Selkoe, D. J. (1992a) Targeting of cell-surface β-amyloid precursor protein to lysosomes: Alternative processing into amyloid-bearing fragments, *Nature 357*, 500–503.

Haass, C., Schlossmacher, M. G., Hung, A. Y., Vigo-Pelfrey, C., Mellon, A., Ostaszewski, B. L., Lieberburg, I., Koo, E. H., Schenk, D., Teplow, D. B. & Selkoe, D. J. (1992b) Amyloid β-peptide is produced by cultured cells during normal metabolism, *Nature 359*, 322–325.

Haass, C. & Selkoe, D. J. (1993) Cellular processing of β-amyloid precursor protein and the genesis of amyloid β-peptide, *Cell 75*, 1039–1042.

Haass, C., Hung, A. Y., Selkoe, D. J. & Teplow, D. B. (1994a) Mutations associated with a locus for familial Alzheimer's disease result in alternative processing of amyloid β-protein precursor, *J. Biol. Chem. 269*, 17741–17748.

Haass, C., Koo, E. H., Teplow, T. P. & Selkoe, D. J. (1994b) Polarized secretion of β-amyloid precursor protein and amyloid β-peptide in MDCK cells, *Proc. Natl Acad. Sci. USA 91*, 1564–1568.

Haass, C., Capell, A., Citron, M., Teplow, D. B. & Selkoe, D. J. (1995) The vacuolar H^+-ATPase inhibitor bafilomycin A1 differentially affects proteolytic processing of mutant and wild-type β-amyloid protein precursor, *J. Biol. Chem. 270*, 6186–6192.

Hardy, J., Houlden, H., Collinge, J., Kennedy, A., Newman, S., Rossor, M., Lilius, L., Winblad, B., Crook, R. & Duff, K. (1993) Apolipoprotein E genotype and Alzheimer's disease, *Lancet 342*, 737–738.

Hardy, J. A. & Higgins, G. A. (1992) Alzheimer's disease: The amyloid cascade hypothesis, *Science 256*, 184–185.

Hendriks, L., van Duijn, C. M., Cras, P., Cruts, M., Van Hul, W., van Harskamp, F., Warren, A., McInnis, M. G., Antonarakis, S. E., Martin, J., Hofman, A. & Van Broeckhoven, C. (1992) Presenile dementia and cerebral haemorrhage linked to a mutation at condon 692 of the β-amyloid precursor protein gene, *Nature Genet. 1*, 218–221.

Hendriks, L., Cras, P., Martin, J.-J. & Van Broeckhoven, C. (1995) Alzheimer's disease and hemorrhagic stroke: Their relationship to βA4 amyloid deposition, in *Alzheimer's disease: Lessons from cell biology* (Kosik, K. S., Christen, Y. & Selkoe, D. J., eds) pp. 37–48, Springer-Verlag, Berlin, Heidelberg.

Heston, L. L., Mastri, A. R., Anderson, V. E. & White, J. (1981) Dementia of the Alzheimer type, *Arch. Gen. Psychiatry 38*, 1085–1090.

Higgens, L. S., Holtzman, D. M., Rabin, J., Mobley, W. C. & Cordell, B. (1994) Transgenic mouse brain histopathology resembles early Alzheimer's disease, *Ann. Neurol. 35*, 598–607.

Hofman, A., Schulte, W., Tanja, T. A., van Duijn, C. M., Haaxma, R., Lameris, A. J., Otten, V. M. & Saan, R. J. (1989) History of dementia and Parkinson's disease in 1st-degree relatives of patients with Alzheimer's disease, *Neurology 39*, 1589–1592.

Jarrett, J. T., Berger, E. P. & Lansbury, P. T. (1993) The carboxy terminus of the -amyloid protein is critical for the seeding of amyloid formation: Implications for the pathogenesis of Alzheimer's disease, *Biochemistry 32*, 4693–4697.

Johnston, J. A., Cowburn, R. F., Norgren, S., Wiehager, B., Benizoles, N., Windblad, B., Vigo-Pelfrey, C., Schenk, D., Lannfelt, L. & O'Neill, C. (1994) Increased β-amyloid release and levels of amyloid precursor protein (APP) in fibroblast cell lines from family members with the Swedish Alzheimer's disease APP670/671 mutation, *FEBS Lett. 354*, 274–278.

Kang, J., Lemaire, H., Unterbeck, A., Salbaum, J. M., Masters, C. L., Grzeschik, K., Multhaup, G., Beyreuther, K. & Müller-Hill, B. (1987) The precursor of Alzheimer's disease amyloid A4 protein resembles a cell-surface receptor, *Nature 325*, 733–736.

Kitaguchi, N., Takahashi, Y., Tokushima, Y., Shiojiri, S. & Ito, H. (1988) Novel precursor of Alzheimer's disease amyloid protein shows protease inhibitory activity, *Nature 331*, 530–532.

Koo, E. H. & Squazzo, S. (1994) Evidence that production and release of β-amyloid protein involves the endocytic pathway, *J. Biol. Chem. 269*, 17386–17389.

LaFerla, F. M., Tinkle, B. T., Bieberich, C. J., Haudenschild, C. C. & Jay, G. (1995) The Alzheimer's Aβ peptide induces neurodegeneration and apoptotic cell death in transgenic mice, *Nature Genet. 9*, 21–30.

Lamb, B. T., Sisodia, S. S., Lawler, A. M., Slunt, H. H., Kitt, C. A., Kearns, W. G., Pearson, P. L., Price, D. L. & Gearhart, J. D. (1993) Introduction and expression of the 400 kilobase precursor protein gene in transgenic mice, *Nature Genet. 5*, 22–29.

Lemaire, G., Salbaum, J. M., Multhaup, G., Kang, J., Bayney, R. M., Unterbeck, A., Beyreuther, K. & Müller-Hill, B. (1989) The $preA4_{695}$ precursor protein of Alzheimer's disease A4 amyloid is encoded by 16 exons, *Nucleic Acids Res. 17*, 517–522.

Levy, E., Carman, M. D., Fernandez-Madrid, I. J., Power, M. D., Lieberburg, I., van Duinen, S. G., Bots, G. T. A. M., Luyendijk, W. & Frangione, B. (1990) Mutation of the Alzheimer's disease amyloid gene in hereditary cerebral hemorrhage, Dutch type, *Science 248*, 1124–1126.

Levy-Lahad, E., Wijsman, E. M., Nemens, E., Anderson, L., Goddard, K. A., Weber, J. L., Bird, T. D., Schellenberg, G. D. (1995a) A familial Alzheimer's disease locus on chromosome 1, *Science 269*, 970–973.

Levy-Lahad, E., Wasco, W., Poorkaj, P., Romano, D. M., Oshima, J., Pettingell, W. H., Yu, C., Jondro, P. D., Schmidt, S. D., Wang, K., Crowley, A. C., Fu, Y.-H., Guenette, Y., Galas, D., Nemmes, E., Wijsman, A. M., Bird, T. D., Schellenberg, G., Tanzi, R. E. (1995b). Candidate gene for the chromosome 1 familial Alzheimer's disease locus, *Science 269*, 973–977.

Luyendijk, W., Bots, G. T. A. M., Vegter-Van der Vlis, M., Went, L. N. & Frangione, B. (1988) Hereditary cerebral haemorrhage caused by cortical amyloid angiopathy, *J. Neurol. Sci. 85*, 267–280.

Martin, L. J., Sisodia, S. S., Koo, E. H., Cork, L. C., Dellovade, T. L., Weideman, A., Beyreuther, K., Masters, C. & Price, D. L. (1991) Amyloid precursor protein in aged nonhuman primates, *Proc. Natl Acad. Sci. USA 88*, 1461–1465.

Maruyama, K., Usami, M., Yamao-Harigaya, W., Tagawa, K. & Ishiura, S. (1991) Mutation of Glu^{693} to Gln or Val^{717} to Ile has no effect on the processing of Alzheimer amyloid precursor protein expressed in COS-1 cells by cDNA transfection, *Neurosci. Lett. 132*, 97–100.

Marx, J. (1992) Alzheimer's debate boils over, *Science 257*, 1336–1338.

Masters, C. L., Simms, G., Weinman, N. A., Multhaup, G., McDonals, B. L. & Beyreuther, K. (1985) Amyloid plaque core protein in Alzheimer disease and Down syndrome, *Proc. Natl Acad. Sci. USA 82*, 4245–4249.

Mattson, M. P., Cheng, B., Culwell, A. R., Esch, F. S., Lieberburg, I. & Rydel, R. E. (1993) Evidence for exito protective and intra neuronal calcium regulating roles for secreted forms of β amyloid precursor protein, *Neuron 10*, 243–254.

Mori, H., Takio, K., Ogawara, M. & Selkoe, D. J. (1992) Mass spectrometry of purified amyloid β protein in Alzheimer's disease, *J. Biol. Chem. 267*, 17082–17086.

Mucke, L., Masliah, E., Johnson, W. B., Ruppe, M. D., Alford, M., Rockenstein, E. M., Forss-Petter, S., Pietropaolo, M., Mallory, M. & Abraham, C. R. (1994) Synaptotrophic effects of human amyloid β-protein precursor in the cortex of transgenic mice, *Brain Res. 666*, 151–167.

Mullan, M., Crawford, F., Axelman, K., Houlden, H., Lilius, L., Winblad, B. & Lannfelt, L. (1992) A pathogenic mutation for probable Alzheimer's disease in the APP gene at the N-terminus of β-amyloid, *Nature Genet. 1*, 345–347.

Murrell, J., Farlow, M., Ghetti, B. & Benson, M. D. (1991) A mutation in the amyloid precursor protein associated with hereditary Alzheimer's disease, *Science 254*, 97–99.

Olson, M. I. & Shaw, C. M. (1969) Presenile dementia and Alzheimer's disease in mongolism, *Brain 92*, 147–156.

Oltersdorf, T., Fritz, L. C., Schenk, D. B., Lieberburg, I., Johnson-Wood, K. L., Beattie, E. C., Ward, P. J., Blacher, R. W., Dovey, H. F. & Sinha, S. (1989) The secreted form of the Alzheimer's amyloid precursor protein with the Kunitz domain is protease nexin-II, *Nature 341*, 144–147.

Pearson, B. E. & Choi, T. K. (1993) Expression of the human β-amyloid precursor protein gene from a yeast artificial chromosome in transgenic mice, *Proc. Natl Acad. Sci. USA 90*, 10578–10582.

Perry, T. A., Torres, E., Czech, C., Beyreuther, K., Richards, S. & Dunnett, S. B. (1995) Cognitive and motor function in transgenic mice carrying excess copies of the 695 and 751 amino acid isoforms of the amyloid precursor protein gene, *Alz. Res. 1*, 5–14.

Ponte, P., Gonzalez-De Whitt, P., Shilling, J., Miller, J., Hsu, D., Greenberg, B., Davis, K., Wallace, W., Lieberburg, I., Fuller, F. & Cordell, B. (1988) A new A4 amyloid mRNA contains a domain homologous to serine proteinase inhibitors, *Nature 331*, 525–527.

Price, D. L. & Sisodia, S. S. (1994) Cellular and molecular biology of Alzheimer's disease and animal models, *Annu. Rev. Med. 45*, 435–446.

Quon, D., Wang, Y., Catalano, R., Scardina, J. M., Murakanmi, K. & Cordell, B. (1991) Formation of β-amyloid protein deposits in brains of transgenic mice, *Nature 352*, 239–241.

Rocca, W. A., Hofman, A., Brayne, C., Breteler, M. M. B., Clarke, M., Copeland, J. R. M., Dartiques, J., Engedal, K., Hagnell, O., Heeren, T. J., Jonker, C., Lindesay, J., Lobo, A., Mann, A. H., Mölsä, P. K., Morgan, K., O'Connor, D. W., da Silva Droux, A., Sulkava, R., Kay, D. W. K., Amaducci, L. & for the EURODEM-Prevalence Research Group (1991) Frequency and distribution of Alzheimer's disease in Europe: A collaborative study of 1980–1990 prevalence findings, *Ann. Neurol. 30*, 381–390.

Rogaev, E. I., Sherrington, R., Rogeava, E. A., et al. (1995) Familial Alzheimer's disease in kindreds with missense mutations in a gene on chromosome 1 related to the Alzheimer's disease type 3 gene, *Nature 376*, 775–778

Roher, A. E., Lowenson, J. D., Clarke, S., Woods, A. S., Cotter, R. J., Gowing, E. & Ball, M. J. (1993) β-amyloid-(1–42) is a major component of cerebrovascular amyloid deposits: Implications for the pathology of Alzheimer disease, *Proc. Natl Acad. Sci. USA 90*, 10836–10840.

Salbaum, J. M., Weidemann, A., Lemaire, H., Masters, C. L. & Beyreuther, K. (1988) The promoter of Alzheimer's disease amyloid A4 precursor gene, *EMBO J. 7*, 2807–2813.

Sambamurti, K., Shioi, J., Anderson, J. P., Pappolla, M. A. & Robakis, N. K. (1992) Evidence for intracellular cleavage of the Alzheimer's amyloid precursor in PC12 cells, *J. Neurosci. Res. 33*, 319–329.

Sandbrink, R., Masters, C. L. & Beyreuther, K. (1994) APP gene family: Unique age-associated changes in splicing of Alzheimer's βA4 amyloid protein precursor, *Neurobiol. Dis. 1*, 13–24.

Saunders, A. M., Strittmatter, W. J., Schmechel, D., St George-Hyslop, P. H., Pericak-Vance, M. A., Joo, S. H., Rosi, B. L., Gusella, J. F., Crapper-Mclachlan, D. R., Alberts, M. J., Hulette, C., Crain, B., Goldgaber, D. & Roses, A. D. (1993) Association of apolipoprotein E allele E4 with late-onset familial and sporadic Alzheimer's disease, *Neurology 43*, 1467–1472.

Scheuner, D., Bird, T., Citron, M., Lannfelt, L., Schellenberg, G., Selkoe, D. J., Viitanen, M. & Younkin, G. (1995) Fibroblasts from carriers of familial AD linked to Chromosome 14 show increased Aβ production, *Soc. Neurosci. 1*, 589.10.

Selkoe, D. J. (1994) Alzheimer's disease: A central role for amyloid, *J. Neuropath. Exp. Neurol. 53*, 438–447.

Selkoe, D. J. (1995) Physiological production and polarized secretion of the amyloid β-peptide in epithelial cells: A route to the mechanism of Alzheimer's disease, in *Alzheimer's disease: Lessons from cell biology* (Kosik, K. S., Christen, Y. & Selkoe, D. J., eds) pp. 70–77, Springer-Verlag, Berlin.

Seubert, P., Vigo-Pelfrey, C., Esch, F., Lee, M., Dovey, H., Davis, D., Sinha, S., Schlossmacher, M. G., Whaley, J., Swindlehurst, C., McCormack, R., Wolfert, R., Selkoe, D. J., Lieberburg, I. & Schenk, D. (1992) Isolation and quantification of soluble Alzheimer's β-peptide from biological fluids, *Nature 359*, 325–327.

Seubert, P., Oltersdorf, T., Lee, M. G., Barbour, R., Blomquist, C., Davis, D. L., Bryant, K., Fritz, L. C., Galasko, D., Thal, L. J., Lieberburg, I. & Schenk, D. B. (1993) Secretion of β-amyloid precursor protein cleaved at the amino terminus of the β-amyloid peptide, *Nature 361*, 260–263.

Sherrington, R., Rogaev, E. I., Liang, Y., Rogaeva, E. A., Levasque, G., Ikeda, M., Chi, H., Lin, C., Li, G., Holman, K., Tsuda, T., Mar, L., Foncin, J. F., Bruni, A. C., Montesi, M. P., Sorbi, S., Rainero, I., Pinessi, L., Nee, L., Chumakov, I., Pollen, D., Brookes, A., Sansosu, P., Polinsky, R. J., Wasco, W., Da Silva, H. A. R., Haines, J. L., Pericak-Vance, M. A., Tanzi, R. E., Roses, A. D., Fraser, P. E., Rommens, J. M. & St George-Hyslop, P. (1995) Cloning of a gene bearing mis-sense mutations in early-onset familial Alzheimer's disease, *Nature 375*, 754–760.

Shoji, M., Golde, T. E., Ghiso, J., Cheung, T. T., Estus, S., Shaffer, L. M., Cai, X., McKay, D. M., Tintner, R., Frangione, B. & Younkin, S. G. (1992) Production of the Alzheimer amyloid β protein by normal proteolytic processing, *Science 258*, 126–129.

Sisodia, S. S. (1992) β-amyloid precursor protein cleavage by a membrane-bound protease, *Proc. Natl Acad. Sci. USA 89*, 6075–6079.

Small, D. H., Nurcombe, V., Reed, G., Clarris, H., Moir, R., Beyreuther, K. & Masters, C. L. (1994) A heparin-binding domain in the amyloid protein precursor of Alzheimer's disease is involved in the regulation of neurite outgrowth, *J. Neurosci. 14*, 2117–2127.

Smith, R. P., Higuchi, D. A. & Broze, G. J. (1990) Platelet coagulation factor XIa-inhibitor, a form of Alzheimer's amyloid precursor protein, *Science 248*, 1126–1128.

Song, X.-H, Suzuki, N., Bird, T., Peskind, E., Schellenberg, G. & Younkin, S. (1995) Plasma amyloid β protein (Aβ) ending at Aβ42(43) is increased in cariers of familials AD (FAD) linked to chromosome 14. *Soc. Neurosci. 1*, 589.11

St George-Hyslop, P. H., Tanzi, R. E., Polinsky, P. J., Haines, J. L., Nee, L., Watkins, P. C., Myers, R. H., Feldman, R. G., Pollen, D., Drachman, D., Growdon, J., Bruni, A., Foncin, J., Salmon, D., Frommelt, P., Amaducci, L., Sorbi, S., Piacentini, S., Steward, G. D., Hobbs, W. J., Conneally, P. M. & Gusella, J. F. (1987) The genetic defect causing familial Alzheimer's disease maps on chromosome 21, *Science 235*, 885–890.

St George-Hyslop, P. H., Haines, J. L., Farrer, L. A., Polinsky, R., Van Broeckhoven, C., Goate, A., Crapper-Mclachlan, D. R., Orr, H., Bruni, A. C., Sorbi, S., Rainero, I., Foncin, J. F., Pollen, D., Cantu, J., Tupler, R., Voskresenskaya, N., Mayeux, R., Growdon, J., Fried, V. A., Myers, R. H., Nee, L., Backhovens, H., Martin, J., Rossor, M., Owen, M. J., Mullan, M., Percy, M. E., Karlinsky, H., Rich, S., Heston, L., Montesi, M., Mortilla, M., Nacmias, N., Gusella, J. F., Hardy, J. A. & other members of the FAD Collaborative Study group (1990) Genetic linkage studies suggest that Alzheimer's disease is not a single homogeneous disorder, *Nature 347*, 194–197.

Strittmatter, W. J., Saunders, A. M., Schmechel, D., Pericak-Vance, M., Enghild, J., Salvesen, G. S. & Roses, A. D. (1993a) Apolipoprotein E: High-avidity binding to β-amyloid and increased frequency of type 4 allele in late-onset familial Alzheimer disease, *Proc. Natl Acad. Sci. USA 90*, 1977–1981.

Strittmatter, W. J., Weisgraber, K. H., Huang, D., Dong, L.-M., Salvesen, G. S., Pericak-Vance, M., Schmechel, D., Saunders, A. M., Goldgaber, D. & Roses, A. D. (1993b) Binding of human apolipoprotein E to synthetic amymloid β peptide: Isoform-specific effects and implications for late-onset Alzheimer disease, *Proc. Natl Acad. Sci. USA 90*, 8098–8102.

Suzuki, N., Cheung, T. T., Cai, X., Odaka, A., Otvos Jr, L., Eckman, C., Golde, T. E. & Younkin, S. G. (1994) An increased percentage of long amyloid β protein secreted by familial amyloid β protein precursor (βAPP_{717}) mutants, *Science 264*, 1336–1340.

Tagliavini, F., Giaccone, G., Frangione, B. & Bugiani, O. (1988) Preamyloid deposits in the cerebral cortex of patients with Alzheimer's disease and nondemented individuals, *Neurosci. Lett. 93*, 191–196.

Tamaoka, A., Odaka, A., Ishibashi, Y., Usami, M., Sahara, N., Suzuki, N., Nukina, N., Mizusawa, H., Shoji, S., Kanazawa, I. & Mori, H. (1994a) APP 717 missence mutation affects the ratio of amyloid β-protein species (Aβ1-42/43 and Aβ1-40) in familial Alzheimer's disease brain, *J. Biol. Chem. 269*, 32721–32724.

Tamaoka, A., Kondo, T., Odaka, A., Sahara, N., Sawamura, N., Ozawa, K., Suzuki, N., Shoji, S. & Mori, H. (1994b) Biochemical evidence for the long-tail form Aβ1-42/43 of amyloid β protein as a seed molecule in cerebral deposits of Alzheimer's disease, *Biochem. Biophys. Res. Commun. 205*, 834–842.

Tanzi, R. E., Gusella, J. F., Watkins, P. C., Bruns, G. A. P., St George-Hyslop, P., Van Keuren, M. L., Patterson, D., Pagan, S., Kurnitt, D. M. & Neve, R. L. (1987) Amyloid β protein gene: cDNA, mRNA distribution, and genetic linkage near the Alzheimer locus, *Science 235*, 880–884.

Tanzi, R. E., McClatchey, A. I., Lamperti, E. D., Villa-Komaroff, L., Gusella, J. F. & Neve, R. L. (1988) Protease inhibitor domain encoded by an amyloid protein precursor mRNA associated with Alzheimer's disease, *Nature 331*, 528–530.

Tanzi, R. E., St George-Hyslop, P. H. & Gusella, J. F. (1989) Molecular genetic approaches to Alzheimer's disease, *Trends Nuerosci. 12*, 152–158.

Timmers, W. F., Tagliavini, F., Haan, J. & Frangione, B. (1990) Parenchymal preamyloid and amyloid deposits in the brain of patients with hereditary cerebral hemorrhage with amyloidosis – Dutch type, *Neurosci. Lett. 118*, 223–226.

Van Broeckhoven, C., Haan, J., Bakker, E., Hardy, J. A., Van Hul, W., Wehnert, A., Vegter-Van der Vlis, M. & Roos, R. A. C. (1990) Amyloid β protein precursor gene and hereditary cerebral hemorrhage with amyloidosis (Dutch), *Science 248*, 1120–1122.

Van Broeckhoven, C. (1995a) Molecular genetics of Alzheimer disease: Identification of genes and gene mutations, *Eur. Neurol. 35*, 8–19.

Van Broeckhoven, C. (1995b) Presenilins and Alzheimer disease, *Nature Genet. 11*, 230–232.

van Duijn, C. M., de Knijff, P., Cruts, M., Wehnert, A., Havekes, L. M., Hofman, A. & Van Broeckhoven, C. (1994) Apolipoprotein E4 allele in a population-based study of early-onset Alzheimer's disease, *Nature Genet. 7*, 74–78.

van Duijn, C. M., de Knijff, P., Wehnert, A., De Voecht, J., Bronzova, J. B., Havekes, L. M., Hofman, A. & Van Broeckhoven, C. (1995) The apolipoprotein E $\varepsilon 2$ allele is associated with an increased risk of early-onset Alzheimer's disease and a reduced survival, *Ann. Neurol. 37*, 605–610.

van Duinen, S. G., Castaño, E. M., Prelli, F., Bots, G. T. A. B., Luyendijk, W. & Frangione, B. (1987) Hereditary cerebral hemorrhage with amyloidosis in patients of Dutch origin is related to Alzheimer disease, *Proc. Natl Acad. Sci. USA 84*, 5991–5994.

Van Nostrand, W. E., Wagner, S. L., Suzuki, M., Choi, B. H., Farrow, J. S., Geddes, J. W., Cotman, C. W. & Cunningham, D. D. (1989) Protease Nexin-II, a potent antichymotrypsin, shows identity to amyloid β-protein precursor, *Nature 341*, 546–549.

Van Nostrand, W. E., Schmaier, A. H., Farrow, J. S. & Cunningham, D. D. (1990) Protease nexin-II (amyloid β-protein precursor): A platelet α-granule protein, *Science 248*, 745–748.

Wisniewski, K. E., Dalton, A. J., Crapper-Mclachlan, D. R., Wen, G. Y. & Wisniewski, H. M. (1985) Alzheimer's disease in Down's syndrome: Clinicopathologic studies, *Neurology 35*, 957–961.

Wisniewski, T., Ghiso, J. & Frangione, B. (1991) Peptides homologous to the amyloid protein of Alzheimer's disease containing a glutamine for glutamic acid substitution have accelerated amyloid fibril formation, *Biochem. Biophys. Res. Commun. 3*, 1247–1254.

Wisniewski, T. & Frangione, B. (1992) Apolipoprotein E: A pathological chaperone protein in patients with cerebral and systemic amyloid, *Neurosci. Lett. 135*, 235–238.

Yamaguchi, H., Nakazato, Y., Hirai, S., Shoji, M. & Harigaya, Y. (1989) Electron micrograph of diffuse plaques, *Am. J. Pathol. 135*, 593–597.

Yanker, B. A., Dawes, L. R., Fischer, S., Villa-Komaroff, L., Oster-Granite, M. L. & Neve, R. L. (1989) Neurotoxicity of a fragment of the amyloid precursor associated with Alzheimer's disease, *Science 245*, 417–420.

Yoshikai, S., Sasaki, H., Doh-ura, K., Furuya, H. & Sakaki, Y. (1990) Genomic organization of the human amyloid β-protein precursor gene, *Gene 87*, 257–263.

Yoshikawa, K., Aizawa, T. & Hayashi, Y. (1992) Degeneration *in vitro* of post-mitotic neurons overexpressing the Alzheimer amyloid protein precursor, *Nature 359*, 64–67.

Zheng, H., Jiang, M., Trumbauer, M. E., Sirinathsinghji, D. J. S., Hopkins, R., Smith, D. W., Heavens, R. P., Dawson, G. R., Boyce, S., Conner, M. W., Stevens, K. A., Slunt, H. H., Sisodia, S. S., Chen, H. Y. & Van der Ploeg, L. H. T. (1995) β-Amyloid precursor protein-deficient mice show reactive gliosis and decreased locomotor activity, *Cell 81*, 525–531.

Eur. J. Biochem. 237, 323–343 (1996)

Review

The regulation of enzymes involved in chlorophyll biosynthesis

Steffen REINBOTHE and Christiane REINBOTHE

Institute for Plant Sciences, Department of Genetics, Swiss Federal Institute of Technology Zurich (ETH), Switzerland

(Received 26 September/10 November 1995) – EJB 95 1581/0

All living organisms contain tetrapyrroles. In plants, chlorophyll (chlorophyll *a* plus chlorophyll *b*) is the most abundant and probably most important tetrapyrrole. It is involved in light absorption and energy transduction during photosynthesis. Chlorophyll is synthesized from the intact carbon skeleton of glutamate via the C_5 pathway. This pathway takes place in the chloroplast. It is the aim of this review to summarize the current knowledge on the biochemistry and molecular biology of the C_5-pathway enzymes, their regulated expression in response to light, and the impact of chlorophyll biosynthesis on chloroplast development. Particular emphasis will be placed on the key regulatory steps of chlorophyll biosynthesis in higher plants, such as 5-aminolevulinic acid formation, the production of Mg^{2+}-protoporphyrin IX, and light-dependent protochlorophyllide reduction.

Keywords: chlorophyll biosynthesis; chloroplast development; C_5-pathway enzymes; enzymology; gene expression.

Tetrapyrroles play an essential role in all living organisms. They are involved in various metabolic processes, such as energy transfer, signal transduction, and catalysis (Dailey, 1990; Senge, 1993). Both linear and cyclic tetrapyrroles are highly reactive; they are able to form radicals upon exposure to light and oxygen. Cyclic tetrapyrroles in particular, such as heme, additionally contain metals as central ions that can undergo reversible changes in their redox state. Due to the remarkable reactivity of all tetrapyrroles, there is in living organisms a substantial danger that uncontrolled chemical reactions may occur and ultimately lead to damage of cellular and subcellular structures. For example, animals and humans impaired in various steps of heme biosynthesis suffer from a large number of disorders, termed porphyrias (Nordman and Deybach, 1990). All living organisms thus must have evolved strategies to neutralize the potentially harmful tetrapyrrole compounds. In most cases, this task is achieved by tightly controlling their actual concentrations. By binding to a variety of different proteins, the level of free tetrapyrroles within a cell is kept to a minimum. In addition, tetrapyrrole synthesis and degradation are carefully adjusted to the cellular requirements, reflecting the different needs under varying environmental conditions.

In plants, at least three structurally and functionally distinct classes of tetrapyrroles can be distinguished, Mg^{2+}-porphyrins, $Fe^{2+/3+}$-porphyrins, and phycobilins (Fig. 1; Beale and Weinstein, 1990). Mg^{2+}-porphyrins, such as chlorophylls (Fig. 1A), are the most abundant tetrapyrrole compounds. As part of the photosynthetic apparatus, the magnesium porphyrin rings of the chlorophylls *a* and *b* (Chls *a* and *b*) are involved in light trapping and energy transduction.

$Fe^{2+/3+}$-porphyrins (hemes) constitute the second class of tetrapyrroles. Protoheme (Fig. 1B) in particular functions as a redox-active cofactor or prosthetic group in many proteins (Padmanaban et al., 1989; Dailey, 1990). For example, protoheme is bound to the various cytochromes of the plastidic and mitochondrial electron transport chains, and to soluble enzymes, such as catalase and peroxidase. Apart from these functions, (proto)-heme and its derivatives have also been shown to regulate a variety of cellular processes in animal and bacterial systems, such as transcription (Guarente and Mason, 1983), translation (Chen et al., 1989; Joshi et al., 1995), and post-translational protein translocation (Lathrop and Timko, 1993).

Phytochromobilin (Fig. 1C) belongs to the third class of plant tetrapyrroles, the phycobilins. In contrast to chlorophylls

Correspondence to S. Reinbothe, Institute for Plant Sciences, Department of Genetics, Swiss Federal Institute of Technology Zurich (ETH), ETH-Zentrum, Universitätsstrasse 2, CH-8092 Zurich, Switzerland

Fax: +41 1632 1081.

Abbreviations. Chl, chlorophyll; Chlide, chlorophyllide; ELIP, early light-induced protein; Pchlide, protochlorophyllide; POR, NADPH: protochlorophyllide oxidoreductase.

Enzymes. Glutamyl-tRNA synthetase (glutamate–tRNA ligase) (EC 6.1.1.17); glutamate-1-semialdehyde aminotransferase (glutamate-1-semialdehyde 2,1-aminomutase) (EC 5.4.3.8); 5-aminolevulinic acid dehydratase (porphobilinogen synthase) (EC 4.2.1.24); porphobilinogen deaminase (hydroxymethylbilane synthase) (EC 4.3.1.8); uroporphyrinogen III synthase (EC 4.2.1.75); uroporphyrinogen III decarboxylase (EC 4.1.1.37); coproporphyrinogen III oxidase (EC 1.3.3.3); protoporphyrinogen oxidase (EC 1.3.3.4); ferrochelatase (protoheme: ferro-lyase) (EC 4.99.1.1); *S*-adenosyl-L-methionine: magnesium-protoporphyrin *O*-methyltransferase (EC 2.1.1.11); NADPH protochlorophyllide oxidoreductase (EC 1.3.1.33).

Note. There appears to be some confusion in the literature about the use of the term heme. In most textbooks, heme is used to define the Fe^{2+}-protoporphyrin chromophore of hemoproteins. According to the recommendations of NC-IUBMB, however, the term heme should be used to collectively refer to iron porphyrin coordination complexes independently of whether the redox state of the iron atom is Fe^{2+} or Fe^{3+} and regardless of the protein to which the pigment chromophore is bound (Moss, 1988). We therefore prefer to use the term protoheme for Fe^{2+}-protoporphyrin IX not bound to protein in this review.

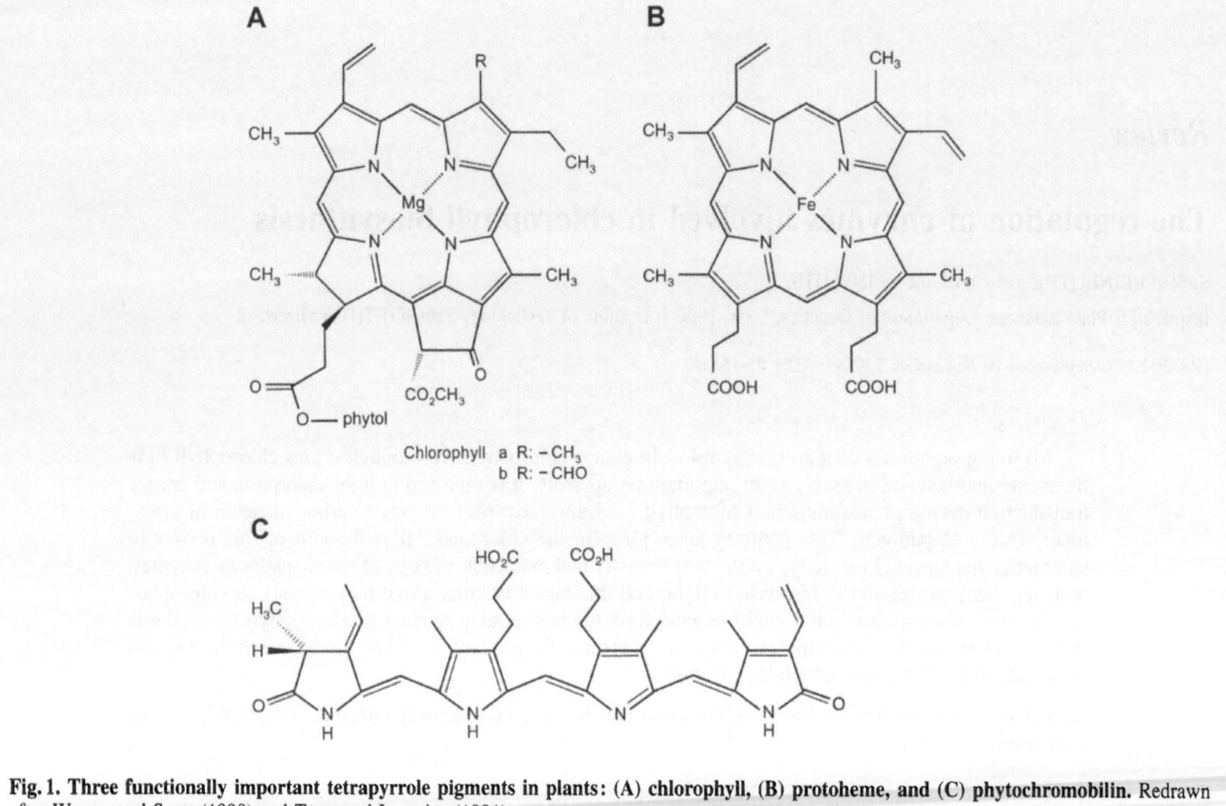

Fig. 1. Three functionally important tetrapyrrole pigments in plants: (A) chlorophyll, (B) protoheme, and (C) phytochromobilin. Redrawn after Warren and Scott (1990) and Terry and Lagarias (1991).

and hemes, phycobilins are linear pigments. Phytochromobilin is covalently bound to the phytochrome apoprotein (Lagarias and Rapoport, 1980), and this holocomplex is known to act as a photoreceptor that controls various developmental processes (Furuya, 1993; Kendrick and Kronenberg, 1994).

In plants and many bacteria, all of the different tetrapyrroles seem to originate from a common biosynthetic pathway (Beale and Weinstein, 1990). As shown in Fig. 2, this pathway, which starts at glutamate and thus has been termed the C_5 pathway, comprises numerous metabolites and a variety of different reactions. In most cases, the enzymes catalyzing these reactions have been identified. In the last few years, considerable progress has been made in elucidating the plant enzymes, cloning their genes, and studying their regulated expression in response to light. A complete picture should emerge as to how the individual enzymes operate, how their activity may be regulated, and how their expression can affect the flow of metabolites through the C_5 pathway, reflecting the varying needs for tetrapyrroles in response to different environmental conditions. Although one might feel prompted to summarize all of the different exciting novel aspects and recent developments in the field of chlorophyll biosynthesis, this review will mainly concentrate on those steps that have been shown to be key regulatory sites of the pathway. Special emphasis will be placed on our own recent results obtained for the NADPH:protochlorophyllide oxidoreductase (POR), which is the key enzyme of chlorophyll biosynthesis in angiosperms. POR is unique in its ability to catalyze the only light-requiring step of the C_5 pathway (Griffiths, 1975; Apel et al., 1980). For a comprehensive treatise of all of the other enzymes of this pathway, the reader is referred to the excellent papers of Rüdiger and Schoch (1988), Beale and Weinstein (1990), Kannangara et al. (1994), and von Wettstein et al. (1995).

Morphological, cellular, and molecular events during the light-induced greening of etiolated angiosperm seedlings

Morphological and cellular changes in angiosperm seedlings upon illumination. When angiosperm seedlings are grown in the dark, they exhibit an etiolated phenotype. In dicotyledonous plant species, the seedling has a long and thin hypocotyl which is topped by an apical hook with small, closely apposed cotyledons. In the light, the phenotype of the seedling is quite distinct: the hypocotyl is short, and the cotyledons are unfolded and in many species expanded. In monocotyledonous plants, the phenotypic differences between dark-grown and light-grown seedlings are much less pronounced. However, in both monocotyledonous and dicotyledonous plant species, light is the key factor for the development of photosynthetically active plants. In particular, light controls plant, leaf, and plastid development (Kendrick and Kronenberg, 1994).

Chloroplasts are derived from undifferentiated progenitors, termed proplastids (Kirk and Tilney-Basset, 1967, 1978; Thomson and Whatley, 1980). These small, spherical organelles are in most cases inherited maternally by the plant zygote. After germination, proplastids are either maintained in an undifferentiated state in meristematic cells of the plant, or develop into other plastid types (Kirk and Tilney-Basset, 1978). In the light, chloroplasts are formed in a process that is coupled to the production of mesophyll cells. Changes in the number, size, and composition of the proplastids occur during mesophyll cell development. Superimposed on these events, leaf development proceeds (Mullet, 1988).

When angiosperm seedlings are germinated in the dark, their cotyledons contain another plastid type, the etioplast (Kirk and Tilney-Basset, 1967, 1978; Thomson and Whatley, 1980).

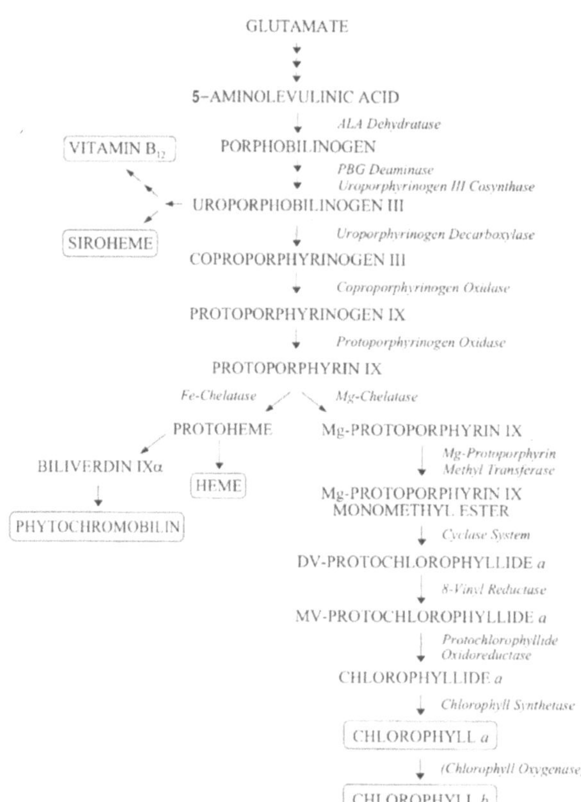

Fig. 2. The biosynthetic pathway of tetrapyrroles in plants and nonphotosynthetic bacteria. In plants, the main products of the C_5 pathway are Chls *a* and *b*, heme, and phytochromobilin. In contrast, nonphotosynthetic bacteria cannot synthesize chlorophylls and phytochromobilin but form heme, siroheme and vitamin B_{12}. Intermediates of the pathway are printed in upper case letters, whereas enzymes catalyzing the different reactions are printed in italics. The nature of the last enzyme of the Mg-branch of the C_5 pathway, chlorophyll oxygenase, is still hypothetical. The abbreviations used are as follows: DV, 3,8-divinyl; MV, 3-monovinyl(protochlorophyllide); PBG, porphobilinogen. Redrawn and modified after Matringe et al. (1994) and Beale and Weinstein (1990).

Within this organelle, the prolamellar body is a highly ordered, paracrystalline structure, which is composed of lipids and proteins (Ryberg and Sundqvist, 1991). The most abundant protein appears to have a molecular mass of approximately 36 kDa (Høyer-Hansen and Simpson, 1977; Santel and Apel, 1981; Ikeuchi and Murakami, 1983). This polypeptide represents POR (Dehesh and Ryberg, 1985; Shaw et al., 1985; Ryberg and Dehesh, 1986). As part of the prolamellar body, POR forms a ternary complex together with NADPH and protochlorophyllide (Pchlide) (Griffiths, 1974, 1975, 1978). Upon illumination, POR photoreduces Pchlide to chlorophyllide (Chlide) (Griffiths, 1978; Mapleston and Griffiths, 1980; Apel et al., 1980; Santel and Apel, 1981) and, simultaneously, the prolamellar body begins to break (Virgin et al., 1963; Henningsen, 1970; Henningsen et al., 1993). Kahn (1968) proposed that photoconversion of Pchlide and transformation of the prolamellar body may represent a single, photodependent process. As a result of this process, chloroplast differentiation is induced, as observed by the formation of the thylakoids (Virgin et al., 1963; Henningsen, 1970). The assembly of the photosynthetic apparatus within the thylakoid membranes leads to the visible greening of the seedling (Thorne, 1971).

Nuclear gene expression in response to light. During the rapid build-up phase of the photosynthetic apparatus, chlorophylls and nuclear-encoded chlorophyll-*a/b*-binding proteins must be synthesized coordinately. Several different steps control this process. At the level of light perception, red/far-red light-absorbing phytochromes (Quail, 1991; Furuya, 1993) and blue light-absorbing photoreceptors (Kaufman, 1993; Short and Briggs, 1994), as well as other less well characterized photoreceptors (Senger, 1984) have been identified. Phytochrome through its ability to reversibly photointerconvert between red and far-red light-absorbing forms, P_r and P_{fr}, sets in motion a complex signal-transduction cascade, the details of which are only poorly understood (Chory, 1993; Deng, 1994; Elich and Chory, 1994). Light absorption by the chromophore (phytochromobilin) is thought to induce a protein conformational change that initiates processes such as heterotrimeric guanine-nucleotide-binding-protein activation, changes in cytosolic Ca^{2+} concentration, and calmodulin activation (Neuhaus et al., 1993; Bowler et al., 1994). At the end of this Ca^{2+}-dependent and cyclic-GMP-dependent signal-transduction pathway, nuclear genes such as *rbcS*, encoding the small subunit of ribulose-1,5-bisphosphate carboxylase/oxygenase, *lhca* and *lhcb*. encoding the light-harvesting chlorophyll-*a/b*-binding proteins of photosystems I and II, respectively, are activated and transcribed (Harpster and Apel, 1985; Tobin and Silverthorne, 1985; Thompson and White, 1991). *Trans*-acting factors bind to *cis*-regulatory elements in the promoters of light-regulated genes (Gilmartin et al., 1990; Schindler and Cashmore, 1990).

In addition to light-induced genes, such as *rbcS* and *lhca/b*, there are also a few genes whose expression is negatively regulated by light (Harpster and Apel, 1985; Thompson and White; 1991). It has been shown that phytochrome depresses transcription of its own gene, *phyA* (Lissemore and Quail, 1988; Bruce and Quail, 1990; Dehesh et al., 1990; Bruce et al., 1991), and also the transcription of *porA*, the gene encoding POR (Batschauer and Apel, 1984; Mösinger et al., 1985). As a result of illumination, the inherently unstable *phyA* and *porA* mRNAs rapidly decline (Apel, 1981; Colbert et al., 1985; Colbert, 1988). Superimposed on these processes, rapid proteolytic degradation drastically lowers the actual concentrations of the PHYA and PORA proteins (Kay and Griffiths, 1983; Häuser et al., 1984; Shanklin et al., 1989). While the low amount of the PHYA holoprotein is consistent with its role as a photoreceptor and key regulator of photomorphogenesis in plants (Kendrick and Kronenberg, 1994), the minute amounts of PORA have long been a matter of dispute (Holtorf et al., 1995). PORA is virtually absent both from rapidly greening and fully light-adapted plants, in which continuous photoreduction of Pchlide to Chlide must occur to sustain chlorophyll synthesis. However, recent results have solved this apparent paradox, as they have shown that there is a second Pchlide-reducing enzyme, termed PORB, which is constitutively expressed in etiolated, illuminated, and fully green barley plants (Holtorf et al., 1995).

Post-translational transport of cytosolic precursor proteins into the plastids. Most plastid proteins are nuclear gene products and thus must be transported post-translationally into the organelle. Numerous steps have been identified that are involved in the targeting of the precursor proteins to the chloroplasts and their different compartments (Keegstra et al., 1989, 1995; Archer and Keegstra, 1990; de Boer and Weisbeek, 1991; Reinbothe et al., 1995b). These include cytosolic synthesis of the precursors with N-terminal extensions, referred to as the chloroplast transit peptides, energy-dependent binding of the precursor proteins to receptors on the outer plastid surface, and translocation of the proteins across the outer and inner plastid envelope membranes. During or shortly after membrane passage, the N-terminal chloroplast transit peptides are cleaved off by the leader peptidase. Prior to, during, and after translocation, complex protein/protein interactions between the precursor polypeptides and cytosolic, membrane and stromal factors, respectively, occur. In addition, enzymes bind cofactors, prosthetic groups, and substrates, to attain their final conformations and activities. In the case of multimeric enzymes, protomers have to assemble into the active holocomplexes. Post-import routing is required for proteins that are destined for either the thylakoid membranes or the thylakoid lumen.

Thus far, the post-translational import pathway by which the various precursor proteins enter the organelle has been considered to be a house-keeping function of the plant cell. However, our recent results obtained for PORA of barley (Reinbothe et al., 1995a, b and unpublished results) demonstrate that protein translocation can also be a specific regulatory step during light-induced chloroplast development.

Plastid gene expression. Within the plastid compartment, both transcriptional and post-transcriptional steps control the light-induced transformation of etioplasts to chloroplasts (Mullet, 1988, 1993; van Grinsven and Kool, 1988; Reinbothe et al., 1993). Differences in the absolute rates of transcription initiation in chloroplasts versus etioplasts, based on differences in the promoter strengths of individual genes and limiting RNA polymerase levels, have been demonstrated to be determinants for mRNA accumulation. Post-transcriptional steps that affect mRNA abundances include intron splicing, 5′ and 3′ transcript processing, as well as preferential mRNA stabilization (Gruissem, 1989). Translational regulation was also implicated as being involved in the control of plastid protein formation (Kreuz et al., 1986; Laing et al., 1988; Danon and Mayfield, 1991; Eichacker et al., 1992).

Plastid-derived factors that control nuclear gene expression. Nuclear gene expression is controlled not only by light operating via phytochrome and other photoreceptors but also by factors originating from the plastid (Oelmüller, 1989; Taylor, 1989; Rajasekhar, 1991; Susek and Chory, 1992). The existence of such factors, which are often collectively referred to as the plastid signal, was implicated based on the following findings.

When angiosperm seedlings are treated with norflurazon [4-chloro-5-(methylamino)-2-(*α,α,α*-trifluoro-*m*-tolyl)-3(2*H*)-pyridazinone], they become photobleached under normal daylight conditions or high white light fluence rates. This effect is due to inhibition of carotenoid biosynthesis and subsequent photooxidation of chlorophyll in the absence of the quenching pigments (Telfer et al., 1990, 1991; De Las Rivas et al., 1993). As a consequence, further chloroplast development does not proceed. Transcripts encoding plastid proteins, such as the small subunit of ribulose-1,5-bisphosphate carboxylase and the light-harvesting chlorophyll-*a/b*-binding proteins, as well as mRNAs for peroxisomal and other proteins related to plastid function rapidly disappear (Reiss et al., 1983; Mayfield and Taylor, 1984; Harpster et al., 1984; Oelmüller and Mohr, 1986; Oelmüller et al., 1986a, 1988; Oelmüller and Schuster, 1987) because their nuclear genes are no longer transcribed (Simpson et al., 1986; Stockhaus et al., 1987; Ernst and Schefbeck, 1988; Hess et al., 1994). Similar to norflurazon, nuclear mutations that cause carotenoid deficiency and thus ultimately lead to photooxidation of the plastid compartment negatively influence the expression of the nuclear *rbcS* and *lhca/b* genes (Mayfield and Taylor, 1984; Batschauer et al., 1986; Burgess and Taylor, 1987). For example, in carotenoid-deficient *albina-f*[17] mutants of barley neither *rbcS* nor *lhcb* transcripts accumulate (Rapp and Mullet, 1991). Collectively, these results demonstrate that the expression of photosynthetic and other nuclear genes is controlled also by signals derived from the plastid. Their release appears to be dependent on the integrity of the plastid compartment.

Numerous attempts have been made to identify the plastid factor. Inhibitor experiments with chloramphenicol showed that plastid protein synthesis is not required for the generation of the plastid factor (Oelmüller et al., 1986b; Hess et al., 1994). By contrast, treatment of leaves from barley and other plant species with tagetitoxin, an inhibitor of prokaryotic-type RNA polymerases (Mathews and Durbin, 1990), completely blocked accumulation of *rbcS* and *lhcb*, but not of actin transcripts (Rapp and Mullet, 1991). Plastid transcription, but not plastid protein synthesis, thus seems to be necessary for the specific expression of nuclear genes during the early stages of chloroplast development.

A participation of RNA in the control of nuclear gene expression has long been proposed (Bradbeer et al., 1979; Parthier, 1982). Although intriguing, this notion is based on the assumption that RNA could be exported from the plastid to the cytosol. Although data obtained by Schoelz and Zaitlin (1989) suggest an exchange of RNA between plastids and their environment, as observed by import of tobacco mosaic virus RNA into tobacco chloroplasts *in vivo*, there is no direct evidence for RNA export from the plastid (Susek and Chory, 1992). Thus, an indirect involvement of RNA in the production and/or liberation of the plastid factor seems to be more likely. Because $tRNA^{Glu}$ is involved in driving tetrapyrrole synthesis (see below), it has been proposed that a pigment intermediate in the C_5 pathway could accomplish the function of the plastid factor (Johanningmeier and Howell, 1984).

The assembly of the photosynthetic apparatus. During the assemby of the photosynthetic apparatus, chlorophylls and the nuclear-encoded and plastid-encoded chlorophyll-binding proteins must interact in a highly coordinated manner (Bogorad, 1967; von Wettstein et al., 1971; Boardman et al., 1978). The accumulation of the different chlorophyll-protein complexes during thylakoid membrane biogenesis appears to be a sequential process that has been suggested to start with the accumulation of the plastid-encoded chlorophyll-binding proteins (Falbel and Staehelin, 1994). First, the reaction centers, comprising Chl *a* and mostly the plastid-encoded apoproteins, such as D1 and D2 of photosystem II (the *psbA* and *psbD* gene products), the P700A and P700B proteins of photosystem I (the *psaA* and *psaB* gene products), and the antenna polypeptides associated with photosystem II, CP43 and CP47 (the *psbC* and *psbB* gene products), are formed. Shortly thereafter, Chl *b* is synthesized and together with Chl *a* bound to the nuclear-encoded, post-translationally imported light-harvesting chlorophyll-*a/b*-binding proteins. Although this sequential process has been documented for different plants species including barley (Anadan et al., 1993; Dreyfuss and Thornber, 1994a, b; Sigrist and Staehelin, 1994), the precise mechanism of its control is only poorly understood (Mattoo et al., 1989; Nitschke and Rutherford, 1991; Goldbeck, 1993).

Previous results suggested that *de novo* synthesis of Chl *a* may be the key factor for the whole assembly pathway. Eichacker et al. (1990, 1992) proposed that nascent Chl *a* may be required for translation of the P700, CP43, CP47, and D2 apoproteins. Different lines of evidence appeared to support this conclusion. First, pulse-labeling studies had shown that the synthesis of these and other apoproteins was depressed in the dark, but occurred in the light (Klein and Mullet, 1986, 1987; Klein et al., 1988a). Secondly, plastid polysomes contained approximately equal amounts of *psaA-psaB* and *psbA* mRNAs both in the dark and in the light (Klein et al., 1988b). Thirdly, light-induced chlorophyll protein accumulation was similar to the time course of photoreduction of Pchlide to Chlide (Castefranco and Beale, 1981; Vierling and Alberte, 1983; Klein and Mullet, 1986). Finally, Chl-*a*-deficient barley mutants were unable to accumulate chlorophyll apoproteins in the light (Klein et al., 1988a). Eichacker et al. (1992) thus concluded that the inability to synthesize Chl *a* in the dark may be the cause for the depression of translation of the chlorophyll apoproteins in barley etioplasts. Seemingly in favor of this explanation were findings which had shown that ribosomes pause during translation of the D1 apoprotein (Kim et al., 1991). However, more refined experiments disproved the possibility of regulation at the level of translational elongation, as they demonstrated that CP43, P700, and D1 protein accumulation was solely due to an increase in apoprotein stability, which was caused by their binding to Chl *a* (Mullet et al., 1990; Kim et al., 1994). Similar to these findings, Chls *a* and *b* have been shown to stabilize the nuclear-encoded chlorophyll-*a/b*-binding proteins (Apel and Kloppstech, 1980; Bennett, 1981; Preiss and Thornber, 1995). Furthermore, oligomerization of chlorophyll-*a/b*-binding proteins is an essential step in the assembly pathway of the photosynthetic complexes (Hobe et al., 1994; Dreyfuss and Thornber, 1994b; Boekema et al., 1995). In addition to Chls *a* and *b*, the different chlorophyll-protein complexes also bind carotenoids (Krauss et al., 1993; Lee and Thornber, 1995), which quench Chl triplet states and thus help prevent the formation of harmful oxygen radicals and singlet oxygen (Siefermann-Harms, 1987). There appears to be an excess of nuclear-encoded and plastid-encoded chlorophyll apoproteins that bind free pigment molecules whenever they are synthesized (Kim et al., 1994).

Biosynthesis of chlorophylls, hemes, and other tetrapyrroles

Steps, intermediates, and enzymes involved in the different branches of the C_5 pathway. *The three-step reaction leading to the formation of 5-aminolevulinic acid.* The C_5-pathway generates many different compounds. This is due to the fact that there are several branch points at which this pathway can diverge. The first committed precursor of all tetrapyrroles is 5-aminolevulinic acid (Fig. 2). Originally, this conclusion was based on the following experiments. When angiosperm seeds are germinated in the dark, they form yellow leaves that are devoid of chlorophyll. If such seedlings are incubated with 5-aminolevulinic acid, they turn greenish however (Granick, 1959; Sisler and Klein, 1963; Nadler and Granick, 1970; Castelfranco et al., 1974). This apparent greening is due to the uncontrolled accumulation of Pchlide. Similarly, isolated plastids accumulate excess Pchlide when fed with 5-aminolevulinic acid in the dark (Kannangara and Gough, 1977; Fuesler et al., 1984). Because the reduction of Pchlide to Chlide is a strictly light-dependent process (Griffiths, 1975, 1978), no chlorophyll accumulation can be observed in dark-grown seedlings. Collectively, these experiments show (a) that all of the enzymes necessary for the conversion of 5-aminolevulinic acid to Pchlide must be present in dark-grown seedlings, and (b) that the entire pathway is likely to operate in the plastids.

5-Aminolevulinic acid is formed from the intact carbon skeleton of glutamate in a three-step reaction involving a tRNA (Fig. 3). Much progress has been made in characterizing this interesting sequence of reactions. Glutamyl-tRNA synthetase is the first enzyme of 5-aminolevulinic acid synthesis (Huang et al., 1984) (Fig. 3). It catalyzes the Mg^{2+}-dependent and ATP-dependent activation of glutamate by esterification at its α-carboxyl group to either the 2′ or 3′ OH of a specific plastid tRNA, $tRNA^{Glu}$ (Kannangara et al., 1988; von Wettstein, 1990; Jahn et al., 1992). This reaction is indistinguishable from that normally occurring during protein biosynthesis (Schimmel and Söll, 1979), and the activated glutamate is used for both protein and chlorophyll biosynthesis. While glutamyl-tRNA synthetase is likely to be encoded in nuclear DNA (Kannangara et al., 1988), the gene for $tRNA^{Glu}$ is located in the plastid DNA (Sugiura, 1992). The nucleotide sequence of $tRNA^{Glu}$ of barley (Schön et al., 1986) is very similar to that of other plant species and algae (Kannangara et al., 1988; Jahn et al., 1992). Universally conserved bases that are required for binding of $tRNA^{Glu}$ to glutamyl-tRNA synthetase have been identified (von Wettstein, 1990; Jahn et al., 1992; Rogers and Söll, 1993).

Glutamyl-tRNA reductase is the first unique enzyme in the pathway leading to 5-aminolevulinic acid. It catalyzes the NADPH and Mg^{2+}-dependent reduction of glutamyl-$tRNA^{Glu}$ to yield glutamate 1-semialdehyde with the release of free tRNA that can be reused for aminoacylation by glutamyl-tRNA synthetase (Fig. 3). Two reaction mechanisms have been proposed (Kannangara et al., 1988; Chen et al., 1990). Glutamyl-tRNA reductases purified from different sources exhibit a significant variability in their catalytic activities, subunit compositions, and molecular masses (Jahn et al., 1992; Pontoppidan and Kannangara, 1994). Their amino acid sequences display a much higher evolutionary conservation than expected from the enzyme parameters, however (Pontoppidan and Kannangara, 1994; Ilag et al., 1994). Glutamyl-tRNA reductase is likely to be a nuclear-encoded plastid protein (Ilag et al., 1994).

Investigations on the tRNA specificity of glutamyl-tRNA reductase have not been conclusive thus far. For example, glutamyl-$tRNA^{Glu}$ of *Escherichia coli* could serve as substrate of the enzyme from *Arabidopsis thaliana* (Ilag et al., 1994) but was not recognized by the purified barley protein (Pontoppidan and Kannangara, 1994). Barley glutamyl-tRNA reductase also did not use cytosolic glutamyl-$tRNA^{Glu}$ in a homologous *in vitro* system (Peterson et al., 1988), which demonstrates its high tRNA specificity. Base-specific requirements imposed by glutamyl-tRNA reductase on $tRNA^{Glu}$ have recently been determined for *Euglena gracilis* (Stange-Thomann et al., 1994). In a yellow mutant of this alga, a single point mutation in $tRNA^{Glu}$ was found to uncouple chlorophyll and plastid protein synthesis (Stange-Thomann et al., 1994). The $tRNA^{Glu}$ could still be charged by glutamyl-tRNA synthetase but was no longer able to participate in the glutamyl-tRNA reductase reaction. There appear to be unique bases also in higher plant $tRNA^{Glu}$ that may serve as recognition (identity) elements for glutamyl-tRNA synthetase, glutamyl-tRNA reductase, and/or the elongation factor EF-Tu (Jahn et al., 1992; Stange-Thomann et al., 1994).

Glutamate-1-semialdehyde aminotransferase catalyzes the final step in 5-aminolevulinic acid synthesis, the transamination of glutamate 1-semialdehyde (Fig. 3). Glutamate-1-semialdehyde aminotransferases have been purified from diverse sources and their catalytic mechanisms have been studied in considerable detail (Avissar and Beale, 1989a,b; Grimm et al., 1989; Nair et al., 1991; Smith et al., 1991a,b; Pugh et al., 1992). It is the generally accepted view that glutamate-1-semialdehyde aminotransferase converts its substrate, (*S*)-glutamate 1-semialdehyde, by a ping-pong bi-bi mechanism via diaminovalerate to 5-aminolevulinic acid, using pyridoxamine 5′-phosphate as an amino group donor (Smith et al., 1991a,b). This reaction is strongly inhibited by 3-amino-2,3-dihydrobenzoic acid (gabaculine) (Friedmann et al., 1987).

A few glutamate-1-semialdehyde aminotransferase cDNAs have been isolated (Grimm, 1990; Höfgen et al., 1994; Ilag et al., 1994; Matters and Beale, 1994). Their encoded products seem to be very similar to one another and appear to display also a high degree of sequence similarity to the *hemL* gene product of

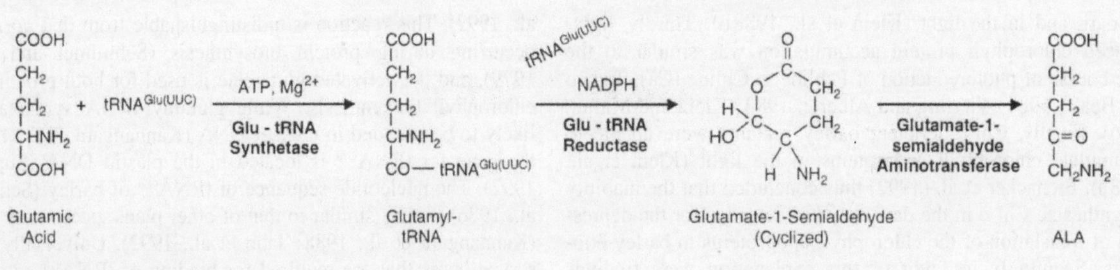

Fig. 3. The three-step reaction of 5-aminolevulinic acid (ALA) synthesis. Redrawn after Beale and Weinstein (1990).

E. coli and other bacteria, which synthesize 5-aminolevulinic acid via the C_5 pathway (Grimm, 1990; Sangwan and O'Brian, 1993; Ilag et al., 1994; Matters and Beale, 1994). Plant glutamate-1-semialdehyde aminotransferase is a nuclear gene product (Grimm, 1990; Sangwan and O'Brian, 1993; Ilag et al., 1994). Expression of antisense glutamate-1-semialdehyde aminotransferase RNA has been shown to cause a remarkable variety of chlorophyll deficiencies in transgenic tobacco plants (Höfgen et al., 1994).

Steps from 5-aminolevulinic acid to protoporphyrin IX. As a next step in the C_5 pathway, two molecules of 5-aminolevulinic acid are condensed to yield porphobilinogen in a reaction that is catalyzed by 5-aminolevulinic acid dehydratase, also known as porphobilinogen synthase (Shemin and Russell, 1953; Granick, 1954). In the active cleft of the bacterial enzyme, which is supposed to be very similar to the plant enzyme, two molecules of 5-aminolevulinic acid are placed staggered side by side, and by the removal of two molecules of water two new carbon bonds are formed, giving rise to the pyrrole ring of porphobilinogen (Kannangara, 1990; von Wettstein, 1990). Two subsequent reactions convert four molecules of porphobilinogen into hydroxymethylbilane (catalyzed by porphobilinogen deaminase, also known as hydroxymethylbilane synthase; Bogorad and Granick, 1953), followed by ring closure and simultaneous isomerization of the pyrrole ring IV, leading to uroporphyrinogen III. The latter reaction is catalyzed by uroporphyrinogen III cosynthase (Battersby, 1988). At uroporphyrinogen III, there is a first branch point in the pathway at which, in bacteria, siroheme and vitamin b_{12} diverge (Fig. 2). However, uroporphyrinogen III can also be converted into coproporphyrinogen, followed by two successive oxidations, which eventually lead to protoporphyrin IX. In the first reaction, the acetic acid side chains of uroporphyrinogen III are shortened to methyl groups by decarboxylation. Then, two propionic acid side chains are 'trimmed' into vinyl groups, followed by oxidation to establish the system of conjugated double bonds in protoporphyrin IX. The enzymes catalyzing these reactions, uroporphyrinogen III decarboxylase, coproporphyrinogen oxidase, and protoporphyrinogen IX oxidase have been characterized in some detail (Beale and Weinstein, 1990; Madsen et al., 1993; Smith et al., 1993), and some of their mRNAs been cloned from tobacco and barley (Kruse et al., 1995a,b; Mock et al., 1995).

Protoporphyrinogen IX oxidase is the target of photobleaching diphenyl ether herbicides, such as acifluorfen (Matringe and Scalla, 1987; Kouji et al., 1989; Lydon and Duke, 1988). Due to the inhibition of the enzyme activity, protoporphyrinogen IX accumulates *in vivo*. After its export from the plastids and non-specific oxidation by herbicide-insensitive oxygenases in the plasma membrane (Lehnen et al., 1990; Jacobs and Jacobs, 1993), the resulting protoporphyrin IX undergoes rapid photooxidation. In the presence of O_2 and light, protoporphyrin IX forms highly reactive oxygen radicals and singlet oxygen that damage membrane structures and ultimately lead to cell death (Jacobs and Jacobs, 1993).

The Fe-branch of the C_5 pathway. At protoporphyrin IX, the C_5 pathway diverges a second time, one route being directed to the synthesis of protoheme and phytochromobilin, and the other route giving rise to the formation of chlorophyll. Ferrochelatase is the enzyme that catalyzes the insertion of Fe^{2+} into the tetrapyrrole ring of protoporphyrin IX to give protoheme.

Since the pioneering work of Jones (1967, 1968) and Porra and Lascelles (1968), ferrochelatase has not received much interest. However, this enzyme was recently reinvestigated (Matringe et al., 1994), and a few ferrochelatase cDNA clones have now become available from different plant sources (Miyamoto et al., 1994; Smith et al., 1994). All plant ferrochelatases characterized thus far are monomeric, ATP-independent proteins that seem to resemble their bacterial counterparts in both their physicochemical and catalytic properties, as well as amino acid sequences (Matringe et al., 1994; Miyamoto et al., 1994; Smith et al., 1994).

The final products of the Fe-branch of the C_5 pathway in higher plants are protoheme and phytochromobilin. Protoheme can either be incorporated into the plastidic hemoproteins, such as cytochromes *f* and b_6 of the cytochrome b_6f complex (Phillips and Gray, 1983; Howe and Merchant, 1992, 1995; Huang et al., 1994), or be exported from the plastids to the cytosol (Thomas and Weinstein, 1990), where it serves different functions (Dailey, 1990). It is not known whether exclusively the plastid-derived protoheme, or an earlier plastid precursor of protoheme, such as protoporphyrinogen IX, can be transported into mitochondria (Beale and Weinstein, 1990; Jacobs and Jacobs, 1984, 1993).

Part of the plastid-localized protoheme must also be converted to biliverdin IXα to finally yield phytochromobilin (Terry and Lagarias, 1991). After export to the cytosol, this linear tetrapyrrole pigment spontaneously combines with the phytochrome apoprotein to form the functional phytochrome holocomplex (Terry and Lagarias, 1991; Li and Lagarias, 1994).

The Mg-branch of the C_5 pathway. Mg^{2+}-chelatase catalyzes the first committed step in the biosynthetic pathway of chlorophyll, the insertion of Mg^{2+} into protoporphyrin IX (Castelfranco et al., 1979; Pardo et al., 1980; Fuesler et al., 1984). This reaction is ATP-dependent (Pardo et al., 1980; Fuesler et al., 1984; Walker and Weinstein, 1991a, b). In contrast to ferrochelatase, which is a single polypeptide, Mg^{2+}-chelatases of cucumber and pea plastids require for activity both soluble and membrane-associated components (Walker and Weinstein, 1991a, b; 1994), although experiments reported by Lee et al. (1992) questioned this point at least for cucumber.

Pioneering studies of Gibson et al. (1995), performed with *Rhodobacter sphaeroides*, a photosynthetic purple bacterium that synthesizes bacteriochlorophyll (Bchl), a compound that is structurally and functionally related to Chl *a* (Beale and

Weinstein, 1990), have identified three components, termed BchH, BchI and BchD, required for Mg^{2+}-chelatase activity. BchH has been proposed to bind protoporphyrin IX prior to the insertion step of Mg^{2+}, while BchI contains an ATP-binding site (Gibson et al., 1995). Gene sequences related to *bchH* and *bchI* of *R. sphaeroides* (Bauer et al., 1993) occur in higher plants, such as *Antirrhinum majus* (Hudson et al., 1993) and *Arabidopsis thaliana* (Koncz et al., 1990). Functional disruption of one allele of the single nuclear gene in *Antirrhinum* by transposon tagging (used to isolate the gene) has been shown to lead to reduced chlorophyll levels and to an altered chloroplast ultrastructure (Hudson et al., 1993). In *Arabidopsis*, the product of the *ch42* gene, a homolog of *bchI*, is found in the stroma (Koncz et al., 1990), which suggests that it may function as a soluble component of the Mg^{2+}-chelatase complex (Hudson et al., 1993).

After chelation to give Mg^{2+}-protoporphyrin IX, a series of reactions follows (Beale and Weinstein, 1990). First, Mg^{2+}-protoporphyrin IX is esterified to Mg^{2+}-protoporphyrin IX monomethyl ester by a methyl transferase. Then, the isocyclic ring of the macrocycle is formed. This reaction requires O_2, NADPH, and some membrane-associated enzymes. The final product of the Mg-branch appears to be 3,8-divinyl Pchlide *a*.

The heterogeneity in the Pchlide pool. There is a remarkable heterogeneity in the Pchlide pool in dark-grown and light-grown plants (Belanger and Rebeiz, 1980; Hanamoto and Castelfranco, 1983; Carey and Rebeiz, 1985; Carey et al., 1985). Work from Rebeiz's group suggests that the last steps of chlorophyll biosynthesis may proceed through a multi-branched pathway in which six main branches with four minor esterification routes and two major monovinyl and divinyl free-acid branches occur. This proposal was based on findings that, depending on the plant species investigated and the experimental conditions used, either the 3-monovinyl or 3,8-divinyl forms of Pchlide *a* predominated (Tripathy and Rebeiz, 1988). Later studies, however, showed that esterified forms of only 3-monovinyl but not of 3,8-divinyl Pchlide *a* could be detected in most plants, with geranylgeraniol, dihydrogeranylgeraniol, tetrahydrogeranylgeraniol, and phytol being the alcohol moieties (McEwen and Lindsten, 1992). Similar to these findings, only the phytol esters of 3-monovinyl Pchlide *b* could be isolated from diverse plant species (Shedbalkar et al., 1991). Collectively, these findings do not seem to be in favor of the original model proposed by Rebeiz's group; they rather suggest the operation of a linear pathway similar to that originally proposed by Granick (1950) (also Fig. 2). However, not only can 3,8-divinyl Pchlide *a* be reduced at the vinyl group at C8 to 3-monovinyl Pchlide *a* but it can also be oxidized at the methyl group at C7 to give 3,8-divinyl Pchlide *b* (Fig. 4). The esterification appears to be a subsequent, multistep process whose initial reaction can use either form of 3-monovinyl Pchlide.

Taking these and other findings (Whyte and Griffiths, 1993) into account, the above-mentioned variability in the monovinyl and divinyl Pchlide (*a* and *b*) levels in different plants can thus solely be explained by species-specific differences in 8-vinyl reductase and 7-methyl oxidase activities. For example, conditions favouring active 8-vinyl reductase would account for the accumulation of 3-monovinyl Pchlide *a* in preference to 3,8-divinyl Pchlide *b*. However, it remains to be demonstrated that 8-vinyl reductase can accept either form of 3,8-divinyl Pchlide as its substrate and that 7-methyl oxidase activity indeed exists in plants.

Light-dependent reduction of Pchlide to Chlide catalyzed by POR. Protochlorophyllide oxidoreductase is the key enzyme of chlorophyll synthesis in angiosperms. It catalyzes the *trans*-reduction of ring D in the tetrapyrrole ring system of Pchlide to yield Chlide (Rüdiger and Schoch, 1988; Schoch et al., 1995), as shown in Fig. 5. POR proteins have been purified and characterized from wheat, barley, pea, and a few other plant species (Schulz and Senger, 1993). Also, there are cDNA clones available for POR proteins from both angiosperms, such as barley (Schulz et al., 1989; Holtorf et al., 1995), wheat (Teakle and Griffiths, 1993), oat (Darrah et al., 1990) (Fig. 6), pea (Spano et al., 1992b), *Arabidopsis* (Benli et al., 1991), and gymnosperms, such as pine (Spano et al., 1992a; Forreiter and Apel, 1993). In cyanobacteria, such as *Synechocystis* sp. PCC 6803, a homolog of the plant-type, light-dependent *por* genes has recently been identified (Suzuki and Bauer, 1995). All POR enzymes characterized thus far share a high degree of conservation in their amino acid sequences (Suzuki and Bauer, 1995). In barley, two closely related POR proteins, termed PORA and PORB, exist (Holtorf et al., 1995; Fig. 6 this review).

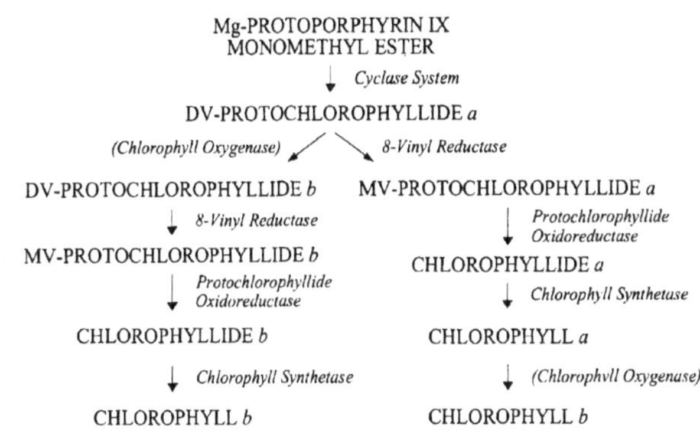

Fig. 4. Alternative routes for the late steps of Chl *a* and Chl *b* synthesis. Abbreviations and further explanations are as described in the legend of Fig. 2.

The reaction catalyzed by POR requires NADPH as a cosubstrate as well as light (Griffiths, 1975, 1978; Apel et al., 1980; Holtorf et al., 1995; Reinbothe et al., 1995a,b). The Pchlide specificity of POR is largely unknown. However, detergent-treated POR enzyme preparations from wheat etioplasts convert 3-monovinyl Pchlide *a* but not the esterified forms of this compound to Chlide (Schoch et al., 1995), which confirms previous results reported by Griffiths (1980). The wheat enzyme also converts Zn^{2+}-protopheophorbide *b*, which is the Zn^{2+} analog of Pchlide *b*, to Zn^{2+}-pheophorbide (Schoch et al., 1995). This finding suggests that POR of wheat etioplasts might exhibit no strict specificity with regard to the 3-monovinyl forms of either Pchlide *a* or *b*. However, as described above for Pchl *a*, the esterified forms of Pchlide *b* (i.e. Pchl *b*) could not be photoreduced by this enzyme preparation (Schoch et al., 1995).

According to the model proposed by Oliver and Griffiths (1982), the observed changes in the spectral forms of Pchlide can be attributed to different steps in enzyme catalysis. As a first step, a photoactive ternary POR-Pchlide-NADPH complex is formed. Such a complex is spectroscopically detectable in prolamellar bodies of wheat and barley etioplasts (Griffiths, 1975, 1978; Oliver and Griffiths, 1982), in greening barley leaves (Franck and Strzalka, 1992), and can also be reconstituted *in vitro* with either detergent-solubilized enzymes, such as those from wheat or barley etioplasts (Griffiths, 1978; Apel et al., 1980; Santel and Apel, 1981; Oliver and Griffiths, 1982; Schoch et al., 1995), or with the cDNA-encoded, *in-vitro*-synthesized PORA and PORB precursor proteins from barley (Holtorf et al., 1995; Reinbothe et al., 1995a,b).

Wilks and Timko (1995) proposed that NADPH binds to a region of the polypeptide which is similar to that of short-chain alcohol dehydrogenases (Baker, 1994). Two conservative resi-

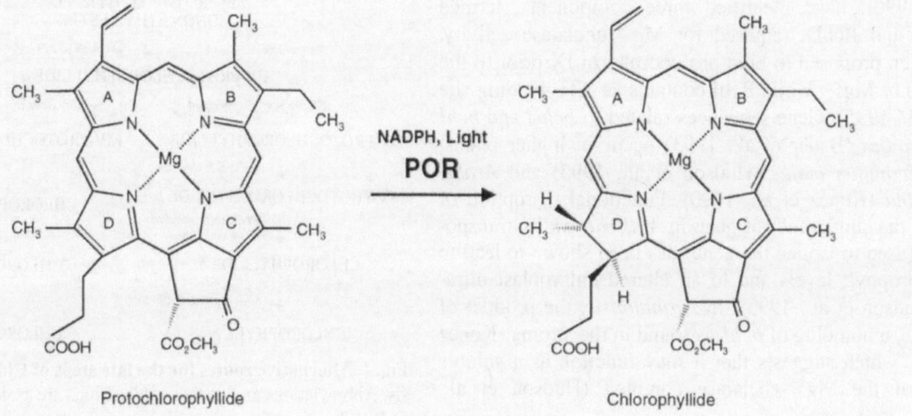

Fig. 5. Light-dependent and NADPH-dependent reduction of Pchlide to Chlide catalyzed by NADPH:protochlorophyllide oxidoreductase (POR).

dues (Tyr275 and Lys279) were suggested to be part of the active site of the pea enzyme. In *Rhodobacter capsulatus* strains that are impaired in their capability of converting Pchlide to Chlide (due to mutations in *bchB*, *bchL*, and/or *bchN*), overexpression of the POR enzyme of pea conferred light dependency on Pchlide reduction (Wilks and Timko, 1995). Bacterial strains expressing mutant pea enzymes, in which Tyr275 or Lys279 was replaced by other amino acids, were unable to convert Pchlide to Chlide, however. In addition to these residues, two or three of the evolutionarily conserved cysteine residues have been implicated as being involved in either the binding or subsequent conversion of Pchlide (Griffiths, 1975; Apel et al., 1980; Oliver and Griffiths, 1981; Dehesh et al., 1986a; Teakle and Griffiths, 1993).

The photoactive ternary POR-Pchlide-NADPH complex undergoes a series of spectral changes during catalysis in the light (Griffiths, 1978; Oliver and Griffiths, 1982; Schulz and Senger, 1993). While the Chlide formed first appears to remain attached to the enzyme, $NADP^+$ has been suggested to be rapidly displaced by fresh NAPDH. However, the final release of Chlide appears to be a relatively slow process and thus may account for the Shibata shift, a spectral change of $Chlide_{684-690}$ to $Chlide_{672-680}$ (Shibata, 1957; Griffiths, 1978; Oliver and Griffiths, 1982). After the final release of Chlide, the regenerated POR-NADPH complex has been proposed to be reused in a subsequent reaction cycle (Oliver and Griffiths, 1982).

Recent results obtained for the cDNA-encoded PORA precursor protein of barley confirm that Chlide is not easily released from the enzyme after catalysis in the light (Reinbothe et al., 1995a,b). Chlorophyllide remained tightly bound to pPORA during gel filtration on Sephadex G-25 and could only be released by denaturation (heat, SDS; Reinbothe et al., 1995a,b). More strikingly, Chlide rendered pPORA susceptible for degradation by plastid proteases (Reinbothe et al., 1995d). pPORA complexed with Chlide was found to be more susceptible to attack by chloroplast proteases than the free enzyme devoid of NADPH and Pchlide. Supporting the view that Chlide destabilizes the enzyme, the mature PORA after its import into isolated barley chloroplasts was rapidly proteolytically degraded in the light (Reinbothe et al., 1995b,e). In the dark, the imported enzyme remained stable, however (Reinbothe et al., 1995e). Under the latter conditions, a protease-resistant ternary PORA-NADPH-Pchlide complex was produced. Similar results were obtained for pPORB both *in vitro* and *in organello* (Reinbothe et al., 1995e).

Light-independent routes of chlorophyllide synthesis. In contrast to angiosperms, most green algae, gymnosperms, and photosynthetic bacteria are able to synthesize chlorophyll in the absence of light (Bogorad, 1950; Suzuki and Bauer, 1992; Schulz and Senger, 1993). The capability of synthesizing chlorophyll in the dark has been attributed to the operation of a light-independent pathway of Pchlide reduction. At least a few of the enzymes involved in this pathway and their genes have been identified in cyanobacteria, such as *Plectonema boryanum* (Fujita et al., 1991), in green algae, such as *Chlamydomonas reinhardtii* (Roitgrund and Mets, 1990; Suzuki and Bauer, 1992), in conifers, such as pine (Lidholm and Gustafsson, 1991), and in liverwort (Kohchi et al., 1988; Ohyama et al., 1988; Fujita et al., 1989). One of the genes studied in detail is the algal *chlL* (Suzuki and Bauer, 1992) and its counterpart in the plastid genome of liverwort, *frxC* (Sugiura, 1992). A similar *frxC* locus has also been suggested to occur in the plastid DNA of conifers (Lidholm and Gustafsson, 1991) but it appears to be absent from the plastid DNAs from tobacco and rice (Sugiura, 1992). All *chlL* (*frxC*) genes studied thus far encode proteins that are related to the *bchL* gene product of *R. capsulatus*, which is involved in the synthesis of bacteriochlorophyll *a*. In addition to the *chlL* (*frxC*) gene product, several other components participate in light-independent Pchlide reduction in *R. capsulatus* (Taylor et al., 1983; Youvan et al., 1984; Zsebo and Hearst, 1984; Yang and Bauer, 1990) and *C. reinhardtii* (Ford and Wang, 1980a,b; Choquet et al., 1992; Suzuki and Bauer, 1992; Li et al., 1993). In *R. capsulatus*, *bchL* and *bchN* encode two of these components (Yang and Bauer, 1990; Bauer et al., 1993; Bollivar et al., 1994). There is also evidence suggesting that some of them may form multimeric complexes related to the nitrogenase holoenzyme. In these complexes, the *bchL* (*chlL*) gene product is thought to be the major, structural component (Suzuki and Bauer, 1992). None of the *bch* (*chl*, *frx*) genes appear to share sequence similarity to the highly conserved genes encoding POR, the light-dependent Pchlide-reducing enzyme of angiosperms (see above).

In gymnosperms, both light-dependent and light-independent versions of the Pchlide-reducing enzymes seem to exist (Selstam et al., 1987; Ou et al., 1990; Spano et al., 1992a; Forreiter and Apel, 1993). In mountain pine (*Pinus mugo*), Forreiter and Apel (1993) detected two immunologically related POR polypeptides using an antibody raised against a barley pPORA–β-galactosidase fusion protein expressed in *E. coli* (Schulz et al., 1989). These two pine proteins must thus be related to the POR proteins found in angiosperms. Sequence analyses of two partial cDNA

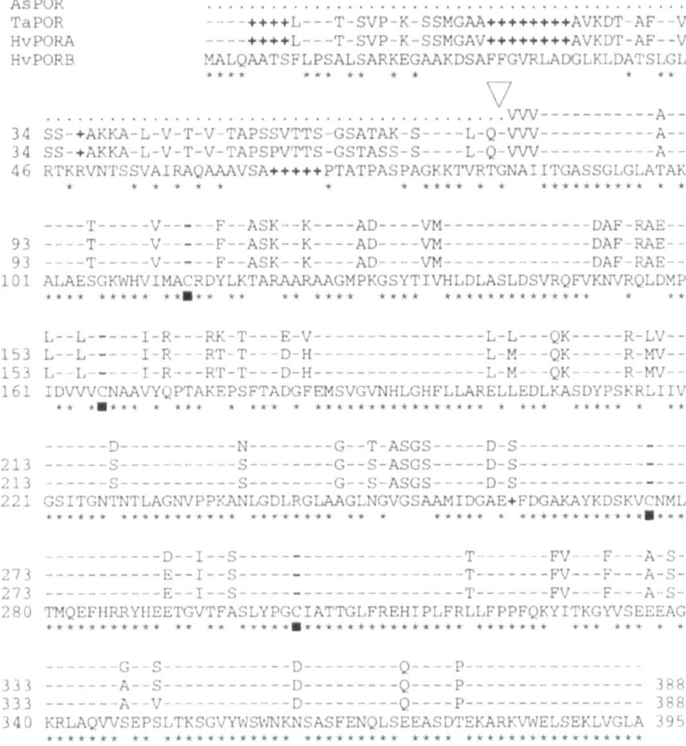

Fig. 6. Amino acid sequence alignment of POR enzymes from oat (*Avena sativa*, As), wheat (*Triticum aestivum*, Ta), and barley (*Hordeum vulgare*, Hv). All sequences refer to that of the barley PORB protein whose amino acid sequence is shown on the bottom line. Identical residues found in all of the different sequences (*) and gaps introduced during sequence alignment (+) are indicated. The positions of the four evolutionarily conserved cysteine residues (■) and the region in the oat POR sequence for which no amino acid sequence could be given (●), because the cDNA available is incomplete, are also indicated. The arrowhead (▽) indicates the putative cleavage site between the transit peptide and the mature part of the PORA of barley (HvPORA), as determined by Schulz et al. (1989). Courtesy of H. Holtorf.

clones support this conclusion, as they showed that the deduced amino acid sequences share an identity of approximately 85% to one another and are also closely related to those of the angiosperm POR enzymes (Forreiter and Apel, 1993). The two pine POR proteins and their mRNAs are differentially regulated by light (Forreiter and Apel, 1993), by analogy to PORA and PORB of angiosperm species such as barley (Holtorf et al., 1995). One pine POR protein is constitutively expressed both in the dark and in the light, whereas the other rapidly disappears upon illumination (Forreiter and Apel, 1993). Both pine POR proteins also differ in their subchloroplast localizations: the 36-kDa polypeptide is present in the prolamellar body of etioplasts and the 38-kDa polypeptide is associated with the thylakoids of chloroplasts (Forreiter and Apel, 1993). Although direct evidence is still lacking, it has been discussed that the constitutively expressed POR protein may be involved in the light-independent route of Pchlide reduction (Forreiter and Apel, 1993). It is tempting to speculate that this polypeptide might interact with plastid-encoded proteins, such as the *frxC* (Lidholm and Gustafsson, 1991) and/or *gidA* (Roitgrund and Mets, 1990) gene products.

Final steps in the C_5 pathway: synthesis of Chls a and b. Chlorophylls *a* and *b* differ only by the substituent at C7: a methyl group in Chl *a* and a formyl group in Chl *b* (Fig. 1). It is generally believed that Chl *b* is formed from Chl *a* by oxidation using molecular oxygen as the precursor of the 7-formyl group (Schneegurt and Beale, 1992; Porra et al., 1993) (Fig. 4). It has been proposed that at least three different enzymes might catalyze this reaction: a mixed-function oxygenase of unusual reaction specificity, a dioxygenase, or a combination of two enzymes, a monooxygenase plus a dehydrogenase (Rüdiger and Schoch, 1988; Porra et al., 1993, 1994). Porra et al. (1994) further proposed that oxidation of the methyl group occurs in a very hydrophobic location within the thylakoid membranes and speculated that Chl *a* already bound to the pigment-protein complexes may serve as the substrate. In contrast, Shlyk (1971) had assembled extensive evidence that Chl *b* may be formed from easily extractable Chl *a*.

Either Chl *a* or an earlier precursor of Chl *a* has been suggested to be the substrate for this oxidation step. For example, Oelze-Karow and Mohr (1978) proposed that 3-monovinyl Chlide *a* instead of Chl *a* may be the precursor of Chl(ide) *b*. Shedbalkar et al. (1991) demonstrated that 3-monovinyl Pchlide *b* is widely distributed in green plants, suggesting that 3-monovinyl Pchlide *a* (or 3,8-divinyl Pchlide *a*) may be the earliest precursor for formyl group formation. A quite distinct view is held by Ito et al. (1994) who showed that cucumber etioplast extracts have the capacity to convert Chl *b* to Chl *a*. However, the physiological relevance of this finding is only poorly understood.

Compartmentation of the C_5-pathway enzymes. As summarized in Fig. 7, all steps beginning at glutamate and ending at either heme or chlorophyll have been shown to take place in the plastid (Kannangara and Gough, 1977; Fuesler et al., 1984). The initial steps of 5-aminolevulinic acid formation, as well as the steps leading to protoporphyrinogen IX are likely to occur in the stroma (Smith and Rebeiz, 1979; Kannangara, 1990; Kruse et al., 1995a; Mock et al., 1995). However, all subsequent steps

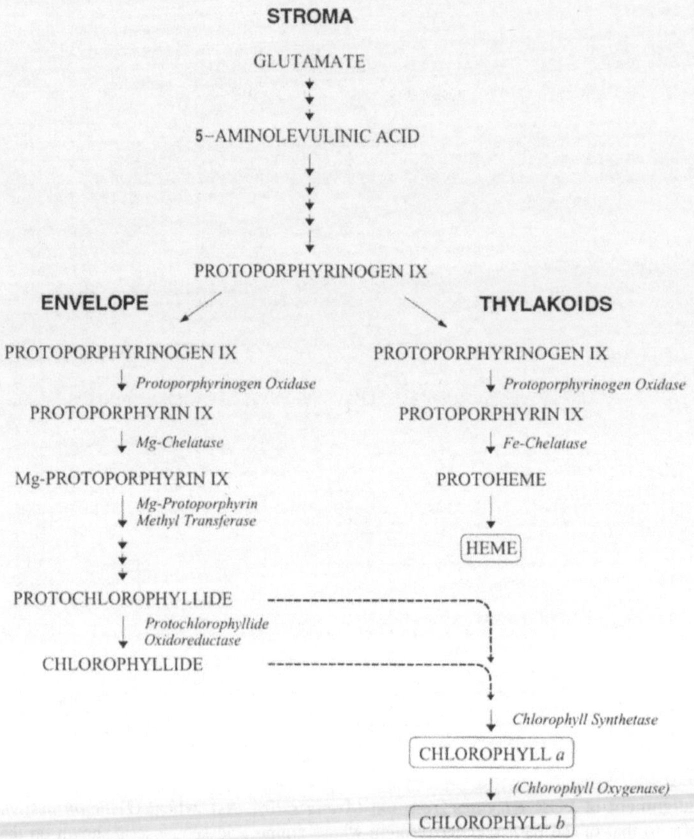

Fig. 7. Compartmentation of the early and late steps of tetrapyrrole synthesis in plastids. Enzymes and intermediates are as in Fig. 2. The dashed lines (- - -) indicate that either Pchlide or Chlide must be transported from the plastid envelope to the thylakoids.

are thought to be catalyzed by membrane-bound enzymes (Castelfranco and Beale, 1983). Unexpectedly, a few of the enzymes and intermediates of the Mg-branch of Chl synthesis could be detected in highly purified envelopes from spinach chloroplasts. Results from Douce's group suggest that the entire sequence of reactions starting at protoporphyrinogen IX and which finally yield Pchlide may occur in the plastid envelope (Pineau et al., 1986, 1993; Joyard et al., 1990, 1991; Matringe et al., 1992). Indeed, a major part of protoporphyrinogen IX oxidase activity is associated with the plastid envelope (Matringe et al., 1992). Mg^{2+}-chelatase contains at least one (envelope) membrane-associated component. Even the later step of light-dependent Pchlide reduction has been shown to occur in isolated plastid envelope membranes (Joyard et al., 1990; Pineau et al., 1993). Both Pchlide and POR are present in isolated envelopes of spinach chloroplasts (Joyard et al., 1990; Pineau et al., 1993). They can form a photoactive POR-NADPH-Pchlide complex that converts Pchlide to Chlide upon illumination (Pineau et al., 1993). However, all of the later steps of chlorophyll synthesis, such as esterification of Chlide *a* with phytol or earlier alcohol precursors, including their reduction, as well as the final step of Chl *a* oxygenation to give Chl *b* are supposed to take place in the thylakoid membranes (Block et al., 1980). This spatial separation of the early and late steps of chlorophyll synthesis implies that either Chlide or an earlier precursor of the C_5 pathway, such as Pchlide, must be transported from the envelope to the thylakoid membranes.

The Fe-branch of protoheme synthesis seems to be confined to the thylakoids (Fig. 7). Ferrochelatase activity is associated exclusively with thylakoids (Matringe et al., 1994). A major part of total chloroplast protoporphyrinogen IX oxidase activity is present in the thylakoids (Matringe et al., 1992). Consistent with these findings, protoporphyrin IX required for Fe^{2+}-chelatase action seems to originate primarily from oxidation of thylakoid-bound protoporphyrinogen IX (Jones, 1968). The final product of the Fe-branch, protoheme, binds to the different components of the thylakoid cytochrome b_6f complex (Phillipps and Gray, 1983). However, another part of protoheme is likely to be released from the thylakoids to the stroma, where the pigment becomes attached to either soluble *c*-type cytochromes (Bhaya and Castelfranco, 1985), or to other proteins, such as catalase (Dailey, 1990). In addition, protoheme export from the plastids to the cytosol has been reported (Thomas and Weinstein, 1990).

Regulation of the metabolite flow through the C_5 pathway. *Feedback inhibition of 5-aminolevulinic acid formation in the dark.* There are only a few steps in the C_5 pathway that have been shown to be regulated by light. The earliest control point is that of 5-aminolevulinic acid formation. When angiosperm seedlings are germinated in the dark, the only intermediate that accumulates to detectable levels is Pchlide. Because 5-aminolevulinic acid feeding drastically increases the level of this pigment, it has been concluded that 5-aminolevulinic acid synthesis is normally restricted in the dark. One explanation put forth is that an end product of the different branches of the C_5 pathway, such as Pchlide or protoheme, may depress 5-aminolevulinic acid formation in the dark (Beale and Weinstein, 1990). In favor of this view, (proto)heme has been shown to inhibit glutamyl-RNA synthetase purified from *C. reinhardtii* (Chang et al., 1990). In contrast, glutamyl-RNA synthetase of *Synechocystis* sp. PCC 6803 was insensitive to (proto)heme inhibition (Rieble and Beale, 1991), raising questions about the general physiological significance of the observed inhibitory effect. It would be difficult to imagine how glutamyl-tRNA^{Glu}-dependent plastid protein synthesis could proceed if glutamyl-RNA synthetase activity was inhibited. For a similar reason, Pchlide inhibition of glutamyl-RNA synthetase, reported for *Scenedesmus obliquus* (Dörnemann et al., 1989), would likely have a strong impact on translation of plastid proteins. Therefore, we prefer to consider feedback inhibition of glutamyl-RNA synthetase by either (proto)heme or Pchlide not to be a specific site of regulation of the metabolite flow through the C_5 pathway, in agreement with Jahn et al. (1992).

The next reaction that might be subject to feedback control is that catalyzed by glutamyl-tRNA reductase. Previous results of Rieble and Beale (1991) and Pontoppidan and Kannangara (1994) showed that (proto)heme inhibits glutamyl-tRNA reductases purified to homogeneity from *Synechocystis* sp. PCC 6803 and barley, respectively. 5-Aminolevulinic acid synthesis was also repressed by (proto)heme administered to extracts of *Chlorella vulgaris* (Weinstein et al., 1993) and *E. coli* (Javor and Febre, 1992), whereas the purified bacterial glutamyl-tRNA reductase appeared to be insensitive to (proto)heme inhibition (Jahn et al., 1991). Thus, one can conclude that some additional factors must be present *in vivo* to confer (proto)heme sensitivity on glutamyl-tRNA reductase of *E. coli* (Jahn et al., 1991). Interestingly, (proto)heme inhibition of 5-aminolevulinic acid synthesis in cell-free extracts of *Chlorella* could be enhanced by glutathione (Weinstein et al., 1993).

Light-dependent transcriptionally and post-transcriptionally controlled expression of 5-aminolevulinic-acid-forming enzymes. As an alternative possibility, it has been proposed that the level of tRNA^{Glu} may restrict chlorophyll synthesis in the dark. However, the transcription of the tRNA^{Glu} gene was not changed in *C. reinhardtii* (Jahn, 1992) nor was the level of glutamyl-tRNA^{Glu} in barley affected by light (Berry-Lowe, 1987). In *Euglena gracilis*, a distinct situation can be observed as light increases the levels of both tRNA^{Glu} and the early C_5-pathway enzymes (Mayer and Beale, 1990). This coordinated induction seems to ensure that rapid consumption of glutamyl-tRNA^{Glu} in chlorophyll synthesis does not depress translation of plastid mRNAs, in particular those which contain an above average frequency of glutamate codons, such as mRNAs for chlorophyll-binding proteins (Reinbothe and Parthier, 1990).

Because neither the levels of glutamyl-RNA synthetase nor of tRNA^{Glu} appear to change in etiolated angiosperm seedlings upon illumination (above), formation of the glutamyl-tRNA^{Glu} cannot be the limiting factor for 5-aminolevulinic acid formation in the dark. This leaves the possibility that the levels of glutamyl-tRNA reductase and glutamate-1-semialdehyde aminotransferase may restrict the accumulation of 5-aminolevulinic acid. Previous studies had provided evidence that 5-aminolevulinic acid synthesis is stimulated in the light (Huang and Castelfranco, 1989), and that this response is mediated by phytochrome (Huang et al., 1989). Recent studies of Ilag et al. (1994) confirm and extend these findings as they show that indeed the mRNAs for glutamyl-tRNA reductase and glutamate-1-semialdehyde aminotransferase are up-regulated in etiolated *Arabidopsis* seedlings upon exposure to light. However, the elevation in the mRNA abundances was found to be too low to account for the net increase in chlorophyll synthesis (Ilag et al., 1994). In contrast, the glutamate-1-semialdehyde aminotransferase mRNA level in etiolated barley seedlings remained constant during illumination (Grimm, 1990), whereas glutamate-1-semialdehyde aminotransferase activity increased drastically in the light (Kannangara et al., 1978). Collectively, these findings imply that post-transcriptional steps may be involved in regulating light-induced 5-aminolevulinic acid formation. One might speculate that glutamate-1-semialdehyde aminotransferase and/or glutamyl-tRNA reductase transcripts could contain sequences within their 5′-untranslated regions that favour their translation in the light. Protoheme, which has been proposed to be an endogenous inhibitor of 5-aminolevulinic acid synthesis in developing cucumber chloroplasts (Castelfranco and Zeng, 1991), could control such post-transcriptional steps. For example, protoheme could affect the availability of free Fe^{2+}, which is known to influence translation of numerous proteins, including ferritin, transferrin, and, more remarkably, 5-aminolevulinic acid synthase in bacteria (May et al., 1990). However, thus far no iron-responsive elements could be detected in the 5′-untranslated leaders of either glutamyl-tRNA reductase or glutamate-1-semialdehyde aminotransferase mRNAs from *Arabidopsis* (Ilag et al., 1994) and barley (Grimm, 1990).

In *Arabidopsis* seedlings that are grown in dark/light cycles, the changes in glutamyl-tRNA reductase and glutamate-1-semialdehyde aminotransferase mRNA levels are much more pronounced than those in etiolated seedlings (Ilag et al., 1994). Both mRNAs are hardly detectable at the end of the dark (night) period but rapidly accumulate in the light. By analogy, glutamate-1-semialdehyde aminotransferase mRNA and glutamate-1-semialdehyde aminotransferase activity are strongly induced by light in dark/light-synchronized cultures of *C. reinhardtii* (Matters and Beale, 1994).

Light induction of porphobilinogen deaminase mRNA. The general increase in 5-aminolevulinic acid synthesis in etiolated angiosperm seedlings upon illumination is accompanied by an increased flow of metabolites through the common path of protoheme and chlorophyll synthesis. Remarkably, not the first enzyme of this pathway, 5-aminolevulinic acid dehydratase (Boese et al., 1991; He et al., 1994), but rather the second enzyme, porphobilinogen deaminase is subject to light induction (Smith, 1986; Spano and Timko, 1991; Witty et al., 1993). The increase in enzyme activity (Smith, 1986; Spano and Timko, 1991) is consistent with the approximately 20–30-fold increase of the steady-state porphobilinogen deaminase mRNA level (Witty et al., 1993).

Light control of ferrochelatase expression. Fe^{2+}-chelatase, which catalyzes the first committed step in the biosynthetic pathway of protoheme, has also been shown to be a site at which the flow of protoporphyrin IX through the C_5 pathway is specifically regulated by light. Ferrochelatase mRNA is induced approximately fivefold in etiolated *Arabidopsis* seedlings upon illumination (Smith et al., 1994). Because the level of protoheme does not change substantially in such seedlings, as discussed by Beale and Weinstein (1990), it has been proposed that the increase in Fe^{2+}-chelatase mRNA might reflect an increase in the requirement for protoheme to be incorporated into the photosynthetic membrane complexes (Smith et al., 1994). The lower increase in Fe^{2+}-chelatase mRNA level, as compared to the light-induced increase in porphobilinogen deaminase mRNA, suggests that the major part of protoporphyrin IX is channelled through the Mg-branch of the C_5 pathway and thus gives rise to chlorophyll formation.

Mg^{2+}-chelatase expression in response to light. Mg^{2+}-chelatase is the first unique enzyme in the biosynthetic pathway of chlorophyll. Its activity is low in etiolated seedlings but increases in the light. Feedback inhibition of the enzyme activity by either Chlide or Pchlide, as proposed by Beale and Weinstein (1990), does not account for this effect (Walker and Weinstein,

1991b). Instead, changes in Mg^{2+}-chelatase activity are thought to result from alterations in its subunit composition, consisting of both soluble and membrane parts.

Unexpectedly, the mRNA level for one of these components, Olive (Oli, which is the homolog of bacterial BchH), was found to be depressed by light in *Antirrhinum majus* seedlings grown under 12-h dark/12-h light cycles (Hudson et al., 1993). The abundance of *oli* transcripts declined progressively during the light period but increased in the night. The *oli* and *bchH* gene products are thought to bind protoporphyrin IX (Hudson et al., 1993; Gibson et al., 1995), while the actual Mg^{2+} insertion step seems to be catalyzed by other components of the enzyme complex. It is therefore conceivable that the *oli* gene product may lower the flow of protoporphyrin IX through the Mg-branch of the C_5 pathway in the dark. In the light, however, tight binding of the pigment to Oli would impair the flow of protoporphyrin IX through the next steps of the pathway and thus could prevent rapid conversion of protoporphyrin IX to chlorophyll. The low expression level of Oli in the light thus would be consistent with the need to release Mg^{2+}-protoporphyrin IX from the Mg^{2+}-chelatase complex during stages of rapid chlorophyll formation, such as in etiolated seedlings exposed to light, or in seedlings grown under alternating dark/light cycles. The alternative explanation that sequestration of protoporphyrin IX into the Mg^{2+}-chelatase complex would lower the level of protoheme and, in turn, would relieve feedback inhibition of 5-aminolevulinic acid synthesis by this compound seems to be rather unlikely, because the Mg^{2+}- and Fe^{2+}-chelatases utilize functionally different pools of protoporphyrin IX. It would be difficult to reconcile such a mechanism with the low level of 5-aminolevulinic acid synthesis in the dark. Although the actual function of Oli is still unknown, it does not completely block the flow of protoporphyrin IX to Pchlide.

Regulation of *por* gene expression. *Two distinct* por *genes in barley and their differential expression in response to light.* Another control point, which occurs a few steps later, is that of Pchlide reduction. In angiosperm seedlings, the enzyme catalyzing this light-dependent reaction, POR, exhibits several interesting features. First, high amounts of the protein are present in etioplasts of dark-grown seedlings, whereas light-grown seedlings contain only traces of POR (Apel, 1981; Batschauer and Apel, 1984; Griffiths et al., 1985; Mösinger et al., 1985; Forreiter et al., 1990; Benli et al., 1991). Secondly, the catalytic activity of POR directly depends on light (Griffiths, 1975, 1978; Apel et al., 1980; Forreiter et al., 1990); however, it has not yet been determined how light operates in this enzyme catalysis. Thirdly, during catalysis, POR is rapidly inactivated and subsequently degraded (Kay and Griffiths, 1983; Häuser et al., 1984; Forreiter et al., 1990).

In our present studies, three different control points of *por* gene expression have been investigated. First, as assessed at the mRNA and protein levels, there are two *por* genes in barley that are differentially expressed in response to light. While the expression of the *porA* gene is depressed by light (Apel, 1981; Batschauer and Apel, 1984; Holtorf et al., 1995), the *porB* gene is active in both etiolated and illuminated seedlings (Holtorf et al., 1995). *porB* mRNA concentration undergoes a diurnal rhythm that seems to provide a way to adjust the synthesis of the PORB polypeptide to the varying needs of fully green plants for chlorophyll during the day/night periods (Holtorf et al., 1995). Unlike *porA*, the *porB* gene is not under the control of phytochrome (Holtorf et al., 1995).

Secondly, the plastid import pathways for PORA and PORB are different. As shown in Fig. 8 for *Arabidopsis thaliana*, the precursor of PORA is taken up by isolated plastids only in the

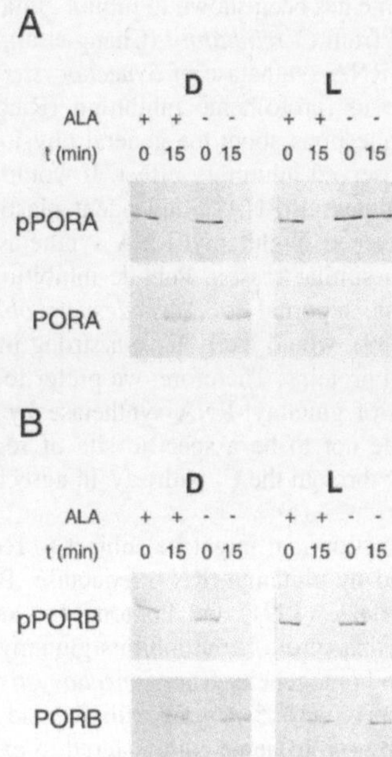

Fig. 8. Post-translational import of *in-vitro*-synthesized pPORA and pPORB of barley into isolated chloroplasts of *A. thaliana* grown in a 10-h-light/14-h-dark regime. Chloroplasts were isolated from *Arabidopsis* plants harvested at the end of the dark (D) or at the end of the light (L) period and incubated either in the dark with 5-aminolevulinic acid to induce intraplastidic Pchlide formation (+) or left untreated (−). Chloroplasts harvested at the end of the dark period reaccumulate Pchlide without additional 5-aminolevulinic acid feeding and thus can import pPORA during a subsequent dark incubation, while chloroplasts harvested at the end of the light period do not. Their transport competence for pPORA can be restored by exogenous 5-aminolevulinic acid-derived Pchlide, however. In contrast to pPORA, pPORB is sequestered by both Pchlide-containing and Pchlide-free chloroplasts.

presence of its substrate, Pchlide. Similar observations have been made in previous and recent experiments with isolated plastids from barley (Reinbothe et al., 1995a, b and unpublished results), pea, and wheat (Reinbothe, C., unpublished results). By contrast, the plastid import pathway of pPORB does not seem to require Pchlide (Fig. 8). Protochlorophyllide synthesized in the plastid envelope (Joyard et al., 1990, 1991) triggers the actual translocation step of pPORA and is likely to operate in close proximity to the plastid import apparatus (Reinbothe et al., 1995a, b and unpublished results).

In etioplasts, the endogenous level of Pchlide is sufficient for the translocation of pPORA, whereas during the light-induced greening, the Pchlide concentration declines drastically (Sundqvist, 1974; Reinbothe et al., 1995e) and soon reaches a level that is too low to allow the import of the precursor protein (Reinbothe et al., 1995a, e). Fully developed chloroplasts are not able to import pPORA. However, their import capacity can be restored by raising their endogenous Pchlide concentration experimentally, either by internalizing exogenously added Pchlide or by feeding 5-aminolevulinic acid in the dark to the isolated organelles (Reinbothe et al., 1995a, b). Similarly, excessive Pchlide accumulation occurring in chloroplasts of darkened *tigrina* o^{34} mutant seedlings (Gough and Kannangara, 1979; Casadoro et al., 1983) restores the transport competence for pPORA. The Pchlide requirement of import appears to be specific for pPORA, because the precursors of other nuclear-encoded plastid proteins, such as the small subunit of ribulose-1,5-bisphosphate carboxylase/oxygenase and PORB, can be imported into both

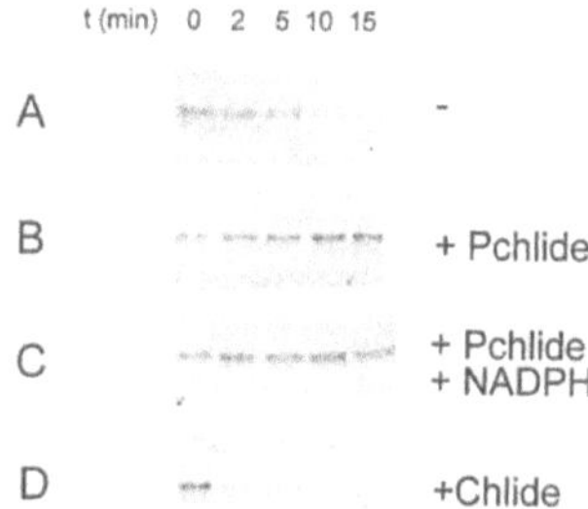

Fig. 9. Protease sensitivities of different PORA-pigment complexes. pPORA-pigment complexes were generated by incubating the *in-vitro*-synthesized pPORA of barley in the dark with either Pchlide (B) or Pchlide plus NADPH (C, D). Control assays were left unsupplemented (A). Half of the assay containing the pPORA-Pchlide-NADPH complexes was exposed to light to induce Chlide formation (D), whereas the other half was kept in the dark (C). Thereafter, the different incubation mixtures were supplemented with a protease preparation from lysed barley chloroplasts and incubated in the dark for the indicated time intervals before trichloroacetic acid was added. The autoradiograms show the levels of pPORA remaining in the trichloroacetic-acid-insoluble fractions after protease treatment.

chloroplasts and etioplasts (Reinbothe et al., 1995a, e). Moreover, the Pchlide-controlled import pathway for pPORA does not seem to reflect a general developmental regulation of the plastid import apparatus that might depend on the age or the type of the plastids, as proposed previously for the import of other nuclear-encoded plastid proteins (Chitnis et al., 1986, 1987, 1988; Halpin et al., 1989; Dahlin and Cline, 1991). Hence, one may conclude that pPORA is likely to enter the plastids by a nonstandard pathway, which supposedly differs from that used by all of the other cytosolic precursor proteins studied thus far.

Thirdly, both POR enzymes are similar to each other in that they exhibit strikingly different protease sensitivities when loaded with their substrates or products (Reinbothe et al., 1995d, e). When incubated with NADPH and Pchlide, they form ternary POR-Pchlide-NADPH complexes that are well protected against plastid proteases, as determined *in vitro* with the cDNA-encoded precursor proteins (Fig. 9) and *in organello* with the post-translationally imported, mature proteins (Reinbothe et al., 1995e). Upon illumination, the ternary (p)POR-pigment complexes photoreduce their Pchlide to Chlide and simultaneously become susceptible toward degradation by plastid proteases (Reinbothe et al., 1995d). The POR-degrading protease is not detectable in etioplasts but is rapidly induced during the light-dependent greening of etiolated seedlings (Reinbothe et al., 1995d, e). Its appearance is not due to light activation of preexisting proteases but requires protein synthesis *de novo* at cytoplasmic ribosomes (Reinbothe et al., 1995d). Partial characterization demonstrates that the POR-degrading protease activity is metal- and ATP-dependent and likely composed of several constituents, comprising both aspartic acid-type and cysteine-type proteases (Reinbothe et al., 1995d).

The putative functions of PORA and PORB. At first glance, PORA and PORB appear to be identical proteins. Their amino acid sequences are closely related, and they both are light-dependent and NADPH-dependent enzymes that reduce Pchlide to Chlide (Holtorf et al., 1995). Because their reaction specificities are identical, one may assume that also their substrate requirements could be the same. However, there is a large heterogeneity in the plastidic Pchlide pool, which suggests that PORA's and PORB's substrate specificities might also be different. For example, several forms of Pchlide can be distinguished by spectroscopic techniques. While the bulk of Pchlide present in etioplasts appears to be associated with PORA and forms the photoactive complex, which rapidly converts Pchlide to Chlide upon illumination, there are several other spectral forms of Pchlide that are less well characterized (Ryberg and Sundqvist, 1991). Furthermore, structurally different forms of Pchlide, such as Pchlide *a* and Pchlide *b*, or the esterified or nonesterified forms of either compound, have been detected within various plastid compartments, such as the envelope, the stroma and the internal membranes of etioplasts and chloroplasts (Ryberg and Sundqvist, 1991). Because the functions of most of these different Pchlide forms are largely unknown, one cannot exclude that certain members of the Pchlide family could be the preferred substrates for either PORA or PORB and thus could give rise to distinctive routes of Chlide synthesis. Studies with detergent-solubilized POR enzyme preparations from wheat or barley etioplasts (Schoch et al., 1995) cannot explicitly answer the question on the substrate specificities of PORA and PORB, because such preparations always contain a mixture of both POR enzymes (Holtorf et al., 1995).

If PORA and PORB represent functionally identical proteins, which display indistinguishable substrate and reaction specificities, one would have to ask why higher plants, and in particular angiosperms such as barley, contain two different versions of the same polypeptide. Do these proteins differ with regard to the timing and location at which they are active? If so, which different functions may be proposed for each protein? In our recent studies, these questions have been addressed. The results summarized in this review show that PORA and PORB are strikingly different in their expression patterns and post-translational plastid import pathways. While PORA is active only during the very early stages of the light-induced greening of etiolated barley seedlings, PORB persists during the transition from etiolated to light growth and is also abundant in green plants. These differences strongly suggest that the two POR enzymes should be able to accomplish distinct functions during chloroplast development.

PORA: enzyme, pigment scavenger, or plastid factor? For PORA, several different functions may be considered. First, PORA could play a role for the assembly of the photosynthetic apparatus during the early stages of the light-induced greening of angiosperm seedlings. Consistent with such a proposal is the finding that the *porA* gene is already active in etiolated seedlings, that large amounts of *porA* mRNA accumulate in the dark, that the precursor of PORA can be taken up by isolated etioplasts, and that there is PORA protein already present in the prolamellar bodies of these organelles. Together with Pchlide and NADPH, PORA forms the photoactive complex that, upon illumination, photoreduces Pchlide to Chlide. Although the light-induced reduction of Pchlide to Chlide catalyzed by PORA seems to be the cause for the rapid disintegration of the prolamellar body, it is as yet undetermined whether the enzyme has to repeatedly proceed through the catalytic cycle to accomplish its function or is used only once. Interestingly, Chlide remains tightly bound to (p)PORA and renders the protein susceptible toward attack by plastid proteases. Although the POR-degrading protease appears in the light after the disintegration of the prolamellar body has been completed, one may speculate that the proteolytic degradation of PORA could be an essential step leading toward the formation of the photosynthetic apparatus. Particularly interesting in this context is the observation that low-molecular-mass fragments could recently be detected during the degradation of the radioactively labeled pPORA *in vitro* (Reinbothe et al., 1995d). Although we do not know yet whether these fragments contain Chlide and whether they are normally also produced *in vivo*, one may assume that such Chlide-binding peptides might operate as vehicles for the transport of freshly formed Chlide molecules to the developing thylakoid membranes and their subsequent integration into the photosynthetic complexes. While similar proposals had already been made in previous studies (Santel and Apel, 1981; Kay and Griffiths, 1983; Häuser et al., 1984), no experimental approach has yet been available to address this important point.

As a second possibility, one may assign PORA a function as a scavenger of the hazardous pigment Pchlide. This proposal is based on our recent results that the import of PORA into the plastids depends on its substrate, Pchlide. Presumably by binding to pPORA, Pchlide triggers the actual translocation step (Reinbothe et al., 1995a and unpublished results). This could reduce the risk of free excited pigment molecules interacting with molecular oxygen during illumination to form oxygen radicals and singlet oxygen which could cause damage of the plastid compartment (Tripathy and Chakraborty, 1991; Chakraborty and Tripathy, 1992). The fact that the mature PORA enzyme remains stable in the light after its *in vitro* import into etioplasts and etiochloroplasts isolated from 30-min-illuminated barley seedlings (Reinbothe et al., 1995e) seems to be consistent with such an idea. Up to the stage when transcription of the *lhca/b* genes has not yet reached its maximal rate and a massive synthesis of light-harvesting chlorophyll-*a/b*-binding proteins thus could not have occurred, excess Chl(ide) would need to be neutralized. This could easily be achieved by its binding to PORA. At a later stage, when the bulk of light-harvesting chlorophyll-*a/b*-binding proteins would have been synthesized, Chlide produced by the degradation of PORA could then be integrated into the light-harvesting complexes. By lowering the actual concentrations of both Pchlide and Chlide, PORA could accomplish the proposed function as a scavenger of tetrapyrrole pigments during the transition from etiolated to light growth.

If such a photoprotective mechanism operated *in vivo* and in the plant, one may expect that the temporal patterns of expression of the light-harvesting chlorophyll-*a/b*-binding proteins and of PORA should at least overlap. However, the *porA* gene is rapidly switched off shortly after the beginning of illumination, the *porA* mRNA level declines soon thereafter, and the plastid import of pPORA is impaired by the rapid lowering of the pool size of Pchlide. At a time point when most PORA protein has disappeared from the developing etiochloroplasts, accumulation of the light-harvesting chlorophyll-*a/b*-binding proteins reaches its maximum rate (Kloppstech et al., 1984). Thus, it seems difficult to reconcile the expression pattern of PORA with the proposed photoprotective function during the early stages of chloroplast development. However, there are findings that may offer an explanation for this apparent paradox.

There are proteins that are likely to cooperate with PORA. One class of these photoprotective proteins appears to be related to the light-harvesting chlorophyll-*a/b*-binding proteins (Grimm and Kloppstech, 1987; Grimm et al., 1989). These polypeptides have been termed early light-induced proteins (ELIPs; Kloppstech et al., 1984). In etiolated barley and pea seedlings, their expression reaches a maximum in mRNA and protein concentrations after approximately 2−4 h illumination (Kloppstech et al., 1984; Meyer and Kloppstech, 1984; Grimm and Kloppstech, 1987). At that time, all of the Pchlide originally present in the photoactive PORA-Pchlide complex of the prolamellar bodies has become photoreduced to Chlide, the POR-degrading protease has reached its full activity and, thus, the PORA protein begins to break down rapidly. It is tempting to speculate that Chlide-containing peptides originating from the degradation of the PORA-Chlide complexes could bind to the ELIPs and thus could transfer the pigment to these proteins. Although ELIPs have been proposed to represent pigment-free proteins (Grimm et al., 1989), their binding capacity for the Chlide released during the degradation of PORA has not been investigated thus far. Previous studies have suggested that ELIPs might also be able to bind carotenoids (Lers et al., 1991), which are known to quench triplet states of chlorophyll (Siefermann-Harms, 1987). By analogy to carotenoids present in the light-harvesting complexes of the thylakoid membranes (Siefermann-Harms, 1987), the ELIP-associated carotenoids could bring about non-radiative dissipation of excess excitation energy by reacting with triplet Chlide or singlet oxygen. Although not yet proven, these findings collectively suggest that PORA and ELIPs could be part of a common photoprotective mechanism which would help prevent that free excited Pchlide and Chl(ide) molecules accumulate and cause photooxidative damage of the plastid compartment during the very early stages of chloroplast development.

As a third and very speculative function, one may propose that PORA could play a role in controlling nuclear gene expression during the establishment of the photosynthetic apparatus. By analogy to PHYA, the key regulatory factor operating during photomorphogenesis in higher plants, PORA exhibits characteristics that are reminiscent of a photoreceptor (Fradkin and Domanski, 1994). As discussed previously, both PHYA and PORA are chromoproteins. Although their spectral properties are different, they both contain tetrapyrrole pigments as the chromophore. Different spectral forms of PHYA and PORA can be distinguished. In the case of PHYA, striking phenotypic effects have been correlated with the expression of this protein. For PORA, a functional analysis of its putative role as a photoreceptor is still lacking. However, PORA has at least been suggested to be somehow correlated with the plastid factor (Batschauer et al., 1986).

Remarkably, the expression pattern of the *porA* gene appears to be inversely correlated with that of the *lhca/b* genes. For example, in etiolated barley seedlings, the *porA* gene is active, while the *lhca/b* genes are not yet transcribed. However, conditions favouring *lhca/b* transcription, such as exposure of etiolated seedlings to light, correlate with a decline in *porA* gene expression (e.g. Batschauer et al., 1986). Furthermore, in seedlings of *A. thaliana* that are grown under far-red light conditions, the PORA protein and its mRNA are absent, while the *lhcb* gene is active and its mRNA transcribed (Runge, 1995, as cited in Holtorf et al., 1995). However, no light-harvesting structures accumulate, because their stabilizing pigments are absent. In carotenoid-deficient *albina* mutants or norflurazon-treated, etiolated seedlings of barley, accumulation of the *lhcb* mRNA is impaired in the light, while that of the *porA* transcript is maintained. However, in chlorophyll-deficient *xantha* mutants of barley, which are blocked after protoporphyrin IX or Mg^{2+}-protoporphyrin IX, the accumulation of the *lhcb* mRNA occurs almost normally when such seedlings are transferred from the dark to the light (Batschauer et al., 1986). Collectively, these findings imply that if any of the reactions leading to chlorophyll formation is involved in the control of *lhcb* mRNA accumulation, it should be one between Pchlide and the esterification step of Chlide.

Pioneering studies of Johanningmeier and Howell (1984) suggested that intermediates in the C_5 pathway, such as protoporphyrin IX, Mg^{2+}-protoporphyrin IX monomethyl ester, or Pchlide could be the signals that inhibit *cab* (*lhcb*) mRNA accumulation in the dark. In the light, due to rapid consumption of the pigments during chlorophyll synthesis, the block of *lhca/b* gene expression would be relieved, and the *lhca/b* mRNAs could accumulate. Johanningmeier and Howell (1984) further proposed that the plastid envelope might be the site from which these pigments are released to the cytosol. The identification of most of the aforementioned pigments and of many of the enzymes necessary for converting protoporphyrinogen IX to

Pchlide in isolated envelope membranes of spinach chloroplasts seems to confirm this proposal. In addition, intermediates and products of the C_5 pathway, such as protoporphyrinogen IX (Lehnen et al., 1990), phytochromobilin (Terry and Lagarias, 1991) and (proto)heme (Thomas and Weinstein, 1990), have been demonstrated to be exportable from the plastids to the cytosol. One may therefore conclude that other intermediates of the C_5 pathway, such as Pchlide, might also be able to leave the plastid, and that either the pigments themselves or pigment-protein complexes could act either directly or indirectly, e.g. as signal transducers or transcription factors, on nuclear genes such as *lhca/b* and *rbcS*.

Recent experiments from our laboratory may add some evidence in support of such a model. They show that pPORA bound to the plastid envelope *in vivo* is able to interact not only with endogenously produced, 5-aminolevulinic-acid-derived Pchlide but also with exogenously administered Pchlide *in vitro* (Reinbothe et al., unpublished results). While the interaction with intraplastidic Pchlide triggers vectorial translocation of pPORA into the plastid and finally gives rise to the mature enzyme, the interaction with exogenously applied Pchlide causes the release of the precursor protein to the cytosolic side of the envelope (Reinbothe et al., unpublished results). Both processes can compete with each other if intraplastidic and extraplastidic Pchlide are offered simultaneously (Reinbothe et al., unpublished results). However, a member of the family of heat shock proteins (70-kDa protein; Hsc70) present in wheat germ and barley leaf extracts prevents the release of pPORA by Pchlide *in vitro* (Reinbothe et al., unpublished results). If such a pigment-induced cytosolic release of envelope-bound pPORA would also occur *in vivo*, either Pchlide or the pPORA-Pchlide complex could exert a regulatory function on nuclear *lhca/b* gene expression. However, the operation of such a mechanism would require some plastid-derived Pchlide to be present in the cytosol and the expression level of Hsc70 to be low. Although isolated envelope membranes from spinach chloroplasts have been shown to contain Pchlide, it is as yet undetermined whether this pigment can be released to the cytosol, and which factors might favour such a pigment export. At the moment, there is no experimental approach available to detect the expected traces of Pchlide in the cytosol *in vivo*. It is also as yet unknown whether Hsc70 is present always in excess *in vivo* to prevent the interaction between pPORA and Pchlide in the cytosol.

One may further speculate that the cytosolic pPORA-Pchlide complex could operate in transcriptional control. For example, transcription of *lhca/b* genes could be depressed in the dark, whereas that of the *porA* gene could be maintained. Upon illumination, the general decrease in the plastidic level of Pchlide would be expected to affect also the level of pigment in the cytosol. As a result, most of the cytosolically synthesized PORA precursor molecules would be sequestered at the plastid envelope, and a decrease in the cytosolic concentration of the pPORA-Pchlide complex would thus be observed. As a consequence, transcription of the *lhca/b* genes could be activated, while transcription of the *porA* gene would decline. The detection of pPORA at the outer envelope of etiochloroplasts (Reinbothe et al., unpublished results) and of an immunoreactive, POR-related protein in the cytosol of illuminated plants (Dehesh et al., 1986b) may support the notion that a pPORA-pigment complex could play the regulatory role of the plastid factor in the control of nuclear gene expression during light-induced chloroplast development. However, this hypothetical model needs careful investigation in future studies.

PORB: the 'true' Pchlide-reducing enzyme? The cumulative negative effect of the substrate-controlled import of pPORA, the rapid proteolytic degradation of freshly formed PORA-Chlide complexes and the light-induced decline in *porA* mRNA concentration leads to a rapid and selective disappearance of PORA from developing chloroplasts soon after the beginning of illumination. Because massive chlorophyll accumulation starts only after the breakdown of PORA has been completed, there must be another Pchlide-reducing enzyme that drives chlorophyll synthesis in illuminated plants.

PORB is a likely candidate for such a Pchlide-reducing enzyme. PORB is strikingly different from PORA in several aspects. First, the *porB* gene is active in both dark-grown and light-grown plants, and *porB* mRNA persists also during the transition from etiolated to light growth (Holtorf et al., 1995). Secondly, the cytosolic precursor protein of PORB is taken up by both isolated chloroplasts and etioplasts (Reinbothe et al., 1995e). While etioplasts contain spectroscopically detectable levels of Pchlide, the concentration of this pigment appears to be at the limit of detection in the chloroplast. One might therefore conclude that pPORB during its import into the chloroplasts does not interact with Pchlide. However, PORB imported into isolated chloroplasts in the dark is stable and thus is likely to form a ternary PORB-Pchlide-NADPH complex. In the light, this complex converts Pchlide to Chlide, and simultaneously, the resulting PORB-Chlide complex becomes susceptible toward proteolytic degradation (Reinbothe et al., 1995e).

Taking the different protease sensitivities of the PORB-Pchlide-NADPH and PORB-Chlide complexes into account, one may predict that PORB should have two different fates in chloroplasts. While PORB-Pchlide-NADPH complexes should be able to cross the stroma and to reach the thylakoids in the dark, PORB-Chlide complexes formed in the light would be destined for proteolytic degradation. Although it is as yet undetermined whether these two different fates of PORB have a physiological significance or not, one might speculate that two different routes of Chlide formation could be initiated, one of which would start in the stroma and the other at the thylakoids. Both possibilities would be consistent with a general function of PORB in sustaining chlorophyll synthesis in green plants but could differ in that they supply distinctive Chlide molecules. For example, the operation of these two postulated pathways of Chlide formation could explain why structurally identical chlorophyll molecules appear in different spectral forms in the various chlorophyll-protein complexes (e.g. Jennings et al., 1993). The operation of the postulated thylakoid targeting pathway of PORB could also offer an explanation of how the Pchlide synthesized in the plastid envelope (Pineau et al., 1986, 1993; Joyard et al., 1990, 1991) can reach the thylakoids. In the dark, PORB could transfer the envelope-synthesized Pchlide to the thylakoid membranes. Upon illumination, the enzyme would then photoreduce its Pchlide to Chlide, presumably followed by proteolytic degradation of the thylakoid-localized PORB-Chlide complex. Consistent with this proposal, Pchlide and, in some cases, Chlide have been detected in the thylakoid membranes of isolated spinach chloroplasts (Pineau et al., 1993).

If PORB would be imported into chloroplasts in the light, it should be degraded in the stroma because the POR-degrading protease would immediately attack the PORB-Chlide complexes. Based on this and the aforementioned findings, pronounced diurnal fluctuations in PORB protein concentration should be expected to take place in plants grown in dark/light cycles. However, such fluctuations have not been observed in previous experiments: the concentration of PORB protein remained rather constant during the day and during the night (Holtorf et al., 1995). One possible explanation for this unexpected finding might be that the observed oscillations in *porB* mRNA concentration during a dark/light cycle with a maximum during the day and a minimum during the night (Holtorf et al.,

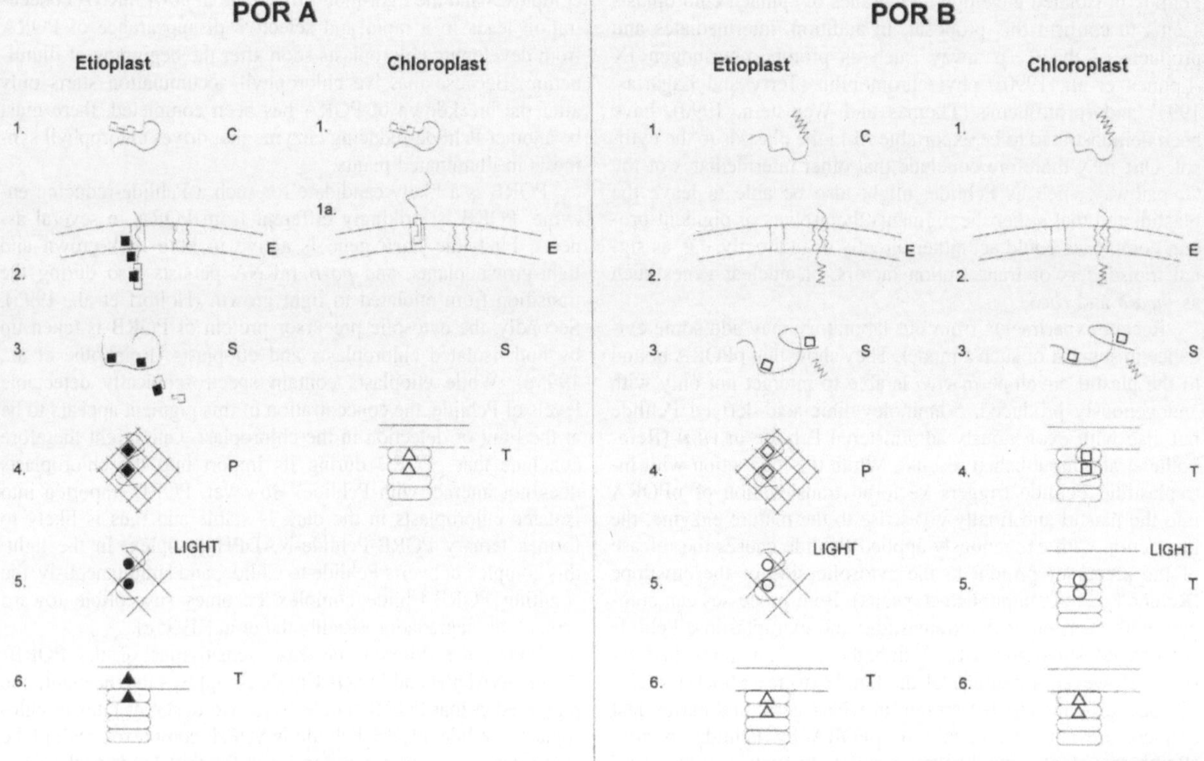

Fig. 10. Working model to illustrate the different post-translational plastid import pathways and postulated functions of PORA and PORB. After synthesis in the cytosol (C) (step 1), the different POR precursor proteins interact with the plastid envelope (E) (step 2). The actual translocation step by which pPORA crosses the plastid envelope membranes depends on its substrate, Pchlide. By virtue of a Pchlide-binding domain in the transit peptide (⊏⊐), pPORA is translocated through a specific translocation machinery in the plastid envelope and thereby binds the pigment (■). After removal of the transit peptide in the stroma (S) (step 3), the mature PORA with Pchlide and NADPH attached to it is targeted to the prolamellar body (P) of the etioplast (step 4). In the dark, the PORA-Pchlide-NADPH complex is stable against plastid proteases. Upon illumination, however, the photoactive ternary complex photoreduces Pchlide to Chlide (●) and, simultaneously, the prolamellar body begins to break (step 5). Superimposed on these effects, a plastid protease activity is induced that degrades the PORA-Chlide complex (not shown). The Chlide released in turn is converted to chlorophyll (▲), which is eventually incorporated into the photosynthetic chlorophyll-protein complexes of the developing thylakoids (step 6). In contrast to etioplasts, chloroplasts present in light-grown plants do not import pPORA because the level of Pchlide is too low to saturate the requirements for import. pPORA can still bind to the plastid surface (step 1a) but is not translocated across the envelope membranes. Unlike pPORA, the import of pPORB occurs virtually in the absence of Pchlide and thus can be observed with both etioplasts and chloroplasts (steps 1 and 2). pPORB is likely to enter the plastids through a translocation machinery that is different from that used by pPORA. The transit peptide of pPORB (WWW) thus would not have to contain a Pchlide-binding domain, although the mature PORB does bind the pigment (□) during import (step 3). Pchlide bound to PORB might either be structurally or functionally identical to the pigment bound to PORA or, as indicated, might be different. In the dark, the resulting ternary PORB-Pchlide-NADPH complex is similarly targeted to the prolamellar body of the etioplast and to the thylakoids of the chloroplast (steps 4). At either place, the photoreduction of Pchlide to Chlide (○) could take place (steps 5), followed by conversion of Chlide to chlorophyll (△) and its subsequent binding to the photosynthetic chlorophyll-protein complexes (steps 6). In etioplasts, chlorophyll could be used for the assembly of the photosynthetic apparatus, whereas in chloroplasts, the pigment could replace those chlorophyll molecules that become damaged during photosynthesis.

1995) could give rise to enhanced rates of PORB synthesis which would compensate for the preferential degradation of the protein in the light. It is also interesting to note that the POR-degrading protease activity is ATP-dependent (Reinbothe et al., 1995d) and thus may be modulated by dark/light treatments. In plants grown under alternating dark/light cycles, light-driven ATP synthesis during photosynthesis would presumably activate this protease during the day, whereas the activity should decline during the night when ATP has been consumed. However, this ATP effect would likely be counteracted by changes in the stromal pH during the day and the night periods. The POR-degrading protease with its optimum in activity around pH 6.5 is suspected to be most active at the beginning of illumination but would become inactivated during the day, followed by its recovery throughout the night.

Summary and perspectives

The results summarized in this review show that chloroplast development in etiolated angiosperm seedlings, such as barley, is a strictly light-dependent, complex process. POR catalyzes the only light-requiring step in chlorophyll synthesis and thus is presumed to play a key regulatory role for the establishment of the photosynthetic apparatus. In barley, two distinct POR enzymes exist; both are light-dependent and NADPH-dependent Pchlide-reducing enzymes. While PORA appears to be active only during the transition from etiolated to light growth, PORB seems to be present in etiolated, illuminated and also in light-adapted plants. One may therefore conclude that PORA and PORB could differ in supplying either functionally or structurally different Chlide molecules. For example, in light-adapted plants, there is a continuous need to replace those chlorophyll molecules that became damaged during the steps of photosynthetic light trapping and energy transduction (Mattoo et al., 1989; Nitschke and Rutherford, 1991; Goldbeck, 1993). Also, the size and composition of the light-harvesting antenna of the photosystems and thus the actual chlorophyll concentration in the thylakoid membranes must continuously be adapted to different light intensities (Melis, 1991). Chlorophyll molecules needed for these purposes could originate from the operation of PORB. By contrast, chlorophyll that is to be incorporated into either light-harvesting complexes or photosynthetic reaction centers during the very early stages of the light-induced greening of etiolated seedlings, could be produced by PORA (plus PORB).

We further assume that the transport machinery involved in pPORA translocation could differ structurally from that required for import of pPORB and also from previously characterized plastid import complexes in that it might contain, in addition to previously identified membrane proteins (Hirsch et al., 1994; Kessler et al., 1994; Schnell et al., 1994) and lipids (Kerber and Soll, 1992; van't Hof et al., 1993), enzymes involved in chlorophyll (Matringe et al., 1992) and carotenoid biosynthesis (Markwell et al., 1992). Particularly interesting in this context is the observation that the target enzyme of norflurazon, phytoene desaturase (Chamovitz et al., 1991), as well as other enzymes involved in carotenoid biosynthesis are localized in the plastid envelope (Joyard et al., 1991). This finding may provide additional evidence for the importance of the plastid envelope in transmitting signals to the cytosol. Whether pPORA and/or its substrate, Pchlide, may be released from the plastids and in turn could play the role of the plastid factor remains to be investigated.

Because the precursors of PORA and PORB are most divergent in their N-termini, it is likely that Pchlide dependency on plastid import is conferred on PORA through the chloroplast transit peptide. One may speculate that the chloroplast transit peptide of pPORA might contain a Pchlide-binding domain that interacts with the pigment during the step of translocation (Fig. 10). Such a pigment-binding domain would then have to be absent from the transit peptide of PORB. This postulated difference also implies that the PORA and PORB precursors enter the plastids through different translocation machineries. This separation would ensure that pPORA does not block the import of other cytosolic precursor proteins under conditions of Pchlide limitation. Future studies will therefore concentrate on the characterization of the different chloroplast transit peptides in pPORA and pPORB and their functions during plastid import.

The detection of a POR-degrading protease activity suggests that Chlide-containing peptide fragments originating from the breakdown of the two POR enzymes could play an essential role for the assembly and maintenance of the photosynthetic apparatus. It will be of interest to identify such POR-derived peptide fragments, to see whether they bind Chlide, and to determine their amino acid sequences to develop tools for the engineering of mutated POR enzymes in which these pigment-binding motifs have been altered. We hope that these approaches will provide information on the actual role of the two POR enzymes during chloroplast development in angiosperms.

We wish to thank Prof. K. Apel for his stimulating interest in this work. We are grateful to Dr D. Rubli for the expert preparation of all of the figures.

REFERENCES

Anadan, S., Morishige, D. T. & Thornber, J. P. (1993) *Plant Physiol. (Bethesda) 101*, 227–236.

Apel, K. & Kloppstech, K. (1980) *Planta (Heidelb.) 150*, 426–430.

Apel, K., Santel, H.-J., Redlinger, T. E. & Falk, H. (1980) *Eur. J. Biochem. 111*, 251–258.

Apel, K. (1981) *Eur. J. Biochem. 120*, 89–93.

Archer, E. K. & Keegstra, K. (1990) *J. Bioenerg. Biomembr. 22*, 789–810.

Avissar, Y. J. & Beale, S. I. (1989a) *Plant Physiol. (Bethesda) 89*, 852–859.

Avissar, Y. J. & Beale, S. I. (1989b) *J. Bacteriol. 171*, 2919–2924.

Baker, M. E. (1994) *Biochem. J. 300*, 605–607.

Batschauer, A. & Apel, K. (1984) *Eur. J. Biochem. 143*, 593–597.

Batschauer, A., Mösinger, E., Kreuz, K., Dörr, I. & Apel, K. (1986) *Eur. J. Biochem. 154*, 625–634.

Battersby, A. R. (1988) *J. Nat. Prod. 51*, 629–642.

Bauer, C. E., Bollivar, D. W. & Suzuki, J. Y. (1993) *J. Bacteriol. 175*, 3919–3925.

Beale, S. I. & Weinstein, J. D. (1990) in *Biosynthesis of heme and chlorophylls* (Dailey, H. A., ed.) pp. 287–391, McGraw-Hill, New York.

Belanger, F. C. & Rebeiz, C. A. (1980) *J. Biol. Chem. 255*, 1266–1272.

Benli, M., Schulz, R. & Apel, K. (1991) *Plant Mol. Biol. 16*, 615–625.

Bennett, J. (1981) *Eur. J. Biochem. 118*, 61–70.

Berry-Lowe, S. (1987) *Carlsberg Res. Commun. 52*, 197–210.

Bhaya, D. & Castelfranco, P. A. (1985) *Proc. Natl Acad. Sci. USA 82*, 5370–5374.

Block, M. A., Joyard, J. & Douce, R. (1980) *Biochim. Biophys. Acta 631*, 210–219.

Boardman, N. K., Anderson, J. M. & Goodchild, D. J. (1978) *Curr. Top. Bioenerg. 8*, 35-109.

Boekema, E., Hankamer, B., Bald, D., Kruip, J., Nield, J., Boonstra, A. F., Barber, J. & Rögner, M. (1995) *Proc. Natl Acad. Sci. USA 92*, 175–179.

Boese, Q. F., Spano, A. J., Li, J. & Timko, M. P. (1991) *J. Biol. Chem. 266*, 17060–17066.

Bogorad, L. (1950) *Bot. Gaz. 111*, 221–241.

Bogorad, L. (1967) in *Biosynthesis and morphogenesis in plastids* (Goodwin, T. W., ed.) pp. 615–631, Academic Press, New York.

Bogorad, L. & Granick, S. (1953) *Proc. Natl Acad. Sci. USA 39*, 1176–1188.

Bollivar, D. W., Suzuki, J. Y., Beatty, J. T., Dobrowolski, J. M. & Bauer, C. E. (1994) *J. Mol. Biol. 237*, 622–640.

Bowler, C., Neuhaus, G., Yamagata, H. & Chua, N.-H. (1994) *Cell 77*, 73–81.

Bradbeer, J. W., Atkinson, Y. E., Börner, T. & Hagemann, R. (1979) *Nature 279*, 816–817.

Bruce, W. B. & Quail, P. H. (1990) *Plant Cell 2*, 1081–1089.

Bruce, W. B., Deng, X.-W. & Quail, P. H. (1991) *EMBO J. 10*, 3015–3024.

Burgess, D. G. & Taylor, W. C. (1987) *Planta (Heidelb.) 170*, 520–527.

Carey, E. E. & Rebeiz, C. A. (1985) *Plant Physiol. (Bethesda) 79*, 1–6.

Carey, E. E., Tripathy, B. C. & Rebeiz, C. A. (1985) *Plant Physiol. (Bethesda) 79*, 1059–1063.

Casadoro, G., Høyer-Hansen, G., Kannangara, G. & Gough, S. (1983) *Carlsberg Res. Commun. 48*, 95–129.

Castelfranco, P. A., Rich, P. M. & Beale, S. I. (1974) *Plant Physiol. (Bethesda) 53*, 615–618.

Castelfranco, P. A., Weinstein, J. D., Schwarcz, S., Pardo, A. D. & Wezelman, B. E. (1979) *Arch. Biochem. Biophys. 192*, 592–598.

Castelfranco, P. A. & Beale, S. I. (1981) in *The biochemistry of plants* (Hatch, M. D. & Boardman, N. K., eds) pp. 375–421, Academic Press, New York.

Castelfranco, P. A. & Beale, S. I. (1983) *Annu. Rev. Plant Physiol. 34*, 241–278.

Castelfranco, P. A. & Zeng, X. (1991) *Plant Physiol. (Bethesda) 97*, 1–6.

Chakraborty, N. & Tripathy, B. C. (1992) *Plant Physiol. (Bethesda) 98*, 7–11.

Chamovitz, D., Pecker, I. & Hirschber, J. (1991) *Plant Mol. Biol. 16*, 967–974.

Chang, T.-E., Wegmann, B. & Wang, W.-Y. (1990) *Plant Physiol. (Bethesda) 93*, 1641–1649.

Chen, J.-J., Yang, J. M., Petryshyn, R., Kowower, N. & London, I. M. (1989) *J. Biol. Chem. 264*, 9559–9564.

Chen, M. W., Jahn, D., O'Neill, G. P. & Söll, D. (1990) *J. Biol. Chem. 265*, 4058–4063.
Chitnis, P. R., Harel, E., Kohorn, B. D., Tobin, E. M. & Thornber, P. (1986) *J. Cell Biol. 102*, 982–988.
Chitnis, P. R., Nechushtai, R. & Thornber, P. (1987) *Plant Mol. Biol. 10*, 3–11.
Chitnis, P. R., Morishige, D. T., Nechushtai, R. & Thornber, P. (1988) *Plant Mol. Biol. 11*, 95–107.
Choquet, Y., Rahire, M., Girard-Bascou, J., Erickson, J. & Rochaix, J.-D. (1992) *EMBO J. 11*, 1697–1704.
Chory, J. (1993) *Trends Genet. 9*, 167–172.
Colbert, J. T. (1988) *Plant Cell Environ. 11*, 305–318.
Colbert, J. T., Hershey, H. P. & Quail, P. H. (1985) *Plant Mol. Biol. 5*, 91–102.
Dahlin, C. & Cline, K. (1991) *Plant Cell 3*, 1131–1140.
Dailey, H. A. (1990) *Biosynthesis of heme and chlorophylls*, McGraw-Hill, New York.
Danon, A. & Mayfield, S. P. Y. (1991) *EMBO J. 10*, 3993–4001.
Darrah, P. M., Kay, S. A., Teakle, G. R. & Griffiths, W. T. (1990) *Biochem. J. 265*, 789–798.
de Boer, D. & Weisbeek, P. (1991) *Biochim. Biophys. Acta 1071*, 221–253.
Dehesh, K. & Ryberg, M. (1985) *Planta (Heidelb.) 164*, 396–399.
Dehesh, K., Klaas, M., Häuser, I. & Apel, K. (1986a) *Planta (Heidelb.) 169*, 162–171.
Dehesh, K., van Cleve, B., Ryberg, M. & Apel, K. (1986b) *Planta (Heidelb.) 169*, 172–183.
Dehesh, K., Bruce, W. B. & Quail, P. H. (1990) *Science 250*, 1397–1399.
De Las Rivas, J. D. L., Shipton, C., Ponticos, M. & Barber, J. (1993) *Biochemistry 32*, 6944–6950.
Deng, X.-W. (1994) *Cell 76*, 423–426.
Dörnemann, D., Kotzabasis, K., Richter, P. & Senger, H. (1989) *Bot. Acta 102*, 112–115.
Dreyfuss, B. W. & Thornber, J. P. (1994a) *Plant Physiol. (Bethesda) 106*, 829–839.
Dreyfuss, B. W. & Thornber, J. P. (1994b) *Plant Physiol. (Bethesda) 106*, 841–848.
Eichacker, L., Soll, J., Lauterbach, P., Rüdiger, W., Klein, R. R. & Mullet, J. E. (1990) *J. Biol. Chem. 265*, 13566–13571.
Eichacker, L., Paulsen, H. & Rüdiger, W. (1992) *Eur. J. Biochem. 205*, 17–24.
Elich, T. D. & Chory, J. (1994) *Plant Mol. Biol. 26*, 1315–1327.
Ernst, D. & Schefbeck, K. (1988) *Plant Physiol. (Bethesda) 88*, 255–258.
Falbel, T. G. & Staehelin, L. A. (1994) *Plant Physiol. (Bethesda) 104*, 639–648.
Ford, C. & Wang, W.-Y. (1980a) *Mol. & Gen. Genet. 179*, 259–263.
Ford, C. & Wang, W.-Y. (1980b) *Mol. & Gen. Genet.* 180, 5–10.
Forreiter, C., van Cleve, B., Schmidt, A. & Apel, K. (1990) *Planta (Heidelb.) 183*, 126–132.
Forreiter, C. & Apel, K. (1993) *Planta (Heidelb.) 190*, 536–545.
Fradkin, L. & Domanski, V. (1994) *Plant Sci. 98*, 135–140.
Franck, F. & Strzalka, K. (1992) *FEBS Lett. 309*, 73-77.
Friedmann, H. C., Thauer, R., Gough, S. P. & Kannangara, C. G. (1987) *Carlsberg Res. Commun. 52*, 363–371.
Fuesler, T. P., Castelfranco, P. A. & Wong, Y.-S. (1984) *Plant Physiol. (Bethesda) 74*, 928–933.
Fujita, Y., Takahashi, Y., Kohchi, T., Ozeki, H., Ohyama, K. & Matsubara, H. (1989) *Plant Mol. Biol. 13*, 551–561.
Fujita, Y., Takahashi, Y., Shonai, F., Ogura, Y. & Matsubara, H. (1991) *Plant Cell Physiol. 32*, 1093–1106.
Furuya, M. (1993) *Annu. Rev. Plant Physiol. Plant Mol. Biol. 44*, 617–645.
Gibson, L. C. D., Willows, R. D., Kannangara, C. G., von Wettstein, D. & Hunter, C. N. (1995) *Proc. Natl Acad. Sci. USA 92*, 1941–1944.
Gilmartin, P. M., Sarokin, L., Memelink, J. & Chua, N.-H. (1990) *Plant Cell 2*, 369–378.
Goldbeck, J. H. (1993) *Proc. Natl Acad. Sci. USA 90*, 1642–1646.
Gough, S. P. & Kannangara, C. G. (1979) *Carlsberg Res. Commun. 44*, 403–416.
Granick, S. (1950) *Harvey Lect. 44*, 220–245.
Granick, S. (1954) *Science 120*, 1105–1106.
Granick, S. (1959) *Plant Physiol. (Bethesda) 34*, 18.
Griffiths, W. T. (1974) *FEBS Lett. 49*, 196–200.
Griffiths, W. T. (1975) *Biochem. J. 152*, 623–635.
Griffiths, W. T. (1978) *Biochem. J. 174*, 681–692.
Griffiths, W. T. (1980) *Biochem. J. 186*, 267–278.
Griffiths, W. T., Kay, S. A. & Oliver, R. P. (1985) *Plant Mol. Biol. 4*, 13–22.
Grimm, B. & Kloppstech, K. (1987) *Eur. J. Biochem. 167*, 493–499.
Grimm, B., Kruse, E. & Kloppstech, K. (1989) *Plant Mol. Biol. 13*, 583–593.
Grimm, B., Bull, A., Welinder, K. G., Gough, S. P. & Kannangara, C. G. (1989) *Carlsberg Res. Commun. 54*, 67–79.
Grimm, B. (1990) *Proc. Natl Acad. Sci. USA 87*, 4169–4173.
Gruissem, W. (1989) *Cell 56*, 161–170.
Guarente, L. & Mason, T. (1983) *Cell 32*, 1279–1286.
Halpin, C., Musgrove, J. E., Lord, J. M. & Robinson, C. (1989) *FEBS Lett. 258*, 32–34.
Hanamoto, C. M. & Castelfranco, P. A. (1983) *Plant Physiol. (Bethesda) 73*, 79–81.
Harpster, M. H., Mayfield, S. P. & Taylor, W. C. (1984) *Plant Mol. Biol. 3*, 59–71.
Harpster, M. & Apel, K. (1985) *Physiol. Plant. 64*, 147–152.
Häuser, I., Dehesh, K. & Apel, K. (1984) *Arch. Biochem. Biophys. 228*, 577–586.
He, Z.-H., Li, J., Sundqvist, C. & Timko, M. P. (1994) *Plant Physiol. (Bethesda) 106*, 537–546.
Henningsen, K. W. (1970) *J. Cell. Sci. 7*, 587–621.
Henningsen, K. W., Boynton, J. E. & von Wettstein, D. (1993) *Royal Dan. Acad. Sci. Lett. Biol. Skrifter 42*, 1–349.
Hess, W. R., Müller, A., Nagy, F. & Börner, T. (1994) *Mol. & Gen. Genet. 242*, 305–312.
Hirsch, S., Muckel, E., Heemeyer, F., von Heijne, G. & Soll, J. (1994) *Science 266*, 1989–1992.
Hobe, S., Prytulla, S., Kühlbrandt, W. & Paulsen, H. (1994) *EMBO J. 13*, 3423–3429.
Höfgen, R., Axelsen, K., Kannangara, C. G., Schüttke, I., Pohlenz, H.-D., Willmitzer, L., Grimm, B. & von Wettstein, D. (1994) *Proc. Natl Acad. Sci. USA 91*, 1726–1730.
Holtorf, H., Reinbothe, S., Reinbothe, C., Bereza, B. & Apel, K. (1995) *Proc. Natl Acad. Sci. USA 92*, 3254–3258.
Howe, G. & Merchant, S. (1992) *EMBO J. 11*, 2789–2801.
Howe, G. & Merchant, S. (1995) *Mol. & Gen. Genet. 246*, 156–165.
Høyer-Hanson, G. & Simpson, D. J. (1977) *Carlsberg Res. Commun. 42*, 379–389.
Huang, D.-D., Wang, W.-Y., Gough, S. & Kannangara, C. G. (1984) *Science 225*, 1482–1484.
Huang, L. & Castelfranco, P. A. (1989) *Plant Physiol. (Bethesda) 90*, 996–1002.
Huang, L., Bonner, B. A. & Castelfranco, P. A. (1989) *Plant Physiol. (Bethesda) 90*, 1003–1008.
Huang, D., Everly, R. M., Cheng, R. H., Heymann, J. B., Schägger, H., Sled, V., Ohnishi, T., Baker, T. S. & Cramer, A. (1994) *Biochemistry 33*, 4401–4409.
Hudson, A., Carpenter, R., Doyle, S. & Coen, E. S. (1993) *EMBO J. 12*, 3711–3719.
Ikeuchi, M. & Murakami, S. (1983) *Plant Cell Physiol. 24*, 71–80.
Ilag, L. L., Kumar, A. & Söll, D. (1994) *Plant Cell 6*, 265–275.
Ito, H., Takaichi, S., Tsuji, H. & Tanaka, A. (1994) *J. Biol. Chem. 269*, 22034–22038.
Jacobs, J. M. & Jacobs, N. J. (1984) *Biochem. Biophys. Res. Commun. 123*, 1157–1164.
Jacobs, J. M. & Jacobs, N. J. (1993) *Plant Physiol. (Bethesda) 101*, 1181–1187.
Jahn, D., Michelsen, U. & Söll, D. (1991) *J. Biol. Chem. 266*, 2542–2548.
Jahn, D. (1992) *Arch. Biochem. Biophys. 298*, 505–513.
Jahn, D., Verkamp, E. & Söll, D. (1992) *Trends Biochem. Sci. 17*, 215–218.
Javor, G. T. & Febre, E. F. (1992) *J. Bacteriol. 174*, 1072–1075.
Jennings, R. C., Bassi, R., Garlaschi, F. M., Dainese, P. & Zucchelli, G. (1993) *Biochemistry 32*, 3203–3210.
Johanningmeier, U. & Howell, S. (1984) *J. Biol. Chem. 259*, 13541–13549.

Jones, O. T. G. (1967) *Biochem. Biophys. Res. Commun. 28*, 671–674.

Jones, O. T. G. (1968) *Biochem. J. 107*, 113–119.

Joshi, B., Morley, S. J., Rhoads, R. E. & Pain, V. (1995) *Eur. J. Biochem. 228*, 31–38.

Joyard, J., Block, M., Pineau, B., Albrieux, C. & Douce, R. (1990) *J. Biol. Chem. 265*, 21 820–21 827.

Joyard, J., Block, M. & Douce, R. (1991) Eur. J. Biochem. 199, 489–509.

Kahn, A. (1968) *Plant Physiol. (Bethesda) 43*, 1781–1785.

Kannangara, C. G. & Gough, S. P. (1977) *Carlsberg Res. Commun. 42*, 441–457.

Kannangara, C. G., Gough, S. P. & von Wettstein, D. (1978) in *Chloroplast development* (Akoyunoglou, G. & Argyroudi-Akoyunoglou, J. H., eds) pp. 147–160, Elsevier, Amsterdam.

Kannangara, C. G., Gough, S. P., Bruyant, P., Hoober, J. K., Kahn, A. & von Wettstein, D. (1988) *Trends Biochem. Sci. 13*, 139–143.

Kannangara, C. G. (1990) in *Cell culture and somatic genetics of plants. The molecular biology of plastids and mitochondria* (Bogorad, L. & Vasil, I. K., eds) vol. 7, pp. 301–329, Academic Press, New York.

Kannangara, C. G., Andersen, R. V., Pontoppidan, B., Willows, R. & von Wettstein, D. (1994) *Ciba Found. Symp. 180*, 3–25.

Kaufman, L. S. (1993) *Plant Physiol. (Bethesda) 102*, 333–337.

Kay, S. A. & Griffiths, W. T. (1983) *Plant Physiol. (Bethesda) 72*, 229–236.

Keegstra, K., Olsen, L. J. & Theg, S. M. (1989) *Annu. Rev. Plant Physiol. Plant Mol. Biol. 40*, 471–501.

Keegstra, K., Bruce, B., Hurley, M., Li, H.-M. & Perry, S. (1995) *Physiol. Plant. 93*, 157–162.

Kendrick, R. E. & Kronenberg, G. H. (1994) *Photomorphogenesis in plants*, Nijhoff, Dordrecht.

Kerber, B. & Soll, J. (1992) *FEBS Lett. 306*, 71–74.

Kessler, F., Blobel, G., Patel, H. A. & Schnell, D. J. (1994) *Science 266*, 1035–1039.

Kim, J., Klein, P. G. & Mullet, J. E. (1991) *J. Biol. Chem. 266*, 14931–14938.

Kim, J., Eichacker, L., Rüdiger, W. & Mullet, J. E. (1994) *Plant Physiol. (Bethesda) 104*, 907–916.

Kirk, J. T. O. & Tilney-Basset, R. A. E. (1967) *The plastids*, Freeman, London.

Kirk, J. T. O. & Tilney-Basset, R. A. E. (1978) *The plastids: their chemistry, structure, growth and inheritance*, Elsevier, Amsterdam/New York.

Klein, R. R. & Mullet, J. E. (1986) *J. Biol. Chem. 261*, 11 138–11 145.

Klein, R. R. & Mullet, J. E. (1987) *J. Biol. Chem. 262*, 4341–4348.

Klein, R. R., Gamble, P. G. & Mullet, J. E. (1988a) *Plant Physiol. (Bethesda) 88*, 1246–1256.

Klein, R. R., Mason, H. S. & Mullet, J. E. (1988b) *J. Cell Biol. 106*, 289–301.

Kloppstech, K., Meyer, G., Bartsch, K., Hundrieser, J. & Link, G. (1984) in *Compartments in algal cells and their interaction* (Wiessner, W., Robinson, D. & Starr, R. C., eds) pp. 36–46, Springer, Berlin.

Kohchi, T., Shirai, H., Fukuzawa, H., Sano, T., Komano, T., Umesono, K., Inokuchi, H., Ozeki, H. & Ohyama, K. (1988) *J. Mol. Biol. 203*, 353–372.

Koncz, C., Mayerhofer, R., Koncz-Kalman, S., Nawrath, C., Reiss, B., Redei, G. P. & Schell, J. (1990) *EMBO J. 9*, 1337–1346.

Kouji, H., Masuda, T. & Matsunaka, S. (1989) *J. Pestic. Sci. 13*, 495–499.

Krauss, N., Hinrich, W., Witt, I., Fromme, P., Pritzkow, W., Dauter, Z., Betzel, C., Wilson, K. S., Witt, H. T. & Saenger, W. (1993) *Nature 361*, 326–331.

Kreuz, K., Dehesh, K. & Apel, K. (1986) *Eur. J. Biochem. 159*, 459–467.

Kruse, E., Mock, H.-P. & Grimm, B. (1995a) *Planta (Heidelb.) 196*, 796–803.

Kruse, E., Mock, H.-P. & Grimm, B. (1995b) *EMBO J. 14*, 3712–3720.

Lagarias, J. C. & Rapoport, H. (1980) *J. Am. Chem. Soc. 102*, 4821–4828.

Laing, W., Kreuz, K. & Apel, K. (1988) *Planta (Heidelb.) 176*, 269–276.

Lathrop, J. T. & Timko, M. P. (1993) *Science 259*, 522–525.

Lee, A. I.-C. & Thornber, J. P. (1995) *Plant Physiol. (Bethesda) 107*, 565–574.

Lee, H. J., Ball, M. D., Parkham, R. & Rebeiz, C. A. (1992) *Plant Physiol. (Bethesda) 99*, 1134–1140.

Lehnen, L. P., Sherman, T. D., Becerril, J. M. & Duke, S. O. (1990) *Pestic. Biochem. Physiol. 37*, 239–248.

Lers, A., Levy, H. & Zamir, A. (1991) *J. Biol. Chem. 266*, 13698–13705.

Li, L. & Lagarias, J. C. (1994) *Proc. Natl Acad. Sci. USA 91*, 12535–12539.

Li, J., Goldschmidt-Clermont, M. & Timko, M. P. (1993) *Plant Cell 5*, 1817–1829.

Lidholm, J. & Gustafsson, P. (1991) *Plant Mol. Biol. 17*, 787–798.

Lissemore, J. & Quail, P. H. (1988) *Mol. Cell. Biol. 8*, 4840–4850.

Lydon, J. & Duke, S. O. (1988) *Pestic. Biochem. Physiol. 31*, 74–83.

Madsen, O., Sandal, L., Sandal, N. N. & Marcker, K. A. (1993) *Plant Mol. Biol. 23*, 35–43.

Mapleston, R. E. & Griffiths, W. T. (1980) *Biochem. J. 189*, 125-133.

Markwell, J., Bruce, B. D. & Keegstra, K. (1992) *J. Biol. Chem. 267*, 13933–13937.

Mathews, D. E. & Durbin, R. D. (1990) *J. Biol. Chem. 265*, 493–498.

Matringe, M. & Scalla, R. (1987) *Proceedings British Crop Protection Conference B 9*, 981–988.

Matringe, M., Camadro, J.-M., Block, M. A., Joyard, J., Scalla, R., Labbe, P. & Douce, R. (1992) *J. Biol. Chem. 267*, 4646–4651.

Matringe, M., Camadro, J.-M., Joyard, J. & Douce, R. (1994) *J. Biol. Chem. 269*, 15010–15015.

Matters, G. L. & Beale, S. I. (1994) *Plant Mol. Biol. 24*, 617–629.

Matto, A. K., Marder, J. B. & Edelman, M. (1989) *Cell 56*, 241–246.

May, B. K., Bhasker, C. R., Bawden, M. J. & Cox, T. C. (1990) *Mol. Biol. Med. 7*, 405–421.

Mayer, S. M. & Beale, S. I. (1990) *Plant Physiol. (Bethesda) 94*, 1365–1375.

Mayfield, S. P. & Taylor, W. C. (1984) *Eur. J. Biochem. 144*, 79–84.

McEwen, B. & Lindsten, A. (1992) *Physiol. Plant. 84*, 343–350.

Melis, A. (1991) *Biochim. Biophys. Acta 1058*, 87–106.

Meyer, G. & Kloppstech, K. (1984) *Eur. J. Biochem. 138*, 201–207.

Miyamoto, K., Tanaka, R., Teramoto, H., Masuda, T., Tsuji, H. & Inokuchi, H. (1994) *Plant Physiol. (Bethesda) 195*, 769–770.

Mösinger, E., Batschauer, A., Schäfer, E. & Apel, K. (1985) *Eur. J. Biochem. 147*, 137–142.

Mock, H.-P., Trainotti, L., Kruse, E. & Grimm, B. (1995) *Plant Mol. Biol. 28*, 245–256.

Moss, G. P. (1988) *Eur. J. Biochem. 178*, 277–328.

Mullet, J. E. (1988) *Annu. Rev. Plant Physiol. Plant Mol. Biol. 39*, 475–502.

Mullet, J. E., Klein, P. G. & Klein, R. R. (1990) *Proc. Natl Acad. Sci. USA 87*, 4038–4042.

Mullet, J. E. (1993) *Plant Physiol. (Bethesda) 103*, 309–313.

Nadler, K. & Granick, S. (1970) *Plant Physiol. (Bethesda) 46*, 240–246.

Nair, S. P., Harwood, J. L. & John, R. A. (1991) *FEBS Lett. 283*, 4–6.

Neuhaus, G., Bowler, C., Kern, R. & Chua, N.-H. (1993) *Cell 73*, 937–952.

Nitschke, W. & Rutherford, W. (1991) *Trends Biochem. Sci. 16*, 241–245.

Nordman, Y. & Deybach, J.-C. (1990) in *Biosynthesis of heme and chlorophylls* (Dailey, H. A., ed.) pp. 491–512, McGraw-Hill, New York.

Oelmüller, R. & Mohr, H. (1986) *Planta (Heidelb.) 167*, 106–113.

Oelmüller, R., Dietrich, G., Link, G. & Mohr, H. (1986a) *Planta (Heidelb.) 169*, 260–266.

Oelmüller, R., Levitan, I., Gergfeld, R., Rajasekhar, V. K. & Mohr, H. (1986b) *Planta (Heidelb.) 168*, 482–492.

Oelmüller, R. & Schuster, C. (1987) *Planta (Heidelb.) 172*, 60–70.

Oelmüller, R., Schuster, C. & Mohr, H. (1988) *Planta (Heidelb.) 174*, 75–83.

Oelmüller, R. (1989) *Photochem. Photobiol. 40*, 229–239.

Oelze-Karow, H. & Mohr, H. (1978) *Photochem. Photobiol. 27*, 189–193.

Ohyama, K., Fukuzawa, H., Kohchi, T., Sano, T., Shirai, H.. Umesono, K., Shiki, Y., Takeuchi, M., Chang, Z., Aota, S., Inokuchi, H. & Ozeki, H. (1988) *J. Mol. Biol. 203*, 281–298.

Oliver, R. P. & Griffiths, W. T. (1981) *Biochem. J. 195*, 93–101.

Oliver, R. P. & Griffiths, W. T. (1982) *Plant Physiol. (Bethesda) 70*, 1019–1025.

Ou, K., Parker, N. & Adamson, H. Y. (1990) *Photosynth. Res. 23*, 89–94.

Padmanaban, G., Venkateswar, V. & Rangarajan, P. N. (1989) *Trends Biochem. Sci. 14*, 492–496.

Pardo, A., Chereskin, B. M., Castelfranco, P., Franceschi, V. R. & Wezelman, B. (1980) *Plant Physiol. (Bethesda) 65*, 956–960.

Parthier, B. (1982) *Biochem. Physiol. Pflanz. 177*, 283–317.

Peterson, D., Schön, A. & Söll, D. (1988) *Plant Mol. Biol. 11*, 293–299.

Phillipps, A. L. & Gray, J. C. (1983) *Eur. J. Biochem. 137*, 553–560.

Pineau, B., Dubertret, G., Joyard, J. & Douce, R. (1986) *J. Biol. Chem. 261*, 9210–9215.

Pineau, B., Gerard-Hirne, C., Douce, R. & Joyard, J. (1993) *Plant Physiol. (Bethesda) 102*, 821–828.

Pontoppidan, B. & Kannangara, C. G. (1994) *Eur. J. Biochem. 225*, 529–537.

Porra, R. J. & Lascelles, J. (1968) *Biochem. J. 108*, 343–348.

Porra, R. J., Schäfer, W., Cmiel, E., Katheder, I. & Scheer, H. (1993) *FEBS Lett. 323*, 31–34.

Porra, R. J., Schäfer, W., Cmiel, E., Katheder, I. & Scheer, H. (1994) *Eur. J. Biochem. 219*, 671–679.

Preiss, S. & Thornber, J. P. (1995) *Plant Physiol. (Bethesda) 107*, 709–717.

Pugh, C. E., Harwood, J. L. & John, R. A. (1992) *J. Biol. Chem. 267*, 1584–1588.

Quail, P. H. (1991) *Annu. Rev. Genet. 25*, 389–409.

Rajasekhar, V. K. (1991) *Biochem. Physiol. Pflanz. 187*, 257–271.

Rapp, J. C. & Mullet, J. E. (1991) *Plant Mol. Biol. 17*, 813–823.

Reinbothe, S. & Parthier, B. (1990) *FEBS Lett. 265*, 7–11.

Reinbothe, S., Krauspe, R. & Parthier, B. (1990) *Planta (Heidelb.) 181*, 176–183.

Reinbothe, S., Reinbothe, C., Heintzen, C., Seidenbecher, C. & Parthier, B. (1993) *EMBO J. 12*, 1505–1512.

Reinbothe, S., Runge, S., Reinbothe, C., van Cleve, B. & Apel, K. (1995a) *Plant Cell 7*, 161–172.

Reinbothe, S., Reinbothe, C., Runge, S. & Apel, K. (1995b) *J. Cell. Biol. 129*, 299–308.

Reinbothe, C., Apel, K. & Reinbothe, S. (1995d) *Mol. Cell. Biol. 15*, 6206–6212.

Reinbothe, S., Reinbothe, C., Holtorf, H. & Apel, K. (1995e). *Plant Cell 7*, 1933–1940.

Reiss, T., Bergfeld, R., Link, G., Thien, W. & Mohr, H. (1983) *Planta (Heidelb.) 159*, 518–528.

Rieble, S. & Beale, S. (1991) *J. Biol. Chem. 266*, 9740–9745.

Rogers, K. & Söll, D. (1993) *Biochemistry 32*, 14210–14219.

Roitgrund, C. & Mets, L. (1990) *Curr. Genet. 17*, 147–153.

Rüdiger, W. & Schoch, S. (1988) in *Plant pigments* (Goodwin, T. W., ed.) pp. 1–9, Academic Press, New York.

Ryberg, M. & Dehesh, K. (1986) *Physiol. Plant. 66*, 616–624.

Ryberg, M. & Sundqvist, C. (1991) in *Chlorophylls* (Scheer, H., ed.) pp. 587–612, CRC press, Boca Raton.

Sangwan, I. & O'Brian, M. R. (1993) *Plant Physiol. (Bethesda) 102*, 829–834.

Santel, H. J. & Apel, K. (1981) *Eur. J. Biochem. 120*, 95–103.

Schimmel, P. & Söll, D. (1979) *Annu. Rev. Biochem. 48*, 601–648.

Schindler, U. & Cashmore, A. R. (1990) *EMBO J. 11*, 3415–3427.

Schneegurt, M. A. & Beale, S. I. (1992) *Biochemistry 31*, 11677–11683.

Schnell, D. J., Kessler, F. & Blobel, G. (1994) *Science 266*, 1007–1012.

Schoch, S., Helfrich, M., Wiktorsson, B., Sundqvist, C., Rüdiger, W. & Ryberg, M. (1995) *Eur. J. Biochem. 229*, 291–298.

Schoelz, J. E. & Zaitlin, M. (1989) *Proc. Natl Acad. Sci. USA 86*, 4496–4500.

Schön, A., Krupp, G., Gough, S., Berry-Lowe, S., Kannangara, C. G. & Söll, D. (1986) *Nature 322*, 281–284.

Schulz, R., Steinmüller, K., Klaas, M., Forreiter, C., Rasmussen, S., Hiller, C. & Apel, K. (1989) *Mol. & Gen. Genet. 217*, 355–361.

Schulz, R. & Senger, H. (1993) in *Pigment-protein complexes in plastids: synthesis and assembly* (Ryberg, M. & Sundqvist, C., eds) pp. 179–218, Academic Press, New York.

Selstam, E., Widell, A. & Johansson, L. (1987) *Physiol. Plant. 70*, 209–214.

Senge, M. O. (1993) *Photochem. Photobiol. 57*, 189–206.

Senger, H. (1984) *Blue light effects in biological systems*. Springer, Berlin.

Shanklin, J., Jabben, M. & Vierstra, R. D. (1989) *Biochemistry 28*, 6028–6034.

Shaw, P., Henwood, J., Oliver, R. P. & Griffiths, W. T. (1985) *Eur. J. Cell Biol. 39*, 50–55.

Shedbalkar, V. P., Ioannides, I. M. & Rebeiz, C. A. (1991) *J. Biol. Chem. 266*, 17151–17157.

Shemin, D. & Russell, C. S. (1953) *J. Am. Chem. Soc. 75*, 4873–4874.

Shibata, A. (1957) *J. Biochem. (Tokyo) 44*, 147–173.

Shlyk, A. A. (1971) *Annu. Rev. Plant Physiol. 22*, 169–184.

Short, T. W. & Briggs, W. R. (1994) *Annu. Rev. Plant Physiol. Plant Mol. Biol. 45*, 143–171.

Siefermann-Harms, D. (1987) *Biochim. Biophys. Acta 811*, 325–355.

Sigrist, M. & Staehelin, L. A. (1994) *Plant Physiol. (Bethesda) 104*, 135–145.

Simpson, J., van Montagu, M. & Herrera-Estrella, L. (1986) *Science 233*, 34–38.

Sisler, E. C. & Klein, W. H. (1963) *Physiol. Plant. 16*, 315–322.

Smith, A. G. (1986) in *Regulation of chloroplast differentiation* (Akoyunoglou, G. A. & Senger, H., eds) pp. 179–218, Academic Press, New York.

Smith, A. G., Marsh, O. & Elder, G. H. (1993) *Biochem. J. 292*, 503–508.

Smith, A. G., Santana, M. A., Wallace-Cook, A. D., Roper, J. M. & Labbe-Rois, R. (1994) *J. Biol. Chem. 269*, 13405–13413.

Smith, B. B. & Rebeiz, C. A. (1979) *Plant Physiol. (Bethesda) 63*, 227–231.

Smith, M. A., Grimm, B., Kannangara, C. G. & von Wettstein, D. (1991a) *Proc. Natl Acad. Sci. USA 88*, 9775–9779.

Smith, M. A., Kannangara, C. G., Grimm, B. & von Wettstein, D. (1991b) *Eur. J. Biochem. 202*, 749–757.

Spano, A. J. & Timko, M. P. (1991) *Biochim. Biophys. Acta 1076*, 29–36.

Spano, A. J., He, Z. & Timko, M. P. (1992a) *Mol. & Gen. Genet. 236*, 86–95.

Spano, A. J., He, Z., Michel, H., Hunt, D. F. & Timko, M. P. (1992b) *Plant Mol. Biol. 18*, 967–972.

Stange-Thomann, N., Thomann, H.-U., Lloyd, A., Lyman, H. & Söll, D. (1994) *Proc. Natl Acad. Sci. USA 91*, 7947–7951.

Stockhaus, J., Eckes, P., Rocha-Sosa, M., Schell, J. & Willmitzer, L. (1987) *Proc. Natl Acad. Sci. USA 84*, 7943–7949.

Sugiura, M. (1992) *Plant Mol. Biol. 19*, 149–168.

Sundqvist, C. (1974) *Physiol. Plant. 30*, 143–147.

Susek, R. E. & Chory, J. (1992) *Aust. J. Plant Physiol. 19*. 387–399.

Suzuki, J. Y. & Bauer, C. E. (1992) *Plant Cell 4*, 929–940.

Suzuki, J. Y. & Bauer, C. E. (1995) *Proc. Natl Acad. Sci. USA 92*, 3749–3753.

Taylor, D. P., Cohen, S. N., Clark, W. G. & Marrs, B. L. (1983) *J. Bacteriol. 154*, 580–590.

Taylor, W. C. (1989) *Annu. Rev. Plant Physiol. Plant Mol. Biol. 40*, 211–233.

Teakle, G. R. & Griffiths, W. T. (1993) *Biochem. J. 296*, 225–230.

Telfer, A., He, W.-Z. & Barber, J. (1990) *Biochim. Biophys. Acta 1017*, 143–151.

Telfer, A., De Las Rivas, J. & Barber, J. (1991) *Biochim. Biophys. Acta 1060*, 106–114.

Terry, M. J. & Lagarias, J. C. (1991) *J. Biol. Chem. 266*, 22215–22221.

Thomas, J. & Weinstein, J. D. (1990) *Plant Physiol. (Bethesda) 94*, 1414–1423.

Thompson, W. F. & White, M. J. (1991) *Annu. Rev. Plant Physiol. Plant Mol. Biol. 42*, 423–466.

Thomson, W. W. & Whatley, J. M. (1980) *Annu. Rev. Plant Physiol. 31*, 375–394.

Thorne, S. (1971) *Biochim. Biophys. Acta 226*, 113–127.

Tobin, E. M. & Silverthorne, J. (1985) *Annu. Rev. Plant Physiol. 36*, 569–593.

Tripathy, B. C. & Rebeiz, C. A. (1988) *Plant Physiol. (Bethesda) 87*, 89–94.

Tripathy, B. C. & Chakraborty, N. (1991) *Plant Physiol. (Bethesda) 96*, 761–767.

van Grinsven, M. Q. J. M. & Kool, A. J. (1988) *Plant Mol. Biol. Rep. 6*, 213–239.

van't Hof, R., van Klompenburg, W., Pilon, M., Kozubek, A., de Korte-Kool, G., Demel, R. A., Weisbeek, P. J. & de Kruijff, B. (1993) *J. Biol. Chem. 268*, 4037–4042.

Vierling, E. & Alberte, R. S. (1983) *J. Cell. Biol. 97*, 1806–1814.

Virgin, H. I., Kahn, A. & von Wettstein, D. (1963) *Photochem. Photobiol. 2*, 83–91.

von Wettstein, D., Henningsen, K. W., Boynton, J. E., Kannangara, C. G. & Nielsen, O. F. (1971) in *Autonomy and biogenesis of mitochondria and chloroplasts* (Boardman, N. K., Linnane, A. W. & Smillie, R. M. C., eds) pp. 205–223, North-Holland Publ. Co., Amsterdam/London.

von Wettstein, D. (1990) in *Plant molecular biology* (Herrmann, R. G. & Larkins, B., eds) pp. 449–459, Plenum Press, New York.

von Wettstein, D., Simpson, D. & Kannangara, C. G. (1995) *Plant Cell 7*, 1039–1057.

Walker, C. J. & Weinstein, J. D. (1991a) *Proc. Natl Acad. Sci. USA 88*, 5789–5793.

Walker, C. J. & Weinstein, J. D. (1991b) *Plant Physiol. (Bethesda) 95*, 1189–1196.

Walker, C. J. & Weinstein, J. D. (1994) *Biochem. J. 299*, 277–284.

Warren, M. J. & Scott, A. I. (1990) *Trends Biochem. Sci. 15*, 486–491.

Weinstein, J. D., Howell, R. W., Leverette, R. D., Grooms, S. Y., Brigola, P. S., Mayer, S. M. & Beale, S. I. (1993) *Plant Physiol. (Bethesda) 101*, 657–665.

Whyte, B. J. & Griffiths, W. T. (1993) *Biochem. J. 291*, 939–944.

Wilks, H. M. & Timko, M. P. (1995) *Proc. Natl Acad. Sci. USA 92*, 724–728.

Witty, M., Wallace-Cook, A. D. M., Albrecht, H., Spano, A. J., Michel, H., Shabanowitz, J., Hunt, D. F., Timko, M. P. & Smith, A. G. (1993) *Plant Physiol. (Bethesda) 103*, 139–147.

Yang, Z. & Bauer, C. E. (1990) *J. Bacteriol. 172*, 5001–5010.

Youvan, D. C., Bylina, E. J., Alberti, M., Begusch, H. & Hearst, J. E. (1984) *Cell 37*, 949–957.

Zsebo, K. M. & Hearst, J. E. (1984) *Cell 37*, 937–947.

Eur. J. Biochem. 237, 519–531 (1996)

Review

Oxygenic photosynthesis

Electron transfer in photosystem I and photosystem II

Jonathan H. A. NUGENT

Department of Biology, Darwin Building, University College London, UK

(Received 5 October 1995) – EJB 95 1621/0

Photosystems I and II drive oxygenic photosynthesis. This requires biochemical systems with remarkable properties, allowing these membrane-bound pigment-protein complexes to oxidise water and produce NAD(P)H. The protein environment provides a scaffold in the membrane on which cofactors are placed at optimum distance and orientation, ensuring a rapid, efficient trapping and conversion of light energy. The polypeptide core also tunes the redox potentials of cofactors and provides for unidirectional progress of various reaction steps. The electron transfer pathways use a variety of inorganic and organic cofactors, including amino acids. This review sets out some of the current ideas and data on the cofactors and polypeptides of photosystems I and II.

Keywords: photosystem I; photosystem II; photosynthesis; oxygen evolution; electron transfer.

Photosynthesis is the basis for most life on earth. Oxygenic photosynthesis by plants, algae and cyanobacteria in particular has had major effects on the planet, which include providing the foundation for the evolution of higher organisms.

Scientifically, the study of photosynthesis presents many interesting challenges for all aspects of biochemistry, from molecular biology to biophysics. It covers topics such as the investigation of the genetic systems involved, their control and evolution; the structure, biosynthesis, assembly and posttranslational modification of the proteins and polypeptide complexes; light trapping; the nature of the electron and proton transfer pathways and the photoactivation, regulation, inhibition and maintenance of the whole system under a variety of conditions.

A feature of this subject is the involvement of many laboratories in a variety of disciplines and the huge literature base. The objective of this article is to present an overview of photosystem I (PSI) and photosystem II (PSII). These are the reaction centre complexes responsible for the initial steps in the conversion of light energy into biochemical products by oxygenic photosynthetic organisms. The two outstanding properties of these complexes are that PSII uses water as a terminal electron donor thereby producing molecular oxygen and protons, and that PSI can directly drive the reduction of NAD(P). However, there are many other areas of biochemical interest within PSI and PSII. Many techniques such as femtosecond laser spectroscopy and magnetic resonance spectroscopy have given information on the cofactors, energy transfer and electron transfer events. The increasing application of molecular biology has given a range of mutations to investigate the properties of polypeptides and their interaction with cofactors. In addition, structural information is being provided by X-ray and electron crystallography. Therefore, we are experiencing a flood of exciting new data which is gradually unlocking the secrets of oxygenic photosynthesis.

The review will concentrate on recent results surrounding the electron transfer reactions and try to give the present state of understanding. References cannot be comprehensive, and are mainly restricted to recent books [1–5], papers and reviews. I refer the reader to these for further detail and to find many of the original papers plus details of techniques used.

Oxygenic photosynthesis

In oxygenic photosynthetic organisms the capture of light energy and its conversion into biochemical intermediates occurs in membrane-bound protein complexes. There are several of these complexes, three electron transfer complexes (PSI, PSII and cytochrome b_6f), light-harvesting pigment complexes (LHC) and ATP synthase, which together produce ATP and NAD(P)H for biosynthesis. The possible role for another membrane-bound complex, NAD(P)H-plastoquinone oxidoreductase, is still under investigation. In eukaryotes, both chloroplast and nuclear genetic systems are involved in forming these complexes.

Both PSI and PSII can be divided into two parts. First, the LHC, which binds the bulk of the light-absorbing pigments. Se-

Correspondence to J. Nugent, Department of Biology, Darwin Building, University College London, Gower Street, London, WC1E 6BT, United Kingdom

Fax: +44 0171 380 7096.

Abbreviations. PSII, photosystem II; PSI, photosystem I; Chl, chlorophyll; Pheo, pheophytin; Cyt, cytochrome; LHC, light-harvesting complex; P700, PSI primary electron donor; A_0, PSI chlorophyll electron acceptor; A_1, PSI phylloquinone electron acceptor; Fe-S_X, Fe-S_B, Fe-S_A and Fe-$S_{A/B}$; PSI iron-sulfur centres X, B and A; P680, PSII primary electron donor; Pheo I, PSII pheophytin electron acceptor I; Qa, the first PSII plastoquinone electron acceptor; Qb, the second PSII plastoquinone electron acceptor; Y_Z, D1 Tyr161 ($Y_Z^{\cdot}$ when oxidised); Y_D, D2 Tyr161 ($Y_D^{\cdot}$ when oxidised); Cyt b_{559}, cytochrome b_{559}; WOC, water-oxidising complex; PC, plastocyanin; Fd, ferredoxin; E_m, midpoint redox potential; E_{m7}, midpoint redox potential at pH 7.

Enzymes. H^+-translocating ATP synthase (EC 3.6.1.34); NAD(P)H-plastoquinone oxidoreductase (EC 1.6.5.3).

cond, a core complex, which contains some light-harvesting pigment, but also binds the cofactors necessary for the photochemical reaction and forward electron transfer. This core complex in each photosystem will be the focus of this review.

Light absorption and electron transfer

The fundamental event of photosynthesis is the absorption of light and the transfer of energy by pigment molecules [6]. The most important pigments in oxygenic photosynthesis are chlorophyll (Chl) and carotenoids. In cyanobacteria, phycobilins are also important pigments in the LHC.

Absorption of a photon by the pigment causes an electronic transition from a ground state to a higher-energy excited singlet state. This absorption is extremely rapid and the energy is then released by several competing mechanisms, i.e. thermal relaxation, fluorescence, delayed light emission (luminescence), phosphorescence or energy transfer via neighbouring pigments to the reaction centre complex. In PSI and PSII, absorption of a photon by antenna pigments in the LHC is mostly followed by the rapid transfer of excitation energy to the reaction centre, where a photochemical charge separation occurs. This initial event must be stabilised by a further separation of the charges using an electron transfer chain. It is the optimised organisation of the light-harvesting pigment-protein and the cofactors in the reaction centre which allows high trapping efficiency and limits loss of energy by other singlet relaxation mechanisms. The LHC also allows the system to cope with a wide variety of light intensities and wavelengths, thereby increasing the overall efficiency. The recent solution of the structure of LHC II has revealed important structural data about light-harvesting and the binding of pigments to membrane proteins [7].

The essential step for energy trapping is the photoinduced charge separation between the primary donor (P) and the primary acceptor (A). In both PSI and PSII, the primary donors are Chl molecules. On excitation by light energy, P becomes a powerful reactant (P*) and interacts with an acceptor (A) resulting in the primary electron transfer event:

$$P + A \xrightarrow{\text{light}} P^* + A \longrightarrow P^+ + A^- .$$

These events occur using light in the visible region between the ionising (high-energy) ultraviolet and the (low-energy) infrared regions.

The reaction centre Chl of PSII, termed P680 due to its bleaching near 680 nm, is photo-oxidised and the electron transferred to the membrane pool of plastoquinone. P680 is re-reduced by the water-oxidising complex (WOC) resulting in oxygen evolution. PSII is therefore a water-plastoquinone oxidoreductase. Analogous events with similar rates of forward electron transfer occur in PSI, the net result being the reduction via ferredoxin (Fd) or flavodoxin of NAD(P) and oxidation of plastocyanin (PC) or cytochrome (Cyt) *c* by the reaction centre Chl, P700.

In PSI and PSII, both inorganic and organic electron transfer cofactors are utilised, with PSII also having amino acid electron transfer components. The initial part of each electron transfer chain is specifically bound to the core complex and allows fast activationless electron transfer. Later in the process, mobile electron carriers allow protonation reactions and transfer of reductant out of the reaction centre complex. An important feature of reaction centres is the organisation of cofactors to produce vectorial electron transfer across the membrane. The vectorial electron transfer, linked to proton transfer, generates an electrochemical potential across the membrane which drives processes such as the synthesis of ATP by ATP synthase.

Electron transfer events in PSI and PSII are linked by the Cyt b_6f complex [8–10] which is a membrane-bound protein

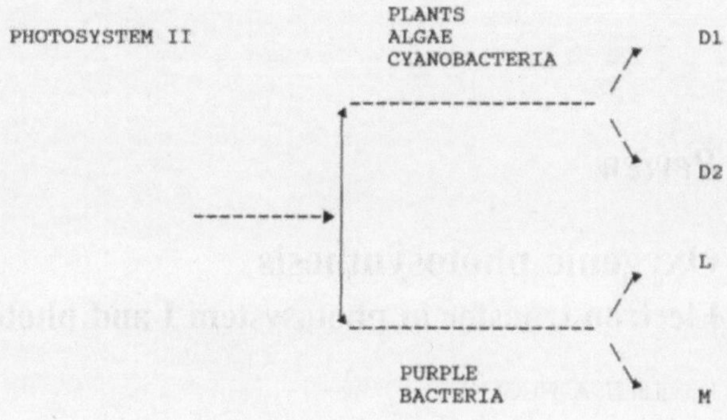

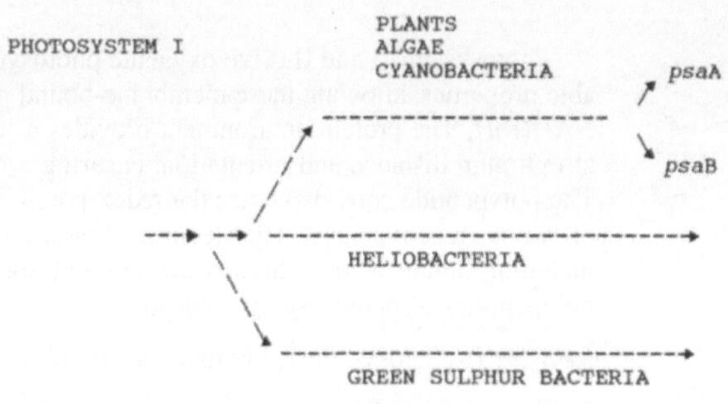

Fig. 1. Simplified evolutionary relationships between the polypeptides of photosynthetic reaction centres. D1 and D2 form the PSII reaction centre, L and M form the purple bacterial reaction centre. They are related through a common ancestral homodimer. *psaA* and *psaB* gene products (PsaA and PsaB) form the PSI reaction centre. They are almost equally related to both heliobacterial and green sulfur bacterial reaction centres. These homodimeric bacterial reaction centres are only distantly related.

complex containing cytochromes and a Rieske iron-sulfur centre. The b_6f complex catalyses the oxidation of plastoquinol and the reduction of plastocyanin or Cyt *c*. This complex is also involved in cyclic electron flow around PSI [11]. Turnover of PSI and PSII is regulated to prevent imbalances in light energy transfer and electron flow [12, 13]. One of the mechanisms used is phosphorylation of LHC and other polypeptides [13].

Evolution

All known types of eukaryotic and prokaryotic photosystems share structural, functional and organisational features. However, their evolutionary relationships are still difficult to elucidate [14–17]. The purple bacterial reaction centre is clearly related to PSII, whilst the reaction centres of green sulfur bacteria and heliobacteria are related to PSI (Figs 1 and 2).

Some authors have further concluded that both PSI and PSII have evolved from a common ancestor, based on their overall structural and functional similarity. This conclusion is based on the fact that the PSI core complex, depleted of its peripheral polypeptides and light-harvesting Chl, contains a cofactor content similar to that of the core of purple bacterial and PSII photosystems. PSI also appears to have the same approximate C_2 symmetry, with two symmetry-related branches of cofactors as observed in purple bacteria and an electron transport chain containing a chlorin primary donor, an intermediary acceptor, and a quinone acceptor. A major difference between the core of

```
         P
HB DF...H......H.T...L.R.....K.S........K......FPCLGPAYGGTC..S..DQ
CB DW...H......L.S...L.R.....H.S........K.. ...FPCLGPVYGGTC..S..DQ
aA DF...H......H.T...L.K.....R.S.....    K......FPCDGPGRGGTC..S..DH
aB DF...H......H.T...L.K.....R.S.....    K......FPCDGPGRGGTC..S..DA
   -----------------------                     | FE-Sx  |      -->
             TM                                                 TM


        P
HB ..AHFIW..........RG...E....     ..WAH.............   ..LS....K..G...F
CB ..GHLVW..........RG...E..........WLG............  . .LT....K..G...Y
aA ..AHFVW..........RG...E......    WAH..........     ...LS....R..G...Y
aB ..GHLVW..........RG...E......    WAH.................LS....R..G...F
   ----------------                                         ------>
         TM                                                    TM
```

Fig. 2. Simplified amino acid sequence alignment of C-terminal regions of PsaA and PsaB polypeptides with those of the related green sulfur bacteria and heliobacteria. Key showing organism and amino acid numbering (total amino acids given in parentheses): HB, *Heliobacillus mobilis* 387–449 and 535–595 (609) [19]; CB, *Chlorobium limicola* 481–542 and 617–680 (730) [18]; aA, *Spinacea oleracea* (spinach) *psaA* 530–589 and 672–730 (750) [20]; aB, *Spinacea oleracea* (spinach) *psaB* 516–575 and 651–713 (734) [20]; TM, predicted transmembrane helix; FE-S_X, highly conserved region (including cysteines) binding iron-sulfur centre X; P labels the two most likely alternatives for P700 ligands. See text for discussion.

PSI and PSII is the existence in PSI of three iron-sulfur centres (Fe-S), termed Fe-S_A, Fe-S_B and Fe-S_X (see below). However the position of Fe-S_A and Fe-S_B on a peripheral 9-kDa polypeptide indicates that it could be a later addition. The water-oxidising system of PSII would also have been a later addition. Fe-S_X could be analagous to the non-haem iron of PSII and purple bacteria.

This argument has merit, but there is also the possibility that PSI and PSII-type reaction centres arose from separate events. The similarities observed could be due to the fact that these membrane complexes both contain proteins with many transmembrane helices, that there are a limited number of biological cofactors with the required properties and that vectorial electron flow is needed to establish the electrochemical gradient. The latter two factors may result in similar sets of cofactors spanning the membrane.

Photosystem I

The PSI core complex consists of >14 polypeptides. Most cofactors are located on the heterodimer encoded by the *psaA* and *psaB* genes. A distinctive feature of PSI is the low redox potential of the electron transfer chain, which allows reduction of NAD(P) on the cytoplasmic (stromal) side of the membrane. Overall PSI allows about 45% conversion of the 1.77 eV of 700-nm light energy (as measured by the midpoint redox potentials (E_m) of PC and Fd). PSI is also involved in cyclic electron flow [11]. The properties of the cofactors will be summarised followed by those of the core polypeptides (see [2–4, 20–24] for reviews).

The primary donor P700. P700 derives its name from the observation of a bleaching maximum near 700 nm upon oxidation of the PSI reaction centre Chl. The E_m of $P700^+/P700$ is about +450 mV placing the excited form, $P700^*/P700$, at about −1300 mV. Measurements of the ground state P700, excited state $P700^*$, cation $P700^+$ and triplet $P700^T$ have been made. Most of the recent spectroscopic data suggest that in the ground state P700 is a Chl dimer, but in $P700^+$ the unpaired electron is distributed asymmetrically with most spin density on one Chl subunit. The two Chl *a* in P700 are oriented almost perpendicular to the membrane [25–27] and are located towards the lumenal side of the membrane (the lumen is a space bounded by the thylakoid membrane).

The electron acceptors A_0 and A_1. The E_m of the primary acceptor A_0^-/A_0 is estimated to be −1000 mV. A_0 has the properties of a monomeric Chl *a*. Recent data indicate the existence of a second Chl *a* reduced by prolonged illumination at 230 K under reducing conditions [28]. This second Chl species may arise from a cofactor on the other branch of cofactors in the heterodimeric core. In summary, the evidence strongly supports the model of a primary event involving a Chl dimer interacting with a monomeric Chl primary electron acceptor. The primary photochemistry, i.e. formation of the $P700^+/A_0^-$ radical pair, occurs in <30 ps. The E_m of the secondary acceptor A_1^-/A_1 is estimated at about −800 mV. A_1 is a phylloquinone (vitamin K_1) molecule, reduced in a few tens of picoseconds and acting between A_0 and iron-sulfur centre Fe-S_X [29–33]. However an exact model of the electron transfer pathway has yet to be established because of the unusually low redox potential at which this quinone has to operate. The environment of the quinone imposed by the protein must be special in order to decrease the operating redox potential of phylloquinone by several hundred millivolts. A_1 can become double reduced under certain conditions [28]. Two phylloquinones are associated with PSI, presumably with one in each proposed symmetry-related branch of cofactors. However, either the second phylloquinone also has a redox potential below A_1, or electron transfer to it is decreased by another mechanism. Under some conditions it may be possible for this second quinone to act as a route for electrons out of the reaction centre for cyclic electron transfer.

The bound iron-sulfur centres. There are three iron-sulfur centres bound to the PSI reaction centre complex: Fe-S_X, bound to the core complex proteins PsaA and PsaB, and Fe-S_A and Fe-S_B both bound to PsaC. Fe-S_X is situated between A_1 and Fe-S_A/Fe-S_B (Fe-$S_{A/B}$). Fe-S_X is reduced in <250 ns [30, 33, 34] and is a single 4Fe-4S centre associated with the heterodimeric core polypeptides. Amino acid sequence information indicates binding by two cysteine residues from each core polypeptide (Fig. 2). It has a E_m of about −730 mV which means it has a distorted or unusual environment. Fe-$S_{A/B}$ are also 4Fe-4S iron-sulfur centres but located on PsaC, a protein related to bacterial 4Fe-4S ferredoxins. The presence of Fe-$S_{A/B}$ is essential for forward electron transfer to $NADP^+$ [35]. The E_m values of the two iron-sulfur centres are not identical but are both close to −550 mV, so the route for electrons from Fe-S_X to Fd is not clear and depends on the relative position of each iron-sulfur centre (see below).

Pathway of electron transport in PSI. In summary the most likely electron transfer chain (Fig. 3) is

$$\text{P700 } A_0 \rightarrow A_1 \rightarrow \text{Fe-}S_X \rightarrow \text{Fe-}S_{A/B} \rightarrow \text{Ferredoxin}.$$

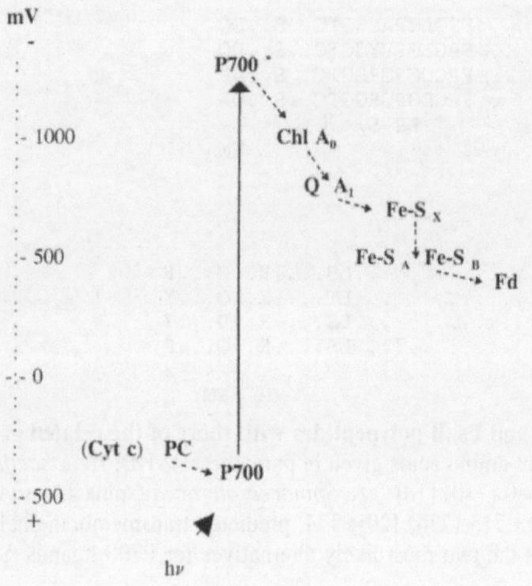

Fig. 3. Diagram showing the main cofactors of PSI, the pathway of electrons through the reaction centre and the probable redox potential relationship between the cofactors. Absorption of light energy (*hν*) by P700 leads to a photochemical charge separation. Electron flow occurs from plastocyanin (PC) to ferredoxin (Fd). Electron transfer from P700 occurs across the membrane to reduce the iron-sulfur centres. PC donates an electron to $P700^+$. Key: P700, primary electron donor; P700*, excited state of P700; A_0, Chl electron acceptor; A_1, phylloquinone (Q); Fe-S_X, Fe-S_A and Fe-S_B, are iron-sulfur centres; Cyt *c*, cytochrome *c* is an alternative electron donor.

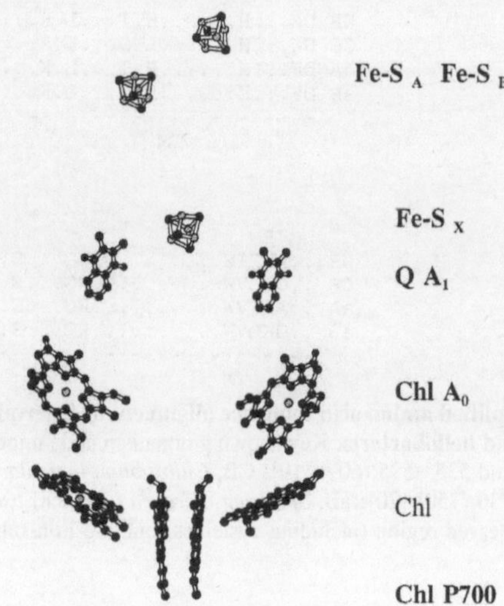

Fig. 4. Diagrammatic view of the PSI cofactors. The view is in the plane of the membrane. Fe-S_A, Fe-S_B and Fe-S_X are represented as cubic structures at the top of the diagram, in positions assigned by X-ray crystallography [36]. The rest of the diagram is assumed from the electron transfer sequence and positions relative to the purple bacterial reaction centre. These cofactors are oriented for easy viewing and are not intended to show specific positional information. Below the iron-sulfur centres are the phylloquinones and then two chlorophylls facing the viewer, one of which may be A_0. The P700 dimer is shown aligned vertically, edge towards the viewer and it is assumed that a pair of accessory chlorophyll molecules either side of P700, each at an angle to the dimer, may be present between P700 and A_0. The diagram was constructed using the program Molscript [37]. For clarity, the cofactors are shown without long sidechains.

Other pathways may operate under particular conditions. The cofactor content and the close relationship between PsaA and PsaB suggest that a symmetry-related branched pathway exists as mentioned above (Fig. 4). However, it is not known if electron transfer is bi- or uni-directional. Asymmetry in the reaction centre may be increased by asymmetry in cofactor binding, post-translational modifications of one polypeptide, or by the binding of the smaller polypeptide subunits.

The core polypeptide composition of PSI. Most PSI proteins are small, <20 kDa, and only PsaA, PsaB and PsaC are known to bind cofactors (Table 1) [20−24, 38, 39]. The key polypeptides of PSI and PSII are thought to be synthesised using membrane-bound ribosomes with cotranslational membrane insertion and incorporation of cofactors such as Chl *a*. Absence of key cofactors or polypeptides results in destabilisation, the degradation of polypeptides and loss of the complex. In eukaryotic species, a further degree of complexity is introduced by the presence of nuclear-encoded subunits, synthesised on cytoplasmic ribosomes and imported into the chloroplast where they are processed and directed to their site of action [40]. Table 1 shows the core polypeptides of PSI with identified functions. These polypeptides show a high degree of sequence conservation among cyanobacteria, algae and green plants, indicating a similar overall structure. In addition to the electron carriers, there are bound antenna Chl, many of which are bound to PsaA and PsaB.

The PSI complex can be isolated from the membrane by treatment with detergents such as Triton X-100. It can then be depleted of its peripheral polypeptides by ion-exchange chromatography and chaotrope extraction. The function of the various polypeptides is currently being studied using biochemical and molecular biology techniques (see [24] for recent review). Cyanobacterial species and the green alga *Chlamydomonas* [41] have been the organisms used for mutagenesis studies, as plants are less suitable being obligate phototrophs.

Table 1. Core PSI subunits and their possible functions. In the Gene column, C and N denote chloroplast and nuclear-encoded in plants and algae. Other PSI polypeptides (PsaG−PsaN) are not shown.

Component	Gene	Mass	Cofactors	Function/ properties
		kDa		
PsaA	*psaA* (C)	83	Chl *a*, P700	light-harvesting,
PsaB	*psaB* (C)	82.5	A_0, A_1, Fe-S_x and β-carotene	electron transfer from P700 to Fe-S_x
PsaC	*psaC* (C)	9	Fe-S_A, Fe-S_B	electron transfer from Fe-S_x to ferredoxin
PsaD	*psaS* (N)	15−18	none	ferredoxin reduction/binding
PsaE	*psaE* (N)	8−11	none	ferredoxin binding/cyclic electron flow
PsaF	*psaF* (N)	17	none	$P700^+$ reduction?

PSI has been crystallised and X-ray structural studies are in progress [36, 42]. The iron-sulfur centres and many helical regions have been identified, as well as the sites of many bound Chl molecules.

PsaA and PsaB. The genes *psaA* and *psaB* encode two high-molecular-mass polypeptides of about 82 kDa, (PsaA and PsaB). They bind redox components P700, A_0, A_1, and Fe-S_X. The polypeptides have a high similarity to each other (about 45%) and show the highest conservation of residues between species in the C-terminal sequence, where two conserved cysteines are located (Fig. 2). Both polypeptides are hydrophobic and each has an estimated 9–11 transmembrane helices. The identification of Fe-S_X as a 4Fe-4S iron-sulfur centre, requiring a total of four cysteine ligands, shows that it must be an inter-polypeptide redox centre located between PsaA and PsaB. The presence of Fe-S_X is essential for rebinding the Fe-$S_{A/B}$ protein, PsaC, to the core complex, suggesting that it plays a role in maintaining PSI structure. The large number of conserved histidine, asparagine and glutamine residues found in PsaA and PsaB are probable ligands to the antenna Chl *a*. P700 was initially proposed to be bound by a conserved histidine in the predicted transmembrane helix prior to the Fe-S_X ligands in the sequence. Mutagenesis experiments now suggest the ligand is in a later transmembrane segment (Fig. 2) [41–44]. The binding sites for other cofactors are unknown, so that the location of most cofactors in Fig. 4 assumes an analogy to the purple bacterial reaction centre which may not prove correct.

PsaC. Site-directed mutagenesis studies have shown that Cys14 and Cys51 are involved in ligation of Fe-S_B and Fe-S_A respectively. The structural relationship to the bacterial 4Fe-4S ferredoxins has allowed the identification of all the ligands to Fe-$S_A$and Fe-S_B. Cys11, Cys14, Cys17 and Cys58 are proposed to ligate Fe-S_B, whilst Cys21, Cys48, Cys51 and Cys54 ligate Fe-S_A [45]. X-ray structural studies suggest a Fe-S_A to Fe-S_B distance of 1.2 nm with distance of 1.5 nm and 2.2 nm, respectively, to Fe-S_X [36], suggesting electron transfer from Fe-S_X to Fe-S_B via Fe-S_A. PsaC is necessary for the stable association of the PsaD, PsaE and PsaL proteins and is thought to bind near Fe-S_X [45–48].

PsaD and PsaE. PsaD is nuclear-encoded in algae and plants and is required for the stable binding of PsaC. The polypeptide from different species shows a high degree of similarity, but not as great as found in the cofactor binding subunits. It has a net positive charge and is implicated in the stable assembly of PSI, correct binding of PsaC, Fd binding and Fd reduction [49, 50]. Another role may be protection of the iron-sulfur centres of PsaC from oxidation. It is thought to lack transmembrane helices and is bound to or near PsaC. In the absence of PsaD, the reconstitution of PsaC is reversible and PSI has a photoinduced EPR signal indicative of an altered orientation of the iron-sulfur centres. The addition of PsaD to the reconstitution system restores normal properties. Cross-linking and antibody labelling studies have also shown a close interaction between PsaC, PsaD, PsaE and Fd. This led to the proposal that PsaD and PsaE play a role in mediation of electron flow from Fe-$S_{A\backslash B}$ to $NADP^+$.

The PsaE polypeptide has a high degree of similarity between species in the C-terminal amino acid sequence. The lack of a transit sequence in cyanobacteria and its hydrophilic sequence suggests it to be an extrinsic polypeptide. The two-dimensional NMR structure reveals an antiparallel β-sheet structure [51]. Mutants deficient in PsaE show autotrophic growth rates equal to the wild type but they are unable to grow heterotrophically in the presence of 3-(3,4-dichlorophenyl)-1,1-dimethylurea. This may mean that PsaE is not involved in forward electron transfer to $NADP^+$ but may have a role in cyclic electron transfer. An interaction with Fd occurs, so that a role in the optimisation of forward electron transfer is also possible [52–55].

PsaF. The amino acid sequence shows a protein of about 17 kDa having a mixture of charged amino acids and some hydrophobic regions. PsaF had been considered to be an extrinsic polypeptide but its hydrophobic nature and resistance to *n*-butanol or chaotrope extraction suggests that it may be membrane-bound. The function of PsaF is also unclear. It has been proposed to be involved in plastocyanin docking and in the reduction of $P700^+$ by plastocyanin but there has been conflicting data. Other studies have shown that it is not necessary for cyt *c* docking to PSI and some mutants lacking PsaF show no defect in photoautotrophic growth [56, 57].

PsaG to PsaN. These low-molecular-mass subunits have been identified, but proposals for their functions and position within the complex are in general rather speculative [24]. However, work with the nuclear-encoded PsaL has revealed a role in the formation of PSI trimers in cyanobacteria [58, 59].

Photosystem II

The proteins and cofactors of the PSII complex have been extensively investigated. A major interest has been the quest to understand the mechanisms of both electron transfer and water oxidation. This has lead to the development of a great variety of PSII preparations, each with particular properties of stability, purity and activity [1–5, 61]. Water oxidation is one of the more labile properties of isolated PSII [61]. As with PSI, various preparations can be made using detergent treatments to strip polypeptides from the complex. Extrinsic proteins (33, 23 and 16 kDa) can be selectively removed by high-ionic-strength treatments. Study of electron transfer has also benefited from the determination of the X-ray structure of the related purple bacterial reaction centre and the application of molecular biological techniques [1–6, 41, 60–62]. Whilst this research has brought a great advance in our knowledge of PSII, there is still much to be understood.

The PSII complex contains >20 polypeptides. It consists of an outer LHC of Chl *a/b* binding proteins (including LHC II [7]) plus an inner antenna of membrane-bound Chl *a* binding proteins termed CP47 and CP43. This inner antenna is tightly associated with the membrane proteins, D1, D2 and Cyt b_{559}, which bind electron transfer components. The PSII reaction centre is more complex than PSI, with more electron transfer cofactors and a variety of electron donor and acceptor pathways available.

Pathway of electron transport in PSII. The initial charge separation is from P680 to a pheophytin termed I (Pheo I; Figs 5 and 6). This probably involves the assistance of an accessory Chl, as in the purple bacterial reaction centre. From Pheo I, the electron is transferred to Qa and then Qb, which are both plastoquinones. Qb accepts two electrons and takes up two protons from the cytoplasmic (stromal) side of the membrane before being released from the complex. Oxidised P680 is reduced by electrons from the WOC via Y_Z (D1 Tyr161) releasing protons on the lumenal side of the membrane. The energy of a singlet exciton at 680 nm is about 1.83 eV. The conversion of energy in the PSII complex is therefore about 45% (based on the E_m of the water/molecular oxygen and Qb/QbH_2 couples). Other cofactors provide alternative pathways. Y_D, Cyt b_{559} and Chl Z are alternative electron donors to oxidised P680. Cyt b_{559} can also be photoreduced suggesting a cyclic pathway could operate perhaps involving Chl Z (Fig. 5). The E_m of the non-haem iron is low-

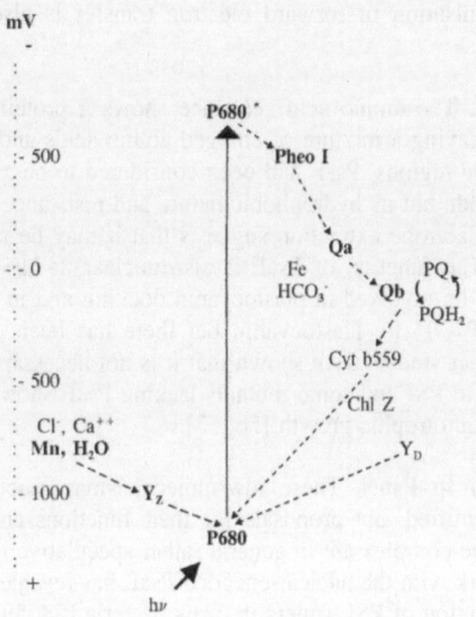

Fig. 5. Diagram showing the main cofactors of PSII, the pathway of electrons through the reaction centre and the probable redox potential relationship between the cofactors. Electron flow occurs from water to Qb. Absorption of light energy (*hv*) by P680, producing P680*, leads to a photochemical charge separation between P680 and Pheo I. Water donates electrons to $P680^+$ via Y_Z (D1 Tyr161). The water-oxidising complex (WOC) involves Mn, plus possibly Ca^{2+} and Cl^- cofactors, oxygen evolution requiring four turnovers of the reaction centre. Y_D (D2 Tyr161), chlorophyll Z (Chl Z) and cytochrome b_{559} (Cyt b_{559}) are alternative electron donors under certain conditions. A non-haem iron (Fe) is located between Qa and Qb and bicarbonate (HCO_3^-) also binds in this region. Qb picks up two electrons and two protons, transferring these to the membrane plastoquinone pool (PQ). Key: P680, the primary chlorophyll electron donor; P680*, excited state of P680; Pheo, pheophytin electron acceptor I; Qa and Qb, primary and secondary plastoquinones.

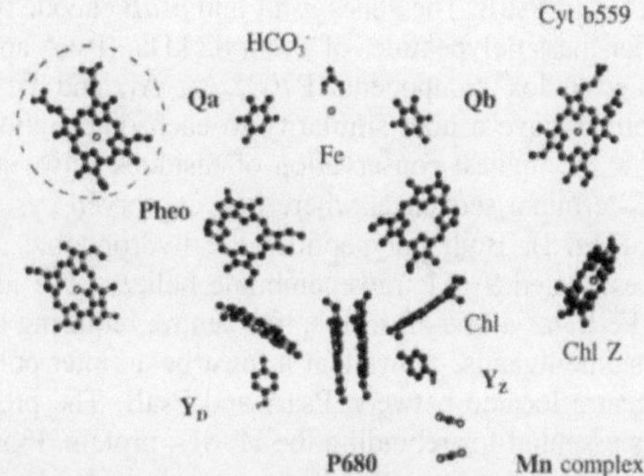

Fig. 6. Diagrammatic view of PSII cofactors. The view is in the plane of the membrane. The presumed P680 dimer is shown edge towards the viewer and it is assumed that a pair of accessory chlorophyll may be present either side of this dimer, between P680 and the pheophytins (shown facing the viewer). Above these are two plastoquinones, Qa and Qb, either side of a ball representing the non-haem iron. Bicarbonate is shown bound at or near to the non-haem iron. Near the P680 dimer are the redox-active tyrosine residues Y_D and Y_Z, with two Mn dimers close to Y_Z. At the edges of the complex are a pair of chlorophyll molecules in a position given by binding to His118 of D1 and D2. Two haems are shown for either one or two (circled) Cyt b_{559}. The existence of some of these cofactors is in doubt, they are oriented for easy viewing and are not intended to show specific positional information. The diagram is based on the available data plus molecular modelling of D1 and D2 [80] (Fig. 8). This is discussed further in the text. The diagram was constructed using the program Molscript [37] as Fig. 4.

ered by comparison with the purple bacterial reaction centre and when oxidised, the non-haem iron can act as an electron acceptor. In the following sections, electron transfer components from P680 to Qb plus Cyt b_{559} and Chl Z are discussed first. This is followed by sections covering water oxidation and electron transfer to P680.

The primary donor P680. $P680^+$/P680 has the highly oxidising E_m of about +1150 mV compared to about +830 mV for Chl *a in vitro*. This high oxidation potential allows the sequential removal of electrons from the WOC, resulting in water oxidation. P680 is only slightly red-shifted from other pigments and therefore forms a shallow trap for excitation energy, with the redox potential of the primary charge separated state being close to the P680 excited state. This has important consequences, among which is an equilibration with antenna pigments resulting in a trapping time which is a function of antenna size [6]. This may allow the antenna to store excitons and act as a buffer to increase the efficiency of charge separation.

The primary charge separation in PSII takes place in a few picoseconds [2–4, 6]. Arguments about the actual rate remain, but this fast rate indicates similar distances between the primary donor and acceptor to those found in purple bacteria [1–4]. The recombination of the radical pair occurs in a few nanoseconds.

The high redox potential and rapid reactions of P680 pose experimental problems. It is especially difficult to study P680 optically because several PSII pigments have overlapping optical transitions near 680 nm. The nature of P680 may also vary depending on whether the ground state, P680, excited state, P680*, triplet state, $P680^T$, or radical cation, $P680^+$, is analyzed. $P680^T$ arises from a charge recombination between $P680^+$ and Pheo I^- (for a review of the triplet state see [63]).

The structure and/or environment of P680 must be different to other types of reaction centre, which have much lower E_m values. The debate as to whether P680 consists of one or more Chl *a* molecules is still not resolved [64–69]. The sequence similarity between D1/D2 and the L/M proteins of purple bacteria, plus the stoichiometry of PSII reaction centre core pigments (6 Chl *a*/2 Pheo *a*/2 β-carotene) argues for a 'purple bacterial' arrangement with P680 as a Chl *a* dimer. In contrast, some experimental data indicate that P680 is a monomer, perhaps in a position equivalent to an accessory Chl in the purple bacterial reaction centre. Other data indicate a monomeric Chl weakly coupled to one or more other pigments.

A consensus from all these results appears to favour P680 being a weakly interacting Chl *a* dimer or multimer. The weak exciton splitting and localisation of the paramagnetic states may be due to an increased separation between the pigments forming P680 compared to purple bacterial reaction centres (from about 0.75 nm to >1 nm centre to centre). A different Chl orientation relative to the purple bacterial structure may be present, perhaps involving an accessory Chl. The reduced coupling and increased monomeric character may be important in achieving the high redox potential for which at present there is no obvious explanation. It would be appropriate for $P680^+$ to reside mainly on a Chl bound to D1, favouring rapid electron transfer by Y_Z and slowing electron transfer by Y_D.

The primary acceptor pheophytin, I. Of the two Pheo *a* bound to the reaction centre, only Pheo I is involved in primary charge separation, indicating an asymmetry which favours one branch of cofactors as in purple bacteria. The E_m of Pheo I/Pheo I^- is below -450 mV. Recent studies have suggested that under some circumstances electron transfer from pheophytin may be used to directly reduce NAD(P) [70]. This is highly controversial and electron transfer to Qa occurs under normal aerobic conditions.

The electron acceptors Qa and Qb. Qa is reduced by Pheo I^-, 200–400 ps after the initial charge separation. Redox titrations determined the E_m at pH 7.0 (E_{m7}) of the Qa/Qa^- couple to be around 0 mV. However, Qa may function at its pK_a and have a working redox potential of approximately −100 mV. Recent work has shown the redox potential of Qa to be sensitive to the integrity of the complex [71–73]. Qa transfers the electron to the secondary quinone Qb, which is a two-electron carrier. In the doubly reduced protonated form, QbH_2 dissociates from PSII and is replaced by a plastoquinone [1–4]. Qa and Qb are identical plastoquinones, so their behaviour can only result from differences in their protein environments.

The rate of electron transfer from Qa^- depends on the reduced state of Qb. A value of 100–200 µs is observed for the first reduction but this is increased to 300–500 µs for the reduction of the semiquinone to the quinol. These relatively slow kinetics are possible because $P680^+$ is reduced by Y_Z on the nanosecond time scale, faster than the $P680^+/Qa^-$ backreaction which has a half-time of 150–200 µs. This stabilises Qa^-, allowing the slower transfer to Qb.

In purple bacteria, the Qb semiquinone is stabilised in the binding site by protonation of the reaction centre polypeptide. Reduction of Qb^- is believed to be coupled to protonation of the doubly reduced quinone. In purple bacteria, the protons arrive in the binding site through a chain of protonatable side chains [3]. In PSII, a comparable proton transfer system probably exists and may include bound bicarbonate.

The Qb binding site is known to be the site of action for several classes of herbicide which act by blocking electron transfer beyond Qa. The herbicides are thought to act by competing with Qb, Qb^- or QbH_2 for the binding pocket, but in PSII some may have additional effects by interfering with bicarbonate binding. Electron transfer between Qa and Qb is also inhibited by some anions which are thought to displace bicarbonate from its binding site (for review of bicarbonate in PSII see Govindjee in [3]). Both the Qa and Qb semiquinones have characteristic EPR signals, which are broadened due to interaction with the nearby non-haem iron. The binding of bicarbonate affects the EPR properties of Qa, Qb, and interactions between them [74, 75].

The non-haem iron. The E_{m7} of the Fe^{2+}/Fe^{3+} couple was determined to be +400 mV with a pH dependence of 60 mV/pH over pH 6–8.5, indicating that reduction is associated with proton binding. Therefore, the non-haem iron acts as a single electron carrier under oxidising conditions. It is not thought to be directly involved in normal electron flow as the E_m of plastoquinol/plastosemiquinone is too low for Qb semiquinone to oxidise Fe^{2+}. The non-haem iron is not vital for Qa to Qb electron transfer as this process is not prevented by removal of the iron. The EPR signal of the oxidised non-haem iron has been used as a sensitive probe of the electron acceptor complex [76, 77]. Changes in the EPR spectrum occur when the Qb binding site is occupied either by herbicides or quinone analogues, and occur when bicarbonate binding is affected. This has led to proposals that bicarbonate provides a single or bidentate ligand to the non-haem iron. Bicarbonate may therefore have a dual role, in the protonation of Qb and as a non-haem iron ligand.

Cytochrome b_{559}. Cyt b_{559} is closely associated with PSII [78], as it is present in purified reaction centres. The cytochrome exhibits a number of redox forms (+50 to +350 mV). The low-redox-potential haem is normally oxidised in the dark by the ambient redox potential. The higher-potential forms can act as electron donors to $P680^+$, when PSII is illuminated at 77 K. These forms are lost during PSII purification or by treatments causing structural changes to PSII, suggesting that a special haem environment is present *in vivo*. The reasons for the different redox potential forms are unknown. Cyt b_{559} is thought to be involved in cyclic electron flow around PSII, because it has been shown to be both photooxidised by $P680^+$ and photoreduced by plastoquinol.

Chlorophyll Z. In PSII samples where Cyt b_{559} is already oxidised, illumination at cryogenic temperatures (below 150 K) generates a Chl *a* cation. Brudvig and coworkers [79] suggest that Cyt b_{559} oxidation by $P680^+$ occurs via this Chl. Their model proposes that the cytochrome reduces Chl^+ and accepts electrons from reduced plastoquinone in a cyclic flow. The reaction centre is thought to bind six Chl molecules. The oxidised Chl could be either an accessory Chl close to P680 or a Chl proposed to be bound to D2 His118 or D1 His118 [64–66, 79, 80]. Recent measurements suggest that it may be the latter [79].

Photoinhibition. Photochemical reactions, especially in the presence of oxygen, are difficult to control. Therefore, photoinhibition of PSII occurs through photochemical damage caused by exposure to high light conditions. PSI is not as vulnerable, probably due to the less oxidising electron donor pathway and protection of the electron acceptors from reaction with oxygen. *In vitro* studies have identified two major forms of PSII photoinhibition [81, 82]. In the first, reduction of the electron acceptors occurs, increasing the population of the $P680^+$ Pheo I^- state. This leads to the formation of a Chl triplet state which can react with oxygen to form singlet oxygen and cause damage to P680 and protein. Reaction between semiquinone states and oxygen may also occur. The second mechanism involves exceeding the ability of electron donors to reduce the highly oxidising $P680^+$. $P680^+$ lifetime increases which leads to damage to the reaction centre. Once damaged, D1 is triggered for degradation.

PSII possesses mechanisms designed to protect PSII from the deleterious effects of light. These dissipate excess excitation energy, minimising photodamage. A cyclic flow around PSII via Cyt b_{559} [79, 83, 84] and modulation of the redox potential of Qa [71, 72] could be part of such a protective function. A 150-mV positive shift in Qa redox potential occurs after procedures which affect binding of Ca and Mn cofactors. This alters the charge recombination route, decreasing backreaction through Chl triplet states and inhibiting forward electron flow to Qb [71–73]. This mechanism would also protect the reaction centre from photodamage during assembly of a functioning complex. Other defensive procedures include alteration in antenna size and dissipation of excitation energy from the antenna (xanthophyll cycle) [12].

Phosphorylation of LHC causes it to uncouple from PSII [13], thereby reducing the absorption cross section. Phosphorylation of LHC is triggered by reduction of the PQ pool in the membrane (i.e. excess PSII activity), which activates the kinase involved. PSII polypeptides including D1 and the *psbH* gene product can also be phosphorylated but the role for this is not clear. Further mechanisms alter PSII activity, sensing saturation of electron transfer by detection of decreased pH in the thylakoid lumen [72]. Degradation and resynthesis of polypeptides is costly to the organism and the mechanisms to protect PSII help minimise this. However, damage to PSII may be the inevitable price the organism has to pay for the advantages the system brings.

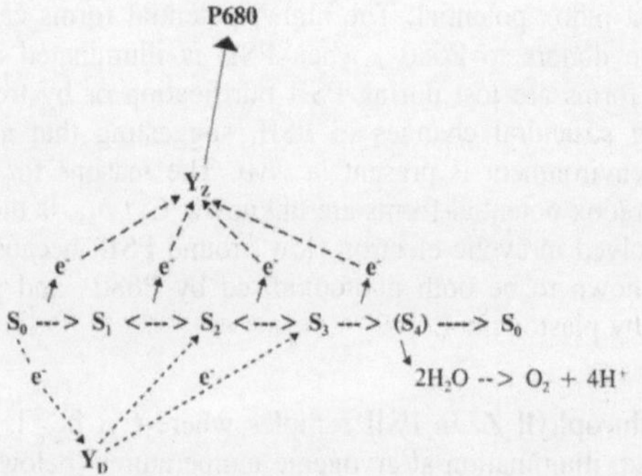

Fig. 7. The S-states of the water-oxidising complex (WOC). Electrons are removed sequentially by $P680^+$ via Y_Z. The S-state number indicates the number of oxidising equivalents stored. On reaching S_4, oxygen is released and the cycle reset. The steps at which water may be bound, oxidised and protons released are discussed in the text. Y_D can slowly oxidise S_0 or can be oxidised by S_2 and S_3. This mechanism helps maintain S_1 as the dark stable state.

The water-oxidising complex and Mn cluster. There are few undisputed facts about this part of PSII (see [2, 61] for recent reviews). In the absence of clear information there has been much controversy and speculation. However, there has been recent progress on some central questions regarding the location, structure and mechanism of the WOC. The oxidation of water to release molecular oxygen is a four-electron transfer process. The turnover of the PSII reaction centre is a one-electron process and the function of the WOC is to act as a system for the accumulation of oxidising equivalents generated by the reaction centre. During its turnover, the WOC passes through different redox states termed S states, S_0-S_4, electrons being removed from S_0 to S_4 and O_2 being evolved at S_4 (Fig. 7). At least two, probably four, Mn atoms are actively involved in the process. The Mn cluster appears to act both as catalytic site and a device for accumulation of oxidising equivalents. Water is a substrate and electron donor, with oxygen, electrons and protons as products. The point in the cycle where water binding occurs is not known. Two alternative mechanisms are mainly discussed, either binding and oxidation at S_3 to S_0 in a concerted reaction, or binding to lower S states with formation of intermediates through deprotonation or oxidation of water [85]. Recent studies strongly suggest that there is a heterogeneity in the binding of the two water molecules. They bind at different sites or different S states [86].

The Mn valence states remain to be resolved, the S_2 state having $3Mn^{3+} + 1Mn^{4+}$ or $1Mn^{3+} + 3Mn^{4+}$. The S_2 and S_3 states are unstable with short half-lives at room temperature, decaying back to S_1 (approximate half-lives of $S_2/Qa^- = 1$ s and $S_2/Qb^- = 30$ s, with similar times for S_3 which decays via S_2). The lifetimes of S_2 and S_3 increase at lower temperatures, recombination being very slow at 200 K. The lifetimes are also pH-dependent and increase when Qa and Qb are oxidised. The S_4 state has a lifetime of about 1 ms, spontaneously decaying to S_0. The dark stable state is not S_0, but S_1. After a few minutes in the dark, about 75% of PSII is in S_1 and 25% in S_0. On dark adaptation, the S_0 is slowly oxidised to S_1 by $Y_D^{\bullet}$.

The oxidation events in S-state turnover have been studied but there are conflicting interpretations. Two popular schemes have either Mn oxidised during each of the S-state transitions [87] or Mn oxidation on some transitions with S_2 to S_3 and possibly S_0 to S_1 not involving Mn oxidation. Two EPR signals, termed the multiline signal and the $g = 4.1$ signal, have been identified as arising from the S_2 state. These have been used as one of the main probes of the WOC and its turnover [61]. The production of these EPR signals support a Mn valence change, to give a Mn^{3+}/Mn^{4+} dimer, on the S_1 to S_2 step; studies also indicate a role for bound chloride in controlling the EPR signal observed. The oxidation of Mn on the S_1 to S_2 step is supported by X-ray spectroscopy [88].

S_1 and S_2 are the most readily accessible redox states of the WOC. They have been extensively studied by X-ray spectroscopy [89]. These studies suggest a structure in S_2 comparable with model compounds containing a Mn^{3+}/Mn^{4+} dimer. Further analysis indicates a Mn−Mn distance of 0.27 nm with Mn coordination by O or N ligands. The data suggest one or two dimeric μ-oxo-bridged Mn pairs are present. Both Cl^- and Ca^{2+} ions have been shown to be required for maximum rates of oxygen evolution [61]. It has been proposed, but not yet proven, that Ca^{2+} may be associated with one of the Mn dimers and that Cl^- may be a Mn ligand [89−92]. The range of Mn models supported by the present data is discussed in [93].

In summary, studies indicate that the intact complex contains two inequivalent Mn dimers. One Ca^{2+} and one Cl^- ion may be associated with one of the dimers, but more studies are required to confirm this. The roles played by calcium and chloride ions are unclear but the data is consistent with the view that they are required for the functional integrity of the complex [94−96].

New EPR signals attributed to the WOC were reported in PSII where oxygen evolution was inhibited by calcium or chloride depletion procedures or by treatment with acetate or ammonia [96−99]. In some preparations a modified multiline signal is observed and assigned to a modified form of the normal S_2 state. In other cases, chloride depletion or acetate treatment, an EPR-silent modified S_2 state appears to be formed. Illumination and freezing to 77 K of these inhibited preparations resulted in the loss of any multiline EPR signal and the generation of a broad signal near $g = 2$. Originally termed the S3 signal, this is now assigned to a S_2X^+ state, where X has been assigned to histidine [97], tyrosine (Y_Z) [99] or a radical formed from water [85]. The S_2X^+ interpretation of the S3 signal has been used to support the lack of oxidation of Mn on the S_2 to S_3 step, but the relationship to normal turnover of the WOC and native S_3 state is not known. Recent studies in the author's laboratory suggest that similar states can be trapped during S-state turnover in uninhibited samples. Another recent study has indicated that X is Y_Z [100]. This data allow the position of the Mn cluster to be suggested, from the proposed interaction of Y_Z and S_2, as within 1 nm of Y_Z [98−100]. This, if confirmed, would eliminate models suggesting much greater distances [101]. The hypothesis that Y_Z is close to the WOC has also lead to models suggesting how Y_Z and the Mn cluster interact in the mechanism of water oxidation [102, 103].

Tyrosine residues D and Z. $Y_D^{\bullet}$ and $Y_Z^{\bullet}$ can be observed by EPR as signals of similar lineshape. Oxidation of tyrosine releases the phenoxyl proton to produce the neutral radical. There are several kinetic forms at physiological temperatures arising from either $Y_Z^{\bullet}$ or $Y_D^{\bullet}$. The E_m of the neutral radical is one of the lowest among common amino acids. Therefore, these radicals will not easily oxidise their protein environments. $Y_Z^{\bullet}$ and $Y_D^{\bullet}$ can be distinguished by their microwave power saturation characteristics which suggest that Y_Z is closer to the WOC.

D1 Tyr161 (Y_Z), acts as an intermediate electron carrier between P680 and the WOC [1−4, 61]. It may increase the efficiency of charge separation by acting as a buffer between P680 and the WOC during the relatively slow multiple electron chemistry of water oxidation. As discussed above, Y_Z has been increasingly thought to be involved in the WOC itself. Electron donation by Y_Z to $P680^+$ is faster than alternative donors such

Table 2. Core PSII subunits and their possible functions. The protein mass is that in spinach. In the Gene column, N denotes nuclear-encoded in plants and algae.

Protein	Gene	Mass	Cofactors	Function/ properties
		kDa		
D1 (PsbA)	*psbA*	39	P680,Pheo,Qa,	electron trans-
D2 (PsbD)	*psbD*		Qb,Fe,Y_Z,Y_D,	fer from Mn to
			Mn?, Chl *a*	Qb
			and β-carotene	
CP47 (PsbB)	*psbB*	56	Chl *a*	light harvesting
CP43 (PsbC)	*psbC*	52	Chl *a*	light harvesting
Cyt b_{559}			Cyt b_{559}	cyclic electron
α (PsbE)	*psbE*	9		flow?
β (PsbF)	*psbF*	4		
33-kDa (PsbO)	*psbO* (N)	27	none	WOC/Mn stability

as Y_D, Chl Z or Cyt b_{559}, which only function when donation from the WOC through Y_Z is blocked. Y_Z is oxidised by $P680^+$ on the nanosecond time scale, but the exact rate is S-state-dependent, increasing in the higher S states (20–40 ns S_0, S_1; 50–250 ns S_3, S_4). The E_m of the $Y_Z/Y_Z^{\bullet}$ couple is estimated at +950–1000 mV [104]. $Y_Z^{\bullet}$ reduction by the WOC is again S-state-dependent; 30–250 µs, 50–100 µs, 200–400 µs and 1.0–1.5 ms respectively for the four S-state transitions between S_0 and S_4 [61]. The reduction of $Y_Z^{\bullet}$ during the S_3 to S_4 transition is the rate-limiting step in water oxidation.

D2 Tyr161 (Y_D) can also be oxidised by $P680^+$. It does not appear to have an essential role in electron flow, it is relatively stable in the oxidised state and is also able to slowly undergo redox reactions with the lower-oxidation states of the WOC (Fig. 7). During dark adaptation, the WOC slowly relaxes to the S_1 state either by advancement from S_0 with electron donation to $Y_D^{\bullet}$ or by deactivation from S_3 and S_2 as Y_D is oxidised. Therefore, the E_m of $Y_D/Y_D^{\bullet}$ should be between that of S_0/S_1 and S_1/S_2. The E_m has been estimated to be +750 mV from a study of equilibrium constants [104]. A number of electron donors compete to reduce $P680^+$ and the oxidation of Y_D may be unfavourable if a series of charge equilibria exist or if, as suggested above, the charge on P680 is located on the D1 Chl, increasing the distance to Y_D. It has been suggested that Y_D is involved in the photoactivation of the WOC, the light-driven process of Mn^{2+} oxidation and complex assembly. It may also be required to oxidise Mn in S_0, maintaining the integrity of the Mn cluster.

Protons and proton gradients. The electron transfer system is designed to generate a proton gradient across the membrane, which can then be used to drive energy-requiring processes in the cell. The electron transfer is vectorial and sites of proton uptake (e.g. Qb) and release (e.g. WOC) are located on opposite sides of the membrane (Figs 6 and 8). The Cyt b_6f complex also contributes to the proton gradient [8–11].

The protons from water are released into the lumen. The pattern of proton release during water oxidation is difficult to follow because of the buffering capacity of the protein. For many years a fixed stoichiometry was assumed for each S-state turnover. However, recent data has allowed a reappraisal of proton release and new ideas to appear [105–107]. Proton release appears to occur upon each oxidation of Y_Z.

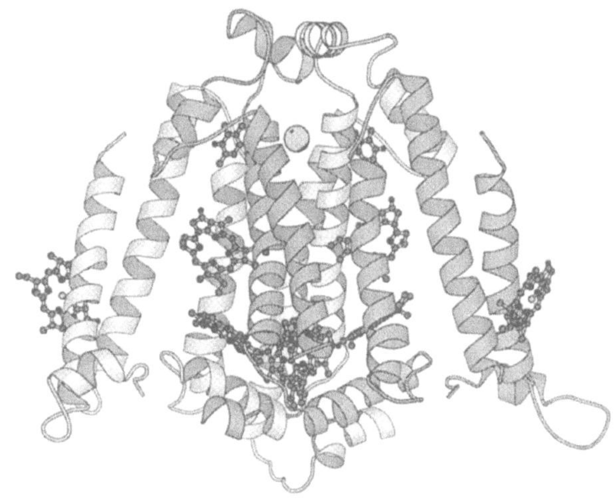

Fig. 8. Illustration from the three-dimensional model of the PSII reaction centre [80]. The view is in the plane of the membrane D1 helices are lightly shaded and D2 is more heavily shaded. The cofactors, a P680 dimer, accessory chlorophylls, pheophytins, plastoquinones and non-haem iron, are shown without long sidechains.

The core polypeptide composition of PSII. PSII consists of a reaction centre containing chloroplast-encoded polypeptides D1 (PsbA), D2 (PsbD), Cyt b_{559} (PsbE +PsbF) and PsbI (4.8-kDa *psbI* gene product). This can be isolated by detergent treatments of PSII complexes. Attached to this are CP47 (PsbB), CP43 (PsbC) and the 33 kDa extrinsic polypeptide (PsbO) (Table 2), making a minimal water-oxidising unit. D1 is synthesized as a precursor polypeptide which is then post-translationally processed to remove several amino acids at the C-terminus. In eukaryotic species, the nuclear-encoded polypeptides, such as the 33-kDa polypeptide, are synthesised with pre-sequences which allow chloroplast uptake, processing and delivery to their site of action. D1 and D2 are thought to form a 1:1 complex with twofold symmetry analogous to the L/M complex of the purple bacterial reaction centre. This has enabled three-dimensional models of D1/D2 to be built based on the X-ray structure of the L and M polypeptides (Fig. 8) [80, 108].

D1 and D2. The core polypeptides D1 and D2 bind most of the electron transfer cofactors. They are hydrophobic polypeptides, each with five predicted membrane spans as in the related L and M polypeptides of purple bacteria. They together bind six Chl *a* molecules, which include P680 and Chl Z. They also bind two Pheo *a* including Pheo I, plus Qa, Qb, β-carotene, non-haem iron and probably provide ligands for the Mn of the WOC. Two amino acids, D1 Tyr161 (Y_Z) and D2 Tyr161 (Y_D), act as electron transfer components. Other factors such as bicarbonate, calcium and chloride may also be bound by D1 and D2.

A Chl dimer is assumed to be ligated by His198 of D1 and D2, which are conserved residues analogous to the ligands found in the purple bacterial reaction centre. In mutants of purple bacteria a number of additional hydrogen bonds have been shown to modulate the redox potential and electron spin distribution of the primary donor [109, 110]. From the models [80, 108] the number of predicted hydrogen bonds is clearly not enough to explain the increase in redox potential of P680. Two conserved prolines (D1 Pro279 and D2 Pro276) are predicted to lie within helix V of each polypeptide. Proline residues are associated with helix bending, which would give a larger cavity and allow for greater lateral separation of the Chl molecules than in purple bacteria. It is unlikely that this larger pocket would accommodate a dimeric structure with one Chl tilted at a steep angle to the other but may allow other conserved residues e.g. Gln187 to play a role. It is asssumed from the electron transfer rate and the number of Chl bound to D1/D2 that the accessory Chl are present as in purple bacteria (Fig. 6). The histidine ligation of these Chl molecules is not conserved from purple bacteria. From the structure of the purple bacterial reaction centre and early site-directed mutagenesis experiments, it was thought that a Mg^{2+}-histidine ligand may be a prerequisite for a bound Chl. However, the situation is now less clear cut as other ligands including water can be used [7]. The lack of histidine ligation to accessory Chl may be part of changes to protect these Chl from oxidation by P680.

Pheo I appears to be bound as in purple bacteria with a hydrogen bond to D1 E130. The non-haem iron is bound by D1 His215, D1 His272, D2 His215 and D2 His269, the D1 and D2 loops between these residues providing the binding site for Qa and Qb. The D1 and D2 loops differ from the corresponding regions in purple bacteria except for D2 Trp254 which may assist electron transfer from Pheo I to Qa. Bicarbonate binds to the non-haem iron or nearby. This could involve D2 Lys265 and D2 Arg266.

Spectroscopic evidence now supports the models in showing that the sidechains of tyrosine residues Y_Z and Y_D are located about 3.5 nm apart, between the lumenal ends of the transmembrane helices 3 and 4 of each polypeptide [111]. The nearest point to the central pair of Chl for Y_Z and Y_D is about 0.8 nm. Hydrogen bonding to Y_D may occur with D2 His190 and/or D2 Gln165 [112]. In D1, interactions of Y_Z with D1 Gln165, D1 Asp170 and D1 His190 are possible. The less hydrophobic environment of Y_Z is contrasted with that of Y_D, which is located in a very hydrophobic pocket.

The D1/D2 heterodimer probably binds at least part of the Mn cluster mainly using D1 ligands. The location of the Mn complex is unknown but it is likely to be located on the lumen side of the membrane. D1 residues predicted to be close to Y_Z and capable of forming ligands to metal ions include Gln165, Ser167, Asp170, Glu189, and His190. Asp170 is a possible binding site for Mn [113]. It has also been proposed that the C-terminus portion of the D1 protein contributes to the Mn binding site since it contains conserved His, Asp and Glu residues and correct post-translational processing to give the terminal residue Ala344 required for Mn complex assembly [114–116]. Although amino acid oxygens are expected to provide most Mn ligands, histidine, probably His190, His332 or His337, has been shown to provide at least one nitrogen ligand [117]. Residues from D2, Glu69 [118] or other PSII polypeptides such as CP47, could provide a small number of Mn ligands. This could give a site for the Mn complex within 1 nm of Y_Z. Other arrangements are possible, some of the residues discussed may bind Ca^{2+}, which may alter the position of the cluster.

Other core polypeptides.

CP47 and CP43 bind Chl *a* and form the inner light-harvesting complex. These polypeptides are also important in stabilising the reaction centre core and modify quinone binding. CP47, especially a sequence forming a large hydrophilic loop, is involved in stabilising the WOC [119].

The Mn complex appears to be shielded and stabilised by the 33-kDa extrinsic polypeptide. Deletion by mutation or biochemical removal of this protein still allow oxygen evolution, but result in an unstable Mn complex. Two other extrinsic polypeptides found in eukaryotes, termed the 16-kDa and 23-kDa polypeptides, further stabilise the Mn complex and modify the Ca^{2+} and Cl^- requirement for optimal oxygen evolution rates.

Both the α and β polypeptide forming Cyt b_{559} have only a single histidine residue, located within a hydrophobic domain. It is thought that the haem cross-links the two different subunits and is positioned towards the Qa/Qb side of the membrane [120]. There are either one or two molecules of Cyt b_{559}/PSII complex (Fig. 6) [121]. The function and location of two small polypeptides found in core complexes, PsbI and the recently discovered PsbW [122], are not known. Further discoveries could increase the list of PSII polypeptides beyond those presently identified (chloroplast-encoded PsbA–PsbF, PsbH–PsbN and nuclear-encoded PsbO–PsbT and PsbW).

I thank the UK Biotechnology and Biological Science Research Council (BBSRC) for their financial support. I thank Mike Evans, Peter Heathcote, Johannes Messinger, Marilyn Nugent, Stephen Rigby, Sandra Turconi and the referees for their comments. I thank the many colleagues who sent preprints and recent papers. I apologise for only being able to mention a fraction of the work in this area.

REFERENCES

1. Hall, D. O. & Rao K. K. (1994) *Photosynthesis*, 5th edn, Cambridge University Press, Cambridge.
2. Ort, D. R. & Yocum, C. F. (1995) *Oxygenic photosynthesis: The light reactions*, Kluwer Academic Publishers, Dordrecht.
3. Deisenhofer, J. & Norris, J. R. (1993) *The photosynthetic reaction center vols. 1 and 2*, Academic Press, San Diego.
4. Barber, J. (1992) *Topics in photosynthesis, vol 11, The photosystems: Structure, function and molecular biology*, Elsevier Science Publishers, Amsterdam.
5. Amesz, J. & Hoff, A. J. (1995) *Biophysical techniques in photosynthesis*, Kluwer Academic Publishers, Dordrecht.
6. van Grondelle, R., Dekker, J. P., Gillbro, T. & Sundström, V. (1994) Energy transfer and trapping in photosynthesis, *Biochim. Biophys. Acta 1187*, 1–65.
7. Kühlbrandt, W., Wang, D. N. & Fujiyoshi, Y. (1994) Atomic model of plant light-harvesting complex by electron crystallography, *Nature 367*, 614–621.
8. Anderson, J. M. (1992) Cytochrome b_6f complex: dynamic molecular organisation, function and acclimation, *Photosynth. Res. 34*, 341–357.
9. Hope, A. B. (1993) The chloroplast cytochrome *b*/*f* complex: a critical focus on function, *Biochim. Biophys. Acta 1143*, 1–22.
10. Cramer, W. A., Martinez, S. E., Huang, D., Tae, G.-S., Everly, R. M., Heymann, J. B., Cheng, R. H., Baker, T. S. & Smith, J. L. (1994) Structural aspects of the cytochrome b_6f complex; structure of the lumen-side domain of cytochrome *f*, *J. Bioenerg. Biomem. 26*, 31–47.
11. Bendall, D. S. & Manasse, R. S. (1995) Cyclic photophosphorylation and electron transport, *Biochim. Biophys. Acta 1229*, 23–38.
12. Horton, P. & Ruban, A. V. (1992) Regulation of photosystem II, *Photoynth. Res. 34*, 375–385.
13. Allen, J. F. (1992) Protein phosphorylation in the regulation of photosynthesis, *Biochim. Biophys. Acta 1098*, 275–335.
14. Nitschke, W. & Rutherford, A. W. (1991) Photosynthetic reaction centres: variations on a common structural theme, *Trends Biochem. Sci. 16*, 241–245.
15. Blankenship, R. E. (1992) Origin and early evolution of photosynthesis, *Photosynth. Res. 33*, 91–111.
16. Golbeck J. H. (1993) Shared thematic elements in photochemical reaction centers, *Proc. Natl Acad. Sci. USA 90*, 1642–1646.
17. Baltscheffsky, H. (1994) *Origin and evolution of biological energy conversion*, VCH Publishers, New York.
18. Büttner, M., Xie, D. L., Nelson, H., Pinther, W., Hauska, G. & Nelson, N. (1992) Photosynthetic reaction center genes in green sulfur bacteria and photosystem 1 are related, *Proc. Natl Acad. Sci. USA 89*, 8135–8139.

19. Liebl, U., Mockensturm-Wilson, M., Trost, J. T., Brune, D. C., Blankenship, R. E. & Vermaas, W. (1993) Single core polypeptide in the reaction center of photosynthetic bacterium *Heliobacillus mobilis*: Structural implications and relations to other photosystems, *Proc. Natl Acad. Sci. USA 90*, 7124–7128.
20. Golbeck, J. H. & Bryant, D. A. (1991) photosystem 1, *Curr. Top. Bioenerget. 16*, 83–177.
21. Golbeck, J. H. (1992) Structure and function of photosystem I, *Annu. Rev. Plant Physiol. Plant Mol. Biol. 43*, 293–324.
22. Lockau, W. & Nitschke, W. (1993) photosystem I and its bacterial counterparts, *Physiol. Plant. 88*, 372–381.
23. Golbeck, J. H. (1994) photosystem I in cyanobacteria in *The molecular biology of cyanobacteria* (Bryant, D., ed.) pp. 179–220, Kluwer Academic Publishers, Dordrecht.
24. Chitnis, P. R., Xu, Q., Chitnis, V. P. & Nechustai, R. (1995) Function and organisation of photosystem I polypeptides, *Photosynth. Res. 44*, 23–40.
25. Rutherford, A. W. & Sétif, P. (1990) Orientation of P700, the primary electron donor of photosystem I, *Biochim. Biophys. Acta 1019*, 128–131.
26. Käss. H., Bittersmann-Weidlich, E., Andréasson, L.-E., Bönigk, B. & Lubitz, W. (1995) ENDOR and ESEEM of the ^{15}N-labelled radical cations of chlorophyll *a* and the primary donor P700 in photosystem I, *Chem. Phys. 194*, 419–432.
27. Davis, I. H., Heathcote, P., MacLachlan, D. J. & Evans, M. C. W. (1993) Modulation analysis of the electron spin echo signals of *in vivo* oxidised primary donor ^{14}N chlorophyll centres in bacterial, P870 and P960 and plant photosystem I, P700 reaction centres, *Biochim. Biophys. Acta 143*, 183–189.
28. Heathcote, P., Hanley, J. A. & Evans, M. C. W. (1993) Double reduction of A_1 abolishes the EPR signal attributed to A_1^-: Evidence for C_2 symmetry in the photosystem I reaction centre, *Biochim. Biophys. Acta 1144*, 54–61.
29. Sétif, P. & Brettel, K. (1993) Forward electron transfer from phylloquinone-A_1 to iron-sulfur centers in spinach photosystem I, *Biochemistry 32*, 7846–7854.
30. Van der Est, A., Bock, C. H., Golbeck, J. H., Brettel, K., Sétif, P. & Stehlik, D. (1994) Electron transfer from the acceptor A_1 to the iron-sulfur centers in photosystem I as studied by transient EPR spectroscopy, *Biochemistry 33*, 11789–11797.
31. Hastings, G., Kleinherenbrink, F. A. M., Lin, S. & Blankenship, R. E. (1994) Time resolved fluorescence and absorption spectroscopy of photosystem I, *Biochemistry 33*, 3185–3192.
32. Hastings, G., Kleinherenbrink, F. A. M., Lin, S., McHugh, T. F. & Blankenship, R. E. (1994) Observation of the reduction and reoxidation of the primary electron acceptor in photosystem I, *Biochemistry 33*, 3193–3200.
33. Moenne-Loccoz, P., Heathcote, P., Maclachlan, D. J., Berry, M. C., Davis, I. H. & Evans, M. C. W. (1994) The path of electron transfer in photosystem I: direct evidence of forward electrontransfer from A_1 to Fe-S_X, *Biochemistry 33*, 10037–10042.
34. Leibl, W., Toupance, B. & Breton, J. (1995) Photoelectric characterisation of forward electron transfer to iron-sulfur centers in photosysytem I, *Biochemistry 34*, 10237–10244.
35. Hanley, J. A., Kear, J., Bredenkamp, G., Li, G., Heathcote, P. & Evans, M. C. W. (1992) Biochemical evidence for the role of the bound iron-sulfur centres A and B in NADP reduction by photosystem I, *Biochim. Biophys. Acta 1099*, 152–156.
36. Krauss, N., Hinrichs, W., Witt, I., Fromme, P., Pritzkow, W., Dauter, Z., Betzel, C., Wilson, K. S., Witt, H. T. & Saenger, W. (1993) Three dimensional structure of system 1 of photosynthesis at 6Å resolution, *Nature 361*, 326–331.
37. Kraulis, P. J. (1991) Molscript: a program to produce both detailed and schematic plots of protein structures, *J. Appl. Crystallog. 24*, 946–950.
38. Ikeuchi, M. (1992) Subunit proteins of photosystem 1, *Plant Cell Physiol. 33*, 669–676.
39. Zilber, A. L. & Malkin, R. (1992) Organisation and topology of photosystem I subunits, *Plant Physiol 99*, 901–911.
40. Mant, A., Nielsen, V. S., Knott, T. G., Moller, B. L. & Robinson, C. (1994) Multiple mechanisms of targeting of photosystem I subunits F, H, K, L, and N into and across the thylakoid lumen, *J. Biol. Chem. 269*, 27303–27309.
41. Webber, A. N., Bingham, S. E. & Lee, H. (1995) Genetic engineering of thylakoid protein complexes by chloroplast transformation in *Chlamydomonas reinhardtii, Photosynth. Res. 44*, 191–205.
42. Fromme, P., Schubert, W-D. & Krauss, N. (1994) Structure of photosystem I: Suggestions on the docking sites for plastocyanin, ferredoxin and the coordination of P700, *Biochim. Biophys. Acta 1187*, 99–105.
43. Cui, L., Bingham, S. E., Kuhn, M., Käss, H., Lubitz, W. & Webber, A. N. (1995) Site directed mutagenesis of conserved histidines in the helix VIII domain of PsaB impairs assembly of the photosystem I reaction center without altering the spectroscopic properties of P700, *Biochemistry 34*, 1549–1558.
44. Rodday, S. M., Webber, A. N., Bingham, S. E. & Biggins, J. (1995) Evidence that the F_X domain in photosystem I interacts with the subunit PsaC: Site directed changes in PsaB destabilize the subunit interaction in *Chlamydomonas reinhardtii, Biochemistry 34*, 6328–6334.
45. Zhao, J., Li, N., Warren, P. V., Golbeck, J. H. & Bryant, D. A. (1992) Directed conversion of a cysteine to an aspartate leads to the assembly of a [3Fe-4S] cluster in PsaC of photosystem 1, *Biochemistry 31*, 5093–5099.
46. Mannan, R. M., Pakrasi, H. B. & Sonoike, K. (1994) The PsaC protein is necessary for the stable association of the PsaD, PsaE and PsaL proteins in the photosystem 1 complex, *Arch. Biochem. Biophys. 315*, 68–73.
47. Yu, L., Bryant, D. A. & Golbeck, J. H. (1995) Evidence for a mixed ligand [4Fe-4S] cluster in the C14D mutant of PsaC. Altered reduction potentials and EPR spectral properties of the F_A and F_B clusters on rebinding to the P700-F_X core, *Biochemistry 34*, 7861–7868.
48. Yu, J., Smart, L. B., Jung, Y-S., Golbeck, J. H. & McIntosh, L. (1995) Absence of the PsaC subunit allows assembly of the photosystem I core but prevents the binding of PsaD and PsaE in *Synechocystis* sp. PCC 6803, *Plant Mol. Biol.* 29, 331–342.
49. Xu, Q., Armbrust, T. S., Guikema, J. A. & Chitnis, P. R. (1994) Organisation of photosystem I Polypeptides: A structural interaction between the PsaD and PsaL subunits, *Plant Physiol. 106*, 1057–1063.
50. Xu, Q., Jung, Y. S., Chitnis, V. P., Guikema, J. A., Golbeck, J. H. & Chitnis, P. R. (1994) Mutational analysis of photosystem I polypeptides in *Synechocystis* sp. PCC 6803. Subunit requirements for the reduction of $NADP^+$ mediated by ferredoxin and flavodoxin, *J. Biol. Chem. 269*, 21512–21518.
51. Falzone, C. J., Kao, Y-H., Zhao, J., Bryant, D. A. & Lecomte, T. T. J. (1994) The three dimensional solution structure of PsaE from the cyanobacterium *Synechococcus* sp. strain PCC 7002: A photosystem I protein that shows structural similarity with SH3 domain, *Biochemistry 33*, 6052–6062.
52. Strotmann, H. & Webber, A. N. (1993) On the function of PsaE in chloroplast photosystem I, *Biochim. Biophys. Acta 1143*, 204–210.
53. Yu, L., Zhao, J., Mühlenhoff, U., Bryant, D. A. & Golbeck, J. H. (1993) PsaE is required for *in vivo* cyclic electron flow around photosystem 1 in the cyanobacterium *Synechococcus* sp. PCC 7002, *Plant Physiol. 103*, 171–180.
54. Rousseau, F., Sétif, P. & Lagoutte, B. (1993) Evidence for the involvement of PSI-E subunit in the reduction of ferredoxin by photosystem 1, *EMBO J. 12*, 1755–1765.
55. Sonoike, K., Hatanaka, H. & Katoh, S. (1993) The *psaE* gene product has a role to promote interaction between the terminal electron acceptor and ferredoxin, *Biochim. Biophys. Acta 1141*, 52–57.
56. Hatanaka, H., Sonoike, K., Hirano, M. & Katoh, S. (1993) Is the *psaF* gene product required for oxidation of cytochrome *c*-553? *Biochim. Biophys. Acta 1141*, 45–51.
57. Xu, Q., Yu, L., Chitnis, V. P. & Chitnis, P. R. (1994) Function and organisation of photosystem I in a cyanobacterial mutant strain that lacks PsaF and PsaJ subunits, *J. Biol. Chem. 269*, 3205–3211.
58. Kruip, J., Boekema, E. J., Bald, D., Boonstra, A. F. & Rögner, M. (1993) Isolation and structural characterisation of monomeric and trimeric photosystem 1 complexes from the cyanobacterium *Synechocystis* sp. PCC 6803, *J. Biol. Chem. 268*, 23353–23360.

59. Chitnis, V. P. & Chitnis, P. R. (1993) PsaL subunit is required for the formation of photosystem 1 trimers in the cyanobacterium *Synechocystis* sp PCC 6803, *FEBS Lett 336*, 330–334.
60. Diner, B. A., Nixon, P. J. & Farchaus, J. W. (1991) Site directed mutagenesis of photosynthetic reaction centres, *Curr. Opin. Struct. Biol. 1*, 546–554.
61. Debus, R. J. (1992) The manganese and calcium ions of photosynthetic oxygen evolution, *Biochim. Biophys. Acta 1102*, 269–352.
62. Vermaas, W. F. J. (1993) Molecular-biological approaches to analyze photosystem II, *Annu. Rev. Plant Physiol. Plant Mol. Biol. 44*, 457–482.
63. Budil, D. E. & Thurnauer, M. C. (1991) The chlorophyll triplet state as a probe of structure and function in photosynthesis, *Biochim. Biophys. Acta 1057*, 1–41.
64. Van Gorkom, H. J. & Schelvis, J. P. M. (1993) Kok's oxygen clock: What makes it tick? The structure of P680 and consequences of its oxidising power, *Photosynth. Res. 38*, 297–301.
65. Rigby, S. E. J., Nugent, J. H. A. & O'Malley, P. J. (1994) ENDOR and special triple resonance studies of chlorophyll cation radicals in photosystem 2, *Biochemistry 33*, 10043–10050.
66. Schelvis, J. P. M., Van Noort, P. I., Aartsma, T. J. & Van Gorkom, H. J. (1994) Energy transfer, charge separation and pigment arrangement in the reaction center of photosystem II, *Biochim. Biophys. Acta 1184*, 242–250.
67. Durrant, J. R., Klug, D. R., Kwa, S. L. S., Van Grondelle, R., Porter, G. & Dekker, J. P. (1995) A multimer model for P680, the primary electron donor of photosystem II, *Proc. Natl Acad. Sci. USA 92*, 4798–4802.
68. Chang, H.-C., Jankowiak, R., Reddy, N. R. S., Yocum, C. F., Picorel, R., Seibert, M. & Small, G. J. (1994) On the question of the chlorophyll *a* content of the photosystem II reaction center, *J. Phys. Chem. 98*, 7725–7735.
69. Hillmann, B., Brettel, K., van Mieghem, F., Kamlowski, A., Rutherford, A. W. & Schlodder, E. (1995) Charge recombination reactions in photosystem II. 2. Transient absorbance difference spectra and their temperature dependence, *Biochemistry 34*, 4814–4827.
70. Greenbaum, E., Lee, J. W., Tevault, C. V., Blankinship, S. L. & Mets, L. J. (1995) CO_2 fixation and photoevolution of H_2 and O_2 in a mutant of *Chlamydomonas* lacking photosystem I, *Nature 376*, 438–441.
71. Krieger, A., Rutherford, A. W. & Johnson, G. N. (1995) On the determination of the redox midpoint potential of the primary quinone electron acceptor, Qa, in photosystem II, *Biochim. Biophys. Acta 1229*, 193–201.
72. Krieger, A., Rutherford, A. W. & Johnson, G. N. (1995) A change in the midpoint potential of the quinone Qa in photosystem II associated with photoactivation of oxygen evolution, *Biochim. Biophys. Acta 1229*, 202–207.
73. Van Mieghem, F., Brettel, K., Hillmann, B., Kamlowski, A., Rutherford, A. W. & Schlodder, E. (1995) Charge recombination reactions in photosystem II. 1. Yields, recombination pathways and kinetics of the primary pair, *Biochemistry 34*, 4798–4813.
74. Hallahan, B. J., Ruffle, S. V., Bowden, S. J. & Nugent, J. H. A. (1991) Identification and characterisation of EPR signals involving Qb semiquinone in plant photosystem II, *Biochim. Biophys. Acta 1059*, 181–188.
75. Bowden, S. J., Hallahan, B. J., Ruffle, S. V., Evans, M. C. W. & Nugent, J. H. A. (1991) Preparation and characterisation of photosystem two core particles with and without bound bicarbonate, *Biochim. Biophys. Acta 1060*, 89–96.
76. Deligiannakis, Y., Petrouleas, V. & Diner, B. A. (1994) Binding of carboxylate ions at the non-heme Fe(II) of PSII. Effects on the $Q_A^-Fe^{2+}$ and the $Q_A^-Fe^{3+}$ EPR spectra and the redox properties of the iron, *Biochim. Biophys. Acta 1188*, 260–270.
77. Petrouleas, V., Deligiannakis, Y. & Diner, B. A. (1994) Binding of carboxylate ions at the non-heme Fe(II) of PSII, *Biochim. Biophys. Acta 1188*, 271–277.
78. Cramer, W. A., Tae, G.-S., Furbacher, P. N. & Bottger, M. (1993) The enigmatic cytochrome b_{559} of oxygenic photosynthesis, *Physiol. Plant. 88*, 705–711.
79. Koulougliotis, D., Innes, J. B. & Brudvig, G. W. (1994) Location of chlorophyll Z in photosystem II, *Biochemistry 33*, 11814–11822.
80. Ruffle, S. V., Donnelly, D., Blundell, T. L. & Nugent, J. H. A. (1992) A three-dimensional model of the photosystem II reaction centre of *Pisum sativum*, *Photosynth. Res. 34*, 287–300.
81. Barber J. (1995) Molecular basis of the vulnerability of photosystem II to damage by light, *Aus. J. Plant Physiol. 22*, 201–208.
82. Aro, E-M., Virgin, I. & Andersson, B. (1993) Photoinhibition of photosystem II. Inactivation, protein damage and turnover, *Biochim. Biophys. Acta 1143*, 113–134.
83. Poulson, M., Samson, G. & Whitmarsh, J. (1995) Evidence that cytochrome b_{559} protects photosystem II against photoinhibition, *Biochemistry 34*, 10932–10938.
84. Barber, J. & de las Rivas, J. (1993) A functional model for the role of cytochrome b_{559} in the protection against donor and acceptor side photoinhibition, *Proc. Natl Acad. Sci. USA 90*, 10942–10946.
85. Kusunoki, M. (1995) EPR evidence for the primary water oxidation step upon the S_2 to S_3 transition in the Joliot-Kok cycle of plant photosystem II, *Chem. Phys. Lett. 239*, 148–157.
86. Messinger, J., Badger, M. & Wydrzynski, T. (1995) Detection of one slowly exchanging substrate water molecule in the S_3 state of photosystem II, *Proc. Natl Acad. Sci. USA 92*, 3209–3213.
87. Ono, T., Noguchi, T., Inoue, Y., Kusunoki, M., Matsushita, T. & Oyanagi, H. (1992) X-ray detection of the period four cycling of the manganese cluster in the photosynthetic water oxidizing enzyme, *Science 258*, 1335–1337.
88. Liang, W., Latimer, M. J., Dau, H., Roelofs, T. A., Yachandra, V. K., Sauer, K. & Klein, M. P. (1994) Correlation between structure and magnetic spin state of the manganese cluster in the oxygen-evolving complex of photosystem II in the S_2 state: Determination by X-ray absorption spectroscopy, *Biochemistry 33*, 4923–4932.
89. Yachandra, V. K., DeRose, V. J., Latimer, M. J., Mukerji, I., Sauer, K. & Klein, M. P. (1993) Where plants make oxygen: A structural model for the photosynthetic oxygen-evolving manganese cluster, *Science 260*, 675–679.
90. Ono, T., Kusunoki, M., Matsushita, T., Oyanagi, H. & Inoue, Y. (1991) Structural and functional modifications of the manganese cluster in calcium depleted S_1 and S_2 states, *Biochemistry 30*, 6836–6841.
91. MacLachlan, D. J., Nugent, J. H. A., Bratt, P. J. & Evans, M. C. W. (1994) The effects of calcium depletion on the O_2 evolving complex in spinach PSII: the S_1^*, S_2^* and S_3^* States and the role of the 17 kDa and 23 kDa extrinsic polypeptides, *Biochim. Biophys. Acta 1186*, 186–200.
92. Dau, H., Andrews, J. C., Roelofs, T. A., Latimer, M. J., Liang, W., Yachandra, V. K., Sauer, K. & Klein, M. P. (1995) Structural consequences of ammonia binding to the manganese center of the photosynthetic oxygen-evolving complex: An X-ray absorption spectroscopy study of isotropic and oriented photosystem II particles, *Biochemistry 34*, 5274–5287.
93. DeRose, V. J., Mukerji, I., Latimer, M. J., Yachandra, V. K., Sauer, K. & Klein, M. P. (1994) Comparison of the manganese oxygen-evolving complex in photosystem II of spinach and *Synechococcus* sp. with multinuclear manganese model compounds by X-ray absorption spectroscopy, *J. Am. Chem. Soc. 116*, 5239–5249.
94. Yocum, C. F. (1991) Calcium activation of photosynthetic water oxidation, *Biochim. Biophys. Acta 1059*, 1–15.
95. Ädelroth, P., Lindberg, K. & Andréasson, L. E. (1995) Studies of Ca^{2+} binding in spinach photosystem II using $^{45}Ca^{2+}$, *Biochemistry 34*, 9021–9027.
96. van Vliet, P., Boussac, A. & Rutherford, A. W. (1994) Chloride-depletion effects in the calcium deficient oxygen-evolving complex of photosystem II, *Biochemistry 33*, 12998–13004.
97. Boussac, A., Zimmermann, J. L., Rutherford, A. W. & Lavergne, J. (1990) Histidine oxidation in the oxygen-evolving photosystem II enzyme, *Nature 347*, 303–306.
98. Baumgarten, M., Philo, J. S. & Dismukes, G. C. (1990) Mechanism of photoinhibition of photosynthetic water oxidation by Cl^- depletion and F^- substitution: Oxidation of a protein residue, *Biochemistry 29*, 10814–10822.
99. Hallahan, B. J., Nugent, J. H. A., Warden, J. T. & Evans, M. C. W. (1992) Investigation of the origin of the "S3" EPR signal

from the oxygen-evolving complex of photosystem 2: The role of tyrosine Z, *Biochemistry 31*, 4562–4573.

100. Gilchrist, M. L., Ball, J. A., Randall, D. W. & Britt, R. D. (1995) Proximity of the manganese cluster of photosystem II to the redox active tyrosine Y_Z, *Proc. Natl Acad. Sci. USA 92*, 9545–9549.
101. Un, S., Brunel, L-C., Brill, T. M., Zimmermann, J.-L. & Rutherford, A. W. (1994) Angular orientation of the stable tyrosyl radical within photosystem II by high field 245 GHz electron paramagnetic resonance, *Proc. Natl Acad. Sci. USA 91*, 5262–5266.
102. Hoganson, C. W., Lydakis-Simantiris, N., Tang, X.-S., Tommos, C., Warncke, K., Babcock, G. T., Diner, B. A., McCracken, J. & Styring, S. (1996) A hydrogen-atom abstraction model for the function of Y_Z in photosynthetic oxygen evolution. *Photosynth. Res.*, in the press.
103. Tommos, C., Tang, X.-S., Warncke, K., Hoganson, C. W., Styring, S., McCracken, J., Diner, B. A. & Babcock, G. T. (1995) Spin density distribution, conformation and hydrogen-bonding status of the redox active tyrosine Y_Z from multiple electron magnetic resonance spectroscopies: Implications for photosynthetic oxygen evolution, *J. Am. Chem. Soc. 117*, 10325–10335.
104. Vass, I. & Styring, S. (1991) pH-dependent charge equilibria between tyrosine D and S-states in photosystem II, *Biochemistry 30*, 830–839.
105. Lavergne, J. & Junge, W. (1993) Proton release during the redox cycle of the water oxidase, *Photosynth. Res. 38*, 279–296.
106. Rappaport, F., Blanchard-Desce, M. & Lavergne, J. (1994) Kinetics of electron transfer and electrochromic change during the redox transitions of the photosynthetic oxygen-evolving complex, *Biochim. Biophys. Acta 1184*, 178–192.
107. Bogershausen, O. & Junge, W. (1995) Rapid proton transfer under flashing light at both functional sides of dark-adapted photosystem II particles, *Biochim. Biophys. Acta 1230*, 177–185.
108. Svensson, B., Vass, I., Cedergren, E. & Styring, S. (1990) Structure of donor side components in photosystem II predicted by computer modelling, *EMBO J. 9*, 2051–2059.
109. Lin, X., Murchison, H. A., Nagarajan, V., Parson, W. W., Allen, J. P. & Williams, J. C. (1994) Specific alteration of the oxidation potential of the electron donor in reaction centres from *Rhodobacter sphaeroides*, *Proc. Natl Acad. Sci. USA 91*, 10265–10269.
110. Rautter, J., Lendzian, F., Schulz, C., Fetsch, A., Kuhn, M., Lin, X., Williams, J. C., Allen J. P. & Lubitz, W. (1995) ENDOR studies of the primary donor cation radical in mutant reaction centers of *Rhodobacter sphaeroides* with altered hydrogen-bond interactions, *Biochemistry 34*, 8130–8143.
111. Koulougliotis, D., Tang, X.-S., Diner, B. A. & Brudvig, G. W. (1995) Spectroscopic evidence for the symmetric location of tyrosines D and Z in photosystem II, *Biochemistry 34*, 2850–2856.
112. Tang, X.-S., Chisholm, D. A., Dismukes, G. C., Brudvig, G. W. & Diner, B. A. (1993) Spectroscopic evidence from site-directed mutants of *Synechocystis* PCC 6803 in favor of a close interaction between histidine 189 and redox-active tyrosine 160, both of polypeptide D2 of the photosystem II reaction center, *Biochemistry 32*, 13742–13748.
113. Nixon, P. J. & Diner, B. A. (1992) Aspartate 170 of the photosystem II reaction center polypeptide D1 is involved in the assembly of the oxygen evolving manganese cluster, *Biochemistry 31*, 942–948.
114. Nixon, P. J., Trost, J. T. & Diner, B. A. (1992) Role of the carboxy terminus of polypeptide D1 in the assembly of a functional water-oxidising manganese cluster in photosystem II of the cyanobacterium *Synechocystis* sp. PCC 6803: Assembly requires a free carboxyl group at C-terminal position 344, *Biochemistry 31*, 10859–10871.
115. Chu, H.-A., Nguyen, A. P. & Debus, R. J. (1995) Amino acid residues that influence the binding of manganese or calcium to photosystem II. 1. The lumenal interhelical domains of the D1 polypeptide, *Biochemistry 34*, 5839–5858.
116. Chu, H.-A., Nguyen, A. P. & Debus, R. J. (1995) Amino acid residues that influence the binding of manganese or calcium to photosystem II. 2. The carboxy terminal domain of the D1 polypeptide, *Biochemistry 34*, 5859–5882.
117. Tang, X.-S., Diner, B. A., Larsen, B. S., Gilchrist, M. L., Lorigan G. A. & Britt, R. D. (1994) Identification of histidine at the catalytic site of the photosynthetic oxygen-evolving complex, *Proc. Natl Acad. Sci. USA 91*, 704–708.
118. Vermaas, W. F. J., Charité, J. & Shen, G. (1990) Glu-69 of the D2 protein in photosystem II is a potential ligand to Mn involved in photosynthetic oxygen evolution, *Biochemistry 29*, 5325–5332.
119. Gleiter, H. M., Haag, E., Shen, J.-R., Eaton-Rye, J., Seeliger, A. G., Inoue, Y., Vermaas, W. F. J. & Renger, G. (1995) Involvement of the CP47 protein in stabilisation and photoinactivation of a functional water-oxidising complex in the cyanobacterium *Synechocystis* sp. PCC 6803, *Biochemistry 34*, 6847–6856.
120. Tae, G.-S. & Cramer, W. A. (1994) Topography of the heme prosthetic group of cytochrome b-559 in the photosystem II reaction center, *Biochemistry 33*, 10060–10068.
121. Buser, C. A., Diner, B. A. & Brudvig, G. W. (1992) Re-evaluation of the stoichiometry of cytochrome b559 in photosystem II and thylakoid membranes, *Biochemistry 31*, 11441–11448.
122. Lorkovíc, Z. J., Schröder, W. P., Pakrasi, H. B., Irrgang, K.-D., Herrmann, R. G. & Oelmüller, R. (1995) Molecular characterisation of PsbW, a nuclear-encoded component of the photosystem II reaction center complex in spinach, *Proc. Natl Acad. Sci. USA 92*, 8930–8934.

Eur. J. Biochem. *238*, 1–27 (1996)

Review

Connections with connexins: the molecular basis of direct intercellular signaling

Roberto BRUZZONE[1], Thomas W. WHITE[2] and David L. PAUL[3]

[1] Unité de Neurovirologie et Régénération du Système Nerveux, Institut Pasteur, Paris, France
[2] Département de Morphologie, Centre Médical Universitaire, Genève, Switzerland
[3] Department of Neurobiology, Harvard Medical School, Boston MA, USA

(Received 27 November 1995/30 January 1996) – EJB 95 1951/0

Adjacent cells share ions, second messengers and small metabolites through intercellular channels which are present in gap junctions. This type of intercellular communication permits coordinated cellular activity, a critical feature for organ homeostasis during development and adult life of multicellular organisms. Intercellular channels are structurally more complex than other ion channels, because a complete cell-to-cell channel spans two plasma membranes and results from the association of two half channels, or connexons, contributed separately by each of the two participating cells. Each connexon, in turn, is a multimeric assembly of protein subunits. The structural proteins comprising these channels, collectively called connexins, are members of a highly related multigene family consisting of at least 13 members. Since the cloning of the first connexin in 1986, considerable progress has been made in our understanding of the complex molecular switches that control the formation and permeability of intercellular channels. Analysis of the mechanisms of channel assembly has revealed the selectivity of inter-connexin interactions and uncovered novel characteristics of the channel permeability and gating behavior. Structure/function studies have begun to provide a molecular understanding of the significance of connexin diversity and demonstrated the unique regulation of connexins by tyrosine kinases and oncogenes. Finally, mutations in two connexin genes have been linked to human diseases. The development of more specific approaches (dominant negative mutants, knockouts, transgenes) to study the functional role of connexins in organ homeostasis is providing a new perception about the significance of connexin diversity and the regulation of intercellular communication.

Keywords: gap junction; channel; connexon; intercellular communication.

A hallmark of multicellularity is the coordinate response of groups of cells against external stimuli, thereby adapting more rapidly to the surrounding medium. Moreover, cells undergoing deleterious changes need to be isolated from the majority to preserve the integrity of the group. In either case, a system is required to allow cells to review and shape the functional state of their neighbors by exchanging signaling molecules. Such a system of communication must also be rapidly modulated to continuously adapt to the immediate needs of the group of coupled cells. These features are met by specialized structures, the intercellular channels, usually collected in distinct regions of the plasma membrane called gap junctions (Fig. 1). The terms 'gap junction channel' and 'intercellular channel' are, therefore, equivalent and define the same structure. These channels control a unique form of communication in that the exchange of molecules is direct and does not involve secretion into the extracellular space. Intercellular channels appear to be present in virtually all multicellular organisms, from mesozoa to humans (Revel, 1988). The remarkable functional conservation of direct cell-to-cell coupling throughout the animal kingdom, however, is not matched at the molecular level of the structural protein components. Thus, intercellular channels in vertebrates are made of connexins, a family of highly related proteins (Beyer et al., 1990; Bennett et al., 1991; Hall et al., 1993; Kanno et al., 1995), whereas the identity of homologous proteins in invertebrates is still debated (Finbow and Pitts, 1993; Ryerse, 1993; Barnes, 1994). In mammals, almost all cell types express intercellular channels at some point during their development, although they are absent in a few fully differentiated cells such as skeletal myocytes, some neurons, most circulating blood cells and spermatozoa. Cell-to-cell signaling can be studied by measuring the flux of either radiolabeled ions and organic molecules (metabolic coupling) or of dyes (dye coupling) (Fig. 2). The presence of intercellular channels can also be determined by recording the passage of current between coupled cells which permits a precise quantitation of junctional conductance. These basic methods have now been applied to study the properties of individual connexins in suitable expression systems such as pairs of *Xenopus*

Correspondence to R. Bruzzone, Unité de Neurovirologie et Régénération du Système Nerveux, Institut Pasteur, 25, rue du Docteur Roux, F-75724 Paris Cedex 15, France

Fax: +33 1 4061 3421.

Abbreviations. Cx, connexin, with the molecular mass in kDa as specified; r, rat; ch, chicken; xen, *Xenopus*; b, bovine; h, human; V_j, transjunctional voltage; $V_{i\text{-}o}$, inside-out voltage; I_j, junctional current; G_j, junctional conductance; v-*src*, viral *src* tyrosine kinase; EGF, epidermal growth factor; PDGF, platelet-derived growth factor; CMT, Charcot-Marie-Tooth disease, with the genetic forms as specified.

Enzymes. Protein kinase C or calcium/phospholipid-dependent protein kinase, protein kinase A or cAMP-dependent protein kinase, cGMP-dependent protein kinase, calcium/calmodulin-dependent protein kinase II and mitogen-activated protein kinase (EC 2.7.1.37); phospholipase A2, or phosphatidylcholine 2-acylhydrolase (EC 3.1.1.4).

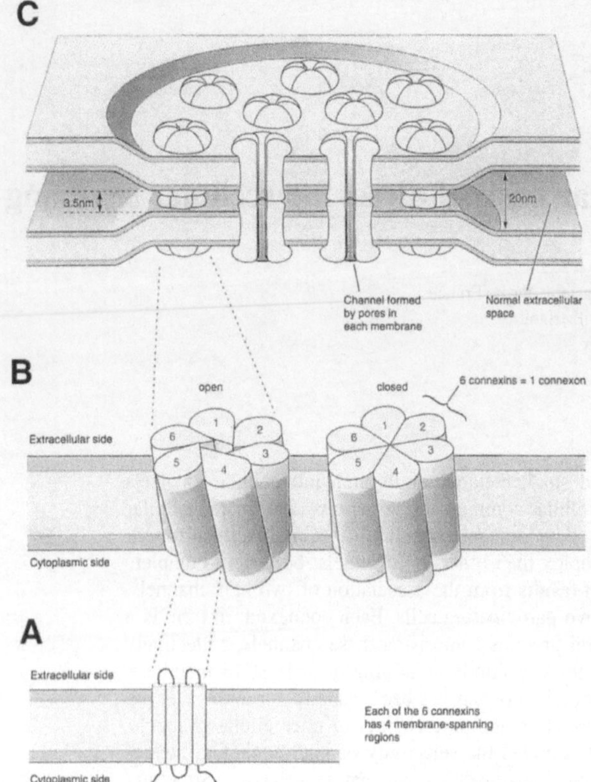

Fig. 1. Schematic view of the molecular steps leading to the formation of intercellular channels. The protein subunits, the connexins (A), oligomerize in a hexameric structure, the connexon (B), that is transported to the plasma membrane. Connexons from adjacent cells interact to form complete intercellular channels (C) that become clustered in specialized membrane regions, the gap junctions. Reproduced with minor modifications from Kandel et al. (1991; p. 130) by copyright permission of Appleton & Lange, Norwalk CT.

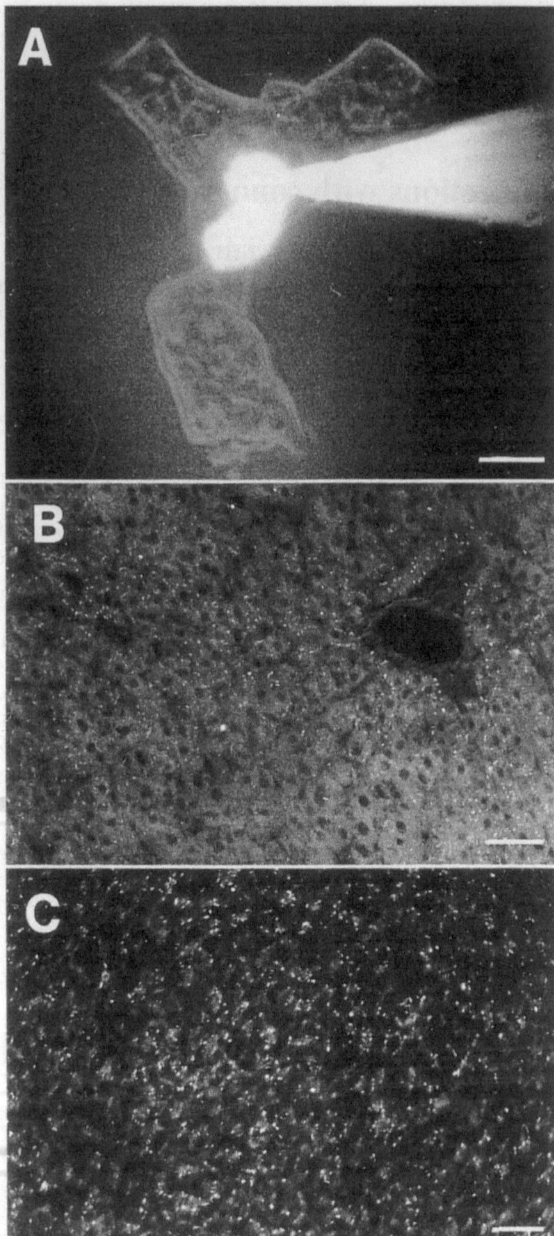

Fig. 2. Functional assay of cell–cell communication and specific patterns of connexin expression. (A) Intercellular channels allow the passage of ions and small molecules between coupled cells, as shown in this fluorescent view of cultured pancreatic β-cells injected with Lucifer Yellow. Following the impalement of a single cell and iontophoretic injection, the dye diffuses to a number of neighboring cells, an example of the ability of communicating cells to exchange a wide range of signaling molecules. Other cells of this cluster are not coupled, an example of the communication gradients that can occur between adjacent cells. (B, C) The immunohistochemical localization of Cx26 and Cx32 in frozen sections of rat liver illustrates the unique and overlapping patterns of connexin expression. Thus, there is a striking specificity in the distribution of immunoreactive spots (Traub et al., 1989; Zhang et al., 1989). The affinity-purified anti-Cx26 antiserum (Goliger and Paul, 1994) shows punctate staining in a restricted area of the hepatic lobule, around the periportal fields (B). In contrast, a mouse monoclonal antibody against Cx32 (Goodenough et al., 1988) decorates uniformly hepatocytes across the entire lobule, with the typical spotty appearance of gap junctions (C). Scale bars are 25 µm in A and 10 µm in B and C.

oocytes, transfected cell lines, or reconstitution into lipid bilayers. The widespread expression and the molecular diversity of connexins has made it a challenging task to assign specific roles for intercellular channels and determine what are the signals exchanged. Intercellular channels participate in the regulation of diverse functions, including contraction of cardiac and smooth muscle (Page and Shibata, 1981; Miller et al., 1989; Spray and Burt, 1990; Huizinga et al., 1992; De Mello, 1994), transmission of neuronal signals at electrotonic synapses (Furshpan and Potter, 1959; Auerbach and Bennett, 1969; Sotelo and Llinás, 1972; Bennett and Verselis, 1992) and metabolic cooperation in development and avascular organs (Larsen and Wert, 1988; Goodenough, 1992). Permeability of intercellular channels to second messengers may also regulate secretion by both the exocrine and endocrine pancreas (Meda et al., 1984; Bruzzone and Meda, 1988) and plays a critical role in pattern formation during development (Caveney, 1985; Guthrie and Gilula, 1989; Kidder, 1992; Warner, 1992), oncogenic transformation and control of cell growth (Yamasaki, 1990; Loewenstein and Rose, 1992; Hotz-Wagenblatt and Shalloway, 1993). It should be acknowledged, however, that most of our understanding of the biological function of intercellular channels is based on correlative studies, rather than direct experimental evidence and that the precise roles of connexins in diverse tissue functions are not well defined. There are, however, clear indications that research on connexins is entering a new promising phase. Recent work has shaken two of the postulates on which our understanding of intercellular channels was based. First, intercellular channels are no longer to be considered as passive conduits allowing the free passage of ions and small molecules in nonspecific fashion. Second, the establishment of communication between cells has proved to be neither passive nor promiscuous but a process that is controlled by the compatibility among connexins. Moreover, the discovery of connexin mutations in human disease and the targeting of specific connexin genes is illustrating a new spectrum of functional implications for intercellular channels. Several articles have reviewed in great detail the early anatomical, biochemical, physiological and pharmacological characterization of intercellular channels (Bennett and Goodenough, 1978; Peracchia, 1980; Hertzberg et al., 1981; Loewenstein, 1981; Sheridan and Atkinson, 1985; Spray and Bennett, 1985; Hertzberg and Johnson, 1988). The aim of this article is to give a brief, but comprehensive review of the progress that has occurred over the past decade, after the cloning of the first connexin started the molecular analysis of the composition and regulation of intercellular channels.

Molecular characterization of intercellular channels

Structure of the gap junction channel. The gap junction is a highly specialized organelle consisting of clustered channels whose unique design permits the direct intercellular exchange of ions and molecules through central aqueous pores. In thin-section electron microscopy, the gap junction was originally characterized as a close apposition of the outer leaflets of the plasma membranes of two neighboring cells which remain separated only by a 2–4-nm space or gap (Robertson, 1963; Benedetti and Emmelot, 1965; Revel and Karnovsky, 1967). Freeze-fracture analysis revealed that this structure is composed of a plaque-shaped region of the plasma membrane containing an array of packed membrane particles on the P fracture face of the cell membrane (Chalcroft and Bullivant, 1970; Goodenough and Revel, 1970; Raviola et al., 1980). The original model of a complete intercellular channel developed by Makowski et al. (1977) is still valid. These authors demonstrated that, unlike other membrane channels, the gap junction channel possesses a unique structure, spanning the plasma membranes of two adjacent cells. Each cell contributes one half of the channel, called the connexon. Two connexons interact in the extracellular space to form the complete gap junction channel, allowing for direct communication between the cytoplasm of the participating cells. Like other membrane channels, each connexon is formed by the oligomerization of the structural protein subunits, termed connexins (Fig. 1). Immunogold labeling of freeze-fracture replicas of hepatocytes has proved that connexins are indeed localized in classical gap junction plaques and that smaller aggregates are also observed (Fujimoto, 1995). A stoichiometry of six protein subunits in each connexon is presently accepted (Unwin and Zampighi, 1980; Caspar et al., 1988; Makowski, 1988; Sosinsky et al., 1988; Yeager and Gilula, 1992; Cascio et al., 1995). More recently, the structure of the liver gap junction channel has been examined by atomic force microscopy. This new technique provides a novel way to image biological structures by tracing out their surface contour at high resolution (Hoh et al., 1991 b, 1993; Revel et al., 1992; Lal et al., 1995). Using this approach gap junctions show a dense packing of connexon-like structures with a center-to-center distance of about 9.4 nm, a value consistent with previous studies. On the extracellular surface of some but not all connexons, atomic force microscopy reveals a small depression, presumably representing the pore. The average pore size in these images is 3.8 nm at the presumptive site of inter-connexon contact, a figure significantly greater than the 2 nm previously reported. Currently, the mouth of the channel on the cytoplasmic aspect has not been resolved. Though powerful, atomic force microscopy has failed to provide the degree of resolution necessary to visualize connexon stoichiometry, as both pentagonal and hexagonal images have been observed (Hoh et al., 1993). The possibility of a pentameric arrangement of connexons has also been suggested in preliminary studies that used chemical cross-linking of connexons isolated from cell lines expressing Cx43 (Musil, 1994). Clearly, the refined structure of the channel awaits the biochemical isolation of a pure connexon population and the production of crystals to achieve near atomic resolution.

The connexin family of proteins. The diversity of the connexin family was first suggested by biochemical analysis of subcellular fractions enriched in gap junctions. These studies identified preponderant polypeptide bands that, depending on the tissue of origin and the method of preparation, exhibited a molecular mass ranging between 16–70 kDa (Henderson et al., 1979; Finbow et al., 1980; Kensler and Goodenough, 1980; Manjunath et al., 1982; Gros et al., 1983; Nicholson et al., 1985, 1987; Buultjens et al., 1988; Kistler et al., 1988). N-terminal sequencing of these polypeptides by Edman degradation confirmed that these proteins were different but in some cases shared some degree of similarity. The debates over the identity of the protein constituent of intercellular channels were put to an end by the cloning of the major rat hepatic gap junction protein (Paul, 1986) and of the homologous murine and human sequences (Heynkes et al., 1986; Kumar and Gilula, 1986). The rat probe was used to screen under low stringency hybridization a cDNA library from rat heart and the second member of the connexin family was identified (Beyer et al., 1987). Additional connexin genes were identified by two strategies: reduced stringency hybridization with probes consisting of full-length cDNA, and PCR amplification using degenerate oligonucleotide primers corresponding to the most conserved domains among previously characterized connexins. Both approaches were successful because of the significant overall homology among connexins and the absence of introns within coding regions, which would have complicated the analysis of PCR products from genomic templates. The rapid addition of new connexin genes brought up the issue of nomenclature. At the 1987 Gap Junction meeting held in Asilomar, it was decided to adopt the system suggested by Beyer et al. (1987). Thus, current nomenclature distinguishes connexins on the basis of species of origin and appends the molecular mass predicted by cloned DNA sequences to the family name connexin (Cx). For example, the 43-kDa protein first identified in myocardial gap junctions is termed Cx43. Connexin homologues from different organisms can be distinguished with a suitable identifying prefix (e.g. r for rat, ch for chicken, xen for *Xenopus*, b for bovine, h for human, etc.). Connexins with very similar molecular mass have been identified, leading to the use of one decimal figure in some cases (rCx30.3, rCx31.1). This system permits the unambiguous identification of each member but is not always sufficiently illustrative. For example, clear functional homologues across species often have very different molecular mass. Despite its limitations, this convention is currently used by the vast majority of researchers. An alternative nomenclature, based on the identification of two major subclasses (α or type II and β or type I) has also been proposed (Risek et al., 1990; Bennett et al., 1991). However, the specific functional and/or structural criteria for distinguishing classes have not been defined.

The connexin family consists, in rodents, of at least 13 members, for which many homologous genes have been identified in

Table 1. Cloned vertebrate connexins. References are listed in the order of species, from rodent to human.

Connexin from				References
rodent	chicken	frog	human	
Cx26			Cx26	Zhang and Nicholson, 1989; Lee et al., 1992
Cx30				Dahl, E. and Willecke, K., personal communication
Cx30.3				Hennemann et al., 1992a
Cx31				Hoh et al., 1991a
Cx31.1				Haefliger et al., 1992
Cx32		Cx30	Cx32	Paul, 1986; Gimlich et al., 1988; Kumar and Gilula, 1986
Cx33				Haefliger et al., 1992
Cx37		Cx38	Cx37	Willecke et al., 1991b; Ebihara et al., 1989;
			Cx41	Yoshizaki and Patiño, 1995; Reed et al., 1993
Cx40	Cx42		Cx40	Haefliger et al., 1992; Beyer, 1990; Kanter et al., 1994
Cx43	Cx43	Cx43	Cx43	Beyer et al., 1987; Musil et al., 1990a; Gimlich et al., 1990; Fishman et al., 1990
Cx45	Cx45		Cx45	Hennemann et al., 1992b; Beyer, 1990; Kanter et al., 1994
Cx46	Cx56			Paul et al., 1991; Rup et al., 1993
Cx50	Cx45.6		Cx50	White et al., 1992; Jiang et al., 1994; Church et al., 1995

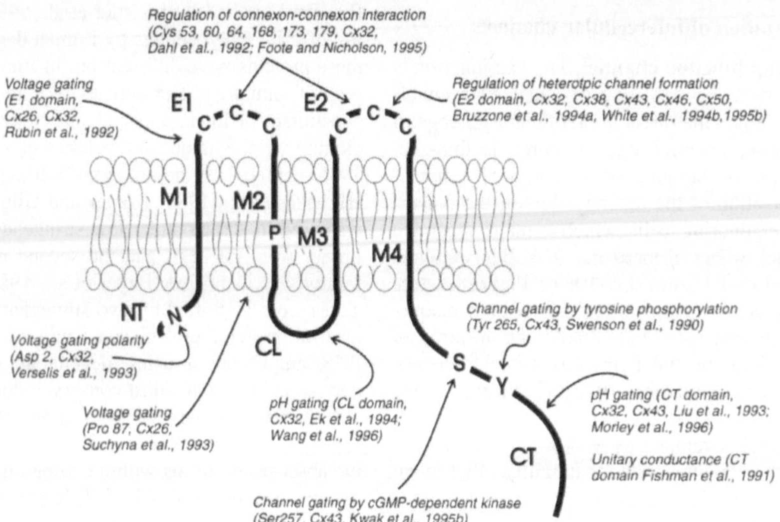

Fig. 3. Schematic representation and topology relative to the plasma membrane of a generic connexin. Hydropathy plots predict four membrane-spanning regions (M1–M4), two extracellular loops (E1 and E2) with characteristically spaced cysteine residues and three cytoplasmic portions, the amino-terminal (NT) and carboxy-terminal (CT) domains and the central cytoplasmic loop (CL). The variations of the molecular mass among connexins are accounted for by the different length of the CL and CT sequences. The putative role of the different domains is inferred from the results of structure/function studies with point mutations and chimeras with swapped domains. In some cases, multiple interdomain interactions are needed to confer distinct properties to each connexin.

different vertebrates species (Table 1). Genetic linkage is observed between pairs of connexins which are relatively dissimilar in amino acid sequence, e.g. Cx26/Cx46 and Cx31.1/Cx37 (Haefliger et al., 1992). This relationship suggests the occurrence of a gene duplication event relatively early in the evolution of the connexin family. A subsequent duplication of these linked genes would result in a pair of tandem genes, consistent with experimental observations. Sequence analysis indicates a significant amount of overall similarity (35–65%) and, in some cases (i.e. Cx43), a remarkable degree of conservation across species (Kumar and Gilula, 1992; Paul et al., 1993; Beyer and Veenstra, 1994). In addition, hydropathy plots predict that all connexins share a common sequence of structural motifs, as schematically represented in Fig. 3. Each connexin contains four transmembrane (M1–M4), two extracellular (E1 and E2) and three cytoplasmic regions. This model has been validated for Cx26, Cx32 and Cx43 by experiments with limited proteolysis, site-specific antibodies and introduction of artificial glycosylation sites (Zimmer et al., 1987; Goodenough et al., 1988; Hertzberg et al., 1988; Milks et al., 1988; Yancey et al., 1989; Evans et al., 1992; Dahl et al., 1994; Falk et al., 1994; Zhang and Nicholson, 1994). Both extracellular loops contain three cysteine residues in a characteristically conserved sequence (CX_6CX_3C in E1; CX_4CX_5C in E2). The only exception to this rule is rCx31, which shows a different pattern, CX_5CX_5C in E2 (Hoh et al., 1991a). It has been shown that the cysteine residues form intramolecular, but not intermolecular, disulfide bonds linking the two extracellular loops of Cx32 and Cx43 (Dupont et al., 1989; Rahman and Evans, 1991; John and Revel, 1991; Foote and Nicholson, 1995). This event occurs in the endoplasmic reticulum, possibly cotranslationally, and it is reasonable to speculate that this structural feature is common to all connexins.

Sequence analysis has not identified consensus sequences for asparagine-linked glycosylation in either extracellular domain, a highly unusual finding for integral membrane proteins. It is likely that addition of large carbohydrate moieties would make connexin oligomerization more difficult and would negatively interfere with head-to-head interactions between connexons. Predicted transmembrane segments and the very short cytoplasmic portion at the N-terminus are also well conserved, whereas the major cytoplasmic domains are unique in both sequence and length (Beyer et al., 1990; Willecke et al., 1991a; Kumar and Gilula, 1992; Paul et al., 1993). The membrane-spanning domains appear to be the critical determinants of the correct topogenic fate of connexins (Leube, 1995). Based on the topological arrangement of connexins discussed above, the interactions between the two connexons should occur through the extracellular regions. The putative wall of the pore is represented by the association of the third, amphipatic, transmembrane domain from each subunit, thereby lining the portion of the channel spanning the bilayer. It is worth noting that parts of the pore wall are formed by extracellular domains in the gap. This model was derived by analogy with other channels, since the assigned pore-lining α-helices share identical alignments of small polar and large hydrophobic residues. It has been proposed that these similarities underlie the existence of a common design in building the wall around an aqueous pore and further suggest that the molecular mechanisms by which channels open and close may obey general rules (Unwin, 1989). On the basis of the channel-forming ability of a peptide corresponding to a putative portion of the second extracellular domain of Cx32, Dahl and coworkers (1994) have proposed an alternative model, in which amino acid residues 147–167 would be part of a pore loop arranged as a β-barrel. This unit could be functionally homologous to the pore regions that are emerging as common structural features of several classes of ion channels (Jan and Jan, 1994; Shen and Pfanninger, 1994; MacKinnon, 1995). A critical test of this hypothesis, that is conferring distinct electrical properties by swapping this domain between connexins, will determine whether the current thinking of connexin topology needs to be modified. A rigorous structural analysis awaits a breakthrough in the crystallization of gap junction channels.

Table 2. Patterns of expression of cloned rodent connexins. The tissue distribution of rodent connexins is by no means complete, but is presented here to illustrate its complex and overlapping nature. In some cases, only data from RNA analysis are available, without indication of the cellular distribution. Note that expression of a given connexin in certain cell types may only occur transiently, during development or physiological and pathological conditions.

Connexin	Tissue and cellular distribution
Cx26	alveolar cells of lactating mammary gland, chorionic villi, decidual cells, ependyma, hepatocytes, intestine, keratinocytes, leptomeninges, myometrium, pancreatic acinar cells, pinealocytes, visceral yolk sac
Cx30.3	kidney, preimplantation blastocyst, skin
Cx31	Harderian gland, keratinocytes, kidney, preimplantation blastocyst, trophoectoderm
Cx31.1	keratinocytes, preimplantation blastocyst, squamous epithelia
Cx32	alveolar cells of lactating mammary gland, hepatocytes, neurons, oligodendrocytes, pancreatic acinar cells, proximal kidney tubules, Schwann cells, thyroid follicular cells, visceral yolk sac
Cx33	testes (Sertoli cells)
Cx37	cortical neuroblasts, endothelium, heart, keratinocytes, stomach, testes
Cx40	conductive myocardium (His bundle, Purkinje fibers), endothelium, preimplantation blastocyst
Cx43	astrocytes, cardiac and smooth muscle, endothelium, ependyma, fibroblasts, keratinocytes, lens and corneal epithelium, leptomeninges, leucocytes, Leydig cells, macrophages, myoepithelial cells of mammary gland, osteocytes, ovarian granulosa, pancreatic β-cells, preimplantation blastocyst, Sertoli cells, thyroid follicular cells, trophoblast giant cells
Cx45	embryonic brain, heart, intestine, kidney, lung, preimplantation blastocyst, skin
Cx46	heart, kidney, lens fibers, Schwann cells
Cx50	atrioventricular valves, corneal epithelium, lens fibers

Tissue-specific expression of connexins. Connexins are expressed in overlapping patterns, with most tissues expressing more than one connexin type. Northern blot hybridization indicates that each connexin gene has its own distinct pattern of expression. On the one side of the spectrum, Cx43 is widely expressed in several organs by many cell types, whereas, on the opposite side, Cx30.3, Cx31 (Tucker and Barajas, 1994), Cx31.1 and Cx33 display a very restricted distribution (Table 2). Another noticeable feature is that certain combinations of connexins are often coexpressed in a given organ or cell type. For example, many epithelia express Cx32 and Cx26 (Nicholson et al., 1987; Traub et al., 1989; Meda et al., 1993; Stutenkemper et al., 1992; Kojima et al., 1995) (Fig. 2) and, similarly, mRNA for Cx37 and Cx40 are frequently found in the same organ, although their relative level of expression differs significantly (Haefliger et al., 1992; Hennemann et al., 1992c; Delorme et al., 1995). Studies with fluorescence and electron microscopy have demonstrated that multiple connexins are often present in the same cell type, and can even be in the same gap junctional plaque (Nicholson et al., 1987; Traub et al., 1989; Paul et al., 1991; Ruangvoravat and Lo, 1992; Sainio et al., 1992; Gourdie et al., 1993; Winterhager et al., 1993; Risek et al., 1994). These findings have been corroborated by electrophysiological analysis (Spray et al., 1991). The size of the connexin family greatly complicates the interpretation of how multiple connexins interact in a tissue to coordinate function. The lens, where there are only two differentiated cell types which express three connexins, appears as one of the least complicated systems in which to study intercellular communication (Goodenough, 1992; Donaldson et al., 1995; Kistler et al., 1995). In contrast, a stratified squamous epithelium, such as skin, utilizes at least five different connexins in a complex and developmentally regulated pattern of expression that varies as a function of the differentiated state of the cells (Risek et al., 1992; Butterweck et al., 1994a; Goliger and Paul, 1994, 1995; Salomon et al., 1994). Current descriptions of connexin distribution are far from complete. Progress in this area of research is needed, in order to test the specific roles of connexins in any given experimental system. It is likely that the use of genetic approaches to the study of intercellular communication will contribute essential information on the presence and regulated expression of connexin genes in different tissues.

In addition to overlap of expression, the levels of connexins may also vary as a function of physiological state and in response to various receptor-activated signaling pathways, toxic agents, cellular proliferation and transformation (Spray et al., 1987; Traub et al., 1987; Fitzgerald et al., 1989; Neveu et al., 1990, 1995; Meda et al., 1991; Miyashita et al., 1991; Lee et al., 1992; Pepper and Meda, 1992; Rosenberg et al., 1992; Kren et al., 1993; Pluciennik et al., 1994; Rosendaal et al., 1994; Chen et al., 1995; Yoshizaki and Patiño, 1996). These two features strongly suggest that expression of connexin proteins is under transcriptional control. One example of exquisite time-dependent regulation is the dramatic increase in both the number of gap junctions and Cx43 mRNA and protein levels in myometrium just prior to parturition, at a time when smooth muscle cells need to acquire a coordinated contractility that results in delivery of the fetus (Garfield et al., 1977; Risek et al., 1990; Lye et al., 1993; but see also Lang et al., 1991). This regulation is clearly tissue-specific since myocardial Cx43 is not modulated during labor or following steroid treatment (Risek et al., 1990; Lang et al., 1991). Post-transcriptional mechanisms may also regulate connexin levels. For example, hepatocytes normally express both Cx26 and Cx32. During rat hepatocarcinogenesis (Fitzgerald et al., 1989; Neveu et al., 1990, 1994) and following ligation of the common bile duct (Fallon et al., 1995) protein levels do not change in parallel with mRNA levels. Thus, connexins can be differentially regulated in the same cell type and the same connexin expressed in different cell types can be regulated differently. How is tissue-specific expression of connexins achieved? Connexin genes have a common architecture, with a single intron of variable length that separates a large exon containing the entire coding sequence from the 5′-untranslated region (Miller et al., 1988; Bai et al., 1993; Sullivan et al., 1993). It is generally accepted that the interaction of DNA-binding proteins with specific sequences in the promoter region are largely responsible for transcriptional regulation of eukaryotic genes (Johnson and McKnight, 1989). While several pharmacological treatments have been correlated with changes of mRNA levels coding for different connexins (Sáez et al., 1989b; Rogers et al., 1990; Mehta et al., 1992; Pepper and Meda, 1992; Petrocelli and Lye, 1993; Risek et al., 1995), the characterization of regulatory elements within the promoter region of connexin genes is just beginning. The promoter region of the Cx43 gene contains a TATA box and AP-1 and AP-2 sites. In addition, a series of half-palindromic estrogen response elements is present in this region (Sullivan et al., 1993; De Leon et al., 1994; Yu et al., 1994). Deletion constructs have also located a positive and a negative regulatory element within 100 base pairs upstream of the transcription start site. Mutational analysis has further confirmed the involvement of these two *cis*-regulatory elements in the transcriptional control of Cx43 by mouse myometrial cells (Chen et al., 1995). The Cx32 promoter lacks a TATA box and uses alternative mechanisms to initiate transcription (Bai et al., 1993). Three DNA-binding proteins have been identified in nuclear extracts of hepatoma cell lines; only one binding activity was recovered, however, from nuclei of normal rat liver cells (Bai et al., 1995). These experiments have mapped the basal promoter region to nucleotides between -179 and -134, immediately upstream of the first exon and have identified a likely candidate for liver-specific expression of Cx32. More recently, it has been reported that splicing of the Cx32 gene gives rise to three alternative promoters that appear to be activated in a cell-type-specific manner (Neuhaus et al., 1995; Söhl et al., 1996). For example, the alternative exon1B transcript is predominantly expressed in the sciatic nerve and developmentally regulated in similar manner to other myelin genes, whereas low levels are detected in liver (Söhl et al., 1996). The molecular dissection of the steps involved in tissue-specific connexin expression may help identify targets for a pharmacological modulation of the function of intercellular channels.

The making of intercellular channels

Oligomerization and intracellular transport. The process leading to the formation of a functionally competent intercellular channel consists of a sequence of steps, whose regulatory mechanisms are beginning to emerge. First, connexins must oligomerize into connexons before reaching the plasma membrane, as it is well documented that quality control mechanisms lead to sequestration of misfolded and incompletely assembled protein subunits in the endoplasmic reticulum and degradation of the retained proteins (Hurtley and Helenius, 1989). Musil and Goodenough (1993) have combined sucrose gradient velocity sedimentation with chemical cross-linking to analyze the assembly of newly synthesized Cx43 into a connexon and locate its site of occurrence. The success of this approach depended on the isolation of a pure connexon fraction, which was achieved in cell lines as well as oocytes injected with cRNA encoding Cx43. Several experimental approaches indicate that assembly of connexons takes place entirely after exit from the endoplasmic reticulum and strongly suggest that the site of Cx43 oligomerization is the *trans*-Golgi network (Musil and Goodenough, 1993). Although extremely well documented, this finding is without precedent. Multimeric assembly has been shown to occur in the endoplasmic reticulum in all other cases studied (Hurtley and Helenius, 1989). Why are connexins an exception to the rule? Once formed, a connexon must remain closed to prevent equilibration between cytosolic and lumenal compartments. A premature oligomerization into connexons, associated with the appearance of double bilayer structures resembling gap junctions within the endoplasmic reticulum, has been described in some systems overexpressing connexins (Kumar and Gilula, 1992). Opening of those channels would perturb cytosolic-lumenal gradients. Thus, a consequence of this delayed assembly is that connexons are being made in an organelle whose vesicles have an average diameter of 50−100 nm. Considering that the estimated connexon length is approximately 8 nm and that the extracellular gap between two connexons does not exceed 3 nm, the size of transport vesicles would provide a physical barrier to accidental pairing of two connexons and the formation of stable structures (see for a comprehensive discussion: Musil, 1994).

What is the molecular composition of a connexon in living cells? Is it made up of only one connexin type as in a homomeric channel, or can different connexins assemble together into a heteromeric channel? This question is particularly relevant, in view of the clear demonstration that most cell types express at least two different connexins and of the precedents of heteromeric voltage- and ligand-gated channels (Jan and Jan, 1990; Hollmann and Heinemann, 1994; Catterall, 1995). To rigorously address this issue, it was essential to develop a system where pure connexons could be isolated from cells programmed to express one or two connexin types at will. This has been achieved by infection of insect cells with baculoviruses coding for Cx32 and Cx26 (Stauffer et al., 1991; Buehler et al., 1995; Stauffer, 1995). Gel filtration of purified homomeric connexons composed of either Cx32 or Cx26 has demonstrated that they run with clearly distinguishable relative mobility, even if they have been mixed for several hours before chromatography. When the Sf9 insect cell line is coinfected with both types of baculovirus, the connexons produced lose this independent behavior and Cx26 now runs together with Cx32. These results have been taken as evidence that Cx32 and Cx26 form a heteromeric connexon, although the stoichiometry of this phenomenon could not be evaluated. As discussed above, Cx32 and Cx26 are frequently found in the same cell type, so that extrapolating this observation from recombinant connexins to living tissues seems justified (Stauffer, 1995). The finding of heteromeric connexons has been confirmed by Jiang and Goodenough (1996), who have investigated the interaction of two connexins *in vivo*. Gap junctional plaques were solubilized from chick lens to produce isolated single connexon-rich fractions that were analyzed by sucrose gradient centrifugation. Biochemical studies of these fractions have shown that Cx45.6 and Cx56, both expressed by lens fiber cells and localized within the same clusters of channels (Jiang et al., 1995) can be coimmuneprecipitated, suggesting the presence of heteromeric connexons *in vivo* (Jiang and Goodenough, 1996). A similar situation may also occur in the cell-to-cell channels of bovine lens fibers that express Cx44 and Cx50 (Konig and Zampighi, 1995). The implications of these studies are far reaching. First, the heteromeric connexons may be endowed with distinct properties that affect the ability to transfer signaling molecules between cells. Such a situation would enormously complicate the task of dissecting the physiological roles of intercellular channels. Alternatively, homomeric connexons may be the preferred form of connexin association, with heteromeric connexons being a fortuitous event, due to leakage in the assembly line. Striking evidence of the separate association of individual connexins with distinct plasma membrane regions has been reported in polarized thyroid epithelial cells (Guerrier et al., 1995). In addition, the composition of connexons *in vivo* has been studied by a combination of electron microscopy and image analysis of liver gap junctions and split junctions. The results obtained are consistent with the interpretation that Cx26 and Cx32 oligomerize only as homomeric connexons (Sosinsky, 1995). Heteromeric connexons could represent, therefore, an exception with no measurable consequences, except in the case of loss of function due to the incorporation of functionally incompetent subunit(s). The solution given to this long-standing issue is likely to raise more questions than can be answered now; for example, are there rules of compatibility governing the coassembly of connexins and defining new, meaningful subclasses? Are post-translational modifications dictating the fate of connexins into homomeric or heteromeric connexons? Is cellular targeting of both populations of connexons regulated differently? Research along these directions should prove fruitful in the near future.

The final destination of a connexon is the gap junction. However, newly-assembled connexons are not likely to move directly to gap junction plaques from *trans*-Golgi membranes. Recent studies using cell-surface biotinylation suggest that connexon are present in regions outside cell-to-cell contacts (Musil and Goodenough, 1991). Before this biochemical demonstration, the existence and functional competence of connexons in the non-junctional plasma membrane had long been debated (Epstein et al., 1977; Loewenstein, 1981). More recently, a provocative report has implicated Cx43 as the ATP^{4-}-sensitive channel present in macrophages (Beyer and Steinberg, 1991) and electrophysiological evidence obtained in single retinal cone horizontal cells has conclusively demonstrated the existence, functional state and gating properties of connexons, or half-intercellular channels (DeVries and Schwartz, 1992; Malchow et al., 1993). The identity of the connexins involved, however, was not determined. Similar findings were generated by expression of either rat Cx46, bovine Cx44 or chicken Cx56 in *Xenopus* oocytes, which resulted in the indiscriminate opening of non-junctional connexons, followed by cell lysis (Paul et al., 1991; Gupta et al., 1994; Ebihara et al., 1995). Cell viability could be restored by raising the concentration of extracellular calcium (Ebihara and Steiner, 1993), which altered the voltage sensitivity of the half-channel and prevented it from opening (Ebihara and Steiner, 1993; White et al., 1994a,b; Gupta et al., 1994; Ebihara et al., 1995). Together, biochemical and physiological analysis suggest that connexons are first transported to the plasma membrane outside junctional areas where they are functionally competent to switch between the open and closed states. The plasma membrane pool may then be used to feed the clustering of channels into junctional plaques. A time-dependent incorporation of connexons from each cell of a pair into complete intercellular channels has been experimentally observed with bovine Cx44, and interpreted as indicative of connexon preference for docking with another connexon rather than remaining an isolated half-channel (Gupta et al., 1994). The regulatory signals involved in the intracellular journey of connexins are poorly understood. Laird et al. (1995) have shown that a phosphorylated form of Cx43 transiently resident in the endoplasmic reticulum/Golgi compartment may represent a pool available for the assembly of new connexin channels. Conceivably, changes in the rate of assembly and/or intracellular transport to the plasma membrane could alter the strength of junctional coupling. This possibility has been raised by showing that prolonged treatment of mouse mammary tumor cells with cAMP increases intercellular communication by modifying the cellular distribution of Cx43, such that a greater proportion is utilized for channels formation (Atkinson et al., 1995). In addition, elevating the levels of cAMP induces clustering of connexons into gap junctional plaques and formation of open channels (Wang and Rose, 1995). In contrast, progesterone appears to inhibit the trafficking of Cx43 to the plasma membrane of myometrial cells in pregnant rats, thus suggesting that an altered rate of transport, targeting and assembly of intercellular channels, but not synthesis *per se*, may contribute to charges in the extent of myometrial coupling that occur during pregnancy (Hendrix et al., 1995).

Docking and gating. Making a connexon is only half the job. Implicit in the structure of the intercellular channel is that, first, two connexons from adjacent cells must recognize each other and align properly. Intercellular channels are defined as homotypic when connexons contributed by each cell are composed of the same connexin, or heterotypic, when each connexon is formed by a different connexin. Inter-connexon binding depends on non-covalent interactions (Ghoshroy et al., 1995) and must be sufficiently strong to keep the channel sealed to the extracellular space. Second, this recognition must be followed by the opening of the two interacting half-channels. This process could be compared to the interaction between an agonist with its receptor, where both functions are taken over by the same molecule. According to Dahl and coworkers (1992), determinants of intercellular channel formation can be divided into two broad categories: (a) extrinsic factors, that are not part of the channel itself, and (b) intrinsic factors, that are represented by regions within the connexin molecule. Extrinsic determinants include, but may not be limited to, adhesion molecules, lectins, lipids, calcium and hydrogen ions. The involvement of cell adhesion molecules in the establishment of gap-junction-mediated intercellular communication had been originally deduced from experiments using a mouse sarcoma cell line (S180). These cells synthesize abundant Cx43 but fail to communicate unless they are transfected with E-cadherin. Thus, the presence of connexin alone is not sufficient to induce communication but specific cell-adhesion molecules are required (Mège et al., 1988). Conversely, signal transduction through gap junction channels may modulate adhesion. Lee et al. (1987) have shown that, in mouse embryos, one blastomere injected with anti-Cx32 antibodies at the two-cell stage continues to divide normally, but the progeny fail to compact at the eight-cell stage. Several other studies have confirmed these observations and extended them to propose a complex relationship between cell adhesion and the establishment of intercellular communication (Keane et al., 1988; Musil et al., 1990b; Jongen et al., 1991). Using both anti-connexin and anti-N-cadherin antibodies, Meyer et al. (1992) have postulated a bidirectional signaling between cell adhesion molecules and connexins in Novikoff hepatoma cells. Similar conclusions can be inferred from studies in the early *Xenopus* embryo, as dominant negative inhibition of either cadherin-based cell adhesion or connexin-mediated intercellular communication generates comparable phenotypes (Detrick et al., 1990; Kintner, 1992; Heasman et al., 1994; Paul et al., 1995). A mechanical and structural link between cadherins and connexins has been strengthened by the unexpected localization of Cx32 (Bergoffen et al., 1993) and E-cadherin (Fannon et al., 1995) at the Schmidt-Lantermann incisures of Schwann cells, between plasma membrane wraps of the same cell. These observations are consistent with a model in which cadherin-based intercellular adhesion and intercellular communication via gap junctions provide mutual homeostatic feedback signals.

Information on other extrinsic determinants is more limited. Incubation of oocytes with lectins before pairing increases both the levels and the rate of formation of junctional coupling. This effect, however, does not appear to be analogous to the action of cell-adhesion molecules (*vide supra*), since it occurred under conditions that did not cause agglutination of cell surfaces (Levine et al., 1991). Instead, the authors suggested that the lectins promoted the removal of steric hindrances. Similarly, it has been reported that both removal of sugar moieties from the cell surface with N-glycosidase F and tunicamycin treatment, which blocks N-linked glycosylation, are associated with increased rate of intercellular channel assembly (Dahl et al., 1992; Wang and Mehta, 1995). Extracellular calcium concentrations show opposite effects on channel formation in oocyte pairs, where low calcium favors coupling (Dahl et al., 1992), and cultured cells, where high calcium increases the extent of dye transfer (Jongen et al., 1991). The latter effect appears to be mediated through the expression of a calcium-dependent cell adhesion molecule, E-cadherin. Specific lipids, such as cholesterol and low-density lipoprotein, also appear to be required for and modulate the efficiency of channel assembly. Both inhibitory as well as stimulatory effects have been reported, depending upon the concentrations used (Malewicz et al., 1990; Meyer et al., 1990). Finally, alkaline extracellular pH promotes junctional communication between oocyte pairs (Dahl et al., 1991), presumably by interacting with intrinsic determinants (cysteine residues on the extracellular loops of connexins, see below). Further work is needed to define the role played by extrinsic determinants other than adhesion molecules in the development and dynamic regulation of the competence to establish intercellular channels.

How do connexons recognize their partners in adjacent cells, what are the determinants intrinsic to the connexin molecule? First, we should consider the simplest case, that of a homotypic channel, where both connexons are composed of the same connexin. From a conceptual standpoint it seems obvious that connexons composed of identical connexins should functionally interact (self-recognition). On the basis of the proposed topology of connexins with respect to the plasma membrane and of the high degree of conservation among connexins, the search for intrinsic determinants has focused on the role of the two extracellular domains (Fig. 3). The involvement of these regions in docking the two hemichannels has been inferred by three separate sets of experiments. First, the formation of coupling between pairs of oocytes programmed to express Cx32 is greatly reduced by incubating the cells before pairing with synthetic peptides corresponding to either the first or second extracellular loop. These results imply that connexons must be assembled and properly inserted in the plasma membrane of single, isolated oocytes and suggest that the synthetic peptide have functioned as receptor antagonists. The inhibition of channel formation appears to be specific, since another peptide sequence derived from a cytoplasmic portion of Cx32 is ineffective (Dahl et al., 1992). Second, replacement of any of the conserved cysteines (present in the extracellular loops of all known vertebrate connexins) with serines and changes in the spacing of the first and third cysteine of each loop result in loss of function of the Cx32 mutants. Because these substitutions do not interfere with the targeting of connexins to the plasma membrane, where they should be found as connexons, it can be deduced that the pattern of cysteines is essential for docking and/or opening of the intercellular channels (Dahl et al., 1992; Foote and Nicholson, 1995). Finally, monovalent antibodies against the extracellular domains of Cx43 specifically block gap junction formation between reaggregating Novikoff cells (Meyer et al., 1992). The binding site has not been mapped precisely, but studies with several synthetic peptides representing sequences of Cx32 indicate that it involves portions of both extracellular loops (Dahl et al., 1995). The possible roles of other connexin domains have not been studied in detail, but it appears that partial truncation of the C-terminal cytoplasmic domain does not abolish the ability of both Cx32 and Cx43 to form junctional channels (Fishman et al., 1991; Werner et al., 1991; Dunham et al., 1992; Rabadan-Diehl et al., 1994). In contrast, deletion of 26 amino acids at the C-terminus of Cx45 results in functional loss due to mistargeting and intracellular retention of the mutant connexin (Hertlein et al., 1995). Swapping of corresponding domains, e.g. C-terminal tails, extracellular loops, has also revealed that some connexins are not composed of functionally exchangeable segments and suggest that multiple regions contribute to the final docking and/or gating steps (Bruzzone et al., 1994a; Wang et al., 1996).

Because of the primary sequence similarity of their extracellular domains (see: *The connexin family of proteins*), all connexins are expected to have similar and stable configuration of their extracellular loops. Similar configurations would play an important role in favoring the head-to-head pairing of connexons each composed of a distinct connexin, thus promoting the formation of heterotypic channels. Early studies documented, with a few exceptions (Fentiman et al., 1976; Pitts and Burk, 1976; Kettenmann et al., 1983; Mesnil et al., 1987), the occurrence of coupling between cells derived from different vertebrate organisms and tissues (Michalke and Loewenstein, 1971; Epstein and Gilula, 1977). It was generally accepted that intercellular communication was a permissive phenomenon occurring each time there were no physical barriers between contacting cells (Hertzberg et al., 1981). The discovery of the multiplicity of connexins, together with the availability of functional expression systems has allowed the rigorous testing of whether connexins actual do speak a common language. While the initial reports showed that some connexins readily form heterotypic channels (Swenson et al., 1989; Werner et al., 1989; Barrio et al., 1991), it has now been established that heterotypic channel formation is a process regulated by compatibility among connexins (Bruzzone et al., 1993; White et al., 1994b, 1995b; Elfgang et al., 1995). A systematic analysis of the ability of adjacent cells expressing different connexins to communicate has brought up extreme examples of selectivity (Table 3). Thus, Cx31 is functional only in homotypic configuration, but not in heterotypic combination with six other connexins: Cx26, Cx32, Cx37, Cx40, Cx43, Cx45 (Elfgang et al., 1995). Cx40 is also highly restricted in its ability to make heterotypic channels, functionally interacting with Cx37, but failing to do so when paired with Cx32, Cx43, Cx46 and Cx50. In contrast, Cx46 interacts well with all connexins tested except Cx40 (White et al., 1995b). Intercellular channel formation appears, therefore, to be dominated by the rules of compatibility, a characteristic displayed by all members of the connexin family and reproduced in different experimental systems (Elfgang et al., 1995; White et al., 1995b). A special case of compatibility is self-discrimination, that is the inability or reduced affinity to form homotypic channel, while retaining functional competence in some heterotypic configurations. This

Table 3. Connexins exhibit selective compatibility in heterotypic intercellular channel formation. These data summarize the results of Elfgang et al. (1995), in transfected HeLa cells, and those of Hennemann et al. (1992a, c), Bruzzone et al. (1993) and White et al. (1994a,b, 1995b), in pairs of *Xenopus* oocytes. Italic characters denote the results obtained in oocyte pairs only; bold characters refer to results obtained in HeLa cells only; normal characters indicate the configurations that were tested in both experimental systems. Functional combinations can be identified because they develop either similar conductance levels or allow the passage of microinjected Lucifer Yellow in both homotypic and heterotypic configurations. Non-functional combinations neither develop conductance that exceed that of antisense-treated control oocyte pairs nor allow the transfer of microinjected Lucifer Yellow between HeLa cells. The question mark denotes that both Cx31.1 and Cx33 are unable to form homotypic channels between paired *Xenopus* oocytes (Hennemann et al., 1992a; Bruzzone et al., 1994a); n.t., not tested.

Connexin	Combinations	
	functional	non-functional
26	26, 32, *46*, *50*	**31**, *31.1*, **37**, 40, 43, **45**
30.3	*30.3*	n.t.
31	**31**	**26**, **32**, **37**, **40**, **43**, **45**
31.1	?	*26*, *31.1*, *32*, *43*
32	26, 32, *46*, *50*	**31**, 37, 40, 43, **45**
33	?	*33*, *37*, *43*
37	37, 40, 43, **45**	**26**, **31**, 32, *33*
40	37, 40, **45**	26, **31**, 32, 43, *46*, *50*
43	37, 43, **45**, **46**	26, **31**, 32, *33*, 40, *50*
45	**37**, **40**, **43**, **45**	**26**, **31**, **32**
46	*26*, *32*, *43*, *46*, *50*	*40*
50	*26*, *32*, *46*, *50*	*40*, *43*

peculiar behavior has been observed only with a chimera (Bruzzone et al., 1994a), but it could represent an additional regulatory feature allowing transient communication between groups of cells.

Selective communication could explain the observation of communication compartments which are characterized by clusters of cells that are permeable to dyes within, but not between, adjacent groups, even in the absence of obvious anatomical boundaries (Lo and Gilula, 1979; de Laat et al., 1980; Warner and Lawrence, 1982; Weir and Lo, 1982; Meda et al., 1983; Blennerhasset and Caveney, 1984; Chanson et al., 1991; LoTurco and Kriegstein, 1991; Laird et al., 1992; Yuste et al., 1992; Serras et al., 1993). An example of compartmentalization is seen in the cardiovascular system, where multiple connexins are expressed in heart, smooth muscle and endothelial cells of the arterial wall (Larson et al., 1990; Pepper et al., 1992; Wiens et al., 1995; see for a review: Beyer and Veenstra, 1994; Severs, 1995). The Purkinje fibers of the cardiac conduction system express largely Cx40, while the working myocardium displays Cx43 predominantly (Beyer et al., 1989; Bastide et al., 1993; Gourdie et al., 1993; Kanter et al., 1993; Gros et al., 1994; Dahl et al., 1995; Delorme et al., 1995; van Kempen et al., 1995). These two connexins are incompatible and do not form heterotypic junctions in expression systems (Bruzzone et al., 1993; Elfgang et al., 1995). Thus, the potential for undesired excitation of myocardial cells is minimized along the length of the conducting fibers. A subset of the fibers at terminal branches, however, expresses Cx43 allowing the coordinated propagation of stimulus to precisely defined regions of the myocardium (Bastide et al., 1993; Gros et al., 1994). The patterns of compatibility between connexins and changes in their expression levels provide a simple mechanism to segregate or promote communication in dynamic fashion. This regulation of signals between different cell types, could have profound biological consequences in physiological and pathological situations such as hypertension, ischemia and atherosclerosis (Bény and Connat, 1992; Segal and Bény, 1992; Bastide et al., 1993; Polacek et al., 1993; Blackburn et al., 1995; Gabriels and Paul, 1995; Jara et al., 1995; Little et al., 1995a).

There are no criteria to predict the pattern of compatibility of any given connexin. Available data obtained in paired *Xenopus* oocytes indicate that connexin selectivity is not simply based on group identity. Although Cx40, Cx46 and Cx50 are all group-α connexins, Cx40 does not interact with either Cx46 or Cx50. Furthermore, Cx32 and Cx26, which are group β, both readily interact with group-α Cx46 and Cx50 (Table 3). Domain swapping experiments have investigated the molecular mechanisms allowing connexins to select a compatible partner. To determine the relative importance of each extracellular domain in the process of discrimination, several chimeras, whose extracellular sequences were derived from two connexins with different patterns of compatibility, have been tested. The ability to form intercellular channels follows the patterns specified by the identity of the second extracellular domain (Bruzzone et al., 1994a; White et al., 1994b, 1995b) (Fig. 3). Whether all connexins discriminate compatible partners on the basis of sequences contained in the second extracellular loop remains to be determined. Altogether these experiments indicate that the formation of a cell-to-cell channel results from the interplay of multiple extrinsic and intrinsic determinants. It remains unclear how the extracellular loops of partnered connexins interact (see for a thorough discussion of the mirror versus staggered model: Peracchia et al., 1994). More detailed studies mapping the minimum number of residues that specify selective recognition will provide a closer definition of the molecular basis for selective interaction between connexins.

Docking and gating of connexons occurs quickly and may be regulated in dynamic fashion. Following initiation of cell-to-cell contact, both intercellular communication and the appearance of gap junctions occur within minutes (Loewenstein et al., 1978; Chow and Poo, 1984; Rook et al., 1988, 1990; Churchill et al., 1993; Bukauskas et al., 1995a). In addition, treatment of re-aggregating Novikoff cells with phorbol esters drastically reduces the formation of functional intercellular channels, without affecting the plasma membrane concentration of Cx43, the only known connexin in this cell system (Lampe, 1994). Phorbol esters are powerful activators of protein kinase C (Kikkawa and Nishizuka, 1986) which, in turn, increases the phosphorylation of Cx43 (Sáez et al., 1993). Because connexins reach the plasma membrane only after oligomerization into connexons (see: *Oligomerization and intracellular transport*), the simplest explanation for these data is that phorbol esters interfere with the docking and/or gating steps of channel formation, although the perturbation of cell adhesion systems cannot be ruled out. At the moment it is not clear whether protein-kinase-C-dependent phosphorylation of Cx43 is the signal regulating connexon–connexon interactions (Lampe, 1994). Initial studies suggest that glycyrrhetinic acid, a drug used to block junctional communication (Davidson and Baumgarten, 1988), also may interfere with the docking between connexons of adjacent cells (Goldberg et al., 1996).

Physiology of connexins

Various assays, reproduced in scores of laboratories, have substantiated the claim that connexins are necessary to form functionally competent intercellular channels. The two most popular systems employed to study connexins are the *Xenopus* oocyte assay (Dahl et al., 1987), where two cells are paired after the injection of specific RNA, and communication-deficient cell lines, where connexin DNA can be stably transfected (Eghbali et al., 1990). Although both techniques have some drawbacks (for example *Xenopus* oocytes express endogenous connexins and none of the recipient cell lines is totally deficient in intercellular communication), in general, they have led to qualitatively similar conclusions. Moreover, conditions can be manipulated, e.g. the injection of antisense oligonucleotides to inhibit the endogenous connexin in *Xenopus* oocytes (Barrio et al., 1991), in order to attribute the recorded properties to the connexin under study. All connexins have been found to form functional channels, with the exception of rCx31.1, rCx33 and Cx32.7 (Hennemann et al., 1992a; Bruzzone et al., 1994a, 1995), the latter having been identified in the ovary of teleosts (Yoshizaki et al., 1994). Detailed investigations have led to the elucidation of some of the elementary properties of connexins and to a new interpretation of the functioning of intercellular channels.

Ionic selectivity and size permeability. Intercellular channels have been previously described as non-selective conduits allowing the passage of ions and molecules with a maximal diameter of approximately 1.5 nm and up to 1–2 kDa of molecular mass (Simpson et al., 1977). Although a few studies had already revealed some selectivity based on charge and size (Brink and Dewey, 1980; Ransom and Kettenmann, 1990; Peinado et al., 1993), it is only during the past year that those historical assumptions on the lack of selectivity have been challenged. First, it has been shown that connexins display a variable degree of ionic selectivity. Thus, Veenstra et al. (1994a,b) have estimated that the anion-to-cation permeability is 0.12 for chCx45 and 0.43 for hCx37. It is clear from these values that connexins do not form ion-selective channels in the classical sense. The selectivity of other voltage-gated channels, such as sodium or potassium, for a given ion is several orders of magnitude greater than that of connexins. Nonetheless, this parameter may not be without consequences. In fact, the marked cationic selectivity of Cx45 most likely accounts for the different permeability of two tracer molecules of similar limiting dimensions, 2′,7′-dichlorofluorescein and 6-carboxyfluorescein, that are commonly used to assess cell coupling (Veenstra et al., 1994a). Similarly, the kinetics of exchange between coupled cells of many intracellular messengers, which are negatively charged, may be connexin-dependent. Next came the finding that rCx45 channels are not permeable to Lucifer Yellow, whereas rCx43 passes the dye to many neighboring cells (Steinberg et al., 1994). Although the two connexins were expressed in different cell lines, these data suggest that size-dependent permeability is also a connexin-specific feature. Interestingly, a switch in connexin gene expression in differentiating keratinocytes is accompanied by selective changes in channel permeability that may operate to control the differentiation process (Brissette et al., 1994). In a more comprehensive analysis, Elfgang et al. (1995) have examined seven murine connexins in the same cell type and have established that size selectivity is a general property of intercellular channels which is exquisitely dependent on the type of connexin. Channels composed of Cx26, Cx37, Cx40, Cx43 and Cx45 transfer propidium iodide (molecular mass = 414 Da, positively charged) and ethidium bromide (molecular mass = 394 Da, positively charged), although with slightly different efficiencies. In contrast, Cx31 and Cx32 are impermeable to those tracers but readily exchange Lucifer Yellow (molecular mass = 443 Da, negatively charged) and DAPI (molecular mass = 279 Da, positively charged). These differences are independent of the levels of macroscopic conductance, which were comparable for all experimental conditions. Finally, it has been suggested that even expression of multiple connexins may control the permeability of gap junction channels through complex intermolecular interactions (Koval et al., 1995). Although these reports do not provide information on the structural basis of ionic and size-selectivity, they demonstrate that it matters very much to a group of communicating cells which connexins are expressed. Multiple connexins may have evolved to provide specific forms of intercellular communication tailored to different functional needs. While additional factors besides charge and size may regulate the passage of molecules through connexin channels, selectivity in this size range may permit discrimination among known second messengers. Gap junctions have been demonstrated to be permeable to cyclic nucleotides, calcium ions and inositol trisphosphate (Lawrence et al., 1978; Murray and Fletcher, 1984; Dunlap et al., 1987; Sáez et al., 1989a; Sandberg et al., 1992). It is currently unknown whether different connexins make channels that are selectively permeable to these signaling molecules and how this mechanism may be operational *in vivo*, but its consequences on tissue homeostasis are potentially important. An interesting possibility is that heterotypic channels generate communication pathways endowed with different size permeability, ionic selectivity and sensitivity to intracellular messengers. A recent study has provided experimental support to this hypothesis, by demonstrating that intercellular channels between AII amacrine and cone bipolar cells, presumably heterotypic, are impermeable to molecules that can diffuse through gap junction channels between AII amacrine cell, presumably homotypic (Mills and Massey, 1995). Furthermore, a striking unidirectionality of cell–cell communication has been observed, with dye movement occurring from astrocytes to oligodendrocytes (Robinson et al., 1993) and from endothelial to smooth muscle cells (Little et al., 1995b) but rarely in the opposite direction. Together, these results suggest the existence of polarized routes of cell–cell signaling, since certain intracellular messengers possess charge characteristics similar to those of the dyes utilized in the above-mentioned studies.

Gating of connexins: transjunctional voltage. The functional state of intercellular channels can be altered in several experimental ways that have provided the essential information on their gating mechanisms. Gating is defined as the reversible transition between open and closed conformations of pre-formed channels. Gating can be the result of non-covalent (e.g. electrostatic, van der Waals' interactions) or covalent (e.g. phosphorylation) modifications of the channel structure. Furthermore, the responsiveness of channels to gating can be modulated by several pharmacological agents as well as extracellular and intracellular messengers. As is the case with conventional ion channels (e.g. sodium, potassium, calcium), the conductance of connexins is affected by a difference of potential between the cells. In a classical study, Furshpan and Potter (1959) described the electrophysiological behavior of the junctions between giant axons of the crayfish nerve cord. These electrotonic synapses showed asymmetrical voltage-dependence, a phenomenon called rectification (Fig. 4). In other words, depolarization of the pre-junctional neuron increases junctional conductance allowing orthodromic conduction, whereas depolarization of the post-junctional neuron closes junctional channels, preventing antidromic conduction. Due to its structure, the gap junction channel can be subjected to voltages established between the coupled cells (transjunctional voltage, V_j) and between the cytoplasm and the extracellular medium (inside-outside voltage, $V_{i\text{-}o}$). Most vertebrate junctions display only V_j-dependence, which is manifested as symmetric changes of junctional currents with equal polarizations on either side and of either sign, although rat Cx26 (Barrio et al., 1991), rCx43 (White et al., 1994b) and *Xenopus* Cx30 (Jarillo et al., 1995) are also sensitive to $V_{i\text{-}o}$ (see for a review of this aspect of connexin physiology: Verselis and Bargiello, 1991). The kinetics and steady-state properties of voltage dependence demonstrated that conductance/voltage relationships differ enough among members of the connexin family to indicate that a wide range of voltage gating behaviors exists (reviewed by Bennett and Verselis, 1992; Nicholson et al., 1993; Beyer and Veenstra, 1994; White et al., 1995a) (Fig. 5). Where is the voltage sensor of connexins located? In the case of other voltage-dependent channels, a common consensus sequence which may underlie their voltage sensitivity has been identified. This motif, called S4, consists of an amphipatic α-helix with positively charged amino acids every fourth residue (Armstrong, 1992; Catterall, 1995). Analysis of connexin sequences, however, has failed to reveal a region similar to S4, thereby implying that the molecular mechanisms which mediate voltage-induced closure of gap junction channels must be different. Rubin et al. (1992a,b) have prepared chimeras by substituting defined amino acid segments of Cx32 with the corresponding ones of Cx26, seeking to identify those sequences that can interconvert their voltage dependence. None of the chimeras display changes that can account for the differences in the calculated gating charges of the parent molecules, a result suggesting that the voltage sensor is not a localized structure but arises from interactions between domains that line the channel and others that do not. Furthermore, the analysis of heterotypic channels has conclusively demonstrated that, in most cases, novel gating properties result from interactions with a different connexon partner (Barrio et al., 1991; Rubin et al., 1992a,b; Nicholson et al., 1993; Bruzzone et al., 1994a; White et al., 1994a,b, 1995a; Donaldson et al., 1995; Chen and DeHaan, 1996). This scenario indicates that connexons do not maintain their electrical fingerprints and that voltage-dependent closure of heterotypic channels containing a given connexin is dramatically influenced by which partner connexin is contributed by the adjacent cell (Fig. 6). Interestingly, certain heterotypic combinations exhibit asymmetric voltage dependence (Swenson et al., 1989; Barrio et al., 1991; White et al., 1994a, 1995b), thus suggesting that

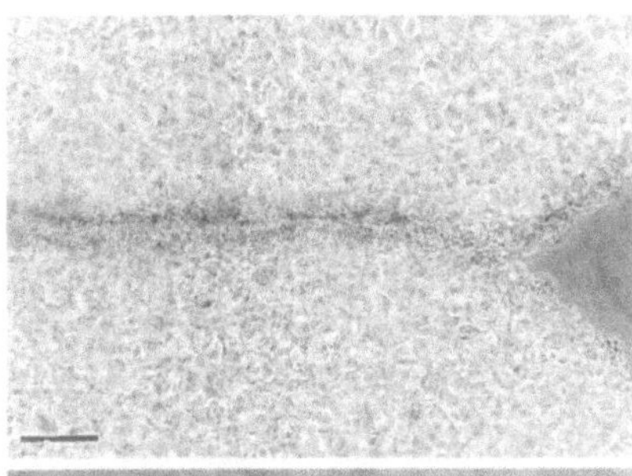

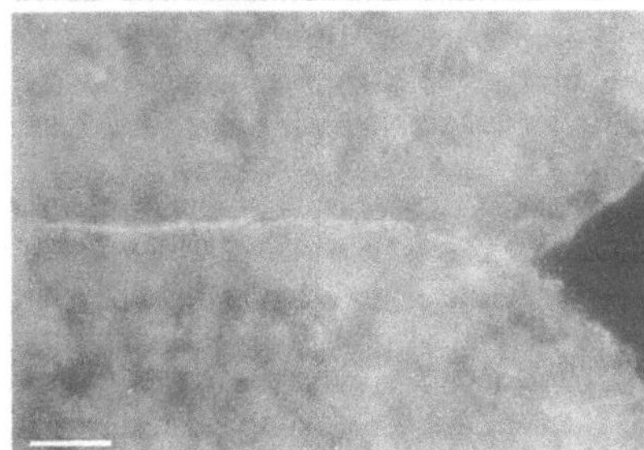

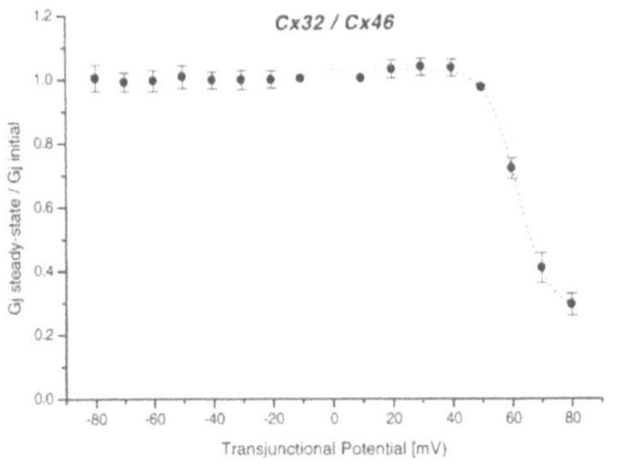

Fig. 4. Analysis of the molecular properties of connexins in programmed expression systems. (A) Schematic representation of the steps involved in the functional expression of connexins using the paired oocyte expression system (Dahl et al., 1987). (1) The coding sequence of a connexin cDNA is subcloned into the expression vector SP64T and synthetic RNA is produced by *in vitro* transcription. (2) Purified RNA is then microinjected into the vegetal pole of defolliculated stage V-VI oocytes. (3) Following removal of the vitelline envelope, oocytes are manipulated to form pairs with the vegetal poles apposed. The functional properties of intercellular channels are assessed using a double voltage clamp procedure which enables junctional conductance to be directly quantitated. V = voltage electrodes; I = current electrodes. (4) The two cells are clamped to the same voltage (V) and alternating symmetrical depolarizing pulses of various amplitudes are imposed. The current (I) supplied to the cell not stepped is equal in magnitude but opposite in sign to the junctional current (I_j); junctional conductance is then calculated by dividing the junctional current by the transjunctional potential (Harris et al., 1981; Spray et al., 1981). (B, C) Oocyte pairs injected with connexin RNA express connexin protein and concentrate it at the region of cell-to-cell apposition. A phase-contrast view (B) of a paraffin section of an oocyte pair displays the cell-to-cell interface and a portion of the cytoplasm of both cells filled with yolk platelets. A fluorescence photomicrograph (C) of the same section probed with an anti-Cx32 antibody (Goodenough et al., 1988) shows the accumulation of the specific immunostaining at the region of cell-to-cell apposition. Scale bar is 20 µm. (D) Heterotypic Cx32/Cx46 channels, expressed in oocyte pairs, illustrate distinct connexin properties and provide an example of rectification. Plots describe the relationship of transjunctional potential to junctional conductance. Transjunctional potential is defined as positive for depolarization of the Cx46 side relative to the Cx32 side and vice versa. The data of junctional conductance (G_j) are calculated as the ratio of steady-state (measured at the end of 30-s pulses) to initial (measured 10 ms after the imposition of a polarizing step) conductance. Conductance shows no voltage dependence for relative positivity of the Cx32 side, whereas it declines sharply for relative positivity of the Cx46 side (redrawn from White et al., 1995b). Thus, channels close asymmetrically in response to voltage. This behavior is equivalent to a unidirectional propagation of electrotonic coupling, as in rectifying synapses (Furshpan and Potter, 1959; Auerbach and Bennett, 1969; Jaslove and Brink, 1986; Giaume et al., 1987).

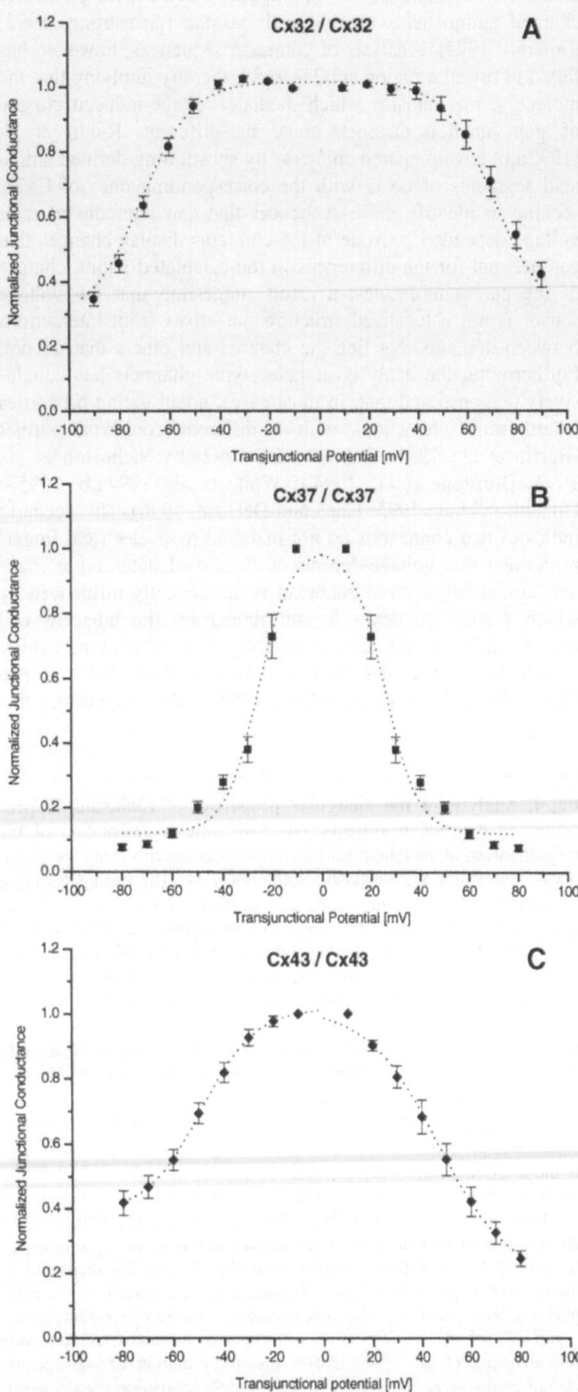

Fig. 5. Voltage-dependent gating of Cx32, Cx37 and Cx43 expressed in pairs of *Xenopus* oocytes. (A–C) Plots describe the relationship of transjunctional voltage to steady-state junctional conductance, normalized to the values obtained at ±10 mV. To better visualize the individual differences, the best fits to a Boltzmann equation (Spray et al., 1981) are displayed (dotted lines). In the case of Cx32/Cx32 pairs (A), potentials greater than ±40 mV represent the threshold to elicit channels closure, whereas in the case of Cx37/Cx37 pairs (B) the same transjunctional voltage inhibits junctional conductance by approximately 80%. Finally, channels composed of Cx43 (C) show an intermediate sensitivity and an asymmetrical voltage dependence. Adapted by selecting representative data from previous publications (Bruzzone et al., 1994a; White et al., 1994b).

heterotypic channels represent the molecular mechanism underlying rectification at electrotonic synapses (Jaslove and Brink, 1986; Giaume et al., 1987).

Attempts to understand the molecular basis of the unpredicted behavior of heterotypic channels have just started and proved to be extremely difficult. A mutational analysis (Suchyna et al., 1993) has lead to the suggestion that a proline in the second transmembrane domain, which is conserved among all members of the connexin family, may act as a transduction element between the voltage sensor and the voltage gate of intercellular channels. In addition. the first two amino acids in E1 may confer the rectifying behavior observed in heterotypic Cx32/Cx26 intercellular channels (Verselis et al., 1994). These authors have proposed that rectification occurs because connexons composed of Cx26 and Cx32 close in response to voltages of opposite polarities, although this assumption has not been verified experimentally. Compelling evidence for an alternative explanation has been deduced from single channel data. These experiments indicate that unitary conductance of the two connexons is differentially modulated by transjunctional potential and shows voltage-dependent rectification (Bukauskas et al., 1995b). Preliminary results also suggest that rectifying channels result from the heterotypic assembly of connexins with different permeability (Cao et al., 1995; Suchyna et al., 1995). The bottom line is that docking of the two connexons contributes to the resulting gating properties. In fact, even polarity of voltage gating does not appear to be an intrinsic characteristic of each connexin, as the polarity of voltage gating for Cx46 connexons reverses when they become incorporated into an intercellular channel (White et al., 1994a). Research in this area needs now to provide the molecular answers to these findings.

Voltage-dependent gating of ion channels underlies common cellular functions, such as transmission of information between neurons, excitation–contraction coupling of muscle cells and stimulus-secretion coupling of certain secretory cells. Although gating by voltage represents a general property of gap junctions, its functional role remains speculative. One consequence of voltage sensitivity of gap junctions could be to transiently uncouple communicating cells. Cell-specific alterations in resting potential would cause transjunctional voltages which close gap junctions and isolate adjacent cells (Harris et al., 1983). For example, in *Xenopus* embryos, cells in the dermatome and myotome layers are electrically coupled within, but not between, layers and the average resting potentials of cells in these two adjacent layers differ by 40 mV (Blackshaw and Warner, 1976).

Unitary conductance, main states and substates. Understanding the physiological properties of a channel requires the analysis of single-channel activity. For intercellular channels this step has been accomplished by applying a dual whole-cell recording method with patch electrodes to high-resistance cell pairs (Neyton and Trautmann, 1985; Veenstra and DeHaan, 1986). In this situation, the opening or closing of an intercellular channel is detected as simultaneous current steps of opposite direction in each cell, whereas other membrane channels cause current steps that occur independently. As soon as connexin diversity was appreciated, it was postulated that the primary sequence differences would confer distinct electrical behavior to the channels they composed. Early studies, using homologous pairs of various cell types, revealed the existence of channels with distinct unitary conductance (Neyton and Trautmann, 1985; Rook et al., 1988; Somogy and Kolb, 1988; Spray and Burt, 1990; Giaume et al., 1991a; Eckert et al., 1993). Analysis of intercellular channels present between pairs of communication-deficient cell lines transfected with one connexin at a time determined that the values of unitary conductance of different con-

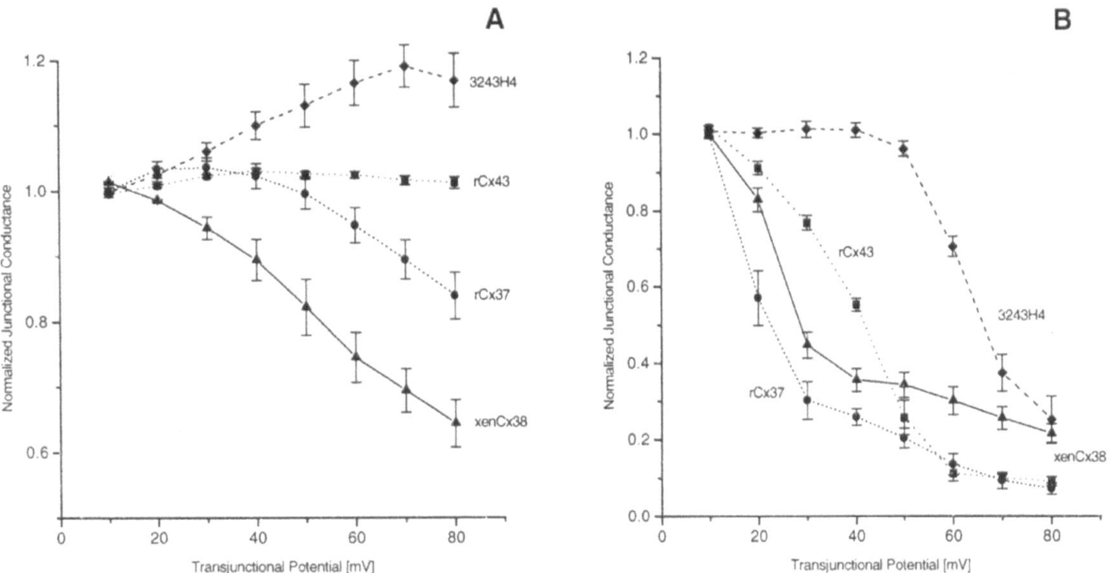

Fig. 6. The physiological properties of *Xenopus* Cx38 (xenCx38) are modified by the identity of the connexin partners expressed in oocyte pairs. Plots describe the relationship of transjunctional potential to initial (A) and steady-state (B) junctional conductance, normalized to the values obtained at ±10 mV. Initial conductance was measured 5−10 ms, whereas steady-state conductance was measured 30 s after the imposition of a polarizing step. It is evident that there is no electrical fingerprint for xenCx38, as voltage dependence is differently affected in the case of homotypic channels (xenCx38, ▲——▲) and heterotypic combinations (rat Cx37, ●- - -●; rat Cx43, ■······■; 3243H4, ◆- - -◆). For example, in combination with the chimera 3243H4 (Bruzzone et al., 1994a) polarizing steps greater than 50 mV represent the threshold to inhibit steady-state conductance, whereas in combination with rCx37 the same transjunctional potential closes approximately 80% of the channels. Adapted, in part, by selecting representative data from a previous publication (Bruzzone et al., 1994a).

nexins range between 30−300 pS (Eghbali et al., 1990; Fishman et al., 1990; Veenstra et al., 1992, 1994a,b; Traub et al., 1994; Buehler et al., 1995; Bukauskas et al., 1995a; Moreno et al., 1995). Careful examination of some recordings, however, suggested a more complex situation: the presence of multiple conductance states of single channels (Rook et al., 1988; Moore et al., 1991; Chen and DeHaan, 1992, 1993a,b; Veenstra et al., 1992; Wang et al., 1992; Ramanan et al., 1993; Donaldson et al., 1994; Pérez-Armendariz et al., 1994). It has now been shown that Cx37, Cx40, Cx43 and Cx45 all form channels with different substates (Moreno et al., 1994a, 1995; Veenstra et al., 1994a,b; Bukauskas et al., 1995a). In the case of five connexins there appears to be no correlation between the limits of molecular size permeability and ionic selectivity, on the one hand, and the values of the main state of unitary conductance, on the other hand (Veenstra et al., 1995). The next effort should be directed at establishing whether, for any given connexin, the reduced conductance of substates is associated with changes in size permeability, charge selectivity and sensitivity to gating mechanisms. Such a mechanism would represent a powerful and dynamic means to modulate channel function and create transient metabolic and/or ionic compartments. These findings may lead to a re-evaluation of whether gap junction channels close in an all-or-none or in a graded fashion. Verselis et al. (1986) found that permeability is linearly related to conductance for transjunctional voltages of opposite signs and concluded that channel closure is an all-or-none phenomenon. In contrast, the effective diameter of intercellular channels was reported to vary under the influence of agents that change macroscopic conductance (Rose et al., 1977).

Chemical and biochemical gating of connexins: cations, phosphorylation, etc. In the case of intercellular channels, there are no known substances that increase the open state probability, whereas closure of channels, or uncoupling, can be observed in response to diverse treatments. The first chemical inhibitors of junctional coupling to be described were pH (Turin and Warner, 1977) and calcium (Rose and Loewenstein, 1976). The issue of pH sensitivity and its molecular basis have been examined in a series of studies using pairs of *Xenopus* oocytes expressing different connexins. All connexins exhibit a variable degree of sensitivity to experimental treatments inducing intracellular acidification (Werner et al., 1991; Liu et al., 1993; Ek et al., 1994; White et al., 1994b), although the actual intracellular pH was measured only in one study (Liu et al., 1993). In particular, comparison of junctional conductance versus pH has demonstrated that, in the same cellular environment, Cx43 and *Xenopus* Cx38 are more sensitive than Cx32 to acidification (Werner et al., 1991; Liu et al., 1993). Experiments with truncated constructs and site-directed mutagenesis have shown that the length and primary sequence of the middle cytoplasmic loop and carboxy-terminal tail play a critical role in this form of chemical gating. First, Cx43 mutants in which histidine at position 95 is replaced with acidic or uncharged residues are less susceptible to acidification, whereas the opposite is true when the same histidine is substituted with a basic residue (Ek et al., 1994). These results agree with the hypothesis that pH-gating of Cx43 may depend on protonation of histidine residues present in the middle cytoplasmic loop (Spray and Burt, 1990). In addition, domain swapping between rat Cx32 and *Xenopus* Cx38 indicates that the greater pH-sensitivity of *Xenopus* Cx38 with respect to rat Cx32 can be mimicked by a chimera in which the middle cytoplasmic loop of Cx32 has been replaced with the corresponding region of Cx38 (Wang et al., 1996). Second, deletion of the C-terminal 125 amino acid residues of Cx43 generates a mutant with a C-terminal tail of equivalent length to Cx32 and, concomitantly, decreases its pH sensitivity to values that are similar to those of Cx32 (Liu et al., 1993). Finally, independent expression of the

C-terminal tail of Cx43 (residues 259–382) rescues the pH-sensitivity of a Cx43 truncated at amino acid 257 and also enhances the susceptibility of Cx32 to acidification (Morley et al., 1996). Together, these data indicate the middle and C-terminal cytoplasmic loops are major determinants of chemical gating (Fig. 3) and are compatible with the existence of a particle–receptor interaction between two separate domains, similar to the original ball-and-chain hypothesis of voltage inactivation (see for a review: Armstrong, 1992; Catterall, 1995). In spite of this progress, it is not clear whether pH gating is a genuinely independent phenomenon or is secondary to elevation of cytosolic calcium, as suggested by Peracchia and coworkers. There is evidence that buffering of cytosolic calcium with calcium chelators antagonizes, at least partially, pH-dependent uncoupling (Lazrak and Peracchia, 1993; Peracchia et al., 1996). Here again, the structural diversity of connexins may help prove or disprove this hypothesis. In this context, it will be interesting to establish whether C-terminal sequences of other connexins can impart to truncated mutants a novel pH-dependent behavior mimicking that of their corresponding full-length proteins.

The principal questions regarding the effect of an elevation of cytosolic calcium on intercellular channels are the precise range of concentrations to achieve closure of channels and whether the effect is direct on the channel itself, or dependent on connexin identity. For years, most studies indicated that the concentrations of calcium required to block intercellular channels were, in general, far higher than those observed in physiological conditions (Spray and Bennett, 1985). A greater sensitivity to calcium has also been reported, estimating the threshold calcium level at which a depression of junctional conductance becomes significant around 400–500 nM, as measured with a calcium-sensitive microelectrode in cardiac myocytes (Dahl and Isenberg, 1980) or with a fluorescent indicator in Novikoff hepatoma cells and cultured bovine lentoids (Lazrak and Peracchia, 1993; Crow et al., 1994). The principal connexin expressed by these cell types is Cx43, which suggests that calcium-sensitivity of a connexin is an intrinsic parameter, independent of the cellular environment. The sensitivity of Cx43 is within physiological concentrations, as most cell types have a resting calcium concentration around 100 nM and display calcium transients and oscillations reaching micromolar concentrations in response to physiological stimuli (Pietrobon et al., 1990). Thus, calcium-induced modulation of cell coupling could be a component of other calcium-dependent cellular activities. A direct effect of calcium seems unlikely, as connexins do not possess consensus sequences for putative calcium-binding sites. Based on its ability to bind to Cx32 (Hertzberg and Van Eldik, 1987) and the presence of possible binding sites on connexins, the ubiquitous calcium-receptor protein, calmodulin, has been considered a logical mediator of the calcium effect (reviewed by Peracchia et al., 1994). This hypothesis is strongly supported by the observation that antisense depletion of endogenous calmodulin markedly reduces the sensitivity of junctional communication to elevated cytosolic calcium in *Xenopus* oocyte pairs (Peracchia et al., 1996).

Another possibility is that calcium gating results from the activation of specific kinases and subsequent connexin phosphorylation. Protein phosphorylation on either serine, threonine and, more recently tyrosine, is a well-characterized mechanism by which extracellular signals can significantly alter the function of ion channels (Hille, 1992). Early studies had suggested the presence of a simple, dual, regulatory system: stimulation of coupling by protein kinase A, the cAMP-dependent kinase (Flagg-Newton et al., 1981; Wiener and Loewenstein, 1983; Sáez et al., 1986) and inhibition of coupling by protein kinase C, the calcium/phospholipid-dependent protein kinase (Yotti et al., 1979; Dotto et al., 1989; Brissette et al., 1991; Berthoud et al., 1993). Depending on the experimental system used, however, the effects of activators and inhibitors of cellular kinases can vary (Piccolino et al., 1984; De Mello, 1986; Neyton and Trautmann, 1986; Chanson et al., 1988; McMahon et al., 1989; Giaume et al., 1991b; see for a review: Loewenstein, 1985; Musil and Goodenough, 1990; Stagg and Fletcher, 1990; Sáez et al., 1993). As it has become clear that cells express a distinct set of connexins, those studies suggest that gating of intercellular channels by phosphorylation could be kinase- and connexin-specific. The first direct evidence of connexin phosphorylation was obtained in hepatocytes for Cx32 (Sáez et al., 1986; Traub et al., 1987). Using synthetic peptides in a biochemical assay, it has been shown that a serine at position 233 is the primary site of Cx32 phosphorylation by protein kinase A, whereas protein kinase C phosphorylates the same residue as well as several others (Sáez et al., 1990). Site-directed mutagenesis of both Ser233 and Ser240, however, does not affect the ability of the mutant Cx32 to form channels in oocyte pairs (Werner et al., 1991), an observation suggesting that constitutive phosphorylation is not necessary for Cx32 to acquire functional competence. Nevertheless, the increase in phosphorylation of Cx32 in hepatocytes treated with cAMP correlates temporally with an increase in junctional conductance, supporting the notion that protein-kinase-A-dependent phosphorylation has a role in the dynamic regulation of cell-to-cell communication in response to extracellular signals (Sáez et al., 1986). Except for Cx26, that has neither consensus sequences for kinases, nor is phosphorylated by protein kinase A, protein kinase C and calcium/calmodulin-dependent protein kinase II (Sáez et al., 1990) all other connexins which have been directly examined are phosphoproteins (Cx40, Cx43, Cx45, Cx46, Cx50 and chCx56; Musil et al., 1990a; Laird et al., 1991; TenBroek et al., 1992, 1994; Jiang et al., 1993; Berthoud et al., 1994; Butterweck et al., 1994b; Laing et al., 1994; Delorme et al., 1995). Cx43 is differentially phosphorylated in a tissue-specific manner (Kadle et al., 1991) and this constitutive phosphorylation has been has been correlated to the communication competence of some cell lines (Musil et al., 1990a,b; Oh et al., 1993; Oyamada et al., 1994). Sequencing of phosphorylated peptides has shown that seryl residues at positions 368 and 372 are phosphorylated by protein kinase C, whereas protein kinase A does not appear to have a direct effect on Cx43 (Sáez et al., 1993). Other consensus sites have been identified (Swenson et al., 1990; Kanemitsu and Lau, 1993) and it will be important to understand whether they can be phosphorylated independently of one another or whether a hierarchical order is maintained.

The most exciting news is coming from studies that are elucidating the basic mechanism of connexin gating by phosphorylation. Takens-Kwak and Jongsma (1992) were the first to suggest that pharmacological activation of kinases coincided with a shift in the unitary conductance of single channels recorded between neonatal cardiac myocytes. The expression of individual connexins in transfected cell lines has now demonstrated that single channel unitary conductance of human Cx43 shows predominance of states with larger values in response to dephosphorylating treatments (Moreno et al., 1994b). In contrast, the decrease in macroscopic conductance induced by cGMP-dependent phosphorylation of rat Cx43 is the result of an increase in the relative frequency of the lowest conductance state (Kwak et al., 1995b). Interestingly, cGMP treatment neither phosphorylates nor affects unitary conductance of human Cx43. Comparison of human and rat sequences of Cx43, that differ only in nine amino acid residues, has led Kwak et al. (1995b) to propose that Ser257 is a plausible site for cGMP-dependent phosphorylation of rat Cx43. The biological implications of these studies could be very significant. If changes in unitary conductance also influence channel permeability and size-selectivity, then the opposite shifts observed after connexin phosphorylation/dephosphorylation would represent a mechanism ideally suited to either restrict or facilitate intercellular coupling in dynamic fashion. The road to a conclusive demonstration is still long, but initial hints that this may be the case are emerging. Thus, examination of dye coupling and single channel conductance of Cx26, Cx43 and Cx45 suggests the existence of a connexin-specific regulation under similar phosphorylating treatments and of a positive correlation between frequency of lower conductance states and decreased permeability (Kwak et al., 1995a). This specificity may explain why intercellular transfer of signals between AII amacrine cells, presumably through homotypic channels, is blocked by elevation of cAMP levels, whereas channels between AII amacrine and cone bipolar, probably heterotypic, are inhibited by cGMP, but are insensitive to cAMP (Mills and Massey, 1995). In addition, connexin phosphorylation may actually lead to an opposite modulation of electrical and metabolic coupling. Activation of protein kinase C decreases the size limit of signals that can be exchanged through intercellular channels, whereas the open probability of the channel increases, as shown by the effect on junctional conductance (Kwak et al., 1995c).

The issue of tyrosine phosphorylation in the modulation of channel signaling is more recent. Rapid inhibition of intercellular communication was observed after transfection of viral *src* tyrosine kinase (v-*src*) or overexpression of the cellular protooncogene product pp60$^{c\text{-}src}$ (Atkinson et al., 1981; Chang et al., 1985; Azarnia et al., 1988). This inhibition did not result in altered gap junction morphology, suggesting that reduced permeability may be due to a direct effect on the channel proteins (Atkinson and Sheridan, 1988). Furthermore, mutations affecting the tyrosine kinase ability of the v-*src* oncogene decreased proportionally its inhibitory action (Azarnia et al., 1988; Filson et al., 1990). A direct causal relationship between tyrosine phosphorylation and the inhibition of communication has since been established by showing that Cx43 is specifically phosphorylated on tyrosine in the presence of v-*src*, and this tyrosine phosphorylation is correlated with inhibited communication (Crow et al., 1990; Filson et al., 1990; Swenson et al., 1990; Goldberg and Lau, 1993). In the paired *Xenopus* oocyte assay, the effects of v-*src* are connexin-specific: Cx43 intercellular channels are completely inhibited by v-*src* while channels composed of Cx32 are unaffected. In addition, mutation of a single tyrosine residue to phenylalanine in the carboxyl terminus of Cx43 results in mutant channels which function normally, but are resistant to v-*src*-mediated closure (Swenson et al., 1990). The effect of v-*src* is not likely dependent on a secondary tyrosine kinase since purified pp60$^{v\text{-}src}$ can phosphorylate Cx43 *in vitro* (Loo et al., 1995). Growth factors that bind to receptors with tyrosine kinase activity also modulate the extent of junctional communication. Thus, a similar reduction of cellular coupling has been observed after stimulation of appropriate cell lines with epidermal growth factor (EGF) or platelet-derived growth factor (PDGF) (Maldonado et al., 1988; Madhukar et al., 1989; Lau et al., 1992). The kinetics, dose-dependence and reversibility of this effect are consistent with the closure of existing channels, rather than an alteration of rates of assembly or disassembly of the gap junctions. Unlike v-*src*, EGF stimulation induces specific phosphorylation of Cx43 only on serine residues, and this may be mediated by a signal cascade that includes mitogen-activated protein kinase (Lau et al., 1992; Kanemitsu and Lau, 1993). The inhibition of communication by PDGF is also independent of tyrosine phosphorylation of Cx43 (Pelletier and Boynton, 1994). PDGF stimulation activates a number of proteins involved in signal transduction (Kaplan et al., 1990), and inhibition of communication by PDGF may occur by multiple pathways. Hepatocyte growth factor/scatter factor, another ligand with a tyrosine kinase receptor, also inhibits dye coupling in a mouse keratinocyte cell line without tyrosine phosphorylation of Cx43 (Moorby et al., 1995). In contrast, basic fibroblast growth factor (bFGF) induces a dose-dependent stimulation of Cx43 mRNA and protein expression in cultured cardiac fibroblasts, which is paralleled by an increased intercellular coupling (Doble and Kardami, 1995). While all these experiments are consistent with the hypothesis that connexins are direct or indirect targets of the action of growth factors, considerable detail is missing from our understanding of how oncogenes and growth factors control the permeability of intercellular channels.

The conductance of intercellular channels is also regulated by several lipophilic substances that have been frequently used to uncouple cells and study single channel activity (Johnston et al., 1980; Délèze and Hervé, 1983; see for a review: Burt, 1991; Chanson and Spray, 1995). Interest in this mode of regulation stems from two observations. First, arachidonic acid, as well as non-esterified fatty acids and long-chain acylcarnitines that accumulate during myocardial ischemia, are capable of closing intercellular channels in cardiac myocytes (Burt, 1989; Fluri et al., 1990; Massey et al., 1992; Hirschi et al., 1993; Wu et al., 1993) as well as other cell types (Giaume et al., 1989; Lazrak et al., 1994). Second, coupling is inhibited between cardiac myocytes during experimentally induced ischemia (Wu et al., 1993). In addition, arachidonic acid, an intracellular messenger produced after phospholipase A2 activation, feeds a metabolic cascade leading to the production of a group of compounds, the eicosanoids, that participate in cell signaling (Axelrod et al., 1988). One such derivative, anandamide, is released from neurons and activates cannabinoid receptors. It has now been shown that anandamide is a potent inhibitor of dye and electrical coupling between astrocytes isolated from the striatum (Venance et al., 1995b). The mechanism of action on communication is unknown but is not mediated through the cannabinoid receptor. Together, these studies provide compelling evidence that lipophilic compounds regulate intercellular channels, possibly during pathophysiological conditions.

Functional roles of connexins: from early developmental stages to human disease

Because gap junctions are found nearly everywhere in animal tissues, their involvement in the regulation of diverse cellular functions has been postulated (Yamasaki, 1990; Huizinga et al., 1992; Loewenstein and Rose, 1992; Warner, 1992; Hall et al., 1993; Beyer and Veenstra, 1994; Kanno et al., 1995; Kistler et al., 1995). In the past few years, the association of connexin mutations and human diseases have raised new and intriguing questions about connexin diversity and functions.

In most organisms, communication between cells is established at early cleavage stages and is likely to participate in coordinating the activity of dividing cells and in pattern formation (Palmer and Slack, 1970; Guthrie, 1984; Guthrie and Gilula, 1989; Nishi et al., 1991; Kidder, 1992; Warner, 1992; Valdimarsson and Kidder, 1995). In mouse embryos, three connexins genes are transcribed already by the four-cell stage and mRNA for three more accumulate beginning with the eight-cell stage (Davies et al., 1996). Several studies have illustrated the consequences of blockade of intercellular channels on morphogenesis and pattern formation (Fraser et al., 1987; Lee et al., 1987; Bevilacqua et al., 1989; Busa and Gimlich, 1989; Olson et al., 1991; Paul et al., 1995; but see also Guthrie et al., 1988). As development proceeds, connexins are expressed in complex and overlapping patterns whose significance is still unclear (Paul,

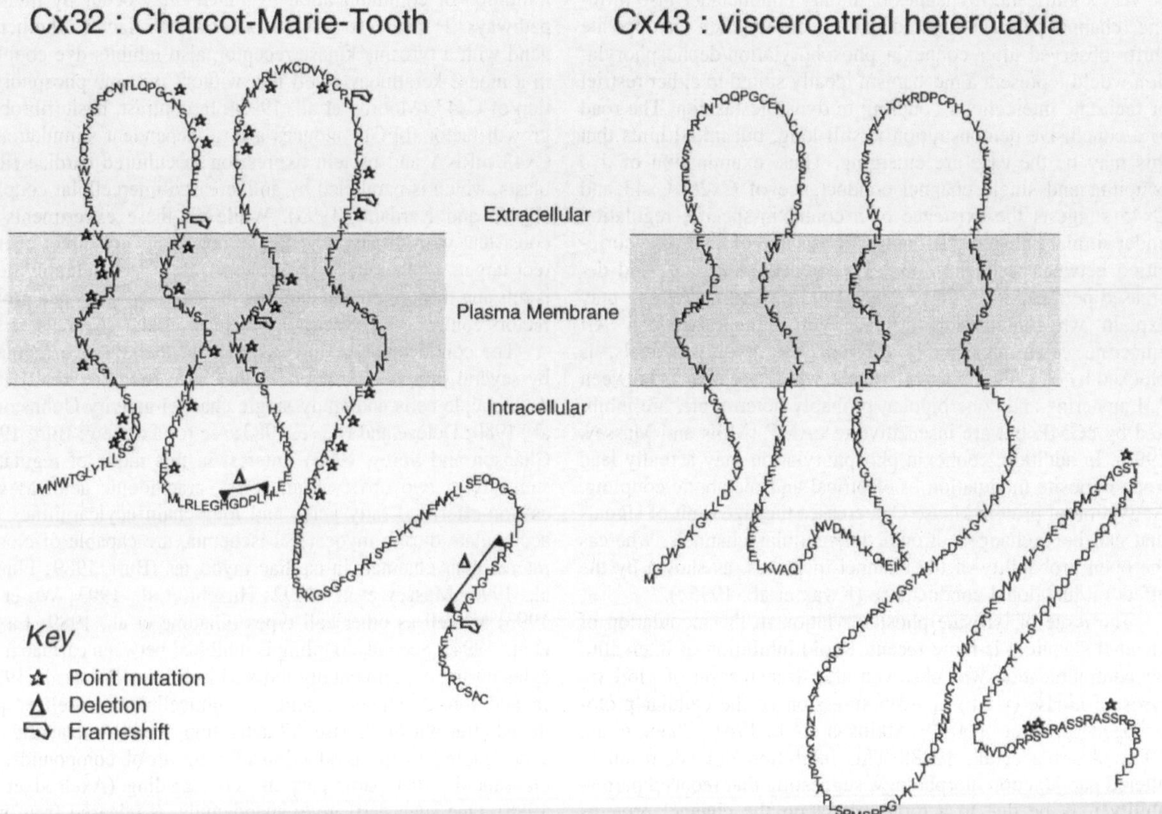

Fig. 7. Diagrams of Cx32 and Cx43 showing the mutations of patients with X-linked Charcot-Marie-Tooth disease (Cx32) and visceroatrial heterotaxia (Cx43). Mutations of Cx32 are scattered throughout the entire sequence: 4 in NT; 6 in M1; 4 in E1; 5 in M2; 3 in CL; 5 in M3; 9 in E2; 6 in CT (Deschênes et al., 1996). In contrast, the four mutations of Cx43 are clustered between amino acid residues 326−373 of the CT region (Britz-Cunningham et al., 1995).

1985; Beyer et al., 1989; Gimlich et al., 1990; Risek and Gilula, 1991; Valdimarsson et al., 1991; Risek et al., 1992; Ruangvoravat and Lo, 1992; Sainio et al., 1992; Dealy et al., 1994; Goliger and Paul, 1994; Green et al., 1994; Grümmer et al., 1994; Jiang et al., 1995; Pozzi et al., 1995). The consequences of this program are the establishment of boundaries or gradients of communication between contacting cells (Lo and Gilula, 1979; Kam and Hodgins, 1982; Warner and Lawrence, 1982; Weir and Lo, 1982; Meda et al., 1983; Blennerhasset and Caveney, 1984; Serras and van den Biggelaar, 1987; Kalimi and Lo, 1988, 1989; Chanson et al., 1991; Laird et al., 1992; Martinez et al., 1992; Serras et al., 1993). Gradients or boundaries of communication would result in distinct signals being sensed by cells and suggest the possibility that gene expression programs be individually regulated within a group of cells. The separate expression of incompatible connexins would be a simple way to achieve compartmentalization without affecting cell-to-cell contacts. For example, within the gastrulating mouse embryo, Cx43 expression is detected in embryonic cells derived from the inner cell mass, whereas Cx31 is expressed in extraembryonic cells derived from the trophoectoderm lineage (Dahl et al., 1996). As Cx31 and Cx43 are incompatible connexins in cultured cells (Elfgang et al., 1995), this segregation correlated with the communication boundaries demonstrated between those compartments (Kalimi and Lo, 1989).

Patterns of communication between cells are often dynamically regulated during development. The vertebrate brain presents a good example of such regulation. Cortical neuroblasts exhibit spontaneous elevations of cytosolic calcium that are synchronous in groups of clonally related cells. It is likely that this synchronization is achieved via gap junctional communication. Cellular coupling within these clusters, dubbed neuronal domains, is abundant during early development, but is gradually eliminated as chemical synapses are established (Vaney, 1991; Yuste et al., 1992, 1995; Peinado et al., 1993; Penn et al., 1994; Wong et al., 1995). Katz and coworkers have proposed that coupling through connexins may control the development of the complex functional architecture of the adult brain, providing a kind of first draft that guides the formation of synapses and the emergence of cortical circuits (Kandler and Katz, 1995; Katz, 1995). Interestingly, neuronal coupling shows size-selectivity, as revealed by the use of neurobiotin (molecular mass = 323 Da) versus Lucifer Yellow (molecular mass = 443 Da) (Peinado et al., 1993). Which connexins are involved? The biochemical characterization and topographical map of connexins in the vertebrate brain is far from complete (Dermietzel et al., 1989; Yamamoto et al., 1990; Dupont et al., 1991; Dermietzel and Spray, 1993; Fisher and Micevych, 1993; Belliveau and Naus, 1995), but selectivity of the exchanged signals is likely to be determined by the connexin type.

Two recent studies have demonstrated the importance of a regulated expression of specific connexin genes and of their functional competence for cardiac embryogenesis. In the first, mice lacking Cx43 were generated by targeted gene ablation (Reaume et al., 1995). Homozygous mice survive to term but die shortly after birth. The only obvious anatomical defect in these animals is a gross enlargement of the conus overlying the right ventricular outflow tract of the heart. In addition, the conus is filled with intraventricular septae leading to interconnected or blind-ended chambers. Thus, obstruction of the pulmonary artery appears to cause neonatal cyanosis and subsequent death. The findings of Reaume et al. (1995) indicate that, although Cx43 is expressed from the start of zygotic transcription and widely distributed in many organs, abnormal morphogenesis is only observed in the heart. Furthermore, Cx43 may be unimportant for myocardial contraction, presumably because other connexins are present. The cardiac defect in Cx43 knockout mice is reminiscent of some forms of stenosis of the pulmonary artery in humans. This exciting possibility has been explored by Britz-Cunningham et al. (1995), who hypothesize that Cx43 mutations lead to functional or developmental abnormalities of the heart. Mutations in Cx43 are found in patients with visceroatrial heterotaxia, but not with other familial cardiac malformations. Visceroatrial ataxia is a genetically heterogeneous syndrome characterized by complex defects of laterality and/or transposition of the great arteries. The common feature with Cx43 knockout mice is stenosis or atresia of the pulmonary artery, which suggests that Cx43 mutations define a subclass of these cardiac malformations. Only the carboxyl-terminal tail from Cx43 of those patients has been sequenced and mutations are clustered in a region which contains several phosphorylation consensus sequences (Fig. 7). Five out of six patients show a single base substitution converting Ser364 to Pro. As Cx43 has been shown to undergo multiple serine phosphorylations that regulate assembly, docking and gating, it is likely that perturbation of either one or all of these processes represents the molecular basis of visceroatrial defects. Thus, transfection of communication-deficient cells with a Cx43 mutant bearing the Ser364→Pro substitution partially restores intercellular communication, but its responses to kinase activation are markedly different from that of wild-type Cx43 (Britz-Cunningham et al., 1995). Although further work is needed to prove a direct causal link, these findings suggest that it is not the functional competence of Cx43 *per se* that is affected by this mutation, but the dynamic modulation of gating properties. Together, the results of knockout mice and visceroatrial heterotaxia indicate that Cx43 may not be necessary for synchronizing heart beats, but is essential for the development of normal cardiac architecture. How alterations of Cx43-based communication, absolute in knockouts or more subtle in human syndromes, lead to the observed anatomical defects is unclear. To address this issue, it will be essential to establish when the defects begin and whether changes in connexin gene expression and cell coupling are present at the same time in the myocardium or other tissues.

The idea that intercellular communication controls the passage of growth regulatory signals, one of the first functions that have been postulated for these channels (Loewenstein and Kanno, 1966; Furshpan and Potter, 1968), has been supported by several independent lines of evidence. On the one hand, tumor promoters, oncogenes and growth factors modulate channel permeability and loss of intercellular communication is a common feature of transformed cells. On the other hand, restoration of cellular coupling is associated with a diminished incidence of tumorigenesis (Yamasaki, 1990; Rogers et al., 1990; Lau et al., 1992; Loewenstein and Rose, 1992; Hossain et al., 1993; Trosko et al., 1993). A straightforward approach has been to characterize the patterns of connexin expression and communication competence of tumor cells, then transfect them with one connexin and check tumorigenicity *in vitro* and *in vivo*. Results have been variable, but in general have confirmed the negative correlation between junctional communication and tumor growth rate (Eghbali et al., 1991; Zhu et al., 1991; Mehta et al., 1991; Naus et al., 1992; Rose et al., 1993; but see also Mesnil et al., 1995, who have found evidence of connexin specificity in the restoration of normal cell growth *in vivo*). An established feature of transformed cells is their lack of contact-dependent growth inhibition. As early studies have already established, normal cells can pass positive signals capable of arresting the growth of transformed cells (Loewenstein and Kanno, 1966; Stoker et al., 1966), a critical issue is that of communication between tumoral and normal cells. Selective lack, reduction and differential regulation of the coupling between transformed and control cells have been demonstrated *in vitro* (Enomoto and Yamasaki, 1984; Mehta et al., 1986; Mesnil et al., 1987, 1993; Mehta and Loewenstein, 1991; Krutovskikh et al., 1994). Alternatively, communication between metastatic tumor cells and vascular endothelium may not be beneficial, but actually help tumor cells cross the endothelium and invade other organs (el-Sabban and Pauli, 1991).

On the basis of the ample evidence that bi-directional signaling between cell adhesion molecules and connexins is critical for the establishment and maintenance of cell-to-cell contact and communication (see above: *Docking and gating*), it has been proposed that both cadherins and connexins are candidate tumor suppressor genes (Yamasaki, 1990; Lee et al., 1992; Birchmeier, 1995). Lee et al. (1992) have divided tumor suppressor genes into two classes: class I, in which the gene is altered at the DNA level by loss or mutation, and class II, in which only expression of the gene is abolished. While mutations in the E-cadherin gene have been found in 50% of diffuse-type gastric carcinomas (Becker et al., 1994), a similar evidence has not been reported for connexins. The transcriptional inhibition of Cx26 and Cx43 in tumor-derived human mammary epithelial cells raises the possibility that these connexins are tumor suppressor genes of class II (Lee et al., 1992). In addition, the differential expression of Cx26 and Cx43 during the cell cycle and their distinct response to pharmacological treatments suggest that the contribution to the malignant potential of tumor cells may be connexin-specific. Thus, the expansion of the connexin family poses the same question: are all connexins sensitive to oncogenes and tumor promoters? Recently, a systematic analysis has demonstrated that individual connexins transfected in HeLa cells respond differentially to tumor-promoters and anti-promoting chemicals (Mazzoleni et al., 1996). In addition, when injected into nude mice, the tumorigenicity of the transfected cell lines is dependent on the connexin species expressed. Thus, transfection with Cx26, but not with Cx40 and Cx43 markedly inhibits cellular growth, in spite of similar extent of intercellular communication (Mesnil et al., 1995). These experiments suggest that the simple measurement of cellular coupling may give only a partial view of the role of connexins in growth control.

A central issue to be addressed is: what are the signals exchanged through intercellular channels? Signals may be nutrients, as cell nutrition may proceed through intercellular channels for cells remotely located from blood supply. Evidence for metabolic cooperation and correction of somatic defects, leading to metabolic improvement, has been provided (Gilula et al., 1972; Subak-Sharpe et al., 1969; Pitts and Simms, 1977; Ledbetter et al., 1986; Hobbie et al., 1987; Puschel and Jungermann, 1988). Likewise, second messengers can diffuse through gap junctions and induce a biological response in the coupled cells (Lawrence et al., 1978; Murray and Fletcher, 1984; Dunlap et al., 1987; Sáez et al., 1989a; Sandberg et al., 1992). The need of coordinated cellular activity, sharing information through intercellular channels, is well illustrated by the finding that stimulus/response coupling in some systems is modulated by the extent of cell communication. A good experimental model is provided by the pancreas, whose function is to synthesize, store and release secretory products in response to appropriate stimuli. In the case of insulin-producing β-cells, that are electrically excitable, it has long been argued that intercellular coupling equalizes their activity and improves their performance. In particular, the close parallel between the rate of insulin secretion and the extent of β-cell coupling indicates that intercellular channels are essential for normal activity of insulin-producing cells (Meda et al., 1984, 1990; Bruzzone and Meda, 1988; Mears et al., 1995). Stimulation of secretion in response to glucose and other stimuli is associated with an increase in extent and strength of intercellular coupling. Thus, the final secretory output depends on the coordinated activity of numerous β-cells that, individually, function in a heterogeneous manner (Bosco and Meda, 1996). Interestingly, inhibition of intercellular communication has been found to be associated with both an increase in basal release of insulin and a decrease in glucose responsiveness, the two secretory defects characteristic of β-cells in type II diabetes (Portha, 1985). Furthermore, several β-cell lines with abnormal glucose responsiveness do not express Cx43 and are not coupled. After correction of these defects by stable transfection of Cx43, those cell lines show a markedly elevated insulin gene expression and restoration of glucose sensitivity (Vozzi et al., 1995). The opposite occurs in the case of the enzyme-releasing exocrine cells. Communication appears to depress the secretory efficiency and stimulation with many secretagogues results in the inhibition of gap junction channels (Iwatsuki and Petersen, 1978; Neyton and Trautmann, 1986; Bruzzone et al., 1987; Meda et al., 1987). Why do exocrine and endocrine cells respond so differently? Basal coupling between β-cells is relatively poor. They use electrical excitability and increase in coupling levels, therefore, to recruit more actively secreting cells. On the other hand, exocrine cells are normally very well coupled and thus need to reduce, rather than expand, the number of cells that respond to a given stimulus so that appropriate levels of enzymes are released (see for a review: Bruzzone, 1990; Chanson and Spray, 1995).

Accumulating evidence indicates that calcium is part of the intercellular signaling pathway. Previous studies have suggested that both calcium and inositol trisphosphate pass through gap junction channels (Sáez et al., 1989a; Sanderson et al., 1990; Charles et al., 1991; Christ et al., 1992; Sandberg et al., 1992; Xia and Ferrier, 1992; Takeda et al., 1995). Moreover, calcium waves that spread across communicating cells and coordination of calcium oscillations are exquisitely dependent on the permeability of intercellular channels (Charles et al., 1991; Boitano et al., 1992; Stauffer et al., 1993). It appears that movement of inositol trisphosphate, rather than calcium itself, diffusing from activated cells represents the molecular basis of this phenomenon (Boitano et al., 1992; Sandberg et al., 1992; Charles et al., 1993; Sneyd et al., 1994; Hansen et al., 1995). Calcium waves among hepatocytes have now been shown to occur in the intact liver, following vasopressin stimulation (Nathanson et al., 1995; Robb-Gaspers and Thomas, 1995). Interestingly, the direction of calcium waves is modulated by the vasopressin concentration. Thus, at low doses, waves initiate from the periportal region and propagate across the hepatic lobule to the pericentral region. In contrast, at high vasopressin doses, the direction is reversed, initiating in the pericentral region and spreading to the periportal region (Robb-Gaspers and Thomas, 1995). The coordination of this signaling pathway is likely to have important consequences for liver function. Calcium waves may contribute to the secretion and direct the flow of bile. Furthermore, because calcium transients stimulate glycogenolysis and gluconeogenesis, the spatial distribution of waves may permit recruitment of cells that are not directly stimulated by vasopressin (Nathanson et al., 1995; Robb-Gaspers and Thomas, 1995).

A series of studies on the chief cell components of the central nervous system (neurons, astrocytes and oligodendrocytes) is providing new insight into the physiological role of intercellular channels and the molecular nature of the signals involved. Inhibition of neuronal growth and communication of neuronal death during development of anterior pagoda neurons in the leech is signaled by calcium waves that cross gap junction channels (Wolszon et al., 1994a,b). Since these cells are not endowed with inositol-trisphosphate-sensitive calcium stores, it is reasonable to propose that direct passage of calcium through intercellular channels signals cell death. In addition, opening of normally closed intercellular channels among neurons may explain the long-distance synchronization that occurs during spreading depression, a condition characterized by extensive neuronal depolarization and prolonged hypoexcitability (Herreras et al., 1994). In keeping with this hypothesis, propagation of a wave of spreading depression is inhibited by agents that depress coupling but not those affecting neuronal excitability (Herreras et al., 1994). Finally, electrical coupling may serve as an amplifier that transduces weaker inputs into larger responses by synchronizing the population of already active fibers and promoting the recruitment of new ones (Pereda et al., 1995).

Glial cells provide another interesting experimental model to study the role of intercellular channels. The finding that astrocytes and oligodendrocytes express distinct connexins in a developmentally and regionally regulated fashion suggests that communication exhibits some degree of specificity in glia (Dermietzel et al., 1991; Giaume et al., 1991b; Sontheimer et al., 1991; Batter et al., 1992; von Blankenfeld et al., 1993; Belliveau and Naus, 1994, 1995; Lee et al., 1994; Giaume and Venance, 1995; Venance et al., 1995a). Astrocytes have been classically assigned a metabolic and trophic role to support neurons in a sort of servant/master relationship (Bignami and Dahl, 1994). Thus, the need for intercellular coupling among astrocytes has been interpreted in the context of removal and buffering of extracellular potassium in order to maintain neuronal excitability (Newmann, 1985; Mugnaini, 1986) as well as supply of glucose metabolites (Giaume and Venance, 1995). However, recent studies suggest that astrocytes may have communication skills that complement those of neurons. Glutamate application as well as release from glutamatergic neurons induce calcium transients that are propagated through intercellular channels between astrocytes (Cornell-Bell et al., 1990; Charles et al., 1991). These studies suggest that, by releasing glutamate, neurons can trigger a signaling response that activates groups of astrocytes. Astrocytes in turn can excite neurons by sending a signal through intercellular channels that leads to increased neuronal calcium levels (Charles, 1994; Nedergard, 1994). Since modulation of neuronal calcium has both short- and long-term effects on excitability, the implications of this bi-directional signaling are that astrocytes may directly participate in information processing in the central nervous system. These novel pathways of glial-neuronal communication may also include oligodendrocytes, the myelin-forming cells of the central nervous system. Oligodendrocytes are coupled to each other and to astrocytes in culture (Kettenmann et al., 1983; Ransom and Kettenmann, 1990; Robinson et al., 1993; von Blankenfeld et al., 1993; Venance et al., 1995a), but the identity of the connexins involved and the role intercellular channels could play in oligodendrocyte homeostasis remain poorly understood.

The critical function of connexins in Schwann cells, that form myelin in the peripheral nervous system, has been clearly demonstrated by the association of mutations in Cx32 with the X-linked form of Charcot-Marie-Tooth disease (Bergoffen et al., 1993; Fairweather et al., 1994; Ionasescu et al., 1994, 1995; Bone et al., 1995). Charcot-Marie-Tooth disease (CMT) is the most common inherited peripheral neuropathy, with an estimated incidence of about 1:2500. From the clinical standpoint, the disease is characterized by progressive weakness of the distal leg and intrinsic hand muscles, development of *pes cavus*, absent or diminished deep tendon reflexes and variable sensory loss (Harding, 1995). The major form of CMT (CMT1) is a demyelinating neuropathy associated with decreased nerve conduction velocity. CMT1 is genetically heterogeneous, with a similar dominantly inherited disease produced by defects of the gene encoding PMP22 (CMT1A), the gene encoding P_0 (CMT1B), or the gene encoding Cx32 (CMTX) (see for a review: Patel, 1993; Chance and Fischbeck, 1994; Suter and Snipes, 1995). A total of 42 different Cx32 mutations in CMTX families have been reported so far (Deschênes et al., 1996). Analysis of the Cx32 gene in patients with CMTX shows that mutations are distributed throughout the molecule, with all domains affected except the fourth transmembrane domain (Fig. 7). The association of mutations in a connexin gene with a demyelinating disorder was unexpected, since morphologically recognizable gap junctions are extremely rare between myelinating Schwann cells, or during Wallerian degeneration and subsequent nerve regeneration (Tetzlaff, 1982). In keeping with these observations, the presence of Cx32 protein in Schwann cells has not been demonstrated at the sites of cell-to-cell contact but, rather, at the paranodal regions and Schmidt-Lantermann incisures (Bergoffen et al., 1993). Such a distribution is incompatible with the formation of intercellular channels between Schwann cells as in orthodox gap junctions, but suggests that Cx32 forms channels within an individual Schwann cell between turns of myelin (Fig. 8). This would greatly shorten the diffusion pathway for the transfer of ions, second messengers and metabolites from the perinuclear to the periaxonal region of Schwann cells, through the compact myelin wraps. How do Cx32 mutations lead to the development of CMTX? It has been speculated that mutations in Cx32 interfere with the diffusion of messenger molecules and nutrients (Bergoffen et al., 1993; Bruzzone et al., 1994b; Deschênes et al., 1996; Paul, 1995). Although the same signals may travel through the cytoplasmic spirals, this pathways is most likely several orders of magnitude slower than the putative radial pathway through intracellular junctions. Potentially, this could interfere with the ability of Schwann cells to respond to the normal glial–neuron interactions that are critical for the proper maintenance of myelin sheaths (Doyle and Colman, 1993). To determine the functional consequence of Cx32 mutation, four mutants have been studied in the paired *Xenopus* oocyte assay. In three cases, two missense mutations (Arg142→Trp and Glu186→Lys) and a frameshift at position 175 resulting in a premature stop codon, a complete loss of channel activity is observed. Since the constructs are efficiently synthesized and normally targeted to the site of cell apposition, these mutations most likely affect channel docking or gating but not intercellular trafficking (Bruzzone et al., 1994b). In contrast, the fourth mutant, a deletion of most of the C-terminal domain, forms channels with voltage and pH gating properties indistinguishable from those of wild-type (Rabadan-Diehl et al., 1994). Preliminary evidence suggests that this mutation is also functional in transfected, cultured mammalian cells, as assessed by dye coupling (Omori et al., 1995). Thus, the carboxyl tail may control a critical function (for example: size selectivity, permeability) that is not related either to the establishment of communication or to basic gating properties. Testing the remaining mutations may prove useful to understand the pathophysiological basis of the disease and reveal new aspects of the structure/function features of connexins. One of the most perplexing issues of CMTX is the exquisite restriction of the phenotype associated with Cx32 mutations, in view of the wide expression pattern of this connexin in many cells, including

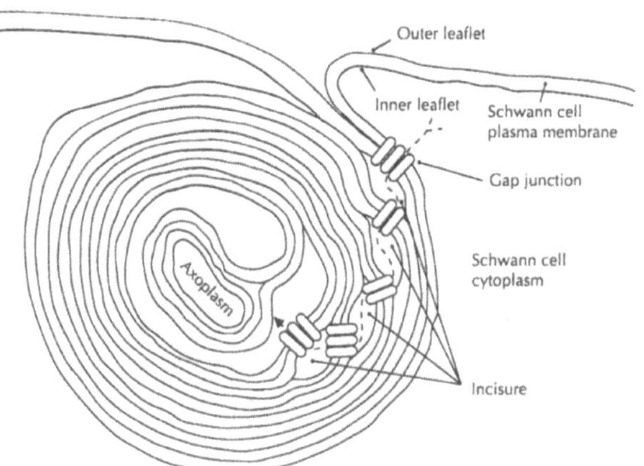

Fig. 8. Schematic representation of the hypothetical role of Cx32 as intracellular channel in myelinating Schwann cells. Cx32 is excluded from compact myelin and is mainly localized to the paranodal regions and incisures of Schmidt-Lantermann. Connexons composed of Cx32 could pair to form a complete channel between adjacent myelin wraps and create preferential communicating domains within the cytosol of a Schwann cell. Thus, these channels would provide a greatly shortened diffusional pathway (dashed lines) for the transfer of ions, second messengers and metabolites from the perinuclear to the periaxonal region of Schwann cells, through the compact myelin layers. Reproduced from Paul (1995) by copyright permission of Current Biology Ltd.

oligodendrocytes (Paul et al., 1993; Spray and Dermietzel, 1995). In the myelin-forming cells of the peripheral nervous system the expression of Cx32 message and protein is developmentally regulated in parallel to that of other myelin-related genes, suggesting the presence of common transcriptional mechanisms (Scherer et al., 1996). It is tempting to speculate that in cells other than Schwann cells the presence of alternative connexins may compensate for the loss of function due to Cx32 mutations. This may imply that myelinating Schwann cells either express only Cx32 or that the other connexins are unable to substitute Cx32 because of different physiological properties or targeting to distinct plasma membrane areas. Analysis of rat peripheral nerve by reverse-transcriptase polymerase chain reaction indicates the presence of transcripts for Cx26, Cx32, Cx33, Cx37 and Cx46 whereas Cx30.3, Cx31, Cx31.1 and Cx40 were not detected (Bone et al., 1996). In addition, cultured, non-myelinating Schwann cells express connexins with biophysical properties that distinguish them from Cx32 (Chanson et al., 1993) under conditions in which Cx32 levels are extremely low (Chandross et al., 1995). Whether the same situation occurs *in vivo*, in myelinating cells, has not been demonstrated. At present one cannot formally exclude the alternative explanation that Cx32 is only essential for communication in Schwann cells. The critical evaluation of these possibilities requires the establishment of an animal model that recreates the phenotype of CMTX. Cx32 knockout mice have been generated, but are phenotypically normal and fertile under caged conditions (Willecke, K., personal communication). It remains possible that more subtle abnormalities and a demyelinating condition will be revealed by longer-term observations. In addition, some CMTX Cx32 mutants exhibit dominant negative activity when coexpressed in oocyte pairs with Cx26, a connexin that is often present in combination with Cx32 *in vivo* (Bruzzone et al., 1994b) and preliminary evidence suggests that three other missense mutations exert dominant negative inhibition in transfected HeLa cells (Omori et al., 1995). As a knockout is a null mutation that eliminates the gene product, reproducing the dominant negative phenotype may require the generation of transgenic animals expressing *bona fide* CMTX mutations.

Perspectives and conclusions

Exciting progress during the past few years has allowed a giant leap in our understanding of the elementary properties of intercellular channels. This knowledge has lead to new hypotheses about the roles of connexins in tissue homeostasis. While it is not clear whether all mammalian connexin genes have been found, we anxiously await the identification of connexin homologues in invertebrates to move beyond pure electrophysiological studies (Verselis and Bargiello, 1991; Gho, 1994). In the meantime, genetic studies may start in zebrafish, where suitable experimental models and members of the connexin family are being characterized (McMahon, 1994; McMahon and Brown, 1994; Essner et al., 1995; Marics et al., 1995). Crystallographic studies of the channel should solve many of the current structural and biophysical questions, such as the location of the pore and of inter-connexon interactions and conformational changes associated with different gating mechanisms. As it has been the case for the nicotinic acetylcholine receptor (Beroukhim and Unwin, 1995), comparison of the open and closed states will provide molecular insight into the mechanisms of channel gating and may allow the development of a rational drug design to target specifically connexin channels. In most cases, voltage-gated ion channels are expressed as a specific protein complex containing the principal, channel-forming, subunit associated with one or more ancillary subunits that can dramatically influence the resulting gating behavior (Isom et al., 1994). This has not been shown for intercellular channels, although initial evidence suggests that interactions between connexins and other proteins, such as small GTP-binding proteins, bFGF and calmodulin, may occur (Hertzberg and Van Eldik, 1987; Kardami et al., 1991; Doucet et al., 1992). Could this represent an additional mechanism to selectively modify the physiological properties of connexins? We need also to test whether expression of incompatible connexins are the molecular basis of communication compartments. Progress will be accomplished by selectively altering the communication properties of defined groups of cells (Goodenough and Musil, 1993). Successful approaches have relied on the use of specific antibodies, which may probe the functional and regulatory role of different domains of the channel protein (Lal et al., 1993; Becker et al., 1995; Bastide et al., 1996), as well as antisense oligonucleotides (Bevilacqua et al., 1989; Moore and Burt, 1994) and dominant negative mutants (Paul et al., 1995; Sullivan and Lo, 1995). A more straightforward strategy to perturb intercellular communication *in vivo* is to generate knockout mice for the various connexins. The first three null mutations of connexin genes obtained thus far exhibit three distinct phenotypes. Ablation of the Cx43 gene results in cardiac malformations and perinatal death (Reaume et al., 1995), Cx26 knockout is embryonically lethal at day 11 post-coitus, whereas homozygous Cx32 knockout mice appear normal and fertile (Willecke, K., personal communication). Specific hypotheses can be tested by producing conditional transgenes that either express dominant negative mutants in order to disrupt signaling loops, or create illicit communications between normally separated compartments. The resulting phenotypes should provide a much needed framework to re-interpret connexin functional roles. Clearly, the most significant discovery in recent years has been the association of connexin defects with genetic and congenital diseases. It is worth exploring the possibility that malfunctioning of intercellular channels may contribute to other pathological conditions, such as synchronization of neuronal firing during seizure-like events (Naus et al., 1991; Perez-Velazquez et al., 1994; Lee et al., 1995) or arrhythmia in Chagas' disease (Campos de Carvalho et al., 1994). In the world of cellular networks, we have begun to understand the hardware, represented by the connexins and their basic language of compatibility and selectivity. Much work is ahead to decipher the various software used, that is the nature of the exchanged messages.

Special thanks are due to Dan Goodenough for his continuous support, insight and interest in our work, to Yan-Hua Chen, Joe Gabriels, Jeff Goliger, Jean Jiang, Linda Musil and Alex Simon for many stimulating discussions and to Danielle Gomès for help with the preparation of Fig. 2. We express our gratitude to Luis Barrio, Eric Beyer, Linda Bone, Marc Chanson, Bob DeHaan, Mario Delmar, Sue Deschênes, Cynthia Foote, Christian Giaume, Daniel Gros, Ross Johnson, Jerry Kidder, Giovanna Mazzoleni, Marc Mesnil, Paolo Meda, Christian Naus, Bruce Nicholson, Camillo Peracchia, Magali Théveniau-Ruissy, Laurent Venance, Cristina Vozzi, Robert Weingart and Klaus Willecke for communicating their results prior to publication. Because an exhaustive bibliographic survey of the connexin field was beyond the scope of this article, we have chosen to focus on the most recent papers and apologize to those colleagues whose contributions could not be discussed. The authors are supported by grants from the *Institut Pasteur*, the *Association française contre les myopathies*, the National Institutes of Health (GM-37551) and the *Fonds national suisse de la recherche scientifique* (32-43'086.95).

REFERENCES

Armstrong, C. M. (1992) *Physiol. Rev. 72* (*suppl.*), S5–S13.

Atkinson, M. M., Menko, A. S., Johnson, R. G., Sheppard, J. R. & Sheridan, J. D. (1981) *J. Cell Biol. 91*, 573–578.

Atkinson, M. M. & Sheridan, J. D. (1988) *Am. J. Physiol. 255*, C674–C683.

Atkinson, M. M., Lampe, P. D., Lin, H. H., Kollander, R., Li, X.-R. & Kiang, D. T. (1995) *J. Cell Sci. 108*, 3079–3090.

Auerbach, A. A. & Bennett, M. V. L. (1969) *J. Gen. Physiol. 53*, 211–237.

Axelrod, J., Burch, R. M. & Jelsema, C. L. (1988) *Trends Neurosci. 11*, 117–123.

Azarnia, R., Reddy, S., Kmiecik, T. E., Shalloway, D. & Lowenstein, W. R. (1988) *Science 239*, 398–401.

Bai, S., Spray, D. C. & Burk, R. D. (1993) *Biochim. Biophys. Acta 1216*, 197–204.

Bai, S., Schoenfeld, A., Pietrangelo, A. & Burk, R. D. (1995) *Mol. Cell. Biol. 15*, 1439–1445.

Barnes, T. M. (1994) *Trends Genet. 10*, 303–305.

Barrio, L. C., Suchyna, T., Bargiello, T., Xu, L. X., Roginski, R. S., Bennett, M. V. L. & Nicholson, B. J. (1991) *Proc. Natl Acad. Sci. USA 88*, 8410–8414.

Bastide, B., Neyses, L., Ganten, D., Paul, M., Willecke, K. & Traub, O. (1993) *Circ. Res. 73*, 1138–1149.

Bastide, B., Jarry-Guichard, T., Briand, J. P., Délèze, J. & Gros, D. (1996) *J. Membr. Biol.*, in the press.

Batter, D. K., Corpina, R. A., Roy, C., Spray, D. C., Hertzberg, E. L. & Kessler, J. A. (1992) *Glia 6*, 213–221.

Becker, D. L., Evans, W. H., Green, C. R. & Warner, A. (1995) *J. Cell Sci. 108*, 1455–1467.

Becker, K. F., Atkinson, M. J., Reich, U., Becker, I., Nekarda, H., Siewert, J. R. & Höfler, H. (1994) *Cancer Res. 54*, 3845–3852.

Belliveau, D. J. & Naus, C. C. G. (1994) *Glia 12*, 24–34.

Belliveau, D. J. & Naus, C. C. G. (1995) *Dev. Neurosci. 17*, 81–96.

Benedetti, E. L. & Emmelot, P. (1965) *J. Cell Biol. 26*, 299–305.

Bennett, M. V. L. & Goodenough, D. A. (1978) *Neurosci. Res. Prog. Bull. 16*, 373–486.

Bennett, M. V. L., Barrio, L. C., Bargiello, T. A., Spray, D. C., Hertzberg, E. & Sáez, J. C. (1991) *Neuron 6*, 305–320.

Bennett, M. V. L. & Verselis, V. K. (1992) *Semin. Cell Biol. 3*, 29–47.

Bény, J.-L. & Connat, J.-L. (1992) *Circ. Res. 70*, 49–55.

Bergoffen, J., Scherer, S. S., Wang, S., Oronzi Scott, M., Bone, L. J., Paul, D. L., Chen, K., Lensch, M. W., Chance, P. F. & Fischbeck, K. H. (1993) *Science 262*, 2039–2042.

Beroukhim, R. & Unwin, N. (1995) *Neuron 15*, 323–331.

Berthoud, V. M., Rook, M. B., Traub, O., Hertzberg, E. L. & Sáez, J. C. (1993) *Eur. J. Cell Biol. 62*, 384–396.
Berthoud, V. M., Cook, A. J. & Beyer, E. C. (1994) *Invest. Ophthalmol. Vis. Sci. 35*, 4109–4117.
Bevilacqua, A., Loch-Caruso, R. & Erickson, R. P. (1989) *Proc. Natl Acad. Sci. USA 86*, 5444–5448.
Beyer, E. C., Paul, D. L. & Goodenough, D. A. (1987) *J. Cell Biol. 105*, 2621–2629.
Beyer, E. C., Kistler, J., Paul, D. L. & Goodenough, D. A. (1989) *J. Cell Biol. 108*, 595–605.
Beyer, E. C. (1990) *J. Biol. Chem. 265*, 14439–14443.
Beyer, E. C., Paul, D. L. & Goodenough, D. A. (1990) *J. Membr. Biol. 116*, 187–194.
Beyer, E. C. & Steinberg, T. H. (1991) *J. Biol. Chem. 266*, 7971–7974.
Beyer, E. C. & Veenstra, R. D. (1994) in *Handbook of membrane channels* (Peracchia, C., ed.) pp. 379–401, Academic Press, London.
Bignami, A. & Dahl, D. (1994) *Glial cells in the central nervous system and their reaction to injury*, R. G. Landes Co., Austin TX.
Birchmeier, W. (1995) *BioEssays 17*, 97–99.
Blackburn, J. P., Peters, N. S., Yeh, H. I., Rothery, S., Green, C. R. & Severs, N. J. (1995) *Arterioscler. Thromb. Vasc. Biol. 15*, 1219–1228.
Blackshaw, S. E. & Warner, A. E. (1976) *J. Physiol. (Lond.) 255*, 209–230.
Blennerhasset, M. G. & Caveney, S. (1984) *Nature 309*, 361–364.
Boitano, S., Dirksen, E. R. & Sanderson, M. J. (1992) *Science 258*, 292–295.
Bone, L. J., Dahl, N., Lensch, M. W., Chance, P. F., Kelly, T., Le Guern, E., Magi, S., Parry, G., Shapiro, H., Wang, S. & Fischbeck, K. H. (1995) *Neurology 45*, 1863–1866.
Bone, L. J., Scherer, S. S., Paul, D. L., Fischbeck, K. H. & Balice-Gordon, R. J. (1996) *J. Cell. Biochem.*, in the press.
Bosco, D. & Meda, P. (1996) *Adv. Exp. Med. Biol.*, in the press.
Brink, P. R. & Dewey, M. M. (1980) *Nature 285*, 101–102.
Brissette, J. L., Kumar, N. M., Gilula, N. B. & Dotto, G. P. (1991) *Mol. Cell. Biol. 11*, 5364–5371.
Brissette, J. L., Kumar, N. M., Gilula, N. B., Hall, J. E. & Dotto, G. P. (1994) *Proc. Natl Acad. Sci. USA 91*, 6453–6457.
Britz-Cunningham, S. H., Shah, M. M., Zuppan, C. W. & Fletcher, W. H. (1995) *N. Engl. J. Med. 332*, 1323–1329.
Bruzzone, R., Trimble, E. R., Gjinovci, A., Traub, O., Willecke, K. & Meda P. (1987) *Pancreas 2*, 262–271.
Bruzzone, R. & Meda, P. (1988) *Eur. J. Clin. Invest. 18*, 444–453.
Bruzzone, R. (1990) *Gastroenterology 99*, 1157–1176.
Bruzzone, R., Haefliger, J.-A., Gimlich, R. L. & Paul, D. L. (1993) *Mol. Biol. Cell 4*, 7–20.
Bruzzone, R., White, T. W. & Paul, D. L. (1994a) *J. Cell Sci. 107*, 955–967.
Bruzzone, R., White, T. W., Scherer, S. S., Fischbeck, K. H. & Paul, D. L. (1994b) *Neuron 13*, 1253–1260.
Bruzzone, R., White, T. W., Yoshizaki, G., Patiño, R. & Paul, D. L. (1995) *FEBS Lett. 358*, 301–304.
Buehler, L. K., Stauffer, K. A., Gilula, N. B. & Kumar, N. M. (1995) *Biophys. J. 68*, 1767–1775.
Bukauskas, F. F., Elfgang, C., Willecke, K & Weingart, R. (1995a) *Biophys. J. 68*, 2289–2298.
Bukauskas, F. F., Elfgang, C., Willecke, K. & Weingart, R. (1995b) *Pflügers Arch. 429*, 870–872.
Burt, J. M. (1989) *Am. J. Physiol. 256*, C913–C924.
Burt, J. M. (1991) in *Biophysics of gap junctions* (Peracchia, C., ed.) pp. 75–93, CRC Press, Boca Raton FL.
Busa, W. B. & Gimlich, R. L. (1989) *Dev. Biol. 132*, 315–324.
Butterweck, A., Elfgang, C., Willecke, K. & Traub, O. (1994a) *Eur. J. Cell Biol. 65*, 152–163.
Butterweck, A., Gergs, U., Elfgang, C., Willecke, K. & Traub, O. (1994b) *J. Membr. Biol. 141*, 247–256.
Buultjens, T. E. J., Finbow, M. E., Lane, N. J. & Pitts, J. D. (1988) *Cell Tissue Res. 251*, 571–580.
Campos de Carvalho, A. C., Masuda, M. O., Tanowitz, H. B., Wittner, M., Goldenberg, R. C. S. & Spray, D. C. (1994) *J. Cardiovasc. Electrophysiol. 5*, 686–698.
Cao, F., Eckert, R., Elfgang, C., Hülser, D., Willecke, K. & Nicholson, B. (1995) in *Proceedings of the 1995 Gap Junction Conference*, Île des Embiez, France.
Cascio, M., Kumar, N. M., Safarik, R. & Gilula, N. B. (1995) *J. Biol. Chem. 270*, 18643–18648.
Caspar, D. L. D., Sosinsky, G. E., Tibbitts, T. T., Phillips, W. C. & Goodenough, D. A. (1988) *Modern Cell Biol. 7*, 117–133.
Catterall, W. A. (1995) *Annu. Rev. Biochem. 64*, 493–531.
Caveney, S. (1985) *Annu. Rev. Physiol. 47*, 319–335.
Chalcroft, J. P. & Bullivant, S. (1970) *J. Cell Biol. 47*, 49–60.
Chance, P. F. & Fischbeck, K. H. (1994) *Hum. Mol. Genet. 3*, 1503–1507.
Chandross, K. J., Chanson, M., Spray, D. C. & Kessler, J. A. (1995) *J. Neurosci. 15*, 262–273.
Chang, C. C., Trosko, J. E., Kung, H. J., Bombick, D. & Matsumura, F. (1985) *Proc. Natl Acad. Sci. USA 82*, 5360–5364.
Chanson, M., Bruzzone, R., Spray, D. C., Regazzi, R. & Meda, P (1988) *Am. J. Physiol. 255*, C609–C704.
Chanson, M., Orci, L. & Meda, P. (1991) *Am. J. Physiol. 261*, G28–G36.
Chanson, M., Chandross, K. J., Rook, M. B., Kessler, J. A. & Spray, D. C. (1993) *J. Gen. Physiol. 102*, 925–946.
Chanson, M. & Spray, D. C. (1995) in *Pacemaker activity and intercellular communication* (Huizinga, J. D., ed.) pp. 51–72, CRC Press, Boca Raton FL.
Charles, A. C., Merrill, J. E., Dirksen, E. R. & Sanderson, M. J. (1991) *Neuron 6*, 983–992.
Charles, A. C., Dirksen, E. R., Merrill, J. E. & Sanderson, M. J. (1993) *Glia 7*, 134–145.
Charles, A. C. (1994) *Dev. Neurosci. 16*, 196–206.
Chen, S. C., Pelletier, D. B., Ao, P. & Boynton, A. L. (1995) *Cell Growth Diff. 6*, 681–690.
Chen, Y.-H. & DeHaan, R. L. (1992) *J. Membr. Biol. 127*, 95–111.
Chen, Y.-H. & DeHaan, R. L. (1993a) *J. Membr. Biol. 136*, 125–134.
Chen, Y.-H. & DeHaan, R. L. (1993b) *Prog. Cell Res. 3*, 97–103.
Chen, Y.-H. & DeHaan, R. L. (1996) *Am. J. Physiol. 270*, C276–C285.
Chen, Z.-Q., Lefebvre, D., Bai, X.-H., Reaume, A., Rossant. J. & Lye, S. J. (1995) *J. Biol. Chem. 270*, 3863–3868.
Chow, I. & Poo, M.-m. (1984) *J. Physiol. (Lond.) 346*, 181–194.
Christ, G. J., Moreno, A. P., Melman, A. & Spray, D. C. (1992) *Am. J. Physiol. 263*, C373–C383.
Church, R. L., Wang, J. H. & Steele, E. (1995) *Curr. Eye Res. 14*, 215–221.
Churchill, D., Coodin, S., Shivers, R. R. & Caveney, S. (1993) *J. Cell Sci. 104*, 763–772.
Cornell-Bell, A. H., Finkbeiner, S. M., Cooper, M. S. & Smith, S. J. (1990) *Science 247*, 470–473.
Crow, D. S., Beyer, E. C., Paul, D. L., Kobe, S. S. & Lau, A. F. (1990) *Mol. Cell. Biol. 10*, 1754–1763.
Crow, J. M., Atkinson, M. M. & Johnson, R. G. (1994) *Invest. Ophtalmol. Vis. Sci. 35*, 3332–3341.
Dahl, E., Winterhager, E., Traub, O. & Willecke, K. (1995) *Anat. Embryol. 191*, 267–278.
Dahl, E., Winterhager, E., Reuss, B., Traub, O., Butterweck, A. & Willecke, K. (1996) *J. Cell Sci 109*, 191–197.
Dahl, G. & Isenberg, G. (1980) *J. Membr. Biol. 53*, 63–75.
Dahl, G., Miller, T., Paul, D., Voellmy, R. & Werner, R. (1987) *Science 236*, 1290–1293.
Dahl, G., Levine, E., Rabadan-Diehl, C. & Werner, R. (1991) *Eur. J. Biochem. 197*, 141–144.
Dahl, G., Werner, R., Levine, E. & Rabadan-Diehl, C. (1992) *Biophys. J. 62*, 172–182.
Dahl, G., Nonner, W. & Werner, R. (1994) *Biophys. J. 67*, 1816–1822.
Davidson, J. S. & Baumgarten, I. M. (1988) *J. Pharmacol. Exp. Ther. 246*, 1104–1107.
Davies, T. C., Barr, K. J., Jones, D. H., Zhu, D. & Kidder, G. M. (1996) *Dev. Genet.*, in the press.
Dealy, C. N., Beyer, E. C. & Kosher, R. A. (1994) *Dev. Dyn. 199*, 156–167.
de Laat, S. W., Tertoolen, L. G. J., Dorresteijn, A. W. C. & van den Biggelaar, J. A. M. (1980) *Nature 287*, 546–548.
De Leon, J. R., Buttrick, P. M. & Fishman, G. I. (1994) *J. Mol. Cell. Cardiol. 26*, 379–389.
Délèze, J. & Hervé, J. C. (1983) *J. Membr. Biol. 74*, 203–215.
Delorme, B., Dahl, E., Jarry-Guichard, T., Marics, I., Briand, J.-P., Willecke, K., Gros, D. & Théveniau-Ruissy, M. (1995) *Dev. Dyn. 204*, 358–371.

De Mello, W. C. (1986) *Biochim. Biophys. Acta 888*, 71–79.

De Mello, W. C. (1994) *Prog. Biophys. Mol. Biol. 61*, 1–35.

Dermietzel, R., Traub, O., Hwang, T. K., Beyer, E., Bennett, M. V. L., Spray, D. C. & Willecke, K. (1989) *Proc. Natl Acad. Sci. USA 86*, 10148–10152.

Dermietzel, R., Hertzberg, E. L., Kessler, J. A. & Spray, D. C. (1991) *J. Neurosci. 11*, 1421–1432.

Dermietzel, R. & Spray, D. C. (1993) *Trends Neurosci. 16*, 186–192.

Deschênes, S. M., Bone, L. J., Fischbeck, K. H. & Scherer, S. S. (1996) in *The biology of gap junctions* (Spray, D. C. & Dermietzel, R., eds) R. G. Landes Co., Austin TX, in the press.

Detrick, R. J., Dickey, D. & Kintner, C. R. (1990) *Neuron 4*, 493–506.

DeVries, S. H. & Schwartz, E. A. (1992) *J. Physiol. (Lond.) 445*, 201–230.

Doble, B. W. & Kardami, E. (1995) *Mol. Cell. Biochem. 143*, 81–87.

Donaldson, P. J., Roos, M., Evans, E., Beyer, E. & Kistler, J. (1994) *Invest. Ophtalmol. Vis. Sci.* 35, 3422–3428.

Donaldson, P. J., Dong, Y., Roos, M., Green, C., Goodenough, D. A. & Kistler, J. (1995) *Am. J. Physiol. 269*, C590–C600.

Dotto, G. P., El-Fouly, M., Nelson, C. & Trosko, J. E. (1989) *Oncogene 4*, 637–641.

Doucet, J.-P., Pierce, G. N., Hertzberg, E. L. & Tuana, B. S. (1992) *J. Biol. Chem. 267*, 16503–16508.

Doyle, J. P. & Colman, D. R. (1993) *Curr. Opin. Neurobiol. 5*, 779–785.

Dunham, B., Liu, S., Taffet, E., Trabka-Janik, E., Delmar, M., Petryshyn, R., Zheng, S., Perzova, R. & Vallano, M. L. (1992) *Circ. Res. 70*, 1233–1243.

Dunlap, K., Takeda, K. & Brehm, P. (1987) *Nature 325*, 60–62.

Dupont, E., El Aoumari, A., Briand, J. P., Fromaget, C. & Gros, D. (1989) *J. Membr. Biol. 108*, 247–252.

Dupont, E., El Aoumari, A., Fromaget, C., Briand, J.-P. & Gros, D. (1991) *Eur. J. Biochem. 200*, 263–270.

Ebihara, L., Beyer, E. C., Swenson, K. I., Paul, D. L. & Goodenough, D. A. (1989) *Science 243*, 1194–1195.

Ebihara, L. & Steiner, E. (1993) *J. Gen. Physiol. 102*, 59–74.

Ebihara, L., Berthoud, V. M. & Beyer, E. C. (1995) *Biophys. J. 68*, 1796–1803.

Eckert, R., Dunina-Barkovskaya, A. & Hülser, D. F. (1993) *Pflügers Arch. 424*, 335–342.

Eghbali, B., Kessler, J. A. & Spray, D. C. (1990) *Proc. Natl Acad. Sci. USA 87*, 1328–1331.

Eghbali, B., Kessler, J. A., Reid, L. M., Roy, C. & Spray, D. C. (1991) *Proc. Natl Acad. Sci. USA 88*, 10701–10705.

Ek, J. F., Delmar, M., Perzova, R. & Taffet, S. M. (1994) *Circ. Res. 74*, 1058–1064.

Elfgang, C., Eckert, R., Lichtenberg-Fraté, H., Butterweck, A., Traub, O., Klein, R. A., Hülser, D. F. & Willecke, K. (1995) *J. Cell Biol. 129*, 805–817.

el-Sabban, M. E. & Pauli, B. U. (1991) *J. Cell Biol. 115*, 1375–1382.

Enomoto, T. & Yamasaki, H. (1984) *Cancer Res. 44*, 5200–5203.

Epstein, M. L. & Gilula, N. B. (1977) *J. Cell Biol. 75*, 769–787.

Epstein, M. L., Sheridan, J. D. & Johnson, R. G. (1977) *Exp. Cell Res. 104*, 25–30.

Essner, J. J., Johnson, R. G. & Hackett, P. B. (1995) in *Proceedings of the 1995 Gap Junction Conference*, Île des Embiez, France.

Evans, W. H., Carlile, G., Rahman, S. & Torok, K. (1992) *Biochem. Soc. Trans. 20*, 856–861.

Fairweather, N., Bell, C., Cochrane, S., Chelly, J., Wang, S., Mostacciuolo, M. L., Monaco, A. P. & Haites, N. E. (1994) *Hum. Mol. Genet. 3*, 29–34.

Falk, M. M., Kumar, N. M. & Gilula, N. B. (1994) *J. Cell Biol. 127*, 343–355.

Fallon, M. B., Nathanson, M. H., Mennone, A., Saez, J. C., Burgstahler, A. D. & Anderson, J. M. (1995) *Am. J. Physiol. 269*, C1186–C1194.

Fannon, A. M., Sherman, D. L., Ilyina-Gragerova, G., Brophy, P. J., Friedrich, V. L. Jr & Colman, D. R. (1995) *J. Cell Biol. 129*, 189–202.

Fentiman, I., Taylor-Papadimitriou, J. & Stoker, M. (1976) *Nature 264*, 760–762.

Filson, A. J., Azarnia, R., Beyer, E. C., Loewenstein, W. R. & Brugge, J. S. (1990) *Cell Growth & Differ. 1*, 661–668.

Finbow, M. E., Yancey, S. B., Johnson, R. G. & Revel, J.-P. (1980) *Proc. Natl Acad. Sci. USA 77*, 970–974.

Finbow, M. E. & Pitts, J. D. (1993) *J. Cell Sci. 106*, 463–472.

Fisher, R. S. & Micevych, P. E. (1993) *Prog. Cell Res. 3*, 141–148.

Fishman, G. I., Spray, D. C. & Leinwand, L. A. (1990) *J. Cell Biol. 111*, 589–598.

Fishman, G. I., Moreno, A. P., Spray, D. C. & Leinwand, L. A. (1991) *Proc. Natl Acad. Sci. USA 88*, 3525–3529.

Fitzgerald, D. J., Mesnil, M., Oyamada, M., Tsuda, H., Ito, N. & Yamasaki, H. (1989) *J. Cell. Biochem. 41*, 97–102.

Flagg-Newton, J. L., Dahl, G. & Loewenstein, W. R. (1981) *J. Membr. Biol. 63*, 195–121.

Fluri, G. F., Rüdisüli, A., Willi, M., Rohr, S. & Weingart, R. (1990) *Pflügers Arch. 417*, 149–156.

Foote, C. I. & Nicholson, B. J. (1995) in *Proceedings of the 1995 Gap Junction Conference*, Île des Embiez, France.

Fraser, S. E., Green, C. R., Bode, H. R. & Gilula, N. B. (1987) *Science 237*, 49–55.

Fujimoto, K. (1995) *J. Cell Sci. 108*, 3443–3449.

Furshpan, E. J. & Potter, D. D. (1959) *J. Physiol. (Lond.) 145*, 289–235.

Furshpan, E. J. & Potter, D. D. (1968) *Curr. Top. Dev. Biol. 3*, 289–235.

Gabriels, J. E. & Paul, D. L. (1995) in *Proceedings of the 1995 Gap Junction Conference*, Île des Embiez, France.

Garfield, R. E., Sims, S. M. & Daniel, E. E. (1977) *Science 198*, 958–959.

Gho, M. (1994) *J. Physiol. (Lond.) 481*, 371–383.

Ghoshroy, S., Goodenough, D. A. & Sosinsky, G. E. (1995) *J. Membr. Biol. 146*, 15–28.

Giaume, C., Kado, R. T. & Korn, H. (1987) *J. Physiol. (Lond.) 386*, 91–112.

Giaume, C., Randriamampita, C. & Trautmann, A. (1989) *Pflügers Arch. 413*, 273–279.

Giaume, C., Fromaget, C., El Aoumari, A., Cordier, J., Glowinski, J. & Gros, D. (1991a) *Neuron 6*, 133–143.

Giaume, C., Marin, P., Cordier, J., Glowinski, J. & Prémont, J. (1991b) *Proc. Natl Acad. Sci. USA 88*, 5577–5581.

Giaume, C. & Venance, L. (1995) *Persp. Dev. Neurobiol. 2*, 335–345.

Gilula, N. B., Reeves, O. R. & Steinbach, A. (1972) *Nature 235*, 262–265.

Gimlich, R. L., Kumar, N. M. & Gilula, N. B. (1988) *J. Cell Biol. 107*, 1065–1073.

Gimlich, R. L., Kumar, N. M. & Gilula, N. B. (1990) *J. Cell Biol. 110*, 597–605.

Goldberg, G. S. & Lau, A. F. (1993) *Biochem. J. 295*, 735–742.

Goldberg, G. S., Moreno, A. P., Bechberger, J. F., Hearn, S. S., Shivers, R. R., MacPhee, D. J., Zhang, Y.-C. & Naus, C. C. G. (1996) *Exp. Cell Res. 222*, 48–53.

Goliger, J. A. & Paul, D. L. (1994) *Dev. Dyn. 200*, 1–13.

Goliger, J. A. & Paul, D. L. (1995) *Mol. Biol. Cell 6*, 1491–1501.

Goodenough, D. A. & Revel, J.-P. (1970) *J. Cell Biol. 45*, 272–290.

Goodenough, D. A., Paul, D. L. & Jesaitis, L. (1988) *J. Cell Biol. 107*, 1817–1824.

Goodenough, D. A. (1992) *Semin. Cell Biol. 3*, 49–58.

Goodenough, D. A. & Musil, L. S. (1993) *J. Cell Sci. Suppl. 17*, 133–138.

Gourdie, R. G., Green, C. R., Severs, N. J., Anderson, R. H. & Thompson, R. P. (1993) *Circ. Res. 72*, 278–289.

Green, C. R., Bowles, L., Crawley, A. & Tickle, C. (1994) *Dev. Biol. 161*, 12–21.

Gros, D. B., Nicholson, B. J. & Revel, J.-P. (1983) *Cell 35*, 539–549.

Gros, D., Jarry-Guichard, T., Ten Velde, I., de Maziere, A., van Kempen, M. J. A., Davoust, J., Briand, J. P., Moorman, A. F. M. & Jongsma, H. J. (1994) *Circ. Res. 74*, 839–851.

Grümmer, R., Chwalisz, K., Mulholland, J., Traub, O. & Winterhager, E. (1994) *Biol. Reprod. 51*, 1109–1116.

Guerrier, A., Fonlupt, P., Morand, I., Rabilloud, R., Audebet, C., Krutovskikh, Gros, D., Rousset, B. & Munari-Silem, Y. (1995) *J. Cell Sci. 108*, 2609–2617.

Gupta, V. K., Berthoud, V. M., Atal, N., Jarillo, J. A., Barrio, L. C. & Beyer, E. C. (1994) *Invest. Ophthalmol. Vis. Sci. 35*, 3747–3758.

Guthrie, S. C. (1984) *Nature 311*, 149–151.

Guthrie, S. C., Turin, L. & Warner, A. E. (1988) *Development 103*, 769–785.

Guthrie, S. C. & Gilula, N. B. (1989) *Trends Neurosci. 12*, 12–16.
Haefliger, J.-A., Bruzzone, R., Jenkins, N. A., Gilbert, D. J., Copeland, N. J. & Paul, D. L. (1992) *J. Biol. Chem. 267*, 2057–2064.
Hall, J. E., Zampighi, G. A. & Davis, R. M. (eds) (1993) *Prog. Cell Res.*, vol. 3, Elsevier Science Publishers, Amsterdam.
Hansen, M., Boitano, S., Dirksen, E. R. & Sanderson, M. J. (1995) *J. Cell Sci. 108*, 2583–2590.
Harding, A. E. (1995) *Brain 118*, 809–818.
Harris, A. L., Spray, D. C. & Bennett, M. V. L. (1981) *J. Gen. Physiol. 77*, 95–117.
Harris, A. L., Spray, D. C. & Bennett, M. V. L. (1983) *J. Neurosci. 3*, 79–100.
Heasman, J., Ginsberg, D., Geiger, B., Goldstone, K., Pratt, T., Yoshida-Noro, C. & Wylie, C. (1994) *Development 120*, 49–57.
Henderson, D., Eibl, H. & Weber, K. (1979) *J. Mol. Biol. 132*, 193–218.
Hendrix, E. M., Myatt, L., Sellers, S., Russel, P. T. & Larsen, W. J. (1995) *Biol. Reprod. 52*, 547–560.
Hennemann, H., Dahl, E., White, J. B., Schwarz, H.-J., Lalley, P. A., Chang, S., Nicholson, B. J. & Willecke, K. (1992a) *J. Biol. Chem. 267*, 17225–17233.
Hennemann, H., Schwarz, H.-J. & Willecke, K. (1992b) *Eur. J. Cell Biol. 57*, 51–58.
Hennemann, H., Suchyna, T., Lichtenberg-Fraté, H., Jungbluth, S., Dahl, E., Schwarz, H.-J., Nicholson, B. J. & Willecke, K. (1992c) *J. Cell Biol. 117*, 1299–1310.
Herreras, O., Largo, C., Ibarz, J. M., Somjen, G. G. & Martín del Río, R. (1994) *J. Neurosci. 14*, 7087–7098.
Hertlein, B., Butterweck, A., Willecke, K. & Traub, O. (1995) in *Proceedings of the 1995 Gap Junction Conference*, Île des Embiez, France.
Hertzberg, E. L. & Van Eldik, L. J. (1987) *Methods Enzymol. 139*, 445–455.
Hertzberg, E. L., Lawrence, T. S. & Gilula, N. B. (1981) *Annu. Rev. Physiol. 43*, 479–491.
Hertzberg, E. L., Disher, R. M., Tiller, A. A., Zhou, Y. & Cook, R. G. (1988) *J. Biol. Chem. 263*, 19105–19111.
Hertzberg, E. L. & Johnson, R. G. (eds) (1988) *Modern Cell Biol.*, vol. 7, Alan R. Liss, New York.
Heynkes, R., Kozjek, G., Traub, O. & Willecke, K. (1986) *FEBS Lett. 205*, 56–60.
Hille, B. (1992) *Ionic channels of excitable membranes*, Sinauer Associates, Sunderland MA.
Hirschi, K. K., Minnich, B. N., Moore, L. K. & Burt, J. M. (1993) *Am. J. Physiol. 265*, C1517–C1526.
Hobbie, L., Kingsley, D. M., Kozarsky, K. F., Jackman, R. W. & Krieger, M. (1987) *Science 235*, 69–73.
Hoh, J. H., John, S. A. & Revel, J.-P. (1991a) *J. Biol. Chem. 266*, 6524–6531.
Hoh, J., Lal, R., John, S., Revel, J.-P. & Arnsdorf, M. (1991b) *Science 253*, 1405–1408.
Hoh, J. H., Sosinky, G. E., Revel, J.-P. & Hansma, P. K. (1993) *Biophys. J. 65*, 149–163.
Hollmann, M. & Heinemann, S. (1994) *Annu. Rev. Neurosci. 17*, 31–108.
Hossain, M. Z., Zhang, L.-X. & Bertram, J. S. (1993) *Prog. Cell Res. 3*, 301–309.
Hotz-Wagenblatt, A. & Shalloway, D. (1993) *Crit. Rev. Oncogen. 4*, 541–548.
Huizinga, J. D., Liu, L. W. C., Blennerhassett, M. G., Thuneberg, L. & Molleman, A. (1992) *Experientia 48*, 932–941.
Hurtley, S. M. & Helenius, A. (1989) *Annu. Rev. Cell Biol. 5*, 277–307.
Ionasescu, V., Searby, C. & Ionasescu, R. (1994) *Hum. Mol. Genet. 3*, 355–358.
Ionasescu, V., Searby, C., Ionasescu, R. & Meschino, W. (1995) *Neuromusc. Disord. 5*, 297–299.
Isom, L. L., De Jongh, K. S. & Catterall, W. A. (1994) *Neuron 12*, 1183–1194.
Iwatsuki, N. & Petersen, O. H. (1978) *J. Cell Biol. 79*, 533–545.
Jan, L. Y. & Jan, Y. N. (1990) *Trends Neurosci. 13*, 415–419.
Jan, L. Y. & Jan, Y. N. (1994) *Nature 371*, 119–122.
Jara, P. I., Boric, M. P. & Sáez, J. C. (1995) *Proc. Natl Acad. Sci. USA 92*, 7011–7015.
Jarrillo, J. A., Barrio, L. C. & Gimlich, R. L. (1995) *Prog. Cell Res. 4*, 399–402.
Jaslove, S. W. & Brink, P. R. (1986) *Nature 323*, 63–65.
Jiang, J. X., Paul, D. L. & Goodenough, D. A. (1993) *Invest. Ophtalmol. Vis. Sci. 34*, 3558–3565.
Jiang, J. X., White, T. W., Goodenough, D. A. & Paul, D. L. (1994) *Mol. Biol. Cell 5*, 363–373.
Jiang, J. X., White, T. W. & Goodenough, D. A. (1995) *Dev. Biol. 168*, 649–661.
Jiang, J. X. & Goodenough, D. A. (1996) *Proc. Natl Acad. Sci. USA 93*, 1287–1291.
John, S. A. & Revel, J.-P. (1991) *Biochem. Biophys. Res. Commun. 178*, 1312–1318.
Johnson, P. F. & McKnight, S. L. (1989) *Annu. Rev. Biochem. 58*, 799–839.
Johnston, M. F., Simon, S. A. & Ramón, F. (1980) *Nature 286*, 498–500.
Jongen, W. M. F., Fitzgerald, D. J., Asamoto, M., Piccoli, C., Slaga, T. J., Gros, D., Takeichi, M. & Yamasaki, H. (1991) *J. Cell Biol. 114*, 545–555.
Kadle, R., Zhang, J. T. & Nicholson, B. J. (1991) *Mol. Cell. Biol. 11*, 363–369.
Kalimi, G. H. & Lo, C. W. (1988) *J. Cell Biol. 107*, 241–255.
Kalimi, G. H. & Lo, C. W. (1989) *J. Cell Biol. 109*, 3015–3026.
Kam, E. & Hodgins, M. B. (1982) *Development 114*, 389–393.
Kandel, E. R., Schwartz, J. H. & Jessell, T. M. (1991) *Principles of neural science*, Appleton & Lange, Norwalk CT.
Kandler, K. & Katz, L. C. (1995) *Curr. Opin. Neurobiol. 5*, 98–105.
Kanemitsu, M. Y. & Lau, A. F. (1993) *Mol. Biol. Cell 4*, 837–848.
Kanno, Y., Katakoa, K., Shiba, Y. & Shimazu, T. (eds) (1995) *Prog. Cell Res.*, vol. 4, Elsevier Science Publishers, Amsterdam.
Kanter, H. L., Laing, J. G., Beyer, E. C., Green, K. G. & Saffitz, J. E. (1993) *Circ. Res. 73*, 344–350.
Kanter, H. L., Saffitz, J. E., Beyer, E. C. (1994) *J. Mol. Cell. Cardiol. 26*, 861–868.
Kaplan, D. R., Morrison, D. K., Wong, G., McCormick, F. & Williams, L. T. (1990) *Cell 61*, 125–133.
Kardami, E., Stoski, R. M., Doble, B. W., Yamamoto, T., Hertzberg, E. L. & Nagy, J. I. (1991) *J. Biol. Chem. 266*, 19551–19557.
Katz, L. C. (1995) *Semin. Dev. Biol. 6*, 117–125.
Keane, R. W., Mehta, P. P., Rose, B., Honig, L. S., Loewenstein, W. R. & Rutishauer, U. (1988) *J. Cell Biol. 106*, 1307–1319.
Kensler, R. W. & Goodenough, D. A. (1980) *J. Cell Biol. 86*, 755–764.
Kettenmann, H., Orkand, R. K. & Schachner, M. (1983) *J. Neurosci. 3*, 506–516.
Kidder, G. M. (1992) *Dev. Genet. 13*, 319–325.
Kikkawa, U. & Nishizuka, Y. (1986) *Annu. Rev. Cell Biol. 2*, 149–178.
Kintner, C. (1992) *Cell 69*, 225–236.
Kistler, J., Christie, D. & Bullivant, S. (1988) *Nature 331*, 721–723.
Kistler, J., Evans, C., Donaldson, P., Bullivant, S., Bond, J., Eastwood, S., Roos, M., Dong, Y., Gruijters, T. & Engel, A. (1995) *Microsc. Res. Tech. 31*, 347–356.
Kojima, T., Mitaka, T., Paul, D. L., Mori, M. & Mochizuki, Y. (1995) *J. Cell Sci. 108*, 1347–1357.
Konig, N. & Zampighi, G. (1995) *J. Cell Sci. 108*, 3091–3098.
Koval, M., Geist, S. T., Westphale, E. M., Kemendy, A. E., Civitelli, R., Beyer, E. C. & Steinberg, T. H. (1995) *J. Cell Biol. 130*, 987–995.
Kren, B. T., Kumar, N. M., Wang, S. Q., Gilula, N. B. & Steer, C. J. (1993) *J. Cell Biol. 123*, 707–718.
Krutovskikh, V. A., Mazzoleni, G., Mironov, N., Omori, Y., Aguelon, A.-M., Mesnil, M., Berger, F., Partensky, C. & Yamasaki, H. (1994) *Int. J. Cancer 56*, 87–94.
Kumar, N. M. & Gilula, N. B. (1986) *J. Cell Biol. 103*, 767–776.
Kumar, N. M. & Gilula, N. B. (1992) *Semin. Cell Biol. 3*, 3–16.
Kwak, B. R., Hermans, M. M. P., De Jonge, H. R., Lohmann, S. M., Jongsma, H. J. & Chanson, M. (1995a) *Mol. Biol. Cell 6*, 1707–1719.
Kwak, B. R., Sáez, J. C., Wilders, R., Chanson, M., Fishman, G. I., Hertzberg, E. L., Spray, D. C. & Jongsma, H. J. (1995b) *Pflügers Arch. 430*, 770–778.
Kwak, B. R., van Veen, T. A. B., Analbers, L. J. S. & Jongsma, H. J. (1995c) *Exp. Cell Res. 220*, 456–463.
Laing, J. C., Westphale, E. M., Engelmann, G. L. & Beyer, E. C. (1994) *J. Membr. Biol. 139*, 31–40.

Laird, D. W., Puranam, K. L. & Revel, J.-P. (1991) *Biochem. J. 273*, 67–72.

Laird, D. W., Yancey, S. B., Bugga, L. & Revel, J.-P. (1992) *Dev. Dyn. 195*, 153–161.

Laird, D. W., Castillo, M. & Kasprzak, L. (1995) *J. Cell Biol. 131*, 1193–1203.

Lal, R., Laird, D. W. & Revel, J.-P. (1993) *Pflügers Arch. 422*, 449–457.

Lal, R., John, S. A., Laird, D. W. & Arnsdorf, M. F. (1995) *Am. J. Physiol. 268*, C968–C977.

Lampe, P. D. (1994) *J. Cell Biol. 127*, 1895–1905.

Lang, L. M., Beyer, E. C., Schwartz, A. L. & Gitlin, J. D. (1991) *Am. J. Physiol. 260*, E787–E793.

Larsen, W. J. & Wert, S. E. (1988) *Tissue & Cell 20*, 809–848.

Larson, D. M., Haudenschild, C. C. & Beyer, E. C. (1990) *Circ. Res. 66*, 1074–1080.

Lau, A. F., Kanemitsu, M. Y., Kurata, W. E., Danesh, S. & Boynton, A. L. (1992) *Mol. Biol. Cell 3*, 865–874.

Lawrence, T. S., Beers, W. H. & Gilula, N. B. (1978) *Nature 272*, 501–506.

Lazrak, A. & Peracchia, C. (1993) *Biophys. J. 65*, 2002–2012.

Lazrak, A., Peres, A., Giovannardi, S. & Peracchia, C. (1994) *Biophys. J. 67*, 1052–1059.

Ledbetter, M. L. S., Young, G. J. & Wright, E. R. (1986) *Am. J. Physiol. 250*, C306–C313.

Lee, S., Gilula, N. B. & Warner, A. E. (1987) *Cell 51*, 851–860.

Lee, S. H., Kim, W. T., Cornell-Bell, A. H. & Sontheimer, H. (1994) *Glia 11*, 315–325.

Lee, S. H., Magge, S., Spencer, D. D., Sontheimer, H. & Cornell-Bell, A. H. (1995) *Glia 15*, 195–202.

Lee, S. W., Tomasetto, C., Paul, D., Keyomarsi, K. & Sager, R. (1992) *J. Cell Biol. 118*, 1213–1221.

Leube, R. E. (1995) *J. Cell Sci. 108*, 883–894.

Levine, E., Werner, R. & Dahl, G. (1991) *Am. J. Physiol. 261*, C1025–C1032.

Little, T. L., Beyer, E. C. & Duling, B. R. (1995a) *Am. J. Physiol. 268*, H729–H739.

Little, T. L., Xia, J. & Duling, B. R. (1995b) *Circ. Res. 76*, 498–504.

Liu, S., Taffet, S., Stoner, L., Delmar, M., Vallano, M. L. & Jalife, J. (1993) *Biophys. J. 64*, 1422–1433.

Lo, C. W. & Gilula, N. B. (1979) *Cell 18*, 411–422.

Loewenstein, W. R. & Kanno, Y. (1966) *Nature 209*, 1248–1249.

Loewenstein, W. R., Kanno, Y. & Socolar, S. J. (1978) *Nature 274*, 133–136.

Loewenstein, W. R. (1981) *Physiol. Rev. 61*, 829–913.

Loewenstein, W. R. (1985) *Biochem. Soc. Symp. 50*, 43–58.

Loewenstein, W. R. & Rose, B. (1992) *Semin. Cell Biol. 3*, 59–79.

Loo, L. W. M., Berestecky, J. M., Kanemitsu, M. Y. & Lau, A. F. (1995) *J. Biol. Chem. 270*, 12751–12761.

LoTurco, J. J. & Kriegstein, A. R. (1991) *Science 257*, 563–566.

Lye, S. J., Nicholson, B. J., Mascarenhas, M., MacKenzie, L. & Petrocelli, T. (1993) *Endocrinology 132*, 2380–2386.

MacKinnon, R. (1995) *Neuron 14*, 889–892.

Madhukar, B. V., Oh, S. Y., Chang, C. C., Wade, M. & Trosko, J. E. (1989) *Carcinogenesis 10*, 13–20.

Makowski, L., Caspar, D. L. D., Phillips, W. C. & Goodenough, D. A. (1977) *J. Cell Biol. 74*, 629–645.

Makowski, L. (1988) *Adv. Cell Biol. 2*, 119–158.

Malchow, R. P., Qian, H. & Ripps, H. (1993) *J. Neurosci. Res. 35*, 237–245.

Maldonado, P. E., Rose, B. & Lowenstein, W. R. (1988) *J. Membr. Biol. 106*, 203–210.

Malewicz, B., Kumar, V. V., Johnson, R. G. & Baumann, W. J. (1990) *Lipids 25*, 419–427.

Manjunath, C. K., Goings, G. E. & Page, E. (1982) *Biochem. J. 205*, 189–194.

Marics, I., Bricaud, O., Jarillo, J. A., Castro, C., Gros, D. & Barrio, L. C. (1995) in *Proceedings of the Congress of the European Developmental Biology Organisation*, p. 135.

Martinez, S., Geijo, E., Sánchez-Vives, M. V., Puelles, L. & Gallego, R. (1992) *Development 116*, 1069–1076.

Massey, K. D., Minnich, B. N. & Burt, J. M. (1992) *Am. J. Physiol. 263*, C494–C501.

Mazzoleni, G., Camplani, A., Telò, P., Pozzi, A., Tanganelli, S., Elfgang, C., Willecke, K. & Ragnotti, G. (1996) *Comp. Biochem. Physiol.*, in the press.

McMahon, D. G., Knapp, A. G. & Dowling, J. E. (1989) *Proc. Natl Acad. Sci. USA 86*, 7639–7643.

McMahon, D. G. (1994) *J. Neurosci. 14*, 1722–1734.

McMahon, D. G. & Brown, D. R. (1994) *J. Neurophysiol. 72*, 2257–2268.

Mears, D., Sheppard, N. F. Jr, Atwater, I. & Rojas, E. (1995) *J. Membr. Biol. 146*, 163–176.

Meda, P., Michaels, R. L., Halban, P. A., Orci, L. & Sheridan, J. D. (1983) *Diabetes 32*, 858–868.

Meda, P., Perrelet, A. & Orci, L. (1984) *Modern Cell Biol. 3*, 131–196.

Meda, P., Bruzzone, R., Chanson, M., Bosco, D. & Orci, L. (1987) *Proc. Natl Acad. Sci. USA 84*, 4901–4904.

Meda, P., Bosco, D., Chanson, M., Giordano, E., Vallar, L., Wollheim, C. & Orci, L. (1990) *J. Clin. Invest. 86*, 759–768.

Meda, P., Chanson, M., Pepper, M., Giordano, E., Bosco, D., Traub, O., Willecke, K., El Aoumari, A., Gros, D., Beyer, E. C., Orci, L. & Spray, D. C. (1991) *Exp. Cell Res. 192*, 469–480.

Meda, P., Pepper, M. S., Traub, O., Willecke, K., Gros, D., Beyer, E., Nicholson, B., Paul, D. & Orci, L. (1993) *Endocrinology 133*, 2371–2378.

Mège, R. M., Matsuzaki, F., Gallin, W. J., Goldberg, J. I., Cunningham, B. A. & Edelman, G. M. (1988) *Proc. Natl Acad. Sci. USA 85*, 7274–7278.

Mehta, P. P., Bertram, J. S. & Loewenstein, W. R. (1986) *Cell 44*, 187–196.

Mehta, P. P., Hotz-Wagenblatt, A., Rose, B., Shalloway, D. & Loewenstein, W. R. (1991) *J. Membr. Biol. 124*, 207–225.

Mehta, P. P. & Loewenstein, W. R. (1991) *J. Cell Biol. 113*, 371–379.

Mehta, P. P., Yamamoto, M. & Rose, B. (1992) *Mol. Biol. Cell 3*, 839–850.

Mesnil, M., Fraslin, J. M., Piccoli, C., Yamasaki, H. & Guguen-Guillouzo, C. (1987) *Exp. Cell Res. 173*, 524–533.

Mesnil, M., Oyamada, M., Fitzgerald, D. J., Jongen, W. M. F., Krutovskikh, V. & Yamasaki, H. (1993) *Prog. Cell Res. 3*, 311–316.

Mesnil, M., Krutovskikh, V., Piccoli, C., Elfgang, C., Traub, O., Willecke, K. & Yamasaki, H. (1995) *Cancer Res. 55*, 629–639.

Meyer, R. A., Lampe, P. D., Malewicz, B., Baumann, W. J. & Johnson, R. G. (1990) *J. Cell Sci. 96*, 231–238.

Meyer, R. A., Laird, D. W., Revel, J.-P. & Johnson, R. G. (1992) *J. Cell Biol. 119*, 179–189.

Michalke, W. & Lowenstein, W. R. (1971) *Nature 232*, 121–122.

Milks, L. C., Kumar, N. M., Houghten, R., Unwin, N. & Gilula, N. B. (1988) *EMBO J. 7*, 2967–2975.

Miller, T., Dahl, G. & Werner, R. (1988) *Biosci. Rep. 8*, 455–464.

Miller, S. M., Garfield, R. E. & Daniel, E. E. (1989) *Am. J. Physiol. 256*, C130–C141.

Mills, S. L. & Massey, S. C. (1995) *Nature 377*, 734–738.

Miyashita, T., Takeda, A., Iwai, M. & Shimazu, T. (1991) *Eur. J. Biochem. 196*, 37–42.

Moorby, C. D., Stoker, M. & Gherardi, E. (1995) *Exp. Cell Res. 219*, 657–663.

Moore, L. K., Beyer, E. C. & Burt, E. C. (1991) *Am. J. Physiol. 260*, C975–C981.

Moore, L. K. & Burt, J. M. (1994) *Am. J. Physiol. 267*, C1371–C1380.

Moreno, A. P., Rook, M. B., Fishman, G. I. & Spray, D. C. (1994a) *Biophys. J. 67*, 113–119.

Moreno, A. P., Saez, J. C., Fishman, G. I. & Spray, D. C. (1994b) *Circ. Res. 74*, 1050–1057.

Moreno, A. P., Laing, J. G., Beyer, E. C. & Spray, D. C. (1995) *Am. J. Physiol. 268*, C356–C365.

Morley, G. E., Taffet, S. M. & Delmar, M. (1996) *Biophys. J. 70*, 1294–1302.

Mugnaini, E. (1986) in *Astrocytes, development, morphology, and regional specialization of astrocytes* (Fedoroff, S. & Vernadakis, A., eds) vol. I, pp. 329–371, Academic Press, New York.

Murray, S. A. & Fletcher, W. H. (1984) *J. Cell Biol. 98*, 1710–1719.

Musil, L. S., Beyer, E. C. & Goodenough, D. A. (1990a) *J. Membr. Biol. 116*, 163–175.

Musil, L. S., Cunningham, B. A., Edelman, G. M. & Goodenough, D. A. (1990b) *J. Cell. Biol. 111*, 2077–2088.

Musil, L. S. & Goodenough, D. A. (1990) *Curr. Opin. Cell Biol. 2*, 875–880.

Musil, L. S. & Goodenough, D. A. (1991) *J. Cell Biol. 115*, 1357–1374.

Musil, L. S. & Goodenough, D. A. (1993) *Cell 74*, 1065–1077.

Musil, L. S. (1994) in *Molecular mechanisms of epithelial cell junctions: from development to disease* (Citi, S., ed.) pp. 173–194, R. G. Landes Co., Austin TX.

Nathanson, M. H., Burgstahler, A. D., Mennone, A., Fallon, M. B., Gonzalez, C. B. & Sáez, J. C. (1995) *Am. J. Physiol. 269*, C167–C171.

Naus, C. C. G., Bechberger, J. F. & Paul, D. L. (1991) *Exp. Neurol. 111*, 198–203.

Naus, C. C. G., Elisevich, K., Zhu, D., Belliveau, D. J. & Del Maestro, R. (1992) *Cancer Res. 52*, 4208–4213.

Nedergard, M. (1994) *Science 263*, 1768–1771.

Neuhaus, I. M., Dahl, G. & Werner, R. (1995) *Gene 158*, 257–262.

Neveu, M. J., Hully, J. R., Paul, D. L. & Pitot, H. C. (1990) *Cancer Commun. 2*, 21–31.

Neveu, M. J., Hully, J. R., Babcock, K. L., Hertzberg, E. L., Nicholson, B. J., Paul, D. L. & Pitot, H. C. (1994) *J. Cell Sci. 107*, 83–95.

Neveu, M. J., Hully, J. R., Babcock, K. L., Vaughan, J., Hertzberg, E. L., Nicholson, B. J., Paul, D. L. & Pitot, H. C. (1995) *Hepatology 22*, 202–212.

Newmann, E. A. (1985) *Trends Neurosci. 8*, 156–159.

Neyton, J. & Trautman, A. (1985) *Nature 317*, 331–335.

Neyton, J. & Trautman, A. (1986) *J. Exp. Biol. 124*, 93–114.

Nicholson, B., Gros, D., Kent, S. B. H., Hood, L. E. & Revel, J.-P. (1985) *J. Biol. Chem. 260*, 6514–6517.

Nicholson, B., Dermietzel, R., Teplow, D., Traub, O., Willecke, K. & Revel, J.-P. (1987) *Nature 329*, 732–734.

Nicholson, B. J., Suchyna T., Xu, L. X., Hammernick, P., Cao, F. L., Fourtner, C., Barrio, L. & Bennett, M. V. L. (1993) *Prog. Cell Res. 3*, 3–13.

Nishi, M., Kumar, N. M. & Gilula, N. B. (1991) *Dev. Biol. 146*, 117–130.

Oh, S. Y., Dupont, E., Madhukar, B. V., Briand, J.-P., Chang, C.-C., Beyer, E. & Trosko, J. E. (1993) *Eur. J. Cell Biol. 60*, 250–255.

Olson, D. J., Christian, J. L. & Moon, R. T. (1991) *Science 252*, 1773–1776.

Omory, Y., Mesnil, M., Mironov, N. & Yamasaki, H. (1995) in *Proceedings of the 1995 Gap Junction Conference*, Île des Embiez, France.

Oyamada, M., Kimura, H., Oyamada, Y., Miyamoto, A., Ohshika, H. & Mori, M. (1994) *Exp. Cell Res. 212*, 351–358.

Page, E. & Shibata, Y. (1981) *Annu. Rev. Physiol. 43*, 431–441.

Palmer, J. F. & Slack, C. (1970) *J. Embryol. Exp. Morphol. 24*, 535, 553.

Patel, P. I. (1993) *Curr. Opin. Genet. Dev. 3*, 438–444.

Paul, D. L. (1985) in *Gap junctions* (Bennett, M. V. L. & Spray, D. C., eds) pp. 107–122, Cold Spring Harbor Laboratory Press, Cold Spring Harbor NY.

Paul, D. L. (1986) *J. Cell Biol. 103*, 123–134.

Paul, D. L., Ebihara, L., Takemoto, L. J., Swenson, K. I. & Goodenough, D. A. 1991) *J. Cell Biol. 115*, 1077–1089.

Paul, D. L., Bruzzone, R. & Haefliger, J.-A. (1993) *Prog. Cell Res. 3*, 15–20.

Paul, D. L. (1995) *Curr. Opin. Cell Biol. 7*, 667–672.

Paul, D. L., Yu, K., Bruzzone, R., Gimlich, R. L. & Goodenough, D. A. (1995) *Development 121*, 371–381.

Peinado, A., Yuste, R. & Katz, L. C. (1993) *Neuron 10*, 103–114.

Pelletier, D. B. & Boynton, A. L. (1994) *J. Cell. Physiol. 158*, 427–434.

Penn, A. A., Wong, R. O. L. & Shatz, C. J. (1994) *J. Neurosci. 14*, 3805–3815.

Pepper, M. S. & Meda, P. (1992) *J. Cell. Physiol. 153*, 196–205.

Pepper, M. S., Montesano, R., El Aoumari, A., Gros, D., Orci, L. & Meda, P. (1992) *Am. J. Physiol. 262*, C1246–C1257.

Peracchia, C. (1980) *Int. Rev. Cytol. 66*, 81–146.

Peracchia, C., Lazrak, A. & Peracchia, L. L. (1994) in *Handbook of membrane channels* (Peracchia, C., ed.) pp. 361–377, Academic Press, London.

Peracchia, C., Wang, X., Li, L. & Peracchia, L. L. (1996) *Pflügers Arch. 431*, 379–387.

Pereda, A. E., Bell, T. D. & Faber, D. S. (1995) *J. Neurosci. 15*, 5943–5955.

Pérez-Armendariz, E. M., Romano, M. C., Luna, J., Miranda, C., Bennett, M. V. L. & Moreno, A. P. (1994) *Am. J. Physiol. 267*, C570–C580.

Perez-Velazquez, J. L., Valiante, T. A. & Carlen, P. L. (1994) *J. Neurosci. 14*, 4308–4317.

Petrocelli, T. & Lye, S. J. (1993) *Endocrinology 133*, 284–290.

Piccolino, M., Gerschenfeld, H. M. & Neyton, J. (1984) *J. Neurosci. 4*, 2477–2488.

Pietrobon, D., Di Virgilio, F. & Pozzan T. (1990) *Eur. J. Biochem. 193*, 599–622.

Pitts, J. D. & Burk, R. R. (1976) *Nature 264*, 762–764.

Pitts, J. D. & Simms, J. W. (1977) *Exp. Cell Res. 104*, 153–163.

Pluciennik, F., Joffre, M. & Délèze, J. (1994) *J. Membr. Biol. 139*, 81–96.

Polacek, D., Lal, R., Volin, M. V. & Davies, P. F. (1993) *Am. J. Pathol. 142*, 593–606.

Portha, B. (1985) *Endocrinology 117*, 1735–1741.

Pozzi, A., Risek, B., Kiang, D. T., Gilula, N. B. & Kumar, N. M. (1995) *Exp. Cell Res. 220*, 212–219.

Puschel, G. P. & Jungermann, K. (1988) *Eur. J. Biochem. 175*, 187–191.

Rabadan-Diehl, C., Dahl, G. & Werner, R. (1994) *FEBS Lett. 351*, 90–94.

Rahman, S. & Evans, W. H. (1991) *J. Cell Sci. 100*, 567–578.

Ramanan, S. V., Manivannan, K., Mathias, R. T. & Brink, P. R. (1993) *Prog. Cell Res. 3*, 121–125.

Ransom, B. R. & Kettenmann, H. (1990) *Glia 3*, 258–266.

Raviola, E., Goodenough, D. A. & Raviola, G. (1980) *J. Cell Biol. 87*, 273–279.

Reaume, A. G., de Sousa, P. A., Kulkarni, S., Langille, B. L., Zhu, D., Davies, T. C., Juneja, S. C., Kidder, G. M. & Rossant, J. (1995) *Science 267*, 1831–1834.

Reed, K. E., Westphale, E. M., Larson, D. M., Wang, H.-Z., Veenstra, R. D. & Beyer, E. C. (1993) *J. Clin. Invest. 91*, 997–1004.

Revel, J.-P. & Karnovsky, M. J. (1967) *J. Cell Biol. 33*, C7–C12.

Revel, J.-P. (1988) *Modern Cell Biol. 7*, 135–149.

Revel, J.-P., Hoh, J. H., John, S. A., Laird, D. W., Puranam, K. & Yancey, S. B. (1992) *Semin. Cell Biol. 3*, 21–28.

Risek, B., Guthrie, S., Kumar, N. & Gilula, N. B. (1990) *J. Cell Biol. 110*, 269–282.

Risek, B. & Gilula, N. B. (1991) *Development 113*, 165–181.

Risek, B., Klier, F. G. & Gilula, N. B. (1992) *Development 116*, 639–651.

Risek, B., Klier, F. G. & Gilula, N. B. (1994) *Dev. Biol. 164*, 183–196.

Risek, B., Klier, F. G., Phillips, A., Hahn, D. W. & Gilula, N. B. (1995) *J. Cell Sci. 108*, 1017–1032.

Robb-Gaspers, L. D. & Thomas, A. P. (1995) *J. Biol. Chem. 270*, 8102–8107.

Robertson, J. D. (1963) *J. Cell Biol. 19*, 201–221.

Robinson, S. R., Hampson, E. C. G. M., Munro, M. N. & Vaney, D. I. (1993) *Science 262*, 1072–1074.

Rogers, M., Berestecky, J. M., Hossain, M. Z., Guo, H. M., Kadle, R., Nicholson, B. J. & Bertram, J. S. (1990) *Mol. Carcinogen. 3*, 335–343.

Rook, M. B., Jongsma, H. J. & van Ginneken, A. C. G. (1988) *Am. J. Physiol. 255*, H770–H782.

Rook, M. B., de Jonge, B., Jongsma, H. J. & Masson-Pévet, M. A. (1990) *J. Membr. Biol. 118*, 179–192.

Rose, B. & Loewenstein, W. R. (1976) *J. Membr. Biol. 28*, 87–119.

Rose, B., Simpson, I. & Loewenstein, W. R. (1977) *Nature 267*, 625–627.

Rose, B., Mehta, P. P. & Loewenstein, W. R. (1993) *Carcinogenesis 14*, 1073–1075.

Rosenberg, E., Spray, D. C. & Reid, L. M. (1992) *Eur. J. Cell Biol. 59*, 21–26.

Rosendaal, M., Green, C. R., Rahman, A. & Morgan, D. (1994) *J. Cell Sci. 107*, 29–37.

Ruangvoravat, C. P. & Lo, C. W. (1992) *Dev. Dyn. 194*, 261–281.

Rubin, J. B., Verselis, V. K., Bennett, M. V. L. & Bargiello, T. A. (1992a) *Proc. Natl Acad. Sci. USA 89*, 3820–3824.

Rubin, J. B., Verselis, V. K., Bennett, M. V. L. & Bargiello, T. A. (1992b) *Biophys. J. 62*, 183–195.

Rup, D. M., Veenstra, R. D., Wang, H.-Z., Brink, P. R. & Beyer, E. C. (1993) *J. Biol. Chem. 268*, 706–712.

Ryerse, J. S. (1993) *Cell Tissue Res. 274*, 393–403.

Sáez, J. C., Spray, D. C., Nairn, A. C., Hertzberg, E. L., Greengard, P. & Bennett, M. V. L. (1986) *Proc. Natl Acad. Sci. USA 83*, 2473–2476.

Sáez, J. C., Conner, J. A., Spray, D. C. & Bennett, M. V. L. (1989a) *Proc. Natl Acad. Sci. USA 86*, 2708–2712.

Sáez, J. C., Gregory, W. A., Watanabe, T., Dermietzel, R., Hertzberg, E. L., Reid, L., Bennett, M. V. L. & Spray, D. C. (1989b) *Am. J. Physiol. 257*, C1–C11.

Sáez, J. C., Nairn, A. C., Czernik, A. J., Spray, D. C., Hertzberg, E. L., Greengard, P. & Bennett, M. V. L. (1990) *Eur. J. Biochem. 192*, 263–273.

Sáez, J. C., Nairn, A. C., Czernik, A. J., Spray, D. C. & Hertzberg, E. L. (1993) *Prog. Cell Res. 3*, 275–281.

Sainio, K., Gilbert, S. F., Lehtonen, E., Miyuki, N., Kumar, N. M., Gilula, N. B. & Saxén, L. (1992) *Development 115*, 827–837.

Salomon, D., Masgrau, E., Vischer, S., Ullrich, S., Dupont, E., Sappino, P., Saurat, J.-H. & Meda, P. (1994) *J. Invest. Dermatol. 103*, 271–289.

Sandberg, K., Ji, H., Iida, T. & Catt, K. J. (1992) *J. Cell Biol. 117*, 157–167.

Sanderson, M. J., Charles, A. C. & Dirksen, E. R. (1990) *Mol. Biol. Cell 1*, 585–596.

Scherer, S. S., Deschênes, S. M., Xu, Y.-T., Grinspan, J. G., Fischbeck, K. H. & Paul, D. L. (1995) *J. Neurosci. 15*, 8281–8294.

Segal, S. & Bény, J.-L. (1992) *Am. J. Physiol. 263*, H1–H7.

Serras, F. & van den Biggelaar, J. M. (1987) *Dev. Biol. 120*, 132–138.

Serras, F., Fraser, S. & Chuong, C.-M. (1993) *Development 119*, 85–96.

Severs, N. J. (1995) *Histol. Histopathol. 10*, 481–501.

Shen, N. V. & Pfanninger, P. J. (1994) in *Handbook of membrane channels* (Peracchia, C., ed.) pp. 5–16, Academic Press, London.

Sheridan, J. D. & Atkinson, M. M. (1985) *Annu. Rev. Physiol. 47*, 337–353.

Simpson, I., Rose, B. & Loewenstein, W. R. (1977) *Science 195*, 294–296.

Sneyd, J., Charles, A. C. & Sanderson, M. J. (1994) *Am. J. Physiol. 266*, C293–C302.

Söhl, G., Gillen, G., Bosse, F., Gleichmann, M., Müller, H. W. & Willecke, K. (1996) *Eur. J. Cell Biol. 69*, 267–275.

Somogy, R. & Kolb, H.-A. (1988) *Pflügers Arch. 412*, 54–65.

Sontheimer, H., Minturn, J. E., Black, J. A., Waxman, S. G. & Ransom, B. R. (1991) *Proc. Natl Acad. Sci. USA 87*, 9833–9837.

Sosinsky, G. E., Jésior, J. C., Caspar, D. L. D. & Goodenough, D. A. (1988) *Biophys. J. 53*, 709–722.

Sosinsky, G. (1995) *Proc. Natl Acad. Sci. USA 92*, 9210–9214.

Sotelo, C. & Llinás, R. (1972) *J. Cell Biol. 53*, 271–289.

Spray, D. C., Harris, A. L. & Bennett, M. V. L. (1981) *J. Gen. Physiol. 77*, 77–94.

Spray, D. C. & Bennett, M. V. L. (1985) *Annu. Rev. Physiol. 47*, 281–303.

Spray, D. C., Fujita, M., Sáez, J. C., Choi, H., Watanabe, T., Hertzberg, E., Rosenberg, L. C. & Reid, L. M. (1987) *J. Cell Biol. 105*, 541–551.

Spray, D. C. & Burt, J. M. (1990) *Am. J. Physiol. 258*, C195–C205.

Spray, D. C., Chanson, M., Moreno, A. P., Dermietzel, R. & Meda, P. (1991) *Am. J. Physiol. 260*, C513–C527.

Spray, D. C. & Dermietzel, R. (1995) *Trends Neurosci. 18*, 256–262.

Stagg, R. B. & Fletcher, W. H. (1990) *Endocrine Rev. 11*, 302–325.

Stauffer, K. A., Kumar, N. M., Gilula, N. B. & Unwin, N. (1991) *J. Cell Biol. 115*, 141–150.

Stauffer, K. A. (1995) *J. Biol. Chem. 270*, 6768–6772.

Stauffer, P. L., Zhao, H., Luby-Phelps, K., Moss, R. L., Star, R. A. & Muallem, S. (1993) *J. Biol. Chem. 268*, 19769–19775.

Steinberg, T. H., Civitelli, R., Geist, S. T., Robertson, A. J., Hick, E., Veenstra, R. D., Wang, H.-Z., Warlow, P. M., Westphale, E. M., Laing, J. & Beyer, E. C. (1994) *EMBO J. 13*, 744–750.

Stoker, M. G. P., Shearer, M. & O'Neil, C. O. (1966) *J. Cell Sci. 1*, 297–310.

Stutenkemper, R., Geisse, S., Schwartz, H. J., Look, J., Traub, O., Nicholson, B. J. & Willecke, K. (1992) *Exp. Cell Res. 201*, 43–54.

Subak-Sharpe, H., Burk, R. R. & Pitts, J. D. (1969) *J. Cell Sci. 4*, 353–382.

Suchyna, T., Xu, L. X., Gao, F., Fourtner, C. R. & Nicholson, B. J. (1993) *Nature 365*, 847–849.

Suchyna, T. S., Kuruvilla, H., Nitsche, J. M., Chilton, M., Veenstra, R. & Nicholson, B. J. (1995) in *Proceedings of the 1995 Gap Junction Conference*, Île des Embiez, France.

Sullivan, R., Ruangvoravat, C., Joo, D., Morgan, J., Wang, B. L., Wang, X. K. & Lo, C. W. (1993) *Gene 130*, 191–199.

Sullivan, R. & Lo, C. W. (1995) *J. Cell Biol. 130*, 419–429.

Suter, U. & Snipes, G. J. (1995) *Annu. Rev. Neurosci. 18*, 45–75.

Swenson, K. I., Jordan, J. R., Beyer, E. C. & Paul, D. L. (1989) *Cell 57*, 145–155.

Swenson, K. I., Piwnica-Worms, H., McNamee, H. & Paul, D. L. (1990) *Mol. Biol. Cell 1*, 989–1002.

Takeda, M., Nelson, D. J. & Soliven, B. (1995) *Glia 14*, 225–236.

Takens-Kwak, B. R. & Jongsma, H. J. (1992) *Pflügers Arch. 422*, 198–200.

TenBroek, E., Arneson, M., Jarvis, L. & Louis, C. (1992) *J. Cell Sci. 103*, 245–257.

TenBroek, E., Johnson, R. G. & Louis, C. F. (1994) *Invest. Ophthalmol. Vis. Sci. 35*, 215–228.

Tetzlaff, W. (1982) *J. Neurocytol. 11*, 839–858.

Traub, O., Look, J., Paul, D. & Willecke, K. (1987) *Eur. J. Cell Biol. 43*, 48–54.

Traub, O., Look, J., Dermietzel, R., Brümmer, D., Husler, D. & Willecke, K. (1989) *J. Cell Biol. 108*, 1039–1051.

Traub, O., Eckert, R., Lichtenberg-Fraté, H., Elfgang, C., Bastide, B., Scheidtmann, K. H., Hülser, D. F. & Willecke, K. (1994) *Eur. J. Cell Biol. 64*, 101–112.

Trosko, J. E., Madhukar, B. V. & Chang, C. C. (1993) *Life Sci. 53*, 1–19.

Tucker, M. A. & Barajas, L. (1994) *Exp. Cell Res. 213*, 224–230.

Turin, L. & Warner, A. (1977) *Nature 270*, 56–57.

Unwin, P. N. T. & Zampighi, G. (1980) *Nature 283*, 545–549.

Unwin, N. (1989) *Neuron 3*, 665–676.

Valdimarsson, G., De Sousa, P. A., Beyer, E. C., Paul, D. L. & Kidder, G. M. (1991) *Mol. Reprod. Dev. 30*, 18–26.

Valdimarsson, G. & Kidder, G. M. (1995) *J. Cell Sci. 108*, 1715–1722.

Vaney, D. I. (1991) *Neurosci. Lett. 125*, 187–190.

van Kempen, M. A., Ten Velde, I., Wessels, A., Oosthoek, P. W., Gros, D., Jongsma, H. J., Moorman, A. F. M. & Lamers, W. H. (1995) *Microsc. Res. Tech. 31*, 420–436.

Veenstra, R. D. & DeHaan, R. L. (1986) *Science 233*, 972–974.

Veenstra, R. D., Wang, H.-Z., Westphale, E. M. & Beyer, E. C. (1992) *Circ. Res. 71*, 1277–1283.

Veenstra, R. D., Wang, H.-Z., Beyer, E. C. & Brink, P. R. (1994a) *Circ. Res. 75*, 483–490.

Veenstra, R. D., Wang, H.-Z., Beyer, E. C., Ramanan, S. V. & Brink, P. R. (1994b) *Biophys. J. 66*, 1915–1928.

Veenstra, R. D., Wang, H.-Z., Beblo, D. A., Chilton, M. G., Harris, A. L., Beyer, E. C. & Brink, P. R. (1995) *Circ. Res. 77*, 1156–1165.

Venance, L., Cordier, J., Monge, M., Zalc, B., Glowinski, J. & Giaume, C. (1995a) *Eur. J. Neurosci. 7*, 451–461.

Venance, L., Piomelli, D., Glowinski, J. & Giaume, C. (1995b) *Nature 376*, 590–594.

Verselis, V., Bennett, M. V. L., Spray, D. C. & White, R. L. (1986) *Science 234*, 461–464.

Verselis, V. K. & Bargiello, T. A. (1991) in *Biophysics of gap junctions* (Peracchia, C., ed.) pp. 117–129, CRC Press, Boca Raton, FL.

Verselis, V. K., Ginter, C. S. & Bargiello, T. A. (1994) *Nature 368*, 348–351.

von Blankenfeld, G., Ransom, B. R. & Kettenmann, H. (1993) *Glia 7*, 322–328.

Vozzi, C., Ullrich, S., Charollais, A., Philippe, J., Orci, L. & Meda, P. (1995) *J. Cell Biol. 131*, 1561–1572.

Wang, H.-Z., Li, J., Lemanski, L. F. & Veenstra, R. D. (1992) *Biophys. J. 63*, 139–151.

Wang, X., Li, L., Peracchia, L. L. & Peracchia, C. (1996) *Pflügers Arch.*, in the press.

Wang, Y. & Mehta, P. P. (1995) *Eur. J. Cell Biol. 67*, 285–296.

Wang, Y. & Rose, B. (1995) *J. Cell Sci. 108*, 3501–3508.

Warner, A. E. & Lawrence, P. A. (1982) *Cell 28*, 243–252.

Warner, A. E. (1992) *Semin. Cell Biol. 3*, 81–91.

Weir, M. P. & Lo, C. W. (1982) *Proc. Natl Acad. Sci. USA 79*, 3232–3235.

Werner, R., Levine, E., Rabadan-Diehl, C. & Dahl, G. (1989) *Proc. Natl Acad. Sci. USA 86*, 5380–5384.

Werner, R., Levine, E., Rabadan-Diehl, C. & Dahl, G. (1991) *Proc. R. Soc. Lond. B Biol. Sci. 243*, 5–11.
White, T. W., Bruzzone, R., Goodenough, D. A. & Paul, D. L. (1992) *Mol. Biol. Cell 3*, 711–720.
White, T. W., Bruzzone, R., Paul, D. L. & Goodenough, D. A. (1994a) *Nature 371*, 208–209.
White, T. W., Bruzzone, R., Wolfram, S., Paul, D. L. & Goodenough, D. A. (1994b) *J. Cell Biol. 125*, 879–892.
White, T. W., Bruzzone, R. & Paul, D. L. (1995a) *Kidney Int. 48*, 1148–1157.
White, T. W., Paul, D. L., Goodenough, D. A. & Bruzzone, R. (1995b) *Mol. Biol. Cell 6*, 459–470.
Wiener, E. C. & Loewenstein, W. R. (1983) *Nature 305*, 433–435.
Wiens, D., Jensen, L., Jasper, J. & Becker, J. (1995) *Anat. Rec. 241*, 541–553.
Willecke, K., Hennemann, H., Dahl, E., Jungbluth, S. & Heynkes, R. (1991a) *Eur. J. Cell Biol. 56*, 1–7.
Willecke, K., Heynkes, R., Dahl, E., Stutenkemper, R., Hennemann, H., Jungbluth, S., Suchyna, T. & Nicholson, B. J. (1991b) *J. Cell Biol. 114*, 1049–1057.
Winterhager, E., Grümmer, R., Jahn, E., Willecke, K. & Traub, O. (1993) *Dev. Biol. 157*, 399–409.
Wolszon, L. R., Gao, W.-Q., Passani, M. B. & Macagno, E. R. (1994a) *J. Neurosci. 14*, 999–1010.
Wolszon, L. R., Rehder, V., Kater, S. B. & Macagno, E. R. (1994b) *J. Neurosci. 14*, 3437–3448.
Wong, R. O. L., Chernjavsky, A., Smith, S. J. & Shatz, C. J. (1995) *Nature 374*, 716–718.
Wu, J., McHowat, J., Saffitz, J. E., Yamada, K. A. & Corr, P. B. (1993) *Circ. Res. 72*, 879–889.
Yamamoto, T., Ochalski, A., Hertzberg, E. L. & Nagy, J. I. (1990) *Brain Res. 508*, 313–319.
Yamasaki, H. (1990) *Carcinogenesis 11*, 1051–1058.
Yancey, S. B., John, S. A., Ratneshwar, L., Austin, B. J. & Revel, J.-P. (1989) *J. Cell Biol. 108*, 2241–2254.
Yeager, M. & Gilula, N. B. (1992) *J. Mol. Biol. 223*, 929–948.
Yoshizaki, G., Patiño, R. & Thomas, P. (1994) *Biol. Reprod. 51*, 493–503.
Yoshizaki, G. & Patiño, R. (1995) *Mol. Reprod. Dev. 42*, 7–18.
Yotti, L. P., Chang, C.-C. & Trosko, J. E. (1979) *Science 206*, 1089–1091.
Yu, W., Dahl, G. & Werner, R. (1994) *Proc. R. Soc. Lond. B 255*, 125–132.
Yuste, R., Peinado, A. & Katz, L. C. (1992) *Science 257*, 665–669.
Yuste, R., Nelson, D. A., Rubin, W. W. & Katz, L. C. (1995) *Neuron 14*, 7–17.
Xia, S.-L. & Ferrier, J. (1992) *Biochem. Biophys. Res. Commun. 186*, 1212–1219.
Zhang, J.-T. & Nicholson, B. J. (1989) *J. Cell Biol. 109*, 3391–3401.
Zhang, J.-T. & Nicholson, B. J. (1994) *J. Membr. Biol. 139*, 15–29.
Zhu, D., Caveney, S., Kidder, G. M. & Naus, C. C. G. (1991) *Proc. Natl Acad. Sci. USA 88*, 1883–1887.
Zimmer, D. B., Green, C. R., Evans, W. H. & Gilula, N. B. (1987) *J. Biol. Chem. 262*, 7751–7763.

Eur. J. Biochem. 238, 297–307 (1996)

Review

DNA-replication fidelity, mismatch repair and genome instability in cancer cells

Asad UMAR and Thomas A. KUNKEL

Laboratory of Molecular Genetics, National Institute of Environmental Health Sciences, North Carolina, USA

(Received 15 December 1995/26 February 1996) – EJB 95 2058/0

It has been suggested that an early event in the multistep progression of a normal cell to a tumor cell could be a defect that leads to an elevated mutation rate, thus providing a pool of mutants upon which selection could act to yield a tumor. Such a mutator phenotype could result from a defect in any of several DNA transactions, including those that determine the DNA replication error rate or the ability to correct replication errors. Recent evidence for the latter is the mutator phenotype observed in tumor cells of patients having a hereditary form of colon cancer. These patients have a germline mutation in genes required for post-replication DNA mismatch repair. A second mutation arises somatically, yielding a greatly elevated mutation rate due to an inability to correct DNA replication errors. This connection between cancer, DNA replication errors and defective mismatch repair is the subject of this review, wherein we consider the key steps and principles for high fidelity replication and how their perturbation results in genome instability.

Keywords: DNA-replication fidelity; mismatch repair; genome instability; mutator phenotype; cancer.

The development of a tumor from a normal cell occurs in multiple steps requiring several changes in genetic information. Given this and the low spontaneous mutation rate in normal cells, it has been suggested that a mutator phenotype may be required for multistage carcinogenesis (Loeb, 1991). Recent evidence consistent with this hypothesis is the genome-wide instability of repetitive DNA sequences in tumors of patients having a hereditary form of colon cancer, hereditary nonpolyposis colorectal carcinoma (HNPCC) (Marra and Boland, 1995). This mutator phenotype, as well as the discovery of several hereditary neurodegenerative diseases characterized by instability in triplet-repeat sequences (Willems, 1994), has spurred great interest in defective replication processes that might yield a high mutation rate, especially in repetitive sequences. The purpose of this article is to review how DNA is replicated with high fidelity and how defects in these processes can lead to a higher than normal mutation rate. This includes our present understanding of the link between defective mismatch repair genes and cancer.

Three major steps determine replication fidelity

To understand the possible origins of mutator phenotypes resulting from replication errors, consider the major steps that operate to determine replication fidelity in both prokaryotes and eukaryotes. Discrimination against errors first occurs as the replication apparatus polymerizes nucleotides at a growing replication fork. If a mistake is made, it may be excised prior to further chain elongation by a 3′→5′ exonuclease. Errors that escape proofreading may be corrected by post-replication mismatch repair, which selectively removes errors from the newly synthesized strand.

The amount of discrimination against base-substitution mutations from these steps has been estimated in *Escherichia coli* by measuring mutation rates in strains selectively disabled in one or more processes (Fig. 1). The wild-type spontaneous mutation rate in the *lacI* reporter gene is $\approx 10^{-10}$ mutations · base pair replicated^{-1} · generation^{-1} (Schaaper, 1993). A mutant lacking methyl directed post-replication mismatch repair has a much higher rate, and analysis of mutational specificity suggests that mismatch repair reduces base substitution rates 20–400-fold, depending on the type of substitution considered. With some simplifying assumptions (Schaaper, 1993), the $\approx 10^{-7}$ mutation rate in the mismatch-repair-defective strain can be considered as the fidelity of chromosomal replication. Analysis of a double mutant lacking mismatch repair and defective in the 3′→5′ exonuclease activity of the ε subunit of the replicative DNA polymerase III holoenzyme suggests that proofreading contributes 40–200-fold to replication fidelity, with the balance (factors of 2×10^5–2×10^6) representing the base selectivity of the replication machinery.

Studies in eukaryotic cells suggest that the contributions of polymerase selectivity, proofreading and mismatch repair to eukaryotic mutation rates may be similar to those in *E. coli*. However, less is known about these processes in eukaryotes, particularly for addition and deletion mutations. This is because past studies have focused more on substitution rates, since mutant cell lines defective in the three processes have only recently become available, and because DNA replication and mismatch

Correspondence to T. A. Kunkel, Laboratory of Molecular Genetics, National Institute of Environmental Health Sciences, Research Triangle Park, North Carolina 27709, USA

Fax: +1 919 541 7613.

Abbreviations. HNPCC, hereditary nonpolyposis colorectal carcinoma; SV40, simian virus 40; HIV-1, type-1 human immunodeficiency virus; RT, reverse transcriptase; m^6G, O^6-methylguanine; MMR, mismatch repair.

Note. Recent reviews of the literature are available for several of the key concepts mentioned here. With apologies to the authors of many of these primary publications, we often refer the reader to these other reviews, in order to limit the number of references and focus on recent work.

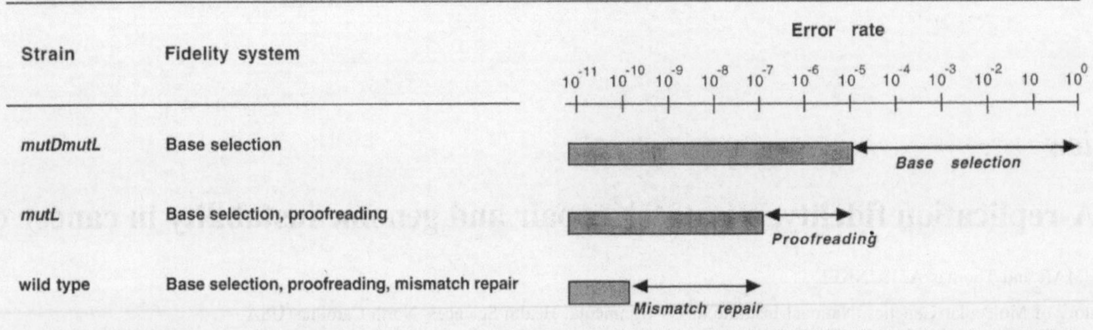

Fig. 1. *E. coli* strains mutants in different replication fidelity systems and their replication error rates. The error rates of DNA replication in each of the strains (indicated by horizontal bars) are estimates of the contributions of the DNA replication fidelity systems to overall genome stability. In a wild-type strain, the serial action of base selectivity, proofreading and mismatch repair reduces the error rate of DNA replication to a low level. In a mismatch-repair-defective phenotype (*mutL*), the error rates represents the efficiency of base selection and proofreading. In a double mutant for mismatch repair and proofreading (*mutDmutL*), the error rates represent the efficiency of base selectivity only.

repair are more highly differentiated in eukaryotes. Before considering in more detail what is known, we first discuss how errors are generated during replication.

Substitution errors by miscoding

The error pathway most relevant to the rates in *E. coli* mentioned above is simple base miscoding, i.e. direct misincorporation of a non-complementary nucleotide leading directly to a base-substitution error. Replication fidelity studies (reviewed in Johnson, 1993 and Roberts and Kunkel, 1995) show that discrimination against errors occurs during dNTP binding to the polymerase · template-primer complex, during a conformational change to position the nucleotide for phosphodiester bond formation, and during catalysis to form the bond. Depending on the polymerase examined, one or more of these steps is disfavored for incorporation of incorrect nucleotides. If an incorrect nucleotide is incorporated, a kinetically slow step after phosphodiester bond formation but prior to release of pyrophosphate and/or the slow rate of incorporation of the next correct nucleotide onto a mispaired terminus provide opportunities for a 3′→5′ exonuclease to remove the error prior to further incorporation.

Microsatellite instability, hereditary non-polyposis colorectal cancer and template-primer misalignment errors

A substantial portion of the eukaryotic genome is comprised of repetitive sequence motifs that are classified according to the length and complexity of the repeat unit. Tandemly repeated mono-, di-, tri-, tetra- or pentanucleotide sequence motifs, so-called microsatellites, are widely distributed on every chromosome in the human genome. Many types of microsatellites are known, differing in the length and nucleotide composition of the repeat unit and in the number of repeat units. Microsatellite sequences are highly unstable in a subset of tumor cells from cancer patients as compared to normal cells. This instability is characterized by both increases and decreases in the number of repeats, which can change by one or more.

The most likely mechanism to explain the origin of the length variations in microsatellite sequences in tumor cells is strand slippage during DNA replication (Kunkel, 1990; Ripley, 1990). If slippage of a template-primer occurs at repetitive sequences during replication, misaligned premutational intermediates can form that are stabilized by adjacent correct base pairs, so long as the unpaired nucleotides are discrete repeat units. This is illustrated (Fig. 2) for a $(CA)_5$ microsatellite. If unpaired nucleotides are in the primer strand, an insertion intermediate results, whereas unpaired nucleotides in the template strand yield a deletion intermediate. These intermediates can be corrected by realignment or 3′→5′ excision of the primer strand prior to continued synthesis. If not corrected during replication, the newly synthesized strand can be repaired after replication is completed by using the parental strand as a template. However, continued polymerization from these intermediates without proofreading or repair yields the insertion or deletion upon subsequent replication of the newly synthesized strand.

Two possibilities (Kunkel, 1990) exist whereby the initiating miscoding or slippage event can lead to an unexpected outcome. An initial misincorporation may be followed by template-primer rearrangement to correctly pair the terminal base with a different template base. Continued polymerization from the misaligned substrate ultimately leads to a frameshift error. Strong evidence for this model has been obtained *in vitro* with normal and damaged substrates (e.g. see Thomas et al., 1995, and references therein). Another possibility is an initial strand slippage followed by correct incorporation, then realignment (e.g. see Bebenek et al., 1993 and references therein). This process forms a terminal mispair that yields a base-substitution.

Eukaryotic DNA replication enzymology

To fully appreciate the opportunities for generating errors during replication of the 6×10^9-nucleotide human genome, consider the complexity of eukaryotic replication enzymology (for a recent comprehensive review, see DePamphilis, 1996). Eukaryotic cells contain (at least) five template-dependent deoxyribonucleotide-polymerizing enzymes, designated DNA polymerases α, β, γ, δ and ε. Two or more of these DNA polymerases are required for replication of the nuclear genome. Replication is initiated at origins, where proteins bind and open the helix, allowing synthesis of short RNA primers followed by DNA polymerization by polymerase α. Concomitant 5′→3′ replication of the two antiparallel strands requires that a replication fork has both a lagging and a leading strand. The leading strand is replicated continuously while the lagging strand is replicated as a series of fragments. Each fragment is about 250 nucleotides and is initiated from an RNA primer that is subsequently replaced with DNA. The fragments are then ligated.

Although the exact roles of the eukaryotic DNA polymerases on the leading and lagging strands in human cells remain to be established, DNA polymerases α, δ and ε all participate. (Poly-

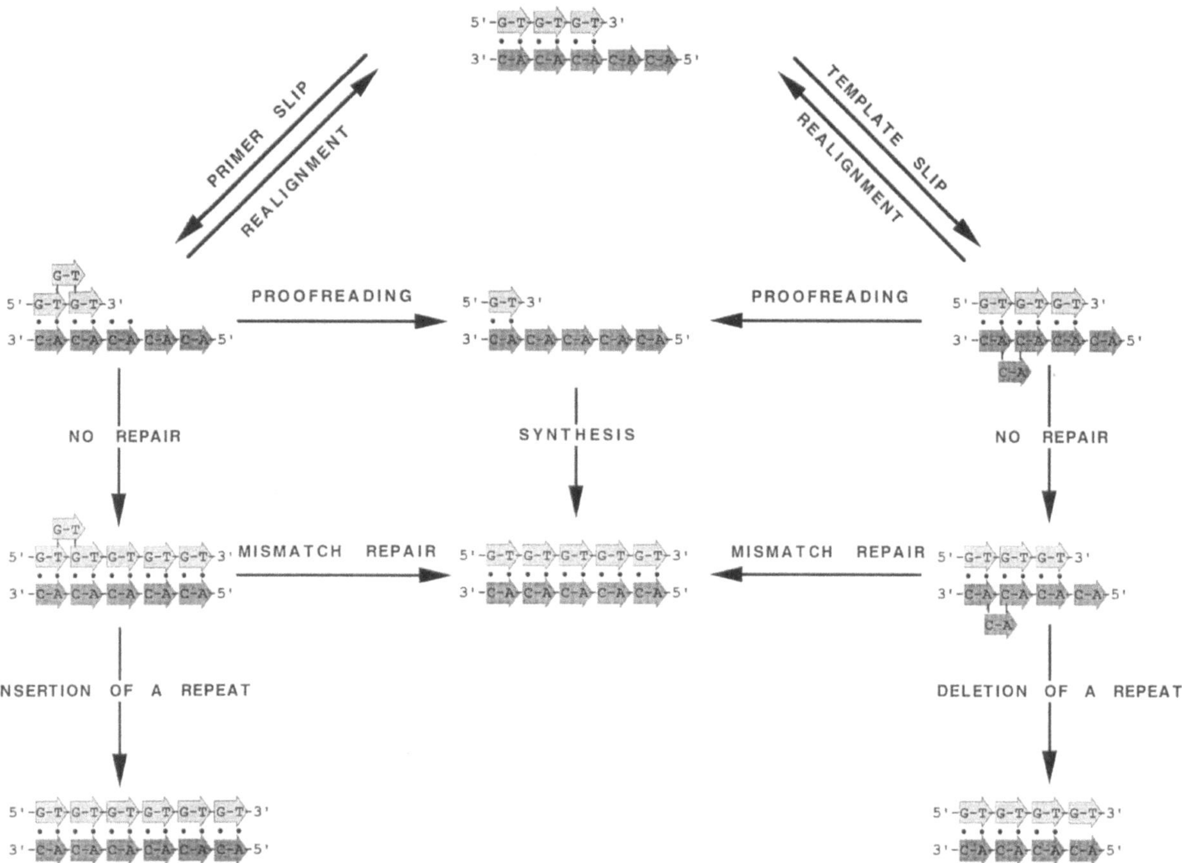

Fig. 2. A schematic flow chart for the instability of repetitive DNA sequences. See text for description.

merase β performs base excision repair and polymerase γ replicates the mitochondrial genome.) Several models have been posited to account for their roles (DePamphilis, 1996). The primase activity associated with polymerase α initiates synthesis at the origin and on the lagging strand, while DNA polymerases δ and ε are thought to perform the bulk of chain elongation on the two strands. Leading-strand replication can occur concomitantly with unwinding of the parental helix, hence little template need be exposed as single-stranded DNA prior to replication. However, synthesis using the complementary-strand template occurs only after exposing an extensive single-stranded region. These enzymological and architectural asymmetries could permit template-primer misalignment errors to occur at different rates during leading and lagging strand replication. DNA synthesis to replace the $\approx$10-nucleotide RNA primer on a 250-nucleotide Okazaki fragment can represent a substantial component of lagging-strand replication if DNA is also excised and replaced in the process. Thus, polymerase misbehavior here could contribute significantly to the mutation rate.

Table 1. Error rates of eukaryotic DNA polymerases α, δ and ε. For a more extensive review of the fidelity of these polymerases, see Roberts and Kunkel (1995). Measurements for polymerase α have been made with polymerase α preparations from several sources. The values given are averages. Reactions for polymerase δ contained PCNA in order to stimulate synthesis by polymerase δ. The replication complex was the SV40 origin-dependent replication in extracts of human or simian cells.

DNA polymerase	Substitutions		One-base deletions
	average error rate ($\times 10^{-6}$)	range of error rates ($\times 10^{-6}$)	average error rate ($\times 10^{-6}$)
Pol α	160	1.2–380	50
Pol δ	~10	≤2.3– 29	18
Pol ε	≤6.7	≤1.0– 19[a]	5.0[a]
Replication complex	≤6.7	≤0.1– ≤6.2	≤1.0

[a] Higher values were from reactions in which proofreading was compromised. Thus, they probably represent a minimal fidelity estimate.

Fidelity of eukaryotic DNA polymerases

Error rates during DNA synthesis *in vitro* by DNA polymerases α, δ and ε (Table 1) are average error rates/detectable nucleotide polymerized for copying a 250-base sequence of the *lacZ* gene in bacteriophage M13mp2 DNA. The wide range of base-substitution error rates illustrates that substitution fidelity depends on the DNA polymerase, the composition of the mispair and the local template-primer sequence.

DNA polymerase α is the least accurate of the three enzymes listed, with an average error rate of $\approx 10^{-4}$. This somewhat lower fidelity is consistent with the fact that most preparations of purified polymerase α lack $3' \rightarrow 5'$ proofreading exonuclease activity and that polymerase α genes lack the three conserved sequence motifs characteristic of such exonucleases. In contrast, DNA polymerases δ and ε both have associated $3' \rightarrow 5'$ exonuclease activity and the coding sequences of the genes do have the characteristic conserved exonuclease motifs. Both enzymes are more accurate than polymerase α, with polymerase ε being the most accurate.

This same order of fidelity applies to the three DNA polymerases for frameshift errors, with polymerase α being the least accurate and polymerase ε being the most accurate (Table 1). For all three polymerases, the most frequently observed error is deletion of a single nucleotide. Frameshift error rates with di-, tri-, tetra- and pentanucleotide repeats have not been reported. The single-base frameshift error rate depends not only on the DNA polymerase, but also on whether the DNA sequence is a homopolymeric repeat and, if so, how many consecutive repeat units are present. For example, the frameshift error rates in homopolymeric runs for polymerase α, T7 DNA polymerase (Kunkel et al., 1994), HIV-1 reverse transcriptase (Bebenek et al., 1995), and T4 and *Pfu* DNA polymerases (Kroutil et al., 1996) all increase as the length of the run increases. Moreover, polymerase frameshift error 'hot spots' at homopolymeric runs are eliminated by interruptions with nonidentical template nucleotides (e.g. Bebenek et al., 1993). These observations are consistent with the hypothesis (Streisinger et al., 1966) that slippage occurs during DNA synthesis to generate misaligned intermediates containing paired termini with unpaired nucleotides in the template or the primer strand, respectively (e.g. Fig. 2 for a dinucleotide repeat). The model predicts that the frameshift error rate during synthesis should increase as the number of consecutive repeats increases because, as the repeat number increases, so does the potential number of stabilizing base pairs and misaligned intermediates and the distance between the polymerase active site and the unpaired nucleotide(s).

The higher frameshift fidelity of polymerase δ and ε as compared to polymerase α is consistent with the idea that some frameshift errors are proofread. Studies directly comparing the frameshift fidelity of exonuclease-proficient DNA polymerases to their exonuclease-deficient derivatives show that proofreading enhances frameshift fidelity for Klenow, T7, T4 and *Pfu* DNA polymerases (see Kroutil et al., 1996 and references therein). However, in all cases, the contribution of proofreading to frameshift fidelity decreases as the length of the homopolymeric run increases. This likely results from protection of the misaligned heteroduplex from exonucleolytic digestion by the correct base pairs possible in a run. Such measurements have yet to be made with the dinucleotide and trinucleotide repeats whose instability is characteristic of HNPCC tumors. However, that exonucleolytic proofreading may contribute less to frameshift fidelity as the number of repeat units in a microsatellite sequence increases may explain the fact that mutations in the proofreading function of replicative DNA polymerases had little effect on the stability of long dinucleotide-repeat sequences in yeast (Strand et al., 1993). The fact that both polymerase selectivity and editing efficiency are repeat-length-dependent has implications for the stability of microsatellites in which the number of consecutive repeats is large. Evidence to date suggests that, for frameshifts, only the third major step for determining replication fidelity, post-replication mismatch repair, is insensitive to repeat number. Thus, the stability of long microsatellites may depend heavily on post-replicative repair of misaligned substrates.

Processivity and microsatellite instability

One property of polymerization relevant to polymerase fidelity during replication of microsatellites is processivity, the number of nucleotides incorporated/polymerase association/dissociation with the template-primer. A relationship between processivity and fidelity (reviewed in Kunkel, 1990) was first suggested by the fact that polymerase α is both more accurate and more processive than is polymerase β. A similar correlation exists with type-1 human immunodeficiency virus reverse transcriptase (HIV-1 RT), which is inaccurate for one-base frameshifts within some but not all template runs. These hot spots for frameshift errors are also template positions where the probability of termination of processive synthesis is high. Changes introduced in the sequences flanking the hot spots increased or decreased frameshift fidelity concomitant with increases or decreases in processive synthesis within the run (Bebenek et al., 1993). Moreover, DNA synthesis by T7 DNA polymerase is highly inaccurate for addition errors in homopolymeric runs when the polymerase operates without thioredoxin (Kunkel et al., 1994), an accessory protein that confers high processivity to the polymerase catalytic subunit. These correlations between low processivity and low frameshift fidelity suggest that the formation and/or utilization of misaligned template-primers is increased during the dissociation-reinitiation phase of a polymerization reaction. Most replication complexes do contain accessory proteins that enhance DNA polymerase processivity (Kelman and O'Donnell, 1995). That the absence of an accessory protein known to confer high processivity to a DNA polymerase affects the fidelity of replication of repetitive DNA offers one possible explanation for microsatellite instability *in vivo*.

Fidelity of the human replication apparatus

Understanding how genomes are stably replicated and how instability may arise to generate diseases is being facilitated by the development of systems that replicate double-stranded DNA *in vitro*. One system for studying human genomic replication utilizes the simian virus 40 (SV40) origin of replication. Circular, double-stranded DNA substrates containing the SV40 origin can be completely replicated by the proteins present in human cells, with only the addition of SV40 T antigen needed to initiate replication at the origin. Host replication proteins can be supplied either by crude extracts of cells grown in culture or by reconstitution with purified proteins prepared from such extracts (Waga and Stillman, 1994).

Using assays that score replication errors in the *lacZ* or *supF* genes (reviewed in Roberts and Kunkel, 1995), SV40 origin-containing DNA is replicated in human HeLa and simian CV-1 cell extracts with high fidelity, with error rates varying from $\leq 6.2\times10^{-6}$ to $\leq 0.1\times10^{-6}$, depending on the substitution or frameshift error considered (Table 1). Since these cell extracts are also competent for mismatch repair (Thomas et al., 1991), these error rates represent the sum of both replication fidelity and any mismatch repair occurring in the extract. However, mismatch repair in the extract only affects error rate determinations by 2–3-fold, suggesting that replication fidelity itself is high.

Since no replication errors are detected from reactions with undamaged DNA and equimolar dNTP concentrations, further understanding of how high replication fidelity is achieved requires perturbation of reaction conditions to generate replication errors. An informative example is the use of unequal dNTP concentrations to force specific misinsertions, then to examine the contribution of proofreading to replication fidelity. Adding an excess of one dNTP generates misinsertions, and editing of these can be manipulated either by increasing the absolute dNTP concentration to stimulate polymerization at the expense of proofreading or by inhibiting proofreading by adding deoxyribonucleoside monophosphate to the replication reaction. Results from both approaches (Roberts and Kunkel, 1995) suggest that proofreading contributes substantially to replication fidelity for base substitution and frameshift errors.

During replication of duplex DNA, errors arising as a result of specific misinsertion events with a dNTP excess also allow

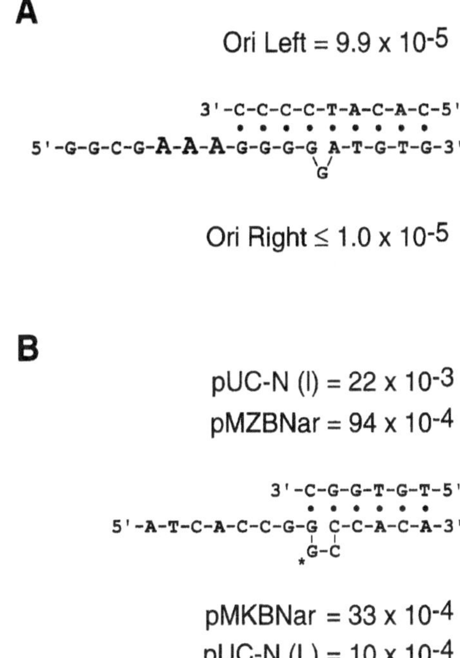

Fig. 3. Mutation frequency differences due to leading-strand and lagging strand synthesis. (A) Fidelity during leading-strand (Ori Right) or lagging-strand (Ori Left) replication in a HeLa cell extract at a run of 5 G · C base pairs (from Roberts et al., 1994). (B) Fidelity during leading-strand (pMKBNar) or lagging-strand (pMKZBNar) replication in a HeLa cell extract (from Thomas et al., 1995) or in *E. coli* [from Veaute and Fuchs, 1993, where pUC-N(L) is leading and pUC-N (l) is lagging] of a substrate containing an *N*-2-acetylaminofluorene-adducted template guanine residue, indicated by an asterisk. The inferred misaligned intermediates are shown.

the strand that templated the misinsertion to be defined. This permits examination of the fidelity of the leading-strand and lagging-strand replication machinery, by comparing replication fidelity with two substrates differing only in the position of the SV40 origin relative to the *LacZ* gene mutational target. Such substrates have been replicated to describe average leading-strand and lagging-strand replication rates for several errors (for review, see Roberts and Kunkel, 1995). The results suggest that the fidelity of the leading-strand and lagging-strand replication apparatus is generally similar for most errors. However some differences have been observed, including 33-fold and 8-fold lower fidelity for G · dTTP and C · dTTP mispairs on the (+) strand by the lagging-strand machinery (Roberts et al., 1994) and several-fold lower lagging-strand fidelity at some template positions during replication of substrates containing randomly introduced ultraviolet-light-induced photoproducts (Thomas et al., 1993).

Of particular interest in considering possible origins of microsatellite instability in cells are studies suggesting that there are differences in nucleotide-deletion error rates between the leading-strand and lagging-strand replication apparatus. For example, frequent loss of a single base pair in a homopolymeric run of G · C base pairs is observed with a substrate containing the replication origin to the left of the mutational target, but not with a substrate containing the replication origin to the left of the mutational target (Fig. 3A). Similarly, differences were observed in two-base-deletion error rates during bypass of a site-specific *N*-2-acetylaminofluorene adduct in a dinucleotide repeat sequence (Fig. 3B). These data suggest that the potential for microsatellite instability in cells may relate to the microsatellite position relative to replication origins. This has also been suggested by studies in *E. coli* demonstrating that the stability of a palindromic sequence (Trinh and Sinden, 1991) and triplet repeat sequences (Kang et al., 1995) depends on their location and orientation relative to the replication origin.

There are several possible explanations for the differences (mentioned above) in leading-strand and lagging-strand replication error rates. One is that mismatch repair in the extract is responsible for the asymmetry. A second is that replication of the two strands is highly asymmetric, providing unequal opportunities to make mistakes. Replication of the two strands may be performed by different DNA polymerases or perhaps the same polymerase but with a different complement of accessory proteins. This could yield differences in misinsertion rates and/or the ability to extend rather than proofread mispaired or misaligned template primers. Replication on the leading strand is highly processive as compared to discontinuous synthesis of Okazaki fragments on the lagging strand, which involves more than one DNA polymerase and/or one or more switches between enzymes to initiate synthesis and to eventually replace RNA primers with DNA. Since both the initiation and completion phases of Okazaki fragment synthesis are specialized relative to the bulk of processive chain elongation, there is unique opportunities for misbehavior.

Post-replication mismatch repair in *E. coli*

Replication errors that escape proofreading may be corrected by post-replication mismatch repair, which selectively removes errors from the newly synthesized strand. In *E. coli*, newly replicated DNA is transiently undermethylated at the N^6 position of adenine residues in dGATC sequences. This provides a signal to target repair to replication errors in the newly synthesized daughter strand (Fig. 4, top; Modrich, 1991, 1994). Correction is initiated by binding of a homodimer (or tetramer) of the 95-kDa product of *MutS* to the mismatch (Fig. 5, top left). A homodimer of the 68-kDa product of *MutL* then forms a complex with MutS, in a reaction that requires ATP but not ATP hydrolysis (Fig. 5, left). This complex activates a latent endonuclease activity associated with the 25-kDa product of *MutH*, which incises the unmethylated DNA strand at the hemimethylated GATC site. An experimentally introduced incision alleviates the requirement for MutH incision. ATP-dependent exonuclease action then generates a gap by excising nucleotides starting at the incision site and proceeding to 100–150 nucleotides beyond the mismatch (Fig. 4). This reaction requires MutS, MutL, DNA helicase II (the product of the *MutU*/*UvrD* gene) and a single-strand-specific exonuclease. Since excision can initiate from an incised dGATC site either 3′ or 5′ to the mismatch, one of three exonucleases can participate, depending on the position of the nick relative to the mismatch. These are exonuclease I (3′→5′ excision) or exonuclease VII or the *RecJ* exonuclease (both 5′→3′ exonucleases). Resynthesis of DNA is catalyzed by DNA polymerase III holoenzyme in the presence of single-strand DNA-binding protein. The nick is then sealed by DNA ligase to complete repair.

Methyl directed *E. coli* mismatch repair corrects most base · base mismatches as well as substrates containing 1–3 unpaired nucleotides. The efficiency of repair varies with the composition of the mismatch (e.g. C · C mispairs are not corrected efficiently) and the sequence surrounding the mismatch. This repair system does not efficiently correct substrates containing more than three unpaired bases (Parker and Marinus, 1992; Carraway and Marinus, 1993), although there is evidence (reviewed in Modrich, 1991) that methylation-independent processing of het-

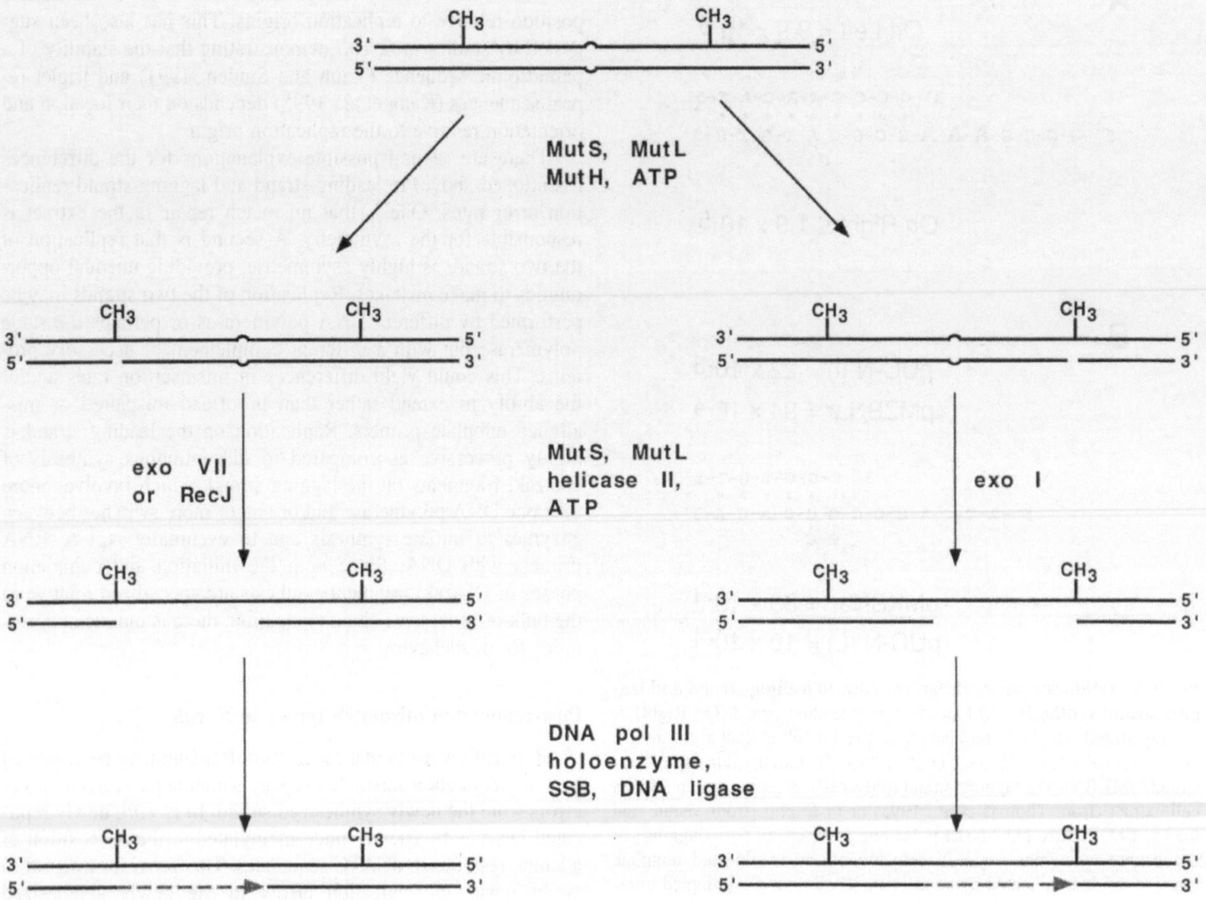

Fig. 4. Methyl-directed mismatch repair in *E. coli*. See text for description.

eroduplexes with loops does occur in *E. coli*. MutS and MutL also interact with recombination intermediates containing mispairs to block recombination between genetically diverged DNA sequences (Worth and Modrich, 1994; for review, Modrich and Lahue, 1996). In addition to the general mismatch repair system, *E. coli* also contains at least two other repair systems for correcting specific mispairs that result from DNA damage, very short patch (VSP) repair (Friedberg et al., 1995) and MutY-dependent repair (Tajiri et al., 1995).

Mismatch repair in human cells

Here we emphasize recent studies of mismatch repair in humans. The reader is also directed to recent reviews (Kolodner and Fishel, 1995; Kolodner, 1995; Modrich and Lahue, 1996) that consider these and additional studies in yeast, *Xenopus* and *Drosophila*. The general mismatch repair (MMR) system in human cells behaves similarly to the *E. coli* methyl directed MMR system. Substrates containing the eight possible base · base mismatches or one to several unpaired bases (Holmes et al., 1990; Thomas et al., 1991; Fang and Modrich, 1993; Fang et al., 1993; Parsons et al., 1993; Umar et al., 1994a) are repaired with an efficiency similar to that in *E. coli*. However, unlike in *E. coli*, some C · C mismatches are well repaired in human cell extracts (Holmes et al., 1990; Fang et al., 1993). Mismatch repair in human cells is strand specific and, although the signal for strand discrimination *in vivo* is unknown, repair *in vitro* can be directed to one strand by an experimentally introduced nick in the DNA substrate (Holmes et al., 1990; Thomas et al., 1991). As in *E. coli*, the human mismatch repair system requires ATP and excision and resynthesis occur between the nick and the mismatch (Holmes et al., 1990; Thomas et al., 1991), which can be separated by as much as 1000 base pairs (hence the designation 'long-patch' mismatch repair). As in *E. coli*, the human mismatch repair system has bidirectional excision capability (Fang and Modrich, 1993). The DNA synthesis step is sensitive to aphidicolin (Holmes et al., 1990; Thomas et al., 1991), suggesting the involvement of a polymerase that also participates in replication, i.e. pol α, δ or ε.

Although the identities of many of the proteins required for excision-resynthesis [helicase, exonuclease(s), ligase, etc.] remain to be established, some of those required for the early steps in the human pathway (Fig. 5) are rapidly emerging from studies showing a connection between cancer and defective mismatch repair. As mentioned above, certain tumor cells exhibit elevated genome-wide mutation rates in short repetitive sequences. Although originally speculated to result from reduced replication fidelity, this phenotype is now thought to result from mutations in genes that result in inactivation of the MMR system. (For an excellent chronological summary of progress through 1994, see Marra and Boland, 1994.) Several candidate mismatch repair genes have now been identified in human cells (Table 2), five of which have been found to be mutant in either tumors or in tumor cell lines.

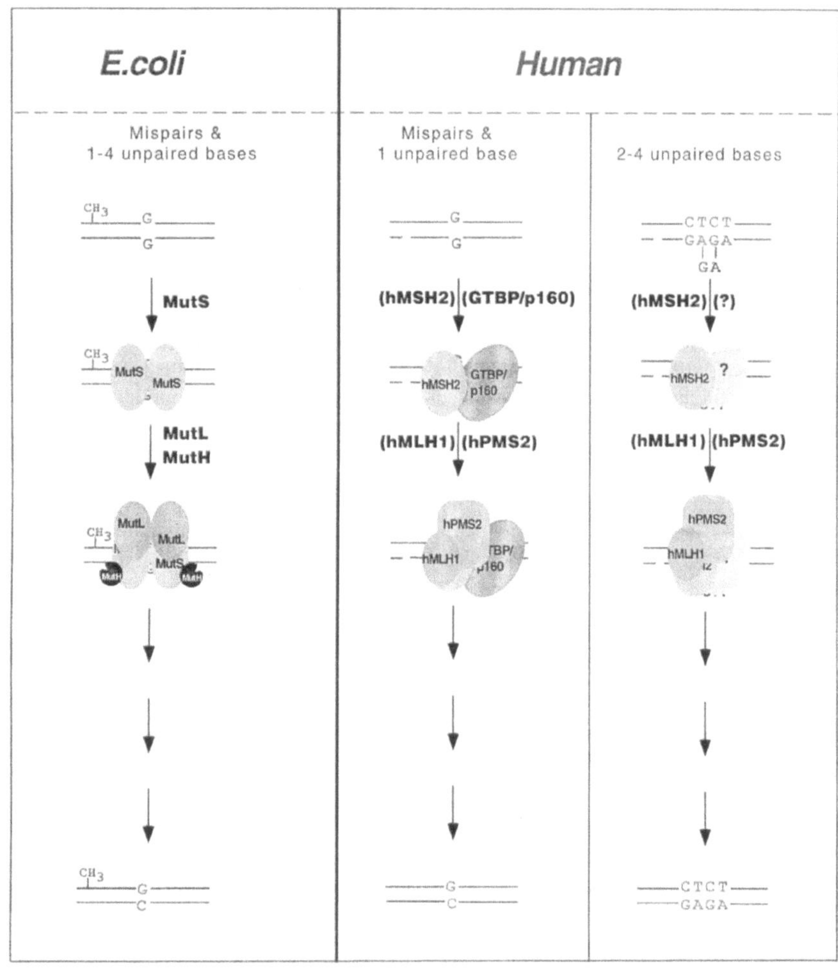

Fig. 5. Comparison of initial steps in mismatch repair in *E. coli* and humans. See text for description.

The MutS homologs

The *hMSH2* gene is so designated because it is one of several genes that shares sequence similarity with the bacterial *MutS* gene (Kolodner and Fishel, 1995). It is located on chromosome 2p and encodes a protein of 105 kDa. Like MutS, the hMSH2 protein binds *in vitro* selectively to DNA containing base · base mispairs and to substrates containing up to 14 extra bases (Fishel et al., 1994a, 1994b). Tumor cell lines that contain mutant *hMSH2* genes are mutators, i.e., they have strongly elevated rates of spontaneous mutations (Branch et al., 1995) and exhibit microsatellite instability (Orth et al., 1994; Shibata et al., 1994; Umar et al., 1994a, 1994b; Risinger et al., 1995a; Boyer et al., 1995). Extracts of such cells deficient in repair of mispaired and unpaired substrates *in vitro* (Umar et al., 1994a, 1994b; Drummond et al., 1995; Risinger et al., 1995; Boyer et al., 1995). Repair activity can be restored to these extracts by a protein preparation designated hMutSα (Drummond et al., 1995), consisting of a heterodimer of hMSH2 and the 160-kDa product of a gene located on chromosome 2p that shares close similarity with the yeast *MSH6* gene. The human gene has been been designated *p160* (Drummond et al., 1995) or *GTBP* (Palombo et al., 1995; Papadopoulos et al., 1995), the latter for G · T-mismatch-binding protein, the property upon which the gene product was first identified (Hughes and Jiricny, 1992). In fact, efficient G · T-mismatch-binding requires both MSH2 and GTBP (Palombo et al., 1995).

Table 2. Genes of interest for mismatch repair in eukaryotics.

Human gene	Other designation	Chromosomal location	Association with HNPCC kinreds
MutS homologs:			
MSH2	–	2p15–16	50–60%
GTBP	*p160/MSH6*	2p16	none
MRP1	*DUG1/MSH3*	5q11–13	none
MutL homologs:			
MLH1	–	3p21–23	20–30%
PMS2	*PMS1*	7p22	two families
PMS2	–	2q31–33	one family

The inability of extracts of *MSH2*-defective cells to repair substrates containing any single-base mismatch or 1–4 unpaired bases strongly suggests that hMSH2 is a central protein in the initial mismatch recognition step of the pathway. Extracts of cells that have mutations in *GTBP* are also deficient in repair of substrates containing mismatches (Kat et al., 1993) or a single unpaired base (Umar et al., 1994a; Drummond et al., 1995). These observations indicate that more than one human gene product is required to fulfill the function of MutS in *E. coli* in correcting single-base mispairs (Fig. 5) and that hMutSα may be the major recognition complex for these substrates. However, the degree of microsatellite instability in *GTBP*-defective cells is not as high as in *hMSH2*-defective or *hMLH1*-defective cells (Shibata et al., 1994; Papadopoulos et al., 1995; Bhattacharyya et al., 1995), and extracts of *GTBP*-defective cells are partially proficient in repair of substrates containing 2−4 unpaired bases (Drummond et al., 1995). Thus, repair of some heteroduplexes may be initiated by hMSH2 alone or when complexed with another protein (Fig. 5, right). Possibilities include the human equivalents of several yeast *MutS* homolgs that have been reported (see Kolodner and Fishel, 1995). Among these is the *yMSH3* gene, whose counterpart in humans has been designated *MRP1* (Shinya and Shimada, 1994) and is located on chromosome 5q (Table 3). While the MRP1 gene product has not been shown to participate in mismatch repair, mutation in the homologous *yMSH3* gene that has little effect on the spontaneous point mutation rate at several loci does strongly elevate the mutation rate for frameshifts (mostly deletions) in dinucleotide sequences (Strand et al., 1995). This implies a repair defect for substrates containing two unpaired bases but not mismatches, a different specificity than for *GTBP* mutation in human cells. Thus, a hypothetical candidate for and hMSH2 partner would be the human homolog of yeast MSH3, perhaps MRP1. Strand et al. (1995) also discuss the possibility that there may be MMR complexes that are specific for correction of leading and lagging-strand replication errors.

The MutL homologs

The *hMLH1*, *hPMS1* and *hPMS2* genes (Table 2) all share close sequence similarity to the bacterial *MutL*. Cells containing mutations in *hMLH1* have greatly elevated rates of spontaneous mutation (Bhattacharyya et al., 1994, 1995; Branch et al., 1995; Eshleman et al., 1995) and cells containing mutations in *hMLH1* or *hPMS2* exhibit microsatellite instability (Parsons et al., 1993; Umar et al., 1994a, 1994b; Orth et al., 1994; Bhattacharyya et al., 1994; Shibata et al., 1994; Liu et al., 1995; Kahn et al., 1995; Risinger et al., 1995b; Boyer et al., 1995). A role for both the *hMLH1* and *hPMS2* gene products in mismatch repair is suggested by observations that extracts made from cells with mutations in these genes are defective in repair of mispairs and unpaired bases (Parsons et al., 1993; Umar et al., 1994a, 1994b; Risinger et al., 1995b; Boyer et al., 1995). Repair activity can be restored to an extract made from a *hMLH1*-defective tumor cell line by a protein fraction designated hMLHα (Li et al., 1995), a heterodimer comprised of hMLH1 (85 kDa) and a second protein (110 kDa), most likely hPMS2. Thus, at least two gene products are required in humans to fulfill the function of one (MutL) in *E. coli*.

A functional role in MMR for *hPMS1* or the other *MutL* homologs that have been reported (Nicolaides et al., 1994; Horii et al., 1994) remains to be established. However, the mere fact that other homologs exist provides for the possibility that mismatch repair may also be functionally differentiated at this step in the pathway.

Other roles for mismatch repair genes

Mismatch repair genes clearly have roles other than correction of normal replication errors. The first human cell line found to be mismatch-repair-defective (Kat et al., 1993) was originally isolated based on its acquired resistance to treatment with *N*-methyl-*N*-nitro-*N*-nitrosoguanidine (Goldmacher et al., 1986). This line has recently been found to have mutations in *GTBP* (Papadopoulos et al., 1995). Since then, several other cell lines with mutations in other mismatch repair genes have been found to be highly resistant to alkylating agents. This includes mouse ES cells containing a homozygous mutation in *MSH2* (de Wind et al., 1995) that is much more MNNG-tolerant than wild-type or heterozygous cells. It also includes human tumor cell lines containing mutations in *hPMS2* (Risinger et al., 1995b) or *hMLH1* (Koi et al., 1994; Branch et al., 1995). The *hMLH1*-defective cell line is reverted to an MNNG-sensitive phenotype by introduction of human chromosome 3 containing a wild-type *hMLH1 gene* (Koi et al., 1994). This is accompanied by restoration of mismatch repair activity and microsatellite stability, reinforcing the model that all three phenotypes of the *hMLH1*-defective cell line result from the mutation in *hMLH1*.

One possible model for this resistance to alkylating agents (see Goldmacher et al., 1986; Karran and Bignami, 1994) is generation of O^6meG · T and O^6meG · C base pairs during replication followed by recognition by the mismatch repair system. Since mismatch repair of the newly synthesized strand would not remove the alkylated template base, leading to repetitive futile repair attempts and eventual lethality. That the mismatch repair system senses DNA damage is further suggested by the observation that the bacterial system responds to damage resulting from ultraviolet irradiation (Feng and Hays, 1995). In the human system, treatment of mismatch repair-proficient cells with 6-thioguanine or *N*-methyl-*N*-nitro-*N*-nitrosoguanidine leads to accumulation of cells in the G2 phase of the cell cycle. No such accumulation is observed with *GTBP*-defective or *hMLH1*-defective cell lines (Kat et al., 1993; Hawn et al., 1995), but accumulation of *hMLH1*-defective cells in G2 is restored upon introduction of chromosome 3 containing the wild-type hMLH1 gene (Hawn et al., 1995). Cumulatively, these observation suggest that the mismatch repair system may play a role as a sensor for genetic damage.

Mismatch repair-defective *mutS* and *mutL* mutants of *E. coli* are also defective in transcription-coupled nucleotide excision repair of ultraviolet photoproducts in the lactose operon (Mellon and Champe, 1995). Similar observations in mismatch-repair-defective human cell lines would suggest that MMR genes might have a role in this very different type of repair that is responsible for correcting a variety of diverse DNA lesions (Selby and Sancar, 1994).

Numerous genetic studies (reviewed in Kolodner, 1995; Modrich and Lahue, 1996) have illustrated that *MutS* and *MutL* genes (or certain of their homologs in eukaryotes) influence recombination processes between DNA strands with less than perfect similarity. This includes long-known effects on gene conversions involving a small number of mismatches as well as effects on homeologous recombination between sequences diverged by several percent. The latter reflects the promiscuous ability of strand transferases to proceed through regions of substantially imperfect similarity. Since mismatch repair genes abort or alter such homeologous recombination (by mechanisms that remain to be elucidated, e.g. see Worth and Modrich, 1994), inactivation of mismatch repair genes provides the potential for significant genetic instability in the form of chromosomal rearrangements. This form of genome instability in a mismatch-repair-defective cell may be highly relevant to tumor progression.

A recent study (Baker et al., 1995) has also indicated a role for a mismatch repair gene in spermatogenesis. Male mice with a null mutant in *PMS2* are infertile, producing only abnormal spermatozoa. Examination of axial element and synaptonemal complex formation during prophase of meiosis I indicated abnormal chromosome synapsis. Thus, this mismatch repair gene is also required for male fertility and normal chromosome synapsis. Interestingly, mice homozygous for mutations in *MSH2* are not reported to be infertile (de Wind et al., 1995; Reitmair et al., 1995).

Mismatch repair and cancer predisposition

As mentioned at the beginning, the development of a tumor requires several mutations and the spontaneous mutation rate in normal cells is low, leading to the hypothesis that an early event in tumorigenesis may be a gene mutation that confers a persistent mutator phenotype (Loeb, 1991). Consistent with this hypothesis is the present link between cancer, microsatellite instability, mutations in mismatch repair genes and defective mismatch repair activity. The initial key to this linkage (for a historical review, see Marra and Boland, 1995) was the observation of thousands of mutations in microsatellite sequences throughout the genome in about 12−15% of sporadic tumors and in most tumors from patients with hereditary non-polyposis cancer, HNPCC (Eshleman and Markowitz, 1995). HNPCC is an autosomal dominant disease in which multiple members of a family get early-onset colon cancer in the absence of polyp formation. HNPCC accounts for several percent of colon cancers and the disease is also associated with frequent tumors in other organs, including the ovary, endometrium, stomach, pancreas, small intestine, skin and urinary tract.

The instability observed in HNPCC tumors is consistent with the idea that the genome wide mutation rate is elevated at an early step in tumorigenesis as a result of a defect in an HNPCC-linked gene, leading to the accumulation of mutations in other genes that provide the selective advantages needed for cells to ultimately give rise to a tumor. The gene defect(s) responsible for HNPCC might be in DNA replication fidelity or in subsequent repair of the heteroduplex intermediates for deletion and addition mutations. Evidence for the latter has rapidly accumulated. Of the HNPCC tumors that exhibit microsatellite instability, over 50% contain mutations in the HNPCC-linked *hMSH2* gene on chromosome 2 and about 20−30% contain mutations in HNPCC-linked *hMLH1* gene on chromosome 3 (Table 3). Analyses of most HNPCC kindreds indicate that normal cells contain a germline defect in one allele, while in tumor cells the remaining wild-type allele is lost or is mutant (reviewed in Kolodner, 1995). These data conform to Knudson's two-hit hypothesis (Knudson, 1971) for genes such as *p53*, *RB* and *APC*, whose functions are lost. Mutations and/or loss of both alleles of *hMSH2* and *hMLH1* have also been found in sporadic tumors (e.g. see Borresen et al., 1995 and references therein) and in mismatch-repair-defective tumor cell lines (see above), where again there is usually no evidence for a wild-type allele. The microsatellite instability in HNPCC is an early event in tumorigenesis that persists (Shibata et al., 1994). The microsatellite instability characteristic of *hMSH2* and *hMLH1* mutant cell lines also uniformly correlates with a greatly elevated spontaneous mutation rates in endogenous genes (see above), as required by the hypothesis. This mutator phenotype obviously places a caveat on studies with cell lines that attempt to establish a cause-effect relationship between a candidate gene and an observed effect, since such cells may contain mutations in other genes that may explain the effect.

A connection between tumorigenesis and other candidate mismatch repair genes is not as strong. Evidence for a germline mutation in the *hPMS1* gene is limited to a single HNPCC family (Nicolaides, 1994) and the *hPMS1* gene product has not yet been shown to be functionally important for mismatch repair. The opposite is true for *GTBP*, where there is substantial evidence for a function in mismatch repair (see above) but no evidence yet for mutations in HNPCC kindreds (Papadopoulos, 1995), although both alleles are mutant in a sporadic colon tumor cell line. While the evidence indicates that hPMS2 functions in mismatch repair (see above), mutations in the *hPMS2* have been described in only two HNPCC families (Nicolaides, 1994; Parsons et al., 1995). Mice containing homozygous null mutations in *PMS2* (Baker et al., 1995), or *MSH2* (de Wind et al., 1995; Reitmair et al., 1995), are viable and develop sarcomas and lymphomas. Tumors and certain tissue samples in these knockout mice exhibit microsatellite instability. However, none of these mutant mice developed colon tumors, suggesting that these mice might not ideally model HNPCC.

HNPCC paitients have also been found to contain *hMLH1* and *hPMS2* mutations that lead to microsatellite instability in both normal and tumor cells (Parsons et al., 1995). Cell lines from these patients lack mismatch repair even though they are detectably mutant in only one of the two copies of the *hMLH1* or *hPMS2* gene. This is in contrast to the usual situation wherein both alleles are mutant and/or lost. The apparent heterozygosity but mismatch repair defect in these cells may thus indicate dominant negative mutations (also see Borresen et al., 1995), as might be anticipated for a repair pathway involving many interacting gene products. These phenotypically repair-defective patients did not develop tumors at an unusually earlier age or higher incidence. This lead the authors to speculate that an additional tissue regenerative stimulus such as that afforded by exposure to a high concentration of mutagens might be required for tumorigenesis.

The idea that a mismatch repair defect leads to an elevated mutation rate that promotes mutations in critical genes needed for tumorigenesis does not account for some of the above observations and the fact that HNPCC individuals do not develop tumors in all tissues with equal probability. This situation reflects our present incomplete understanding of mismatch repair processes in human cells. These seem to be more differentiated than in bacteria with respect to the number of genes involved, their substrate specificity, developmental control and possibly their tissue specificity. The gene defects responsible for a substantial number of HNPCC families are not known, but could be in yet to be discovered genes in the mismatch repair pathway. Here, as in the case of our present understanding of eukaryotic mismatch repair, lessons learned from studies in other organisms, particularly yeast (Strand et al., 1993, 1995; Johnson et al., 1995; Kolodner, 1995), should prove invaluable. Also incomplete is our understanding of the roles of some of these genes in recombination, meiosis, other types of repair, cell-cycle control or other processes. Especially important for understanding tumor progression will be an appreciation of the selective advantages imparted to a cell upon loss of one or both alleles of candidate mismatch repair genes (Bodmer et al., 1994). The present link between microsatellite instability, defective mismatch repair and cancer has greatly stimulated interest in identifying and understanding the relevant genes and their functions, and more broadly, in deciphering other replication, repair or recombination processes that might destabilize the genome and promote tumorigenesis.

REFERENCES

Baker, S. M., Bronner, C. E., Zhang, L., Plug, A. W., Robatzek, M., Warren, G., Elliott, E. A., Yu, J., Ashley, T., Arnheim, N., Flavell, R. A. & Liskay, R. M. (1995) Male mice defective in the DNA mismatch repair gene *PMS2* exhibit abnormal chromosomal synapsis in meiosis, *Cell 82*, 309−319.

Bebenek, K., Abbotts, J., Wilson, S. H. & Kunkel, T. A. (1993) Error-prone polymerization by HIV-1 reverse transcriptase. Contribution of template-primer misalignment, miscoding, and termination probability to mutational hot spots, *J. Biol. Chem. 268*, 10324–10334.

Bebenek, K., Beard, W. A., Casas-Finet, J. R., Kim, H.-R., Darden, T. A., Wilson, S. H. & Kunkel, T. A. (1995) Reduced frameshift fidelity and proceeivity of HIV-1 reverse transcriptase mutants containing alanine substitutions in helix H of the thumb subdomain, *J. Biol. Chem. 270*, 19516–19523.

Bhattacharyya, N. P., Skandalis, A., Ganesh, A., Groden, J. & Meuth, M. (1994) Mutator phenotypes in human colorectal carcinoma cell lines, *Proc. Natl Acad. Sci. USA 91*, 6319–6323.

Bhattacharyya, N. P., Ganesh, A., Phear, G., Richards, B., Skandalis, A. & Meuth, M. (1995) Molecular analysis of mutations in mutator colorectal carcinoma cell lines, *Hum. Molec. Genet. 4*, 2057–2064.

Bodmer, W., Bishop, T. & Karran, P. (1994) Genetic steps in colorectal cancer, *Nature Genet. 6*, 217–219.

Borresen, A.-L., Lothe, R. A., Meling, G. I., Lystad, S., Morrison, P., Lipford, J. R., Kane, M. F., Rognum, T. O. & Kolodner, R. D. (1995) Somatic mutations in the *hMSH2* gene in microsatellite unstable colorectal carcinomas, *Hum. Mol. Genet. 4*, 2065–2072.

Boyer, J. C., Umar, A., Risinger, J. I., Lipford, R., Kane, M., Yin, S., Barrett, J. C., Kolodner, R. D. & Kunkel, T. A. (1995) Microsatellite instability, mismatch repair deficiency and genetic defects in human cancer cell lines, *Cancer Res. 55*, 6063–6070.

Branch, P., Hampson, R. & Karran, P. (1995) DNA mismatch binding defects, DNA damage tolerance, and mutator phenotypes in human colorectal carcinoma cell lines, *Cancer Res. 55*, 2304–2309.

Carraway, M. & Marinus, M. G. (1993) Repair of heteroduplex DNA molecules with multibase loops in *Escherichia coli*, *J. Bacteriol. 175*, 3972–3980.

de Wind, N., Dekker, M., Berns, A., Radman, M. & te Riele, H. (1995) Inactivation of the mouse *Msh2* gene results in mismatch repair deficiency, methylation tolerance, hyperrecombination, and predisposition to cancer, *Cell 82*, 321–330.

DePamphilis, M. (1996) *DNA replication in eukaryotic cells: concepts, enzymes & systems*, Cold Spring Harbor Laboratories, Cold Spring Harbor, New York, in the press.

Drummond, J. T., Li, G.-M., Longley, M. J. & Modrich, P. (1995) Isolation of an hMSH2-p160 heterodimer that restores DNA mismatch repair to tumor cells, *Science 268*, 1909–1912.

Eshleman, J. R., Lang, E. Z., Bowerfind, G. K., Parsons, R., Vogelstein, B., Wilson, J. K. V., Veigl, M. L., Sedwick, W. D. & Markowitz, S. D. (1995) Increased mutation rate at the *hprt* locus accompanies microsatellite instability in colon cancer, *Oncogene 10*, 33–37.

Eshleman, J. R. & Markowitz, S. D. (1995) Microsatellite instability in inherited and sporadic neoplasms, *Current Opin. Oncol. 7*, 83–89.

Fang, W.-H. & Modrich, P. (1993) Human strand-specific mismatch repair occurs by a bidirectional mechanism similar to that of the bacterial reaction, *J. Biol. Chem. 268*, 11838–11844.

Fang, W.-H., Li, G.-M., Longley, M., Holmes, J., Thilly, W. & Modrich, P. (1993) Mismatch repair and genetic stability in human cells, *Cold Spring Harbor Symp. Quant. Biol. 58*, 597–603.

Feng, W. Y., Lee, E. H. & Hays, J. B. (1991) Recombinagenic processing of UV-light photoproducts in nonreplicating phage DNA by the *Escherichia coli* methyl-directed mismatch repair system, *Genetics 129*, 1007–1020.

Fishel, R. & Kolodner, R. D. (1995) Identification of mismatch repair genes and their role in the development of cancer, *Current Opin. Genet. & Dev. 5*, 382–395.

Fishel, R., Ewel, A., Lee, S., Lescoe, M. K. & Griffith, J. (1994a) Binding of mismatched microsatellite DNA Sequences by the human MSH2 protein, *Science 266*, 1403–1405.

Fishel, R., Ewel, A. & Lescoe, M. K. (1994b) Purified human MSH2 protein binds to DNA containing mismatched nucleotides, *Cancer Res. 54*, 5539–5542.

Friedberg, E. C., Walker, G. C. & Siede, W. (1995) in: *DNA repair and mutagenesis*, pp. 396–398, Washington, D. C., A. S. M. Press.

Goldmacher, V. S., Cuzick, R. A. J. & Thilly, W. G. (1986) Isolation and partial characterization of human cell mutants differing in sensitivity to killing and mutation by methylnitrosourea and N-methyl-N'-nitro-N-nitrosoguanidine, *J. Biol. Chem. 261*, 12462–12471.

Hawn, M. T., Umar, A., Carethers, J. M., Marra, G., Kunkel, T. A., Boland, C. R. & Koi, M. (1995) Evidence for a connection between the mismatch repair system and the G2 cell cycle checkpoint, *Cancer Res. 55*, 3721–3725.

Holmes, J., Clark, S. J. & Modrich, P. (1990) Strand-specific mismatch correction in nuclear extracts of human and *Drosophila melanogaster* cell lines, *Proc. Natl Acad. Sci. USA 87*, 5837–5841.

Horii, A., Han, H.-J., Sasaki, S., Shimada, M. & Nakamura, Y. (1994) Cloning, charachterization and chromosomal assignment of the human genes homologous to yeast PMS1, a member of mismatch repair genes, *Biochem. Biophys. Res. Commun. 204*, 1257–1264.

Hughes, M. J. & Jiricny, J. (1992) The purification of a human mismatch-binding protein and association of its associated ATPase and helicase activities, *J. Biol. Chem. 267*, 23876–23882.

Johnson, K. A. (1993) Conformational coupling in DNA polymerase fidelity, *Annu. Rev. Biochem. 62*, 685–713.

Johnson, R. E., Kovvali, G. K., Prakash, L. & Prakash, S. (1995) Requirement of the yeast RTH1 5' to 3' exonuclease for the stability of simple repetitive DNA, *Science 269*, 238–240.

Kahn, S. M., Klein, M. G., Jiang, W., Xing, W. Q., Xu, D. B., Perucho, M. & Weinstein, I. B. (1995) Design of a seletable reporter for the detection of mutations in mammalian simple repeat sequences, *Carcinogenisis 16*, 1223–1228.

Kang, S., Jaworski, A., Ohshima, K. & Well, R. D. (1995) Expansion and deletion of CTG repeats from human disease genes are determined by the direction of replication in *E. coli*, *Nature Genet. 10*, 213–218.

Karran, P. & Bignami, M. (1994) DNA damage tolerance, mismatch repair and genome instability, *BioEssays 16*, 833–839.

Kat, A., Thilly, W. G., Fang, W.-H., Longley, M. J., Li, G.-M. & Modrich, P. (1993) An alkylation-tolerant, mutator human cell line is deficient in strand-specific mismatch repair, *Proc. Natl Acad. Sci. USA 90*, 6424–6428.

Kelman, Z. & O'Donnell, M. (1994) DNA replication: enzymology and mechanisms, *Current Opin. Genes Dev. 4*, 185–195.

Knudson, A. G. (1971) Mutation and cancer: statistical study of retinoblastoma, *Proc. Natl Acad. Sci. USA 68*, 820–823.

Koi, M., Umar, A., Chauhan, D. P., Cherian, S. P., Carethers, J. M., Kunkel, T. A. & Boland, C. R. (1994) Human chromosome 3 corrects mismatch repair deficiency and microsatellite instability and reduces *N'*-methyl-*N*-nitro-*N*-nitrosoguanidine-tolerance in colon tumor cells with a homozygous *hMLH1* mutation, *Cancer Res. 54*, 4308–4312.

Kolodner, R. D. (1995) Mismatch repair: mechanisms and relationship to cancer susceptibility, *Trends Biol. Sci. 20*, 397–401.

Kroutil, L. C., Register, K., Bebenek, K. & Kunkel, T. A. (1996) Exonucleolytic proofreading during replication of repetitive DNA, *Biochemistry 35*, 1046–1053.

Kunkel, T. A. (1990) Misalignment-mediated DNA synthesis errors, *Biochemistry 29*, 8003–8011.

Kunkel, T. A., Patel, S. S. & Johnson, K. A. (1994) Error-prone replication of repeated DNA sequences by T7 DNA polymerase in the absence of its processivity subunit, *Proc. Natl Acad. Sci. USA 91*, 6830–6834.

Li, G.-M. & Modrich, P. (1995) Restoration of mismatch repair to nuclear extracts of H6 colorectal tumor cells by a heterodimer of human MutL homologs, *Proc. Natl Acad. Sci. USA 92*, 1950–1954.

Liu, B., Nicolaides, N. C., Markowitz, S., Willson, J. K. V., Parsons, R. E., Jen, J., Papadopoulos, N., Peltomäki, P., de la Chapelle, A., Hamilton, S. R., Kinzler, K. W. & Vogelstein, B. (1995) Mismatch repair gene defects in sporadic colorectal cancers with microsatellite instability, *Nature Genet. 9*, 48–55.

Loeb, L. A. (1991) Mutator phenotype may be required for multistage carcinogenesis, *Cancer Res. 54*, 5059–5063.

Marra, G. & Boland, C. R. (1995) Hereditary nonpolyposis colorectal cancer (HNPCC): The syndrome, the genes, and historical perspectives, *J. Natl Can. Inst. 87*, 1114–1125.

Mellon, I. & Champe, G. N. (1995) Products of DNA mismatch repair genes *mutS* and *mutL* are required for transcription-coupled nucleotide excision repair of the lactose operon in *Escherichia coli*, *Proc. Natl Acad. Sci. USA 92*, 1292–1297.

Modrich, P. (1991) Mechanisms and biological effects of mismatch repair, *Annu. Rev. Genet. 25*, 229–253.

Modrich, P. (1994) Mismatch repair, genetic stability, and cancer, *Science 266*, 1959–1960.

Modrich, P. & Lahue, R. (1996) Mismatch repair in replication fidelity, genetic recombination, and cancer biology, *Annu. Rev. Biochem.*, in the press.

Nicolaides, N. C., Papadopoulos, N., Liu, B., Wel, Y.-F., Carter, K. C., Ruben, S. M., Rosen, C. A., Haseltine, W. A., Fleishmann, R. D., Fraser, C. M., Adams, M. D., Venter, J. C., Dunlop, M. G., Hamilton, S. R., Peterson, G. M., de la Chapelle, A., Vogelstein, B. & Kinzler, K. W. (1994) Mutations of two *PMS* homologues in hereditary nonpolyposis colon cancer, *Nature 371*, 75–80.

Orth, K., Hung, J., Gazdar, A., Bowcodk, A., Mathis, J. M. & Sambrook, J. (1994) Genetic instability in human ovarian cancer cell lines, *Proc. Natl Acad. Sci. USA 91*, 9495–9499.

Palombo, F., Gallinari, P., Iaccarino, I., Lettieri, T., Hughes, M., D'Arrigo, A., Truong, O., Hsuan, J. J. & Jiricny, J. (1995) GTBP, a 160 kD protein essential for mismatch binding activity in human cells, *Science 268*, 1912–1914.

Papadopoulos, N., Nicolaides, N. C., Liu, B., Parsons, R. E., Lengauer, C., Palombo, F., D'Arrigo, A., Markowitz, S., Willson, J. K. V., Kinzler, K. W., Jiricny, J. & Vogelstein, B. (1995) Mutations of *GTBP* in genetically unstable cells, *Science 268*, 1915–1917.

Parker, B. O. & Marinus, M. G. (1992) Repair of DNA heteroduplexes containing small heterologous sequences in *Escherichia coli*, *Proc. Natl Acad. Sci. USA 89*, 1730–1734.

Parsons, R., Li, G.-M., Longley, M. J., Fang, W.-H., Papadopoulos, N., Jen, J., de la Chapelle, A., Kinzler, K. W., Vogelstein, B. & Modrich, P. (1993) Hypermutability and mismatch repair deficiency in RER^+ tumor cells, *Cell 75*, 1227–1236.

Parsons, R., Li, G.-M., Longley, M., Modrich, P., Liu, B., Berk, T., Hamilton, S. R., Kinzler, K. W. & Vogelstein, B. (1995) Mismatch repair deficiency in phenotypically normal human cells, *Science 268*, 738–740.

Reitmair, A. H., Schmits, R., Ewel, A., Bapat, B., Redston, M., Mitri, A., Waterhouse, P., Mittrücker, H.-W., Wakeham, A., Liu, B., Thomason, A., Griesser, H., Gallinger, S., Ballhausen, W. G., Fishel, R. & Mak, T. W. (1995) *MSH2* deficient mice are viable and susceptible to lymphoid tumours, *Nature Genet. 11*, 64–70.

Ripley, L. S. (1990) Frameshift mutation: Determinants of specificity, *Annu. Rev. Genet. 24*, 189–213.

Risinger, J. I., Umar, A., Boyer, J. C., Evans, A. C., Berchuck, A., Kunkel, T. A. & Barrett, J. C. (1995a) Microsatellite instability in gynecological sarcomas and in uterine sarcoma cell lines defective in mismatch repair activity and mutant in the *hMSH2* gene, *Cancer Res. 55*, 5664–5669.

Risinger, J. I., Umar, A., Barrett, J. C. & Kunkel, T. A. (1995b) A *hPMS2* mutant cell line is defective in strand-specific mismatch repair, *J. Biol. Chem. 270*, 18183–18186.

Roberts, J. D., Izuta, S., Thomas, D. C. & Kunkel, T. A. (1994) Mispair-, site-, and strand-specific error rates during simian virus 40 origin-dependent replication *in vitro* with excess deoxythymidine triphosphate, *J. Biol. Chem. 269*, 1711–1717.

Roberts, J. D., Nguyen, D. & Kunkel, T. A. (1993). Frameshift fidelity during replication of double-stranded DNA in HeLa cell extracts, *Biochemistry 32*, 4083–4089.

Roberts, J. D., Thomas, D. C. &. Kunkel, T. A. (1991) Exonucleolytic proofreading of leading and lagging strand DNA replication errors, *Proc. Natl Acad. Sci. USA 88*, 3465–3469.

Roberts, J. D. & Kunkel, T. A. (1995) Eukaryotic DNA replication fidelity, in *DNA replication in eukaryotic cells: concepts, enzymes & systems* (De Pamphilis, M., ed.) Cold Spring Harbor Laboratories, Cold Spring Harbor, New York, in the press.

Schaaper, R. M. (1993). Base selection, proofreading, and mismatch repair during DNA replication in *Escherichia coli*, *J. Biol. Chem. 268*, 23762–23765.

Selby, C. P. & Sancar, A. (1994) Mechanism of transcription-repair coupling and mutation frequency decline, *Micobiol. Rev. 58*, 317–329.

Shibata, D., Peinado, M. A., Ionov, Y., Malkhosyan, S. & Perucho, M. (1994) Genomic instability in repeated sequences is an early somatic event on colorectal tumorigenesis that persists after transformation, *Nature Genet. 6*, 273–281.

Shinya, E. & Shimada, T. (1994) Identification of two initiator elements in the bidirectional promoter of the human dihydrofolate reductase and mismatch repair protein 1 genes, *Nucleic Acids Res. 22*, 2143–2149.

Strand, M., Prolla, T. A., Liskay, R. M. & Petes, T. D. (1993). Destabilization of tracts of simple repetitive DNA in yeast by mutations affecting DNA mismatch repair, *Nature 365*, 274–276.

Strand, M., Earley, M. C., Crouse, G. F. & Petes, T. D. (1995) Mutations in the *MSH3* gene preferentially lead to deletions within tracts of simple repetitive DNA in *Saccharomyces cerevisiae*, *Proc. Natl Acad. Sci. USA 92*, 10418–10421.

Streisinger, G., Okada, Y., Emrich, J., Newton, T., Tsugita, A.. Terzaghi, E. & Inouye, M. (1966) Frameshift mutations and the genetic code, *Cold Spring Harbor Symp. Quant. Biol. 31*, 77–84.

Tajiri, T., Maki, H. & Sekiguchi, M. (1995) Functional cooperation of MutT, MutM and MutY proteins in preventing mutations caused by spontaneous oxidation of guanine nucleotide in *Escherichia coli*, *Mutation Res. 336*, 257–267.

Thomas, D. C., Roberts, J. D. & Kunkel, T. A. (1991) Heteroduplex repair in extracts of human HeLa cells, *J. Biol. Chem. 266*, 3744–3751.

Thomas, D. C., Nguyen, D. C., Piegorsch, W. W. & Kunkel, T. A. (1993) Relative probability of mutagenic translesion synthesis on the leading and lagging strands during replication of UV-irradiated DNA in a human cell extract, *Biochemistry 32*, 11476–11482.

Thomas, D. C., Veaute, X., Fuchs, R. P. P. & Kunkel, T. A. (1995) Frequency and fidelity of translesion synthesis of site-specific N-2-acetylaminofluorene adducts during DNA replication in a human cell extract, *J. Biol. Chem. 270*, 21226–21233.

Trinh, T. Q. & Sinden, R. R. (1991) Preferential DNA secondary structure mutagenesis in the lagging strand of replication in *E. coli*, *Nature 352*, 544–548.

Umar, A., Boyer, J. C., Thomas, D. C., Nguyen, D. C., Risinger, J. I., Boyd, J., Ionov, Y., Perucho, M. & Kunkel, T. A. (1994a) Defective mismatch repair in extracts of colorectal and endometrial cancer cell lines exhibiting microsatellite instability, *J. Biol. Chem. 269*, 14367–14370.

Umar, A., Boyer, J. C. & Kunkel, T. A. (1994b) DNA loop repair by human cell extracts, *Science 266*, 814–816.

Veaute, X. & Fuchs, R. P. P. (1993) Greater susceptibility to mutations in lagging strand of DNA replication in *Escherichia coli* than in leading strand, *Science 261*, 598–600.

Waga, S. & Stillman, B. (1994) Anatomy of a DNA replication fork revealed by reconstitution of SV40 DNA replication *in vitro*, *Nature 369*, 207–212.

Willems, P. J. (1994) Dynamic mutations hit double figures, *Nature Genetics 8*, 213–215.

Worth, L. Jr, Clark, S., Radman, M. & Modrich, P. (1994) Mismatch rpair proteins MutS and MutL inhibit RecA-catalyzed strand transfer between diverged DNAs, *Proc. Natl Acad. Sci. USA 91*, 3238–3241.

Eur. J. Biochem. *239*, 539–557 (1996)

Review

The emerging three-dimensional structure of a receptor
The nicotinic acetylcholine receptor

Ferdinand HUCHO[1], Victor I. TSETLIN[2] and Jan MACHOLD[1]

[1] Freie Universität Berlin, Institut für Biochemie, Germany
[2] Shemyakin-Ovchinnikov-Institute for Bioorganic Chemistry, Moscow, Russia

(Received 12 February 1996) – EJB 96 0187/0

The nicotinic acetylcholine receptor is the neurotransmitter receptor with the most-characterized protein structure. The amino acid sequences of its five subunits have been elucidated by cDNA cloning and sequencing. Its shape and dimensions (approximately 12.5 nm×8 nm) were deduced from electron-microscopy studies. Its subunits are arranged around a five-fold axis of pseudosymmetry in the order (clockwise) $\alpha_H\gamma\alpha_L\delta\beta$. Its two agonist/competitive-antagonist-binding sites have been localized by photolabelling studies to a deep gorge between the subunits near the membrane surface. Its ion channel is formed by five membrane-spanning (M2) helices that are contributed by the five subunits. This finding has been generalized as the Helix M2 model for the superfamily of ligand-gated ion channels. The binding site for regulatory non-competitive antagonists has been localized by photolabelling and site-directed-mutagenesis studies within this ion channel.

Therefore a three-dimensional image of the nicotinic acetylcholine receptor is emerging, the most prominent feature of which is an active site that combines the agonist/competitive-antagonist-binding sites, the regulatory site and the ion channel within a relatively narrow space close to and within the bilayer membrane.

Keywords: nicotinic acetylcholine receptor; three-dimensional structure; ligand-gated ion channel.

About ten integral membrane proteins have been crystallized, and their three-dimensional structures have been solved recently (Iwata et al., 1995). Among those is still no neurotransmitter receptor. First crystals obtained from acetylcholine receptor preparations were reported (Hertling-Jaweed et al., 1988), but they are still too small for X-ray crystallography. Disappointing as this may be, considerable progress with structural analysis of receptors was achieved. Especially the nicotinic acetylcholine receptor (AChR) can now be described in much detail, and structure/function relationships can be discussed. This is mainly due to virtually unlimited amounts of receptor protein available from the electric tissue of the electric ray *Torpedo* sp., to advances in biochemical and biophysical methodology, and to the exploitation of the combination of recombinant DNA technology with patch clamp electrophysiology. This review intends to draw a picture of a receptor in the stage before X-ray crystallography.

Correspondence to F. Hucho, Freie Universität Berlin, Institut für Biochemie, Thielallee 63, D-14195 Berlin, Germany

Abbreviations. 4TM, four transmembrane model; $5HT_3$, 5-hydroxytryptamine; AChR, acetylcholine receptor; nAChR, nicotinic acetylcholine receptor; αBgt, α-bungarotoxin; βBgt, β-bungarotoxin; DDF, $[^3H]$*p*-(dimethylamino)benzenediazonium fluoroborate; DMT, dimethyl(+)-tubocurarine; dTC, (+)-tubocurarine; FTIR, Fourier-transform infrared; HTX, histrionicotoxin; MBTA, 4-(*N*-maleimido)benzyltrimethylammonium iodide; MIR, main immunogenic region; NCI, non-competitive inhibitors; SCAM, substituted-cysteine-accessible method; ^{125}I-TID, 3-trifluoromethyl-3-(*m*-$[^{125}I]$iodophenyl)diazirine; $TPMP^+$, triphenylmethylphosphonium.

We focus on the structure and functional domains of the AChR from *Torpedo* electric tissue, drawing some arguments from investigations of other nicotinic receptors as well, especially neuronal and higher vertebrate muscle acetylcholine receptors from various species. We attempt to update the picture presented in previous reviews (Barrantes, 1983; Hucho, 1986; Claudio, 1989; Changeux, 1990; Karlin, 1991, 1993; Karlin and Akabas, 1995; Devillers-Thiéry et al., 1993). More recent reviews by the above-mentioned and other authors will be given credit as we go along. We do not include aspects of developmental and molecular biology, and we touch upon its pharmacology and toxicology only in passing. For these topics we would like to refer the reader to specialized reviews (Changeux, 1990; Sargent, 1993; Hucho, 1992; Swanson and Albuquerque, 1992).

We describe receptors as signal converters (Hucho, 1993), whose function is the transduction of extracellular signals into intracellular effects. With the AChR the signal is the neurotransmitter acetylcholine and the intracellular effect is primarily depolarization of the plasma membrane (at least in *Torpedo* electrocytes and in muscle cells; in neuronal cells an increase of the Ca^{2+} concentration seems to be of similar importance). Signal recognition (acetylcholine binding) is converted by the receptor protein into opening of a cation-specific ion channel. The AChR shares this function with other members of the superfamily of ligand gated ion channels, also termed type-I receptors, like the 5-hydroxytryptamine ($5HT_3$) receptors and the anion-specific $GABA_A$ and glycine receptors (Schofield et al., 1987; Betz, 1990). The aim of structural analysis of receptor proteins is to understand how this conversion takes place and how it is regulated. Of special importance for pharmacologists is understanding how specific signal recognition occurs and how agonists and antagonists are discriminated, i.e. what the structure of the binding site for nicotinic effectors may be.

As compared to muscarinic acetylcholine receptors signal transduction through nicotinic receptors is much faster. The mechanism accordingly is entirely different, possibly simpler.

The protein structure

Primary structure. AChR from *Torpedo* is composed of four different polypeptide chains named α (437 amino acids, calculated M_r 50116, apparent M_r in SDS PAGE 40000), β (469 amino acids, calculated M_r 53681, apparent M_r 48000), γ (489 amino acids, calculated M_r 56279, apparent M_r 60000), δ (501 amino acids, calculated M_r 57565, apparent M_r 65000) (see for a review also Changeux, 1990). Cloned and sequenced in the eighties by several laboratories (Noda et al., 1982, 1983a; Claudio et al., 1983) they were found to be the product of different genes (in some cases even localized on different chromosomes) but to be similar in sequence (Fig. 1), hinting at a common evolutionary origin. Over the whole length they show about 18% sequence identity. The *Torpedo* receptor is used by neurochemists as a model for homologous receptors of higher vertebrates because of its conserved structure through evolution (Fig. 2).

Muscle AChR contains several developmental variants: embryonic muscle receptor contains a γ subunit which is postnatally replaced by an ε subunit (Takai et al., 1985; Witzemann et al., 1990). In addition, several variants of the α and γ subunits were detected, the physiological significance of which is not fully understood (Mileo et al., 1995).

More conspicuous is the receptor's significant sequence similarity with the polypeptide chains of the glycine, $GABA_A$, and $5HT_3$ receptors which is the basis for putting them into a common superfamily (Betz, 1990). Cloning similar sequences led to the discovery of several other members of the gene family in vertebrates coding for acetylcholine receptor subunits. Presently a total of nine α (termed $\alpha1-9$) and four β ($\beta1-4$) subunits are known; $\alpha1$ and $\beta1$ occur in *Torpedo* electric tissue and in vertebrate muscle. They are called 'muscle type' receptor subunits, to discriminate them from the neuronal subunits $\alpha2-9$ and $\beta2-4$. Muscle and neuronal sequences are very similar (Fig. 3). Since the β chains, have little in common except being different from the α chains, they are often called 'non-α chains'; γ and δ chains have been found only in electrocyte and muscle receptors. Neuronal receptors are made up of only two types of subunits (α and non-α) (Anand et al., 1991; Sargent, 1993), or they are even homopolymers composed of one type ($\alpha7$). The neuronal subunits are expressed tissue specifically: for example, $\alpha9$ is specifically expressed in cochlear hair cells (Elgoyhen et al., 1994).

In addition to these polypeptide chains which are integral membrane proteins there is one peripheral membrane protein which is present in stoichiometric amounts: originally termed the '43 k protein' it was named rapsyn (from receptor associated protein of the synapse) to differentiate it from other substoichiometric proteins of M_r 43000 (Frail et al., 1988). Rapsyn is proposed to mediate receptor clustering and linkage to the cytoskeleton (Froehner et al., 1990; Phillips et al., 1991, 1993; Apel et al., 1995). Though structurally unrelated it may act similarly to gephyrin of the glycine receptors (Kirsch et al., 1993). Rapsyn has no sequence similarity with any receptor subunit.

Oligosaccharide structure. Description of the covalent structure is not complete without mentioning the carbohydrate moieties. About 20 kDa of the receptor's 290 kDa is carbohydrate. The majority of the AChR's oligosaccharide side chains is of the high-mannose type, the rest is of the sialylated complex type (Nomoto et al., 1986; Poulter et al., 1989; Shoji et al., 1992; Strecker et al., 1994). For a comprehensive discussion of the structural and functional details see Shoji et al. (1992).

The α, β, γ, δ chains contain one, one, four and three potential N-glycosylation sites, respectively (Mattson and Heilbronn, 1975; Vandlen et al., 1979; Lindstrom et al., 1979). Not all of these are actually glycosylated. For example, the potential glycosylation site Asn453 of the γ subunit was shown not to be glycosylated (Poulter et al., 1989). This site, according to the 4TM folding model (see below), is located on the cytoplasmic side. On the other hand, the consensus site at α-Asn141 is glycosylated. The oligosaccharide located here is of the high-mannose type (Poulter et al., 1989). Of the three potential glycosylation sites of the δ chain, Asn70 and Asn208 carry a complex-type and Asn143 a high-mannose type oligosaccharide (Strecker et al., 1994).

Secondary structure, tertiary structure and transmembrane folding

Spectroscopic investigations of the receptor showed that most of the protein is either α-helical or β-structure. In terms of the percentage of the secondary structure components, the data obtained by CD (Moore et al., 1974; Mielke and Wallace, 1988; Wu et al., 1990), Raman (Aslanian et al., 1983; Yager et al., 1984) or Fourier-transform infrared (FTIR) (Fong and McNamee, 1987; Butler and McNamee, 1993; Naumann et al., 1993) are not quite comparable because of the different methods applied and the different receptor preparations used (detergent solubilized versus membrane-bound, native versus reconstituted membranes). Investigations from the author's lab using FTIR spectroscopy of native receptor-rich membranes showed that β-structure predominates (36−43% β-structure, 32−33% α-helical, 14−24% turns and the rest unordered) (Naumann et al., 1993).

From the first cDNA-derived sequences and their hydropathy plots it became evident that all receptor sequences contained four stretches of particular hydrophobicity (Noda et al., 1983b). They were thought to traverse the membrane as α-helices (Finer-Moore and Stroud, 1984). All type-I receptors share this pattern (Fig. 4). They are believed to be composed of a large extracellular N-terminal domain thought to be composed largely from β-structure. This domain precedes the first three membrane-spanning sequences termed M1, M2, and M3. Between M3 and M4 there is, according to this model, a long (109−142 amino acids) connecting loop assumed to extend to the cytoplasm (Fig. 5) and which contains all the receptor's phosphorylation sites (Huganir and Greengard, 1987). Models with three, five, or seven transmembrane sequences, postulated on theoretical grounds or on the basis of epitope mapping using antibodies, are not considered anymore, primarily because both the N- and C-termini are assumed to be located extracellularly. Positioning both ends to the same side of the membrane excludes transmembrane folding patterns with uneven numbers of membrane passings.

From affinity labeling experiments the disposition of the transmembrane sequences relative to the lipid environment was deduced: M4 of all subunits faces the lipids (Giraudat et al., 1985; Blanton and Cohen, 1994). It is the least conserved and most hydrophobic among M1−M4. M2, on the other hand, is not in contact with lipids, rather, it faces the lumen of the ion channel surrounding the AChR's central axis of fivefold pseudosymmetry (Hucho et al., 1986; see below). M1 and M3 are also labeled by polar affinity labels approaching the receptor through

```
                    signal sequence      1
α-SUBUNIT     -24 MILCSYWHVGLVLLLFSCCGLVLG SEHETRLVANLL--ENYNKVIRPVEH  24
β-SUBUNIT     -24 MENVRRMALGLVVMMALALSGVGA SVMEDTLLSVLF--ETYNPKVRPAQT  24
γ-SUBUNIT     -17 MV------LTLLLIICLALEV-RS ENEEGRLIEKLL--GDYDKRIIPAKT  24
δ-SUBUNIT     -21 MGNI---HFVYLLISCLYYSGCSG VNEEERLINDLLIVNKYNKHVRPVKH  26
                  *          ...     .   .    *. *.. *.  ..*.  . *..

α-SUBUNIT     HTHFVDITVGLQLIQLISVDEVNQIVETNVRLRQQWIDVRLRWNPADYGG   74
β-SUBUNIT     VGDKVTVRVGLTLTNLLILNEKIEEMTTNVFLNLAWTDYRLQWDPAAYEG   74
γ-SUBUNIT     LDHIIDVTLKLTLTNLISLNEKEEALTTNVWIEIQWNDYRLSWNTSEYEG   74
δ-SUBUNIT     NNEVVNIALSLTLSNLISLKETDETLTSNVWMDHAWYDHRLTWNASEYSD   76
               .. ... . *.* .*. ..*  . ...** .  .* * ** *....*..

α-SUBUNIT     IKKIRLPSDDVWLPDLVLYNNADGDFAIVHMTKLLLDYTGKIMWTPPAIF  124
β-SUBUNIT     IKDLRIPSSDVWQPDIVLMNNNDGSFEITLHVNVLVQHTGAVSWQPSAIY  124
γ-SUBUNIT     IDLVRIPSELLWLPDVVLENNVDGQFEVAYYANVLVYNDGSMYWLPPAIY  124
δ-SUBUNIT     ISILRLPPELVWIPDIVLQNNNDGQYHVAYFCNVLVRPNGYVTWLPPAIF  126
              *. .*.*.. .* **.** ** ** . ..   ..*.  .* . * *.**.

α-SUBUNIT     KSYCEIIVTHFPFDQQNCTMKLGIWTYDGTKVSI----SPESDR--P---  165
β-SUBUNIT     RSSCTIKVMYFPFDWQNCTMVFKSYTYDTSEVTLQHALDAKGER--EVKE  172
γ-SUBUNIT     RSTCPIAVTYFPFDWQNCSLVFRSQTYNAHEVNLQLSAEEGEA----VEW  170
δ-SUBUNIT     RSSCPINVLYFPFDWQNCSLKFTALNYDANEITMDLMTDTIDGKDYPIEW  176
              .* * * * .**** ***.. .   .*.. ....    .  ..

α-SUBUNIT     ---DLSTFMESGEWVMKDYRGWKHWVYYTCCPDTPYLDITYHFIMQRIPL  212
β-SUBUNIT     IVINKDAFTENGQWSIEHKPSRKNWR----SDDPSYEDVTFYLIIQRKPL  218
γ-SUBUNIT     IHIDPEDFTENGEWTIRHRPAKKNYNWQLTKDDTDFQEIIFFLIIQRKPL  220
δ-SUBUNIT     IIIDPEAFTENGEWEIIHKPAKKNIYPDKFPNGTNYQDVTFYLIIRRKPL  226
                 . ..* *.*.* . . .. *.        .. . .... .*..* **

α-SUBUNIT     YFVVNVIIPCLLFSFLTGLVFYLPTDSGEKM-TLSISVLLSLTVFLLVIV  261
β-SUBUNIT     FYIVYTIIPCILISILAILVFYLPPDAGEKM-SLSISALLAVTVFLLLLA  267
γ-SUBUNIT     FYIINIIAPCVLISSLVVLVYFLPAQAGGQKCTLSISVLLAQTIFLFLIA  270
δ-SUBUNIT     FYVINFITPCVLISFLASLAFYLPAESGEKMST-AISVLLAQAVFLLLTS  275
              ....  * **.*.* *. *...**...*... . .**.**. ..**..

α-SUBUNIT     ELIPSTSSAVPLIGKYMLFTMIFVISSIIITVVVINTHHRSPSTHTMPQW  311
β-SUBUNIT     DKVPETSLSVPIIIRYLMFIMILVAFSVILSVVVLNLHHRSPNTHTMPNW  317
γ-SUBUNIT     QKVPETSLNVPLIGKYLIFVMFVSMLIVMNCVIVLNVSLRTPNTHSLSEK  320
δ-SUBUNIT     QRLPETALAVPLIGKYLMFIMSLVTGVIVNCGIVLNFHFRTPSTHVLSTR  325
              . .*.*. .**.* .*..*.*      ..   .*.*   *.*.** ..

α-SUBUNIT     VRKIFIDTIPNVMFF--------STMKRASKEKQE---------NKIF--  342
β-SUBUNIT     IRQIFIETLPPFLWIQRPVTTPSPDSKPTIISRAN---------DEYFIR  358
γ-SUBUNIT     IKHLFLGFLPKYLGMQLEPSEETPEKPQP---RRRSSFGIMIKAEEYILK  367
δ-SUBUNIT     VKQIFLEKLPRILHMSRADESEQPDWQNDLKLRRSSSVGYISKAQEYFNI  375
              ....*.. .*  . .        ..       .           .. .

α-SUBUNIT     --ADDIDISDISGK-QVTG-EVIFQT------------PLIKNPDVKSAI  376
β-SUBUNIT     KPAGDFVCPVDNARVAVQP-ERLFSEMKW--HLNGLTQPVTLPQDLKEAV  405
γ-SUBUNIT     KPRSELMFEEQKDRHGLKRVNKMTSDIDIGTTVDLYKDLANFAPEIKSCV  417
δ-SUBUNIT     KSRSELMFEKQSERHGL--VPRVTPRIGFGNNNENIAASDQLHDEIKSGI  423
                 ...     ...  .     .                     ..*. .

α-SUBUNIT     EGVKYIAEHMKSDEESSNAAEEWKYVAMVIDHILLCVFMLICIIGTVSVF  426
β-SUBUNIT     EAIKYIAEQLESASEFDDLKKDWQYVAMVADRLFLYVFFVICSIGTFSIF  455
γ-SUBUNIT     EACNFIAKSTKEQNDSGSENENWVLIGKVIDKACFWIALLLFSIGTLAIF  467
δ-SUBUNIT     DSTNYIVKQIKEKNAYDEEVGNWNLVGQTIDRLSMFIITPVMVLGTIFIF  473
              .. ..*..  .. .. ..   .*  .. . *.  . .   .  .**  .*

α-SUBUNIT     AGRLIELSQE-----------------G   437
β-SUBUNIT     LDASHNVPPDNPF--------------A   469
γ-SUBUNIT     LTGHFNQVPEFPFPGDPRKYVP------   489
δ-SUBUNIT     VMGNFNHPPAKPFEGDPFDYSSDHPRCA   501
                   .  ..
```

Fig. 1. Amino acid sequences of the precursors of the four acetylcholine receptor subunits from *Torpedo californica*. The alignment was done with the CLUSTAL computer program. '*', sequence identity; '.', sequence similarity. Consensus length, 528; identity, 96 (18.2%); similarity, 183 (34.7%). The putative transmembrane sequences M1–M4) are shown in bold letters.

```
                      signal sequence      1
α1_TORPEDO   -24 MILCSYWHVGLVLLLFSCCGLVLG- SEHETRLVANLLENYNKVIRPVEHHT   26
α1_HUMAN     -20 MEPWPL----LLLFSLCSAGLVLG- SEHETRLVAKLFKDYSSVVRPVEDHR   26
α1_BOVIN     -20 MEPRPL----LLLLGLCSAGLVLG- SEHETRLVAKLFEDYNSVVRPVEDHR   26
α1_RAT       -20 MELTAV----LLLLGLCSAGTVLG- SEHETRLVAKLFKDYSSVVRPVGDHR   26
α1_CHICK     -20 MELCRV----LLLI-FSAAGPALCY -EHETRLVDDLFREYSKVVRPVENHR   25
                 *         *.*. ..  * .*    *******..*. .*..*.***..*

α1_TORPEDO   HFVDITVGLQLIQLISVDEVNQIVETNVRLRQ------------------   58
α1_HUMAN     QVVEVTVGLQLIQLINVDEVNQIVTTNVRLKQGDMVDLPRPSCVTLGVPL   76
α1_BOVIN     QAVEVTVGLQLIQLINVDEVNQIVTTNVRLKQ------------------   58
α1_RAT       EIVQVTVGLQLIQLINVDEVNQIVTTNVRLKQ------------------   58
α1_CHICK     DAVVVTVGLQLIQLINVDEVNQIVTTNVRLKQ------------------   57
             . * .**********.********.*****.*

α1_TORPEDO   -------QWIDVRLRWNPADYGGIKKIRLPSDDVWLPDLVLYNNADGDFA   101
α1_HUMAN     FSHLQNEQWVDYNLKWNPDDYGGVKKIHIPSEKIWRPDLVLYNNADGDFA   126
α1_BOVIN     -------QWVDYNLKWNPDDYGGVKKIHIPSEKIWRPDLVLYNNADGDFA   101
α1_RAT       -------QWVDYNLKWNPDDYGGVKKIHIPSEKIWRPDVVLYNNADGDFA   101
α1_CHICK     -------QWTDINLKWNPDDYGGVKQIRIPSDDIWRPDLVLYNNADGDFA   100
                    **.* .*.***.****.*.*..**...* **.***********

α1_TORPEDO   IVHMTKLLLDYTGKIMWTPPAIFKSYCEIIVTHFPFDQQNCTMKLGIWTY   151
α1_HUMAN     IVKFTKVLLQYTGHITWTPPAIFKSYCEIIVTHFPFDEQNCSMKLGTWTY   176
α1_BOVIN     IVKFTKVLLDYTGHITWTPPAIFKSYCEIIVTHFPFDEQNCSMKLGTWTY   151
α1_RAT       IVKFTKVLLDYTGHITWTPPAIFKSYCEIIVTHFPFDEQNCSMKLGTWTY   151
α1_CHICK     IVKYTKVLLEHTGKITWTPPAIFKSYCEIIVTYFPFDQQNCSMKLGTWTY   150
             **. **.**..**.* ****************.****.***.****.***

α1_TORPEDO   DGTKVSISPESDRPDLSTFMESGEWVMKDYRGWKHWVYYTCCPDTPYLDI   201
α1_HUMAN     DGSVVAINPESDQPDLSNFMESGEWVIKESRGWKHSVTYSCCPDTPYLDI   226
α1_BOVIN     DGSVVVINPESDQPDLSNFMESGEWVIKESRGWKHWVFYACCPSTPYLDI   201
α1_RAT       DGSVVAINPESDQPDLSNFMESGEWVIKEARGWKHWVFYSCCPNTPYLDI   201
α1_CHICK     DGTMVVINPESDRPDLSNFMESGEWVMKDYRGWKHWVYYACCPDTPYLDI   200
             **. * *.****.****.********.*. ***** * *.***.******

α1_TORPEDO   TYHFIMQRIPLYFVVNVIIPCLLFSFLTGLVFYLPTDSGEKMTLSISVLL   251
α1_HUMAN     TYHFVMQRLPLYFIVNVIIPCLLFSFLTGLVFYLPTDSGEKMTLSISVLL   276
α1_BOVIN     TYHFVMQRLPLYFIVNVIIPCLLFSFLTGLVFYLPTDSGEKMTLSISVLL   251
α1_RAT       TYHFVMQRLPLYFIVNVIIPCLLFSFLTSLVFYLPTDSGEKMTLSISVLL   251
α1_CHICK     TYHFLMQRLPLYFIVNVIIPCLLFSFLTGFVFYLPTDSGEKMTLSISVLL   250
             ****.***.****.**************..********************

α1_TORPEDO   SLTVFLLVIVELIPSTSSAVPLIGKYMLFTMIFVISSIIITVVVINTHHR   301
α1_HUMAN     SLTVFLLVIVELIPSTSSAVPLIGKYMLFTMVFVIASIIITVIVINTHHR   326
α1_BOVIN     SLTVFLLVIVELIPSTSSAVPLIGKYMLFTMVFVIASIIITVIVINTHHR   301
α1_RAT       SLTVFLLVIVELIPSTSSAVPLIGKYMLFTMVFVIASIIITVIVINTHHR   301
α1_CHICK     SLTVFLLVIVELIPSTSSAVPLIGKYMLFTMVFVIASIIITVIVINTHHR   300
             *******************************.***.******.*******

α1_TORPEDO   SPSTHTMPQWVRKIFIDTIPNVMFFSTMKRASKEKQENKIFADDIDISDI   351
α1_HUMAN     SPSTHVMPNWVRKVFIDTIPNIMFFSTMKRPSREKQDKKIFTEDIDISDI   376
α1_BOVIN     SPSTHVMPEWVRKVFIDTIPNIMFFSTMKRPSREKQDKKIFTEDIDISDI   351
α1_RAT       SPSTHIMPEWVRKVFIDTIPNIMFFSTMKRPSRDKQEKRIFTEDIDISDI   351
α1_CHICK     SPSTHTMPPWVRKIFIDTIPNIMFFSTMKRPSRDKPDKKIFAEDIDISEI   350
             *****.** ****.*******.********.*..*....**..*****.*

α1_TORPEDO   SGKQVTGEVIFQTPLIKNPDVKSAIEGVKYIAEHMKSDEESSNAAEEWKY   401
α1_HUMAN     SGKPGPPPMGFHSPLIKHPEVKSAIEGIKYIAETMKSDQESNNAAAEWKY   426
α1_BOVIN     SGKPGPPPMGFHSPLIKHPEVKSAIEGIKYIAETMKSDQESNNAAEEWKY   401
α1_RAT       SGKPGPPPMGFHSPLIKHPEVKSAIEGVKYIAETMKSDQESNNASEEWKY   401
α1_CHICK     SGKQGPVPVNFYSPLTKNPDVKNAIEGIKYIAETMKSDQESSNAADEWKF   400
             ***. .  . * .**.*.*.**.****.***** ****.**.**..***.

α1_TORPEDO   VAMVIDHILLCVFMLICIIGTVSVFAGRLIELSQEG   437
α1_HUMAN     VAMVMDHILLGVFMLVCIIGTLAVFAGRLIELNQQG   462
α1_BOVIN     VAMVMDHILLAVFMLVCIIGTLAVFAGRLIELNQQG   437
α1_RAT       VAMVMDHILLGVFMLVCLIGTLAVFAGRLIELHQQG   437
α1_CHICK     VAMVLDHLLLVIFMLVCIIGTLAVFAGRLIELNQQG   436
             ****.**.** .***.*.***..********* *.*
```

Fig. 2. A selection of AChR α subunits from different species. Information on alignment and symbols as in Fig. 1.

Fig. 3. Sequence similarity between muscle (α1) and neuronal (α7) AChR subunits from rat. For information on the alignment procedure, see the legend for Fig. 1. Vertical lines indicate sequence identity: 172 (37.6%) of the residues are identical.

the lipid phase. Experiments performed mostly with reconstituted receptor/lipid systems have proven the importance of lipids, especially cholesterol and negatively charged lipids, for the structural and functional integrity of the receptor (Fong and McNamee, 1987).

These transmembrane folding patterns including the secondary structures involved are 'predictions'; supporting experimental evidence is scarce. It has been pointed out (Lunt, G., unpublished remark) that the now generally accepted four-transmembrane pattern probably would not even have been predicted, if another receptor, e.g. the $GABA_A$ receptor, had been cloned and sequenced first. The hydropathy plots of $GABA_A$ receptor subunits are much more ambiguous with respect to the four hydrophobic sequences.

This is some of the supporting evidence: The N-termini, though never really identified *in situ*, must be extracellular, because biosynthesis of the subunit chains including signal peptides at the rough endoplasmic reticulum demands this location. The C-terminus of the δ subunit has been shown to be extracellular, by the accessibility of the cysteine residue δ Cys500 next to it (Czajkowski et al., 1989; DiPaola et al., 1989). No comparable experiment was performed for the other subunits. Another line of evidence supporting transmembrane folding models comes from the proven or predicted glycosylation sites. As mentioned above glycosylation at α-Asn141, δ-Asn143, δ-Asn70 and δ-Asn208 was experimentally proven (Poulter et al., 1989; Strecker et al., 1994). These sites therefore are landmarks of extracellular position. Artificial N-glycosylation sites introduced

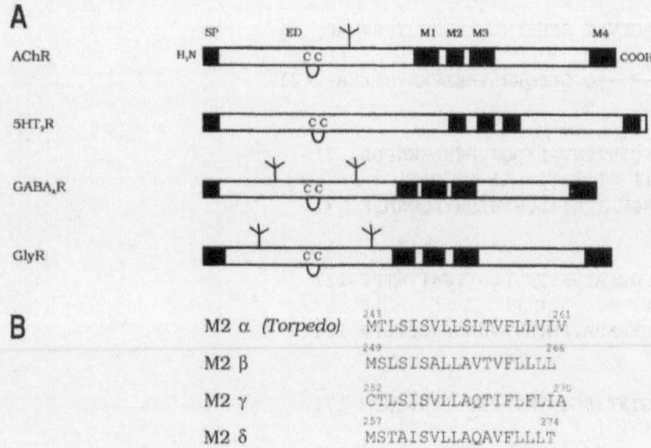

Fig. 4. (A) Schematic representation of the primary sequences of ligand-gated ion channels. (type I-receptors). (B) Amino acid sequences of the putative transmembrane helices M2 of the AChR's four subunits which participate in the formation of the channel wall. SP, signal peptide; M1–M4, putative transmembrane sequences; C C, cystine-loop characteristic for the ligand binding α subunits; Ỵ, oligosaccharide. In B, note the sequence similarity.

by site-directed mutagenesis proved that the N-terminal sequence of the α subunit at least up to residue α207 extends from the extracellular side of the membrane (Chavez and Hall, 1991). Fusion proteins constructed from fragments of the α and δ chains (mammalian muscle) with prolactin were in agreement with the fold predicted on the basis of hydropathy plots (Chavez and Hall, 1992).

While most agree on the four transmembrane model (4TM) the secondary structure of the intramembrane portion of the protein is still in dispute. Most integral membrane proteins, the structure of which has been solved, span the membrane via α-helices (Deisenhofer et al., 1985; Henderson et al., 1990; Iwata et al., 1995), but exceptions, notably the porins forming β-barrels, are known (Parker et al., 1992; Weiss et al., 1991). Indeed, it was suggested that β-barrels are of general importance for ion channels (and transporters) (MacKinnon, 1995). No mixed α-helical and β-structure transmembrane domains have been discovered yet.

Cryo-electron microscopy at 0.9-nm resolution (Unwin, 1993) did detect just one rod, believed to represent an α-helical structure in the intramembrane part of the AChR (see below), and FTIR spectroscopy of a membrane-bound receptor, from which the extramembrane domains had been removed by proteolysis, detected more than 40% β-structure (Görne-Tschelnokow et al., 1994). Furthermore, modeling studies showed that a bundel of 5×4 transmembrane helices is not compatible with the given dimensions (Ortells and Lunt, 1994). The conclusion from these theoretical studies was that M1 and M3 are most likely β-strands.

These considerations are supported in part by affinity labeling studies using the hydrophobic reagent 3-trifluoromethyl-3-(m-[^{125}I]iodophenyl)diazirine (^{125}I-TID) (White et al., 1991; White and Cohen, 1992; Blanton and Cohen, 1994). From the periodicity of the labeled amino acids an α-helical conformation for M2 (and M4) was concluded. Similar conclusions were obtained from electrophysiological studies of native and mutated receptor using the channel blocker QX 222 (Charnet et al., 1990; Lester, 1992), and from photolabeling experiments with chlorpromazine (Revah et al., 1990). On the other hand, cysteine mutants investigated by the accessibility of the Cys residues by thiosulfonate cations (SCAM) showed that possibly not all the residues in the membrane-spanning segments M1 and M2 are in an α-helical conformation (Akabas et al., 1992; Akabas and Karlin, 1995).

The transmembrane fold of active receptor *in situ* remains an enigma especially with the unexplained observations made by several laboratories. As mentioned above, immunologists presented completely different folding patterns (Criado et al., 1985). Furthermore, it was observed that the supposedly intracellular loop connecting M3 and M4 is the primary target of proteases, even immobilized ones, and of iodinization, even with iodogen which supposedly iodinizes primarily surface structures (Moore et al., 1989; Mund, 1994). Part of the loop connecting M3 and M4 (Fig. 5) was proposed to form an amphipathic helix immersed in the membrane (Finer-Moore and Stroud, 1984). The possibility still exists that it forms an intra-membrane structure as was recently shown to exist with the glutamate receptors (Hollmann et al., 1994; Bennett and Dingledine, 1995). Since this loop contains a variety of phosphorylation sites there is no doubt that it is located at least in part or at times intracellularly (Wagner et al., 1991). But given its length and its amphipathic nature there is still room for speculation (and experimenting) as to its localization with respect to the plasma membrane.

```
α-subunit   HHRSPSTHTMPQWVRKIFIDTIPNVMFF--------STMKRASKEKQE--   338
ß-subunit   HHRSPNTHTMPNWIRQIFIETLPPFLWIQRPVTTPSPDSKPTTISRAN--   352
γ-subunit   SLRTPNTHSLSEKIKHLFLGFLPKYLGMQLEPSEETPEKPQP---RRRSS   354
δ-subunit   HFRTPSTHVLSTRVKQIFLEKLPRILHMSRADESEQPDWQNDLKLRRSSS   362
              *.*.** ..  ....*.. .*  .  .        ..    ↑  . ↑↑

α-subunit   -------NKIF----ADDIDISDISGK-QVTG-EVIFQT----------   364
ß-subunit   -------DEYFIRKPAGDFVCPVDNARVAVQP-ERLFSEMKW--HLNGLT   392
γ-subunit   FGIMIKAEEYILKKPRSELMFEEQKDRHGLKRVNKMTSDIDIGTTVDLYK   404
δ-subunit   VGYISKAQEYFNIKSRSELMFEKQSERHGL--VPRVTPRIGFGNNNENIA   410
                    ..↑.  ↑ ...      ...  .

α-subunit   -PLIKNPDVKSAIEGVKYIAEHMKSDEESSNAAEEWKYVAMVIDH   408
ß-subunit   QPVTLPQDLKEAVEAIKYIAEQLESASEFDDLKKDWQYVAMVADR   437
γ-subunit   DLANFAPEIKSCVEACNFIAKSTKEQNDSGSENENWVLIGKVIDK   449
δ-subunit   ASDQLHDEIKSGIDSTNYIVKQIKEKNAYDEEVGNWNLVGQTIDR   455
                 ..*. ... ..*..  .. .. ..  .*  .. .*.
```

Fig. 5. Alignment of the sequences of the predicted cytosolic loops between M3 and M4 of the four subunits of the nAChR from *Torpedo californica*. Bold underlined letters show phosphorylated serine and tyrosine residues. Other symbols as in Fig. 1.

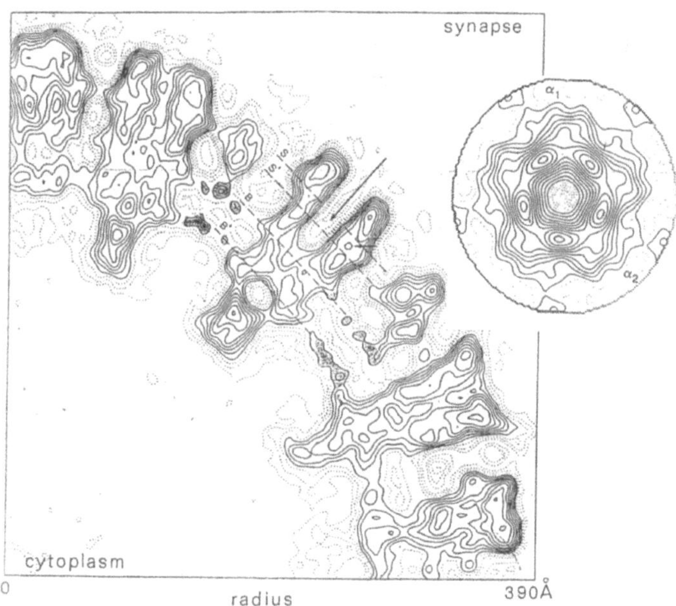

Fig. 6. Electron microscopic image of an AChR-rich membrane, longitudinal section. The long arrow indicates the path presumably taken by the ions; the short arrow points to a specialized region, interpreted as the possible agonist binding site. The nearly rectangular-shaped blob on the cytoplasmic side represents the 43-kDa subunit (rapsyn). Insert: cross-section of an AChR molecule, membrane-spanning domain, after image-averaging (from Unwin, 1993, with kind permission of the author). S S′ and B B′ indicate cross-sections analyzed in more detail in Unwin (1993).

The transmembrane folding is only one aspect of the receptor's tertiary structure. Very little is known concerning the structure of the large extracellular domain. As mentioned above, since during biosynthesis a signal peptidase removes a signal peptide, it contains the N-terminus (Anderson and Blobel, 1981). As also mentioned above, the C-terminus, at least of the *Torpedo* AChR δ chain, was shown to be extracellular too (DiPaola et al., 1989). There are several topological markers which provide information related to the spatial arrangement of the polypeptide chains: the glycosylation sites (see above) are located extracellularly and on the surface of the extracellular domain. The phosphorylation sites define intracellular structures. The main immunogenic region (MIR), shown to comprise amino acids around α48−67 (Tzartos and Lindstrom, 1980; Bellone et al., 1989; Papadouli et al., 1993, and for a review see Conti-Tronconi et al., 1994), was shown to be positioned as expected on the receptor surface (Beroukhim and Unwin, 1995). The amino acid residues which were found to be part of the ligand-binding sites (see below) are located extracellularly though, as we shall see, not necessarily on the surface. Further spatial information will be presented when we discuss the intersubunit contacts within the quaternary structure and the structure and positioning of the functional domains. This paragraph shall end with a summary of the information obtained by high-resolution electron microscopy.

Electron microscopy of AChR. At the resolution presently reached of 0.9 nm (Unwin, 1993), in addition to the overall shape of the molecule, certain features of the tertiary and even the secondary structure emerge. A side view of the molecule at 0.9-nm resolution (Fig. 6) shows its dimensions (Unwin, 1993): The subunits appear as elongated bodies 12.5 nm long, extending about 6.0 nm from the extracellular and 2.0 nm from the cytoplasmic surface of the membrane. Thus, only about 33% of the protein is immersed in the hydrophobic membrane, about 50% is extracellular. A view from the top shows that the five receptor subunits are arranged around a fivefold axis of pseudosymmetry (Fig. 6). The dimensions of the molecule in the membrane plane are 8.0 nm for the diameter of the pentamer and about 2.5 nm of the cylindrical tube enclosed by the subunits which narrows down at the mebrane plane beyond electron microscope resolution and which is believed to contain the receptor's ion channel gate (see below).

The great merit of the electron microscopy applied in these investigations is that it uses quick-frozen receptor-rich membranes embedded in amorphous ice. Among other advantages, this method avoids the artifacts of negative staining. On the other hand the samples used, two-dimensional tubular crystals obtained in sufficient yield at extreme conditions from only one out of ten fish of a certain *Torpedo* species (Unwin, 1993), may warrant some caution concerning the interpretation of the images obtained.

At 0.9-nm resolution not much detail of the secondary and tertiary structures can be seen. Most conspicuous are several 'rods' perpendicular to the membrane which are interpreted to represent α-helices. In the intramembrane part of the molecule each subunit contains one such rod, positioned close to the lumen in the center. They may represent the channel-forming helices M2 (discussed below). No further rods are visible in the transmembrane domain. The absence of any features which hint at secondary structure elements is interpreted as possible β-structures. Additionally, in the extracellular part of the molecule sets of three rods are seen in each subunit, about 3.0 nm above the membrane bilayer. They form a cavity which is more pronounced in the α subunits and which is proposed to represent the ligand-binding site.

The quaternary structure of the nAChR. Interest in the AChR's quaternary structure arises from at least three basic properties, which will be discussed in detail below: (a) the receptor is an allosteric protein (see the reviews by Léna and Changeux, 1993; Galzi and Changeux, 1994), allosterism being based on quaternary structures; (b) the receptor's cation channel is a result of the subunit arrangement; (c) the binding sites for agonists and competitive antagonists are located at subunit interfaces.

In 1973, electron microscopic images showed the negatively stained AChR as a donut-shaped molecule composed of five or six subunits, arranged around a central pore (Nickel and Potter, 1973; Cartaud et al., 1973). In the same year (Biesecker, 1973; Hucho and Changeux, 1973) a pentameric structure of the receptor prepared from the electric eel *Electrophorus electricus* and shortly thereafter for the *Torpedo* receptor (Hucho et al., 1978) was deduced from cross linking experiments. The *Torpedo* AChR is composed of four different polypeptide chains (α, β, γ and δ) with the stoichiometry $\alpha_2\beta\gamma\delta$ (Reynolds and Karlin, 1978; Lindstrom et al., 1979; Raftery et al., 1980). Different subunit compositions (though also forming pentameric complexes; Cooper et al., 1991; Anand et al., 1991) were reported for the neuronal-type AChRs which will not be discussed in this review (for a detailed review see Sargent, 1993).

The five receptor subunits are arranged around a fivefold axis of pseudosymmetry (Brisson and Unwin, 1985). Photoaffinity labeling with the channel blocker chlorpromazine (Heidmann et al., 1983) suggested that all five subunits together form the receptor's high-affinity binding site for non-competitive inhibitors regulating the permeability of the channel (see below).

The nAChR from *Torpedo* occurs in two forms: as a monomer (9.0 S) and as a dimer (13.7 S) (Raftery et al., 1972). The dimeric form predominates in postsynaptic membranes (Suarez-Isla and Hucho, 1977). The dimers are stabilized by a disulfide bridge between the δ subunits of two neighboring receptor monomers (Suárez-Isla and Hucho, 1977; Chang and Bock, 1977; Hucho et al., 1978). The cysteine residues, required for the formation of this S-S linkage, have been shown to be situated at the cytoplasmic side of the δ subunit (Wennogle et al., 1981). The importance of this disulfide bridge is unclear, since muscle AChR does not have it, and since no differences in ligand-binding and channel properties of the monomeric and the dimeric receptor forms could be found (Fels et al., 1982; Anholt et al., 1980). Nevertheless, it helped structural analysis serving as a reference point in the interpretation of electron microscopic images.

Several methods have been employed to deduce the subunit arrangement around the channel in the receptor monomer. For an $\alpha_2\beta\gamma\delta$ heteropentamer six different subunit arrangements (plus their mirror images) are possible. Neighboring α-subunits could be excluded by electron microscopy (Wise et al., 1981) and because they could not be cross-linked to each other (Schiebler et al., 1980). Thus, the two α subunits are separated from each other by a non-α subunit, and further efforts were directed to the identification of this single subunit located between the α subunits.

Taking the δ-δ disulfide as a reference axis and introducing an additional artificial β-β disulfide bridge, the angles between the resulting receptor trimers were examined by electron microscopy (Wise et al., 1981). This study excluded the possibility that the δ subunit is flanked by the two α subunits. Therefore only two possible arrangements are left: α-β-α or α-γ-α.

Several experiments were interpreted in favour of an α-β-α arrangement of the receptor subunits around the channel. Electron diffraction studies of two-dimensional crystals of the nAChR labeled with Fab fragments of subunit-specific antibodies, in combination with wheat germ agglutinin-labeling of *N*-acetylglucosamine residues which are present only on the δ subunit were interpreted as an α-β-α arrangement (Kubalek et al., 1987). Cross-linking experiments with the α-neurotoxin from *Naja naja siamensis* in the presence of (+)-tubocurarine and bromoacetylcholine supported this proposal (Hamilton et al., 1985). The same conclusion was reached by Chatrenet et al. (1990) as an interpretation of the cross-linking pattern obtained with photoactivatable derivatives of an α-neurotoxin from *Naja nigricollis*.

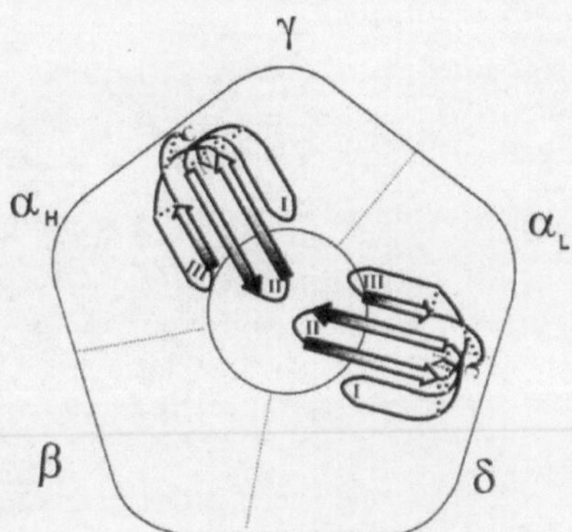

Fig. 7. Quaternary structure of the heteropentameric AChR, including the handedness of subunit arrangement around the fivefold axis of pseudosymmetry. α_H and α_L indicate the high- and low-affinity binding sites for (+) tubocurarine (Machold et al., 1995b).

However, an α-β-α order of the receptor subunits seems to be unlikely and can now be dismissed on the basis of compelling evidence. The α-γ-α order was proposed from angle estimations between the β and δ subunits in artificial receptor trimers (Wise et al., 1981) in combination with angle estimation of the avidin-biotin-labeled α-neurotoxin from *Naja naja siamensis*, bound to nAChR monomers and dimers (Holtzman et al., 1982; Karlin et al., 1983). Further strong support for an α-γ-α order is given by coexpression studies in which pairs of subunits were coexpressed in BC3H-1 cells. It was shown that α and γ chains expressed together yielded a high-affinity pair, α and δ chains a low-affinity pair with respect to antagonist binding, while α and β chains coexpressed gave no antagonist-binding pair at all (Blount and Merlie, 1989; Sine and Claudio, 1991). These experiments are in agreement with the finding that (+)-tubocurarine binds with high affinity to the α/γ interface and with low affinity to the α/δ interface (Neubig and Cohen, 1979; Pedersen and Cohen, 1990), which also excludes the α-β-α arrangement. Examining specific intersubunit contacts, the specificity of subunit association, formation of functional ligand-binding sites and the transport of assembled $\alpha_2\beta\gamma\delta$ pentamers by the use of chimeric subunits, Kreienkamp et al. (1995) also supported the conclusion that the γ subunit stands between the two α subunits.

Finally, to discriminate between the two possible mirror images, the handedness of the circular subunit arrangement was elucidated by labeling with different photoactivatable derivatives of the α-neurotoxin II from *Naja naja oxiana*. The observed labeling is explained only by the clockwise receptor subunit arrangement α_H-γ-α_L-δ-β, as viewed from the synaptic cleft (Fig. 7; Machold et al., 1995b).

The ligand-binding sites: the interface-gorge model

Biochemical and molecular biology data can be summarized into a spatial model of the binding pockets. Initially it was believed that the two ligand-binding sites in the *Torpedo* receptor are on the two α subunits. Karlin and coworkers were the first to identify such sites and to specify amino acid residues of the α subunit sequence by labeling α-Cys192/Cys193 with the affinity reagent 4-(*N*-maleimido)benzyltrimethylammonium (MBTA) iodide (Kao et al., 1984). This and other findings attributed the ligand-binding site to the α-subunit region comprising amino acids 180−200. The corresponding subunit fragments or syn-

Table 1. Identified residues in the ligand-binding sites of nicotinic acetylcholine receptors.

Residue	Subunit $\alpha\beta\gamma\delta$	Receptor (source, species)	Methods: affinity labelling (AL) or mutations/ chimaera (MC)	Comments (Ligand used, etc.)	References
Ser36	δ	mouse muscle	MC	α-conotoxin, M1, high-affinity site	Sine et al., 1995
Lys34	γ	mouse muscle	MC	α-conotoxin, M1, low-affinity site	Sine et al., 1995
Trp54	α7	neuronal α7-5HT$_3$-chimera	MC	ACh, chimera with 5HT$_3$	Corringer et al., 1995
Trp55	γ	*Torpedo*	MC	ACh, dTC	O'Leary et al., 1994
Trp55	γ	*Torpedo*	AL	dTC, high-affinity site	Cohen et al., 1992
Trp57	δ	*Torpedo*	AL	dTC, low-affinity site	Cohen et al., 1992
Trp86	α	*Torpedo*	AL	DDF	Galzi et al., 1990
Tyr93	α	*Torpedo*	AL	DDF	Galzi et al., 1990
Tyr93	α	*Torpedo*	AL	ACh-mustard	Cohen et al., 1991
Ser111	γ	mouse muscle	MC	α-conotoxin M1, low-affinity site	Sine et al., 1995
Tyr113	δ	mouse muscle	MC	α-conotoxin M1, high-affinity site	Sine et al., 1995
Ile116	γ	mouse muscle	MC	DMT, high-affinity site	Sine, 1993
Tyr117	γ	mouse muscle	MC	DMT, high-affinity site	Sine, 1993; Fu and Sine, 1994
Val116	δ	mouse muscle	MC	DMT, low-affinity site	Sine, 1993
Thr117	δ	mouse muscle	MC	DMT, low-affinity site	Sine, 1993
Ile145	γ	mouse muscle	MC	promotes α/γ site formation	Kreienkamp et al., 1995
Lys145	δ	mouse muscle	MC	promotes α/δ site formation	Kreienkamp et al., 1995
Thr150	γ	mouse muscle	MC	promotes α/γ site formation	Kreienkamp et al., 1995
Lys150	δ	mouse muscle	MC	promotes α/δ site formation	Kreienkamp et al., 1995
Tyr149	α	*Torpedo*	AL	DDF	Dennis et al., 1988
Tyr151	α	*Torpedo*	AL	DDF	Galzi et al., 1990
Ser161	γ	mouse muscle	MC	DMT, high-affinity site	Sine, 1993
Lys161	δ	mouse muscle	MC	DMT, low-affinity site	Sine, 1993
Asp180	δ	*Torpedo*	AL	forms the cross-link with the αCys192/Cys193	Czaikowski and Karlin, 1995
Glu182	δ	*Torpedo*	MC	forms the cross-link with the αCys192/Cys193	
Asp180	δ	mouse muscle	MC	affects binding of ACh and βBgt	Czaikowski et al., 1993
Glu189		mouse muscle	MC	affects binding of ACh and αBgt	Czaikowski et al., 1993
Asn187	α	monogoose/ mouse muscle	MC	glycosylation of these residues precludes αBgt binding	Keller et al., 1995
Asn189	α	monogoose/ mouse muscle	MC	glycosylation of these residues precludes αBgt binding	
Tyr190	α	*Torpedo*	AL	[H^3]nicotine	Middleton and Cohen, 1991
Tyr190	α	*Torpedo*	AL	DDF	Dennis et al., 1988
Tyr190	α	*Torpedo*	AL	lophotoxin	Abramson et al., 1989
Cys192	α	*Torpedo*	AL	MBTA	Kao et al., 1984
Cys192	α	*Torpedo*	AL	DDF	Dennis et al., 1988
Cys193	α	*Torpedo*	AL	MBTA	Kao et al., 1984
Cys193	α	*Torpedo*	AL	DDF	Dennis et al., 1988
Tyr198	α	*Torpedo*	AL	[H^3]nicotine	Middleton and Cohen. 1991
Tyr198	α	mouse muscle	MC	DMT, high affinity site	Fu and Sine, 1994

thetic peptides were shown to bind snake venom α-neurotoxins (Neumann et al., 1986; Gershoni et al., 1983). On the other hand, AChR from mongoose and snakes which are resistant to snake α-neurotoxins contain conspicuous mutations in this region (Neumann et al., 1989). Subsequently, a number of residues in this region were labeled by different agonists and antagonists (Table 1): α-Tyr190 by lophotoxin (Abramson et al., 1989); α-Tyr190, α-Cys192 and α-Cys193 by the antagonist [^{3}H]*p*-(dimethylamino)benzenediazonium fluoroborate (DDF) (Dennis et al., 1988); α-Tyr190, α-Cys192 and, most efficiently, α-Tyr198 by the agonist [3H]nicotine (Middelton and Cohen, 1991).

Two characteristic features became obvious early on. First, the ligand-binding site on the α subunit is not confined to a contiguous stretch of the primary structure including amino acids 180–200. In addition to the above-mentioned residues, DDF was found to label α-Tyr86, α-Tyr93, α-Tyr149, and α-Tyr151 (Dennis et al., 1988; Galzi et al., 1990). Moreover, [^{3}H]acetylcholine mustard was cross-linked exclusively to α-Tyr93 (Cohen et al., 1991), thus discovering in the putative cat-

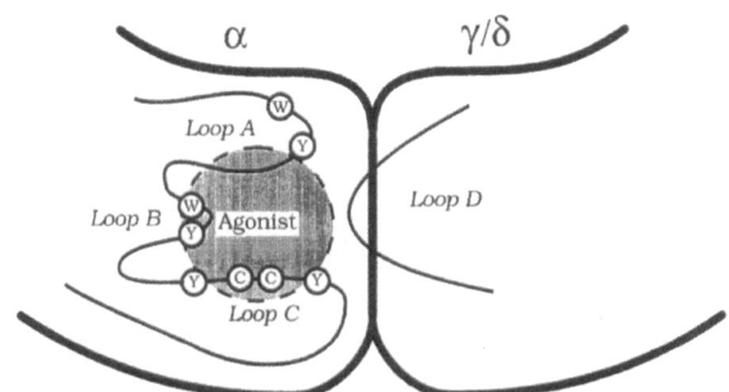

Fig. 8. The three-loop model of the agonist binding site, according to Bertrand and Changeux (1995). Note that a fourth loop extends from the neighboring subunits, indicating the location of the binding sites at subunit interfaces.

ion-binding subsite an aromatic residue, instead of the expected carboxylic group. The cross-linking pattern with DDF was found to depend on the receptor's functional state, the labeling of α-Tyr93 and α-Trp149 becoming more pronounced in the desensitized state (Galzi et al., 1991a). These data provided the basis for the three-loop model of the α subunit-binding site for low-molecular-mass ligands (Fig. 8), which has been demonstrated to be applicable also to the neuronal α7 AChR (Galzi, et al., 1991b; Devillers-Thiéry et al., 1993).

Second, the ligand-binding site is not confined to the α subunits; there are also contributions, direct or indirect (see below), from the neighboring subunits. Considering the AChR subunit composition and pseudosymmetrical arrangement around the central axis (Fig. 7), the two α subunits must have different neighbors. This geometrical non-equivalence of the α subunits provides an explanation for the observed non-equivalence of the two ligand-binding sites, which manifests itself in a two-step dissociation of bound α-neurotoxin (Maelicke and Reich, 1976), in different affinities for antagonists (Neubig and Cohen, 1979), antibodies, or in different accessibility for alkylating reagents (Damle and Karlin, 1978; Dowding and Hall, 1987; Dunn et al., 1993; and a recent review by Bertrand and Changeux, 1995). Agonists like acetylcholine and carbamoylcholine do not show comparably distinct affinities for the two sites.

Photolabeling experiments by Pedersen and Cohen (1990) demonstrated that the high-affinity site for (+)-tubocurarine extends from the α to the γ subunits, whereas the low-affinity site resides at the interface between the α and δ subunits. A similar conclusion has been drawn from the heterologous co-expression of different subunit combinations mentioned above (Blount and Merlie, 1989). Analysis of ligand-binding to the mouse AChR subunit combinations expressed in a fibroblast cell line demonstrated an important role of the γ and δ subunits in shaping the high- and low-affinity binding sites, respectively (Sine and Claudio, 1991). The α-neurotoxin binding site, too, is located at the subunit interfaces (Kreienkamp et al., 1992).

In summary, a large body of data indicates that the binding sites are in fact at the α/γ and α/δ interfaces and not just pockets within the α subunits. Such a localization is favorable for providing an efficient means for converting the neurotransmitter-induced conformational changes at the ligand-binding sites into rearrangements which open the receptor's channel. Interaction between subunits are the basis for allosteric mechanisms (Monod et al., 1965).

A number of amino acid residues in the ligand-binding domains of the γ and δ subunits have been identified (Sine, 1993). γ-Trp55 and δ-Trp57 were found in the (+)-tubocurarine binding sites using photoinduced cross-linking (Cohen et al., 1992). The importance of the γ-Trp55 residue for the high-affinity binding of agonists and antagonists was also demonstrated in mutagenesis experiments (O'Leary et al., 1994). For the homopentameric neuronal α7 receptor it was shown by analysis of an α7−$5HT_3$ chimera (Corriger et al., 1995; Eiselé et al., 1993), that Trp 54 of α7 is important for agonist binding.

Many chimera studies were aimed at defining residues in the extracellular domains of the γ and δ subunits which are essential for the formation of the α/γ and α/δ binding sites (see Sine, 1993; Kreienkamp et al., 1995). With this approach it is not always possible to distinguish whether a residue is essential for the assembly of the oligomeric receptor complex, or whether it is a part of the ligand-binding site. Interestingly, even a portion of the cytoplasmic loop in the α subunit was found to be important for the receptor assembly (Xiao-Mei and Hall, 1994). Mutations of α-Tyr190, α-Tyr198, and α-Asp200 did not significantly change the affinity of the ligand binding site, but interfered with the coupling of ligand-binding to channel opening (O'Leary and White, 1992). Noteworthy, the residues γ-Ile154, γ-Thr150, δ-Lys145 and δ-Lys150, which promote formation of the α/γ and α/δ sites, respectively (Kreienkamp et al., 1995), belong to the same loop which in the α subunit contains α-Tyr149 and α-Tyr151 tagged by DDF.

Another class of compounds which discriminate between the two ligand-binding sites in the nicotinic AChR's are α-conotoxins, short neurotoxic peptides from the marine snails of the *Conus* genera. Their affinity depends not only on a particular α-conotoxin, but even more so on the receptor species: for the *Torpedo* receptor, α-conotoxins M1 and G1 bind 10−100 times more efficiently to the α/γ interface (Hann et al., 1994; Utkin et al., 1994), while M1 shows about 10000-fold preference for the α/δ site in the mouse muscle receptor (Kreienkamp et al., 1994). By combining the chimaera and mutation approaches, residues Lys34, Ser111 and Phe172 of the γ subunit were found to confer low affinity to the α/γ site, whereas Ser36, Tyr113 and Ile178 of the δ subunit are essential for maintaining the high affinity towards α-conotoxin M1 at the α/δ interface (Sine et al., 1995).

The idea of the ligand-binding sites situated at the interfaces of the two α subunits with their γ and δ neighbors at a first glance does not agree with the cryoelectron microscopy data. The three α-helical structures in the extracellular domains of the α subunits, hypothesized to be the ligand-binding sites, are about 1−1.5 nm apart from the neighboring subunits (Unwin, 1993). However, Fu and Sine (1994), based on mutation analysis, came to the conclusion that one quaternary amonium group of dimethyl-(+)-tubocurarine interacts with Tyr198 in the α subunit, whereas another quaternary amonium of the same molecule (separated from the first one by 1.08 nm), is bound by γ-Tyr117. This finding illustrates that bis-quaternary amonium antagonists can bridge the α/γ subunit interface. Along the same line, an elegant experiment by Karlin and coworkers (Czajkowski and Karlin, 1991, 1995; Czajkowski et al., 1993), showed that a cross-linker of 0.9 nm length attached at the reduced Cys192/193 S-S bond of the *Torpedo* AChR α subunit reacted with δ-Asp165, δ-Asp180, and δ-Glu182. The corresponding mutations of the mouse receptor heterologously expressed demonstrated the essential role of δ-Asp180 for the high-affinity acetylcholine binding. Thus, it was concluded that the ligand-binding site does contain a carboxylic group (an anionic subsite not found in the α subunits where only aromatic amino acid residues were found to be labeled by DDF and other photo reagents), and the respective residue δ-Asp180 is at close proximity to α-Cys192/Cys193.

Here it is appropriate to say a word of caution concerning the distinction between the spatial proximity of residues and/or their functional importance. On the one hand, affinity labelling undoubtedly discloses the proximity of a tagged residue to the binding site (with the uncertainties introduced by a spacer arm or other modifications of the structure of the agonist used for the labeling) but does not prove functional importance of the revealed proximity. On the other hand, mutations allow one to pinpoint a residue of possible functional importance. But whether or not those residues are directly involved in ligand binding should be checked in independent experiments, though conclusions based on mutations and functional assays alone, without further structural investigation, may be ambiguous as well. For example, δ-Glu189 was also, on the basis of mutation-analysis, considered as a candidate carboxylate in the binding site, but lost this role in view of cross-linking experiments (Czajkowski and Karlin, 1995). Mutations in M4 had strong effects on the channel properties although, according to all other evidence, M4 is located at the protein/lipid interface, far away from the channel (quoted in Karlin, 1993, p. 306).

The three α-helical rods in the extracellular portions of the α subunits mentioned above make pockets situated at about 3.0 nm above the membrane bilayer. Support for the hypothesis that these pockets are indeed the agonist-binding sites (Unwin, 1993) came from fluorescence studies which gave a similar estimate for the location of low-molecular-mass gonists/antagonist-binding sites (Venezuela et al., 1994). However, distance determination from fluorescence energy transfer depends on many factors which are difficult to control; not surprisingly, the localisation of binding sites for various effectors in AChR has been revised in recent years (Herz et al., 1989; Johnson and Nuss, 1994): Evidence was obtained that one type of binding site for noncompetitive inhibitors is located contrary to previous conclusions outside the transmembrane domain.

The location of a ligand-binding site relative to the membrane surface can be determined by the analysis of cross-links formed by the receptor with different agonists and antagonists. The curarimimetic snake venom α-neurotoxins seem especially promising in this respect. In the section on quaternary structure, we have already discussed the application of photoactivatable derivatives of α-neurotoxins in elucidating the receptor subunit arrangement around the channel axis. Since α-neurotoxins act as competitive antagonists, mapping of their binding sites can also shed light on the neurotransmitter binding site location. We recently found that the *p*-azidosalicylamidoethyl-1,3′-dithiopropyl derivative at Lys25 of neurotoxin II from *Naja naja oxiana* forms a photoinduced cross-link with the δ-Ala268 residue in the upper part of the channel-forming M2 helix (Machold et al., 1995a). Taking into account the known crystal and solution structures of α-neurotoxins (Low and Corfield, 1986; Betzel et al., 1991; Golovanov et al., 1993) and the 1.65 nm length of the photolabel spacer arm, the center of the bound neurotoxin molecule should lie in the extracellular domain at 2.0 nm or closer to the membrane surface (Machold et al., 1995a). Considering further the multiplicity of contacts which the three loops I, II and III of the α-neurotoxin molecule make with the AChR (Tsetlin et al., 1982; Pillet et al., 1993; Fulachier et al., 1994), one comes to the conclusion that α-neurotoxins are bound in three dimensions [comparable to the three-loop model of low-molecular-mass antagonist binding (Devillers-Thiéry et al., 1993)], rather than as flat structures bound to the uppermost portion of the receptor's extracellular domain.

A comparison of the dimensions ($3.5 \times 2.0 \times 1.5$ nm) of α-neurotoxin (Bourne et al., 1985; Betzel et al., 1991; Golovanov et al., 1993), with the diameter of the AChR pore, 2.0–2.5 nm in the upper part and narrowing towards the cytoplasmic side (Unwin, 1993), indicates that it is not very likely that the α-neurotoxin enters through the channel mouth, reaching its binding site from inside, even more so when one considers that two toxin molecules have to find their way to the receptor's two binding sites. It seems that the toxin molecule accommodates itself between the receptor subunits, possibly approaching their binding pockets from the outside, in such a way that the photolabel, attached via a spacer arm to the toxin's central loop II, extends to the channel-forming helix M2.

While the photo cross-link proves the appearance of parts of the toxin on the inside, other parts have been shown to be accessible from the outside. Streptavidin binding to biotinylated toxin was detected by electron microscopy on the periphery of the donut-shaped receptor complex, and fluorescence energy transfer measurements suggested that the distance between toxin molecules bound to neighboring receptor complexes was smaller than between the two toxin molecules bound to the two binding sites within one AChR, an observation again interpreted as indicating binding to the periphery of the donut shape. Extension of one toxin molecule both to the inside and the outside of the receptor's subunit ring would mean that the binding site is actually a tunnel or gorge spanning the wall of the ring-shaped receptor (Fig. 9). This gorge would be situated at the interfaces of the α/γ and α/δ subunits, respectively, less than 2 nm above the membrane, and it would contain the amino acid residues of the antagonist and agonist binding sites discussed above, including the tandem cystein residues α-Cys192/193 and the conspicuous aromatic residues. Such an aromatic interface gorge reminds one of the 2.5–3.0 nm intersubunit tunnel found by X-ray crystallography in the tryptophan synthase $\alpha_2\beta_2$ bienzyme complex, which was shown to play a role in substrate (indole) and information transfer within this allosteric protein (Dunn et al., 1990; Ruvinov et al., 1995).

It is not yet clear whether this pocket, or gorge, exists in the native receptor. Alternatively, the toxin could move apart certain domains of the subunits. The last supposition does not appear improbable, since cryoelectron microscopy unraveled quite pronounced shifts induced by ligand binding. Namely, while a short pulse of acetylcholine (which was applied with the intention to open the AChR channel) induces a small rotation of transmembrane (most probably, M2) helices (Unwin, 1995), a large difference between the electron diffraction maps of the AChR in the resting and desensitized states was interpreted as a marked shift or tilt of the whole γ and δ subunits, the latter moving away from the channel axis (Unwin et al., 1988). There are also some data which might be relevant for the mode of neurotoxin penetration into the AChR. Introduction of glycosylation sites at positions Trp187 and Phe189 of the toxin-sensitive α subunit of the mouse muscle receptor confers a resemblance to the mongoose and snake AChRs, especially the resistance against the α-neurotoxin action (Kreienkamp et al., 1994; Keller et al., 1995). It was found that glycosylation does not reduce the association rate constant for the α-toxin binding to the receptor, but makes the dissociation of the toxin from the receptor complex extremely fast.

In addition to agonist and competitive antagonists, which have overlapping binding sites, and noncompetitive antagonists directly blocking the channel, in recent years the AChR was found to be sensitive to different classes of compounds such as physostigmine (Shaw et al., 1985; Okonjo et al., 1991; Pereira et al., 1993), steroids (Bertrand et al., 1991; Valera et al., 1992; Fernandez-Ballester et al., 1994), or the neuropeptide subtance P (Blanton et al., 1994). The mechanisms and the physiological significance of the modulatory effects of these compounds are still far from clear, whereas, for some of them, the sites of action have already been identified. A photoactivatable derivative of physostigmine labels Lys125 in the extracellular domain of the α subunit (Schrattenholz et al., 1993). The *p*-benzoylphenylalanine derivative of substance P was found to form a cross-link within the δ-M2 helix (Blanton et al., 1994).

Taken together, these data depict the AChR's ligand binding site as an intersubunit gorge near the membrane surface and also near the ion channel, containing conspicuously many aromatic amino acid side chains placed on at least three loops of the α polypeptide chains, and containing negatively charged side chains contributed by the neighboring γ and δ subunits. This gorge seems to span the wall of the channel entrance. Despite much detail, this picture still lacks an answer to the key question of pharmacology: what is the characteristic structural feature explaining the functional difference between agonists and competitive antagonists?

The regulatory domain: the allosteric model

The nicotinic acetylcholine receptor (as any other neurotransmitter receptor) is an allosteric protein (Changeux et al., 1967; Karlin, 1967; Galzi and Changeux, 1994). (a) The dose/response curve of postsynaptic depolarization is sigmoidal, as is

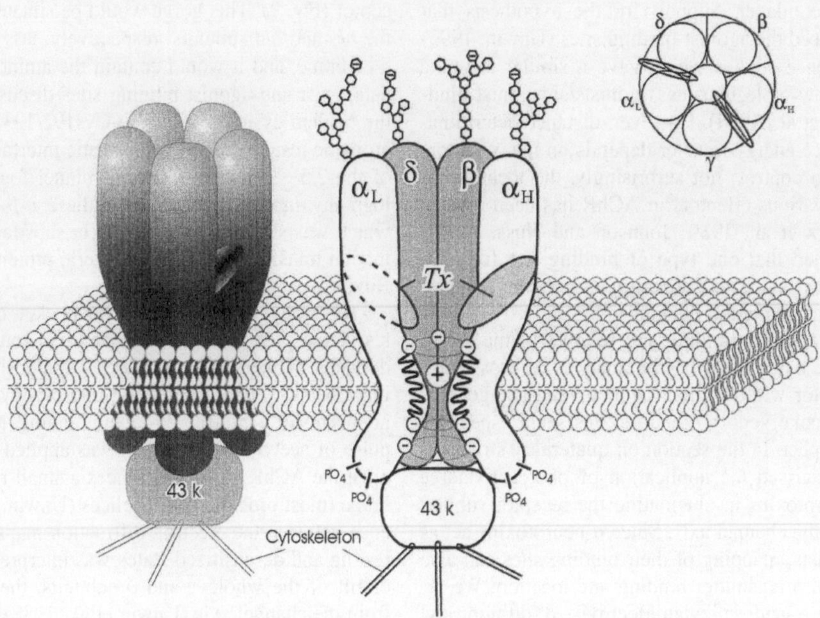

Fig. 9. The interface gorge model of the AChR, indicating the ligand binding site spanning the receptor wall. The figure summarizes the present state of the three-dimensional research on the AChR.

the agonist-binding curve. This indicates that both ligand-binding and channel-opening are cooperative processes (Colquhoun and Sakmann, 1985). (b) The functional domains of the receptor are separate structures, interacting indirectly through conformational changes. Furthermore, there are effector molecules, termed non-competitive inhibitors, acting (exclusively negatively) via regulatory binding sites which are distinct from the agonist-binding sites. (c) The AChR is composed of more than one polypeptide chain, a prerequisite for allosteric regulation (Monod et al., 1965). (d) Finally, the receptor exists in various functional states: resting, active, desensitized, which are in a rapid equilibrium affected by reversibly binding ligands. The desensitized state is characterized by an affinity two orders of magnitude higher for agonists and by a closed channel. It is thought to be an element of synaptic plasticity, though with the AChR it has not been proven to exist under physiological conditions (i.e. without blocking the acetylcholinesterase).

While these features comply with the definition given for allosteric proteins (Monod et al., 1965); others do not, making the AChR an unconventional allosteric protein (Galzi and Changeux, 1994): (a) The heteropentameric pseudosymmetric subunit structure and its non-equivalent agonist-binding α subunits preclude symmetric transitions which describe the saturation curves of, for example, hemoglobin or GAPDH. (b) The allosteric model proposed by Monod and coworkers (Monod et al., 1965) postulates the pre-existence of the various functional states. Allosteric regulation accordingly is described as a shift in the preformed equilibrium. For the nicotinic acetylcholine receptor such pre-existing states have not been proven to exist unequivocally, at least not in concentrations sufficient to describe the kinetics of the receptor's structural transitions after ligand binding (Jackson, 1984).

Neurotransmitter receptors have been described as triune signal converters (Hucho, 1992): They are composed of a signal receiving moiety R (the ligand-binding site), an effector moiety E (in the case of the AChR the ion channel), and a transducer T, which couples R with E (Fig. 10). While with seven-transmembrane (type-II) receptors, T is a separate molecule (a G-protein), with type-I receptors (ligand-gated ion channels) the transducer is an integral feature of the pentameric protein complex, in other words: its allosteric propensity.

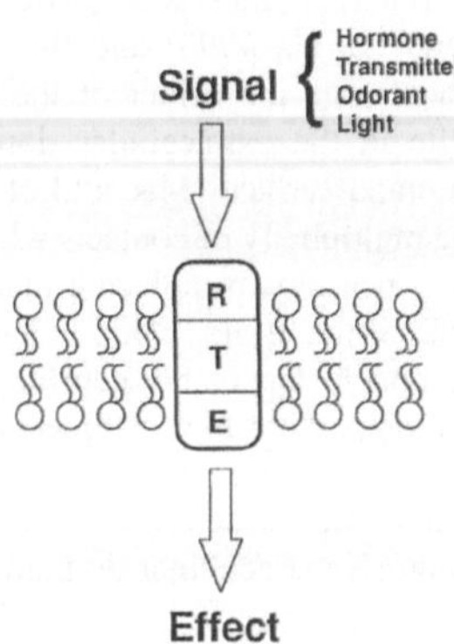

Fig. 10. The triune receptor concept. R, signal receiving (ligand-binding) domain; E, effector (ion channel domain; T, transducing (allosterically coupling) domain.

The structural correlations of the receptor's allosterism are becoming increasingly evident. We have discussed already the importance of the quaternary structure and especially of the subunit interfaces as means for the transduction of the extracellular signals from the ligand-binding site to the gating site of the ion channel.

Another set of information leading to protein structures which are important for the AChR's allosteric regulation is provided by the non-competitive inhibitors (NCIs). NCIs are a structurally heterogeneous group of compounds, which seem to bind to several distinct sites on the receptor. Characteristically, they bind in a voltage-dependent manner to a high-affinity site within the channel, thereby blocking the ion flux sterically in addition to their allosteric action (reviewed by Changeux, 1981). A prototype of these compounds is histrionicotoxin (HTX), an alkaloid from the skin of certain Latin-American frogs. Its allo-

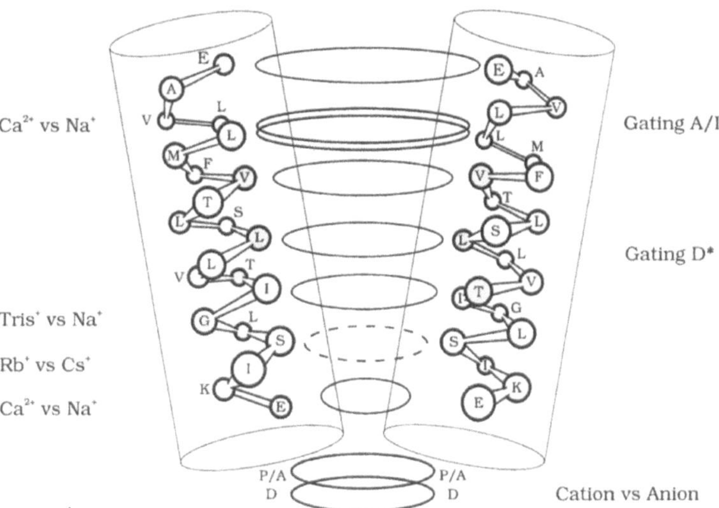

Fig. 11. The stratification of the ion channel (modified from Bertrand et al., 1993). Depicted are M2 sequences from the neuronal α7-subunit. The parameters which are affected by mutations in the respective ring are indicated in the margin. D* indicates a mutation which resulted in a desensitized/channel-open state.

steric action is evidenced by its positive cooperativity with the agonist (acetylcholine) binding site. Because of this action, HTX and other NCIs enhance the rate of desensitization. The first direct identification of a site involved was achieved with chlorpromazine (a voltage-independent NCI). Upon photoactivation it reacted covalently with the Ser262 of the *Torpedo* receptor's δ subunit (Giraudat et al., 1986). As we shall see in the next section, this site was also identified by means of triphenylmethyl phosphonium ($TPMP^+$), another NCI (Oberthür et al., 1986), as part of the channel wall, which led to the helix-M2 model of the receptor's ion channel (Hucho et al., 1986; Hucho and Hilgenfeld, 1989). In addition to sites within the channel NCIs can act at the lipid-protein interface.

A third set of data, besides the quaternary structure and the photolabeling with NCIs, was provided by site-directed mutagenesis, in combination with patch clamp physiology. But the mutants investigated so far exhibit pleiotropic effects (see, for example, Revah et al., 1991). Therefore, and because of the general ambiguity of the interpretation of mutants, which are accessible only to functional and not to structural investigation, it is not possible to assign an allosteric gating function to individual amino acid residues. It must suffice to state that several mutations within the channel-forming transmembrane helix M2 have significant effects on channel gating and desensitization. As pointed out by Karlin (1991), the mere speed of channel gating precludes large structural rearrangements within the receptor protein which, on the other hand, might be the basis of desensitization (channel opening occurs in the 10–100-μs range, while desensitization is 1000-fold slower). Channel opening most likely requires only few side-chain rotations (Hilgenfeld and Hucho, 1988).

In brief, the data relevant to the structural bases of the regulatory (transducing) domain do not hint at one specialized protein structure responsible for the transmission of the agonist binding signal into opening of the channel. Important features seem to be the quaternary structure, especially the localization of the binding sites at the subunit interfaces, and the binding sites for NCIs within the channel. No large-scale re-arrangements of the receptor structure have been observed which could have been interpreted as correlates of allosteric regulation and channel gating.

The ion channel: the helix-M2 model

The hole in the donut revealed by electron microscopy was from the beginning and still is, interpreted as the entrance to the ion channel. Photoaffinity labeling with the channel-blocking noncompetitive antagonist chlorpromazine suggested that the binding site for this compound was located in the axis of pseudosymmetry in the channel (Heidmann et al., 1983). This site was localized on δ-Ser262, a residue of the putative transmembrane helix M2. It was initially interpreted as one of the binding sites of noncompetitive blockers (Giraudat et al., 1986). But photolabeling with the channel blocker $TPMP^+$ (triphenylmethylphosphonium) revealed that the homologous helices M2 of the different subunits are oriented towards this axis and contribute to the wall of the ion channel (Hucho et al., 1986). This proposition was supported by many molecular biological (Imoto et al., 1986), electrophysiological (Dani, 1989; Leonard et al., 1988; Lester, 1992) and biochemical experiments (Giraudat et al., 1987, 1989) and is now widely accepted. It is even extended to the other members of the superfamily of ligand-gated ion channels (type-I receptors) the channel walls of which are also thought to be formed by their homologue M2 helices (summarized e.g. by Betz, 1990).

The channel formed by the five helices M2 is funnel-shaped (Fig. 9) (Hucho and Hilgenfeld, 1989). This can be concluded from the three available diameters at the channel entrance (2.0–2.5 nm, visible by electron microscopy; Unwin, 1993), at the reaction site of the noncompetitive channel blockers (the diameter of $TPMP^+$ as determined by X-ray crystallography is 1.15 nm; McPhail et al., 1971) and at the narrowest part, where the diameter of the largest permeating cation is 0.64 nm (Huang et al., 1978). The funnel-shaped arrangement of the five helices M2 implies that, in the wider upper (extracellular) part, there is room for other membrane-spanning peptides. In the activated (open-channel) state M1 could be labeled by the noncompetitive channel blocker quinacrine azide (DiPaola et al., 1990). Parts of M1 may be candidates to fill the gaps. The residues involved have been identified by SCAM; Akabas et al., 1992) and it was shown that they vary depending on the functional state of the receptor. (A by-product of these experiments was evidence that this part of M1 is not α-helical.)

The ion conductance properties of the channel (channel conductance, ion selectivity, rectification) are determined by three rings of negatively charged amino acid side chains (Fig. 9), one at the C-terminal end of the five M2 helices, one on its cytoplasmic side and one midway between these two, named the outer, the inner, and the intermediate ring, respectively (Imoto et al., 1988). The latter forms the constriction, probably the selectivity filter. The role of the selectivity filter was more specifically assigned to a threonine residue (α-Thr264, in rat muscle) located one helix turn above the intermediate ring (Villaroel and Sakmann, 1992).

The sequence similarity of the five channel-forming helices M2 implies a stratified structure (Bertrand et al., 1993): side chains at homologous positions form rings of distinct physical properties which determine various parameters of the channel: ion selectivity, gating, and cation or anion specificity (Fig. 11).

Two most elegant experiments may be placed at the end of this discussion. In the first, mutations based on a comparison of the anion-selective glycine receptor channel with the cation-selective AChR channel converted the latter into an anion channel (Galzi et al., 1992). Two point mutations within M2 of the neuronal α7 AchR (Val251Thr and Glu237Ala) and the insertion of a Pro (or Ala) residue at the cytoplasmic end of M2, brought this switch about. The replacement of a ring of Val side chains by a Thr ring and the substitution of the intermediate ring of negative charges by a neutral ring seem to be the only chracteristics within the M2 channel relevant for cation versus anion discrimination. The necessary insertion of a Pro (or Ala) points at the importance of the connecting loop between M1 and M2 for the geometry of the channel structure.

The other experiment pertains to the modular architecture of receptors and supports the RTE triune concept mentioned above, especially with respect to R and E. A chimaera was constructed by recombinant DNA techniques which combines features of the neuronal α7 AChR with the serotonine receptor $5HT_3$ (Eiselé et al., 1993). The homopolymeric α7 AChR is permeable for Ca^{2+} and external calcium potentiates the action of acetylcholine allosterically. In contrast, the channel of the $5HT_3$ receptor is blocked by Ca^{2+}. One chimaeric construct was obtained which combines AChR and $5HT_3$ properties: its channel was blocked by calcium ions, but it was activated by acetylcholine and this activation was potentiated by Ca^{2+}. The chimaera contained the N-terminal 201 amino acids of α7 and the C-terminal portion of the $5HT_3$ receptor. During heterologous expression, apparently the agonist-binding domain and the ion channel domain folded independently of the other domains.

To summarize this section, we can state that the AChR's ion channel is located in the center of the donut formed by the five receptor subunits, i.e. in the fivefold axis of pseudosymmetry. Each subunit contributes its homologous transmembrane helix M2 to the channel wall. The channel has the shape of a funnel. The channel properties are determined by three rings of negatively charged amino acid side chains, two of which are placed at the upper and lower end of M2, with the third, the intermediate ring, at the constriction of the channel. Ion selectivity is determined by a Thr residue present close to this constriction. The channel domain folds independently of the other receptor domains.

The active site of the AChR

Activation and channel-opening of the AChR is an all-or-nothing event taking place faster than the time resolution of the patch clamp technique, i.e. within 20–100 µs after agonist binding. As discussed above, this velocity precludes large-scale rearrangements within the receptor protein, which would be necessary for lon-range allosteric interactions between the ligand-binding pocket and the channel gating device.

One could hypothesize that long-range interactions are not necessary because there are no long distances between the functional domains. The cross-link between Lys25 of the α-neurotoxin bound to the receptor and δ-Ala268 of the receptor (Machold et al., 1995a) proves the proximity of R and E: The length of the cross-linker was 1.65 nm. It reached into the channel to δ-Ala268, which is located almost in the middle of M2, three helical turns (more than 1 nm) down from the membrane surface. Even if one includes the length of the Lys side chain (0.7 nm) to which the cross-linker was attached, and even if one assumes that the binding site for low-molecular-mass agonists and antagonists is not located under loop II of the toxin but rather higher up, under its disulfide-stabilized core (about 1 nm), there are only about 2 nm between agonist-binding site and the channel.

All the regulatory NCI binding sites identified so far are within this distance, on the protein domain covering these 2 nm: the site for chlorpromazine (δ-Ser262, Giraudat et al., 1986, and the corresponding positions on the other subunits, Giraudat et al., 1987; Revah et al., 1990), $TPMP^+$ (δ-Ser262, Oberthür et al., 1986, and corresponding positions on other subunits, Hucho et al., 1986), tetracain (δ-Ala268, Cohen, J. B., unpublished observation), the highly lipophilic probe 3-(trifluoromethyl)-3-(*m*-iodophenyl)diazirine [δ-Leu265, δ-Val269 and the corresponding positions on the β subunit, White and Cohen, 1992)], even meproafiden mustard which labeled α-Glu262, located on the extracellular margin of M2 (Pedersen et al., 1992). Furthermore, the mutations affecting allosteric regulation are also found within this range. The conclusion we would like to put forward is therefore as follows. The sites for signal reception (agonist binding), signal transduction (allosterism), and effector (ion channel), i.e. R, T, and E, are confined within an active site with a length of about 2 nm. This site is very close to the membrane surface, which may be the reason for the sensitivity of AChR activity to the composition of the lipid environment.

Conclusion: The nicotinic acetylcholine receptor (*Torpedo*); what we know and what we do not know

What do we know? We know the overall shape and dimensions, the primary sequence of its subunits, part of the secondary structure, overall and in its transmembrane domain, its transmembrane folding and the structure and position of its post-translational modifications. We have first ideas of its tertiary structure with respect to the agonist and antagonist binding sites, the ion channel and its selectivity filter. We know the quarternary structure as far as the subunit arrangement around the fivefold central axis of pseudosymmetry is concerned.

What do we not know? We do not know much about the structure and function of the vast extramembrane domains (beyond the ligand binding pockets, which are small and close to the membrane surface). We do not know the important structural details of the ligand binding pocket which triggers the channel gating upon agonist binding. In other words, we do not know the structural basis of the difference between agonists and antagonists (we don't know the structures involved in channel gating). We do not know the structure and function of the β subunit and of the domains of the other subunits, which are not involved in ligand binding. In summary: though there is much that we do know, we do not know very much and we need a crystal structure.

The work from the authors' laboratories was supported by the *Deutsche Forschungemeinschaft* (SfB 312), the *Ernst-Schering-*

Forschungsgesellschaft, a cooperation grant from the German Federal Ministry BMBF (O310588A), and the *Fonds der Chemischen Industrie*. Figs 9 and 10 were drawn by K. von Etzel, Figs 4, 8 and 11 by C. Kiecker; the sources of the figures are indicated in the legends.

REFERENCES

Abramson, S. N., Li, Y., Culver, P. & Taylor, P. (1989) An analog of lophotoxin reacts covalently with Tyr190 in the α-subunit of the nicotinic acetylcholine receptor, *J. Biol. Chem. 264*, 12666–12672.

Akabas, M. H., Stauffer, D. A., Xu, M. & Karlin, A. (1992) Acetylcholine receptor channel structure probed in cysteine substitution mutants, *Science 258*, 307–310.

Akabas, M. H. & Karlin, A. (1995) Identification of acetyclholine receptor channel-lining residues in the M1 segment of the α-subunit, *Biochemistry 34*, 12496–12500.

Anand, R., Conroy, W., Schoepfer, R. & Lindstrom, J. (1991) Neuronal nicotinic acetylcholine receptors expressed in *Xenopus* oocytes have a pentameric quaternary structure, *J. Biol. Chem. 266*, 11192–11198.

Anderson, D. J. & Blobel, G. (1981) *In vitro* synthesis, glycosylation and membrane insertion of the four subunits of *Torpedo* acetylcholine receptor, *Proc. Natl Acad. Sci. USA 78*, 5598–5602.

Anholt, R., Lindstrom, J. & Montal, M. (1980) Functional equivalence of monomeric and dimeric forms of purified acetylcholine receptors from *Torpedo californica* in reconstituted lipid vesicles, *Eur. J. Biochem. 109*, 481–487.

Apel, E. D., Roberds, S. L., Campbell, K. D. & Merli, J. P. (1995) Rapsyn may function as a link between the acetylcholine receptor and the agrin-binding dystrophin associated glycoprotein-complex, *Neuron 15*, 115–126.

Aslanian, D., Heidmann, T., Negrerie, M. & Changeux, J.-P. (1983) Raman spectroscopy of acetylcholine receptor-rich membranes from *Torpedo marmorata* and of their isolated components, *FEBS Lett. 164*, 393–400.

Barrantes, F. J. (1983) Recent developments in the structure and function of the acetylcholine receptor, *Int. Rev. Neurobiol. 24*, 259–341.

Bellone, M., Tang, F., Millius, R. & Conti-Tronconi, B. M. (1989) The main immunogenic region of the nicotinic acetylcholine receptor. Idenfication of amino acid residues interacting with different antibodies, *J. Immunol. 143*, 3568–3579.

Bennett, J. A. & Dingledine, R. (1995) Topology profile for a glutamate receptor: three transmembrane domains and a channel lining reentrant membrane loop, *Neuron 14*, 373–384.

Beroukhim, R. & Unwin, N. (1995) Three-dimensional location of the main immunogenic region of the acetylcholine receptor, *Neuron 15*, 323–331.

Bertrand, D., Valera, S., Bertrand, S., Ballivet, M. & Rungger, D. (1991) Steroids inhibit nicotinic acetylcholine receptors, *Neuroreport 2*, 277–280.

Bertrand, D., Galzi, J.-L., Devillers-Thiéry, A., Bertrand, S. & Changeux, J.-P. (1993) Stratification of the channel domain in neurotransmitter receptors, *Curr. Opin. Cell. Biol. 5*, 688–693.

Bertrand, D. & Changeux, J.-P. (1995) Nicotinic receptor: an allosteric protein specialized for intercellular communication, *The Neurosciences 7*, 75–90.

Betz, H. (1990) Ligand-gated ion channels in the brain: the amino acid receptor superfamily, *Neuron 5*, 383–392.

Betzel, C., Lange, G., Pal, G. P., Wilson, K. S., Maelicke, A. & Saenger, W. (1991) The refined crystal structure of α-cobrotoxin from *Naja naja siamensis* at 2. 4-Å resolution, *J. Biol. Chem. 266*, 21530–21536.

Biesecker, G. (1973) Molecular properties of the cholinergic receptor purified from *Electrophorus electricus*, *Biochemistry 12*, 4403–4409.

Blanton, M. P. & Cohen, J. B. (1994) Identifying the lipid-protein interface of the *Torpedo* nicotinic acetylcholine receptor: secondary structure implications, *Biochemistry 33*, 2859–2872.

Blanton, M. P., Li, Y. M., Stimson, E. R., Maggio, J. E. & Cohen, J. B. (1994) Agonist-induced photoincorporation of a *p*-benzoylphenylalanine derivative of substance P into membrane spanning region 2 of the *Torpedo* nicotinic acetylcholine receptor delta subunit, *Mol. Pharmacol. 46*, 1048–1055.

Blount, P. & Merlie, J. P. (1989) Molecular basis of the two nonequivalent ligand binding sites of the muscle nicotinic acetylcholine receptor, *Neuron 3*, 349–357.

Bourne, P. E., Sato, A., Corfield, P. W., Rosen, L. S., Birken, S. & Low, B. W. (1985) Erabutoxin b. Initial protein refinement and sequence analysis at 0.140-nm resolution, *Eur. J. Biochem. 153*, 521–527.

Brisson, A. & Unwin, N. (1985) Quaternary structure of the nicotinic acetylcholine receptor, *Nature 315*, 474–477.

Butler, D. H. & McNamee, M. G. (1993) FTIR analysis of nicotinic acetylcholine receptor secondary structure in reconstituted membranes, *Biochim. Biophys. Acta 1150*, 17–24.

Cartaud, J., Benedetti, E. L., Cohen, J. B., Meunier, J.-C. & Changeux, J.-P. (1973) Presence of a lattice structure in membrane fragments rich in nicotinic receptor protein from the electric organ of *Torpedo marmorata*, *FEBS Lett. 33*, 109–113.

Chang, H. W. & Bock, E. (1977) Molecular forms of the acetylcholine receptor. Effects of calcium ions and a sulfhydryl reagent on the occurence of oligomers, *Biochemistry 16*, 4513–4520.

Changeux, J.-P., Thiéry, J.-P., Tung, Y. & Kittel, C. (1967) On the cooperativity of biological membranes, *Proc. Natl Acad. Sci. USA 57*, 335–341.

Changeux, J.-P. (1981) The acetylcholine receptor: an 'allosteric' protein, *Harvey Lect. 75*, 85–254.

Changeux, J.-P. (1990) Functional architecture and dynamics of the nicotinic acetylcholine receptor: a ligand-gated ion channel, *Fidia Res. Found. Neurosi Award Lect. 4*, 21–168.

Charnet, P., Labarca, C., Leonhard, R. J., Vogelaar, N. J., Czyzyk, L., Gouin, A., Davidson, N. & Lester, H. A. (1990) An open-channel blocker interacts with adjacent turns of α-helices in the nicotinic acetylcholine receptor, *Neuron 4*, 87–95.

Chatrenet, B., Tremeau, O., Bontems, F., Goeldner, M. P., Hirth, C. G. & Ménez, A. (1990) Topography of toxin-acetylcholine receptor complexes by using photoactivatable toxin derivatives, *Proc. Natl Acad. Sci. USA 87*, 3378–3382.

Chavez, R. A. & Hall, Z. (1991) The transmembrane topology of the amino terminus of the α subunit of the nicotinic acetylcholine receptor, *J. Biol. Chem. 266*, 15532–15538.

Chavez, R. A. & Hall, Z. (1992) Expression of fusion proteins of the nicotinic acetylcholine receptor from mammalian muscle identifies the membrane-spanning regions in the α and δ subunits, *J. Cell. Biol. 116*, 385–393.

Claudio, T. (1989) Molecular genetics of acetylcholine receptor channels, in *Frontiers in molecular biology: molecular neurobiology volume* (Glover, D. M. & Hames, B. C., eds) pp. 63–142, IRL, Oxford.

Claudio, T., Ballivet, M., Patrick, J. & Heinemann, S. (1983) Nucleotide and deduced amino acid sequences of *Torpedo californica* acetylcholine receptor γ-subunit, *Proc. Natl Acad. Sci. USA 80*, 1111–1115.

Cohen, J. B., Blanton, M. P., Chiara, D. C., Sharp, S. D. & White, B. H. (1992) Structural organization of functional domains of the nicotinic acetylcholine receptor, *J. Cell. Biochem. 16E*, 217-T003.

Cohen, J. B., Sharp, S. D. & Liu, W. S. (1991) Structure of the agonist binding site of the nicotinic acetylcholine receptor: [^{3}H]acetylcholine mustard identifies residues in the cation binding subsite, *J. Biol. Chem. 266*, 23354–23364.

Colquhoun, D. & Sakmann, B. (1985) Fast-events in single-c channel currents activated by acetylcholine and its analogues at the frog neuromuscular endplate, *J. Physiol. (Lond.) 369*, 501–557.

Conti-Tronconi, B. M., McLane, K. E., Raftery, M. A., Grando, S. A. & Protti, M. P. (1994) The nicotinic acetylcholine receptor: structure and autoimmune pathology, *Crit. Rev. Mol. Biol. 29*, 69–123.

Cooper, E., Couturier, S. & Ballivet, M. (1991) Pentameric structure and subunit stoichiometry of a neuronal nicotinic acetylcholine receptor, *Nature 350*, 235–238.

Corringer, P. J., Galzi, J. L., Eisele, J. L., Bertrand, S., Changeux, J. P. & Bertrand, D. (1995) Identification of a new component of the agonist binding site of the nicotinic α7 homooligomeric receptor, *J. Biol. Chem. 270*, 11749–11752.

Criado, M., Hochschwender, S., Sarin, V., Fox, J. L. & Lindstrom, J. (1985) Evidence for unpredicted transmembrane domains in acetylcholine receptor subunits, *Proc. Natl Acad. Sci. USA 82*, 2004–2008.

Czajkowski, C., DiPaola, M., Bodkin, M., Salazar-Jimenez, G., Holtzman, E. & Karlin, A. (1989) The intactness and orientation of acetylcholine receptor-rich membranes from *Torpedo californica* electric tissue, *Arch. Biochem. Biophys. 272*, 412–420.

Czajkowski, C. & Karlin, A. (1991) Agonist binding site of *Torpedo* electric tissue nicotinic acetylcholine receptor. A negatively charged region of the δ subunit within 0.9 nm of the α subunit binding site disulfide, *J. Biol. Chem. 266*, 22603–22612.

Czajkowski, C., Kaufmann, C. & Karlin, A. (1993) Negatively charged amino acid residues in the nicotinic acetylcholine receptor δ subunit that contribute to the binding of acetylcholine, *Proc. Natl Acad. Sci. USA 90*, 6285–6289.

Czajkowski, C. & Karlin, A. (1995) Structure of the nicotinic receptor acetylcholine binding site – identification of acidic residues in the δ subunit within 0.9 nm of the α subunit binding site disulfide, *J. Biol. Chem. 270*, 3160–3164.

Damle, V. N. & Karlin, A. (1978) Affinity labeling of one of the two α-neurotoxin binding sites in acetylcholine receptor from *Torpedo californica*, *Biochemistry 17*, 2039–2045.

Dani, J. (1989) Site-directed mutagenesis and single-channel currents define the ionic channel of the nicotinic acetylcholine receptor, *Trends Neurosci. 12*, 125–128.

Deisenhofer, J., Epp, O., Miki, K., Huber, R. & Michel, H. (1985) Structure of the proteins subunits in the photosynthetic center of *Rhodopseudomonas viridis* at 3 Å resolution, *Nature 318*, 618–624.

Dennis, M., Giraudat, J., Kotzyba-Hibert, F., Goeldner, M., Hirth, C., Chang, J. Y., Lazure, C. Chretien, M. & Changeux, J.-P. (1988) Amino acids of the *Torpedo marmorata* acetylcholine receptor subunit labeled by a photoaffinity ligand for the acetylcholine binding site, *Biochemistry 27*, 2346–2357.

Devillers-Thiéry, A., Galzi, J. L., Eiselé, J. L., Bertrand, S., Bertrand, D. & Changeux, J.-P. (1993) Functional architecture of the nicotinic acetylcholine receptor: a prototype of ligand gated ion channels, *J. Membr. Biol. 136*, 97–112.

DiPaola, M., Czajkowski, C. & Karlin, A. (1989) The sidedness of the COOH terminus of the acetylcholine receptor δ-subunit, *Biol. Chem. 264*, 15457–15463.

DiPaola, M., Kao, P. N. & Karlin, A. (1990) Mapping of the α-subunit site photolabeled by the noncompetitive inhibitor [^{3}H]quinacrine azide in the active state of the nicotinic acetylcholine receptor, *J. Biol. Chem. 265*, 11017–11029.

Dowding, A. J. & Hall, Z. W. (1987) Monoclonal antibodies specific for each of the two toxin-binding sites of the *Torpedo* acetylcholine receptor, *Biochemistry 26*, 6372–6381.

Dunn, M. F., Aguilar, V., Brzovic, P., Drewe, W. F. Jr, Houben, K. F., Leja, C. A. & Roy, M. (1990) The tryptophan synthase bienzyme complex transfers indole between the α and β-sites via a 25–30 Å long tunnel, *Biochemistry 29*, 8598–8607.

Dunn, S. M. J., Conti-Tronconi, B. M. & Raftery, M. A. (1993) A high-affinity site for acetylcholine occurs to the α-γ subunit interface of *Torpedo* nicotinic acetylcholine receptor, *Biochemistry 32*, 8616–8621.

Eiselé, J. L., Bertrand, S., Galzi, J. L., Devillers-Thiéry, A., Changeux, J.-P. & Bertrand, D. (1993) Chimaeric nicotinic-serotonergic receptor combines distinct ligand binding and channel specificities, *Nature 366*, 479–483.

Elgoyhen, A. B., Johnson, D. S., Boulter, J., Vetter, D. E. & Heinemann, S. (1994) Alpha 9: An acetylcholine receptor with novel pharmacological properties in rat cochlear hair cells, *Cell 79*, 705–715.

Fels, G., Wolff, E. K. & Maelicke, A. (1982) Equilibrium binding of acetylcholine to the membrane-bound acetylcholine receptor, *Eur. J. Biochem. 127*, 31–38.

Fernandez-Ballester, S., Castresana, J., Fernandez, A. M., Arrondo, J.-L. R., Ferragut, J. A. & Gonzalez-Ros, J. M. (1994) A role for cholesterol as a structural effector of the nicotinic acetylcholine receptor, *Biochemistry 33*, 4065–4071.

Finer-Moore, J. & Stroud, R. M. (1984) Amphipathic analysis and possible formation of the ion channel in an acetylcholine receptor, *Proc. Natl Acad. Sci. USA 81*, 155–159.

Fong, T. M. & McNamee, M. G. (1987) Stabilization of acetylcholine receptor secondary structure by cholesterol and negatively charged phospholipids in membranes, *Biochemistry 26*, 3871–3880.

Frail, D. E., McLaughlin, L. L., Mudd, J. & Merlie, J. P. (1988) Identification of the mouse muscle 43000 Dalton acetylcholine receptor-associated protein (RAPsyn) by cDNA cloning, *J. Biol. Chem. 263*, 15602–15607.

Froehner, S. C., Luetje, C. W., Scotland, P. B. & Patrick, J. (1990) The postsynaptic 43 K protein clusters muscle nicotinic acetylcholine receptors in *Xenopus* oocytes, *Neuron 5*, 403–410.

Fu, D.-X. & Sine, S. M. (1994) Competitive antagonists bridge the α-γ subunit interface of the acetylcholine receptor through quaternary ammonium-aromatic interactions, *J. Biol. Chem. 269*, 26152–26157.

Fulachier, M. H., Mourier, G., Cotton, J., Servent, D. & Ménez, A. (1994) Interaction of protein ligands with receptor fragments. On the residue of curarimimetic toxins that recognize fragments 128–142 and 185–199 of the α-subunit of the nicotinic acetylcholine receptor, *FEBS Lett. 338*, 331–338.

Galzi, J. L., Revah, F., Black, D., Goeldner, M., Hirth, C. & Changeux, J.-P. (1990) Identification of a novel amino acid α-Tyr93 within the active site of the acetylcholine receptor by photoaffinity labelling: additional evidence for a three loop model of the acetylcholine binding site, *J. Biol. Chem. 265*, 10430–10437.

Galzi, J. L., Revah, F., Bouet, F., Ménez, A., Goeldner, M., Hirth, C. & Changeux, J.-P. (1991a) Allosteric transitions of the acetylcholine receptor probed at the amino acid level with a photolabile cholinergic ligand, *Proc. Natl Acad. Sci. USA 88*, 5051–5055.

Galzi, J. L., Bertrand, D., Devillers-Thiéry, A., Revah, F., Bertrand, S. & Changeux, J.-P. (1991b) Functional significiance of aromatic amino acids from three peptide loops of the α7 neuronal nicotinic receptor site investigated by site-directed mutagenesis, *FEBS Lett. 294*, 198–202.

Galzi, J.-L., Devillers-Thiery, A., Hussy, N., Bertrand, S. & Changeux, J.-P. (1992) Mutations in the ion channel domain of a neuronal nicotinic receptor convert ion selectivity from cationic to anionic, *Nature 359*, 500–505.

Galzi, J.-L. & Changeux, J.-P. (1994) Neurotransmitter-gated ion channels as unconventional allosteric proteins, *Curr. Opin. Struct. Biol. 4*, 554–565.

Gershoni, J. M., Hawrot, E. & Lentz, T. L. (1983) Binding of α-bungarotoxin to isolated α-subunit of the acetylcholine receptor of *Torpedo californica*: quantitative analysis with protein blots, *Proc. Natl Acad. Sci. USA 80*, 4973–4977.

Giraudat, J., Montecucco, C., Bisson, R. & Changeux, J.-P. (1985) Transmembrane topology of acetylcholine receptor subunits probed with photoreactive phospholipids, *Biochemistry 24*, 3121–3127.

Giraudat, J., Dennis, M., Heidmann, T., Chang, J.-Y. & Changeux, J.-P. (1986) Structure of the high affinity binding site for noncompetitive blockers of the acetylcholine receptor: serine 262 of the δ subunit is labeled by [3H]chlorpromazine, *Proc. Natl Acad. Sci. USA 83*, 2719–2723.

Giraudat, J., Dennis, M., Heidmann, T., Haumont, P.-Y., Lederer, F. & Changeux, J.-P. (1987) Structure of the high-affinity binding site for noncompetitive blockers of the acetylcholine receptor: [^{3}H]chlorpromazine labels homologous residues in the beta and delta chains, *Biochemistry 26*, 2410–2418.

Giraudat, J., Galzi, J.-L., Revah, F., Changeux, J.-P., Haumont, P. Y. & Lederer, F. (1989) The noncompetitive blocker [^{3}H]chlorpromazine labels segment M2 but not segment M1 of the nicotinic acetylcholine receptor α-subunit, *FEBS Lett. 253*, 190–198.

Golovanov, A. P., Lomize, A. L., Arseniev, A. S., Utkin, Y. N. & Tsetlin, V. I. (1993) Two-dimensional ^{1}H-NMR study of the spatial structure of neurotoxin II from *Naja naja oxiana*, *Eur. J. Biochem. 213*, 1213–1223.

Görne-Tschelnokow, U., Strecker, A., Kaduk, C., Naumann, D. & Hucho, F. (1994) The transmembrane domains of the nicotinic acetylcholine receptor contain α-helical and β structures, *EMBO J. 13*, 338–341.

Hamilton, S. L., Pratt, D. R. & Eaton, D. C. (1985) Arrangement of the subunits of the nicotinic acetylcholine receptor of *Torpedo californica* as determined by α-neurotoxin cross linking, *Biochemistry 24*, 2210–2219.

Hann, R. M., Pagan, O. R. & Eterovic, V. A. (1994) The α-conotoxins GI and MI distinguish between the nicotonic acetylcholine receptor agonist sites while SI does not, *Biochemistry 33*, 14058–14063.

Heidmann, T., Oswald, R. E. & Changeux, J.-P. (1983) Multiple sites of action for non-competitive blockers on acetylcholine receptor rich membrane fragments from *Torpedo marmota*, *Biochemistry 22*, 3112–3127.

Henderson, R., Baldwin, J. M., Ceska, T. A., Zemlin, F., Beckmann, E. & Downing, K. H. (1990) Model for the structure of bacteriorhodopsin based on high-resolution electron cryomicroscopy, *J. Mol. Biol. 213*, 899–929.

Hertling-Jaweed, S., Bandini, G., Müller-Fahrnow, A., Dommes, V. & Hucho, F. (1988) Rapid preparation of the nicotinic acetylcholine receptor for crystallization in detergent solution, *FEBS Lett. 241*, 29–32.

Herz, J. M., Johnson, D. A. & Taylor, P. (1989) Distance between the agonist and noncompetitive inhibitor sites on the nicotinic acetylcholine receptor, *J. Biol. Chem. 264*, 12439–12448.

Hilgenfeld, R. & Hucho, F. (1988) Properties and problems of the helix-M2 model of the acetylcholine receptor ion channel, *Transport through membranes: carriers, channels, and pumps* (Pullman, A., Jordtner, J. & Pullman, B., eds) pp. 359–367, Kluwer Academic Publishers, Dordrecht.

Hollmann, M., Maron, C. & Heinemann, S. (1994) N-glycosylation site tagging suggests a three transmembrane domain topology for the glutamate receptor GluR1, *Neuron 13*, 1331–1343.

Holtzman, E., Wise, D., Wall, J. & Karlin, A. (1982) Electron microscopy of complexes of isolated acetylcholine receptor, biotinyl-toxin, and avidin, *Proc. Natl Acad. Sci. USA 79*, 310–314.

Huang, L.-Y., Catterall, W. A. & Ehrenstein, G. (1978) Selectivity of cations and nonelectrolytes for acetylcholine activated channels in cultured muscle cells, *J. Gen. Physiol. 71*, 394–410.

Hucho, F. & Changeux, J.-P. (1973) Molecular mass and quaternary structure of the cholinergic receptor protein extracted by detergents from *Electrophorus electricus* electric tissue, *FEBS Lett. 38*, 11–15.

Hucho, F., Bandini, G. & Suarz-Isla, B. A. (1978) The acetylcholine receptor as part of a protein complex in receptor-enriched membrane fragments from *Torpedo californica* electric tissue, *Eur. J. Biochem. 83*, 335–340.

Hucho, F. (1986) The nicotinic acetylcholine receptor and its ion channel, *Eur. J. Biochem. 158*, 211–226.

Hucho, F., Oberthür, W. & Lottspeich, F. (1986) The ion channel of the nicotinic acetylcholine receptor is formed by the homologous helices M II of the receptor subunits, *FEBS Lett. 205*, 137–142.

Hucho, F. & Hilgenfeld, R. (1989) The selectivity filter of a ligand-gated ion channel: The helix-M2 model of the ion channel of the nicotinic acetylcholine receptor, *FEBS Lett. 257*, 17–23.

Hucho, F. (1992) Transmitter receptors – general principles and nomenclature, *Neurotransmitter receptors, New comprehesive biochemistry*, pp. 3–13, Elsevier Science Publishers, Amsterdam.

Huganir, R. L. & Greengard, P. (1987) Regulation of receptor function by protein phosphorylation, *Trends Pharmacol. Sci. 8*, 472–477.

Imoto, K., Busch, C., Sakmann, B., Mishina, M., Konno, T., Nakai, J., Bujo, H., Mori, Y., Fukuda, K. & Numa, S. (1986) Rings of negatively charged amino acids determine the acetylcholine receptor channel conductance, *Nature 335*, 645–648.

Imoto, K., Methfessel, C., Sakmann, B., Mishina, M., Konno, T., Nakai, J., Bujo, H., Fujita, Y. & Numa, S. (1988) Location of a δ-subunit region determining ion transport through the acetylcholine receptor channel, *Nature 324*, 670–674.

Iwata, S., Ostermeier, C., Ludwig, B. & Michel, H. (1995) Structure at 2. 8 Å resolution of cytochrome oxidase from *Paracoccus denitrificans*, *Nature 376*, 660–669.

Jackson, M. B. (1984) Spontaneous openings of the acetylcholine receptor channel, *Proc. Natl Acad. Sci. USA 81*, 3901–3904.

Johnson, D. A. & Nuss, J. M. (1994) The histrionicotoxin sensitive ethidium binding site is located outside of the transmembrane domain of the nicotinic acetylcholine receptor: a fluorescence study, *Biochemistry 33*, 9070–9077.

Kao, P. N., Dwork, A. J., Kaldany, R. R., Silver, M. L., Wideman, J., Stein, S. & Karlin, A. (1984) Identification of the α subunit half-cysteine specifically labeled by an affinity reagent for the acetylcholine receptor binding site, *J. Biol. Chem. 259*, 11662–11665.

Karlin, A. (1967) On the application of a 'plausible model' of allosteric proteins to the receptor for acetylcholine, *J. Theoret. Biol. 16*, 306–320.

Karlin, A., Holtzman, E., Yodh, N., Lobel, P., Wall, J. & Hainfeld, J. (1983) The arrangement of the subunits of the acetylcholine receptor of *Torpedo californica*, *J. Biol. Chem. 258*, 6678–6681.

Karlin, A. (1991) Explorations of the nicotinic acetylcholine receptor, *Harvey Lect. 86*.

Karlin, A. (1993) Structure of nicotinic acetylcholine receptors, *Curr. Opin. Neurobiol. 3*, 299–309.

Karlin, A. & Akabas, M. H. (1995) Toward a structural basis for the function of nicotinic acetylcholine receptors and their cousins, *Neuron 15*, 1231–1244.

Keller, S. H., Kreienkamp, H.-J., Kawanishi, C. & Taylor, P. (1995) Molecular determinants conferring α-toxin resistance in recombinant DNA-derived acetylcholine receptors, *J. Biol. Chem. 270*, 4165–4171.

Kirsch, J., Wolter, I., Triller, A. & Betz, H. (1993) Gephyrin antisense oligonucleotides prevent glycine receptor clustering in spinal neurons, *Nature 366*, 745–748.

Kreienkamp, H.-J., Maeda, R. K., Sine, S. M. & Taylor, P. (1995) Intersubunit contacts governing assembly of the mammalian nicotinic acetylcholine receptor, *Neuron 14*, 635–644.

Kreienkamp, H.-J., Utkin, Y. N., Weise, C., Machold, J., Tsetlin, V. I. & Hucho, F. (1992) Investigation of ligand binding sites of the acetylcholine receptor using photoactivatable derivatives of neurotoxin II from *Naja naja oxiana*, *Biochemistry 31*, 8239–8244.

Kreienkamp, H.-J., Sine, S., Maeda, R. K. & Taylor, P. (1994) Glycosylation sites selectively interfere with α-toxin binding to the nicotinic acetylcholine receptor, *J. Biol. Chem. 269*, 8108–8114.

Kubalek, E., Ralston, S., Lindstrom, J. & Unwin, N. (1987) Location of subunits within the acetylcholine receptor by electron image analysis of tubular crystals from *Torpedo marmorata*, *J. Cell. Biol. 105*, 9–18.

Léna, C. & Changeux, J.-P. (1993) Allosteric modulations of the nicotinic acetylcholine receptor, *Trends Neurosci. 16*, 181–186.

Leonard, R. J., Labarca, C. G., Charnet, P., Davidson, N. & Lester, H. A. (1988) Evidence that the Mmembrane-spanning region lines the ion channel pore of the nicotinic receptor, *Science 242*, 1578–1581.

Lester, H. A. (1992) The permeation pathway of neurotransmitter-gated ion channels, *Annu. Rev. Biophys. Biomol. Struct. 21*, 267–292.

Lindstrom, J., Merlie, J. P. & Yogeeswaran, G. (1979) Biochemical properties of acetylcholine receptor subunits from *Torpedo californica*, *Biochemistry 18*, 4465–4470.

Low, B. W. & Corfield, P. W. R. (1986) Structure/function relationships following initial protein refinement at 0.140-nm resolution, *Eur. J. Biochem. 161*, 579–587.

Machold, J., Utkin, Y., Kirsch, D., Kaufmann, R., Tsetlin, V. & Hucho, F. (1995a) Photolabeling reveals the proximity of the α-neurotoxin binding site to the M2-helix of the ion channel in the nicotinic acetylcholine receptor, *Proc. Natl Acad. Sci. USA 92*, 7282–7286.

Machold, J., Weise, C., Utkin, Y., Tsetlin, V. & Hucho, F. (1995b) The handedness of the subunit arrangement of the nicotinic acetylcholine receptor from *Torpedo californica*, *Eur. J. Biochem. 234*, 427–430.

MacKinnon, R. (1995) Pore loops: An emerging theme in ion channel structure, *Neuron 14*, 889–892.

Maelicke, A. & Reich, E. (1976) On the interaction between cobra α-neurotoxin and the acetylcholine receptor, *Cold Spring Harbor Symp. Quant. Biol.*, 203–210.

Mattson, C. & Heilbronn, E. (1975) The nicotinic acetylcholine receptor: a glycoprotein, *J. Neurochem. 25*, 899–901.

McPhail, A. T., Semenink, G. M. & Chesnut, D. B. (1971) Crystal structures of methyltriphorylarsonium and methyltriphonylphosphonium Bis-(α,αα′α′-tetracyanoquinodimethanmides) {Di-[3, 6-bis(dicyanomethylene)cychmexa-1,4-dienes]} *J. Chem. Soc. (A)*, 2174–2180.

Middelton, R. E. & Cohen, J. B. (1991) Mapping of the acetylcholine binding site of the nicotinic acetylcholine receptor: [^{3}H]nicotine as an agonist photoaffinity label, *Biochemistry 30*, 6987–6997.

Mielke, D. L. & Wallace, B. A. (1988) Secondary structural analysis of the nicotinic acetylcholine receptor as a test of molecular models, *J. Mol. Biol. 263*, 3177–3182.

Mileo, A. M., Monaco, L., Palma, E., Grassi, F., Miledi, R. & Eusebi, F. (1995) Two forms of acetylcholine receptor γ subunit in mouse muscle, *Proc. Natl Acad. Sci. USA 92*, 2686–2690.

Monod, J., Wyman, J. & Changeux, J.-P. (1965) On the nature of allosteric transitions, *J. Mol. Biol. 12*, 88–118.

Moore, W. M., Holladay, L. A., Puett, D. & Brady, R. N. (1974) On the conformation of the acetylcholine receptor protein from *Torpedo nobiliana*, *FEBS Lett. 45*, 145–149.

Moore, C. R., Yates, J. R., Griffin, P. R., Shabanowitz, J., Martino, P. A., Hunt, D. F. & Cafiso, D. S. (1989) Proteolytic fragments of the nicotinic acetylcholine receptor identified by mass spectrometry: implications for receptor topography, *Biochemistry 28*, 9184–9191.

Mund, M. (1994) Kartierung der Protein-Oberfläche des nikotinischen Acetylcholin-Rezeptors durch Jodierung exponierter Tyrosin-Reste, Dissertation, Berlin.

Naumann, D., Schultz, C., Görne-Tschelnokow, U. & Hucho, F. (1993) Secondary structure and temperature behavior of the acetylcholine receptor by Fourier transform infrared spectroscopy, *Biochemistry 32*, 3162–3168.

Neubig, R. R. & Cohen, J. B. (1979) Equlibrium binding of [^{3}H]tubocurarine and [^{3}H]acetylcholine by *Torpedo* postsynaptic membranes: stoichiometry and ligand interactions, *Biochemistry 18*, 5464–5475.

Neumann, D., Barchan, D., Fridkin, M. & Fuchs, S. (1986) Analysis of ligand binding to the synthetic dodecapeptide 185–196 of the acetylcholine receptor α subunit, *Proc. Natl Acad. Sci. USA 83*, 9250–9253.

Neumann, D., Barchan, D., Horowitz, M., Kochva, E. & Fuchs, S. (1989) Snake acetylcholine receptor: Cloning of the domain containing the four extracellular cysteins of the α-subunit, *Proc. Natl Acad. Sci. USA 86*, 7255–7259.

Nickel, E. & Potter, L. T. (1973) Ultrastructure of isolated membranes of *Torpedo* electric tissue, *Brain Res. 57*, 508–517.

Noda, M., Takahashi, H., Tanabe, T., Toyosato, M., Furutani, Y., Hirose, T., Asai, M., Inayama, S., Miyata, T. & Numa, S. (1982) Primary structure of α-subunit precursor of *Torpedo californica* acetylcholine receptor deduced from cDNA sequence, *Nature 299*, 793–797.

Noda, M., Takahshi, H., Tanabe, T., Toyosato, M., Kikyotani, S., Hirose, T., Asai, M., Takashima, H., Inayama, S., Miyata, T. & Numa, S. (1983a) Primary structure of β and δ subunit precursors of *Torpedo californica* acetylcholine receptor deduced from cDNA sequences, *Nature 301*, 251–255.

Noda, M., Takahashi, H., Tanabe, T., Toyosato, M., Kikyotani, S., Furutani, Y., Hirose, T., Takashima, H., Inayama, S., Miyata, T. & Numa, S. (1983b) Structural homology of *Torpedo californica* acetylcholine receptor subunits, *Nature 302*, 528–532.

Nomoto, H., Takahashi, N., Nagaki, Y., Endo, S., Arata, Y. & Hayashi, K. (1986) Carbohydrate structures of acetylcholine receptor from *Torpedo californica* and distribution of oligosaccharides among the subunits, *Eur. J. Biochem. 157*, 233–242.

Oberthür, W., Muhn, P., Baumann, H., Lottspeich, F., Wittmann-Liebold, B. & Hucho, F. (1986) The reaction site of a noncompetitive antagonist in the δ-subunit of the nicotinic acetylcholine receptor, *EMBO J. 5*, 1815–1819.

Okonjo, K. O., Kuhlmann, J. & Maelicke, A. (1991) A second pathway of activation of the *Torpedo* acetylcholine receptor channel, *Eur. J. Biochem. 200*, 671–677.

O'Leary, M. E. & White, M. M. (1992) Mutational analysis of ligand-induced activation of the *Torpedo* acetylcholine receptor, *J. Biol. Chem. 267*, 8360–8365.

O'Leary, M. E., Filatov, G. N. & White, M. M. (1994) Characterization of *d*-tubocurarine binding site of *Torpedo* acetylcholine receptor, *Am. J. Physiol. 266*, C648–C653.

Ortells, M. O. & Lunt, G. G. (1994) The transmembrane region of the nicotinic acetylcholine receptor: is it an all-helix bundle? *Receptors and Channels 2*, 53–59.

Papadouli, I., Sakarello, C. & Tzartos, S. J. (1993) High-resolution epitope mapping and fine antigenic characterization of the main immunogenic region of the acetylcholine receptor, *Eur. J. Bichem. 211*, 227–234.

Parker, M. W., Postma, J. P. M., Pattus, F., Tacker, A. D. & Tsernoglu, D. (1992) Refined structure of the pore forming domain of colicin A at 2.4 Å resolution, *J. Mol. Biol. 224*, 639–657.

Pedersen, S. E. & Cohen, J. B. (1990) *d*-Tubocuarine binding sites are located at α-γ and α-δ subunit interfaces of the nicotinic acetylcholine receptor, *Proc. Natl Acad. Sci. USA 87*, 2785–2789.

Pedersen, S. E., Sharp, S. D., Liu, W. S. & Cohen, J. B. (1992) Structure of the noncompetitive antagonist-binding site of the *Torpedo* nicotinic acetylcholine receptor. [^{3}H]Meproadifen mustard reacts selectively with a-subunit Glu-262, *J. Biol. Chem. 267*, 10489–10499.

Pereira, E. F., Alkondon, M., Tano, T., Castro, N. G., Froes, F. M., Aronstam, R. S., Schrattenholz, A., Maelicke, A. & Albuquerque, E. X. (1993) A novel agonist binding site on nicotinic acetylcholine receptors, *J. Recept. Res. 13*, 413–436.

Phillips, W. D., Maimone, M. M. & Merlie, J. P. (1991) Mutagenesis of the 43-kD postsynaptic protein defines domains involved in plasma membrane targeting an AChR clustering, *J. Cell. Biol. 115*, 1713–1723.

Phillips, W. D., Noakes, P. G., Roberds, S. L., Campell, K. P. & Merlie, J. P. (1993) Clustering and immobilization of acetylcholine receptors by the 43-kD postsynaptic protein: a possible role for dystrophin-related protein, *J. Cell. Biol. 123*, 729–740.

Pillet, L., Tremeau, O., Ducancel, F., Drevet, P., Zinn-Justin, S., Pinkasfeld, S., Boulain, J. C. & Menez, A. (1993) Genetic engineering of snake toxins. Role of invariant residues in the structural and functional properties of a curarimimetic toxin, as probed by site directed mutagenesis, *J. Biol. Chem. 268*, 909–916.

Poulter, L., Earnest, J. P., Stroud, R. M. & Burlingame, A. L. (1989) Structure, oligosaccharide structures, and posttranslationally modified sites of the nicotinic acetylcholine receptor, *Proc. Natl Acad. Sci. USA 86*, 6645–6649.

Raftery, M. A., Hunkapiller, M. W., Strader, C. D. & Hood, L. E. (1980) Acetylcholine receptor: complex of homologous subunits, *Science 208*, 1454–1457.

Raftery, M. A., Schmidt, J. & Clark, D. G. (1972) Specificity of α-bungarotoxin binding to *Torpedo californica* electroplax, *Arch. Biochem. Biophys. 152*, 882–886.

Revah, F., Galzi, J. L., Giraudat, J., Haumont, P.-Y., Lederer, F. & Changeux, J.-P. (1990) The noncompetitive blocker [^{3}H]chlorpromazine labels three amino acids of the acetylcholine subunit: Implication for the α helical organisation of the regions M II and for the structure of the ion channel, *Proc. Natl Acad. Sci. USA 87*, 4675–4679.

Revah, F., Bertrand, D., Galzi, J.-L., Devillers-Thiery, A., Mulle, C., Hussy, N., Bertrand, S., Ballivet, M. & Changeux, J. P. (1991) Mutations in the channel domain alter desensitization of a neuronal nicotinic receptor, *Nature 353*, 846–849.

Reynolds, J. A. & Karlin, A. (1978) Molecular mass in detergent solution of acetylcholine receptor from *Torpedo californica*, *Biochemistry 17*, 2035–2038.

Ruvinov, S. B., Yang, X. J., Parris, K. D., Banik, U., Ahmed, S. A., Miles, S. W. & Sackett, D. L. (1995) Ligand-mediated changes in the tryptophan synthase indole tunnel probed by nile red fluorescence with wild type, mutant, and chemically modified enzymes. *J. Biol. Chem. 270*, 6357–6369.

Sargent, P. B. (1993) The diversity of neuronal nicotinic acetylcholine receptors, *Annu. Rev. Neurosci. 16*, 403–443.

Schofield, P. R., Darlison, M. G., Fujita, N., Burt, D. R., Stephenson, F. A., Rodriguez, H., Rhee, L. M., Ramachandran, J., Reale, V., Glencorse, T. A., Seeburg, P. H. & Barnard, E. A. (1987) Sequence and functional expression of the $GABA_A$ receptor shows a ligand-gated receptor super-family, *Nature 328*, 221–227.

Schrattenholz, A., Godovac-Zimmermann, J., Schäfer, H.-J., Albuquerque, E. X. & Maelicke, A. (1993) Photoaffinity labeling of *Torpedo* acetylcholine receptor by physostigmine, *Eur. J. Biochem. 216*, 671–677.

Shaw, K. P., Aracava, Y., Akaide, A., Daly, J. W., Rickett, D. L. & Albuquerque, E. X. (1985) The reversible cholinesterase inhibitor physostigmine has channel-blocking and agonist effects on the acetylcholine receptor ion-channel complex, *Mol. Pharmacol. 28*, 527–538.

Shoji, H., Takahashi, N., Nomoto, H., Ishikawa, M., Shimada, I., Arata, Y. & Hayashi, K. (1992) Detailed structural analysis of asparagine-linked oligosaccharides of the nicotinic acetylcholine receptor from *Torpedo californica*, *Eur. J. Biochem. 207*, 631–641.

Sine, S. M. & Claudio, T. (1991) γ- and δ-subunits regulate the affinity and cooperativity of ligand binding to the acetylcholine receptor, *J. Biol. Chem. 266*, 19369–19377.

Sine, S. M. (1993) Molecular dissection of subunit interfaces in the acetylcholine receptor: identification of residues that determine curare selectivity, *Proc. Natl Acad. Sci. USA 90*, 9436–9440.

Sine, S. M., Kreienkamp, H.-J., Bren, N., Maeda, R. & Taylor, P. (1995) Molecular dissection of subunit interfaces in the acetylcholine receptor: identification of residues that determine α-conotoxin M1 selectivity, *Neuron 15*, 205–211.

Strecker, A., Franke, P., Weise, C. & Hucho, F. (1994) All potential glycosylation sites of the nicotinic acetylcholine receptor δ-subunit from *Torpedo californica* are utilized, *Eur. J. Biochem. 220*, 1005–1011.

Suárez-Isla, B. A. & Hucho, F. (1977) Acetylcholine receptor: SH group reactivity as indicator of conformational changes and functional states, *FEBS Lett. 76*, 65–69.

Swanson, K. L. & Albuquerque, E. X. (1992) Nicotinic acetylcholine receptors and low molecular mass toxins, *Handbook of Experimental Pharmacology*, vol. 102, ch. 17, Springer-Verlag, Berlin.

Takai, T., Noda, M., Mishina, M., Shimizu, S., Furuntani, Y., Kayano, T., Ikeda, T., Kubo, T., Takahashi, H., Takahashi, T., Kuno, M. & Numa, S. (1985) Cloning sequencing and expression of cDNA for a novel subunit of acetylcholine receptor from calf muscle, *Nature 315*, 761–764.

Tsetlin, V. I., Karlsson, E., Utkin, Y. N., Pluzhnikov, K. A., Arseniev, A. S., Surin, A. M., Kondakov, V. V., Bystrov, V. F., Ivanov, V. T. & Ovchinnikov, Y. A. (1982) Interacting surfaces of neurotoxins and acetylcholine receptor, *Toxicon 20*, 83–93.

Tzartos, S. J. & Lindstrom, J. (1980) Monoclonal antibodies used to probe acetylcholine receptor structure: localization of the main immunogenic region and detection of similarities between subunits, *Proc. Natl Acad. Sci. USA 77*, 755–759.

Unwin, N., Toyoshima, C. & Kubalek, E. (1988) Arrangement of the acetylcholine receptor subunits in the resting and desensitized states, determined by cryoelectron microscopy of crystallized *Torpedo* postsynaptic membranes, *J. Cell. Biol. 107*, 1123–1138.

Unwin, N. (1993) Nicotinic acetylcholine receptor at 9 Å resolution, *J. Mol. Biol. 229*, 1101–1124.

Unwin, N. (1995) Acetylcholine receptor channel imaged in the open state, *Nature 373*, 37–43.

Utkin, Y. N., Kabayashi, Y., Hucho, F. & Tsetlin, V. I. (1994) Relationships between the binding sites for an α-conotoxin and snake venom toxins in the nicotinic acetylcholine receptor from *Torpedo californica*, *Toxicon 32*, 1153–1157.

Valera, S., Ballivet, M. & Bertrand, D. (1992) Progesterone modulates a neuronal nicotinic acetylcholine recptor, *Proc. Natl Acad. Sci. USA 89*, 9949–9953.

Vandlen, J. L., Wu, W. C.-S., Eisenach, J. C. & Raftery, M. A. (1979) Studies of the composition of purified *Torpedo californica* acetylcholine receptor and of its subunits, *Biochemistry 18*, 1845–1854.

Venezuela, C. F., Weign, P., Yguerabide, J. & Johnson, D. A. (1994) Transverse distance between the membrane and the agonist binding sites of the *Torpedo* acetylcholine receptor: A fluorecence study, *Biophys. J. 66*, 674–682.

Villarroel, A. & Sakmann, B. (1992) Threonine in the selectivity filter of the acetylcholine receptor channel, *Biophys. J. 62*, 196–208.

Wagner, K., Edson, K., Heginbotham, L., Post, M., Huganir, R. L. & Czernik, A. J. (1991) Determination of the tyrosine phosphorylation sites of the nicotinic acetylcholine receptor, *J. Biol. Chem. 266*, 23784–23789.

Weiss, M. S., Kreusch, A., Schiltz, E., Nestel, U., Welte, W., Weckesser, J. & Schulz, G. E. (1991) The structure of porin from *Rhodobacter capsulatus* at 1.8 Å resolution, *FEBS Lett. 280*, 379–382.

Wennogle, L. P., Oswald, R., Saitho, T. & Changeux, J.-P. (1981) Dissection of the 66. 000-Dalton subunit of the acetylcholine receptor, *Biochemistry 20*, 2492–2497.

White, B. H., Howard, S., Cohen, S. G. & Cohen, J. B. (1991) The hydrophobic photoreagent 3-(trifluoromethyl)-3-(m[^{125}I]iodophenyl)-diazerine is a novel noncompetitive antagonist of the nicotinic acetylcholine receptor, *J. Biol. Chem. 266*, 21595–21607.

White, B. H. & Cohen, J. B. (1992) Agonist-induced changes in the structure of the acetylcholine receptor M 2 regions revealed by photoincorporation of an uncharged nicotinic non-competitive antagonist, *J. Biol. Chem. 267*, 15770–15783.

Wise, D. S., Wall, J. & Karlin, A. (1981) Relative locations of the beta and delta chains of the acetylcholine receptor determined by electron microscopy of isolated receptor trimer, *J. Biol. Chem. 256*, 12624–12627.

Witzemann, V., Stein, E., Barg, B., Konno, T., Koenen, M., Kues, W., Criado, M., Hofmann, M. & Sakmann, B. (1990) Primary structure and functional expression of the α-, β-, γ- and δ-subunits of the acetylcholine receptor from rat muscle, *Eur. J. Biochem. 194*, 437–448.

Wu, C.-S., Sun, X. H. & Yang, Y. T. (1990) Conformation of acetylcholine receptor in the presence of agonists and antagonists, *J. Protein Chem. 9*, 119–126.

Xiao-Mei, Y. & Hall, Z. W. (1994) A sequence in the main cytoplasmic loop of the α subunit is required for assembly of mouse muscle nicotinic acetylcholine receptor, *Neuron 13*, 247–255.

Yager, P., Chang, E. L., Williams, R. W. & Dalziel, A. W. (1984) The secondary structure of acetylcholine receptor reconstituted in a single lipid component as determined by Raman spectroscopy, *Biophys. J. 45*, 26–28.

Eur. J. Biochem. *240*, 307–313 (1996)

Review

The role of controlled proteolysis in cell-cycle regulation

Andor UDVARDY

Institute of Biochemistry, Biological Research Center of the Hungarian Academy of Sciences, Szeged, Hungary

(Received 28 May 1996) – EJB 96 0783/0

Cyclins and cyclin-dependent kinases are key regulators of the cell cycle. The binding of different cyclins, required to activate the catalytically inactive cyclin-dependent kinases, determines the substrate specificity of the enzymes. Cyclin-dependent-kinase inhibitors have an adverse effect, blocking the catalytic activity of cyclin-activated cyclin-dependent kinases. The cell cycle is a cyclic process of successive transient activation or inactivation of cyclin-dependent kinases by association with different cyclin regulatory subunits or cyclin-dependent kinase inhibitors. As the concentration of cyclin-dependent kinases is fairly constant during the cell cycle and exceeds the total amount of cyclins present in the cell, the exchange of regulatory subunits is determined by the availability of the different cyclins. Transcriptional control of cyclin gene expression is the most decisive factor determining the total amount of different cyclins synthesized. The actual concentration of a cyclin, however, is always the result of an equilibrium between the rates of its synthesis and degradation. While cyclin gene expression has long been known to be cell-cycle controlled, the idea of the rapid destruction of cyclins or cyclin-dependent-kinase inhibitors as an equally important factor contributing to the progress of the cell cycle is more recent. The role of controlled proteolysis in the regulation of cell cycle is discussed in this review. Two general features of this regulation are worth mentioning: cyclin-dependent kinases activated by different cyclin regulatory subunits have a central role both in the transcriptional regulation of their own genes and in the regulated, selective destruction of cyclins or cyclin-dependent kinase inhibitors; transcriptional regulation of cyclin gene expression ensures fine-tuned, continuous changes, and controlled proteolysis generates abrupt, irreversible transitions. The progress of the cell cycle is based on a delicate balance of the these mutual, but opposite regulations.

Keywords: cell cycle; cyclin; cyclin-dependent kinase; proteolysis; ubiquitin.

The cell cycle, a regulatory circuit operating in every eukaryotic cell, is a meticulously ordered series of events controlling the division of cells. Cyclins and cyclin-dependent kinases (CDK) have long been considered to act as key regulators of this process by sequentially phosphorylating a series of substrate proteins promoting the events of the cell cycle. Three basic mechanisms are involved in this regulation: as CDK alone are catalytically inactive, the binding of different cyclins, which can activate CDK, eventually determines the set of substrate proteins phosphorylated; the activity of cyclin-CDK complexes can be modulated by postsynthetic modifications of CDK, or by binding of specific inhibitors; the regulated intracellular localization of cyclin-CDK complexes further restrains and diversifies the spectrum of phosphorylated proteins.

As the concentration of the CDK is usually in excess in comparison with the total amount of cyclins present in the cell, the actual cyclin partner of a CDK will always be determined by the availability of the different cyclins. The cell-cycle-coupled transcriptional regulation of cyclin genes has long been considered to be the key element determining the concentration of cyclins. The idea of the well-timed, selective destruction of cyclins as an equally important factor adjusting the actual concentration of the different cyclins is more recent. The significance of this controlled proteolysis, however, is well-documented today in almost every phase and transition of the cell cycle. This review demonstrates the advances that have occurred during recent years in the elucidation of the role of controlled proteolysis in cell-cycle regulation. To emphasize the significance of the precisely regulated balance of cyclin accumulation and degradation, the most relevant features of the transcriptional control of cyclin gene expression will be briefly discussed. A number of excellent reviews which focus on other aspects of cell cycle regulation have been published recently (Cross, 1995; Deshaies, 1995; Doree and Galas, 1994; Hartwell and Kastan, 1994; Kamb, 1995; Kranenburg et al., 1995; Lees, 1995; Morgan, 1995; Müller, 1995; Nigg, 1995; Pines, 1995; Woollard and Nurse, 1995).

Controlled intracellular proteolysis is an event comprising two distinct consecutive steps: selection of the protein for degradation by ubiquitination, and degradation of the ubiquitinated protein by a specific protease complex. These steps are briefly summarized here; more detailed information can be found in several recent reviews (Ciechanover, 1994; Hochstrasser, 1995; Jentsch and Schlenker, 1995; Jennissen, 1995; Hilt and Wolf,

Correspondence to A. Udvardy, Institute of Biochemistry, Biological Research Center of Hungarian Academy of Sciences, H-6701 Szeged, P.O. Box 521, Hungary

Abbreviations. CDK, cyclin-dependent Kinase; Cln, cyclin; SBF, Swi4/6-dependent cell-cycle box-binding factor; APC, anaphase-promoting complex.

1996). The selectivity of intracellular protein degradation is ensured by an enzyme system which can postsynthetically modify proteins intended for rapid degradation. This modification, the attachment of a multiubiquitin chain, is achieved by a series of enzymatic reactions. The first step in this reaction is the activation of ubiquitin, a small, highly conserved protein, by the ubiquitin-activating enzyme (E_1 enzyme). This activation is ATP dependent and results in the formation of an E_1-ubiquitin thiol ester. In the second step, ubiquitin-conjugating enzymes (E_2 enzymes or ubiquitin-carrier proteins) mediate the transfer of ubiquitin from the E_1-ubiqutin complex to the protein selected for degradation. While this reaction is catalyzed in certain cases by the ubiquitin-conjugating enzyme (E_2 enzyme) alone, the E_3-enzyme (ubiquitin-protein ligase) generally cooperates in this transfer reaction. Several different E_2 enzymes with different substrate specificities have been characterized from all eukaryotic cells studied. The E_2 enzymes can associate to form multiple hetero-oligomeric complexes, and this combinatorial association modifies the substrate specificity of ubiquitination. The association of different E_3 enzymes to ubiquitin-conjugating enzymes further expands the specificity of ubiquitination. In this way, a limited number of different enzymes can generate a wide variety of specificity. The ubiquitination of protein is not irreversible. Several deubiquitinating enzymes have been purified and characterized. The fate of a ubiquitinated protein thus depends on the competitive action of deubiquitinating enzymes and the protease responsible for the degradation of the ubiquitinated proteins. The significance of deubiquitinating enzymes is clearly demonstrated by the observation that the yeast *DOA4* gene, which encodes a deubiquitinating enzyme, is related to the human *tre-2* oncogene (Papa and Hochstrasser, 1993).

Ubiquitinated proteins are degraded by a specific multiprotein protease complex, the 26S protease, which is composed of two regulator complexes and a catalytic core. The catalytic core of the 26S protease is the 20S proteosome, a multicatalytic protease complex which cleaves only non-ubiquitinated proteins adjacent to acidic, basic or hydrophobic amino acids. The proteolytic activity of the 20S proteosome is profoundly altered after an ATP-dependent association with the regulator complexes: the reconstituted 26S protease degrades only ubiquitinated proteins, and this proteolysis is absolutely ATP-dependent.

Role of controlled proteolysis in regulation of G1/S transition

Transition through the G1/S phase depends on the concerted action of G1-specific cyclins (Cln1, -2 and -3 in yeast, cyclin C, -D, -E in higher eukaryotes) and CDK (Cdc28 in yeast or CDK 2,-4,-5 and-6 in higher eukaryotes). Cdc28 is present at a relatively constant level throughout the cell cycle (Wittenberg and Reed, 1988). After mitosis, Cln 3 is the only G1 cyclin present (Tyers et al., 1992). Cln3-Cdc28 triggers transcription of the *CLN1* and *CLN2* genes by phosphorylating the Swi4/6-dependent cell-cycle box binding factor (SBF), a transcription factor responsible for the transcriptional activation of several G1-specific genes (Tyers et al., 1993). The binding of newly synthesized Cln1 and Cln2 to Cdc28 accelerates the phosphorylation of SBF, which further enhances *CLN1* and *CLN2* transcription. This positive-feedback loop (Cross and Tinkelenberg, 1991; Nasmyth and Dirick, 1991), however, does not lead to an unlimited increase in G1-specific cyclins, because the actual concentration of the G1 cyclins is determined by the rates of their synthesis and degradation. All G1 cyclins are fairly unstable, short-lived proteins; this instability may be due to the presence of a sequence in the C-terminal part of these proteins which is rich in the amino acids Pro, Glu, Ser and Thr flanked by basic residues. These sequence motifs, called PEST sequences, have been found in many short-lived proteins (Rogers et al., 1986). It was earlier shown that deletion of the PEST sequences from ornithine decarboxylase stabilized the truncated proteins (Ghoda et al., 1989), while fusion of this PEST sequence to a stable protein destabilized it (Loetscher et al., 1991). The role of PEST sequences in the metabolic stability of G1 cyclins has been extensively studied. All dominant hyperactive alleles of Cln1,- 2 and -3 are truncated at the C-terminal part of these proteins, the PEST sequences being removed (Cross, 1988; Nash et al., 1988; Hadwiger et al., 1989). In the case of Cln3, it was shown that this truncation stabilized the protein (Tyers et al., 1992). Fusion of the C-terminal 178-residue segment of Cln2 to a highly stable truncated form of human thymidine kinase destabilized the enzyme. The half-life of the fusion protein was similar to that of authentic Cln2, indicating that the Cln2 PEST domain contains a signal for rapid protein turnover. The PEST sequence alone, however, is not a self-contained determinant of protein instability, because fusion to human thymidine kinase of a 37-residue segment of Cln2, carrying the most prominent PEST sequence, did not detectably destabilize the fusion protein (Salama et al., 1994).

In vitro experiments revealed that, after binding of Cln2 to its cognate CDK (Cdc28), Cln2 is hyperphosphorylated and multiubiquitinated. This *in vitro* ubiquitination is dependent on the ubiquitin-conjugating enzyme Cdc34 (Goebl et al., 1988). Although the degradation of *in vitro* ubiquitinated Cln2 was not efficient, these reactions were considered to be physiologically significant, because the ubiquitination of Cln2 was positively regulated by its phosphorylation, and Cln2 was shown to be an *in vivo* substrate of Cdc34 (Deshaies et al., 1995).

In vivo experiments with a nested set of Cln3 deletions, in which different portions of the PEST-containing C-terminal tail of the protein were removed, revealed that the stabilization of Cln3 was proportional to the number of PEST sequences deleted (Yaglom et al., 1995). All five PEST sequences contributed to the turnover of Cln3. Fusion of the 85-residue C-terminal tail of Cln3 to β-galactosidase destabilized the hybrid protein *in vivo*. Deletion analysis of this fusion protein suggested the importance of PEST sequences as degradation signals. The fusion protein was phosphorylated *in vivo*. Two separate regions of the C-terminal tail were required for efficient phosphorylation, while the sequence between the phosphorylation sites was essential for degradation. Three independent experimental approaches suggested that the degradation occurs via the ubiquitination pathway: turnover of the fusion protein was strongly inhibited in a strain carrying a mutation in a subunit of the 20S proteosome; strong stabilization was observed in a double-mutant strain in which the attachment of the multiubiquitin chain to the fusion protein was disturbed; and mutants with a temperature-sensitive defect in the Cdc34 ubiquitin-conjugating enzyme exhibited strong stabilization of the fusion protein.

A somewhat different conclusion was drawn from *in vivo* experiments on mutagenized Cln2 derivatives, in which the major phosphorylation sites were destroyed (Lanker et al., 1996). The authors claim that phosphorylation of Cln2 by Cln-Cdc28 kinase is the major signal necessary for its rapid degradation, and specific phosphorylated residues are probably the recognition signals for the protein degradation machinery. The exact mechanism of this proteolysis was not revealed. Thus, the relationship between PEST sequences, phosphorylation sites, ubiquitination by Cdc34 and rapid proteolysis is still not completely resolved. Nevertheless, from a purely regulatory viewpoint, targeting G1 cyclins to degradation by phosphorylation performed by Cln-activated Cdc28 is an effective mechanism self-limiting their concentration during the G1 phase, so creating the perfect balance between transcriptional accumulation and proteolytic elimination.

The regulation of G1-cyclin degradation is coupled to environmental signals. Significant stabilization of the Cln1-β-galactosidase fusion protein was found in a *Saccharomyces cerevisiae* mutant (GRR1), which is involved in glucose uptake, glucose repression and divalent-cation transport. G1-cyclin stabilization is not a direct consequence of nutrient uptake defects, but rather suggests a common regulatory pathway linking nutrient uptake and G1-cyclin-controlled cell division (Barral et al., 1995).

The accumulation of G1 cyclins is absolutely essential for S phase entry, but G1 cyclins alone cannot trigger DNA replication. This follows from the observation that high levels of all three Cln-Cdc28 kinases accumulate in mutations affecting the *CDC4*, *CDC34* and *CDC53* genes, but entry into the S phase is blocked in these mutants (Tyers et al., 1992). As *CDC34* encodes a ubiquitin-conjugating enzyme (Goebl et al., 1988), the destruction of certain protein(s) seemed to be needed for entry into the S phase. Among the six B-type cyclin genes, only *CLB5* and *CLB6* are transcribed in late G1 (Epstein and Cross, 1992; Schwob and Nasmyth, 1993), and they are needed for punctual and proper DNA replication. This suggested their direct involvement in the initiation of DNA replication. Sextuple mutants lacking all six *CLB* genes cannot initiate DNA replication, despite having normal Cln activities (Schwob et al., 1994). This further supported the assumption that not Cln-Cdc28 kinases, but rather Clb-Cdc28 kinases are the initiators of the S phase. As premature expression of Clb5 cannot initiate entry into the S phase in the presence of normal Cln-Cdc28 activity, it was reasonable to suppose that Clb-Cdc28 kinases are kept inactive during the G1 phase. $p40^{SIC1}$ protein (Mendenhall, 1993; Nugroho and Mendenhall, 1994; Donovan et al., 1994) was shown to be responsible for this inactivation (Schwob et al., 1994). The CDK inhibitor $p40^{SIC1}$ is a short-lived protein which accumulates as the cell exits from mitosis, and suddenly disappears shortly before the S phase. During this short period of the cell cycle Clb-Cdc28 activity is completely blocked by $p40^{SIC1}$. Its disappearance is dependent on the ubiquitin-conjugating activity encoded by the *CDC34* gene. This follows from the observation that in Cdc34 mutants the $p40^{SIC1}$ inhibitor is stabilized. In this mutant, no Clb-Cdc28 kinase activity was detected, and cells could not enter the S phase. Because of the rapid degradation of $p40^{SIC1}$ at the end of the G1 phase, Clb-Cdc28 kinases are activated, which is a prerequisite for the initiation of DNA replication. As $p40^{SIC1}$ is a good substrate of Cln2 kinase *in vitro*, and Clb5 kinase activation requires Cln kinase activity, the ubiquitination-dependent degradation of $p40^{SIC1}$ may be initiated at the end of the G1 phase through phosphorylation by Cln-Cdc28 kinase. Inactivation of $p40^{SIC1}$ is the only non-redundant, essential function of Cln cyclins, because an *SIC1* deletion rescued the inviability of Cln1-Cln2-Cln3 triple mutant (Schneider et al., 1996). Phosphorylation of $p40^{SIC1}$ might be required for recognition by Cdc34. It is not known how the degradation of $p40^{SIC1}$ is turned off at the end of mitosis. A human Cdc34 homolog can complement the yeast $Cdc34^{ts}$ mutation, indicating that the Cdc34-encoded ubiquitination pathway is conserved amongst eukaryotes (Plon et al., 1993).

The most relevant features of cell-cycle regulation during the G1 phase can be summarized as follows (Fig. 1). Cln3-Cdc28 kinase activates Cln1 and Cln2 expression by phosphorylating the SBF transcription factor. The newly synthesized Cln1 and Cln2 bind and activate Cdc28 kinase, enhancing phosphorylation of the SBF transcription factor. This positive-feedback loop causes the Cln-Cdc28 activities to increase during the G1 phase. When the accumulation of Cln-Cdc28 kinase activities exceeds a critical threshold level budding and spindle pole body duplication are initiated. Phosphorylation of Cln cyclins by the activated Cln-Cdc28 kinases leads to destabilization of the Cln proteins and the transcriptional accumulation and the ubiquitination-dependent degradation of Cln cyclins attain equilibrium. Cln-Cdc28 kinases alone are not sufficient to initiate DNA replication. At the end of the G1 phase, Cln-Cdc28 kinases, being above the critical threshold concentration, initiate the Cdc34-dependent degradation of the $p40^{SIC1}$ inhibitor, and the reactivated Clb-Cdc28 kinases initiate the DNA replication.

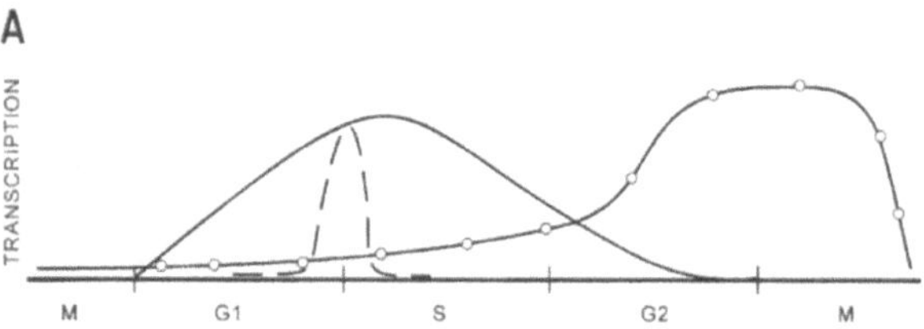

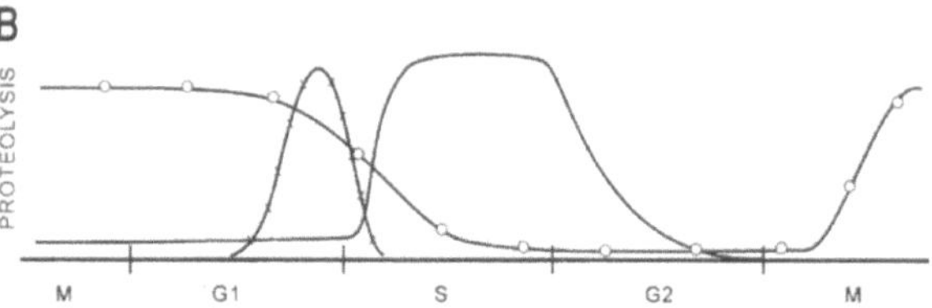

Fig. 1. Cell cycle control at the level of gene expression and proteolysis. (A) The rate of cyclin gene expression during the cell cycle. G1 cyclins (——), G2-type cyclins; Clb 5 and Clb 6; (- - -); Clb 1, 2, 3 and 4; (O—O). (B) The rate of proteolysis of cyclins and CDK inhibitors during the cell cycle. G1 cyclins (——); G2 cyclins (O—O); $p40^{SIC1}$ (X—X).

S/G2 phase switching: replacement of the cyclin components of Cdc28

During the G1 and S phases, Cdc28 is complexed with G1-type cyclins. The lack of G2-type cyclins was previously thought to be due to the very low transcriptional rates of the *CLB1,-2,-3* and-4 genes during G1. Recently, however, it was demonstrated (Amon et al., 1994) that the extensive Clb2 degradation which occurs at the end of mitosis is not restricted to a limited time period following mitosis as previously believed (Hunt et al., 1992) but continues up to the end of the subsequent G1 phase. Thus, the lack of G2-type cyclins during the G1 phase is due to a combined effect of very low transcription and efficient proteolytic degradation. Clb proteolysis at the end of mitosis requires the presence of an N-terminal segment of G2-type cyclins, which is called the destruction box. Deletion of this destruction box stabilizes Clb2 during the G1 phase, indicating that the same pathway is utilized for Clb proteolysis throughout the M and G1 phases. Accumulation of Cln-Cdc28 kinases at the end of the G1 phase inactivates Clb2 proteolysis (Amon et al., 1994), and this is an important precondition of a subsequent Clb build-up. The exact mechanism as to how Cln-Cdc28 inactivates Clb2 proteolysis is not known. The low transcriptional rate, the controlled proteolysis of G2-type cyclins and the Cln-Cdc28 kinase-dependent inactivation of proteolysis ensure that Clb do not accumulate, and do not initiate a new mitosis before attainment of the concentration of Cln-Cdc28 kinases required for budding, spindle pole body duplication and initiation of DNA replication.

The ubiquitin-conjugating enzyme Ubc9 of *Saccharomyces cerevisiae* was reported to be involved in Clb2 degradation during the G1 phase, but the destruction box dependency of this proteolysis has not been demonstrated (Seufert et al., 1995).

The Cln-Cdc28 kinase-dependent inactivation of Clb proteolysis generates a positive-feedback and a negative-feedback loop, which eventually leads to complete replacement of the G1-type cyclins by G2-type cyclins. Clb1/2 autocatalytically activate transcription of the *CLB1* and *CLB2* genes, and this activation requires the presence of active Cdc28 kinase (Amon et al., 1993). This positive feedback loop, which results in a rapid build-up of Clb-Cdc28 kinase activity during the G2 phase, is analogous to the autocatalytic activation of *CLN1/2* transcription which occurs during the G1 phase. Transcription of G1-phase-specific and S-phase-specific genes is regulated by SBF or MBF (MluI cell cycle box binding factor) transcription factors (McKinney and Cross, 1992). SBF is required for Cln1 and Cln2 transcription, while the majority of genes involved in DNA replication are under the control of MBF. Clb1 and Clb2 repress the transcription of SBF-regulated genes, generating a negative-feedback loop which finally eliminates Cln cyclins associated with Cdc28 kinase (Amon et al., 1993). This repression is probably due to the ability of Clb2 to interact physically with the SWI4 protein, a subunit of the SBF transcription factor. The Clb-generated positive-feedback and negative-feedback loops, i.e. the autocatalytic activation of *CLB* gene transcription and the repression of *CLN* gene transcription, is the driving force reshaping the cyclin-*CDK* pattern during the S/G2 transition.

Cyclin proteolysis at the metaphase/anaphase transition

G2-type cyclins which accumulate during the interphase drive cells into mitosis, and after a short lag period trigger the cyclin-destruction machinery. Elimination of mitotic cyclins is a prerequisite for entry to the next G1 phase. Cyclins are abruptly degraded after the anaphase (Luca and Ruderman, 1989; Felix et al., 1990; Ghiara et al., 1991; Hunt et al., 1992; Gallant and Nigg, 1992). The N-terminal 90-amino-acid sequence of cyclin B was shown to be required for degradation (Murray et al., 1989; Glotzer et al., 1991; Luca et al., 1991), because addition of a truncated cyclin B that lacks this N-terminal first 90 amino acids to a *Xenopus* interphase cell extract can activate the maturation promoting factor, indicating induction of the metaphase, but the truncated cyclin B is not degraded in this metaphase extract. Proteolysis of full-length cyclin B, however, was very high in this extract, indicating that the metaphase state persistently maintained by the truncated cyclin B resembles the physiological conditions of the metaphase. Fusion of a segment of the N-terminal part of *Xenopus* cyclin B (amino acids 13–91) to IgG destabilizes it; the fusion protein is degraded in the metaphase extract, but is stable in the interphase extract. Residues 13–91 thus contain all the information required for cell-cycle-dependent proteolysis. Deletion analysis of this N-terminal segment revealed a nine-amino-acid sequence, called the destruction box, which is a determinant of cell cycle-dependent cyclin proteolysis.

Although the destruction box is absolutely required for cell-cycle-dependent proteolysis of all mitotic cyclins, the destruction box alone is not sufficient for cyclin A destruction. Mutants outside the destruction box, affecting the binding of cyclin A to $p34^{cdc2}$ kinase, are indestructible, indicating that this cyclin must be complexed with the cognate CDK for proteolysis (Stewart et al., 1994). $p34^{cdc2}$ binding also appears to be required for the destruction of cyclin B2, but not for that of cyclin B1. The sequential destruction of cyclins A, B and B3 has a different role in mitotic progression. The expression of truncated cyclin variants in *Drosophila* embryos revealed that deletion of the destruction box from cyclin A resulted in a delay in the metaphase, while the same deletion from cyclin B and cyclin B3 resulted in an early or late anaphase arrest, respectively (Sigris et al., 1995).

Cyclin B is ubiquitinated during the metaphase just before the onset of proteolysis. All degradable cyclin derivatives are ubiquitinated, while non-degradable deletion variants are not (Glotzer et al., 1991). Methylated ubiquitin, which interferes with polyubiquitin formation, blocks cyclin A and B proteolysis (Hershko et al., 1991). The substrate specificity of mitotic cyclin ubiquitination, i.e. its strict dependence on the presence of the destruction box, and the limitation of ubiquitination to a narrow period of the cell cycle (Hunt et al., 1992; Amon et al., 1994) are unique features of these reactions, suggesting that a highly specialized enzyme system may be responsible for mitotic cyclin ubiquitination. Early attempts to characterize this ubiquitination system from clam oocytes revealed that four components are required for cyclin ubiquitination: a non-specific ubiquitin-activating enzyme, a cyclin-selective soluble ubiquitin-conjugating activity, a ubiquitin ligase activity present in a particulate fraction, and the 26S protease (Hershko et al., 1994). The particulate fraction was recently identified as a 1500-kDa complex with cyclin-selective ubiquitin ligase activity. This complex, termed the cyclosome, requires the presence of the destruction box for ubiquitination. It is present, but inactive during the interphase and can be activated *in vitro* by Cdc2 kinase (Sudakin et al., 1995). The active form of cyclosome present in mitotic extracts can be inactivated by an okadaic-acid-sensitive phosphatase, and Cdc2 can restore the activity (Lahav-Baratz et al., 1995). The cyclin-selective ubiquitin carrier protein (E2-C) has recently been cloned and sequenced (Aristarkhov et al., 1996). It is a new member of UBC enzymes with unique properties.

Two different, cell-cycle-independent ubiquitin-conjugating enzymes that can support destruction-box-dependent cyclin ubiquitination have been found in mitotic *Xenopus* egg extract (King et al., 1995). One of the ubiquitin-conjugating enzymes was identified as Ubc4, an enzyme responsible for the ubiquitination of many different proteins. Ubcx, the other destruction box-dependent ubiquitin-conjugating enzyme present in *Xenopus* mitotic egg extract has distinct enzymatic properties. Ubcx generates cyclin-ubiquitin conjugates of lower molecular mass than Ubc4 (Yu et al., 1996). The specificity of cyclin ubiquitination was determined by a 20S complex that contained homologs of Cdc27 and Cdc16 proteins (King et al., 1995). These proteins, together with Cdc23, form a complex which is essential for mitosis (Lamb et al., 1994). The 20S complex is probably the *Xenopus* homolog of the clam cyclosome. As the 20S complex can complement recombinant ubiquitin-activating enzyme and Ubc4 in cyclin ubiquitination, and is the only mitosis-specific component of the *Xenopus* extract required, it can be considered to be a cell cycle-regulated cyclin-specific ubiquitin ligase.

Yeast mutants defective in B-type cyclin proteolysis have been identified recently (Irniger et al., 1995). The *CDC16*, *CDC23* and *CSE1* genes were affected in these mutations. The synthetic lethality of Cdc23 and Ubc4 double mutants indicated that ubiquitination is obligatory in this proteolysis. The *CDC16*, *CDC23* and *CSE1* genes were essential for Clb proteolysis during the anaphase and G1, but were not required for the destruction of other unstable proteins. Mitotic cyclin destruction thus utilizes a specialized ubiquitination pathway. The cyclin-dependent 20S ubiquitin ligase complex (King et al., 1995) is most probably a component of this pathway, as Cdc16 is a subunit of the 20S complex. Several observations suggest, however, that the ubiquitination pathway required for cyclin B destruction should be utilized for the elimination of other anaphase-specific proteins. Inhibition of Clb2 proteolysis cannot explain why the Cdc16 and Cdc23 mutants are arrested in metaphase, because the hyperaccumulation of Clb1,-2 and-3 cannot prevent entry to the anaphase (Surana et al., 1993). The metaphase arrest may be due to a failure of the degradation of other proteins required for anaphase separation. This idea is supported by the observation that addition of a N-terminal cyclin B fragment to the *Xenopus* extract inhibited the onset of the anaphase (Holloway et al., 1993). Proteins with destruction-box-like recognition sequences may compete for a component of this destruction machinery present in limited amount. The substrates of this degradation pathway still await identification. Cut2 protein of *Schizosaccharomyces pombe*, which is essential for sister-chromatid separation (Hirano et al., 1986), is selectively degraded by the 20S complex during anaphase (Funabiki et al., 1996). Cut2 may be identical with the hypothetical glue protein, which presumably holds sister chromatids together (Holloway et al., 1993). The kinetochore microtubule motor protein CENP-E (Brown et al., 1994), the mitotic inducer NIMA protein kinase (Pu and Osmani, 1995) and thymidine kinase (Kauffman and Kelly, 1991), which are all stable during interphase and are degraded during mitosis, may also belong to this substrate group. As the ubiquitin ligase activity of the 20S complex is most probably involved in the destruction of all the proteins required for transition to anaphase, the complex is termed the anaphase-promoting complex (APC). It is probably highly conserved, because homologs of Cdc16 and Cdc27 have been detected in humans (Tugendreich et al., 1995), in *S. pombe* (Samejima and Yanagida, 1994) and in *Aspergillus nidulans* (Mirabito and Morris, 1993). APC colocalizes to the centrosome and to the mitotic spindle (Tugendreich et al., 1995). This association may be important for the regulation of anaphase, preventing the premature activation of APC before the chromosomes are properly aligned. Yeast APC may have an important function controlling the DNA replication. DNA overreplication was observed in conditional yeast Cdc16 and Cdc27 mutants (Heichman and Roberts, 1996), suggesting that Cdc16 and Cdc27 proteins may participate in destroying the licensing factor (Blow and Laskey, 1988) during S phase.

The role of ubiquitination in the control of cell-cycle progression is supported by the observation that mutation of the hus5 protein, a novel member of the ubiquitin-conjugating enzymes from *S. pombe*, results in high levels of abortive mitosis (Al-Khodairy et al., 1995). Hus5 is a homolog of Ubc9 ubiquitin-conjugating enzyme. The Lys48→Arg mutation of ubiquitin, which prevents formation of the ubiquitin-ubiquitin linkage, interfering in this way in the multiubiquitination of proteins intended for degradation, produces a cell cycle arrest (Finley et al., 1994). The *Drosophila* cell cycle gene fizzy is required for the normal degradation of cyclins A and B; its mutation causes a metaphase arrest and an enrichment of spindle microtubules (Dawson et al., 1995). fizzy is a homolog of the CDC20 gene of *S. cerevisiae*.

Proteolysis of mitotic cyclins is disturbed in mutations affecting subunits of the 26S protease. Mutations in the ATPase subunits of the regulator complex of the 26S protease result in defective chromosome segregation in *S. pombe* (Gordon et al., 1993), and arrest cell division in G2/metaphase and stabilize ubiquitinated proteins in *S. cerevisiae* (Ghislain et al., 1993; Kominami et al., 1995).

Controlled proteolysis is implicated in the transition when quiescent cells are activated for proliferation. In quiescent cells the cyclin-dependent-kinase inhibitor p27 is present in very high concentration. When cells are induced to enter the cell cycle, the amount of p27 decreases (Firpo et al., 1994; Kato et al., 1994; Halevy et al., 1995). This drop is due to an increased proteolysis of p27, because the amount of p27 mRNA and the rate of protein synthesis are constant during this transition (Halevy et al., 1995). Both *in vivo* and *in vitro* p27 proteolysis occur via the ubiquitin-proteosome pathway (Pagano et al., 1995). The Ubc2 and Ubc3 ubiquitin-conjugating enzymes are responsible for p27 ubiquitination. The ubiquitinating activity is lower in quiescent cells than in proliferating cells, and this accounts for the longer half-life of p27 in quiescent cells. Addition of Ubc2 and Ubc3 led to stimulated p27 degradation in quiescent cells. In this case, the activity of the ubiquitin-dependent proteolytic system was modulated, and this may be an alternative mechanism in cell-cycle regulation.

Future prospects

The rapid progress which has occurred in recent years in elucidating the role of controlled proteolysis in cell-cycle progression will probably continue in three different directions. Great effort will be made to identify cell-cycle proteins which are selectively degraded during certain phases of the cell cycle. The combination of genetic and biochemical methods has been a highly efficient tool identifying cyclins and CDK inhibitors degraded during the G1 and S phases in yeast. The fate of different G1-cyclins, G1-CDK and CDK inhibitors in higher eukaryotes is much less understood, and systematic effort is needed to study the exact timing of their degradation. Another main field of interest will be the identification of different ubiquitin-conjugating enzymes and ubiquitin protein ligases involved in the selective ubiquitination of different cell-cycle proteins intended for rapid degradation. While these enzymes are solely responsible for the substrate specificity of ubiquitination, the exact timing of ubiquitination is probably governed by an elaborate network of phosphorylation events, which coordinates the growth of cell mass, the replication of DNA and the duplication of cell. CDK activated by different cyclin-regulatory subunits are implicated in this regulation, but the elucidation of the exact mechanism by which they influence the different ubiquitination reactions will be one of the greatest challenge of future cell-cycle research.

The author was supported by grants Országos Tudományos Kutatasi Alap 16055 and 12836 from the National Scientific Research Fund. He wishes to thank Dr D. Dudits for a critical reading of the manuscript.

REFERENCES

Al-Khodairy, F., Enoch, T., Hagan, I. M. & Carr, A. M. (1995) The *Schizosaccharomyces pombe* hus5 gene encodes a ubiquitin-conjugating enzyme required for normal mitosis, *J. Cell Sci. 108*, 475–486.

Amon, A., Tyers, M., Futcher, B. & Nasmyth, K. (1993) Mechanism that help the yeast cell cycle clock tick: G2 cyclins transcriptionally activate G2 cyclins and repress G1 cyclins, *Cell 74*, 993–1007.

Amon, A., Irniger, S. & Nasmyth, K. (1994) Closing the cell cycle in yeast: G2 cyclin proteolysis initiated at mitosis persists until the activation of G1 cyclins in the next cycle, *Cell 77*, 1037–1050.

Aristarkhov, A., Eytan, E., Moghe, A., Admon, A., Hershko, A. & Ruderman, J. V. (1996) E2-C, a cyclin-selective ubiquitin carrier protein required for the destruction of mitotic cyclins, *Proc. Natl Acad. Sci. USA 93*, 4294–4299.

Barral, Y., Jentsch, S. & Mann, C. (1995) G1 cyclin turnover and nutrient uptake are controlled by a common pathway in yeast, *Genes & Dev. 9*, 399–409.

Blow, J. J. & Laskey, R. A. (1988) A role for the nuclear envelope in controlling DNA replication within the cell cycle, *Nature 332*, 546–548.

Brown, K. D., Coulson, R. M. R., Yen, T. J. & Cleveland, D. W. (1994) Cyclin-like accumulation and loss of the putative kinetochor motor CENP-E results from coupling continuous synthesis with specific degradation at the end of mitosis, *J. Cell Biol. 125*, 1303–1312.

Ciechanover, A. (1994) The ubiquitin-proteosome proteolytic pathways, *Cell 79*, 13–21.

Cross, F. R. (1988) DAF1, a mutant gene affecting size control, pheromone arrest, and cell cycle kinetics of *Saccharomyces cerevisiae*, *Mol. Cell. Biol. 8*, 4675–4684.

Cross, F. R. & Tinkelenberg, A. H. (1991) A potential positive feedback loop controlling CLN1 and CLN2 gene expression at the start of the yeast cell cycle, *Cell 65*, 875–883.

Cross, F. R. (1995) Starting the cell cycle: what's the point? *Curr. Opin. Cell Biol. 7*, 790–797.

Dawson, I. A., Roth, S. & Artavanis-Tsakonas, S. (1995) The *Drosophila* cell cycle gene fizzy is required for normal degradation of cyclins A and B during mitosis and has homology to the CDC20 gene of *Saccharomyces cerevisiae*, *J. Cell Biol. 129*, 725–737.

Deshaies, R. J. (1995) The self-destructive personality of a cell cycle in transition, *Curr. Opin. Cell Biol. 7*, 781–789.

Deshaies, R. J., Chau, V. & Kirschner, M. (1995) Ubiquitination of the G1 cyclin Cln2 by a Cdc34-dependent pathway, *EMBO J. 14*, 303–312.

Donovan, J. D., Toyn, J. H., Johnson, A. L. & Johnston, L. H. (1994) p40SDB25, a putative CDK inhibitor, has a role in the M/G1 transition in Saccharomyces cerevisiae, *Genes & Dev. 8*, 1640–1653.

Doree, M. & Galas, S. (1994) The cyclin-dependent protein kinases and the control of cell division, *FASEB J. 8*, 1114–1121.

Epstein, C. B. & Cross, F. (1992) CLB5: a novel B cyclin from budding yeast with a role in S phase, *Genes & Dev. 6*, 1695–1706.

Félix, M. A., Labbe, J. C., Doree, M., Hunt, T. & Karsenti, E. (1990) Triggering of cyclin degradation in interphase extracts of amphibian eggs by cdc2 kinase, *Nature 346*, 379–382.

Finley, D., Sadis, S., Monia, B. P., Boucher, P., Ecker, D. J., Crooke, S. T. & Chau, V. (1994) Inhibition of proteolysis and cell cycle progression in a multiubiquitination-deficient yeast mutant, *Mol. Cell. Biol. 14*, 5501–5509.

Firpo, E. J., Koff, A., Solomon, M. J. & Roberts, J. M. (1994) Inactivation of a Cdk2 inhibitor during interleukin2-induced proliferation of human T lymphocytes, *Mol. Cell. Biol. 14*, 4889–4901.

Funabiki, H., Yamano, H., Kumada, K., Nagao, K., Hunt, T. & Yanagida, M. (1996) Cut2 proteolysis required for sister-chromatid separation in fission yeast, *Nature 381*, 438–441.

Gallant, P. & Nigg, E. A. (1992) Cyclin B2 undergoes cell cycle-dependent nuclear translocation and, when expressed as a non-destructable mutant, causes mitotic arrest in HeLa cells, *J. Cell Biol. 117*, 213–224.

Ghiara, J. B., Richardson, H. E., Sugimoto, K., Henze, M., Lew, D. J., Wittenberg, C. & Reed, S. I. (1991) A cyclin B homolog in *S. cerevisiae*: chronic activation of the CDC28 protein kinase by cyclin prevents exit from mitosis, *Cell 65*, 163–174.

Ghislain, M., Udvardy, A. & Mann, C. (1993) *S. cerevisiae* 26S protease mutants arrest cell division in G2/metaphase, *Nature 366*, 358–362.

Ghoda, L., van Daalen Wetters, T., Macrae, M., Ascherman, D. & Coffino, P. (1989) Prevention of rapid intracellular degradation of ODC by a carboxy-terminal truncation, *Science 243*, 1493–1495.

Glotzer, M., Murray, A. W. & Kirschner, M. W. (1991) Cyclin degradation by the ubiquitin pathway, *Nature 349*, 132–138.

Goebl, M. G., Yochem, J., Jentsch, S., McGrath, J. P., Varshavsky, A. & Byers, B. (1988) The yeast cell cycle gene CDC34 encodes a ubiquitin-conjugating enzyme, *Science 241*, 1331–1335.

Gordon, C., McGurk, G., Dillon, P., Rosen, C. & Hastie, N. D. (1993) Defective mitosis due to a mutation in the gene for a fission yeast 26S protease subunit, *Nature 366*, 355–357.

Hadwiger, J. A., Wittenberg, C., de Baross Lopes, M. A., Richardson, H. E. & Reed, S. I. (1989) A family of cyclin homologs that control the G1 phase in yeast, *Proc. Natl Acad. Sci. USA 86*, 6255–6259.

Halevy, O., Novitch, B. G., Spicer, D. B., Skapek, S. X., Rhee, J., Hannon, G. J., Beach, D. & Lassar, A. B. (1995) Correlation of terminal cell cycle arrest of skeletal muscle with induction of p21 by Myo D, *Science 267*, 1018–1021.

Hartwell, L. H. & Kastan, M. B. (1994) Cell cycle control and cancer, *Science 266*, 1821–1828.

Heichman, K. A. & Roberts, J. M. (1996) The yeast CDC16 and CDC27 genes restrict DNA replication to once per cell cycle, *Cell 85*, 39–48.

Hershko, A., Ganoth, D., Pehrson, D., Palazzo, R. & Cohen, L. H. (1991) Methylated ubiquitin inhibits cyclin degradation in clam embryo extracts, *J. Biol. Chem. 266*, 16376–16379.

Hershko, A., Ganoth, D., Sudakin, V., Dahan, A., Cohen, L. H., Luca, F. C., Ruderman, J. V. & Eytan, E. (1994) Components of a system that ligates cyclin to ubiquitin and their regulation by the protein kinase cdc, *J. Biol. Chem. 269*, 4940–4946.

Hilt, W. & Wolf, D. (1996) Proteosomes: destruction as a programme, *Trends Biochem. Sci. 21*, 96–102.

Hirano, T., Funahashi, S., Uemura, T. & Yanagida, M. (1986) Isolation and characterization of *Schizosaccharomyces pombei cut* mutants that block nuclear division but not cytokkinesis, *EMBO J. 5*, 2973–2979.

Hochstrasser, M. (1995) Ubiquitin, proteosomes, and the regulation of intracellular protein degradation, *Cur. Opin. Cell Biol. 7*, 215–223.

Holloway, S. L., Glotzer, M., King, R. W. & Murray, A. W. (1993) Anaphase is initiated by proteolysis rather than by the inactivation of maturation-promoting factor, *Cell 73*, 1393–1402.

Hunt, T., Luca, F. C. & Ruderman, J. (1992) The requirement for protein synthesis and degradation and the control of destruction of cyclins A and B in the meiotic and mitotic cell cycles of the clam embryo, *J. Cell Biol. 116*, 707–724.

Irniger, S., Piatti, S., Michaelis, C. & Nasmyth, K. (1995) Genes involved in sister chromatid separation are needed for B-type cyclin proteolysis in budding yeast, *Cell 81*, 269–277.

Jennissen, H. P. (1995) Ubiquitin and the enigma of intracellular protein degradation, *Eur. J. Biochem. 231*, 1–30.

Jentsch, S. & Schlenker, S. (1995) Selective protein degradation: a journey's end within the proteosome, *Cell 82*, 881–884.

Kamb, A. (1995) Cell cycle regulators and cancer, *Trends Genet. 11*, 136–140.

Kato, J., Matsuoka, M., Polyak, K., Massague, J. & Sherr, C. J. (1994) Cyclic AMP-induced G1 phase arrest mediated by an inhibitor (p27Kip1) of cyclin-dependent kinase 4 activation, *Cell 79*, 487–496.

Kauffman, M. G. & Kelly, T. J. (1991) Cell cycle regulation of thymidine kinase – residues near the carboxy terminus are essential for the specific degradation of the enzyme at mitosis, *Mol. Cell. Biol. 11*, 2538–2546.

King, R. W., Peters, J.-M., Tugendreich, S., Rolfe, M., Hieter, P. & Kirschner, M. W. (1995) A 20S complex containing cdc27 and cdc16 catalyzes the mitosis specific conjugation of ubiquitin to cyclin B, *Cell 81*, 279–288.

Kominami, K., DeMartino, G. N., Moomaw, C. R., Slaughter, C. A., Shimbara, N., Fujimoro, M., Yokosawa, H., Hisamatsu, H., Tanahashi, N., Shimizu, Y., Tanaka, K. & Toh-e, A. (1995) Nin1p, a regulatory subunit of the 26S proteasome, is necessary for activation of Cdc28p kinase of *Saccharomyces cerevisiae*, *EMBO J. 13*, 3105–3115.

Kranenburg, O., van der Eb, A. J. & Zantema, A. (1995) Cyclin-dependent kinases and pRb: regulators of the proliferation-differentiation switch, *FEBS Lett. 367*, 103–106.

Lahav-Baratz, S., Sudakin, V., Ruderman, J. V. & Hershko, A. (1995) Reversible phosphorylation controls the activity of cyclosome-associated cyclin-ubiquitin ligase, *Proc. Natl Acad. Sci. USA 92*, 9303–9307.

Lamb, J. R., Michaud, W. A., Sikorski, R. S. & Hieter, P. A. (1994) Cdc16p, Cdc23p and Cdc27p form a complex essential for mitosis, *EMBO J. 13*, 4321–4328.

Lanker, S., Henar Valdivieso, M. & Wittenberg, C. (1996) Rapid degradation of the G1 cyclin Cln2 induced by CDK-dependent phosphorylation, *Science 271*, 1597–1601.

Lees, E. (1995) Cylin dependent kinase regulation, *Curr. Opin. Cell Biol. 7*, 773–780.

Loetsher, P., Pratt, P. & Rechsteiner, M. (1991) The C terminus of mouse ornithine decarboxylase confers rapid degradation on dihydrofolate reductase, *J. Biol. Chem. 266*, 11213–11220.

Luca, F. C. & Ruderman, J. V. (1989) Control of programmed cyclin destruction in a cell-free system, *J. Cell Biol. 109*, 1895–1909.

Luca, F. C., Shibuya, E. K., Dohrmann, C. D. & Ruderman, J. V. (1991) Both cyclin A-delta60 and B-delta97 are stable and arrest cells in M-phase, but only cyclin B-delta97 turns on cyclin destruction, *EMBO J. 10*, 4311–4320.

McKinney, J. & Cross, F. (1992) A switch-hitter at the start of the cell cycle, *Curr. Biol. 2*, 421–423.

Mendenhall, M. D. M. (1993) An inhibitor of p34CDC28 protein kinase activity from *S. cerevisiae*, *Science 259*, 216–219.

Mirabito, P. M. & Morris, N. R. (1993) BIMA, a TRP containing protein required for mitosis, localizes to the spindle pole body in *Aspergillus nidulans*, *J. Cell Biol. 120*, 959–968.

Morgan, D. O. (1995) Principles of CDK regulation, *Nature 374*, 131–134.

Murray, A. W., Solomon, M. J. & Kirschner, M. W. (1989) Role of cyclin synthesis and degradation in the control of maturation promoting factor activity, *Nature 339*, 280–286.

Müller, R. (1995) Transcriptional regulation during the mammalian cell cycle, *Trends Genet. 11*, 173–178.

Nash, R., Tokiwa, G., Anand, S., Erikson, K. & Futcher, A. B. (1988) The WHI1$^+$ gene of *Saccharomyces cerevisiae* tethers cell division to cell size and is a cyclin homolog, *EMBO J. 7*, 4335–4346.

Nasmyth, K. & Dirick, L. (1991) The role of SWI4 and SWI6 in the activity of G1 cyclins in yeast, *Cell 66*, 995–1013.

Nigg, E. A. (1995) Cyclin-dependent protein kinases: key regulators of the eukaryotic cell cycle, *Bioessays 17*, 471–480.

Nugroho, T. T. & Mendenhall, M. D. M. (1994) An inhibitor of yeast cyclin-dependent protein kinase plays an important role ensuring the genomic integrity of daughter cells, *Mol. Cell. Biol. 14*, 3320–3328.

Pagano, M., Tam, S. W., Theodoras, A. M., Beer-Romero, P., Del Sal, G., Chau, V., Yew, P. R., Draetta, G. F. & Rolfe, M. (1995) Role of the ubiquitin-proteosome pathway in regulating the abundance of the cyclin-dependent kinase inhibitor p27, *Science 269*, 682–685.

Papa, F. R. & Hochstrasser, M. (1993) The yeast DOA4 gene encodes a deubiquitinating enzyme related to a product of the human tre-2 oncogene, *Nature 366*, 313–319.

Pines, J. (1995) Cyclins and cyclin-dependent kinases: a biochemical view, *Biochem. J. 308*, 697–711.

Plon, S. E., Leppig, K. A., Do, H. N. & Groudine, M. (1993) Cloning of the human homolog of the CDC34 cell cycle gene by complementation in yeast, *Proc. Natl Acad. Sci. USA 90*, 10484–10488.

Pu, R. T. & Osmani, S. A. (1995) Mitotic destruction of the cell cycle regulated NIMA protein kinase of *Aspergillus nidulans* is required for mitotic exit, *EMBO J. 14*, 995–1003

Rogers, S., Wells, R. & Rechsteiner, M. (1986) Amino acid sequences common to rapidly degraded proteins: the PEST hypothesis, *Science 234*, 364–368.

Salama, S. R., Hendriks, K. B. & Thorner, J. (1994) G1 cyclin degradation: the PEST motif of yeast Cln2 is necessary, but not sufficient, for rapid protein turnover, *Mol. Cell. Biol. 14*, 7953–7966.

Samejima, I. & Yanagida, M. (1994) Bypassing anaphase by fission yeast cut9 mutation – requirement of cut9$^+$ to initiate anaphase, *J. Cell Biol. 127*, 1655–1670.

Schneider, B. L., Yang, Q.-H. & Futcher, A. B. (1996) Linkage of replication to Start by the Cdk inhibitor Sic1, *Science 272*, 560–562.

Schwob, E. & Nasmyth, K. (1993) CLB5 and CLB6, a new pair of B cyclins involved in S phase and mitotic spindle formation in *S. cerevisiae*, *Genes & Dev. 7*, 1160–1175.

Schwob, E., Böhm, T., Mendenhall, M. D. & Nasmyth, K. (1994) The B-type cyclin kinase inhibitor p40^{SIC1} controls the G1 to S transition in *S. cerevisiae*, *Cell 79*, 233–244.

Seufert, W., Futcher, B. & Jentsch, S. (1995) Role of a ubiquitin-conjugating enzyme in degradation of S- and M-phase cyclins, *Nature 373*, 78–81.

Sigris, S., Jacobs, H., Stratmann, R. & Lehner, C. F. (1995) Exit from mitosis is regulated by Drosophila fizzy and the sequential destruction of cyclins A, B and B3, *EMBO J. 14*, 4827–4838.

Stewart, E., Kobayashi, H., Harrison, D. & Hunt, T. (1994) Destruction of *Xenopus* cyclins A and B2, but not B1, requires binding to p34^{cdc2}, *EMBO J. 13*, 584–594.

Sudakin, V., Ganoth, D., Dahan, A., Helle, H., Hershko, J., Luca, F. C., Ruderman, J. V. & Hershko, A. (1995) The cyclosome, a large complex containing cyclin-selective ubiquitin ligase activity, targets cyclins for destruction at the end of mitosis, *Mol. Biol. Cell 6*, 185–198.

Surana, U., Amon, A., Dowzer, C., McGrew, J., Byers, B. & Nasmyth, K. (1993) Destruction of the CDC28/CLB mitotic kinase is not required for the metaphase to anaphase transition in budding yeast, *EMBO J. 12*, 1969–1978.

Tugendreich, S., Tomkiel, J., Earnshaw, W. & Hieter, P. (1995) CDC27Hs colocalizes with CDC16Hs to the centrosome and mitotic spindle and is essential for the metaphase to anaphase transition, *Cell 81*, 261–268.

Tyers, M., Tokiwa, G., Nash, R. & Futcher, B. (1992) The Cln3-cdc28 kinase complex of S.cerevisiae is regulated by proteolysis and phosphorylation, *EMBO J. 11*, 1773–1784.

Tyers, M., Tokiwa, G., Nash, R. & Futcher, B. (1993) Comparison of the Saccharomyces cerevisiae G1 cyclins: Cln3 may be an upstream activator of Cln1 and Cln2 cyclins, *EMBO J. 12*, 1955–1968.

Wittenberg, C. & Reed, S. I. (1988) Control of yeast cell cycle is associated with assembly/disassembly of the cdc28 protein kinase complex, *Cell 54*, 1061–1072.

Woollard, A. & Nurse, P. (1995) G1 regulation and checkpoints operating around START in fission yeast, *Bioessays 17*, 481–490.

Yaglom, J., Linskens, M. H. K., Sadis, S., Rubin, D. M., Futcher, B. & Finley, D. (1995) p34^{Cdc28}-mediated control of Cln3 cyclin degradation, *Mol. Cell. Biol. 15*, 731–741.

Yu, H., King, R. W., Peters, J.-M. & Kirschner, M. W. (1996) Identification of a novel ubiquitin-conjugating enzyme involved in mitotic cyclin degradation, *Curr. Biol. 6*, 455–466.

Eur. J. Biochem. *240*, 491–507 (1996)

Review

The regulation of human immunodeficiency virus type-1 gene expression

Susan M. KINGSMAN and Alan J. KINGSMAN

Department of Biochemistry, University of Oxford, Oxford, England

(Received 13 March 1996) – EJB 96 0365/0

Despite 15 years of intensive research we still do not have an effective treatment for AIDS, the disease caused by human immunodeficiency virus (HIV). Recent research is, however, revealing some of the secrets of the replication cycle of this complex retrovirus, and this may lead to the development of novel antiviral compounds. In particular the virus uses strategies for gene expression that seem to be unique in the eukaryotic world. These involve the use of virally encoded regulatory proteins that mediate their effects through interactions with specific viral target sequences present in the messenger RNA rather than in the proviral DNA. If there are no cellular counterparts of these RNA-dependent gene-regulation pathways then they offer excellent targets for the development of antiviral compounds. The viral promoter is also subject to complex regulation by combinations of cellular factors that may be functional in different cell types and at different cell states. Selective interference of specific cellular factors may also provide a route to inhibiting viral replication without disrupting normal cellular functions. The aim of this review is to discuss the regulation of HIV-1 gene expression and, as far as it is possible, to relate the observations to viral pathogenesis. Some areas of research into the regulation of HIV-1 replication have generated controversy and rather than rehearsing this controversy we have imposed our own bias on the field. To redress the balance and to give a broader view of HIV-1 replication and pathogenesis we refer you to a number of excellent reviews [Cullen, B. R. (1992) *Microbiol. Rev. 56*, 375–394; Levy, J. A. (1993) *Microbiol. Rev. 57*, 183–394; Antoni, B. A., Stein, S. & Rabson, A. B. (1994) *Adv. Virus Res. 43*, 53–145; Rosen, C. A. & Fenyoe, E. M. (1995) *AIDS (Phila.) 9*, S1–S3].

Keywords: human immunodeficiency virus type 1; Tat; Rev; AIDS.

Human immunodeficiency virus, HIV-1, a retrovirus of the lentivirus subgroup, causes the debilitating and generally fatal disease, AIDS (aquired immune deficiency syndrome). The key pathological feature of AIDS is a gradual but accelerating decline in immune competence culminating in overwhelming infection with one or more other microorganisms which exploit the immunocompromised state. AIDS is a slowly developing disease and overt symptoms may take years to appear. The major target cells for viral infection are the CD4+ cells that are pivotal to the development of humoral and cell mediated immunity and it is now generally considered that virally mediated destruction of these cells is a major contributor to immune breakdown (reviewed in [5]).

Correspondence to S. M. Kingsman, Department of Biochemistry, South Parks Road, Oxford, OX1 3QU, England

Fax: +44 1865 285 304.

Abbreviations. CRS, *cis*-active repression sequence; INS, instability sequence; PBL, peripheral blood lymphocyte; CP, coat protein; LTR, long terminal repeat; NF-κB, nuclear factor κB; PIC, pre-initiation complex; pol II, RNA polymerase II; InR, initiation region; IST, inducer of short transcripts; TAR, Tat-activation-response element; Tat-SF, Tat-stimulating factor; EIAV, equine infectious anaemia virus; TAP, Tat-associated protein; PKR, dsRNA-induced protein kinase; bHLH, basic helix-loop-helix (a class of transcription factors); PRDII, positive-regulatory-domain-binding factor; RRE, Rev-response element; TPA, 12-*O*-tetradecanoyl-phorbol-13-acetate; GR, glucocorticoid receptor; snRNA small nuclear RNA; TBP, TATA-binding protein; TAF, TBP-associated factor; TAK, Tat-associated kinase.

In common with all retroviruses the replication cycle of HIV-1 involves the creation of a double-stranded DNA provirus from an RNA genome and the subsequent integration of the proviral DNA into the host chromosome. The integrated provirus can remain unexpressed latent infection), it can be expressed (productive infection) to produce new virus or replication may be blocked prior to integration (abortive infection). The replication pattern has major implications for pathogenesis because the balance between latent versus productive infection combined with the virus yield/cell will dictate the degree of tissue damage and the progression of infection. In addition, through the action of virally encoded proteins, the physiological state of the host cell may be altered resulting in changes in viral replication kinetics. Understanding the contribution of each type of replication pattern to disease progression and establishing the role of viral and cellular factors in regulating replication are central to designing therapeutic strategies for AIDS.

The patterns of virus replication

The replication pattern adopted by the virus depends upon both the cell type and cell proliferative state. HIV-1 can infect quiescent, proliferating and terminally differentiated CD4+ target cells that are circulating in the blood (peripheral blood lymphocytes, PBL) or lymphatic systems or that are trapped or fixed in the lymph nodes. This includes naive (un-primed) T lymphocytes, i.e. which do not respond to recall antigens, these are essentially quiescent (non-dividing) cells blocked in G0 of the cell cycle, memory cells, i.e primed T lymphocytes which do respond to recall antigens and which divide every 5–6 months, actively dividing T cells undergoing clonal expansion and macrophages and dendritic cells which are terminally differentiated and blocked at the G1/S or G2 phases of the cell cycle. This ability to replicate in non-dividing cells is a major point of difference between the lentiviruses and oncoretroviruses. Within the lymph node, the target cells for HIV replication are in a milieu of cytokines that variously influence intracellular signalling pathways and cellular gene expression [6–8]. The intracellular environment of an activated T cell is therefore not static and different transcription factors and signalling molecules are available at different stages of activation.

When HIV-1 infects naive or quiescent T cells there is an abortive infection where reverse transcription is initiated but it is either not completed [9] or it generates a transcriptionally inert form of proviral DNA [10]. Once the T cell is activated cell division occurs and provirus formation and integration is completed. Latency results if, after a single cell division, the T cell resumes the quiescent state of a memory cell. Subsequent exposure of the memory cell to antigen, cytokines or superinfection with other viruses, e.g. herpes viruses, will then activate the latent genome. In contrast, when terminally differentiated non-dividing cells or actively proliferating T cells are infected, reverse transcription and integration occur within hours and there is no true silent phase after integration, virus is produced continuously until ultimately the cell dies. In the HIV-infected patient it appears that significant viral replication is occurring throughout the disease [11–13]. It is likely however that disease progression to the development of overt AIDS is associated with a reduction in the number of memory T cells and an increase in antigenically activated T cells, which would shift the balance from infection of naive T lymphocytes, which might give rise to latency in the initial stages of the disease, to infection of proliferating T cells at later stages when true viral latency would be less significant. The disease becomes terminal when viral-replication-induced cell death exceeds the capacity for renewal of the T cell population. The significance of the balance between latency and productive infection is still a matter of some debate and the issues are reviewed in [14, 15].

To understand the control of HIV-1 gene expression we must therefore consider factors which can essentially suppress expression to maintain latency, factors that can switch on expression, and factors that can amplify and/or optimise expression during the period within the activated T cell when signalling and regulatory molecules are available. It is not surprising therefore that there appears to be a complex interplay between viral and cellular factors that achieves these expression states. This complexity of interactions may also explain why the area of research into HIV-1 gene expression has generated controversy and conflicting data. Any model system be it a T cell line, a CD4-expressing HeLa cell, a *Xenopus* oocyte, an *in vitro* transcription system or an artificially activated PBL is unlikely to display all of the features of HIV-1 replication in the infected patient. Each system may accentuate just one aspect of the expression cycle which, hopefully, might ultimately assist the dissection of the complexity.

The viral genome and expression cycle

The primate immunodeficiency viruses have more complex genomes than any of the other retroviruses (Fig. 1). There are three genes common to all retroviruses that define the essence of the retroviral life style. These are the *gag*, *pol* and *env* genes. The *gag* gene encodes the core proteins that package the viral genomic RNA. The *pol* gene encodes the enzymatic activities that process precursor proteins (protease) and that copy the RNA into a double-stranded DNA provirus and integrate it into the host cell chromosome (reverse transcriptase, RNase H and integrase). The *env* gene encodes the envelope protein (Env) that is embedded in the host cell membrane which surrounds the core particles and is responsible for docking the virions onto the CD4 molecule and mediating entry by the virus into the cell by membrane fusion. In addition there are two regulatory proteins, Tat and Rev, which, except for a minority of experimental configurations, are essential for viral replication *in vivo* and *in vitro*. Tat, is an activator of transcription in the majority of cell types analysed but, perplexingly, it can also activate translation in *Xenopus* oocytes and in rabbit reticulocyte lysates. Rev regulates the nucleocytoplasmic export of unspliced and partially spliced mRNAs which is essential as HIV-1 relies upon differential splicing to generate the full range of viral proteins. There are four further genes whose products, referred to as accessory factors, are not essential for viral replication *per se* in cultured cells but which are conserved and undoubtedly play key roles in determining pathogenesis *in vivo*. These are *vif*, *vpu*, *vpr* and *nef* (see below; reviewed in [16, 17]).

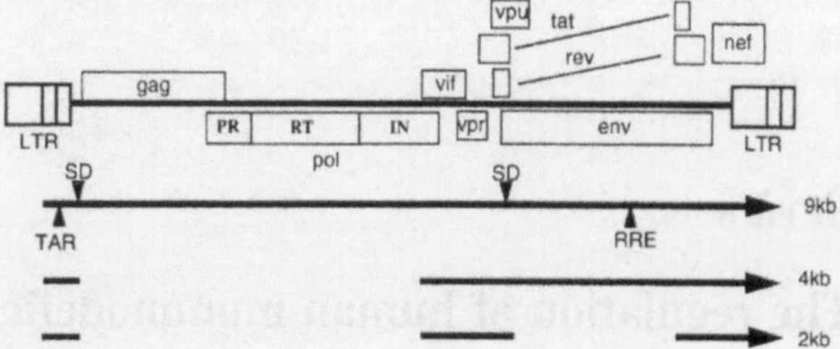

Fig. 1. The HIV-1 genome and major transcripts. LTR, long terminal repeat; PR, protease; IN, integrase; RT, reverse transcriptase. The coding regions are shown as open boxes. The major size classes of transcript are shown and the location of the key *cis* regulatory sites, TAR and RRE, are indicated. The spliced transcripts used fixed splice donor (SD) sites as indicated but there are a variety of acceptor sites, therefore sizes are approximate (see [18]). The Tat and Rev coding sequences are interrupted by an intron.

The HIV-1 genome is transcribed to produce three size classes of mRNA, depending upon the pattern of splicing (Fig. 1). These are full length (9 kb), partially spliced (4–5 kb) and multiply spliced (2 kb). A number of splice donor and acceptor sites have been identified that give HIV-1 the potential to produce more than 30 differentially spliced mRNAs [18]. With the exception of Env, every expressed HIV-1 ORF has a splice acceptor upstream of the initiator AUG, giving at least one unique message/protein. The full-length transcript is, however, multifunctional, as it is used to produce Gag and a Gag–Pol fusion protein and, at later times in infection, it is packaged as the genome into new viral particles. Different mRNAs predominate at different times during the viral replication cycle [19]. Early in infection the multiply spliced 2-kb RNAs predominate and high levels of Nef, Tat and Rev are produced. There is then an increase in partially spliced and unspliced mRNAs with a concomitant decrease in the multiply spliced mRNAs until at the end of the replication cycle the full-length RNA predominates and the virion structural proteins and accessory proteins are synthesised [19]. It appears that the switch from the early phase of replication, marked by the abundance of the short 2-kb transcripts, to the late phase of replication, marked by the presence of 4-kb and full-length transcripts, is in part mediated by the intracellular concentration of Rev which must reach a critical threshold [20]. In chronically infected cells which show low levels of viral replication, induction of viral replication is also associated with an increase in the proportion of unspliced RNA [21] suggesting that latency is also controlled by Rev. There is, however, evidence that limiting concentrations of Tat dictate latency [22] and therefore both of these essential proteins may orchestrate the switch from latency to productive infection.

The role of accessory factors in regulating viral gene expression

The accessory factors Nef, Vif, Vpu and Vpr are all involved in determining the virulence of HIV-1. The functions of these proteins have not been fully characterised and in most cases they seem to have multiple activities (reviewed by [16]).

Vpr (virion protein R) is found in viral particles and a major function of this protein appears to be to facilitate the nuclear import of the viral preintegration complex. It has also been shown to weakly *trans*-activate the viral promoter, possibly via the glucocorticoid-receptor (GR) complex which might interact with a potential GR-binding site located at −276 to −235 in the HIV-1 promoter [24]. Vpr might however have more profound effects on viral replication because it has marked effects on the cell cycle. The expression of Vpr induces growth arrest by blocking cells in the G2 phase of the cell cycle [23]. This might influence the activity of transcription factors.

The Nef (negative factor) protein is produced at all stages during the viral replication cycle and functions to enhance virion infectivity and to regulate T cell functions. It has also been proposed to have a direct inhibitory effect on viral transcription, although this is not always observed. Nef may have an indirect effect on viral gene expression via its effects on cellular physiology. Nef is known to interfere with the T-cell-activation pathway, in some circumstances inhibiting activation and in others stimulating proliferation and apoptosis. It also influences the ability of the T cell to respond to activation signals by down-regulating the CD4 receptor. The effects of Nef on T cell activation may therefore influence the availability and activity of transcription factors.

Vif (virion infectivity factor) is a late gene product that functions during viral assembly to enhance the infectivity of the virions. The requirement for Vif is most marked in primary cell lines. As yet no profound effects on the host cell have been attributed to this protein.

Vpu (virion protein U) is a late gene product expressed from the same mRNA that produces Env. It is an integral phosphoprotein that influences viral release from the cell but also down-modulates CD4 by accelerating internalisation and degradation. This reduction in cell surface expression of CD4 might influence the state of the T cell and therefore indirectly influence gene expression.

In addition to the effects of accessory genes it is also likely that the regulatory protein Tat has an indirect effect on viral gene expression via its effects on cellular genes. Tat can stimulate the expression of various cytokines and has been shown to regulate cell growth, cell migration and apoptosis (reviewed in [25, 26]).

Clearly the interplay between virally encoded accessory and regulatory proteins and cellular factors may be significant in determining the patterns of viral gene expression in different cell types in the infected patient. It should be remembered therefore that most of what we know about HIV-1 gene expression has been derived from studies of model transcription units that lack these accessory factors.

Transcription-control elements in the HIV-1 long terminal repeat (LTR)

When the viral RNA (vRNA) is reverse transcribed to produce the provirus, repeated sequences from the extreme ends of the vRNA (R) and adjacent unique 5′ (U5) and 3′ (U3) sequences are duplicated in such a way as to produce a directly repeated structure with the composition U3 R U5 at each end of the provirus, called the long terminal repeat (LTR). The LTR DNA contains a constellation of *cis*-active sites that mediate the up-regulation and down-regulation of viral transcription (Fig. 2; reviewed in [27]). A key feature of the control of HIV-1 gene expression is that elements located both upstream and downstream of the mRNA start site are required. The RNA starts at the beginning of the R region so that upstream control elements are contained within U3 and downstream elements within R and U5. The LTR can be divided into several functional regions with respect to the control of gene expression. These are, the NRE (the negative-regulatory element), the enhancer, the basal promoter elements, the core promoter, TAR (Tat-activation-response element) and the IST (inducer of short transcripts). Most of these transcription control elements are shared by many promoters of eukaryotic genes, however, TAR and IST appear to be unique to the lentiviruses. Both of these elements are found downstream of the RNA start and it is probably these elements that endow the HIV-1 promoter with the flexibility of gene expression that is required to meet the range of cellular environments faced by this virus. The IST is a DNA element that functions as a positive downstream enhancer that stimulates the production of abortive transcripts. The TAR element is a positive enhancer that stimulates the synthesis of productive transcripts but which is unique in terms of eukaryotic transcription control because it is only functional as an RNA element.

The requirement for Tat for viral replication

In the absence of the virally encoded Tat protein there is little or no gene expression directed by the proviral LTR or by the LTR fused to a reporter gene [28, 29]. The lack of protein synthesis is paralleled by low levels of transcripts. If, however, transcription is analysed more closely, by probing for promoter proximal RNA, then it is clear that initiation has occurred because short RNA species are readily detected. It appears that there is a block to transcription elongation, the polymerase complex is said to be poorly processive. The processivity block is bypassed if Tat protein is cointroduced with the LTR-reporter gene or if Tat protein is injected into, or taken up by, cells. Tat action appears to be conserved between cell types, with the exception of rodent cells which have altered requirements for Tat function (see below) [30] and *Xenopus* oocytes [31] in which Tat functions to stimulate translation rather than transcription.

Activation by Tat is entirely dependent upon the presence of the TAR RNA sequence and it is now known that Tat functions to activate expression by specifically binding to TAR. In the presence of Tat there is a 10−1000-fold increase in the level of reporter protein expression and this is in general parallelled by a corresponding increase in the levels of transcription. Full-length transcripts are readily detected and in many cases the overall level of transcription is increased. Stimulation of transcription is also seen in *in vitro* transcription systems where only transcription elongation can be measured. This stimulation is preferentially inhibited by dichloro-1-β-D-ribofuranosyl-benzimidazole) which is known to inhibit elongation by RNA polymerase II (pol II) [32]. These observations have lead to the view that the primary function of Tat is to increase the processivity of RNA polymerase II rather than to stimulate transcription initiation. The overall increase in transcription initiation that is observed in intact cells is seen as a secondary consequence of increased promoter clearance that results from the enhanced processivity.

In addition to a role in determining polymerase processivity Tat may have other effects on gene expression. A number of

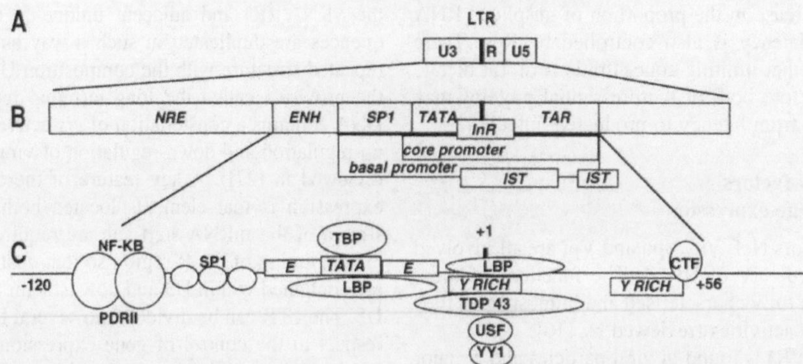

Fig. 2. The LTR is shown as a box (A) that is expanded (B) to show the key functional domains of the DNA sequence. NRE, negative-regulatory element; ENH, enhancer; Sp1, SP1 binding region; TAR, Tat activation region; InR, initiation region; IST, inducer of short transcripts. Part of the LTR is further expanded (C) to show some of the critical protein interactions. The factors that bind to the LTR are NF-κB, nuclear factor κ of B cells; PRDII-BF, positive-regulatory-domain-binding factor; TBP, TATA-binding protein; LBP, leader binding protein; TDP43, 43-kDa TAR DNA protein; USF, upstream stimulatory factor; YY1, a matrix-associated transcription factor. CTF, CAAT transcription factor. E refers to the E box that binds bHLH transcription factors and Y rich refer to the pyrimidine-rich region that probably functions as initiation element.

studies have shown marked increases in transcriptional initiation in addition to effects on elongation (e.g. [33]). Tat may also have a post-transcriptional activity, because in several studies the magnitude of Tat activation when measured at the level of protein production was at least an order of magnitude greater than the effect on transcription [34, 35]. In the *Xenopus* oocyte system Tat only activates post-transcriptionally [30]. The oocyte is, however, somewhat atypical, being a non-dividing cell blocked at G2 in the first meiotic prophase. In this system HIV-1-directed full-length transcripts are produced but they are not translated unless Tat is added. It remains to be seen whether this system is holding any clues as to the requirement for the unusual TAR RNA-dependent transcriptional activation mechanism that is used by the lentiviruses.

Although Tat appears to be essential for ensuring multiple rounds of infection by HIV-1, it is possible, experimentally, to bypass or minimise the Tat requirement. If the level of transcription initiation is increased, for example by the action of cytokines or viral transactivators such as adenovirus E1a/E1b then the probability of some full-length transcripts being produced is increased. This effect may be analogous to the transcription-activation-dependent read-through of termination sites that is observed in the human c-*myc* gene [36]. However in the case of activation with ultraviolet or mitomycin C, particularly if the proviral DNA is integrated into chromosomes, expression is maximal with little or no additional effect of Tat [37]. It has also been possible to produce TAR-less viruses which can replicate to a limited degree in stimulated PBL although these viruses are defective in T cell lines [38]. Although, in general, Tat functions synergistically with other activators (e.g. [33]), the observations of Tat and TAR independence in some systems suggests that a particular constellation of cellular factors may determine the degree of pol II processivity in the HIV-1 transcription unit and will determine the requirement for Tat and TAR. The weight of evidence now suggests that it is the HIV-1 core promoter (see below) that senses these conditions via differential protein interactions and determines both the low processivity of pol II and the ability of the promoter to respond to Tat.

Models for the regulation of transcription elongation

The basic requirements for a regulatory pathway that involves a control of transcription termination are, the presence of a negative component that limits polymerase processivity and a positive component that overcomes this limitation. A good illustration of these principles can be found in the life cycle of bacteriophage λ where gene expression is controlled, in part, by two phage proteins, N and Q that, respectively, prevent premature transcription termination in the early and late operons (reviewed in [39]). Target sites, *nut* and *qut*, for the N and Q proteins are located upstream of specific termination sites. In the case of the N system the *nut* site is a specific RNA sequence that forms a stem-loop structure. N protein binds to *nut* and an array of cellular factors are recruited. The ternary complex then loops forward and contacts the polymerase that is stalled at downstream termination sites and activates it for production of full-length transcripts. The λ Q systems is slightly different in that *qut* is a DNA site that spans the RNA start site. Transcription pauses just downstream of the start and the polymerase is held in a form that is susceptible to the action of Q. When Q protein binds to *qut* the polymerase becomes processive and reads through the pause site. It is possible that Q remains in contact with the polymerase via DNA looping but this is not yet clear. In both these systems the negative determinant is a transcription pause or termination site and the positive element is a specific RNA or DNA sequence that recruits an anti-termination complex. The mechanism of anti-termination involves an upstream signal contacting the currently engaged polymerase and is therefore referred to as a reach-forward model.

In eukaryotes the regulation of transcription termination is less well characterised. The regulation of the *Drosophila hsp70* gene appears to share some features with the λ Q system. In this case the polymerase is paused at about 25 nucleotides downstream of the RNA start and the binding of heat-shock transcription factor to sites in the promoter releases this block. However the negative component also involves an upstream DNA sequence as transcriptional pausing is dependent upon the binding of a GAGA-specific transcription factor [40]. The notion that, in eukaryotic systems, upstream DNA sequences play a significant role in determining polymerase processivity is further supported by studies of the small nuclear (sn)RNA genes. Transcription of these genes is pol II dependent but the precise configuration of the promoter sequences is critical for correct termination of transcription. If for example a heterologous pol II promoter is substituted for the natural snRNA promoter then downstream termination sites are ignored [41]. This implies that the quality of transcription can determine the effectiveness of a transcription terminator. There is a similar situation with the human c-*myc* gene which contains a strong premature transcription-termination signal distal to the promoter, but this is used differently according to the activity of the promoter under different developmental conditions [42]. In the snRNA and human c-*myc* systems the transcription-termination sites are far downstream of the RNA start and so direct contact between the upstream DNA sequences and the polymerase that is paused at the terminators is thought to be unlikely. This has lead to the development of the polymerase-education model. It is proposed that the polymerase which is recruited to the promoter can be 'educated' to increase general processivity and/or to read through specific downstream termination sites. The details of this 'education' are far from clear but presumably it must involve the recruitment of specific cellular anti-termination factors. The key distinction between the two general models that have been proposed to explain the regulation of transcription termination is that in the reach-forward model it is the currently engaged polymerase that is influenced whereas in the polymerase-education model newly recruited polymerases are affected.

In the case of HIV-1 there seems to be an amalgamation of the features of the reach-forward and polymerase-education models to produce a reach-backwards, polymerase-education model. There is little evidence to suggest the presence of specific transcription-termination sites within the HIV-1 transcription unit. Certainly TAR does not function as a termination signal [28, 33]. It appears that the polymerase simply readily dissociates from the template creating a gradient of terminated transcripts as progressively fewer polymerase molecules reach the distal regions of the transcription unit. The prematurely terminated transcripts are rapidly degraded but the exonucleolytic attack is retarded at the base of the TAR stem resulting in the generation of short RNAs that are essentially TAR sequences and that give the appearance of a specific termination event. However, even though there is no obvious transcription terminator, the Tat/TAR system is quite capable of dealing with the presence of strong termination signals within the transcript. For example, in *in vitro* transcription assays, strong artificial terminators placed downstream of TAR are ignored in the presence of Tat [43] and in the c-*myc* transcription unit, when the TAR sequence is placed at the start of the transcribed region, Tat can overcome the natural transcription termination that would otherwise occur [44]. In addition when Tat/TAR is placed downstream of an snRNA promoter the normal transcription-termination signal is ignored [45]. TAR RNA is therefore not a negative element but a positive element that serves to orchestrate anti-termination. This is clearly reminiscent of the λ N/*nut* system and one could envisage the Tat/TAR complex reaching forward to contact the polymerase. However this does not seem to be likely as, in the HIV-1 context, TAR cannot function in isolation but requires upstream DNA sequences, the most important of which appears to be the core promoter (see below). Furthermore, it is critical that TAR is located close to the promoter as the activity is rapidly lost when TAR is moved downstream [46]. This suggests that there is an intimate association between the Tat/TAR complex and the promoter. It is therefore proposed that the Tat/TAR complex does not reach forward to contact the polymerase but rather reaches backwards and educates the incoming polymerase to ensure that it can read through any downstream pause or termination site. This polymerase-education model is further qualified by considerations of how the contact with the promoter might influence processivity. There are two general ideas. A pre-assembled pre-initiation complex might be activated by the Tat/TAR complex, alternatively the Tat/TAR complex might effect a complete remodelling of the pre-initiation complex such that there is *de novo* assembly of a processivity-competent complex (reviewed in [47]). Polymerase education via reach backwards of the Tat/TAR complex is the currently favoured model for the action of Tat/TAR and some progress is now being made towards establishing the mechanistic details (see below). A simple reach-forward model and the reach-backwards, polymerase-education model as they might apply to HIV-1 are outlined in Fig. 3.

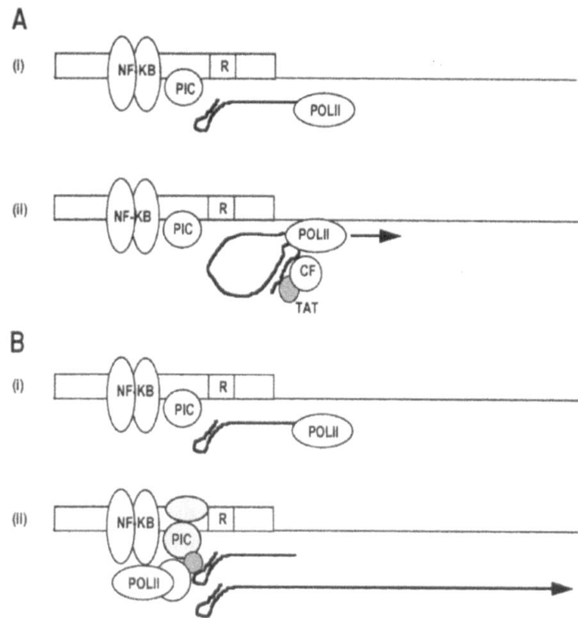

Fig. 3. The mechanism of action of Tat. (A) Reach forward mechanism. (i) The polymerase is paused, NF-κB is bound to the enhancer and the pre-initiation complex (PIC) is engaged. (ii) Nascent TAR RNA recruits Tat and cofactors (CF) and the complex loops forward to activate the polymerase (pol II) to complete the round of transcription. In this model no changes to the PIC are envisaged. (B) Polymerase-education model. (i) As above. (ii) Nascent TAR RNA recruits Tat and cellular factors and the complex reaches back to interact with the PIC. New factors are recruited and the character of the PIC changes to produce a new round of transcription in which the polymerase is fully processive.

The inducer of short transcripts

Any consideration of the control of HIV-1 gene expression at the level of RNA elongation is complicated by the presence of the IST. This is a functionally autonomous element which can be placed downstream of many pol II promoters where it induces the synthesis of short RNAs [45]. These RNAs all terminate at the base of the TAR and are therefore essentially TAR RNAs. The IST spans the initiation site from -5 to $+26$ and a secondary element maps between $+40$ and $+59$. There seems to be agreement that the IST is a positive element that determines the formation of a particular type of pre-initiation complex that generates short transcripts, presumably as a consequence of poor processivity of the polymerase. What is in question however is whether the elongation enhancement by Tat is mediated through an effect on this particular pre-initiation complex and whether the IST is an essential component of the Tat activation system. Mutations have been described which abolish the activity of the IST yet which still allow Tat-dependent stimulation of elongation [48]. These data suggest that the IST-induced initiation complex cannot be acted upon by Tat but rather this complex may be in competition with a Tat activable complex. These ideas are consistent with the finding that short transcripts can still be detected after Tat-mediated activation but their levels are often reduced. These data also suggest that the IST-induced short transcripts are not required *per se* for Tat action. However, it is possible that *in vivo*, where basal levels of transcription are undoubtedly lower than in cultured cells, the IST may be the primary means of creating the TAR RNA target for Tat and is therefore essential to initiate the activation cascade.

An alternative view is that the IST is entirely separate from the Tat activation process but it is essential to generate short TAR RNAs for another purpose. For example, TAR RNA may be mimicking a cellular RNA and serving to regulate viral gene expression by titrating inhibitory factors, by saturating positive factors or by removing Tat from the pool that is available to activate transcription. It is possible that TAR RNA contributes to the control of the intracellular environment that determines latency versus productive infection. Certainly if the levels of TAR RNA are artificially elevated in cells by over expression from a poly(TAR) DNA template or by expressing TAR in pol III transcripts linked to tRNA then subsequent infection by HIV-1 is severely diminished [49, 50]. The TAR RNAs appear to be acting as decoys to remove Tat. Sequestration of Tat may not necessarily be part of a viral gene regulatory pathway but may be involved in regulating the other activities of Tat (reviewed in [25, 26]). To date there has been no analysis of the replication of HIV-1 that is defective in IST. This is largely due to the difficulty of making mutations in IST that do not disrupt other key functions in the LTR and TAR.

Upstream control elements

The upstream control elements in the HIV-1 LTR comprise the NRE, the enhancer and the basal promoter elements which map in the U3 region upstream of the RNA start. The NRE is a dispensable far-upstream region extending back from position −120 (the RNA start is +1). In some studies deletion of this region resulted in a small but significant increase in gene expression and viral replication hence the notion that it contains elements that suppress gene activity [51]. However these negative effects are not consistently observed and there is no clear role for this region in HIV replication and pathogenesis. The NRE does however contain strong consensus binding sites for a number of transcription factors, and *in vivo* footprinting indicates occupancy of these regions (e.g. [52]). These include, TCF-1 (−121 to −150), USF-1 (−159 to −173), NFAT-1, ILF-1 (−216 to −254), GRE (−235 to −276), AP1, COUP-TF, RAR, NRT1/2, T cell factor B, myb and GATA3 (−300 to −350) (reviewed in [53]; see also [54, 55]). In the *Xenopus* oocyte model system point mutations in the −300 to −350 region appear to regulate the translation of the cognate RNA [56]. Given the conservation of the NRE region and the hints at biological function in some experimental systems it seems likely that it plays some role in viral infection.

The enhancer region extends from −82 to −103 and it is essential for any significant level of viral replication to be achieved in the course of a normal infection *in vivo*. It contains two binding sites for the nuclear factor κB (NF-κB)/relB family of transcription factors [57]. A number of other proteins also bind to the enhancer region of the HIV-1 LTR, these include MBP-1 and PRDII-BF-1 (positive-regulatory-domain-binding factor) [58] the roles of which, in regulation, are less well defined than that of the NF-κB/rel family. NF-κB/rel proteins upregulate transcription in response to a variety of stimuli including cytokines, mitogens and other viral activator proteins e.g. adenovirus E1a/E1b. There is also evidence that NF-κB can mediate activation of the LTR by Tat independently of the TAR in astrocytic glial cells where *trans*-activation and viral replication do not require TAR [59]. The NF-κB proteins are major regulators of gene activation in a developing immune response [60] and the predominance of any one member of this family and the relative activity varies during the cellular response. The requirement for NF-κB for HIV-1 replication may depend upon the cellular environment, for example in lymphoblastoid cell lines it is possible to produce replication-competent viruses that lack these upstream sequences [61]. However in isolated human blood lymphocytes and in unstimulated human primary monocyte/macrophages there is little or no activity of the LTR in transient transfections and no viral replication without stimulation of NF-κB activity [62, 63]. It therefore appears that the HIV-1 LTR is essentially inactive in resting T lymphocytes and monocytes/macrophages and that enhancement by NF-κB is a prerequisite for viral replication, whereas in cell lines there is a high level of spontaneous activity of the LTR which is augmented by, but not dependent upon, stimulating NF-κB. Certainly when single cells derived from a clonal cell line carrying an integrated HIV-1 LTR are examined there are a significant number that are displaying very high levels of expression in the absence of stimulation of NF-κB or of the viral *trans*-activator, Tat [64].

The basal promoter elements map to a (G+C)-rich region extending from −46 to −78 and comprise three Sp1 boxes. The sites can be substituted with other basal elements such as ATF, USF or AP1 in some assays of promoter function [65, 66] and replication-competent virus can be produced if the SP1 sites are deleted, provided that NF-κB sites are preserved. However, it is generally accepted that the Sp1 sites are important and function synergistically with the TATA box to ensure maximum activation of the promoter by Tat [67]. Certainly in some assays the precise spatial relationship between the SP1 sites and the TATA box is critical for ensuring optimum responsiveness to Tat [68]. It seems likely that, as discussed for NF-κB, the requirements for SP1 may be more stringent when physiologically relevant cells, rather than cell lines, are studied.

The core promoter

The HIV-1 core promoter elements map between −30 and +50 and appear to be unusually complex. As with the majority of genes transcribed by pol II the HIV-1 core promoter is sufficient to mediate low-level transcription in the absence of additional upstream-DNA-binding proteins. In general, core promoters comprise either a TATA box or the initiation region (InR), a pyrimidine-rich site that spans the RNA start, but the HIV-1 LTR contains both types of element.

The TATA box region. The function of the TATA box is to recruit an array of protein factors and pol II, referred to as the pre-initiation complex (PIC). The initial step is the binding of the highly conserved 38-kDa TATA-binding protein (TBP) which is sufficient to recruit pol II and allow basal transcription. However high-level transcription generally requires the action of upstream-enhancer-binding proteins which can only interact with the TATA box region if additional proteins have been recruited into the PIC. The binding of TBP usually initiates a cascade of additional interactions with at least eight different proteins referred to as TBP-associated factors (TAF). The complex between TBP and the TAF is referred to as TFIID and it is the minimal assembly through which enhancer-binding proteins can function. Once TFIID is formed then additional multicomponent transcription factors are recruited. TFIIB associates with the polymerase which is probably already complexed with TFIIF. The latter facilitates elongation by the polymerase and under certain restricted conditions such as the presence of ATP and a supercoiled template this minimal PIC can process to produce full-length transcripts. Usually, however, additional elongation factors comprising a TFIIE complex with TFIIH are required. TFIIH is a remarkable multisubunit complex that possesses ATPase, helicase, nucleotide excision repair and kinase activities. The kinase phosphorylates the C-terminal domain of RNA polymerase II. This phosphorylation is proposed to facilitate promoter clearance by allowing elongation to proceed without competition from the initiation assembly reaction. Clearly the balance between the formation of a TBP/pol II complex and TFIID/pol II complex dictates the levels of basal versus activated transcription. The recruitment of the different TF into the activated complex dictates the efficiency of promoter clearance as the elongation factors assist the complex to traverse the transcription unit (reviewed in [69, 70]).

The HIV-1 TATA box, although conforming to a consensus is flanked by two conserved palindromic sequences called E boxes which are not commonly found in eukaryotic promoters. These are recognition sites for the bHLH (basic helix-loop-helix) class of transcription factors and two such factors, AP4 and E47 bind to these sites *in vitro*. This binding precludes the interaction of the TATA box with TBP but stabilises the interaction with TFIID. This suggests that protein binding to the E boxes might regulate a switch from basal to activated gene expression [71]. A number of studies have shown that substitution of the HIV-1 TATA box region with the corresponding regions from other promoters can severely diminish the responsiveness of the LTR to activation by Tat [72, 73]. It has also been shown that the core HIV-1 promoter comprising only the TATA box region and downstream elements is responsive to Tat implying that the TATA box is the critical upstream determinant of Tat action [74].

It seems very likely that in the HIV-1 promoter there is the capacity to regulate basal versus activated transcription via effects on the TATA box and the PIC and that Tat interacts specifically with one type of complex. This notion is further supported by the finding that Hela nuclear extracts can be depleted of factors that are required for Tat activation without affecting their capacity to support basal transcription [75].

The initiation region (InR). The HIV-1 InR maps from −2 to +7 and shows partial sequence similarity to other InR but cannot functionally substitute for this region in promoters which are dependent upon InR-mediated initiation [76, 77]. Nonetheless it is pyrimidine rich and interacts with TFII-I, USF and YY1 [76]. This suggests that a PIC distinct from one that forms at the TATA box could form on the InR. A second InR-like sequence maps to +32 to +42 which overlaps critical residues in TAR (see below). USF has been shown to stimulate HIV-1 transcription [78], whereas YY1, which is a nuclear-matrix-associated protein [79] appears to be an inhibitor [80]. Recently, a non-bHLH protein, TDP-43 has been shown to bind to the InR and to block the assembly of transcription complexes that can be activated by Tat [81]. Another protein that appears to control elongation is LBP1. This has a high-affinity binding site between −38 and −16 which overlaps the TATA region and a lower-affinity site at −4 to +21 that overlaps InR. In *in vitro* transcription reactions high concentrations of LBP preclude the binding of TFIID and addition of LBP after PIC formation blocks elongation, suggesting that LBP could be part of the mechanism that restricts processivity [82]. However, despite the elegance of these *in vitro* analyses of core promoter requirements there are several studies that show efficient viral replication which is Tat-dependent despite mutations in InR and in the high-affinity LBP1-binding site [83]. Other proteins also bind in this region, including CTF/NF1 and less well characterised proteins, but no function in transcription or Tat activation has been ascribed to these factors.

The protein interactions at the core promoter can also be influenced by global factors, for example the tumour suppressor protein p53 represses transcription from the HIV-1 LTR whereas mutated derivatives activate expression. p53 can interact with both Sp1 and TFIID and its levels and activity are influenced by factors ranging from ultraviolet light to viral *trans*-activators [84]. It is also clear that chromatin organisation over the initiation region plays an important role in regulating initiation. The activation of expression from the HIV-1 LTR by mitomycin C or ultraviolet light is accompanied by a change in chromatin organisation [85] and a nucleosome (nuc-1) has been mapped to a position immediately after the transcription initiation site and this is disrupted after treatment of cells with TPA (12-*O*-tetradecanoyl-phorbol-13-acetate) or tumour-necrosis factor α [86]. These data suggest that a condensed chromatin structure could limit expression from the HIV-1 LTR.

It seems likely that the interplay between positive and negative factors binding to the core promoter can determine the relative functions of the TATA and InR in assembling a PIC, and that the composition of that complex might be dictated by the constellation of flanking proteins, by chromatin assembly and by global regulators. The character of the PIC then determines the level of basal expression, the degree of requirement for Tat and the degree of responsiveness to Tat.

The Tat/TAR interaction

Despite the presence of an array of upstream activation signals, maximal expression from the HIV-1 LTR is clearly dependent upon the Tat protein. The first requirement for Tat-mediated activation appears to be for Tat to bind to TAR. This interaction has been well characterised by *in vitro* binding studies (reviewed in [87]). Purified recombinant Tat protein or synthetic polypeptide binds as a monomer to a single molecule of TAR with a K_d of 3 nM. It is a matter of some debate as to whether specific TAR-binding cellular factors are required to facilitate this interaction *in vivo* and a number of models to describe the functions of TAR-binding cellular factors have been proposed.

The properties of Tat. HIV-1 Tat is an 86-amino-acid protein the structure of which has yet to be determined; however, genetic analysis has revealed functionally autonomous regions (Fig. 4). Residues 48−57 comprise an (Arg+Lys)-rich basic region that is essential for binding to TAR and which functions as a nuclear/nucleolar targeting signal, residues 32−47 comprise a highly conserved hydrophobic core region, residues 22−31 are cysteine rich and mediate dimerisation although this is now thought not to be significant for Tat function, and the N-terminal amino acids constitute an acidic, proline-rich region that is critical for activation. The arginine at position 52 mediates the specific complex formation between Tat and TAR but there is some sequence flexibility in the rest of the basic region as, for example, the Rev basic region can functionally substitute for the Tat basic region [88]. The last part of exon 1 (residues 57−72) is glutamine rich and although non-essential it appears to augment Tat action in some systems. Residues 72−86 encoded by the second exon are redundant for activation but they contain the conserved RGD motif that mediates binding to integrins to allow exogenous Tat to be taken up by cells. Residues 1−48 can function as an autonomous transcription-activation domain when fused to a heterologous DNA-binding domain. For example, if multiple binding sites for the heterologous binding domain are placed upstream of the HIV-1 basal promoter then this hybrid Tat fusion protein can activate transcription via a pathway that seems to be identical to that used by classical transcription factors such as VP16. The basic region is also functionally au-

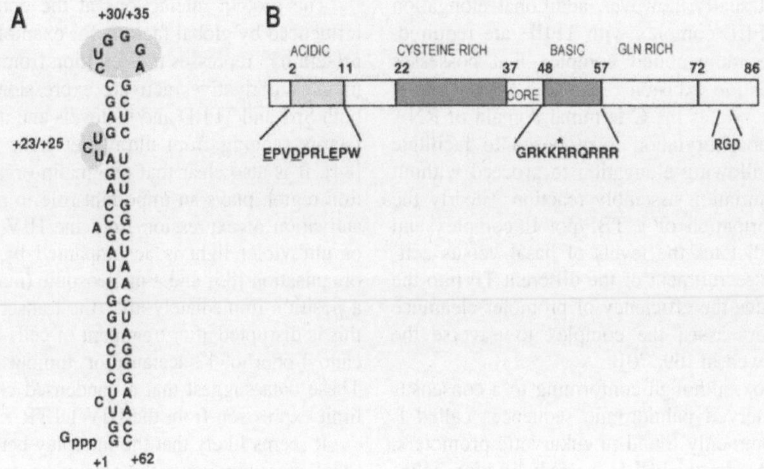

Fig. 4. TAR and Tat. (A) TAR RNA sequence with key bulge and loop regions indicated. (B) Functional domains of Tat, amino acid residues are numbered and key sequences are indicated. Arg52 is indicated by a star. See text for details.

tonomous as it can bind to TAR RNA independently of any other part of Tat. Heterologous activation proteins such as VP16 can be brought to the promoter via an interaction with TAR that is mediated by the Tat RNA-binding domain. In this configuration VP16 can activate the HIV-1 promoter. Tat can therefore be regarded as a two-domain protein consisting of a nucleic-acid-binding domain and a transcription-activation domain. This has obvious superficial similarity to the structure of many transcription factors.

Despite the apparent flexibility of sequence requirements in the basic region, the RNA-binding domain appears to confer some specificity on Tat action such that, for example, HIV-1 Tat fails to activate the analogous TAR region from equine infectious anaemia virus (EIAV, also a lentivirus) and vice versa. However the activation domains of the two Tat proteins are interchangeable such that the EIAV activation domain can be linked to the HIV-1 TAR-binding domain to produce a fully functional protein that will activate the HIV-1 promoter. The EIAV activation domain is much smaller than that of HIV-1 in that it lacks the cysteine-rich region and is essentially the conserved core region. It has been possible to produce a minimal functional lentiviral Tat protein which comprises a 25-amino-acid fusion of the HIV-1 basic region (10 amino acids) and the EIAV activation domain (15 amino acids; [89]). The notion that the basic region has an autonomous structure is consistent with the finding that peptides from this region can bind to TAR, albeit with a lower affinity than full-length Tat and can inhibit the binding of the full-length TAR. A single arginine or arginimide can also interact specifically with TAR and this interaction has been use to develop models for the binding interaction. However in the case of HIV-1 Tat it appears that residues flanking the basic region are also important in conferring the specificity of binding [90]. It is possible that these residues serve to present the TAR structure in the optimum configuration for interacting with the RNA. Recently the structure of EIAV Tat and the minimal EIAV/HIV Tat protein have been determined by NMR spectroscopy [91, 92]. The protein appears to be largely unordered with the exception of the conserved, hydrophobic core domain which appears to acts as a scaffold to anchor the flexible N and C terminal domains. It has been suggested that the basic domain wraps around the core to create a positively charged cleft which can accommodate the RNA. Alternatively it has been suggested that the basic domain forms an α-helix which can interact with the deformed major groove of the TAR RNA [91]. This latter suggestion seems less likely, because mutations that would disrupt a potential helix do not affect Tat binding to TAR [93].

The properties of TAR. The TAR sequence forms a stable secondary structure, somewhat reminiscent of a tRNA; this has been confirmed *in vitro* by chemical and enzyme mapping and by NMR analysis [94, 95] (Fig. 4). Basically the RNA exists as a partially base-paired stem with a critical tripyrimidine bulge located in the upper stem. The stem terminates with a hexanucleotide loop which is also partially base paired although there is some disagreement as to whether the base pairing is dynamic or whether there are preferred stable structures. The bulge and flanking nucleotides induce a pronounced kink in the helix axis of the stem [96]. The key nucleotides in TAR for Tat binding are the uridine residue at position 23 in the tripyrimidine bulge and the G26 · C39 base pair in the upper stem and there is a requirement for a base paired purine at position A27. Nucleotide pairs below the bulge (A22 · U40, G21 · C41) are also important in stabilising Tat binding [90]. Tat recognises the conformation and primary structure of the TAR RNA and probably interacts with the major groove of the A form helix where it is deformed at the tripyrimidine bulge [97]. Critical contacts have been defined as the N^3-H of U23, the N^7 of G26, the N^7 of A26 and the phosphate between A22 and U23 [98]. Tat binds to an RNA with the correct bulge structure with a 1000-fold higher affinity than to an RNA with general double-strand character e.g. tRNA. Some studies have suggested that a binding pocket is formed by an interaction between U23 and A27 · U38, which can stabilise the interaction of the arginine with G26 and the phosphate backbone at residues 22 and 23, this is referred to as the arginine fork model (proposed by [99] and modified by [100]). This model is derived from an analysis of the binding of a single arginimide to TAR and it has been argued that this does not accurately model the interaction with full-length Tat which displays a far higher binding constant and greater specificity [98, 101]. The overall conformation of TAR may be more significant for Tat interaction than the primary structure. This view is consistent with the finding that certain mutations, e.g. at U23 and A27 are not as deleterious as deletion of the bulge or displacement of the bulge to the 3′ side of the loop. These latter mutations cannot be overcome by increasing the concentration of Tat whereas the point mutations can all be compensated for to a degree by excess

Tat, implying that they affect the affinity of the interaction (e.g. [102, 103]).

One of the most perplexing features of the Tat/TAR interaction is that the TAR sequence requirements for Tat binding that have been defined by *in vitro* studies do not seem to be completely mirrored by the sequence requirements for responsiveness to Tat *in vivo*. The Tat/TAR interaction *in vitro* has no requirement for the loop sequences, Tat will bind to a double-stranded TAR RNA provided the bulge and upper stem sequences are preserved (e.g. [98]). However, *in vivo*, any mutation in the hexanucleotide loop (nucleotides 30–35) abolishes Tat-dependent activation [104, 105]. This has been presumed, probably correctly, to be due to an effect on the TAR RNA, but some of these mutations would influence some of the downstream DNA-binding interactions. In addition to point mutations in the loop, perturbations of the spatial relationship between the loop and the bulge region also interfere with Tat action. For example a 3-bp insertion into the upper stem abolishes Tat activation [106]. However the bulge and the loop sequences can be placed on adjacent stem-loop structures and restore Tat activation provided that the bulge is upstream of the loop [107]. These data suggest that the loop sequence in some way cooperates with Tat when it is interacting with the Tat-binding site but in a spatially constrained way. In the *Xenopus* oocyte system where Tat activates the translation of TAR RNA, mutations in the loop and extensions in the upper stem also abolish Tat action. As these latter experiments were performed using only an injected TAR RNA then the loop sequences must be functioning at the RNA level [30]. A different situation exists in rodent cells. These cells are refractory to Tat activation of the HIV-1 LTR but curiously any mutation in the loop restores responsiveness to Tat implying an inhibitory function for this region [108].

There are few clues as to the real function of the TAR RNA loop *in vivo* but there are several hypotheses which account for some of the observations. It has been proposed that the affinity of the Tat/TAR interaction *in vivo* is not sufficiently high to create a stable complex and that a co-binding cellular factor is required [109]. An extension of this idea is that the Tat/cellular factor complex is required not only to bind to TAR but also to create a functional transcriptional activator. Conversely in the *Xenopus* oocyte system it has been possible to activate translation of TAR RNA which has a mutated loop sequence provided that this mutated TAR RNA target is present in great excess over Tat. This has been interpreted as suggesting that Tat can bind to TAR with high affinity in the context of a cell nucleus but that it is normally denied access to the bulge by a competing bulge-binding cellular factor [110]. It is proposed that this factor is removed by the excess loop mutated TAR RNAs. In this model the role of the loop sequence is to bind a cellular factor that displaces the inhibitory protein. To complicate the issue there are experiments that appear to cast doubt upon any specific requirement for TAR RNA to achieve activation by Tat. In one of these studies TAR RNA was replaced with the RNA-binding site for the bacteriophage R17 coat protein (CP) and the activation domain (residues 1–48) of Tat was fused to CP. This Tat–CP fusion protein was capable of binding to the R17 RNA target with high affinity *in vitro* and the hybrid protein could activate an HIV-1 LTR *in vivo* provided that TAR had been replaced with the R17 sequence [46]. This lead to the notion that the only function of TAR was to tether Tat in the vicinity of the promoter. However it has been noted that the efficiency of activation was never as high as the authentic configuration [111] and the hybrid configuration does not appear to function *in vitro*. Clearly there is much still to learn about the properties of TAR RNA *in vivo* and its role in the activation process.

TAR-binding cellular factors. A number of cellular factors have been identified that bind to TAR with some degree of specificity but no clear role for any of these factors in Tat activation or viral infection has been demonstrated. The best characterised factor is TRBP for which a molecular clone is available. TRBP binds preferentially to (G+C)-rich double-stranded RNAs but has no specific sequence requirements and does not require a bulge for binding [112]. It appears to interact with the upper stem of TAR and it can bind the double-stranded RNA induced protein kinase (PKR) to form a PKR/TRBP heterodimer [113]. Overexpression of TRBP marginally increases LTR-directed transcription but has no preferential effect on Tat activation. There are however some intriguing features of TRBP. It can be found in rodent cells which are refractory to Tat action, but it maps to human chromosome 12 which is known to complement the defect in rodent cells. Also the sequence molecular mass of TRBP is 37 kDa yet it appears to behave as a 55-kDa protein in human cells [114]. It is possible therefore that there is a family of TRBP proteins that have different roles in different cell types.

There are a number of proteins that bind to TAR at the bulge region these include TRP2 (a family of 70–110-kDa proteins) which enhances Tat activation in *in vitro* transcription reactions [111] and BBP, a 38-kDa protein that inhibits Tat binding to TAR *in vitro* [115]. Recently a protein with similarity to a yeast ribosomal protein has been found to bind to the TAR bulge region and stimulate *trans*-activation [116]. A 185-kDa protein that specifically binds to the TAR loop has been identified and this binding apparently requires a cofactor. This loop binding complex has some stimulatory effect on Tat activation in *in vitro* transcription reactions [111, 117]. Recently, a molecular clone for this loop-binding protein has been obtained [118]. An 83-kDa loop-binding protein has been found in rodent cells carrying human chromosome 12, and as chromosome 12 restores Tat responsiveness this protein may be a candidate as a positive cofactor of Tat action [119]. A 68-kDa loop-binding factor has been shown to stimulate Tat activation in an *in vitro* transcription system [120]. Several proteins bind to the lower stem of TAR including the La autoantigen [121] and a 140-kDa phosphoprotein [122] and there are at least seven RNA-binding proteins that bind preferentially to TAR after ultraviolet light treatment of cells [123]. Further progress in defining a role for any of these TAR RNA-binding proteins awaits, in many cases, the availability of molecular clones.

Another puzzling feature of TAR is its variable effect on the translation of TAR-containing mRNAs. There is a significant body of work that indicates that stem-loop structures at the 5′ ends of mRNAs are deleterious to translation (e.g. [124]). Consistent with this, in two conventional assays for translational competence, the rabbit reticulocyte lysate and the *Xenopus* oocyte, TAR RNAs are indeed poorly translated [30, 125, 126]. Likewise in the yeast *Saccharomyces cerevisiae*, TAR RNA can be produced but there is no translation (C. Stanway, A. J. Kingsman and S. M. Kingsman, unpublished results). Although there is some evidence that TAR is functioning in *trans* to activate the translational inhibitory DI kinase [127] these data have been contested [128]. The RNA analyses by Kozak [124] would suggest that translation inhibition should be a *cis* effect of the TAR sequence. Intriguingly it has recently been shown that the TAR-mediated translational repression in rabbit reticulocyte lysates can be alleviated by La autoantigen, which binds specifically to TAR [129]. In support of these observations on translation, two early studies of Tat action in mammalian cells indicated that the effect on expression could not be explained solely on the basis of an effect on transcription, implying that there was a translation block that was also overcome by Tat [34, 35]; however, when TAR RNA was microinjected into HeLa cells there was no inhibition of translation [130]. It is possible that any translational inhibitory effects of TAR might be more evident in primary cells but this has yet to be tested. Despite the observations in oocytes, reticulocyte lysates and yeast, the importance of TAR as an inhibitor of translation in viral infection remains enigmatic.

The mechanism of action of Tat

As we have discussed above, the currently favoured model for Tat action is polymerase education via reach backwards where Tat contacts the PIC and effects a switch from non-processive to processive transcription. The first key question is whether Tat is functioning through classical activation pathways that are used by upstream activators to stimulate the PIC. Some studies suggest that this might not be the case. For example VP16, which is considered a classical transcription activator, can function tethered to RNA but activation occurs independently of the orientation of upstream promoter components. In contrast, when Tat functions tethered to RNA the orientation of upstream elements is critical. This indicates that VP16 and Tat are not behaving identically in this downstream configuration. Likewise when Tat functions in artificial constructs as an upstream activator of the HIV-1 promoter then the presence of downstream Sp1 sites is critical whereas there is no such requirement for Sp1 sites for VP16 to activate expression when located in the same upstream site [131]. The notion that Tat functions via cofactors that are different from those used by VP16 is supported by the finding that in a reciprocal saturation experiment, the Tat activation domain can inhibit activation by full-length Tat but has no effect on VP16 mediated activation [132]. It has also recently been shown that a partially reconstituted transcription system that contained all of the TF and TAF required to support SP1 and VP16 activation of the LTR, failed to support Tat activation. Addition of a further nuclear fraction allowed Tat action, this Tat-stimulatory factor (Tat-SF) functioned to stimulate transcription elongation. Tat and Tat-SF stimulated transcription in extracts that contained recombinant TBP and not other TF or TAF which is in marked contrast to SP1 and VP16 which depend upon holoTF-IID for their activity [133]. These *in vitro* systems therefore add weight to the notion that Tat is not simply a misplaced classical activator protein that for some bizarre reason has to function through an RNA element, but that it is working via a non-classical pathway.

It is still not clear whether Tat directly contacts the PIC or is linked via a bridging cofactor. In *in vitro* reactions, a Tat-TBP complex can form and this specifically requires residues 36–50 in Tat [134]. These residues in Tat also play an important auxiliary role in stabilising the binding of Tat to TAR *in vitro* and so one might envisage an initial interaction with TAR followed by an interaction with TBP. The notion that the interaction between Tat and TAR might be transient is supported by tethering experiments where Tat fusion proteins that had a low dissociation rate were less active *in vivo* [135]. Tat also binds strongly to SP1 *in vitro* and a Tat/Sp1 complex can be immunoprecipitated from HIV-infected cells [136]. This interaction could help to stabilise the PIC but as discussed above a specific interaction with SP1 might not be central to Tat activation. There is some evidence that Tat can functionally substitute for elongation factor TFIIF in *in vitro* transcription reactions [137] although this has not always been observed [133]. A candidate for a bridging cofactor is TAP (Tat-associated protein) which binds specifically to the conserved core region of Tat and binds strongly to TFIIB [138]. A number of other novel Tat-binding proteins have been identified that are candidates for cofactors. A Tat affinity column identified a kinase that can phosphorylate the C-terminal domain of the large subunit of pol II [139] which has obvious implications for elongation [140]. This Tat-associated kinase (TAK) is sensitive to dichloro-1-β-D-ribofuranosyl-benzimidazole which preferentially blocks transcription elongation. However, this kinase is a predominantly cytoplasmic protein which does not fit easily with a role in transcription activation and it binds with greater efficiency to a truncated version of Tat containing only the first 48 amino acids. This truncated Tat is inactive in most assays but can function as a *trans*-dominant inhibitor of full-length Tat action implying that it does bind a cellular factor (e.g. [141]). Further characterisation of TAK will be needed before we can conclude that Tat functions via phosphorylating the polymerase. Other Tat-binding proteins have also been detected by affinity chromatography. These include a 36-kDa protein that stimulates Tat action *in vitro* and in rodent cells that are normally refractory to Tat, and a 200-kDa protein that is essential for Tat activation in an *in vitro* transcription system [67]. Yeast two-hybrid screening has identified a novel zinc-finger protein but as yet no specific function for this protein in the Tat activation pathway has been discovered [142]. When a λ expression library was probed with Tat a specific binding protein, TBP1, was identified. TBP1 is a concentration-dependent repressor/activator of HIV-1 transcription and has similarity with the MSS1/SUG1 family of proteins that are normally involved in transcription and cell-cycle control and which stimulate Tat action in some assays [143]. Clearly there is no shortage of potential cofactors for Tat function but detailed mechanistic information has not been obtained for any of them to date.

A tentative description of the Tat activation pathway is therefore that the SP1/holoTFIID complex, perhaps in concert with other core promoter elements, generates high levels of non-processive transcripts that provide the TAR RNA targets for Tat. Once Tat is bound to TAR it instigates a switch in the type of PIC that is formed, and now full-length transcripts are produced. This switch is mediated either via direct binding of Tat to the PIC or via Tat-recruited cofactors.

Very little further progress will be made in unravelling the mechanism of action of Tat until these various cellular factors have been purified and their activities tested in the different *in vivo* and *in vitro* assays for Tat action. A major breakthrough would be the development of a Tat-dependent assay in yeast. This would allow genetic ablation and reconstitution of the activation yielding the identity of the key cofactors. To date, however, all attempts at reconstituting the activation pathway in either *S. cerevisiae* or *Schizosaccharomyces pombe* have been unsuccessful. Further characterisation of the rodent cell system may also be informative as genetic complementation of the Tat activation pathway has been possible by whole chromosome transfer.

Why is the TAR enhancer an RNA?

There are many examples of gene-regulatory circuits in eukaryotic cells that display exquisite temporal and quantitative control of gene activity without requiring an RNA enhancer. This is achieved by concentration-dependent combinatorial interactions between upstream-DNA-binding proteins such that a gene does not have to be on or off but can display degrees of activity. It has been proposed that the presence of nascent TAR RNA allows rapid recruitment of Tat to the PIC but there is no evidence that this is the case. It is not known if prematurely terminated RNAs remain spatially associated with the proviral DNA. It is also not clear why combinatorial upstream interactions coupled with the activation of NF-κB should not be sufficient to produce high levels of transcription at an appropriate time in the T-cell-activation pathway. One possibility is that the window of opportunity for expression in the activated T cell is very narrow and that maximal transcription must be ensured without delay. There is some evidence that Tat stimulation of transcription *in vitro* is only transient, ceasing after 4 h [144]. In this case assembling all of the ingredients at the promoter may be necessary to maximise the activation.

An alternative view is that Tat activation of transcription is an important but secondary activity of Tat. The main function of Tat could be to identify viral mRNAs by assembling a specific complex on the 5′ end. The only evidence that this could be a possibility comes from studies in *Xenopus* oocytes. In this system TAR RNA that is transcribed from the HIV-1 LTR or which is directly injected into the oocyte, is only translated in the presence of Tat. Intriguingly the RNA target and Tat must both be in the nucleus for activation to occur implying the involvement of a nuclear factor. On the whole, the same mutations in TAR and Tat that compromise transcriptional activation also compromise translational activation. The benzodiazapine drug Ro24-7429 that inhibits viral replication via a specific effect on Tat activation of transcription [145] also specifically inhibits Tat activation of translation and genetic analyses in both systems implicate a cellular factor as the target for this drug [146]). Also, dichloro-1-β-D-ribofuranosyl-benzimidazole, a general inhibitor of transcription elongation is a specific inhibitor of Tat activation of transcription in mammalian cells and of Tat activation of translation in *Xenopus* oocytes [147]. These data could be interpreted as suggesting that there are cellular factors common to transcription and translation that are targets for Tat action. There is no evidence that such factors exist and the notion that translation control might be mediated though nuclear factors is not a mainstream idea. However one scenario that might dictate the necessity for a special translation complex to be associated with all viral mRNAs would occur if HIV-1 RNAs failed to reach the ribosomes via the pathway normally travelled by cellular RNAs. There is a possibility that this is the case because recent work has suggested that viral mRNAs may exit from the nucleus via the snRNA pathway and not the mRNA pathway (see below). For the RNA to be recognised for translation it might be necessary to load factors onto the RNA in the nucleus.

While the oocyte data are intriguing and, despite the atypical nature of this cell, should not be dismissed, at present the majority view is that the only role of Tat/TAR is to function as an instant response and amplification signal for transcription. Such a view might however have to be revised once the functions of the key Tat-binding and TAR-binding cellular factors have been defined.

Post-transcriptional regulation

There is clearly significant control of the transcription of the HIV-1 provirus. However elevated levels of transcription alone are not sufficient to ensure high levels of virus replication. There is a second critical control point that operates post-transcriptionally. This control is effected by the virally encoded Rev protein that orchestrates the export of unspliced and partially spliced transcripts. The full-length (9 kb) and singly spliced (≈4 kb) mRNAs all retain a short (≈300 nucleotide) RNA sequence that is the target for Rev action. This is the Rev-response element (RRE) which folds into a complex multiple-stem-loop structure that contains a specific binding site for Rev (Fig. 5). In the absence of Rev the 9-kb and 4-kb mRNAs are retained in the nucleus, whereas in the presence of Rev they are exported into the cytoplasm. Rev functions to promote the expression of intron-containing RNAs in a wide range of experimental systems including yeasts [148] and *Xenopus* oocytes [149]. In general in the HIV-infected cell the overall levels of the different species

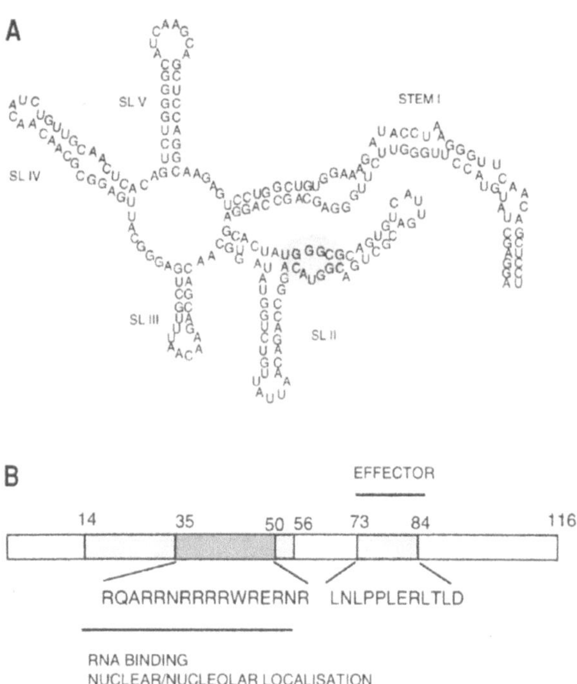

Fig. 5. RRE and Rev. (A) The minimal functional RRE. The initial binding site for Rev on SLIIB is indicated. SL, stem loop. (B) Functional domains of REV. Key sequence are indicated.

of RNA appear to be largely unaffected by Rev and the function of the protein is therefore seen as one of facilitating the nucleocytoplasmic export of intron-containing RNAs rather than of regulating splicing *per se*. However, under some experimental conditions the binding of Rev or Rev-derived peptides to RRE can disrupt the splicing of an intron on the same transcript [150]. There has been some controversy about whether RRE can only function on a mRNA that contains splicing signals. It has been possible to establish assays for Rev in *Xenopus* oocytes by using mRNAs that lack introns, which argues that there is not an absolute requirement for splice sites [149]. However in *S. cerevisiae* the presence of an intron was essential to detect any effect of Rev and, furthermore, mutations in the 5′ splice site and the branch point abolished Rev action, implying that early steps in splicing precede Rev action [148]. It is clear that removal of the introns in the HIV-1 mRNAs is not an efficient process and it may be that *in vivo* commitment to the splicing pathway allows RNA to be retained in the nucleus and the consequent delay in splicing provides an opportunity for Rev to act. In one study the presence of TAR in the transcription unit could substitute for an intron in facilitating the Rev response [51]. It also appears that HIV-1 mRNAs contain sequences, *cis* active repression sequences (CRS), that can promote nuclear retention. These are distributed throughout the mRNA and when transplanted to some heterologous mRNAs render them Rev dependent for expression. No details of the possible mechanism of action of CRS have emerged and their activity is not universally demonstrable, which tends to imply that they may not play a major role in the Rev pathway (Reviewed in [151]). There is also evidence that some HIV-1 transcripts are unstable and this has been mapped to instability sequences (INS) that are dispersed throughout the transcript [152]. These do not seem to be functional in all cell types [153] but certainly in some assays the Rev/RRE interaction is seen as counteracting the effect of INS. In general it seems likely that those RNAs where expression will be stimulated to the highest degree by the Rev/RRE interaction are ones which are otherwise trapped in the nucleus and that there may be several features of an RNA that might contribute to such entrapment.

Rev is a 116-amino-acid nuclear protein that binds directly to RRE. It has two essential domains (Fig. 5). The first domain comprises a short arginine-rich sequence (amino acids 34–50) that is required for binding to RNA and for nuclear localisation. Several amino acids on each flank of this sequence mediate multimerisation of Rev which is required for it to function *in vivo* [154]. The second domain is positioned towards the C-terminus and contains three closely spaced leucine residues which are critical for function. Mutation of any of these abolishes Rev function but has no effect on nuclear localisation or binding to RRE. The two domains of Rev are functionally independent and as observed with Tat they can be linked to heterologous proteins and confer either specific RNA binding or Rev activity on a heterologous RNA. The binding interaction between Rev and RRE is complex as it involves an initial interaction with a high affinity site (stem loop IIB) which is followed by the cooperative binding of additional Rev monomers. About 6–8 proteins bind to the RRE sequence but at high protein concentrations polymerisation can continue along the RNA [155]. The interaction with the high-affinity site is mediated via the recognition of non-Watson-Crick base pairs in a stem-bulge-stem structure [156]. In particular a G · G base pair distorts the sugar-phosphate backbone. NMR analysis of peptide interactions with the high-affinity site indicates that this unusual base pairing is induced during binding which stabilises a conformational change in RRE [157, 158]. The recognition of a bulged stem via basic amino acids is therefore a feature that is shared by Tat and Rev.

Rev is found predominantly in the nucleus and nucleolus but Rev/RRE complexes can be detected in the cytoplasm. If transcription is inhibited then all the Rev moves out of the nucleus and into the cytoplasm [159]. This is a feature of a number of proteins, particularly those involved in packaging RNA such as the hnRNP protein A1, which appear to shuttle between the cellular compartments. The ability of Rev to shuttle is consistent with a role in RNA export and it has now been shown that Rev can specifically bind to a class of nucleoporin-like proteins [160–162]. This binding is mediated by the Rev-activation domain. BSA conjugates with peptides containing the Rev activation domain mediated the specific export of RRE containing mRNAs via a pathway that was distinct from the normal mRNA export pathway [163]. It appears that Rev/RRE complexes may exit the nucleus using components of a pathway that is used to facilitate the export of 5SRNA and snRNAs. A number of other cellular factors have been implicated in Rev action [164]. In particular efficient Rev action in *Xenopus* oocytes was dependent upon the addition of a HeLa cell nuclear factor, eIF-5A, which has previously been implicated in translation control [165].

Cellular RNAs with recognisable introns that are not properly spliced are usually retained within the nucleus [166]. However, high-level transcription which generates a large intranuclear pool of such RNAs can allow leakage to the cytoplasm [167]. High levels of expression of HIV-1 mRNA in some cells can lead to partially spliced RNAs appearing in the cytoplasm in the absence of Rev [168, 169]. However, surprisingly, these RNAs are not translated. This has lead to the notion that Rev is a chaperone that not only facilitates RNA export but that ensures polysome association. Certainly the fact that Rev is a nucleocytoplasmic shuttle protein and is associated with RRE-containing RNA in the cytoplasm is consistent with a chaperone role. The translation-promoting role of Rev has not however been observed in all cellular systems and it may therefore be secondary to the main function of Rev as a nucleocytoplasmic export protein [153].

Much of the data therefore suggests that the major role of Rev is as an RNA export protein that either diverts RNA away from the splicing pathway or rescues it from this pathway. One might expect all viruses that use differential splicing in their life cycle to possess a Rev-like activity. Mason Pfizer Monkey Virus, an oncoretrovirus has an RNA element (cytoplasmic transport element, CTE) that can functionally substitute for RRE in the absence of Rev [170]. This suggests that there are Rev-like cellular factors that orchestrate the export of RNAs that would otherwise be trapped in the nucleus.

Inhibitors of HIV-1 gene expression for prevention and treatment of AIDS

There is little doubt that a human immunodeficiency virus that lacked either Tat or Rev would be replication defective and non-pathogenic. There has therefore been tremendous interest in developing therapeutic strategies for the treatment and prevention of AIDS that involve inhibiting one or both of these critical proteins. Low-molecular-mass inhibitors of Tat and Rev have been identified. For example, a class of benzodiazepine derivatives were found to inhibit Tat activation of the *in vitro* HIV-1 LTR and viral replication, but unfortunately no antiviral effect was seen in patients. This may have been due to the toxicity of these compounds which could have reduced efficacy and have prevented their clinical use (reported in [171]). Compounds related to the aminoglycoside antibiotics interfere with the binding of Rev to the target RRE and inhibit viral replication but to date there is no information on the clinical effectiveness of such compounds [172]. Combinations of inhibitors that target NF-κB and Tat have been tested with considerable success in inhibiting viral replication in cultured cells [173] but clinical data are not yet available. Macromolecules are also being developed as inhibitors of viral gene expression. In particular derivatives of Tat and Rev, referred to as *trans*-dominant inhibitors, that are themselves inactive but that competitively inhibit the normal activity of the proteins are being developed. For example the expression of Rev with a mutation in the activation/nucleoporin binding domain of Rev, the M10 mutation, reduces viral replication in the same cells [174]. Likewise derivatives of Tat that retain the activation domain but lack the RNA-binding motif are inhibitory (e.g. [141]). The expression of antisense oligonucleotides directed to the Tat or Rev transcripts and the expression of single-chain antibodies directed against Tat and Rev are also effective at inhibiting viral replication in cultured cells (e.g. [175]). The overexpression of TAR RNA has also been shown to inhibit viral replication [50]. The key problem with using any of these macromolecular inhibitors in patients is developing methods to deliver them to infected cells. The tools of gene therapy are being applied to this problem and various viral and non-viral delivery systems are being tested (reviewed in [176]). A key development in genetic therapeutic approaches for the treatment of AIDS will be when highly efficient and specific targeting of therapeutic molecules to CD4+ cells can be achieved.

Conclusions

There is exquisite control of HIV-1 gene expression exerted by cellular and viral factors which allows the virus to exploit maximally different target cells and different cell states. This virus is almost unique in mediating the major control of gene expression through RNA targets rather than through DNA targets. There is however still no satisfactory explanation for the requirement for an RNA target for transcriptional activation. This is particularly perplexing as Tat appears to function via the PIC which is the target for conventional upstream-binding activator proteins. Despite attempts to model Tat activation on conventional activation pathways that are mediated by classical DNA-binding transcription factors there are sufficient clues in the data to suggest to us that Tat is functioning via a completely novel pathway, and until we discover this pathway we will not understand the real role of the TAR RNA enhancer. In addition to this RNA-mediated control on mRNA production, the virus orchestrates the use of the transcripts via their interaction with Rev which ensures their export to the cytoplasm.

It is clear that Tat and Rev play a pivotal role in HIV-1 replication and it is intriguing that they occupy the same space in the HIV genome and must therefore have co-evolved. The proteins are remarkably similar in structure, being small, nuclear, two-domain, specific RNA-binding proteins that use very similar strategies for RNA recognition. It has been suggested that the Tat/TAR system is required solely to effect rapid amplification of gene expression, it is possible however that given the intimate genetic relationship of Tat and Rev, there is a greater degree of functional relationship than has so far been suspected. Perhaps HIV-1 ensures that once transcripts are produced they will be chaperoned by Tat and Rev to ensure both efficient export and translation. Consistent with this notion is the observation that the Tat/TAR complex functions as a small ribonucleoprotein particle [177]. The development of models for Tat and Rev action amenable to genetic analysis and the identification of molecular clones for all the putative cofactors will improve our understanding of the regulation of HIV-1 gene expression and will probably lead us to some novel ideas about cellular pathways of gene expression. This understanding will undoubtedly lead to the development of new approaches for the treatment and prevention of AIDS.

The authors' research is supported by grants from the Medical Research Council, the Biotechnology and Biological Sciences Research Council, Glaxo Wellcome and The Wellcome Trust. Thanks go to our colleagues for helpful comments and to Frances Ryan for help in preparing the manuscript.

REFERENCES

1. Cullen, B. R. (1992) Mechanism of action of regulatory proteins encoded by complex retroviuses, *Microbiol. Rev. 56*, 375–394.
2. Levy, J. A. (1993) Pathogenesis of human immunodeficiency virus infection, *Microbiol. Rev. 57*, 183–289.
3. Antoni, B. A., Stein, S. & Rabson, A. B. (1994) Regulation of human immunodeficiency virus infection: implications for pathogenesis, *Adv. Virus Res. 43*, 53–145.
4. Rosen, C. A. & Fenyoe, E. M. (1995) Virology, an overview, *AIDS (Phila.) 9*, S1–S3.
5. Pantaleo, G. & Fauci, A. S. (1994) Tracking HIV during disease progression, *Curr. Opin. Immunol. 6*, 600–604.
6. Ullman, K. S., Northrop, J. P., Verweij, C. L. & Crabtree, G. R. (1990) Transmission of signals from the T lymphocyte antigen receptor to the genes responsible for cell proliferation and immune function: the missing link, *Ann. Rev. Immunol. 8*, 421–452.
7. McLean, A. R. & Michie, C. A. (1995) *In vivo* estimates of division and death rates of human T lymphocytes, *Proc. Natl Acad. Sci. USA 92*, 3707–3711.
8. McCune, J. M. (1995) Viral latency in HIV disease, *Cell 82*, 183–188.
9. Zack, J., Haislip, A. M., Krogstad, P. & Chen, I. S. Y. (1992) Incompletely reverse-transcribed human immunodeficiency virus type 1 genomes in quiescent cells can function as intermediates in the retroviral life cycle, *J. Virol. 66*, 1717–1725.
10. Spina, C. A., Guatelli, J. & Richman, D. D. (1995) Establishment of a stable, inducible form of human immunodeficiency virus type 1 DNA in quiescent CD4 lymphocytes *in vitro*, *J. Virol. 69*, 2977–2988.
11. Pantaleo, G., Graziosi, C., Demarest, J. F., Butini, L., Montroni, M., Fox, C. H. & Orenstein, J. M. (1993) HIV infection is active and progressive in lymphoid tissue during the clinically latent stage of disease, *Nature 362*, 355–358.
12. Ho, D. D., Heumann, A. U., Perelson, A. S., Chen, W., Leonard, J. M. & Markowitz, M. (1995) Rapid turnover of plasma virions and CD4 lymphocytes in HIV-1 infection, *Nature 373*, 123–126.
13. Wei, X., Ghosh, S. K., Taylor, M. E., Johnson, V. A., Emini, E. A., Deutsch, P., Lifson, J. D., Bonhoeffer, S., Nowak, M. A., Hahn, B. H., Saag, M. S. & Shaw, G. M. (1995) Viral dynamics in human immunodeficiency virus type 1 infection, *Nature 373*, 117–122.
14. Coffin, J. M. (1995) HIV population dynamics *in vivo*: implications for genetic variation, pathogenesis, and therapy, *Science 267*, 483–489.
15. Perelson, A. S., Neumann, A. U., Markowitz, M., Leonard, J. M. & Ho, D. D. (1996) HIV-1 dynamcis *in vivo*: virion clearance rate, infected cell life-span, and viral generation time, *Science 271*, 1582–1585.
16. Trono, D. (1995) HIV accessory proteins: leading roles for the supporting cast, *Cell 82*, 189–192.
17. Subbramanian, R. A. & Cohen, E. A. (1994) Molecular biology of the human immunodeficiency virus accessory proteins, *J. Virol. 68*, 6831–6835.
18. Schwartz, S., Felber, B. K., Benko, D. M., Fenyo, E. M. & Pavlakis, G. N. (1990) Cloning and functional analysis of multiply spliced mRNA species of human immunodeficiency virus type 1, *J. Virol. 64*, 2519–2529.
19. Kim, S., Byrn, R., Groopman, J. E. & Baltimore, D. (1989) Temporal aspects of DNA and RNA synthesis during human immunodeficiency virus infection: evidence for differential gene expression, *J. Virol. 63*, 3708–3713.
20. Pomerantz, F. J., Seshamma, T. & Trono, D. (1992) Efficient replication of human immunodeficiency virus type 1 requires a threshold level of Rev: potential implications for latency, *J. Virol. 66*, 1809–1813.
21. Michael, N. L., Morrow, P., Mosca, J., Vahey, M., Burke, D. & Redfield, R. R. (1991) Induction of human immunodeficiency virus type 1 expression in chronically infected cells is associated primarily with a shift in RNA splicing patterns, *J. Virol. 65*, 1291–1303.
22. Adams, M., Sharmeen, L., Kimpton, J., Romeo, J. M., Garcia, J. V., Peterlin, B. M., Groudin, M. & Ermerman, M. (1994) Cellular latency in human immunodeficiency virus-infected individuals with high CD4 levels can be detected by the presence of promoter-proximal transcripts, *Proc. Natl Acad. Sci. USA 91*, 3862–3866.
23. Jowett, J. B. M., Planelles, V., Poon, B., Shah, N. P., Chen, M. L. & Chen, I. S. Y. (1995) The human immunodeficiency virus type 1 vpr gene arrests infected T cells in the G2 + M phase of the cell cycle, *J. Virol. 69*, 6304–6313.
24. Refaeli, Y., Levy, D. N. & Weiner, D. B. (1995) The glucocorticoid receptor type 1I complex is a target of the HIV-1 vpr gene product, *Proc. Natl Acad. Sci. USA 92*, 3621–3625.
25. Chang, H.-K., Gallo, R. C. & Ensoli, B. (1995) Regulation of cellular gene expression and function by the human immunodeficiency virus type 1 tat protein, *J. Biomed. Sci. 2*, 189–202.
26. Li, C. J., Friedman, D. J., Wang, C., Metelev, V. & Pardee, A. B. (1995) Induction of apoptosis in unifected lymphocytes by HIV-1 Tat protein, *Science 268*, 429–431.
27. Jones, K. A. (1993) Tat and the HIV-1 promoter, *Curr. Opin. Cell Biol. 5*, 461–468.
28. Kao, S. Y., Calman, A. F., Luciw, P. A. & Peterlin, B. M. (1987) Anti-termination of transcription within the long terminal repeat of HIV-1 by tat gene product, *Nature 330*, 489–493.
29. Feinberg, M. B., Baltimore, D. & Frankel, A. D. (1991) The role of Tat in the human immunodeficiency virus life cycle indicates a primary effect on transcriptional elongation, *Proc. Natl Acad. Sci. USA 88*, 4045–4049.

30. Alonso, A., Cujec, T. P. & Peterlin, M. (1994) Effects of human chromosome 12 on interactions between tat and TAR of human immunodeficiency virus type 1, *J. Virol. 68*, 6505–6513.
31. Braddock, M., Chambers, A., Wilson, W., Esnouf, M. P., Adams, S. E., Kingsman, A. J. & Kingsman, S. M. (1989) HIV-1 TAT 'activates' presynthesized RNA in the nucleus, *Cell 58*, 269–279.
32. Marciniak, R. & Sharp, P. A. (1991) HIV-1 Tat protein promotes formation of more processive elongation complexes, *EMBO J. 10*, 4189–4196.
33. Laspia, M. F., Rice, A. & Mathews, M. B. (1989) HIV-1 Tat protein increases transcriptional intitation and stabilizes elongation, *Cell 59*, 283–292.
34. Rosen, C. A., Sodroski, J., Goh, W. C., Dayton, A. I., Lippke, J. & Haseltine, W. (1986) Post-transcriptional regulation accounts for the trans-activation of the human T-lymphotropic virus type 1II, *Nature 319*, 555–559.
35. Cullen, B. R. (1986) Trans-activation of human immunodeficiency virus occurs via a bimodal mechanism, *Cell 46*, 973–982.
36. Yankulov, K., Blau, J., Purton, T., Roberts, S. & Bentley, D. L. (1994) Transcriptional elongation by RNA polymerase II is stimulated by transactivators, *Cell 77*, 749–759.
37. Valerie, K., Delers, A., Bruck, C., Thiriart, C., Rosenberg, H., Debouck, C. & Rosenberg, M. (1988) Activation of human immunodeficiency virus type 1 by DNA damage in human cells, *Nature 333*, 78–81.
38. Harrich, D., Hsu, C., Race, E. & Gaynor, R. B. (1994) Differential growth kinetics are exhibited by human immunodeficiency virus type 1 TAR mutants, *J. Virol. 68*, 5899–5910.
39. Greenblatt, J., Nodwell, J. R. & Mason, S. W. (1993) Transcriptional antitermination, *Nature 364*, 401–406.
40. Lee, H.-S., Kraus, K. W., Wolfner, M. F. & Lis, J. T. (1992) DNA sequence requirements for generating paused polymerase at the start of hsp70, *Genes & Dev. 6*, 284–295.
41. Hernandez, N. & Lucito, R. (1988) Elements required for transcription initiation of the human U2 snRNA gene coincide with elements required for snRNA3′ end formation, *EMBO J. 7*, 3125–3134.
42. Spencer, C. A. & Groudine, M. (1990) Transcription elongation and eucaryotic gene regulation, *Oncogene 5*, 777–785.
43. Graeble, M. A., Churcher, M. J., Lowe, A. D., Gait, M. J. & Karn, J. (1993) Human immunodeficiency virus type 1 transactivator protein,tat, stimulates transcriptional read-though of distal terminator sequences *in vitro*, *Proc. Natl Acad. Sci. USA 90*, 6184–6188.
44. Wright, S., Lu, X. & Peterlin, B. M. (1994) Human immunodeficiency virus type 1 tat directs transcription through attenuation sites within the mouse c-myc gene, *J. Mol. Biol. 243*, 568–573.
45. Ratnasabapathy, R., Sheldon, M., Johal, L. & Hernandez, N. (1990) The HIV-1 long terminal repeat contains an unusual element that induces the synthesis of short RNAs from various mRNA and snRNA promoters, *Genes & Dev. 4*, 2061–2074.
46. Selby, M. J., Bain, E. S., Luciw, P. A. & Peterlin, B. M. (1989) Structure, sequence, and position of the stem-loop in TAR determine transcriptional elongation by Tat through the HIV-1 long terminal repeat, *Genes & Dev. 3*, 547–558.
47. Cullen, B. R. (1993) Does HIV-1 tat induce a change in viral initiation rights? *Cell 73*, 417–420.
48. Sheldon, M., Ratnasabapathy, R. & Hernandez, N. (1993) Characterization of the inducer of short transcripts, a human immunodeficiency virus type 1 transcriptional element that activates the sysnthesis of short RNAs, *Mol. Cell. Biol. 13*, 1251–1263.
49. Sullenger, B. A., Gallardo, H. F., E, U. G. & Gilboa, E. (1991) Analysis of transacting response decoy RNA-mediated inhibition of human immunodeficiency virus type 1 transactivation, *J. Virol. 65*, 6811–6816.
50. Lisziewicz, J., Sun, D., Smythe, J., Lusso, P., Lori, F., Louie, A., Markham, P., Rossi, J., Reitz, M. & Gallo, R. C. (1993) Inhibition of human immunodeficiency virus type 1 replication by regulated expression of a polymeric Tat activation response RNA decoy as a strategy for gene therapy in AIDS, *Proc. Natl Acad. Sci. USA 90*, 8000–8004.
51. Lu, X., Heimer, J., Rekosh, D. & Hammarskjoeld, M. L. (1990) U1 small nuclear RNA plays a direct role in the formation of a Rev-regulated human immunodeficiency virus env mRNA that remains unspliced, *Proc. Natl Acad. Sci. USA 87*, 7598–7602.
52. Demarchi, F., D'Agaro, P., Falaschi, A. & Giacca, M. (1993) In vivo footprinting analysis of constitutive and inducible protein-DNA interactions at the long terminal repeat of human immunodeficiency virus type 1, *J. Virol. 67*, 7450–7460.
53. Gaynor, R. B. (1995) Regulation of HIV-1 gene expression by the transactivator protein tat, in *Transacting functions of human retroviruses* (Srinivasan, A. & Vogt, P. K., eds) pp. 51–77, Springer-Verlag, Heidelberg.
54. Fagagna D'Adda di, F., Marzio, G., Gutierrex, M. I., Kang, L. Y., Falaschi, A. & Giacca, M. (1995) Molecular and functional interactions of transcription factor USF with the long terminal repeat of human immunodeficiency virus type 1, *J. Virol. 69*, 2765–2775.
55. Lee, M.-O., Hobbs, P. D., Zhang, X.-K., Dawson, M. I. & Pfahl, M. (1994) A synthetic retinoid antagonist inhibits the human immunodeficiency virus type 1 promoter, *Proc. Natl Acad. Sci. USA 91*, 5632–5636.
56. Gunkel, N., Braddock, M., Thorburn, A. M., Muckenthaler, M., Kingsman, A. J. & Kingsman, S. M. (1995) Promoter control of translation in *Xenopus* oocytes, *Nucleic Acids Res. 23*, 405–412.
57. Nabel, G. & Baltimore, D. (1987) An inducible transcription factor activates expression of human immunodeficiency virus in T-cells, *Nature 326*, 711–713.
58. Seeler, J.-S., Muchardt, C., Suessle, A. & Gaynor, R. B. (1994) Transcription factor PRDII-BF1 activates human immunodeficiency virus type 1 gene expression, *J. Virol. 68*, 1002–1009.
59. Taylor, J. P., Kundu, M. & Khalili, K. (1993) TAR-independent activation of HIV-1 requires the activation domain but not the RNA-binding domain of Tat, *Virology 195*, 780–785.
60. Baeuerle, P. A. & Henkel, T. (1994) Function and activation of NF-κB in the immune system, *Annu. Rev. Immunol. 12*, 141–591.
61. Leonard, J., Parrott, C., Buckler-White, J., Turner, W., Ross, E. K., Martin, M. A. & Rabson, A. B. (1989) The NF-κB binding sites in the human immunodeficiency virus type 1 long terminal repeat are not required for virus infectivity, *J. Virol. 63*, 4919–4924.
62. Alcami, J., Lain de Lera, T., Fogueira, L., Pedraza, M. A., Jacque, J. M., Bachelerie, F., Noriega, A. R., Hay, R. T., Harrich, D., Gaynor, R. B., Virelizier, J. L. & Arenzana-Seisdedos, F. (1995) Absolute dependence on kappa-B responsive elements for initiation and Tat-mediated amplification of HIV transcription in blood CD4 T lymphocytes, *EMBO J. 14*, 1552–1560.
63. Moses, A. V., Ibanez, C., Gaynor, R., Ghazal, P. & Nelson, J. A. (1994) Differential role of long terminal repeat control elements for the regulation of basal and tat-mediated transcription of the human immunodeficiency virus in stimulated and unstimulated primary human macrophages, *J. Virol. 68*, 298–307.
64. White, M., Masuko, M., Amet, L., Elliott, G., Braddock, M., Kingsman, A. J. & Kingsman, S. M. (1995) Real time analysis of the transcriptional regulation of HIV and hCMV promoters in single mammalian cells, *J. Cell Sci. 108*, 441.
65. Southgate, C. M. & Green, M. R. (1991) Tat protein activates transcription from an upstream DNA binding site: implications for tat function, *Genes & Dev. 5*, 2496–2507.
66. Madore, S. J. & Cullen, B. R. (1995) Functional similarities between HIV-1 tat and DNA sequence-specific transcriptional activators, *Virology 206*, 1150–1154.
67. Sune, C. & Garcia-Blanco, M. (1995) Sp1 transcription factor is required for *in vitro* basal and tat-activated transcription from the human immunodeficiency virus type 1 long terminal repeat, *J. Virol. 69*, 6572–6576.
68. Huang, L. M. & Jeang, K.-T. (1993) Increased spacing between Sp1 and TATAA renders human immunodeficiency virus type 1 replication defective: implication for Tat function, *J. Virol. 67*, 6937–6944.
69. Buratowski, S. (1994) The basics of basal transcription by RNA polymerase II, *Cell 77*, 1–3.
70. Tjian, R. & Maniatis, T. (1994) Transcriptional activation: A complex puzzle with few easy pieces, *Cell 77*, 5–8.

71. Ou, S.-H. I., Garcia-Martinez, L. F., Paulssen, E. J. & Gaynor, R. B. (1994) Role of flanking E box motifs in human immunodeficiency virus type 1 TATA element function, *J. Virol. 68*, 7188–7199.
72. Olsen, H. S. & Rosen, C. (1992) Contribution of the TATA motif to Tat-mediated transcriptional activation of human immunodeficiency virus gene expression, *J. Virol. 66*, 5594–5597.
73. Berkhout, B. & Jeang, K.-T. (1992) Functional roles for the TATA promoter and enhancers in basal and Tat-induced expression of the human immunodeficiency virus type 1 long terminal repeat, *J. Virol. 66*, 139–149.
74. Lu, X., Welsh, T. M. & Peterlin, M. (1993) The human immunodeficiency virus type 1 long terminal repeat specifies two different transcription complexes, only one of which is regulated by tat, *J. Virol. 67*, 1752–1760.
75. Sune, C. & Garcia-Blanco, M. A. (1995) Transcriptional trans activation by human immunodeficiency virus type 1 Tat requires specific coactivators that are not basal factors, *J. Virol. 69*, 3098–3107.
76. Roy, A. L., Meisterernst, M., Pognonec, P. & Roeder, R. G. (1991) Cooperative interaction of an initiator binding transcription initiation factor and the helix-loop-helix activator USF, *Nature 354*, 245–248.
77. Zenzie-Gregory, B., Sheridan, P., Jones, K. A. & Smale, S. T. (1993) HIV-1 core promoter lacks a simple initiator element but contains a bipartite activator at the transcription start site, *J. Biol. Chem. 268*, 15823–15832.
78. Du, H., Roy, A. L. & Roeder, R. G. (1993) Human transcription factor USF stimulates transcription through the initiator elements of the HIV-1 and the Ad-ML promoters, *EMBO J. 12*, 501–511.
79. Guo, B., Odgren, P. R., Wijnen, V. A. J. & Last, T. J., Nickerson, J., Penman, S., Lian, J. B., Stein, J. L. & Stein, G. S. (1995) The nuclear matrix protein NMP-1 is the transcription factor YY1, *Proc. Natl Acad. Sci. USA 92*, 10526–10530.
80. Margolis, D. M., Somasundaran, M. & Green, M. R. (1994) Human transcription factor YY1 represses human immunodeficiency virus type 1 transcription and virion production, *J. Virol. 68*, 905–910.
81. Ou, S.-H. I., Wu, F., Harrich, D., Garcia-Martinez, L. F. & Gaynor, R. B. (1995) Cloning and characterization of a novel cellular protein, TDP-43, that binds to human immunodeficiency virus type 1 TAR DNA sequence motifs, *J. Virol. 69*, 3584–3596.
82. Parada, C. A., Yoon, J.-B. & Roeder, R. G. (1995) A novel LBP-1 mediated restruction of HIV-1 transcription at the level of elongation in vitro, *J. Biol. Chem. 270*, 2274–2283.
83. Klaver, B. & Berkhout, B. (1994) Evolution of a disrupted TAR RNA hairpin structure in the HIV-1 virus, *EMBO J. 13*, 2650–2659.
84. Duan, L., Ozaki, I., Oakes, J. W., Taylor, J. P., Khalili, K. & Pomerantz, R. J, (1994) The tumor suppressor protein p53 strongly alters human immunodeficiency virus type 1 replication, *J. Virol. 68*, 4302–4313.
85. Valerie, K. & Rosenberg, M. (1990) Chromatin structure implicated in activation of HIV-1 gene expression by ultraviolet light, *New Biol. 2*, 712–718.
86. Verdin, E., Paras, P. & VanLint, C. (1993) Chromatin disruption in the promoter of human immunodeficiency virus type 1 during transcriptional activation, *EMBO J. 12*, 3249–3259.
87. Gait, M. J. & Karn, J. (1993) RNA recognition by the human immuno-deficiency virus Tat and Rev proteins, *Trends Biochem. Sci. 18*, 255–259.
88. Tao, J. & Frankel, A. D. (1993) Electrostatic interactions modulate the RNA-binding and transactivation specificities of the human immunodeficiency virus and simian immunodeficiency virus tat proteins, *Proc. Natl Acad. Sci. USA 90*, 1571–1575.
89. Derse, D., Carvalho, M., Carroll, R. & Peterlin, B. M. (1991) A minimal lentivirus Tat, *J. Virol. 65*, 7012–7015.
90. Churcher, M. J., Lamont, C., Hamy, F., Dingwall, C., Green, S. M., Lowe, A. D., Butler, P. J. G., Gait, M. J. & Karn, J. (1993) High affinity binding of TAR RNA by the human immunodeficiency virus Type-1 tat protein requires base-pairs in the RNA stem and amino acid residues flanking the basic region, *J. Mol. Biol. 230*, 90–110.
91. Mujeeb, A., Bishop, K., Peterlin, B. M., Turck, C., Parslow, T. G. & James, T. L, (1994) NMR structure of a biologically active peptide containing the RNA-binding domain of human immunodeficiency virus type 1 Tat, *Proc. Natl Acad. Sci. USA 91*, 8248–8252.
92. Willbold, D., Rosin-Arbesfeld, R., Sticht, H., Frank, R. & Rosch, P. (1994) Structure of the equine infectious anemia virus tat protein, *Science 264*, 1584–1587.
93. Tan, R. & Frankel, A. D. (1995) Structural variety of arginine-rich RNA-binding peptides, *Proc. Natl Acad. Sci. USA 92*, 5282–5286.
94. Jaeger, J. A. & Tinoco, I. (1993) An NMR study of the HIV-1 TAR element hairpin, *Biochemistry 32*, 12522–12530.
95. Critchley, A. D., Haneef, I., Cousens, D. J. & Stockley, P. G. (1993) Modeling and solution structure probing of the HIV-1 TAR stem-loop, *J. Mol. Graphics 11*, 92–97.
96. Riordan, F. A., Bhattacharyya, A., McAteer, S. & Lilley, D. M. J. (1992) Linking of RNA helices by bulged bases, and the structure of the human immunodeficiency virus transactivator response element, *J. Mol. Biol. 226*, 305–310.
97. Weeks, K. M. & Crothers, D. M. (1991) RNA recognition by Tat-derived peptides: interaction in the major groove, *Cell 66*, 577–588.
98. Hamy, F., Asseline, U., Grasby, J., Shigenori, I., Pritchard, C., Slim, G., Butler, P. J. G., Karn, J. & Gait, M., J. (1993) Hydrogen-bonding contacts in the major groove are required for human immunodeficiency virus Type-1 tat protein recognition of TAR RNA, *J. Mol. Biol. 230*, 111–123.
99. Calnan, B. J., Biancalan, S., Hudson, D. & Frankel, A. D. (1991) Analysis of argininge-rich peptides from the HIV Tat protein reveals unusual features of RNA-protein recognition *Genes & Dev. 5*, 201–210.
100. Puglisi, J. D., Chen, L., Frankel, A. D. & Williamson, J. R. (1993) Role of RNA structure in arginine recognition of TAR RNA, *Proc. Natl Acad. Sci. USA 90*, 3680–3684.
101. Long, K. S. & Crothers, D. M. (1995) Interaction of human immunodeficiency virus type 1 Tat-derived peptides with TAR RNA, *Biochemistry 34*, 8885–8895.
102. Blanchard, A. D., Powell, R., Braddock, M., Kingsman, A. J. & Kingsman, S. M. (1992) An adenosine at position 27 in the human immunodeficiency virus type 1 trans-activation response element is not critical for transcriptional or translational activation by Tat, *J. Virol. 66*, 6769–6772.
103. Delling, U., Reid, L. S., Barnett, R. W., Ma, M. Y.-X., Climie, S., Summer-Smith, M. & Sonenberg, N. (1992) Conserved nucleotides in the TAR RNA stem of human immunodeficiency virus type 1 are critical for Tat binding and trans activation: model for TAR RNA tertiary structure, *J. Virol. 66*, 3018–3025.
104. Feng, S. & Holland, E. C. (1988) HIV-1 tat transactivation requires the loop sequences within TAR, *Nature 334*, 165–167.
105. Roy, S., Parkin, N. T., Rosen, C., Itovitch, J. & Sonenberg, N. (1990) Structural requirements of transactivation of human immunodeficiency virus type 1 long terminal repeat-directed gene expression by tat: importance of base pairing, loop sequences and bulges in the tat-responsive region, *J. Virol. 64*, 1402–1406.
106. Berkhout, B. & Jeang, K.-T. (1991) Detailed muational analysis of TAR RNA: critical spacing between the bulge and loop recognition domains, *Nucleic Acids Res. 19*, 6169–6176.
107. Braddock, M., Powell, R., Sutton, J., Kingsman, A. J. & Kingsman, S. M. (1994) Orientation specific cis complementation by bulge- and loop mutated human immunodeficiency virus type 1 TAR RNAs, *J. Virol. 68*, 8396–8400.
108. Sutton, J. A., Braddock, M., Kingsman, A. J. & Kingsman, S. M. (1995) Requirement for HIV-1 TAR sequences for Tat activation in rodent cells, *Virology 206*, 690–694.
109. Madore, S. J. & Cullen, B. R. (1993) Genetic analysis of the cofactor requirement for human immunodeficiency virus type 1 Tat function, *J. Virol. 67*, 3703–3711.
110. Braddock, M., Powell, R., Blanchard, A. D., Kingsman, A. J. & Kingsman, S. M. (1993) HIV-1 TAR RNA-binding proteins control Tat activation of translation in Xenopus oocytes, *FASEB J. 7*, 214–222.

111. Sheline, T., Milocco, L. H. & Jones, K. A. (1991) Two distinct nuclear transcription factors recognize loop and bulge residues of the HIV-1 TAR RNA hairpin, *Genes & Dev. 5*, 2508–2520.
112. Gatignol, A., Buckler, C. & Jeang, K.-T. (1993) Relatedness of an RNA-binding motif in human immunodeficiency virus type 1 TAR RNA-binding protein TRBP to human P1/dsI kinase and *Drosophila staufen*, *Mol. Cell. Biol. 13*, 2193–2202.
113. Cosentino, G. P., Venkatesan, S., Serluca, F. C., Green, S. R., Mathews, M. B. & Sonenberg, N. (1995) Double-stranded-RNA dependent protein kinase and TAR RNA-binding protein form homo- and heterodimers *in vivo*, *Proc. Natl Acad. Sci. USA 92*, 9445–9449.
114. Kozak, C. A., Gatignol, A., Graham, K., Jeang, K. T. & McBride, O. W. (1995) Genetic mapping in human and mouse of the locus encoding TRBP, a protein that binds the TAR region of the human immunodeficiency virus (HIV-1), *Genomics 25*, 66–72.
115. Baker, B., Muckenthaler, M., Blanchard, A. D., Vives, E., Braddock, M., Nacken, W., Kingsman, A. J. & Kingsman, S. M. (1994) Identification of a novel HIV-1 RAR RNA bulge binding protein, *Nucleic Acids Res. 22*, 3365.
116. Reddy, T. R., Suhasini, M., Rappaport, J., Looney, D. J., Kraus, G. & Wong-Staal, F. (1995) Molecular cloning and characterization of a TAR-binding nuclear factor from T cells, *AIDS Res. Hum. Retroviruses 11*, 663–669.
117. Wu, F. J., Garcia, D., Sigman, D. & Gaynor, R. (1991) Tat regulates binding of the human immunodeficiency virus transactivating region RNA loop-binding protein TRP-185, *Genes & Dev. 5*, 2128–2140.
118. Wu-Baer, F., Lane, W. S. & Gaynor, R. B. (1995) The cellular factor TRP-185 regulates RNA polymerase II binding to HIV-1 TAR RNA, *EMBO J. 14*, 5995–6009.
119. Hart, C. E., Saltarelli, M. M. J., Galphin, J. C. & Schochetman, G. (1995) A human chromosome 12 associated 83 kilodalton cellular protein specifically binds to the loop region of human immunodeficiency virus type 1 trans-activation response element RNA, *J. Virol. 69*, 6593–6599.
120. Marciniak, R. A., Garcia-Blanco, M. A. & Sharp, P. A. (1990) Identification and characterisation of a Hela nuclear protein that specifically binds to the transactivation response element of human immunodeficiency virus, *Proc. Natl Acad. Sci. USA 87*, 3624–3628.
121. Chang, Y.-N., Kenan, D. J., Keene, J. D., Gatignol, A. & Jeang, K.-T. (1994) Direct interactions between autoantigen La and Human immunodeficiency virus leader RNA, *J. Virol. 68*, 7008–7020.
122. Han, X.-M., Laras, A., Rounseville, M. P., Kumar, A. & Shank, P. R. (1992) Human immunodeficiency virus type 1 tat-mediated trans activation correlates with the phosphorylation state of a cellular TAR RNA stem-binding factor, *J. Virol. 66*, 4065–4072.
123. Carrier, F., Gatignol, A., Hollander, M. C. & Jeang, K.-T. (1994) Induction of RNA-binding proteins in mammalian cells by DNA-damaging agents, *Proc. Natl Acad. Sci. USA 91*, 1554–1558.
124. Kozak, M. (1994) Features in the 5′ non-coding sequences of rabbit alpha and beta-globin mRNAs that affect translational efficiency, *J. Mol. Biol. 235*, 95–110.
125. SenGupta, D. N., Berkhout, B., Gatignol, A., Zhou, A. & Silverman, R. H. (1990) Direct evidence for translational regulation by leader RNA and Tat protein of human immunodeficiency virus type 1, *Proc. Natl Acad. Sci. USA 87*, 7492–7496.
126. Parkin, N. T., Cohen, E. A., Darveau, A., Rosen, C., Haseltine, W. & Sonenberg, N. (1988) Mutational analysis of the 5′ non-coding region of human immunodeficiency virus type 1: effects of secondary structure on translation, *EMBO J. 7*, 2831–2837.
127. Edery, I., Petryshyn, R. & Sonenberg, N. (1989) Activation of double-stranded RNA-dependent kinase (dsl) by the TAR region of HIV-1 messenger RNA: A novel translational control mechanism, *Cell 56*, 303–312.
128. Gunnery, S., Greeen, S. R. & Mathews, M. B. (1990) Tat responsive region of human immunodeficiency virus type 1 stimulates protein synthesis *in vivo* and *in vitro*: relationship between structure and function, *Proc. Natl Acad. Sci. USA 89*, 11556–11561.
129. Svitkin, Y. V., Pause, A. & Sonenberg, N. (1994) La autoantigen alleviates translational repression by the 5′ leader sequence of the human immunodeficiency virus type 1 mRNA, *J. Virol. 68*, 7001–7007.
130. Chin, D. J., Selby, M. J. & Peterlin, B. M. (1991) Human immunodeficiency virus type 1 tat does not transactivate mature transacting responsive region RNA species in the nucleus or cytoplasm of primate cells, *J. Virol. 65*, 1758–1764.
131. Kamine, J., Subramanian, T. & Chinnadurai, G. (1993) Activation of a heterologous promoter by human immunodeficiency virus type 1 Tat requires Sp1 and is distinct from the mode of activation by acidic transcriptional activators, *J. Virol. 67*, 6828–6834.
132. Song, C.-Z., Loewenstein, P. M. & Green, M. (1994) Transcriptional activation *in vitro* by the human immunodeficiency virus type 1 Tat protein: evidence for specific interaction with a coactivator(s), *Proc. Natl Acad. Sci. USA 91*, 9357–9361.
133. Zhou, Q. & Sharp, P. A. (1995) Novel mechanism and factor for regulation by HIV-1 Tat, *EMBO J. 14*, 321–328.
134. Kashanchi, F., Piras, G., Radonovich, M. F., Duvall, J. F., Fattaey, A., Chiang, C. M., Roeder, R. G. & Brady, J. N. (1994) Direct interaction of human TFIID with the HIV-1 transactivator Tat, *Nature 367*, 295–299.
135. Selby, M. J. & Peterlin, M. B. (1990) Transactivation by HIV-1 tat via a heterologous RNA-binding protein, *Cell 62*, 769–776.
136. Jeang, K.-T., Chun, R., Lin, N. H., Gatignol, A., Glabe, C. G. & Fan, H. (1993) *In vitro* and *in vivo* binding of human immunodeficiency virus type 1 Tat protein and Sp1 transcription factor, *J. Virol. 67*, 6224–6233.
137. Kato, H. M., Sumimoto, H., Pognonec, P., Chen, C.-H., Rosen, C. A. & Roeder, R. G. (1992) HIV-1 Tat acts as a processivity factor *in vitro* in conjunction with cellular elongation factors, *Genes & Dev. 6*, 655–666.
138. Yu, L., Zhang, Z., Loewenstein, P. M., Desai, K., Tang, Q., Mao, D., Symington, J. S. & Green, M. (1995) Molecular cloning and characterization of a cellular protein that interacts with the human immunodeficiency virus type 1 Tat transactivator and encodes a strong transcriptional activation domain, *J. Virol. 69*, 3007–3016.
139. Herrmann, C. H. & Rice, A. (1995) Lentivirus Tat proteins specifically associate with a cellular protein kinase, TAK, that hyperphosphorylates the carboxyl-terminal domain of the large subunit of RNA polymerase II: candidate for a Tat cofactor, *J. Virol. 69*, 1612–1620.
140. O'Brien, T., Hardin, S., Greenleaf, A. & Lis, J. T. (1994) Phosphorylation of RNA polymerase II C-terminal domain and transcriptional elongation, *Nature 370*, 75–77.
141. Modesti, N., Garcia, J., Debouck, C., Peterlin, M. & Gaynor, R. (1991) Trans-dominant tat mutants with alterations in the basic domain inhibit HIV-1 gene expression, *New Biol. 3*, 759–768.
142. Fridell, R. A., Harding, L. S., Bogerd, H. P. & Cullen, B. R. (1995) Identification of a novel human zinc finger protein that specifically interacts with the activation domain of lentiviral tat proteins, *Virology 209*, 347–357.
143. Ohana, B., Moore, P. A., Ruben, S. M., Southgate, C. D., Green, M. R. & Rosen, C. A. (1993) The type 1 human immunodeficiency virus Tat binding protein is a transcriptional activator belonging to an additional family of evolutionarily conserved genes, *Proc. Natl Acad. Sci. USA 90*, 138–142.
144. Drysdale, C. M. & Pavlakis, G. N. (1991) Rapid activation and subsequent down-regulation of the human immunodeficiency virus type 1 promoter in the presence of Tat: possible mechanisms contributing to latency, *J. Virol. 65*, 3044–3051.
145. Hsu, M.-C., Dhingra, U., Early, J. V., Holly, M., Keith, D., Nalin, C. M., Richou, A. R., Schutt, A. D., Tam, S. Y., Potash, M. J., Volsky, D. J. & Richman, D. D. (1993) Inhibition of type 1 human immunodeficiency virus replication by a Tat antagonist to which the virus remains sensitive after prolonged exposure *in vitro*, *Proc. Natl Acad. Sci. USA 90*, 6395–6399.
146. Braddock, M., Cannon, P., Muckenthaler, M., Kingsman, A. J. & Kingsman, S. M. (1994) Inhibition of human immunodeficiency virus type 1 Tat-dependent activation for translation in *Xenopus* oocytes by the benzodiazepine Ro24-7429 requires transactivation response element loop sequences, *J. Virol. 68*, 25–33.
147. Braddock, M., Thorburn, A. M., Kingsman, A. J. & Kingsman, S. M. (1991) Blocking of Tat-dependent HIV-1 RNA modification

by an inhibitor of RNA polymerase II processivity, *Nature 350*, 439–441.

148. Stutz, F. & Rosbash, M. (1994) A functional interaction between Rev and yeast pre-mRNA in the nucleus, *EMBO J. 13*, 4096–4104.
149. Fisher, U., Meyer, S., Teufel, M., Heckel, C., Luhrmann, R. & Rautmann, G. (1994) Evidence that HIV-1 Rev directly promotes the nuclear export of unspliced RNA, *EMBO J. 13*, 4105–4112.
150. Chang, D. D. & Sharp, P. A. (1989) Regulation by HIV Rev depends upon recognition of splice sites, *Cell 59*, 789–795.
151. Cullen, B. R. (1995) Regulation of HIV gene expression, *AIDS (Phila.), Supplement A: A Year in Review 9*, S19–S32.
152. Schwartz, S., Felber, B. K. & Pavlakis, G. N. (1992) Distinct RNA sequences in the gag region of human immunodeficiency virus type 1 decrease RNA stability and inhibit expression in the absence of Rev protein, *J. Virol. 66*, 150–159.
153. Malim, M. H. & Cullen, B. R. (1993) Rev and the fate of pre-mRNA in the nucleus: implications for the regulation of RNA processing in eukaryotes, *Mol. Cell. Biol. 13*, 6180–6189.
154. Malim, M. H. & Cullen, B. R. (1991) HIV-1 structural gene expression requires the binding of multiple Rev monomers to the viral RRE: implications for HIV-1 latency, *Cell 65*, 241–248.
155. Heaphy, S., Finch, J., Gait, M. J., Karn, J. & Singh, M. (1991) Human immunodeficiency virus type 1 regulator of virion expression, rev, forms nucleoprotein filaments after binding to a purine-rich 'bubble' located within the rev-responsive region of viral mRNAs, *Proc. Natl Acad. Sci. USA 88*, 7366–7370.
156. Bartel, D. P., Zapp, M. L., Green, M. R. & Szostak, J. W. (1991) HIV-1 rev regulation involves recognition of non-Watson-Crick base pairs in viral RNA, *Cell 67*, 529–536.
157. Battiste, J. L., Tan, R., Frankel, A. D. & Williamson, J. R. (1994) Binding of an HIV Rev peptide to Rev responsive element RNA induces formation of purine-purine base pairs, *Biochemistry 33*, 2741–2747.
158. Peterson, R. D., Bartel, D. P., Szostak, J. W., Horvath, S. J. & Feigon, J. (1994) 1H NMR studies of the high-affinity Rev binding site of the Rev responsive element of HIV-1 mRNA: base pairing in the core binding element, *Biochemistry 33*, 5357–5366.
159. Meyer, B. E. & Malim, M. H. (1994) The HIV-1 Rev trans-activator shuttles between the nucleus and the cytoplasm, *Genes & Dev. 8*, 1538–1547.
160. Fritz, C. C., Zapp, M. L. & Green, M. R. (1995) A human nucleoporin-like protein that specifically interacts with HIV Rev, *Nature 376*, 530.
161. Stutz, F., Neville, M. & Rosbash, M. (1995) Identification of a novel nuclear pore-associated protein as a functional target of the HIV-1 rev protein in yeast, *Cell 82*, 495–506.
162. Bogerd, H. P., Fridell, R. A., Madore, S. & Cullen, B. R. (1995) Identification of a novel cellular cofactor for the Rev/Rex class of retroviral regulatory proteins, *Cell 82*, 485–494.
163. Fischer, U., Huber, J., Boelens, W. C., Mattaj, I. W. & Luhrmann, R. (1995) The HIV-1 rev activation domain is a nuclear export signal that accesses an export pathway used by specific cellular RNAs, *Cell 82*, 475–483.
164. Luo, Y., Yu, H. & Peterlin, B. M. (1994) Cellular protein modulates effects of human immunodeficiency virus type 1 Rev, *J. Virol. 68*, 3850–3856.
165. Ruhl, M., Himmelspach, M., Bahr, G. M., Hammerschmid, F., Jaksche, H., Wolff, B., Aschauer, H., Farrington, G. K., Probst, H., Bevec, D. & Hauber, J. (1993) Eukaryotic initiation factor 5A is a cellular target of the human immunodeficiency virus type 1 rev activation domain mediating trans-activation, *J. Cell Biol. 123*, 1309–1320.
166. Elliott, D., Stutz, F., Lescure, A. & Rosbash, M. (1994) mRNA nuclear export, *Curr. Opin. Genet. Dev. 4*, 305–309.
167. Legrain, P. & Rosbash, M. (1989) Some cis and trans-acting mutants for splicing target pre-mRNA to the cytoplasm, *Cell 57*, 573–583.
168. Arrigo, S. & Chen, I. S. Y. (1991) Rev is necessary for translation but not cytoplasmic accumulation of HIV-1 vif, vpr. and enby vol. pu 2 RNAs, *Genes & Dev. 5*, 808–819.
169. Lawrence, J., Cochrane, A., Johnson, C. V., Perkins, A. & Rosen, C. A. (1991) The HIV-1 rev protein: A model system for coupled RNA transport and translation, *New Biol. 3*, 1220–1232.
170. Bray, M., Prasad, S., Dubay, J. W., Hunter, E., Jeang, K.-T. & Rekosh, D., Hammerskjold, M.-L. (1994) A small element from the Mason-Pfizer monkey virus genome makes human immunodeficiency virus type 1 expression and replication Rev-independent, *Proc. Natl Acad. Sci. USA 91*, 1256–1260.
171. Cupelli, L. A. & Hsu, M.-C. (1995) The human immunodeficiency virus type 1 tat antagonist, Ro 5–3335, predominantly inhibits transcription initiation from the viral promoter, *J. Virol. 69*, 2640–2643.
172. Zapp, M. L., Stern, S. & Green, M. R. (1993) Small molecules that selectively block RNA binding of HIV-1 Rev protein inhibit Rev function and viral production, *Cell 74*, 969–978.
173. Biswas, D. K., Ahlers, C. M., Dezube, B. J. & Pardee, A. B. (1993) Cooperative inhibition of NF-κB and Tat-induced superactivation of human immunodeficiency virus type 1 long terminal repeat, *Proc. Natl Acad. Sci. USA 90*, 11044–11048.
174. Malim, M. G., Bohnlein, S., Hauber, J. & Cullen, B. R. (1989) Functional dissection of the HIV-1 rev trans-activator-derivation of a trans-dominant repressor of rev function, *Cell 58*, 205–214.
175. Mhashilkar, A. M., Bagley, J., Chen, S. Y., Szilvay, A. M., Helland, D. G. & Marasco, W. A. (1995) Inhibition of HIV-1 Tat-mediated transactivation and HIV-1 infection by anti-Tat single chain intrabodies, *EMBO J. 14*, 1542–1551.
176. Lever, A. M. L. (1995) Gene therapy for HIV infection, *Br. Med. Bull. 1995*, 149–166.
177. Pfeifer, K., Bachmann, M., Schroder, H. C., Weiler, B. E., Ugarkovic, D., et al. (1991) Formation of a small ribonucleoprotein particle between tat protein and trans-acting response element in human immunodeficiency virus-infected cells, *J. Biol. Chem. 266*, 14620–14626.

Eur. J. Biochem. 242, 1–19 (1996)

Review

Immunostimulating agents: what next?

A review of their present and potential medical applications

Georges H. WERNER[1] and Pierre JOLLÈS[2]

[1] Institut de Chimie des Substances Naturelles, CNRS, Gif-sur-Yvette, France

[2] Laboratoire de Chimie des Substances Naturelles, URA CNRS no. 401, Muséum National d'Histoire Naturelle, Paris, France

(Received 19 July 1996) – EJB 96 1083/0

Many chemical entities, either from natural sources or prepared by synthesis, are known to exert stimulating activities on various functions of the immune system, such as antibody production, resistance to infections, rejection of malignant cells, etc. In this review, the origin, chemical structures and main activities of several immunostimulants are described, with special emphasis on their present or potential medical usefulness. An attempt is made to envisage the future of this type of pharmacological agents, excluding however from the presentation the endogenous modulators of the immune system (cytokines), the production and activities of which are influenced by the immunostimulants themselves.

Keywords: immunopharmacology; immunostimulant; immunomodulator; adjuvant; biological response modifier.

It is convenient and probably also justified to adopt the apparently finalistic concept that, in the course of evolution, the immune system of Vertebrates arose primarily in response to the threat posed by infectious agents. This immune system is characterized by a learned self-nonself discrimination mechanism that distinguishes it from all other protective strategies [1]. There are marked resemblances, however, between the immune system, which has been called 'a mobile brain' [2] and the central nervous system, and bidirectional interactions between these systems have been recognized: they are the object of intensive investigation, in a discipline called psychoneuroimmunology [3]. The various cell populations of the immune system (macrophages, neutrophils, monocytes, lymphocytes) are organized in a highly complex and intricate network, in which connections are operated through direct interactions between cells and by communication through intercellular chemical signals, the monokines and lymphokines globally called cytokines (interferons, interleukins, transforming growth factors, chemotactic factors). Differentiation and maturation of the bone-marrow-derived precursor cells of the immune system are under the influence of a series of hematopoietic growth factors (colony-stimulating factors).

Whereas this 'sophisticate' immune system has been a major weapon in the struggle between Vertebrates and an enormous diversity of parasitic organisms, it is by no means perfect. Attempts to 'domesticate' the immune system for the benefit of man and domestic animals have been a major endeavour of medicine during the last two centuries: specific vaccines and anti-infectious immunoglobulins illustrate this point. More recently, it was recognized that several pathological conditions are caused by inadequate functioning and regulation of the immune system: immunological deficiencies (congenital or acquired, dramatically exemplified by AIDS) and a significant number of autoimmune diseases, in which an immune response is aberrantly mounted against the individual's own cells or tissues. It is also recognized that certain tumor cells express specific antigens that distinguish them from non-malignant cells, thus giving rise to the hope that immunological intervention can play a role in the treatment of certain cancers. On the other hand, the normal function of the immune system in distinguishing nonself from self represents a serious obstacle to transplantation of tissues and organs from not fully histocompatible donors and this has justified the search for agents that could temporarily and selectively suppress some components of the immune system. Early in this century, it was recognized that the protective efficacy of vaccines could be enhanced by adding to them substances called adjuvants that nonspecifically stimulate the humoral and cell-mediated responses to the antigens [4].

Even earlier, at the end of the last century, Metchnikoff suggested [5] that nonspecific stimulation of the phagocytic cells through the injection of aseptic inflammatory substances could increase patients' resistance against infections and Coley [6] used bacterial culture filtrates to enhance rejection of tumor cells by the patients' immune system. Progressively, therefore, and very rapidly over the last two decades, a new scientific and medical discipline has emerged that deserves the name of immunopharmacology. Immunopharmacological agents or immunomodulators, can conveniently be divided between immunosuppressive drugs and immunostimulating drugs (including adjuvants), which correspond to the therapeutic needs outlined above.

Correspondence to G. H. Werner, Institut de Chimie des Substances Naturelles, CNRS, F-91198 Gif-sur-Yvette, France and P. Jollès, Laboratoire de Chimie des Substances Naturelles, URA CNRS 401, Muséum National d'Histoire Naturelle, 63, rue Buffon, F-75005 Paris, France

Abbreviations. BCG, bacillus Calmette-Guérin; Etn, ethanolamine; LPS, lipopolysaccharide; MPL, monophosphoryl lipid A; MDP and MTP, muramyldipeptide and muramyltripeptide, respectively; Ptd, phosphatidyl.

Note. This Review will be reprinted in *EJB Reviews* 1996, to appear in April 1997.

The major issue facing immunopharmacological research in general is to define from which sources effective immunostimulating, adjuvant and immunosuppressive agents are better obtained: should one exploit the rich arsenal of endogenous molecules (mostly peptides), which serve as communication factors between cells of the immune system or look for immunomodulators (or biological response modifiers, as they are frequently called) among the almost infinite resource of natural or synthetic chemical substances, which can be tested in the laboratory for their possible immunomodulating activities?

As the title of our review implies, we shall restrict our discussion to the so-called immunostimulants, the medicinal purpose of which may be, according to the case, to enhance resistance against infectious agents, to stimulate rejection of cancer cells or to increase the immunogenicity of vaccines. There can be little doubt that several endogenous polypeptides of the immune system (cytokines and related factors) will play a role in various modes of immunostimulating therapy: examples are provided by interferon α in the treatment of chronic hepatitis B virus infection, by interleukin-2 in the treatment of metastatic renal cell carcinoma, and by colony-stimulating factors (GM-CSF and G-CSF) which have been shown to accelerate neutrophil recovery following chemotherapy of malignant diseases, including very intensive therapy used with bone marrow transplantation. Interleukin-12 is showing remarkably broad potential powers against a variety of infectious diseases (AIDS, leishmaniasis, malaria, tuberculosis, schistosomiasis) and possibly some malignant conditions [7]. On the other hand, the therapeutic potential of cytokine manipulation includes also a great variety of illnesses in which a number of cytokines (tumor-necrosis factor, interleukins-1, -4, and -6) are implicated as part of the pathophysiological process: inflammation, rheumatic diseases, AIDS. Antagonists of such cytokines, including monoclonal antibodies against them, soluble receptors, receptor antagonists, are being developed [8–10] and immunomodulation by cytokine antisense oligonucleotides is being contemplated [11]. Direct administration of cytokines or of their antagonists as well as *in vitro* treatment of immune cells that are reinfused in the patient's circulation, and gene transfer (for instance the gene coding for the tumor-necrosis factor) are some of the strategies used in this expanding approach. It must be realized, however, that cytokines are definitely pleiotropic in their activities and that side effects (including fatalities) can be anticipated and are indeed observed in many of the clinical trials involving cytokine or anti-cytokine therapies. In the healthy individual, the various cytokines collaborate efficiently, in spite of their pleiotropism and redundancy, but artificial introduction of additional amounts of a given cytokine may disturb this equilibrium, which explains why cytokine therapy still poses difficult problems of dosage and timing of administration.

Another therapeutic approach, to which we shall devote the main part of this review, is to look for immunostimulating and adjuvant activities in exogenous molecules, either natural (of microbial origin for instance), semi-synthetic or totally synthetic. The possibilities there are unlimited and search for such agents is facilitated by the wide array of *in vitro* and *in vivo* tests that are available for the detection and analysis of immunostimulating activities, such as enhancement of resistance to infections, rejection of transplanted tumors, augmentation of antibody production or delayed type hypersensitivity against various antigens. Of course it should be realized that the activities of the various immunostimulating substances of exogenous origin which we shall review must be, to a large extent, mediated through their interactions (either already demonstrated or still awaiting clarification) with the cytokine network and that therefore there is no contradiction between the direct and the indirect approach. One distinct advantage, however, of the exogenous immunostimulants is that, in contrast to the polypeptides of the cytokine system, several of them are of small molecular mass, synthetic or readily accessible to chemical synthesis, thereby endowed with easily recognized pharmacokinetic characteristics, making it possible to administer them orally, for instance; they may also be stable in dry form at room temperature and precise dosage fractionation may be easier than with the cytokine molecules. Finally, many of them may be cheaper to produce, with simpler analytical controls than biotechnological products. One additional advantage is that, in the case of synthetic or semi-synthetic immunostimulants, there is a wide open possibility of designing chemical analogues with improved activity and benefit: risk ratio.

The decision to pursue biological and clinical investigations of exogenous immunostimulants finds another justification when we look into the area of their counterparts, the immunosuppressive drugs. Whereas the field of pharmacological immunostimulation is still at a very early stage of development, immunosuppressive therapy has become part of everyday medical practice, either to prevent allograft rejection in patients receiving kidney, heart, lung, liver transplants or in treating a variety of autoimmune disorders, such as rheumatoid arthritis, insulin-dependent diabetes, lupus erythematosus, Crohn's disease and probably other conditions in the future [12–14]. With the exception of a monoclonal antibody against the CD-3 marker of T lymphocytes (OKT 3), all the immunosuppressive drugs used in medical practice or undergoing clinical investigation are exogenous molecules, either microbial secondary metabolites (such as cyclosporin, the macrolides rapamycin and tacrolimus, 15-deoxyspergualin, mycophenolate mofetil) or synthetic, such as leflunomide or Brequinar sodium.

We shall now start a detailed review of the various immunostimulating agents and vaccine adjuvants that are either already used in medicine (actually a small number of them) or under preclinical and clinical development. We will make no attempt to exhaustively cover this field and the reader's attention is called to some reviews that have appeared recently [15–18].

Bacteria and bacterial products

Whole bacteria. About 40 years ago, Biozzi et al. [19] showed that inoculation of live bacillus Calmette-Guérin (BCG, the antituberculosis vaccine) induced in the mouse a sustained hyperactivity of the phagocytic cells. BCG-inoculated mice were shown by Dubos and Schaedler [20] to be more resistant than control animals to infection with virulent staphylococci and Biozzi et al. [21] demonstrated that in BCG-inoculated mice there was an inhibition of growth of the Ehrlich ascite tumor. These experimental findings were the basis of the clinical use of BCG for the active immunotherapy of certain leukemias and subsequently of various solid tumors, an approach which was pioneered by Mathé et al. [22]. BCG has been employed, with various degrees of success, in the treatment of malignant diseases, alone or in combination with other therapies (surgery, chemotherapy, radiotherapy) but, at the present time, the most important and clearly beneficial application of BCG is in the treatment of bladder cancer [23, 24], in which live bacillus suspensions are administered intravesically. Bladder cancer is the sixth most common cancer in men and the fourteenth most common cancer in women; in 1985, it accounted for 182000 new cases and 82000 deaths worldwide. The first clinical trials with intravesical BCG showed a remarkable decrease in the rates of recurrence of bladder cancer and in the last 20 years, many tens of thousands of patients have benefited from this therapy, which involves repeated intravesical instillations of 4×10^8 bacilli every week for six weeks, followed by two more at six and twelve months after the beginning of treatment. The most frequent complication is cystitis or more distant infections (controlled by isoniazid administration) and occasional fatalities due to anaphylaxis or sepsis. Progress is being made toward optimization of BCG therapy for bladder cancer [25]. Jackson and James [26] have carefully analyzed the mechanism of action of BCG in immunotherapy of bladder cancer. They have shown that several cytokines are identified in patients' urine following repeated instillation of BCG (interleukins-1, -2, -6, and -8, tumor necrosis factor α, interferon γ) and that actually BCG interacts also directly with the tumor cells and exerts potent effects both on the growth of the tumor cells and on their phenotype. Their observations have led them to postulate a two-stage mechanism for an optimal clinical response to BCG therapy: stage 1, the tumor system; stage 2, the immune system. Simultaneous activation of the immune system and a response of the tumor system are necessary: the tumor cells must be induced to display some molecules (such as intercellular adhesion molecules, like intercellular adhesion molecule-1) which predispose them to cell-mediated cytotoxicity by the stimulated immune system. In conclusion, one of the earliest known immunostimulants, namely live BCG, has a recognized beneficial role in cancer immunotherapy. Indeed, many years before the work of Biozzi, Dubos and Mathé, the immunostimulating activities of tubercle bacilli (mycobacteria) had been experimentally demonstrated and led to the development of Freund's complete adjuvant in 1937 [27]. We shall come back later to the mycobacteria-derived chemically defined substances as important immunostimulants.

Bacterial extracts. While BCG is the only live bacillus clinically employed as an immunostimulant, a significant number of other bacterial products, of various degrees of purification, have found medical applications. One of them is OK 432 (Picibanil), produced by a low virulence strain of human *Streptococcus pyogenes* treated with penicillin G; it is widely used in Japan in cancer immunotherapy, in combination with chemotherapy or radiotherapy. This preparation augments NK (natural killer) cell activity, activates macrophages to exert tumoricidal activity, enhances cytotoxic T lymphocytes and induces the production of various cytokines (interleukin-1, interleukin-2, interferon-γ, tumor necrosis factor, colony-stimulating factor GM-CSF). In Japan, this immunostimulant is part of the management of head and neck cancer, colorectal and gastric cancer, small cell cancer of the lung and brain tumors [28, 29].

It must be emphasized at this point that western medicine in general does not appear to share the optimism of Japanese investigators and clinicians concerning the usefulness of bacterial immunostimulants in the management of malignant diseases (with the exception of BCG in bladder cancer). In Europe, for instance, several preparations of bacterial origin are on the market exclusively as anti-infectious immunostimulants: these include a preparation composed of two glycoprotein fractions (molecular masses of 95 and 350 kDa, respectively) from *Klebsiella pneumoniae* and a preparation consisting of ribosomal and proteoglycan fractions from several common respiratory pathogens.

These bacterial immunostimulants are administered either orally or intranasally (taking advantage of the mucosal immunity, through lymphoid areas present in the nasopharynx): placebo-controlled studies have shown that repeated administrations during the winter period provides a rather modest but statistically significant protection against respiratory infections in children or in elderly subjects (with chronic bronchitis), as evidenced by reduced numbers of infectious episodes, shorter duration and milder symptoms of such episodes, and decreased need to institute antibiotic treatment. Lyophilized lysates of *Haemophilus influenzae*, *Streptococcus pneumoniae*, *Staphylococcus aureus*, *Streptococcus pyogenes*, and *Neisseria catarrhalis* enter into the composition of another orally active anti-infectious immunostimulant, whereas fractions (primarily high-molecular-mass membrane proteins) from *Escherichia coli* are claimed to be an effective immunostimulating agent, always by the oral route, against recurrent urinary infections. Surprisingly, the same preparation exhibits immunomodulating activities that indicate its potential usefulness in the treatment of an autoimmune disease, namely rheumatoid arthritis [30].

Apart from immunopharmacological considerations, it is noteworthy that bacteria of the physiological microflora (digestive tract) exert a natural and permanent stimulus on certain immune functions: Beuth et al. [31] have shown that antibiotic decontamination of laboratory animals resulted in immunodepression; these authors found that certain members of the mouse gastrointestinal microflora (e.g. *Bacteroides*, *Clostridium*, *Lactobacillus*, *Propionibacterium*) actually liberate low-molecular-mass peptides (molecular mass less than 6.5 kDa) which are apparently essential for adequate immune responses of the host.

Chemically defined bacterial products. Of all the bacterial products that exert a stimulating effect on the functions of the immune system through their interaction with macrophages, monocytes and lymphocytes, the most potent ones, are probably the endotoxins, or lipopolysaccharides of gram-negative bacteria. Chemically, lipopolysaccharides (LPS) are heteropolymers consisting of polysaccharides covalently bound to a nitrogen-containing phospholipid called lipid A. LPS exert a wide spectrum of biological activities, lipid A being the active part in most of these effects. LPS preparations and purified lipid A have been shown to stimulate antibody production (LPS are polyclonal activators of B lymphocytes), to enhance the resistance of laboratory animals to bacterial, viral, fungal, and parasitic infections, to exert a necrotizing effect on tumors, all these effects being concomitant with increased production and release of various cytokines, such as interleukin-1, interleukin-6 and tumor-necrosis factor. Bocci [32] has formulated the interesting and plausible hypothesis that minute amounts of LPS constantly liberated from the gram-negative flora of healthy individuals and animals behave as natural immunostimulants, which play a crucial role in protection against pathogenic microorganisms. As the author stated: 'Human intestine, by harboring the bulk of gram-negative bacteria, contains more than enough to kill the host in a few hours, should they enter the circulation (as happens in septic shock). Luckily, this potential bomb rarely harms the host and actually, most of the time, has two beneficial effects. The first is mainly local, on the gut-associated lymphoid tissue, with consequent induction of IgA and a localized production of cytokines, which activate mononuclear cells. The second beneficial effect is due to traces of carrier-bound LPS which, by entering the circulation, reach organs such as liver, lungs, spleen, and bone marrow and elicit focal immunological reactions'. On the other hand, it is quite clear that exogenous administration of LPS or lipid A for therapeutic purposes would be a much too dangerous procedure, in view of the marked and pleiotropic toxicity of these molecules. Takada and Kotani [33] have performed a thorough study of structural requirements of lipid A for endotoxocity and other biological activities, with the purpose of designing novel compounds that would present an adequate balance between endotoxicities and beneficial bioactivities. Such a goal has been partly achieved by Ribi et al. [34] who showed that the minimal structure required for toxicity was a bisphosphorylated diglucosamine moiety, to which long chain fatty acids were attached: this was called diphosphoryl lipid A. Several of the nontoxic or less toxic analogues differed from diphos-

Diphosphoryl-lipid A (DPL)

Monophosphoryl-lipid A (MPL)

MDP

phoryl lipid A by the simple fact that they lacked a phosphate group on the reducing end of the disaccharide: such material was called monophosphoryl lipid A. Bioassays on monophosphoryl lipid A showed that, while it was 1000 times less potent on a molar basis in eliciting toxic and pyrogenic responses, it was comparable to diphosphoryl lipid A (and endotoxin itself) in immunostimulating activities. Monophosphoryl lipid A has been shown to cause regression of certain grafted tumors in laboratory animals, to enhance production of colony-stimulating factors, to enhance resistance against infections, and to confer protection against X-radiation. At the present time, monophosphoryl lipid A (from *Salmonella typhimurium* or from *Salmonella minnesota*) is being marketed as an immunity adjuvant for laboratory animals, in particular within the framework of the production of monoclonal antibodies. Recently, a rather surprising finding was reported by Yao et al. [35], namely that monophosphoryl lipid A could represent a new approach for cardioprotection. Brown et al. [36] had observed that a low dose of endotoxin, given as a 24-h pretreatment, increased endogenous myocardial catalase activity and decreased ischemia-reperfusion injury in isolated rat hearts. Yao et al. performed *in vivo* studies of this phenomenon, by substituting much less toxic monophosphoryl lipid A to the endotoxin used by the previous authors. They found that this glycolipid induced functional protection against myocardial ischemia-reperfusion injury in rabbits and rats and that the same preparation reduced myocardial infarct size in dogs. The mechanisms by which monophosphoryl lipid A produces a cardioprotective effect when administered 24 h prior to ischemia is not clear, but the observation that pretreatment with monophosphoryl lipid A produced a marked decrease in myeloperoxidase activity (an index of polymorphonuclear leukocyte infiltration in the border zone surrounding the necrotic tissue) suggests that an effect of monophosphoryl lipid A on neutrophil function and/or traffic may play a role in the cardioprotective action of this preparation. Neutrophils are a major contributing factor in the pathogenesis of myocardial ischemia-reperfusion injury. In conclusion, monophosphoryl lipid A could be administered as preventive therapy 24 h prior to such procedures as angioplasty and coronary bypass surgery. According to Yao et al. [35], a phase II clinical trial with monophosphoryl lipid A in patients who are having elective coronary artery by-pass surgery is ongoing. As this example shows, through a particular aspect of its mechanisms of action, an immunostimulant may have quite unexpected applications.

Next to the gram-negative bacteria and their lipopolysaccharides, the mycobacteria represent a remarkable source of immunostimulants, as already illustrated by the case of BCG.

Polar glycopeptidolipids extracted from the cell wall of *Mycobacterium chelonae* (a microorganism isolated from the turtle) were shown to exert in the mouse a reversal of the doxorubicin-induced leucopenia, an effect comparable in its intensity to that exerted by the colony-stimulating factor GM-CSF [37]. The same preparation was also shown to enhance resistance of mice to disseminated infection with *Candida albicans*, an experimental model resembling the invasive and life-threatening fungal infections in immunocompromised patients [38].

On the other hand, attempts at fully characterizing the molecular entities responsible for the immunopotentiating activities of mycobacteria (such as those of BCG or of Freund's complete adjuvant) culminated 22 years ago [39] with the identification of *N*-acetylmuramyl-L-alanyl-D-isoglutamine (muramyldipeptide, MDP) as the minimal essential structure required for adjuvant activity.

Many publications have appeared concerning the activities and therapeutic potential of MDP and the numerous chemical analogs that were synthesized, and we shall not attempt to give here a comprehensive review of this field. The reader is referred to an excellent review published six years ago [40]. We shall only deal here with possible clinical applications of the MDP analogs that appear to be particularly interesting, namely murabutide (*N*-acetyl-muramyl-L-alanyl-D-glutaminyl-*n*-butyl ester), temurtide (threonyl-MDP), romurtide [MDP-Lys(18)N^2-(*N*-acetylmuramoyl)-L-alanyl-D-isoglutaminyl)-N^6-stearoyl-L-lysine] and MTP-PtdEtn (muramyltripeptide phosphatidylethanolamine).

Parenteral administration of MDP and its analogs in laboratory animals (mouse, guinea pig, rabbit) induces a complex range of biological effects: increased production of antibodies to various antigens, enhanced resistance to infections, enhanced tumoricidal activity of macrophages, restoration of myelopoiesis, fever, stimulation of slow wave sleep, induction of meningeal inflammation, etc. On the other hand, oral administration of muramylpeptides has been reported to induce certain other biological effects that are not observed following parenteral ad-

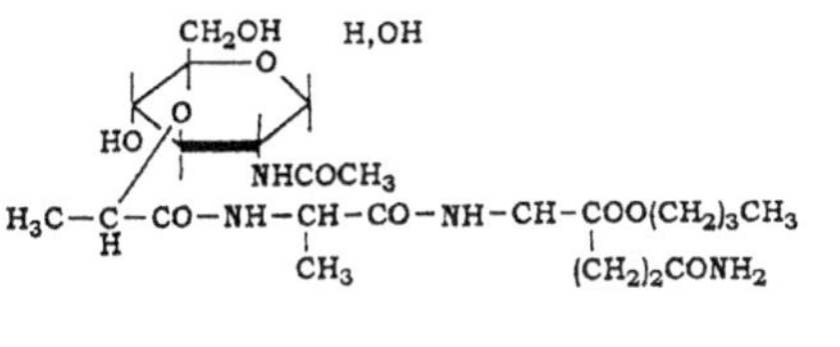

Murabutide

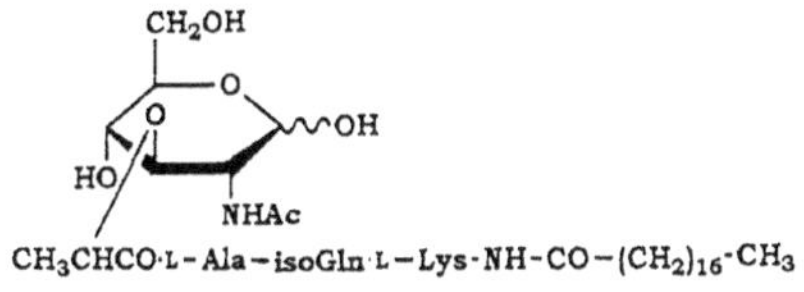

Romurtide

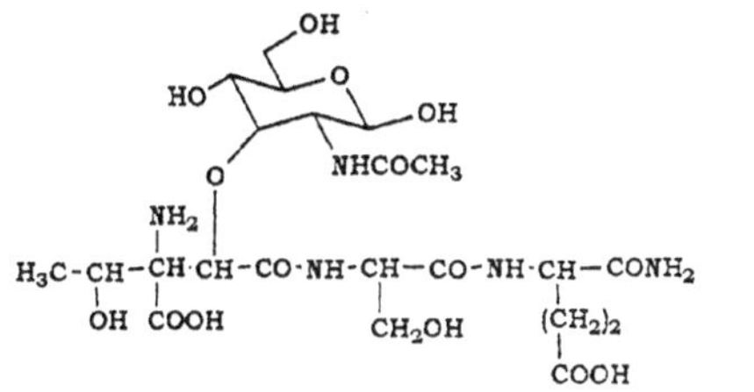

Temurtide

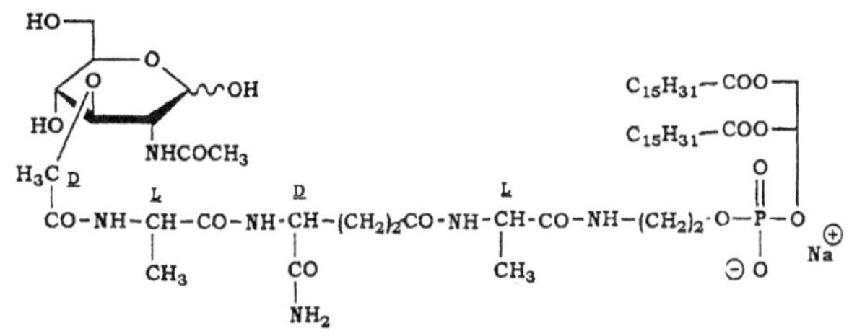

MTP-PtdEtn

ministration, such as a downregulation of anamnestic antigen-specific IgE response. Bahr et al. [41] have reported the efficacy of a lipophilic derivative, *N*-acetyl-Mur-L-Thr-D-isoGln-*sn*-glyceryldipalmitoyl-MDP, incorporated into liposomes, in suppressing polyclonally induced serum IgE levels in anti-IgD-treated mice. The anti-allergic activity of certain muramylpeptides, exerted through their effects on gut-associated lymphoid tissues, may be one of the possible therapeutic applications of this family of compounds. The same authors have reported, on the other hand, that murabutide has been administered (in phase I and phase I/IIa studies) to 200 healthy volunteers and cancer patients: good tolerance was observed at doses up to 200 μg/kg injected subcutaneously. Synergistic activity between murabutide and interferon α was demonstrated *in vitro* and in *in vivo* models (protection of mice against endotoxic shock and against infection with the murine encephalomyocarditis virus) and such results gave support for assessment of a combination therapy between murabutide and interferon-α, in malignancies against which the latter cytokine has already shown some activity. Preliminary studies in healthy volunteers demonstrated excellent tolerance of murabutide when injected in association with interferon-α; interestingly, the incidence of side-effects caused by a high dose of interferon-α (6×10^6 International Units) was reduced upon association with murabutide.

Temurtide is an MDP analog, in which alanine has been replaced with threonine, that was selected from approximately 20 MDP analogs because of its superior adjuvant activity (on both humoral and cell-mediated immunity) and its lack of side effects at adjuvant-active doses (pyrogenicity, induction of adjuvant arthritis or uveitis) and which is now being developed as a vaccine adjuvant, as reviewed by Lidgate and Byars [42]. The dilution vehicle for threonyl-MDP (highly soluble in water) is composed of a finely dispersed metabolizable oil in an aqueous continuous phase. The oil is squalane and the emulsifiers are polysorbate 80 (Tween 80) and a nonionic surfactant (Pluronic L 121). Virtually all of the adjuvant remains in the aqueous phase of the emulsion, which is an oil-in-water emulsion instead of the water-in-oil emulsion of Freund's adjuvant. Studies in laboratory animals have shown that temurtide exerts superior adjuvant activities toward the following antigens: influenza virus hemagglutinin; hepatitis B surface antigen; herpes simplex type 2 virus glycoprotein; Epstein-Barr virus, simian immunodeficiency virus, HIV-1 glycoprotein 120, as well as melanoma and β lymphoma tumor antigens.

Another MDP derivative, romurtide [N^2-(*N*-acetylmuramoyl)-L-alanyl-D-isoglutaminyl)-N^6-stearoyl-L-lysine] is, since 1991, on the pharmaceutical market in Japan [43, 44]. Experimental data have shown this stearoyl-MDP to stimulate resistance of laboratory animals to bacterial and mycotic infections and to be a potent inducer of several cytokines, such as interleukin-1, interleukin-6, tumor-necrosis factor, interferon α and of the colony-stimulating factors G-CSF and GM-CSF. The hematopoietic stimulation observed in monkeys receiving subcutaneous administrations of romurtide is attributable to the augmenting effect of this agent on the production of cytokines, and notably the colony-stimulating factors. In mice, romurtide promoted the recovery from a leucopenic state induced by cyclophosphamide treatment or X-irradiation and in X-irradiated guinea pigs, the same agent enhanced platelet recovery. Clinical trials have been performed in lung cancer patients undergoing chemotherapy with cisplatin, vindesine, and mitomycin C. Subcutaneous treatments with low doses of romurtide (200 μg) led to rapid recovery of white blood cells counts and to recovery of platelet counts, in comparison with a slower recovery in control patients that received chemotherapy alone. A similar restorative effect of romurtide was observed in cancer patients submitted to radiotherapy. Romurtide is now widely utilized in Japan for this therapeutic application. Fever is the most common side effect of romurtide treatment. However, whereas this drug induces the production of several cytokines, the severe toxic manifestations reported following injection of recombinant human interleukin-1, interleukin-6 and tumor-necrosis factor were not observed in the course of treatment with romurtide.

Another synthetic MDP analog of considerable interest is the lipophilic muramyltripeptide phosphatidylethanolamine (MTP-PtdEtn). This lipophilic MDP analog was synthesized to obtain a stable association of the compound with liposomes (because MTP-PtdEtn can be inserted into the phospholipid bilayer of the latter), following the observation that repeated systemic administrations of liposomal MDP, but not of the free form of MDP, caused in mice eradication of pulmonary and lymph node metastases from a subcutaneous murine melanoma [45, 46]. Systemic administration of liposomes containing MTP-PtdEtn has been shown to eradicate spontaneous metastases in several animal tumor models, including dogs with autochthonous osteogenic sarcoma metastases. Indeed, activating macrophages and monocytes to a tumoricidal state is one of the possible immunotherapeutic approaches to the immunological treatment of cancer. After several preclinical and phase I clinical trials in cancer patients had given the necessary information concerning the biological parameters influenced by liposomal MTP-PtdEtn and its side effects as well as acceptable doses, a phase II trial of this agent was initiated in patients with osteosarcoma who developed pulmonary metastases during adjuvant chemotherapy or who presented pulmonary metastases that persisted despite chemotherapy. Liposomal MTP-PtdEtn was infused at a dose of 2 mg/m^2 twice weekly for three months. Histological examination of the tumor nodules recurring after this period showed unique morphological changes marked by peripheral fibrosis surrounding the tumor and inflammatory cell infiltration and/or a change in malignant characteristics from high grade before therapy to low grade after therapy [47]. Killion et al. [48], on the other hand, have shown that in mice, oral administration of lipophilic MTP-PtdEtn prevents the monocytopenia induced by chemotherapy (doxorubicin) or whole body X-irradiation, thus adding another aspect to the therapeutic potential of this agent in cancer patients.

The last three examples have illustrated the possible prophylactic or therapeutic applications of synthetic MDP analogs: potent adjuvant effect on the immunogenicity of vaccines in the case of temurtide, recovery from chemotherapy or radiotherapy-induced leukopenia in the case of romurtide, activation of tumoricidal state of monocytes as well as prevention of chemotherapy or radiotherapy-induced monocytopenia in the case of MTP-PtdEtn. The last two effects take us away from the original medicinal profile of MDP, based on its activities in experimental models, namely an enhancement of the host's resistance to infections. However, as O'Reilly and Zak rightly stress [49], muramylpeptide immunostimulants could also be used clinically to enhance the effectiveness of conventional chemotherapies of bacterial, fungal, parasitic and viral infections, possibly by allowing the use of lower doses of the antimicrobial drugs or shortening the duration of treatment or simply in promoting faster patient recovery through combination of two attacks on the pathogenic microorganisms: the antimicrobial agent and the immune system (phagocytes and lymphocytes). It should be pointed out, however, that in the experimental models of infections that are used to analyze the activities of MDP and its analogs, treatment must be initiated before inoculation of the infectious agent (i.e. prophylactically) to exhibit significant activity, even when associated with post-infection chemotherapy.

The title of the present section is Bacteria and bacterial products and, as is evident from the discussion concerning MDP derivatives, many of the so-called bacterial products do have a structure basically reminiscent of the structure of a natural bacterial product (cell wall constituent, for instance) but are in fact synthetic preparations (MDP analogs such as murabutide, temurtide, romurtide, and MTP-PtdEtn).

Semi-synthetic molecules of bacterial origin. A similar approach was followed by our team about 15 years ago, when we became interested in the immunostimulating activity of crude water-soluble extracts of a *Streptomyces* strain and attempted to identify their active component. A tetrapeptide, L-Ala-D-Glu(L,L-A_2pm(Gly))NH_2 was isolated, but was found to to be inactive in immunostimulation tests. Following a lead provided by our earlier work showing that chemical conjugation with fatty acids of water-soluble fragments from strains of *M. tuberculosis* modified the immunoadjuvant properties of these substances by rendering them active in the absence of mineral oil [50], the inactive tetrapeptide was conjugated with lauric anhydride. The compound thus obtained (lauroyltetrapeptide) exhibited marked *in vitro* and *in vivo* immunopotentiating activities [51, 52], which justified its preparation by total synthesis as well as the synthesis of close to 100 analogs. The biological activities of the synthetic lauroyltetrapeptide and of several analogs have been described in several publications [53–58] and may be summarized as follows: adjuvant activity on antibody production and on delayed type hypersensitivity reactions against various antigens, enhancement of the resistance of mice against bacterial infections (including intracellular microorganisms, like *L. monocytogenes*), increased resistance of mice to the lethal effect of γ-ray irradiation (radioprotective activity), decrease of the amount of hepatic microsomal cytochrome *P*-450 and of the level of CCl_4-induced lipid peroxidation. Among the likeliest mechanisms of action of pimelautide and its analogs, one can quote a stimulation of phagocytosis and of interleukin-1 and tumor-necrosis factor production by macrophages and monocytes, an enhancement of interleukin-2 production by lymphocytes, a stimulation of cytolytic T lymphocytes and the induction of the production of colony-stimulating-factor-like factors in the serum of mice. It was also realized that the synthetic lauroyltetrapeptide is a mixture of two stereoisomers, of which the active component was the molecule containing diaminopimelic acid in its L,L form (pimelautide, RP 44102). The presence of glycine in the molecule is not essential for the immunopharmocological activity of lauroyltetrapeptide, since the lauroyltripeptide N_2-(*N*-(*N*-lauroyl-L-alanyl)-γ-D-glutamyl)-L,L,-2,6-diaminopimelamic acid (RP 56142, trimexautide) exhibited activities comparable to those of the tetrapeptide. Recently, Déprez et al. [59] described the synthesis of analytically pure immunogens, in which a hexadecameric peptide (V3) derived from the principal neutralizing domain of the envelope glycoprotein of HIV-1 was associated with either pimelautide or trimexautide. The *in vivo* (in Balb/c mice) immunogenicity of these compounds (built-in adjuvants) was evaluated according to two criteria: the ability to elicit a T-lymphocyte cytotoxicity and the ability to stimulate antibody response. The results indicated that the trimexautide conjugate was able to induce an efficient and relevant HIV-1-specific T-lymphocyte cytotoxicity response, whereas the pimelautide conjugate stimulated a strong antibody response to the linked viral peptide. Such a chemically defined model of peptide vaccines against HIV-1 may be used to selectively stimulate subpopulations in immunocompetent cells and, obviously, a similar strategy could be used in designing vaccines against other viruses against which both cell-mediated and humoral immune responses are important in terms of protection.

```
                      (L)       (D)
CH3(CH2)10·CO·NH·CH-CO-NH-CH-COOH
                      |         |              (*)
                      CH3       (CH2)2CO·NH-CH-COOH
                                               |
                                               (CH3)3
                                               |
                           H2NCH2-CO·NH-CH-CONH2
                                               (*)
```

Pimelautide

```
                      (L)       (D)
CH3(CH2)10·CO·NH·CH-CO-NH-CH-COOH      (L)
                      |         |
                      CH3       (CH2)2CO·NH-CH-COOH
                                               |
                                               (CH3)3
                                               |
                                          H2N·CH-CONH2
                                               (L)
```

Trimexautide

One should also mention the discovery in 1981 in Japan of the marked immunostimulating activities of compounds belonging to the same general family as the lipopeptides described above, the acyloligopeptides. The parent compound in this case was a natural metabolite of *Streptomyces olivaceogriseus*: D-lactoyl-L-alanyl-γ-D-glutamyl-(L)-*meso*-2,6-diaminopimeloyl-glycine (FK-156), of which several analogs were synthesized [60–63]. One of the most active analogs appears to be the acyltripeptide FK-565: heptanoyl-γ-D-Glu-(L-*meso*-α,ε-A_2pm(L)-

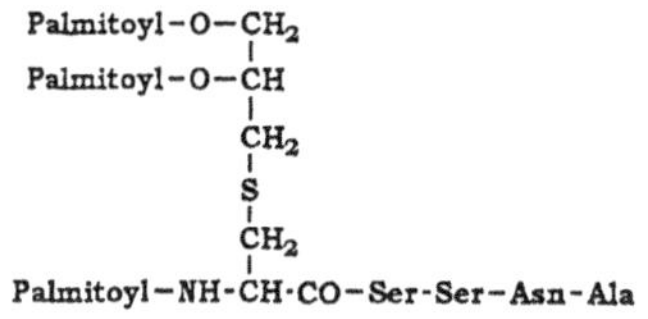

Palmitoyl-O-CH_2
Palmitoyl-O-CH
CH_2
S
CH_2
Palmitoyl-NH-CH-CO-Ser-Ser-Asn-Ala

Palmitoyllipopeptide

Ala-OH. Acyloligopeptides were found to enhance resistance against bacterial or viral infections in animals with either compromised or competent immune systems and FK-565 did exert in murine tumor models an anti-metastatic activity possible related to a stimulation of NK (natural killer) cells [64].

Puri et al. [65] have replaced 2,6-diaminopimelic acid (A_2pm) with lysine and introduced other structural modifications in the lauroyltetrapeptide molecule. One of the compounds thus synthesized is

$$\text{MeAla-D-Glu-NH}_2 \quad \text{Gly-Lys-NH.C}_{12}\text{H}_{25}(n),$$

which was shown to exert in mice a partial protective effect against infection with *Leishmania donovani*. Recently, Sidwell et al. [66] reported that a lipophilic desmuramyl dipeptide analog, the hydrochloride salt of octadecyl-D-alanyl-L-glutamine, enhanced the resistance of mice to the infection with murine cytomegalovirus and exerted a modest protective effect on the infection with mouse-adapted human influenza A virus.

Bessler and his coworkers have concentrated their research efforts on a totally different class of synthetic lipopeptides with immunoadjuvant activities, the chemical structure of which is derived from the lipoprotein of gram-negative bacteria such as *Escherichia coli* [67–70]. One example is provided by the palmitoylipopeptide that is identical to the N-terminus of the bacterial lipoprotein with respect to the *S*-glycerylcysteine and the peptide part, but differs in the composition of the fatty acid part (in this case palmitic acid). These lipopeptides exert potent adjuvant activities, when administered in combination with particulate or soluble antigens to various laboratory animals; they potentiate both the cellular and the humoral responses. These adjuvants, which are devoid of pyrogenicity (like some but not all muramylpeptides), can be coupled covalently to low-molecular-mass haptens (peptides or toxins) and the resulting conjugates are able to elicit high hapten-specific antibody titers in mice and rabbits. Conjugates containing B or T helper cell epitopes constitute novel synthetic vaccines that protect against viral infections by inducing the production of virus-specific antibodies. When coupled to T-lymphocyte cytotoxicity epitopes, the conjugates can induce the appearance of cytotoxic T lymphocytes, which *in vivo* may eliminate virus-infected cells. For instance, a vaccine against foot-and-mouth virus disease was prepared by coupling a lipopeptide adjuvant to suitable VP-1 segments of the viral protein: this preparation gave full protection to guinea pigs against challenge with the homologous virus type. The technique of *in vitro* immunization using a lipopeptide adjuvant can be used for the production of monoclonal antibodies. Lipopeptide-antigen conjugates were also used to produce *in vivo* (mice) and *in vitro* (human or murine cells) HIV-specific antibodies.

Substances of fungal origin

Numerous polysaccharides from various biological origins (yeasts, algae, bacteria, higher plants and especially fungi) have been investigated for anti-tumor and immunomodulating activities. Active polysaccharides have been shown experimentally to exert anti-tumor effects against allogenic, syngenic and even autologous tumors and the general consensus at the present time is that these antitumor activities are mediated to a large extent by stimulating effects on the immune system (macrophages, NK cells, cytokine induction).

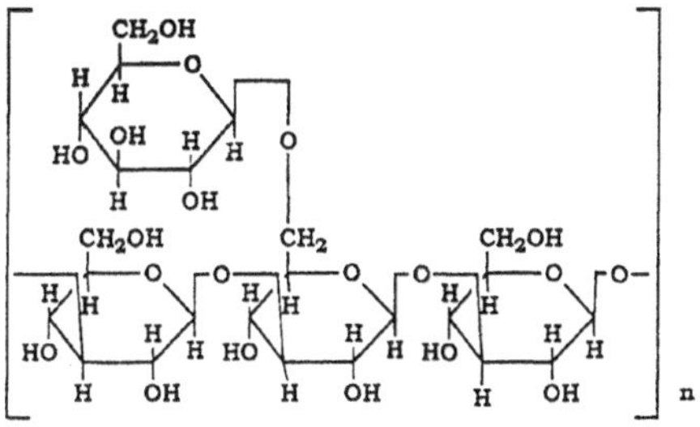

n = 700

Repeating unit of a glucan

Several recent reviews have been devoted to the biochemical and pharmacological characteristics of glucans endowed with immunostimulating activities [71–74]. These glucans consist of a linear backbone of β(1,3)-linked D-glucopyranosyl groups with varying degrees of branching from the C6 position. Reported degrees of branching (number of branches/main chain residues) should be considered average values. Branches are usually only a single glucose residue, although more than one glucose unit may be present in some glucans. The immunostimulating glucans which were first discovered (and are used clinically) in Japan, namely schizophyllan (an extracellular polysaccharide from the culture filtrate of *Schizophyllum commune)* and lentinan (a cell wall glucan from *Lentinus edodes*) present interestingly a triple helical conformation, as shown by viscosity measurements, NMR and X-ray diffraction studies. Triple helical parallel strands of glucan are hydrogen bonded via the C2 hydroxyls, with higher order structures arising through hydrogen bonding between C4 and C6 hydroxyls, the latter being present on the external surface of the triple helix. β-Glucans can also adopt single chain, and single helix forms as assessed by NMR spectroscopy. The physicochemical properties and biological activities of glucans can be modulated by chemical modification; for instance, insoluble yeast glucan was converted to water soluble glucan sulfate, or glucan phosphate, with a significant decrease of its toxicity (hepatosplenomegaly, granuloma formation, formation of microembolisms), thus providing safer compounds for parenteral administration. The immunostimulating activities of glucans have been demonstrated in many experimental models, in addition to those involving tumors: stimulation of resistance to bacterial, fungal and viral infections, stimulation of hematopoiesis (likely through induction of colony-stimulating factors), stimulation of wound healing. Enhancement of the production of various cytokines (interleukin-1, tumor-necrosis factor, interleukin-6) was reported *in vitro* and *in vivo*, as well as stimulation of cytotoxic T lymphocytes. Tumor regression in various animal models can be ascribed to vascular damage to tumor blood flow and to necrosis caused by T cells and local tumor-necrosis factor production. When injected into one of a pair of double-grafted tumors, in mice, soluble scleroglucan (an exopolymer produced by *Sclerotium glucanicum*) remained localized in the injected tumor but exerted shrinking effects on the distal tumor, presumably through stimulation of the immune response [75]. Scleroglucan was also found in the blood, liver and spleen when administered to normal mice, with significant amounts remaining in the liver and spleen for up to four weeks.

Clinical studies in man of various glucans have been performed essentially by Japanese investigators. Lentinan (500 kDa molecular mass, composed of a linear $\beta(1,3)$-glucan with $\beta(1,6)$-linked branches) is being used in association with various chemotherapeutic regimens, in the treatment of gastric, colorectal, and breast cancers. Schizophyllan (450 kDa molecular mass) is used in conjunction with radiotherapy in the management of cervical cancer. Increased disease-free survival and improved quality of life for the patients are reported both for lentinan and schizophyllan, but it does not appear that these interesting results have convinced oncologists outside of Japan about the usefulness of combining this immunostimulating approach with conventional anti-cancer chemotherapy.

In the United States, clinical development of a genetically modified (engineered) glucan from *Saccharomyces cerevisiae* PGG-glucan (Betafectin) is in progress. This compound is a triple-helical $\beta(1-3)\beta(1-6)$-linked glucose polymer with a unique branching structure [76]. Its superior *in vitro* biological activity is reflected in its increased glucan receptor-binding affinity as compared to naturally occurring β-glucan. Through β-glucan receptor binding, PGG activates macrophages, inducing a cascade of interactions mediated by the release of monocyte-derived cytokines. Administration of PGG to mice enhances resistance to bacterial and fungal infections and accelerates recovery from drug-induced neutropenia in mice treated with cyclophosphamide [77]. *In vitro* cells treated with PGG generated significantly increased levels of hydrogen peroxide and the glucan also stimulated nitric oxide production by rat neutrophils. Recently, double-blind placebo-controlled randomized phase II clinical studies examined the safety and efficacy of PGG-glucan in preventing postoperative infection in patients undergoing major thoracic or abdominal surgery [74]. The patients received multiple intravenous doses of PGG (from 0.5 mg/kg to 2.0 mg/kg). Infection incidence and severity was lower in the PGG-glucan-treated patient group, as was the incidence of postoperative antibiotic usage and duration of hospital stay. Two other independent studies reported efficacy of PGG administration to trauma patients: intravenous injections every 12 h decreased the incidence of pneumonia and sepsis in a group of 21 patients, in a randomized, double-blind placebo-controlled trial. Slight increases in interleukin-2 levels, but not tumor-necrosis factor levels, in the blood were transiently observed.

It is interesting to note that several studies have demonstrated the biological activity of glucans administered orally to laboratory animals. For instance, Nicoletti et al. [78] have shown that a glucan extracted from *Candida albicans* enhanced, upon oral administration of low doses to mice, their resistance to systemic infection with either *Staphylococcus* or *Candida albicans* and stimulated interleukin-2 production in the blood. This anti-infectious activity is of prophylactic nature, since treatment of the mice must be initiated ten days before the infectious challenge. The fact that a high-molecular-mass substance can exert immunostimulating activities when administered orally is not altogether surprising in view of what is presently known concerning the existence and activity of a system of mucosal immunity, such as the gut-associated lymphoid tissue and the lymphoid tissue associated with the upper respiratory tract.

Even more complex than the glucans described above, is the protein-bound glucan extracted from the edible mushroom *Coriolus versicolor* and used in clinical oncology in Japan under the name of Krestin, or PSK. Krestin (average molecular mass 100 kDa) is obtained by hot water extraction from the cultured mycelium; it contains 18–38% protein and the main fraction of the polysaccharide part is a β-glucan, with main chain 1–4 bonds and branches at the 3 and 6 positions, in a proportion of one per several residues. This interesting substance has been shown, in various experimental models, to exert significant antitumor activities mediated through stimulation of macrophages, monocytes, NK cells and various T lymphocyte populations; it also enhances the resistance of laboratory animals to bacterial, fungal and viral infections and exerts a chemopreventive effect against carcinogenesis [79]. In models with various methylcholanthrene-induced tumors, there was a good correlation between the antitumor activity and the antigenicity of the tumor. There is clear evidence that Krestin is mostly effective in host-tumor systems, in which tumor-induced immunosuppression can be observed. The compound has also been shown to exert antimetastatic effects [80] and to induce the production of monokines such as tumor-necrosis-factor-α and interleukin-1β. Activities are exerted following various routes of administration, including the oral route.

Krestin is extensively used clinically in Japan, as an anticancer immunotherapeutic agent, in association with conventional anticancer chemotherapy (for instance 5-fluorouracil or mitomycin). Randomized controlled trials of the compound have been performed, following primary tumor resection, in patients with gastric cancer [81] and in colorectal cancer patients [82], comparing the disease-free periods and the overall survival in patients receiving chemotherapy alone and in those treated in addition with Krestin (3 g/day per os for prolonged periods). In both sets of trials, the disease-free period and the five-year survival were significantly increased in the patients receiving the combination in comparison with those treated with chemotherapy alone. Side effects (diarrhea, nausea) were rarely observed. A plausible explanation of the clinical efficacy of Krestin is that this immunostimulant antagonizes humoral immunosuppressive factors that are produced in tumor-bearing hosts: quality of life and resistance to infections are improved and antimetastatic effects may also be observed.

Immunostimulating activities have also been reported for polysaccharides extracted from higher plants and, in some cases, have led to the pharmaceutical use in certain countries of complex extracts of plants such as *Echinacea purpurea* or *Viscum album* [83, 84]. These preparations are administered, generally by the oral route, to young children or to elderly individuals to increase their natural resistance to bacterial or viral infections of the upper respiratory tract; their widespread use in some areas gives indirect indication of their efficacy in this context. The active ingredients in these complex preparations appear to be water-soluble polysaccharides, such as xyloglucan, arabinogalactan and pectin, and probably also glycoproteins as well as flavonoids.

Immunostimulating peptides have, on the other hand, been isolated from soybean, and the following amino acid sequences have been reported for the active peptides: Ala-Glu-Ile-Asn-Met-Pro-Asp-Tyr, Ile-Gln-Gln-Gly-Asn, and Ser-Gly-Phe-Ala-Pro [85].

Substances of mammalian origin

Strictly speaking, the thymus-derived hormones (thymic peptides) and tuftsin (a tetrapeptide derived from the Fc portion of the immunoglobulin molecule) should not be discussed here, since they are not the exogenous substances (to which the present review restricts itself), but can be considered as natural endogenous immunomodulators. They will however deserve brief mention, inasmuch as several such peptides are prepared by chemical synthesis or extracted from bovine thymus and used in various clinical applications.

The nature and therapeutic uses of synthetic thymic peptides have been reviewed fairly recently [17, 18, 86]. The pentapeptide thymopentin (TP-5), corresponding to the Arg-Lys-Asp-Val-

Tyr sequence of thymopoietin has found clinical use in the treatment of autoimmune diseases, such as rheumatoid arthritis. Thymosin $\alpha1$ is a 28-amino-acid recombinant peptide derived from thymosin fraction V: current trials are in cancer immunotherapy and in the treatment of chronic hepatitis B virus infection [87]. Thymic humoral factor is the synthetic octapeptide leucyl-glutamyl-aspartyl-glycyl-prolyl-lysyl-phenylalanyl-leucine; in an open nonrandomized pilot clinical study of patients with immunological defects associated with anticancer chemotherapy or severe infection, thymic humoral factor augmented lymphocyte populations in 70% of the patients, with a trend toward CD4/CD5 ratio normalization [88]. It is also in phase II clinical trials, in combination with azidothymidine, in HIV-seropositive individuals.

Thymomodulin and thymostimulin are purified extracts from calf thymus and are rather extensively used clinically, especially in Italy. Thymomodulin contains a mixture of low-molecular-mass (less than 10 kDa) proteins. It has been claimed to improve the course of chronic infectious diseases, such as chronic bronchitis and recurrent pediatric respiratory infections [89]. Braga et al. have reported restoration of polymorphonuclear leukocyte functions in elderly subjects treated with thymomodulin [90]. Thymostimulin is another bovine thymic extract that has been reported to show efficacy, following repeated intramuscular injections, in the treatment of recurrent respiratory infections in small children [91]. Periti et al. [92] have conducted a prospective controlled multicenter study showing the benefit of chemoimmunoprophylaxis (the antibiotic cefotetan and thymostimulin) in patients undergoing colorectal surgery: abdominal abscesses and respiratory tract infections were significantly less frequent in patients submitted to this chemoprophylactic combination than in those receiving a placebo.

The tetrapeptide tuftsin, Thr-Lys-Pro-Arg, is known to enhance phagocytosis, bactericidal, and tumoricidal activities of macrophages and to enhance the release of interleukin-1 and tumor-necrosis factor; numerous analogs of tuftsin have been synthesized (polytuftsin, glycotuftsin derivatives, O-glycosylated tuftsins), but as far as we know, very little has been achieved so far in terms of delineating the potential clinical usefulness of these molecules, which seems surprising when considering that the biological activities of this natural tetrapeptide were first described almost 30 years ago.

Switching from the thymus and the immunoglobulin molecule, we may turn now to the mammary gland as a source of immunostimulating substances. Casein has been described as a milk protein with diverse biologic consequences [93]: casein fragments exert a rather wide range of immunopharmacological, hematological and neuropharmacological effects. Several years ago, our team observed that enzymic fragments obtained from human casein (maternal milk being frequently and advantageously man's first food) exhibited immunostimulating activities, such as stimulation of phagocytosis by macrophages and enhancement of resistance of mice to a bacterial infection [94]. The purification, sequence, synthesis and further biological activities of an immunostimulating hexapeptide (Val-Glu-Pro-Ile-Pro-Tyr) from human β-casein were described [95]. Furthermore, the presence of some other biologically active short peptides has been detected in human as well as in cow milk caseins, such as Gly-Leu-Phe and Leu-Leu-Tyr [96, 97]. Whether such peptides, released from milk by enzymes in the gut, do play a role in the natural resistance of breastfed infants to infections (in addition to the obvious role of maternal antibodies) and could find therapeutic applications remains to be investigated.

Vitamins

As stated in a comprehensive review on vitamin A status and its relationship to immunity [98], vitamin A (retinol), a fat-soluble vitamin, plays an essential role in several biological processes, many of which concern growth, cellular differentiation and cell-cell or cell-substrate interactions. Vitamin A is important in maintaining the functional integrity of epithelial and mucosal surfaces and in the production of mucous secretions. Clinical and experimental data have shown that vitamin A deficiency is associated with impaired resistance to infections. Vitamin A supplementation can decrease morbidity in preschool children in populations that are at high risk for vitamin A deficiency; some of the benefit of vitamin A is due to restoration of normal epithelial barriers, resulting in improved resistance to respiratory and gastrointestinal infections but stimulation of the immune system must also be taken into account. For instance, moderate increases in dietary vitamin A were reported to enhance host resistance to infection with bacterial or fungal pathogens and experiments in chicks, rats and mice have shown that administration of nontoxic doses of retinol or retinoic acid stimulates phagocytic functions of macrophages and antibody production as well as NK cell activity and antibody-dependent cell-mediated cytotoxicity. Vitamin A can function as an adjuvant in immunization experiments with tetanus toxoid, cholera toxin, pneumococcal polysaccharide or other antigens. Vitamin A and retinoic acid have been also shown to stimulate the rejection of certain immunogenic tumors in mice [99]. Several studies on antibody production and phagocytosis support a role of retinoids in immunostimulation in animals in which vitamin A nutritional status is normal, so that retinoids can be considered as *bona fide* immunostimulating agents, and not just immunorestoring agents in deficient hosts. Even in developed countries, vitamin A dietary intake may in many individuals (especially women using low-fat diets and poorly nourished elderly subjects) be definitely suboptimal and supplementation (at least 3000 IU/day) is advisable to avoid the immunosuppressive effects of this deficiency.

Turning now to another vitamin, it is known that, in addition to its established role as a calcium regulating factor, the active metabolite of vitamin D, namely the sterol 1,25-dihydroxyvitamin D_3, exerts anti-proliferative, predifferentiating and immunosuppressive properties. Lemire has shown [100] that this compound inhibits the production of interleukin-2 and interleukin-12 as well as that of interferon γ and prevents *in vivo* the development of spontaneous or induced models of auto-immunity; Alroy et al. [101] have reported that other cytokines, such as interleukin-4 and GM-CSF (granulocyte-macrophage colony-stimulating factor), seem to be downregulated by vitamin D3 in T lymphocytes, in which promoters have the same binding sites for activators. The question then arises about a possible physiological role of vitamin D in the immune system; in this respect, it is worth noting that suboptimal levels of vitamin D are frequently observed in elderly individuals, due to a deficient diet and insufficient exposure to sunlight.

Vitamin E (α-tocopherol) has also been claimed to be endowed with immunomodulating activities: Shkalar et al. [102] have reported that treatment with vitamin E of hamsters, in which buccal pouches were repeatedly painted with dimethylbenzanthracene, inhibited the formation of the gross tumors observed in the control animals; instead, the small tumors seen in the treated hamsters were densely infiltrated with lymphocytes and macrophages. On the other hand, Wang and Watson [103] have reviewed the available evidence on the immunosuppressive effect of excessive alcohol consumption and the immunoenhancing activities of vitamin E and proposed that supplementation with this vitamin could provide a useful therapeutic approach to enhance the resistance of alcoholics to infections.

Antibiotics

The concept of a possible interaction of clinically effective antimicrobial agents with various functions of the immune system is presently the subject of intensive research, which has been lucidly reviewed by Labro [104]. Obviously antibiotics capable of exerting immunosuppressive effects (such as chloramphenicol and rifampicin) should not be used in immunocompromised patients, whereas antibacterial agents endowed with an immunostimulating activity, in addition to their antimicrobial activity, could represent a most welcome therapeutic strategy in several infectious situations. Apart from such a direct potentiating effect on immune reactions it is clear that the alterations induced in bacterial metabolism and morphology by most antibacterial agents, at concentrations inferior to their minimal inhibitory or bactericidal values, may facilitate phagocytosis by macrophages and polymorphonuclear leukocytes; furthermore, several antibiotics (such as the macrolides) have the ability to penetrate phagocytes in infections caused by intracellular bacteria (such as *Listeria monocytogenes*). On the other hand, a great number of *in vitro* experiments have shown some antibiotics to exert stimulating or inhibitory activities on several functions of phagocytic cells and T and B lymphocytes, including production of cytokines. Whether such immunomodulating activities have a real significance in *in vivo* situations is open to question, inasmuch as the *in vitro* effective concentrations are quite different from the blood levels reached following *in vivo* administration of effective antibacterial doses. Nevertheless, it is now established that some antibacterial agents are indeed truly capable to directly modify the immune responses. As mentioned by Labro [104], cefpimizole was the first molecule to be proposed as a *bona fide* immunomodulating antibiotic, capable of stimulating monocytes to release neutrophil-activating factors. More recently, the oxyimino-amino-2-thiazolyl cephalosporin cefodizime has been shown to enhance and/or restore several immune functions *in vitro* and *ex vivo* and to be effective in various experimental models of infections, in which not only sensitive but antibiotic-resistant bacterial species and immunocompromised as well as normal hosts were used [105]. The mechanisms of the immunostimulating activities of cefodizime are still unclear, but the chemical structure necessary for these activities has been determined as the thiothiazolyl moiety present at position 3 of the cephem nucleus. Indeed, the thiazole ring appears to be able to confer immunomodulating activities to several molecules, as will be shown in the next section of the present review. Let us also mention the observation that ciprofloxacin (a quinolone antibiotic) administered orally to healthy subjects induced the production in their serum of a factor potentiating the chemotactic activity of polymorphonuclear leukocytes [106]. Clarithromycin, a broad spectrum macrolide, administered to healthy volunteers and chronic bronchitis patients, exerted immunostimulating effects, as shown by increased phagocytosis and intracellular killing by their circulating leukocytes [107].

Miscellaneous synthetic molecules

We shall review here a rather large number of synthetic compounds exerting immunostimulating activities (often leading to clinical applications) but the chemical structures of which could not *a priori* allow prediction of such activities, in marked contrast with molecules originating from microbial, fungal or mammalian sources. Several years ago, Georgiev [16] published a comprehensive review on synthetic immunomodulating agents, but we shall restrict the present review to those synthetic compounds exhibiting a sufficiently wide range of *in vivo* immunostimulating activities, without mentioning the many molecules the effects of which on the immune system were documented solely by some *in vitro* tests. Such *in vitro* immunopharmacological systems are so sensitive that many compounds of no practical potential may exert various types of effects that are not confirmed by *in vivo* experiments. We must also stress the fact that many of the compounds from natural sources described in the preceding sections are obtained by chemical synthesis, but their structure is identical to or inspired from the structure of natural molecules.

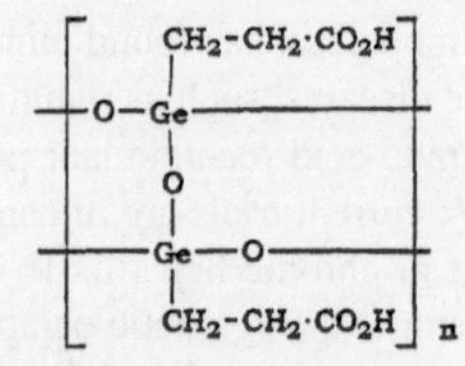

Propagermanium

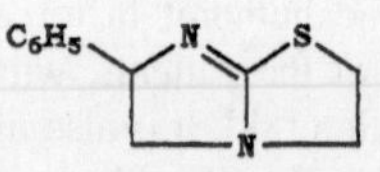

Levamisole

Compounds from mineral chemistry. Compounds from mineral chemistry have been shown to exert immunostimulating activities. For instance, dietary suplementation with selenium (as sodium selenite) restores in mice the age-related decline in immune cell function, probably through an increase in the number of high-affinity interleukin-2 receptors on T lymphocytes [108] and enhances internalization of interleukin-2 [109]. Supplementation with Se *in vivo* or *in vitro* results in an earlier expression of high-affinity interleukin-2 receptors, whereas Se deficiency results in a delayed expression of lower numbers of receptors [110].

Propagermanium, i.e. 3-oxygermylpropionic acid polymer, was shown, upon oral administration in mice, to enhance resistance to infection with herpes simplex virus type 1, through stimulation of cytotoxic T lymphocytes and NK cells and the same compound enhanced interferon production in BCG-sensitized and in influenza-virus-infected mice [111]. Clinical studies are under way, following reassuring toxicological data in laboratory animals and in healthy volunteers [112].

AS-101, or ammonium trichloro(dioxyethylene-*O-O'*)tellurate was developed in Israel and was shown to stimulate the production of a variety of cytokines. Phase I clinical trials in cancer patients showed an enhancement of the secretion of tumor-necrosis factor, interferon and interleukin-2. This compound was shown to have radioprotective activities, when injected into mice prior to sublethal or lethal doses of radiation and to protect mice from hematopoietic cell damage caused by cyclophosphamide [113].

Tetramisole and levamisole. The imidazothiazole tetramisole and its levorotatory isomer levamisole have been known for many years for their marked effectiveness as anthelminthics (especially against nematode infestations, such as ascaridiasis) in man and domestic animals. In 1971, Renoux and Renoux [114] observed that mice immunized with a poorly immunogenic

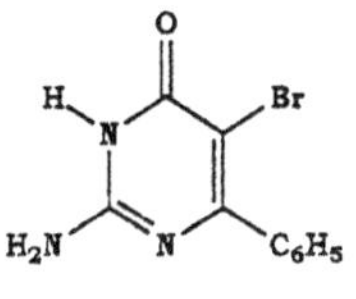

Bropirimine

anti-Brucella vaccine were nevertheless protected against challenge with live Brucella bacilli if they had been treated with levamisole or tetramisole at the time of immunization. It was the first time that a simple synthetic compound was shown to potentiate an immune response, in other words to behave as an adjuvant. The mechanisms of the immunopharmacological acivities of levamisole have been investigated by many immunologists and thus far the picture is still quite complex; stimulation of recruitment and functions of macrophages, monocytes and various classes of T lymphocytes, especially in conditions of suboptimal immune responses, can be referred to when dealing with the global effects of levamisole. But what is more important is that, after many rather inconclusive clinical trials of levamisole in a wide variety of illnesses possibly associated with impaired immune functions (recurrent infections, rheumatoid arthritis, Crohn's disease, leukemia, aphtous stomatitis, etc.), this drug is now used, in combination with the anticancer agent 5-fluorouracil, in the treatment of colorectal cancer in patients with stage C (Duke's classification) disease and the results in terms of prolongation of disease-free interval and overall survival have been so superior to those observed in patients receiving 5-fluorouracil alone, that the Food and Drug Administration has approved the use of levamisole and 5-fluorouracil as the standard therapy for stage C colon carcinoma [115, 116].

The mechanism(s) by which levamisole exerts this beneficial effect are still inadequately elucidated. Kimball [117] states that this agent may be able to promote antitumor responses by increasing the production of cytokines that can be directly cytotoxic to tumors (in synergy with 5-fluorouracil) and that can promote and sustain cell-mediated host responses to cancer cells. In addition, levamisole may serve to render certain tumors more susceptible to immune responses of the host, by increasing the exposure of their surface recognition molecules, but these are mainly hypotheses. On the other hand, recent reports involving the use of levamisole in the treatment of steroid-dependent pediatric nephrotic syndrome demonstrated encouraging results [118], a somewhat paradoxical finding inasmuch as the alternative to corticosteroids for the treatment of this syndrome has generally been immunosuppresive agents, such as cyclophosphamide.

Nucleic acid derivatives. Nucleic acid derivatives provide another source of immunostimulating agents, as exemplified by pyrimidinones, guanosine derivatives, inosine-5'-methylmonophosphate and hypoxanthine derivatives. The immunostimulating properties of pyrimidinones can be summarized as follows [119]: (a) induction of interferon-α, tumor-necrosis factor α and interleukin-2; (b) stimulation of the expression of receptors to interleukin-2; (c) induction of proliferation and differentiation of B lymphocytes; (d) stimulation of macrophage and NK cell activities; (e) restoration of levels and function of T_4 lymphocytes in immunosuppressed mice. All these studies were performed *in vivo*, very few direct effects being demonstrable *in vitro*. It appears that the primary effector cells activated by pyrimidinones are the macrophages and the NK cells. Of all the pyrimidinones synthesized, the most interesting one is bropirimine, which exerts in animal models a wide variety of antiviral

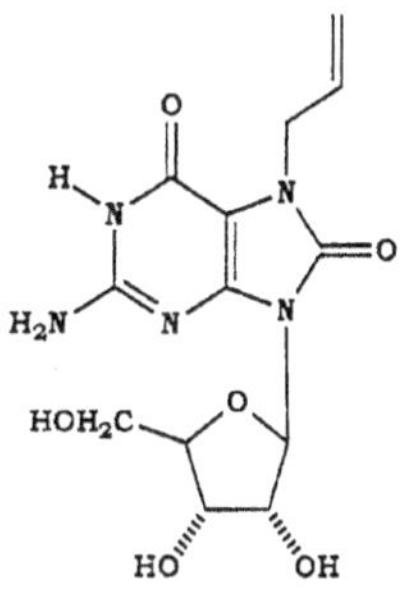

5 (loxoribine)-Syn 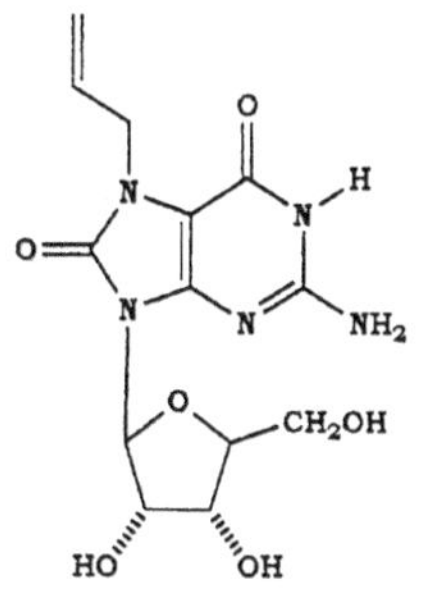 **5 (loxoribine)-Anti**

and antitumor activities. The antiviral activities of this agent (in a large number of animal models) can be explained by its interferon-inducing property as well as the stimulation of NK-cells. The antitumor activity of bropirimine, on the other hand, demonstrated in several syngeneic tumor models in mice and rats, has been linked to its induction of tumor-necrosis factor-α. The most demonstrative antitumor activity of bropirimine (and the pyrimidones in general) has been in models involving its association with chemotherapeutic agents (cyclophosphamide, doxorubicin, mitomycin C, cisplatin, etc.), especially with those exerting no suppressive effects on NK cells functions. As a result of these extensive observations on the antitumor potential of bropirimine, clinical studies have been performed with this agent in bladder cancer patients, to whom the drug has been administered by the oral route [120]. The phase I trial indicated that bropirimine had significant single-agent activity in carcinoma *in situ* of the bladder, a good number of patients demonstrating complete or partial responses.

It appears too early to demonstrate that, in bladder cancer patients, the efficacy of bropirimine approaches that of BCG, administered intravesically: if this was indeed the case, bropirimine might have several advantages over BCG (more convenient route of administration and less side effects); the total daily dose of 4.5 g for three consecutive days each week was well tolerated. Trials of combination of bropirimine with BCG are also under way. Phase II trials also showed possible efficacy of bropirimine in nodular lymphoma and hepatoma, but renal cell carcinoma was unresponsive to this therapy.

Guanosine derivatives have been extensively studied with respect to their immunomodulating activities. Anderson and Capetola [121] described the activities of 7-allyl-8-oxoguanosine which, in various experimental models, exhibited adjuvant-like effects in enhancing antibody production in response to several antigens, even in the absence of T lymphocytes, while being devoid of pyrogenicity and unable to exacerbate autoimmune manifestations. More recently, several related analogs of 7-allyl-8-oxoguanosine (loxoribine) have been synthesized by Chen et al. [122]: 2',3'-ketals of loxoribine display a significant activity on their own, apparently without being cleaved to the free nucleoside.

Hypoxanthine derivatives, such as ST-789, a 9-pentylarginine-hypoxanthine compound, were shown to enhance resistance of immunosuppressed mice to infections and to exert antitumor activities, likely mediated through stimulation of NK cell activity and of production of interleukin-6, a pleiotropic cytokine [123].

Inosine-5'-methyl monophosphate was described as a thymomimetic immunomodulator, acting directly on T lymphocytes (induction of T-cell differentiation markers and interleukin-2 receptors in human pro-thymocytes), increasing T-cell dependent antibody formation and delayed type hypersensitivity (DTH)

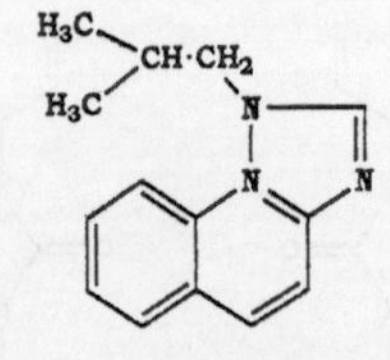

Imiquimod

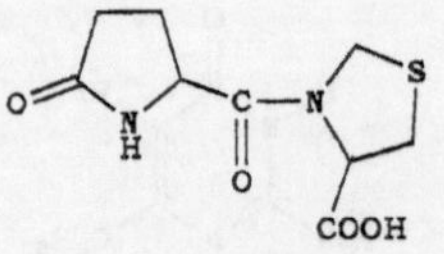

Pidotimod

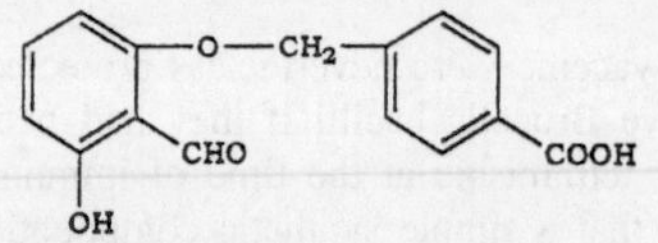

Tucaresol (substituted benzaldehyde)

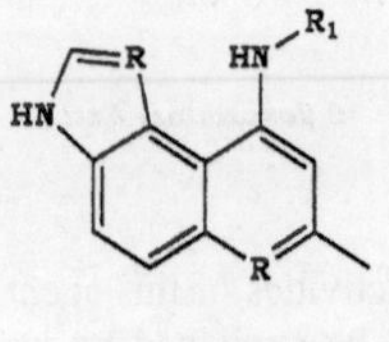

	R_1
a	3-methoxyphenyl
b	4-(cyclohexylmethoxy)phenyl
c	4-butoxyphenyl
d	4-butylphenyl
e	4-cyanophenyl
f	cyclohexyl

Pyrazolo[3,4-f]quinoline derivatives

responses, enhancing resistance of laboratory animals to bacterial and viral infections and exerting antitumor activities [124]; it is active both orally and parenterally.

Unrelated structures. A number of unrelated chemical structures have been shown to provide new leads for the development of promising immunostimulating agents; One of those is imiquimod (R-837 or S-26308), i.e. (1-(2-methyl propyl)-1*H*-imidazo(4,5-c)quinolin 4-amine), which has been found to be an effective antiviral and antitumor agent in animal models, without exerting such activities *in vitro* but acting through the induction of various cytokines, such as interferon α, tumor-necrosis factor-α, interleukin-1 α and β, interleukin-6 and interleukin-8, monocytes being largely responsible for the cytokines produced [125]. Imiquod protects guinea pigs from infection by herpes simplex virus when administered intravaginally, intramuscularly, intraperitoneally, and orally, and reduces recurrences caused in guinea pig by latent infection with herpes simplex virus [126]. The same agent is active prophylactically and therapeutically against cytomegalovirus infection and arbovirus infection in mice. On the other hand, imiquimod has been shown to inhibit the growth of a number of murine tumors, including a tumor induced by a chemical carcinogen. Phase I studies in healthy adult males indicated that imiquimod administered orally induced detectable serum concentrations of interferon α. Imiquimod also exerts adjuvant activities versus a herpes simplex virus type 2 glycoprotein vaccine in guinea pigs [127]. Recently, the cellular requirements for cytokine production in response to imiquimod have been characterized [128]: the cell population responsible for the majority of cytokine release in human peripheral blood monocytes in response to this agent appears to be E rosette-, CD14+, CD 36+, HLA-DR+ monocyte.

Pyrazolo(3,4-f)quinoline derivatives have been reported to exert *in vivo* anti-infectious immunostimulating activities, using a model infection of mice with a pathogenic strain of *Escherichia coli* and structure-activity relationships in this chemical series have been precisely determined [129−131].

Pidotimod, a synthetic peptide (3-L-pyroglutamyl-L-thiazolidine-4-carboxylic acid) deserves special mention in the present review, since it is already on the pharmaceutical market in Italy as an orally effective immunostimulating drug. In human subjects, pidotimod has been shown to stimulate the phagocytic activity of peripheral blood polymorphonuclear cells and the proliferative response of lymphocytes to T-cell mitogens and this effect, together with an enhancement of interferon-γ production, was especially noticeable when the drug was administered orally to elderly individuals [132]. Pidotimod was also shown to stimulate natural killer (NK) cell activity in mice [133] and to stimulate residual B-cell immunity in nude mice [134]. An increased rate of survival in *Streptococcus pneumoniae*-infected rats was demonstrated when the animals were treated intraperitoneally with pidotimod [135]. Studies in 40 young healthy volunteers treated orally with pidotimod showed, in *ex vivo* studies, a significant enhancement of neutrophil functions, such as chemotaxis and phagocytosis [136]. Clinical studies of pidotimod have shown that the drug exerts a prophylactic effect on recurrent acute tonsillitis in children: patients treated with this immunostimulant had a significant reduction in the frequency of acute episodes and in symptomes (cough and dysphagia) as well as recourse to antibiotics, when compared to children receiving a placebo [137]. There is also some evidence that treatment with pidotimod (in this study by the intramuscular route at a dose of 200 mg twice a day) counteracted the transient by significant immunosuppressive effects of surgical interventions [138]. Altogether and based on the numerous experimental and clinical studies performed in Italy, pidotimod can be considered as an effective immunostimulating drug, more particularly in infectious situations.

Schiff base-forming molecules. A quite novel approach to therapeutic potentiation of the immune system has recently been reported by Rhodes et al. [139]. These authors have shown that small Schiff base-forming molecules, such as tucaresol, can substitute for the physiological donor of carbonyl groups (constitutively expressed on antigen-presenting cells) and thus interact with the amines on T-cell receptors, thereby providing a costimulatory signal to T lymphocytes, by activating transport of Na^+ and K^+, leading to phosphorylation of key signalling proteins. Tucaresol enhances CD4 T-cell responses, selectively favoring a Th-1-type profile of cytokine production. *In vivo*, tucaresol potently enhances CD4 T-cell priming and CD8 cytotoxic T-cell priming to viral antigens. Enhancement of CD8 cytotoxic T-cell priming to influenza virus peptide antigens by parenterally administered tucaresol was demonstrated and, at immunopotentiating doses, this agent was shown to be therapeutically effective in a murine model of cytomegalovirus infection as well as in a model of syngeneic tumor growth (murine colon adenocarcinoma). Oral administration of tucaresol to mice caused an increase in production of interleukin-2 and interferon-γ and a decrease in the production of interleukin-4 and interleukin-6.

A distinct therapeutic advantage of small Schiff base-forming molecules over cytokine therapy is likely to be the avoidance of toxic effects mediated through cytokine cascade events. Promotion of a Th-1-like profile of cytokine production and selective enhancement of cell-mediated responses can be therapeuti-

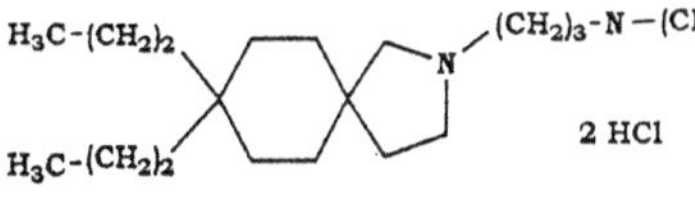

$$H_3C-(CH_2)_2 \;\; (CH_2)_3-N-(CH_3)_2 \;\; H_3C-(CH_2)_2 \;\; 2\,HCl$$

Azaspirane SK and F 105 685

cally favorable against intracellular pathogens (viruses, mycobacteria, protozoal parasites).

Stimulators of nonspecific suppressor cell activity induction. Finally, it is worth mentioning that compounds stimulating the induction of nonspecific suppressor cell activity are capable of exerting interesting immunopharmacological effects: one example is provided by the azaspirane SKF 105685 (*N,N*-dimethyl-8,8-dipropyl-Z-azaspiro(4,5)decane-2-propanamine dihydrochloride). This compound exerts therapeutic activities in animal models of autoimmune disease [140] while, at the same time, stimulating myelopoiesis and enhancing survival from lethal irradiation in mice [141].

The special case of adjuvants

Several compounds described in this review have been reported as exerting a particular type of immunostimulating activity, namely an adjuvant activity, that is the power of enhancing significantly the humoral and/or cell-mediated immune responses to particulate or soluble antigens. This is the case of monophosphoryl lipid A, of muramyldipeptide (MDP) and its derivatives such as murabutide, of temurtide (threonyl-MDP), of MTP-PtdEtn (muramyl-tripeptide phosphatidylethanolamine), of the lipopeptides pimelautide and trimexautide and of other lipopeptides such as palmitoyl-lipopeptides.

An adjuvant activity is of distinct practical interest for the development of human and veterinary vaccines, inasmuch as the highly purified antigens that are presently incorporated in vaccines tend often to be poorly immunogenic *per se*. Furthermore, some individuals are weak responders to certain vaccines, as in the case of hepatitis B vaccine or of influenza vaccines in elderly people.

Adjuvants are also very useful when attempts are made to immunize laboratory animals with various substances to prepare high titer antibodies against these molecules for research purposes. Freund's complete adjuvant (i.e. killed mycobacteria in a water-in-oil emulsion) provided biochemists and immunologists, almost 60 years ago, with a powerful tool, which is still used today and has led to the discovery of the adjuvant activities of muramyldipeptides, but the use of this adjuvant is restricted to laboratory animals, in view of its multiple side effects.

At the present time, the most common adjuvants for human vaccines are aluminium hydroxide and aluminium phosphate, although calcium phosphate is also being used in some vaccines and may have some advantages over aluminium-based adjuvants. Aluminium adjuvants have an extensive record of efficacy and safety and any new type of adjuvant will have to be endowed with distinctive superior properties to be considered for development. The aluminium adjuvants seem to act mainly by depot formation, allowing the slow release of antigen and thereby prolonging the time for interaction between antigen and antigen-presenting cells as well as lymphocytes. Contrary to the other potential adjuvants mentioned in this review, aluminium hydroxide or phosphate and calcium phosphate do not exert the overall immunostimulating activities that the former exhibit.

Several excellent reviews have been devoted recently to the topic of adjuvants for human and veterinary vaccines [142−145] and we refer to them for details. It is, however, important to note that several adjuvants have been shown to selectively modulate the immune response to elicit humoral and/or cellular immune responses. With their use, this immune response can be either an MHC class I or an MCH class II response. An MHC class I response is usually directed against intracellular pathogens (viruses, for instance) leading to the induction of cytotoxic T lymphocytes (CTL); it is normally not observed with protein or peptide antigens, but some adjuvants are capable of promoting this response with the latter. The MHC class II response is usually elicited against protein antigens or inactivated microorganisms and leads to antibody production: most adjuvants are efficient in eliciting a MHC class II response. We have mentioned earlier in the text that two lipopeptides of very close chemical structure, pimelautide and trimexautide, are capable when conjugated with an HIV-1-derived peptide, to stimulate a specific cytotoxic T lymphocyte response in the case of trimexautide and an antibody response, in the case of pimelautide. This example also illustrates the potential usefulness of built-in adjuvants, i.e. chemical conjugation between an antigen and an adjuvant.

Adjuvants can also modulate the immune response to different T-helper cells (Th-1 and Th-2). Stimulation of the Th-1-type response leads to a cell-mediated immune response and production of IgG2a antibodies (in mice), whereas stimulation of the Th-2-type response leads to the production of IgG1 and IgE antibodies. MDP, temurtide, monophosphoryl lipid A stimulate the Th-1-type response; aluminium adjuvants are known to stimulate a Th-2-type response.

As things stand presently in a fast moving field, temurtide (threonyl-MDP) may well be one of the most promising adjuvants for possible use in human vaccines, as indicated above in the section devoted to this compound [42], probably because of its effective formulation in a stable oil-in-water emulsion. Furthermore, temurtide appears to be devoid of other immunostimulating activities beside adjuvanticity: it is nonpyrogenic and does not stimulate resistance to infections. This selectivity of action pleads in favor of a minimal potential of causing side effects. As stated by Gupta et al.[143] the search for vaccine adjuvants is guided by a balance between toxicity and true adjuvanticity and we have seen in this review that many immunostimulants do exert some toxic effects. Lack of toxicity is an essential prerequisite for vaccine adjuvants since most vaccines are administered to young children.

There are several other possible strategies for the design of adjuvants which have not been mentioned in our review on immunostimulants and which we shall briefly describe here, based mainly on the paper by Gupta et al. [145] and a few other publications. Saponin, for instance, isolated from the bark of *Quillaja saponoria*, can be purified and give an adjuvant-active fraction called Quil A, which is used in several veterinary vaccines. But its hemolytic activity and local reactions make it unsuitable for human vaccines. Further purification of Quil A led to the discovery of QS 21, a water-soluble substance with potent adjuvant activities and minimal toxicity, which in mice elicits a Th-1-type response. Clinical studies in man have established the adjuvanticity and acceptable toxicology of QS 21 [146]. Iscom (immune-stimulating complexes) is the name given to noncovalently bound complexes of Quil A adjuvant, cholesterol and amphipathic antigens: in the Iscom, the antigens are attached as multimers to a 40-nm cage-like particle with a built-in adjuvant; the antigens in Iscoms are swiftly transported from the injection site to the draining lymph node. Iscoms, which stimulate both humoral and cell-mediated immune response to amphipathic antigens (hepatitis B virus surface antigen, influenza virus hemagglutinins, herpesvirus glycoproteins, for instance) are now used in veterinary vaccines but have not yet been approved for human vaccines.

Another approach to adjuvanticity of vaccines is the encapsulation of antigens in liposomes; furthermore, substances such as monophosphoryl lipid A or MDP show enhanced adjuvanticity and reduced side effects when encapsulated in liposomes. Adjuvanticity of liposomal formulations is probably due to depot formation at the site of injection and efficient presentation of the antigens to macrophages.

Biodegradable polymer microspheres containing vaccine antigens receive also much attention due to their potential as a vehicle to target the antigen to antigen-presenting cells and for the controlled release of antigens, allowing the reduction of the number of doses for primary immunization. The microspheres are made from various polymers such as poly(lactic)/glycolic acid.

Gjata et al. [147] have described the adjuvant activity of polar glycopeptidolipids from *Mycobacterium chelonae* on the immunogenic and protective effects in mice of an inactivated influenza virus vaccine.

As could be expected, several interleukins have been examined from the point of view of their possible adjuvant activity on vaccines. For instance, Alfonso et al. [148] have described the adjuvant activity of interleukin-12 in the vaccination of mice against the protozoan parasite *Leishmania major*. In leishmaniasis, protection requires the induction of Leishmania-specific CD4+T helper lymphocytes; mice vaccinated with a soluble leishmanial antigen and injected simultaneously with interleukin-12 were resistant to a subsequent normally lethal challenge with the parasite. Duits et al. [149] have shown the adjuvant activity of interleukin-6 entrapped in liposomes and injected into mice immunized with a heat-shock protein used as a model antigen. The multiple, pleiotropic activities of most known interleukins may however represent an obstacle to their routine use as adjuvants for vaccines. A more subtle and probably safer approach, but applicable only to live microorganisms, is engineering to express a particular cytokine, thus enhancing the immune response, as demonstrated very recently by Murray et al. [150] who engineered the BCG strain of tubercle bacillus to express several murine cytokines, such as interleukin-2, interleukin-4, interleukin-6 or interferon-γ. There is an obvious need for a more potent BCG vaccine than the one presently used, in the face of the current threat posed by widespread infection of certain populations with antibiotic-resistant strains of *Mycobacterium tuberculosis*.

We can also mention the demonstration by Dempsey et al. [151] that vaccination of mice with a recombinant model antigen (hen egg lysozyme) fused to the C3 component of complement caused a 10000-fold higher humoral and cell-mediated immune response than when the antigen alone was administered, enabling the authors to conclude that the third complement protein, which is a molecular adjuvant of innate immunity, can profoundly influence an acquired immune response.

Thus far most experimental data on adjuvants and adjuvant activity have been obtained in systems in which both the antigens and the adjuvants are administered by parenteral routes, inasmuch as most human or veterinary vaccines are ordinarily injected either subcutaneously or intramuscularly, with the notable exception of live poliovirus vaccines which are administered orally. Nevertheless, with the present recognition of the importance of mucosal immunity involving lymphoid cells present on diverse mucosal surfaces (gut-associated lymphoid tissue, bronchi-associated lymphoid tissue, Langerhans-dendritic cells in the skin), the trend will be to preferentially stimulate mucosal immunity with vaccines, especially when the primary site of infection in the natural pathological process is indeed a mucosal surface (skin, respiratory tract, digestive tract). So-called mucosal vaccines are probably going to be developed in the near future against respiratory syncytial virus, influenza virus, adenoviruses, measles and rubella viruses, caries-causing streptococci, *Hemophilus pertussis*, *Shigella*, pathogenic *Escherichia coli* strains, *Vibrio cholerae* and possibly HIV-1 and HIV-2. Whether these mucosal vaccines will contain live or inactivated infectious agents, there will be a need for adjuvants (of humoral and cell mediated immunity) which will be active when administered together with the vaccines on a mucosal surface (orally, intranasally). There is therefore a definite need for experimental systems which will be able to demonstrate an immunoadjuvant activity under such conditions of administration.

Furthermore, a novel and likely promising approach to anti-infectious immunization is represented by the DNA vaccines, in which the DNA sequence coding for a particular antigen, involved in protective mechanisms, is injected directly in the host. Whether or not there will be, in such a strategy, a need for adjuvants, to boost the immune response against the endogenously produced antigen, is open to question.

Conclusions and perspectives

Table 1 is a list, hopefully not too incomplete, of the immunostimulants that are being used as such presently (1996) in various clinical applications. Note that only two immunostimulants have become part of a standard therapeutic scheme in several countries: BCG (a live bacillus) for cancer of the bladder, levamisole (a simple synthetic molecule) for colorectal cancer. Nothing that was known about the immunostimulating activities of BCG and of levamisole would have allowed a reasonable guess that these agents would indeed find their special usefulness in those two particular pathologies (although intravesical treatment was a rather obvious choice in the case of bladder cancer). Immunotherapy of cancers with the other complex preparations listed in the table (picibanil, schizophyllan, lentinan, krestin) is performed exclusively in Japan and, for a number of reasons, which are not altogether clear, has not been adopted by Western oncologists and probably never will.

The other possible applications of immunostimulants are in the field of infectious diseases, especially when such infectious episodes occur in individuals whose immune system does not function optimally: young children, elderly persons, patients submitted to major anesthetic and surgical procedures. In this domain, we find various bacterial lysates (some of which may in part behave as vaccines), chemically pure bacterial glycoproteins, plant extracts and only one chemically characterized molecule, namely pidotimod. Also used in this context are the thymic extracts thymomodulin and thymostimulin. At this point, one may however ask the impertinent question: do we really need anti-infectious immunostimulants? There is no doubt that recurrent ear-nose-and-throat infections are a real problem in immunologically immature young children, most of whom finally outgrow this situation. In these children, some of the immunostimulants listed in the table appear to reduce to a certain extent the number and duration of infectious episodes and to allow a less frequent recourse to antibiotics, but these immunostimulants need to be administered rather frequently throughout the cold season. With respect to elderly individuals, the situation is more complex, inasmuch as adequate immune functions are practically preserved in many quite old persons, whereas they decline in others. Here again, the existence of recurrent respiratory (and possibly urinary) infections may justify the use of immunostimulants, as in the case in chronic bronchitis.

On the other hand, it is often claimed that major surgical interventions requiring prolonged and deep anesthesia, exert a

Table 1. Immunostimulants presently used in clinical practice (1996).

Preparation	Therapeutic application	Remarks
1. Bacille Calmette-Guérin (BCG)	bladder cancer	standard therapy
2. *Streptococcus pyogenes* strain (Picibanil)	various solid tumors	in Japan
3. Bacterial lysates (from various respiratory pathogens)	recurrent respiratory infections	in some countries
4. Glycoproteins from *Klebsiella pneumoniae*	recurrent respiratory infections	in some countries
5. *Escherichia coli* lysate	urinary infections	in some countries
6. Extract from *Echinacea purpurea*	recurrent respiratory infections	in Germany
7. Schizophyllan (glucan)	gastric, colorectal, breast cancers (association with chemotherapy)	in Japan
8. Lentinan (glucan)	cervical cancer (association with chemotherapy)	in Japan
9. Krestin, PSK (protein-bound glucan)	gastric and colorectal cancers (association with chemotherapy)	in Japan
10. Thymomodulin } purified extract from 11. Thymostimulin } bovine thymus	chronic or recurrent infections, post-surgical infections	in Italy
12. Levamisole (synthetic)	colorectal cancer (stage C) (in association with 5-fluorouracil)	standard therapy
13. Pidotimod (synthetic)	chronic or recurrent respiratory infections, post-surgical infections	in Italy
14. Romurtide (synthetic)	Recovery from chemotherapy or radiotherapy-induced leukopenia	in Japan

transient immunosuppressive effect, thereby favoring the appearance of severe nosocomial infections. In this case, the association of anti-infectious immunostimulants with wide spectrum antibiotics may become a favorable strategy; further evidence of the usefulness of PGG-glucan in this context is of course awaited with considerable interest. On the other hand, and although such an approach appears to be scientifically founded, the day has probably not yet come when the treatment of infectious diseases in general will consist in an association between an immunostimulant and antibiotics, thereby allowing a reduction in the amount of antibiotic and the duration of its administration. Is it utopian to think that anti-infectious immunostimulants might play a role in the fight against emergent pathogens and/or against antibiotic-resistant microorganisms? Remember at this point that some antibiotics may possess immunomodulating activities *per se*.

If there is one infectious pathology in which therapeutic immunostimulation would clearly appear necessary it is indeed the acquired immunodeficiency syndrome, i.e. AIDS. But, as far as things stand nowadays, the immunopathological mechanisms of this syndrome appear so complex and still insufficiently known, that the use of immunostimulants, either to treat the HIV infection itself or to prevent the occurrence of opportunistic infections, is generally considered much too hazardous (fear of reactivating virus replication, etc.). The results of the various current protocols in which certain cytokines are administered to AIDS patients will probably provide the necessary guidelines in the future.

Turning now to the immunotherapy of cancer, present trends of fundamental and clinical research in cancer immunology may cautiously help to predict the future. Progress continues to be made in characterizing various specific tumor antigens, as summarized recently by Boon [152]. This opens the way to the possible use of therapeutic cancer vaccines and, in such a strategy, the association of a tumor antigen with one or another of the immunological adjuvants mentioned in this review may turn out to be advantageous. Another approach, currently the object of extensive investigation, is to engineer tumor cells to make cytokines [such as interleukin-2 or interleukin-4) and to reinject those engineered cells into the tumor bearing host [153]; the well-defined cytokine-inducing properties of some immunostimulants (such as bropirimine or imiquimod, for instance) might make those useful in such therapeutic schemes.

There is however still room in the future for the direct utilization of an immunostimulant in a more conventional type of cancer immunotherapy, as suggested by the recently reported efficacy of liposomal muramyltripeptide in the treatment of relapsed osteosarcoma [154], justifying an ongoing multicenter phase III clinical trial of this agent in patients with newly diagnosed osteosarcoma. Just as in the case of levamisole and colorectal cancer, liposomal muramyltripeptide looks like an effective immunostimulant against a precise type of malignancy.

To the general question 'do exogenous immunostimulants of the type described in this review have a real future in human medicine?', the answer can be only cautiously affirmative and only for some of them. This will depend, to a large extent, on what will be achieved therapeutically with the endogenous immunomodulators represented by the various cytokines, now extensively investigated in several areas, since it is clear that all known immunostimulants do exert their activities through the *in vivo* induction of one or several cytokines. For instance, romurtide is used clinically in Japan to help cancer patients to recover from chemotherapy or radiotherapy-induced leukopenia: this drug does act indeed through induction of colony-stimulating factors, but its use may be more practical and economical than that of the factors themselves.

It appears also likely that some of the adjuvants described in this review will be used in the future to boost the immune responses to various anti-infectious vaccines.

Finally, it is not impossible to envisage the practical use of certain immunostimulants in therapies that do not appear, at least superficially, to involve the immune system, as exemplified by the cardioprotective activity of monophosphoryl lipid A (MPL). The recognized existence of bidirectional interactions between the neuroendocrine system and the immune system may well constitute the basis of still unknown therapeutic applications of immunomodulating agents. In a very recent paper, Sigel and Rosenbaum [155] have reviewed the evidence showing that many cytokines (such as interferon-γ, colony-stimulating factor GM-CSF, tumor-necrosis factor, interleukin-1, interleukin-6) promote neuronal survival in cultured cell populations, suggesting that some cytokines (and the immunomodulators that are able to induce their production and/or their activity) might ultimately have a potential in the treatment of stroke, trauma and dementia.

One should however always remember the caveat of Lewis Thomas [156]: 'Good applied science in medicine requires a high degree of certainty about the basic factors at hand, and especially about their meaning, and we have not yet reached this point for most of medicine'. Such a cautious view is indeed appropriate when dealing with molecules acting on the immune system.

We wish to thank Mrs J. Rolland for typing the manuscript and Mrs F. Cossart for drawing the chemical formulae.

REFERENCES

1. Langman, R. E. (1989) *The immune system*, Academic Press Inc., New York, p. 310.
2. Fridman, W. H. (1991) *Le cerveau mobile*, Hermann publishers, Paris, p. 216.
3. Black, P. H. (1994) Central nervous system-immune system interactions: psychoneuroendocrinology of stress and its immune consequences, *Antimicrob. Agents Chemother. 38*, 1–6.
4. Ramon, G. (1926) Procédés pour accroître la production des antitoxines, *Ann. Inst. Pasteur (Paris) 40*, 1–10.
5. Metchnikoff, E. (1892) *Leçons sur la pathologie comparée de l'inflammation*, G. Masson publishers, Paris, p. 239.
6. Coley, W. B. (1895) The treatment of inoperable malignant tumors with the toxins of erysipelas and *Bacillus prodigiosus*, *Med. Rec. Ann. 47*, 65–70.
7. Hall, S. S. (1995) IL-12 at the crossroads, *Science 268*, 1432–1434.
8. Henderson, B. & Blake, S. (1992) Therapeutic potential of cytokine manipulation, *Trends Pharmacol. Sci. 13*, 145–152.
9. McIntyre, C. A. & Rees, R. C. (1992) Cytokines and inflammation, *Drug News Perspect. 5*, 207–213.
10. Brakenhoff, J. P. J. (1995) Interleukin-6 receptor antagonists, *Drug News Perspect. 8*, 397–403.
11. Lefebvre d'Hellencourt, C., Diaw, L. & Guenounou, M. (1995) Immunomodulation by cytokine antisense oligonucleotides, *Eur. Cytokine Network 6*, 7–19.
12. Allison, A. C., Lafferty, K. J. & Fliri, H. (1993) Immunosuppressive and antiinflammatory drugs, *Ann. N.Y. Acad. Sci. 696*, 1–421.
13. St Georgiev, V. & Yamaguchi, H. (1993) Immunomodulating drugs, *Ann. N.Y. Acad. Sci. 685*, 1–812.
14. Klingemann, H. G. (1995) Update on immunosuppressive drugs, *Drug News Perspect. 8*, 303–309.
15. St Georgiev, V. (1991) Immunomodulating peptides of natural and synthetic origin, *Med. Res. Rev. 11*, 81–119.
16. St Georgiev, V. (1990) Synthetic immunomodulating agents, *Med. Res. Rev. 10*, 371–409.
17. Chirigos, M. A. (1992) Immunomodulators: current and future development and application, *Thymus 19*, S7–S20.
18. Hadden, J. W. (1993) Immunostimulants, *Immunol. Today 14*, 275–280.
19. Biozzi, G., Benacerraf, F. B., Grumbach, F., Halpern, B. N., Levaditi, J. & Rist, N. (1954) Etude de l'activité granulopéxique du système réticulo-endothelial au cours de l'infection tuberculeuse expérimentale de la souris, *Ann. Inst. Pasteur (Paris) 87*, 291–300.
20. Dubos, R. J. & Schaedler, R. W. (1957) Effects of cellular constituents of mycobacteria on the resistance of mice to heterologous infections. I. Protective effects, *J. Exp. Med. 106*, 703–719.
21. Biozzi, G., Stiffel, C., Halpern, B. N. & Mouton, D. (1959) Effet de l'inoculation du bacille de Calmette-Guérin sur le développement de la tumeur ascitique d'Ehrlich chez la souris, *C.R. Soc. Biol. (Paris) 153*, 987–989.
22. Mathé, G., Amiel, J. L., Schwarzenberg, L., Schneider, M., Cattan, A., Schlumberger, J. L., Hayat, M. & de Vassal, F. (1969) Active immunotherapy for acute lymphoid leukemia, *Lancet 1*, 697–699.
23. Morales, A., Eidinger, D. & Bruce, A. W. (1976) Intercavitary bacillus Calmette-Guérin in the treatment of superficial bladder tumors, *J. Urol. 116*, 180–183.
24. Morales, A. (1984) Long term results and complications of intercavitary bacillus Calmette-Guérin therapy for bladder cancer, *J. Urol. 132*, 457–459.
25. Lamm, D. L. (1995) Advances in the treatment of superficial bladder cancer: optimizing BCG immunotherapy. *Eur. Urol. suppl. 1*, p. 27, Karger, Basel.
26. Jackson, A. M. & James, K. (1994) Understanding the most successful immunotherapy of cancer, *The Immunologist 2*, 208–215.
27. Freund, J., Casals, J. & Hosmer, E. P. (1937) Sensitization and antibody formation after injection of killed tubercle bacilli and paraffin oil, *Proc. Soc. Exp. Biol. Med. 37*, 509–513.
28. Ishida, N. & Hoshino, T. A. (1985) Streptococcal preparation as a potent biological response modifier (OK-432), *Excerpta Med.*, 2nd edn, 1–69.
29. Tsuchiya, I., Kasahara, T., Yamashita, K., Ko, Y. C., Kanazawa, K., Matsushima, K. & Mukaida, N. (1993) Induction of inflammatory cytokines in the pleural effusion of cancer patients after the administration of an immunomodulator OK-432: role of IL-8 for neutrophil infiltration, *Cytokine 5*, 595–603.
30. Robinson, C. (1994) Subreum (OM-89), *Drugs Future 19*, 845–849.
31. Beuth, J., Ko, H. L., Roszkowski, W. & Pulverer, G. (1990) Bacteria of physiological microflora liberate immunomodulating peptides, *Microecol. Ther. 20*, 175–183.
32. Bocci, V. (1992) The neglected organ: bacterial flora has a crucial immunomodulatory role, *Perspect. Biol. Med. 35*, 251–260.
33. Takada, H. & Kotani, S. (1989) Structural requirements of lipid A for endotoxicity and other biological activities, *CRC Crit. Rev. Microbiol. 16*, 477–523.
34. Ribi, E., Amano, K., Cantrell, J. L., Schwartzman, S. M., Parker, R. & Takayama, K. (1982) Preparation and antitumor activity of nontoxic lipid A, *Cancer Immunol. Immunother. 12*, 91–102.
35. Yao, Z., Elliott, G. T. & Gross, G. J. (1994) Monophosphoryl lipid A: a new approach for cardioprotection, *Drug News Perspect. 7*, 96–102.
36. Brown, J. M., Grosso, M. A., Terada, L. S., Whitman, G. J. R., Banerjee, A., White, C. W., Harken, A. H. & Repine, J. E. (1989) Endotoxin pretreatment increases endogenous myocardial catalase activity and decreases ischemia-reperfusion injury of isolated rat hearts, *Proc. Natl Acad. Sci. USA 86*, 2516–2520.
37. Neway, T., Boulouis, H. J., Thibault, D. & Pilet, C. (1992) Activité des glycopeptidolipides polaires de *Mycobacterium chelonae* sur la restauration d'une leucopénie chimio-induite chez la souris, *C.R. Acad. Sci. Paris 315*, 13–19.
38. Lagrange, P. H., Fourgeaud, M., Neway, T. & Pilet, C. (1995) Les glycopeptidolipides polaires mycobactériens augmentent la résistance de la candidose expérimentale, *C.R. Acad. Sci. Paris 318*, 359–365.
39. Ellouz, T., Adam, A., Ciorbaru, R. & Lederer, E. (1974) Minimal structural requirements for adjuvant activity of bacterial peptidoglycan derivatives, *Biochem. Biophys. Res. Commun. 59*, 1317–1325.
40. Baschang, G. (1989) Muramylpeptides and lipopeptides: studies towards immunostimulants, *Tetrahedron 45*, 6331–6360.
41. Bahr, G. M., Darcissac, E., Bevec, D., Dukor, P. & Chedid, L. (1995) Immunopharmacological activities and clinical development of muramylpeptides with particular emphasis on murabutide, *Int. J. Immunopharmacol. 17*, 117–131.
42. Lidgate, D. M. & Byars, N. E. (1995) Development of an emulsion-based muralyldipeptide adjuvant formulation for vaccines, in *Vaccine design, the subunit and adjuvant approach* (Powell, M. F. & Newman, M. J., eds) pp. 313–324, Plenum Press, New York.
43. Azuma, I. (1992) Development of the cytokine inducer romurtide: experimental studies and clinical applications, *Trends Pharmacol. Sci. 13*, 425–428.
44. Namba, K., Otani, T. & Osada, Y. (1994) Enhancement of platelet recovery in X-irradiated guinea pigs by romurtide, a synthetic muramyldipeptide derivative, *Blood 83*, 2480–2488.
45. Fidler, I. J., Sone, S., Fogler, W. E. & Barnes, Z. L. (1981) Eradication of spontaneous metastases and activation of alveolar macrophages by intravenous injection of liposomes containing muramyldipeptide, *Proc. Natl Acad. Sci. USA 78*, 1680–1684.
46. Gay, B., Cardot, J. M., Schnell, C., Van Hoogevest, P. & Gygax, D. (1993) Comparative pharmacokinetics of free muramyltripeptide

phosphatidyl-ethanolamine (MTP-PtdEtn) and liposomal MTP-PE, *J. Pharm. Sci. 82*, 997–1001.

47. Kleinerman, E. S., Raymond, A. K., Bucana, C. D., Jaffe, N., Harris, M. B., Krakoff, I. H., Benjamin, R. & Fidler, I. J. (1992) Unique histological changes in lung metastases of osteosarcoma patients following therapy with liposomal muramyltripeptide, *Cancer Immunol. Immunother. 34*, 211–220.
48. Killion, J. J., Brown, D. R., Wilson, M. R., Lloyd, M. M. & Fidler, I. J. (1994) Prevention of chemotherapy or X-irradiation-induced monocytopenia by oral administration of lipophilic muramyltripeptide, *Oncol. Res. 6*, 357–364.
49. O'Reilly, T. & Zak, O. (1992) Enhancement of the effectiveness of antimicrobial therapy by muramylpeptide immunomodulators, *Clin. Infect. Dis. 14*, 1100–1109.
50. Migliore-Samour, D., Floc'h, F., Maral, R., Werner, G. H. & Jollès, P. (1977) Adjuvant activities of chemically modified water-soluble substances from *Mycobacterium tuberculosis*, *Immunology 33*, 477–484.
51. Migliore-Samour, D., Bouchaudon, J., Floc'h, F., Zerial, A., Ninet, L., Werner, G. H. & Jollès, P. (1979) Propriétés immunostimulantes et adjuvantes d'un lipopeptide de faible poids moléculaire, *C.R. Acad. Sci. Paris 289D*, 473–476.
52. Migliore-Samour, D., Bouchaudon, J., Floc'h, F., Zerial, A., Ninet, L., Werner, G. H. & Jollès, P. (1980) A short lipopeptide, representative of a new family of immunological adjuvants devoid of sugar, *Life Sci. 26*, 883–888.
53. Floc'h, F., Bouchaudon, J., Fizames, C., Zerial, A., Dutruc-Rosset, G. & Werner, G. H. (1984) Lauroyltetrapeptide (R. P. 40639) and related lipopeptides: a novel class of synthetic immunomodulating agents, *Drugs Future 9*, 763–776.
54. Floc'h, F. & Poirier, J. (1988) Immunopotentiating activities of a low molecular mass lipopeptide, R.P. 56142. Studies in infectious models, *Int. J. Immunopharmacol. 10*, 863–873.
55. Floc'h, F., Poirier, J., Fizames, C. & Woehrle, R. (1987) Pimelautide (R. P. 40639) from experimental results to clinical trials: an illustration, in *Immunostimulants now and tomorrow* (Azuma, I. & Jollès, G., eds) pp. 183–204, Japanese Scientific Societies Press, Tokyo and Springer-Verlag, Berlin.
56. Migliore-Samour, D., Delaforge, M., Jaouen, M., Mansuy, D. & Jollès, P. (1989) *In vivo* effects of immunomodulating lipopeptides on mouse liver microsomal cytochromes *P*-450 and on paracetamol-induced toxicity, *Experientia (Basel) 45*, 882–886.
57. Migliore-Samour, D., Bousseau, A., Caillaud, J. M., Naussac, A., Sedqi, M., Ferradini, C. & Jollès, P. (1993) Radioprotective effects of the immunostimulating lauroylpeptide L tri P (R.P. 56142), *Experientia (Basel) 49*, 160–166.
58. Sedqi, M., Delaforge, M., Mansuy, D., Martin, B., Jollès, P. & Migliore-Samour, D. (1995) Immunostimulating lipopeptide, L tri P (R.P. 56142); comparison of the effect on hepatic cytochrome *P*-450 modulation and radioprotection in male and female of three mouse strains, *Experientia (Basel) 51*, 790–798.
59. Deprez, B., Gras-Masse, H., Martinon, F., Gomard, E., Levy, J. P. & Tartar, A. (1995) Pimelautide or trimexautide as built-in adjuvants associated with an HIV-1-derived peptide: synthesis and *in vivo* induction of antibody and virus-specific cytotoxic T-lymphocyte-mediated-response, *J. Med. Chem. 38*, 459–465.
60. Goto, T., Nakahara, K., Nishiura, T., Hashimoto, M., Kino, T., Kuroda, Y., Okuhara, M., Kohsaka, M., Aoki, H. & Imanaka, H. (1982) Studies on a new immunoactive peptide, FK-156 II. Fermentation, extraction and chemical and biological characterization, *J. Antibiot. 35*, 1286–1292.
61. Goto, T. & Aoki, H. (1987) The immunomodulatory activities of acylpeptides, in *Immunostimulants, now and tomorrow* (Azuma, I. & Jollès, G., eds) pp. 99–108, Japanese Scientific Societies Press, Tokyo and Springer-Verlag, Berlin.
62. Kusumi, T., Yamada, A., Cao, M., Tanaka, A., Takenada, H. & Imanishi, J. (1989) Enhancing effects of immunoactive peptide FR 48127 on immunological responses to vaccination by inactivated influenza virus, *Vaccine 7*, 351–356.
63. Mine, Y., Watanabe, Y., Tawara, S., Yokota, Y., Nishida, M., Goto, S. & Kuwahara, S. (1983) Immunoactive peptides, FK-156 and FK-565. III. Enhancement of host defense mechanisms against infection, *J. Antibiot. 36*, 1059–1066.
64. Inamura, N., Nakahara, K., Kito, T., Gotoh, T., Kawamura, I., Aoki, H. & Sone, S. (1985) Activation of tumoricidal properties in macrophages and inhibition of experimentally-induced murine metastases by a new synthetic acyltripeptide, FK-565, *J. Biol. Resp. Mod. 4*, 408–417.
65. Puri, A., Rizvi, S. Y., Haq, W., Guru, P. Y., Kundu, B., Saxena, R. P., Shukla, R., Mathur, K. B. & Saxena, K. C. (1993) Immunostimulant activity of a novel lipopeptide and its protective action against *Leishmania donovani*, *Immunopharmacol. Immunotoxicol. 15*, 539–556.
66. Sidwell, R. W., Smee, D. F., Huffman, J. H., Bailey, K. W., Warren, R. P., Burger, R. A. & Penney, C. L. (1995) Antiviral activity of an immunomodulatory lipophilic desmuramyldipeptide analog, *Antiviral. Res. 26*, 145–159.
67. Reitermann, A., Metzger, J., Wiesmüller, K. H., Jung, G. & Bessler, W. G. (1989) Lipopeptide derivatives of bacterial lipoprotein constitute potent immune adjuvants combined with or covalently coupled to antigen or hapten, *Biol. Chem. Hoppe-Seyler 370*, 343–352.
68. Bessler, W. G., Kleine, B., Biesert, L. & Schlecht, S. D. (1990) Bacterial surface components as immunomodulators, in *Immunotherapeutic prospects of infectious diseases* (Masihi, K. N. & Lange, W., eds) pp. 37–48, Springer-Verlag, Berlin, Heidelberg.
69. Wiesmüller, K. H., Hess, G., Bessler, W. G. & Jung, G. (1990) Diastereomers of tripalmitoyl-*S*-glyceryl-L-cysteinyl carrier adjuvant systems induce different mitogenic and protective immune responses, in *Chirality and biological activity* (Holmstedt, B., Frank, H. & Testa, B., eds) pp. 267–272, Alan R. Liss, New York.
70. Loleit, M., Tröger, W., Wiesmüller, K. H., Jung, G., Strecker, M. & Bessler, W. G. (1990) Conjugates of synthetic lymphocyte-activating lipopeptides with segments from HIV proteins induce protein-specific antibody formation, *Biol. Chem. Hoppe-Seyler 371*, 967–975.
71. Chihara, G. (1984) Immunopharmacology of lentinan and the glucans, *Rev. Immunol. Immunopharmacol. 4*, 85–96.
72. Kraus, J. & Franz, G. (1991) β(1-3) glucans: antitumor activity and immunostimulation, in *Fungal cell wall and immune response* (Latgé, J. P. & Boucias, D., eds) pp. 431–444, Springer-Verlag, Berlin.
73. Trnovec, T. & Hrmova, M. (1993) Immunomodulator polysaccharides: chemistry, disposition and metabolism, *Biopharm. & Drug Dispos. 14*, 187–195.
74. Goldman, R. C. (1993) Biological response modification by β-D-glucans, *Ann. Rep. Med. Chem. 130*, 129–138.
75. Pretus, H. A., Ensley, H. E., McNamee, R. B., Jones, E. L., Browder, I. W. & Williams, D. L. (1991) Isolation, physicochemical characterization and preclinical efficacy evaluation of soluble scleroglucan, *J. Pharmacol. Exp. Ther. 257*, 500–510.
76. Jamas, S., Easson, D. D. Jr & Ostroff, G. R. (1990) PGG, a novel class of macrophage activating immunomodulators, *Abstr. Int. Congress. Infect. Dis.*, no. 698, p. 143.
77. Lagrange, P. H. & Fourgeaud, M. (1991) Enhanced natural resistance against severe disseminated *Candida albicans* infection in mice treated with betafectin, *Int. J. Exp. Clin. Chemother. 4*, 48–55.
78. Nicoletti, A., Nicoletti, G., Ferraro, G., Palmieri, G., Mattaboni, P. & Germogli, R. (1992) Preliminary evaluation of immunoadjuvant activity of an orally administered glucan extracted from *Candida albicans*, *Arzneim.-Forsch./Drug Res. 42*, 1246–1250.
79. Kobayashi, H., Matsunaga, K. & Fujii, M. (1993) PSK as a chemopreventive agent, *Cancer Epidemiol. Biomarkers and Prevention 2*, 271–276.
80. Kobayashi, H., Matsunaga, K. & Fujii, M. (1995) Antimetastatic effects of PSK, a protein-bound polysaccharide obtained from basidiomycetes: an overview, *Cancer Epidemiol. Biomarkers and Prevention 4*, 275–281.
81. Nakazato, H., Koike, A., Saji, S., Ogama, N. & Sakamoto, I. (1994) Efficacy of immunochemotherapy as adjuvant treatment after curative resection of gastric cancer, *Lancet 343*, 1122–1126.
82. Mitomi, T., Tsuchiya, S., Iijima, N., Aso, K., Suzuki, K., Nishiyama, K., Amano, T., Takahashi, T., Murayama, N., Oka, H.,

Oya, K., Noto, T. & Ogawa, N. (1992) Randomized controlled study on adjuvant immuno-chemotherapy with PSK in curatively resected colorectal cancer, *Dis. Colon Rectum 35*, 123–130.
83. Franz, G. (1992) Immunomodulation durch *Echinacea purpurea* und *Viscum album*, *Erfahrungsheilkunde 6*, 401–405.
84. Franz, G. (1993) Risiko ohne Nutzen-Nutzen ohne Risiko, am Beispiel von Immunomodulatoren, *Therapeutikon 7*, 562–568.
85. Chen, J. R., Suetsuna, K. & Yamauchi, F. (1995) Isolation and characterization of immunostimulative peptides from soybean, *Nutritional Biochem. 6*, 310–313.
86. Werner, G. H. (1987) Immunostimulants: the western scene, in *Immunostimulants, now and tomorrow* (Azuma, I. & Jollès, G., eds) pp. 3–39, Springer-Verlag, Berlin, Japan Scientific Societies Press, Tokyo.
87. Perrillo, R. P. & Mason, A. L. (1994) Therapy for hepatitis B virus infection, *Gastroenterol. Clin. North Am. 23*, 581–601.
88. Handzel, Z. T. & Trainin, N. (1990) Immunomodulation of T-cell deficiency in humans by thymic humoral factor, *J. Biol. Response Modif. 9*, 269–278.
89. Fiocchi, A., Borella, E., Riva, E. & Arensi, D. (1986) A double-blind clinical trial for the evaluation of the therapeutic effectiveness of a calf thymus derivative (thymomodulin) in children with recurrent respiratory infections, *Thymus 8*, 331–339.
90. Braga, P. C., Dal Sasso, M., Maci, S., Piatti, G. & Palmieri, R. (1994) Restoration of polymorphonuclear leukocyte function in elderly subjects by thymomodulin, *J. Chemother. 6*, 354–359.
91. De Mattia, D., Decandia, P., Ferrante, P. & Pace, D. (1993) Effectiveness of thymostimulin and study of lymphocyte-dependent antibacterial activity in children with recurrent respiratory infections, *Immunopharmacol. Immunotoxicol. 15*, 447–459.
92. Periti, P., Tonelli, F., Mazzei, T. & Ficari, F. (1993) Antimicrobial chemoimmunoprophylaxis in colorectal surgery with cefotetan and thymostimulin: prospective, controlled multicenter study, *J. Chemother. 5*, 37–42.
93. Miller, M. J. S., Witherly, S. A. & Clark, D. A. (1990) Casein: a milk protein with diverse biologic consequences, *Proc. Soc. Exp. Biol. Med. 195*, 143–159.
94. Jollès, P., Parker, F., Floc'h, F., Migliore-Samour, D., Alliel, P., Zerial, A. & Werner, G. H. (1982) Immunostimulating substances from human casein, *J. Immunopharmacol. 3*, 363–369.
95. Parker, F., Migliore-Samour, D., Floc'h, F., Zerial, A., Werner, G. H., Jollès, J., Casaretto, M., Zahn, H. & Jollès, P. (1984) Immunostimulating hexapeptide from human casein. Amino acid sequence, synthesis and biological properties, *Eur. J. Biochem. 145*, 677–682.
96. Berthou, J., Migliore-Samour, D., Lifchitz, A., Delettré, J., Floc'h, F. & Jollès, P. (1987) Immunostimulating properties and three-dimensional structure of two tripeptides from human and cow caseins, *FEBS Lett. 218*, 55–58.
97. Gattegno, L., Migliore-Samour, D., Saffar, L. & Jollès, P. (1988) Enhancement of phagocytic activity of human monocyte-macrophage cells by immunostimulating peptides from human casein, *Immunol. Lett. 18*, 27–31.
98. Ross, A. C. (1992) Vitamin A status: relationship to immunity and the antibody response, *Proc. Soc. Exp. Biol. Med. 200*, 303–320.
99. Forni, G., Cerruti, S. S., Giovanelli, M., Santoni, A., Martinetto, P. & Vietti, D. (1986) Effect of prolonged administration of low doses of dietary retinoids on cell-mediated immunity and the growth of transplantable tumors in mice, *J. Nat. Cancer Inst. 76*, 527–533.
100. Lemire, J. M. (1995) Immunomodulatory actions of 1,25-di-hydroxyvitamin D3, *J. Steroid Biochem. Mol. Biol. 53*, 599–602.
101. Alroy, I., Towers, T. L. & Freedman, L. P. (1995) Transcriptional repression of the interleukin-2 gene by vitamin D3: direct inhibition of NFATp/AP-1 complex formation by a nuclear hormone receptor, *Mol. Cell. Biol. 15*, 5789–5799.
102. Shkalar, G., Schwartz, J. L., Trickler, D. & Reid, S. (1990) Prevention of experimental cancer and immunostimulation by vitamin E (immunosurveillance), *J. Oral Pathol. Med. 19*, 60–64.
103. Wang, Y. & Watson, R. R. (1994) Ethanol, immune responses and murine AIDS: the role of vitamin E as an immunostimulant and antioxidant, *Alcohol 11*, 75–84.
104. Labro, M. T. (1993) Immunomodulation by antibacterial agents. Is it clinically relevant? *Drugs 45*, 319–328.
105. Labro, M. T. (1990) Cefodizime as a biological response modifier: a review of its *in vivo*, *ex vivo* and *in vitro* immunomodulatory properties, *J. Antimicrob. Chemother. 26*, 37–47.
106. Lianou, P. E., Votta, E. G., Papavassiliou, J. T. & Bassaris, H. P. (1993) *In vivo* potentiation of polymorphonuclear leukocyte function by ciprofloxacin, *J. Chemother. 5*, 223–227.
107. Scaglione, F., Ferrara, F., Dugnani, S., Demartini, G., Triscari, F. & Fraschini, F. (1993) Immunostimulation by clarithromycin in healthy volunteers and chronic bronchitis patients, *J. Chemother. 5*, 228–232.
108. Roy, M., Kiremidjian-Schumacher, L., Wishe, H. I., Cohen, M. W. & Stotzky, G. (1995) Supplementation with selenium restores age-related decline in immune cell function, *Proc. Soc. Exp. Biol. Med. 209*, 369–375.
109. Roy, M., Kiremidjian-Schumacher, L., Wishe, H. I., Cohen, M. W. & Stotzky, G. (1993) Selenium supplementation enhances the expression of interleukin 2 receptor subunits and internalization of interleukin 2, *Proc. Soc. Exp. Biol. Med. 202*, 295–299.
110. Roy, M., Kiremidjian-Schumacher, L., Wishe, H. I., Cohen, M. W. & Stotzky, G. (1991) Effect of selenium on the expression of high affinity interleukin 2 receptors, *Proc. Soc. Exp. Biol. Med. 200*, 36–41.
111. Ishiwata, Y., Suzuki, E., Yokochi, S., Otsuka, T., Tasaka, F., Usuda, H. & Mitani, T. (1994) Studies on the antiviral activity of propagermanium with immunostimulating action, *Arzneim.-Forsch./Drug Res. 44*, 357–361.
112. Hoshi, A. (1993) Proxigermanium (propagermanium) antiviral immunostimulant, *Drugs Future 18*, 905–908.
113. Kalechman, Y., Shani, A., Albeck, M., Sotnik-Barkai, I. & Sredni, B. (1993) Increased DNA repair ability after irradiation following treatment with the immunomodulator AS-101, *Radiation Res. 136*, 197–204.
114. Renoux, G. & Renoux, M. (1971) Effet immunostimulant d'un imidazothiazole dans l'immunisation de la souris contre l'infection par *Brucella abortus*, *C. R. Acad. Sci. Paris D272*, 349–350.
115. Borden, E. C., Davies, E. & Chirigos, M. A. (1982) Interim analysis of a trial of levamisole and 5-fluorouracil in metastatic colorectal carcinoma, in *Immunotherapy of human cancer* (Terry, W. D. & Rosenberg, S. A., eds) pp. 231–235, Excerpta Medica, New York.
116. Laurie, J., Moertal, C., Fleming, T. & Wieland, H. (1989) Surgical adjuvant therapy of large-bowel carcinoma: an evaluation of levamisole and the combination of levamisole and fluorouracil, *J. Clin. Oncol. 7*, 1447–1456.
117. Kimball, E. S. (1993) Experimental modulation of IL-1 production and cell surface molecule expression by levamisole, *Ann. N.Y. Acad. Sci. 685*, 259–268.
118. Kurman, M. R. (1993) Recent clinical trials with levamisole, *Ann. N.Y. Acad. Sci. 685*, 269–277.
119. Wierenga, W. (1993) Overview on the chemistry and immunomodulating properties of novel pyrimidinones, *Ann. N.Y. Acad. Sci. 685*, 296–300.
120. Sarosdy, M. F. (1993) Bropirimine in bladder cancer: clinical studies, *Ann. N.Y. Acad. Sci. 685*, 301–308.
121. Anderson, D. W. & Capetola, R. J. (1990) Immunoactive properties of RWJ 21757, a substituted guanosine with adjuvant activity, in *Gamete interaction, prospects for immunocontraception* (Alexander, N. J., ed.) pp. 523–548, Wiley-Liss, Inc., New York.
122. Chen, R., Goodman, M. G., Argenteri, D., Bell, S. C., Levelle, E. B., Come, J., Goodman, J. H. & Klaubert, D. H. (1994) Guanosine derivatives as immunostimulants: discovery of loxoribine, *Nucleosides & Nucleotides 131*, 551–562.
123. De Simone, C., Famularo, G., Foresta, P., Albertoni, C. & Arrigoni Martelli, E. (1993) *In vivo* antitumor activity of the hypoxanthine derivatives St 789 and ST 689, *Ann. N.Y. Acad. Sci. 685*, 344–346.
124. Sosa, M., Saha, A., Giner-Sorolla, A., Hadden, E. & Hadden, J. W. (1993) Immunopharmacologic properties of inosine 5′-methylmonophosphate (MIMP), *Ann. N.Y. Acad. Sci. 685*, 458–463.
125. Testerman, T. L., Gerster, J. F., Imbertson, L. M., Reiter, M. J., Miller, R. L., Gibson, S. J., Wagner, T. L. & Tomai, M. A. (1995)

Cytokine induction by the immunomodulators imiquimod and S-27609, *J. Leukocyte Biol. 58*, 365–372.

126. Bernstein, D. I. & Harrison, C. J. (1989) Effects of the immunomodulating agent R 837 on acute and latent herpes simplex virus type 2 infections, *Antimicrob. Agents Chemother. 33*, 1511–1515.
127. Bernstein, D. I., Miller, R. L. & Harrison, C. J. (1993) Adjuvant effects of imiquimod on a herpes simplex virus type 2 glycoprotein vaccine in guinea pig, *J. Infect. Dis. 167*, 731–735.
128. Gibson, S. J., Imbertson, L. M., Wagner, T. L., Testerman, T. L., Reiter, M. J., Miller, R. L. & Tomai, M. A. (1995) Cellular requirements for cytokine production in response to the immunomodulators imiquimod and S-27609, *J. Interferon and Cytokine Res. 15*, 537–545.
129. Moyer, M. P., Weber, F. H., Gross, J. L., Isaac, J. W. & Saint Fort, R. (1992) The synthesis and identification of 4,6-diaminoquinoline derivatives as potent immunostimulants, *Bioorganic & Med. Chem. Lett. 2*, 1589–1594.
130. Moyer, M. P., Weber, F. H. & Gross, J. L. (1992) Structure-activity relationships of imidazo(4,5-f)quinoline partial structures and analogs. Discovery of pyrazolo(3,4-f)quinoline derivatives as potent immunostimulants, *J. Med. Chem. 35*, 4595–4601.
131. Moyer, M. P., Weber, F. H., Canning, P. C., Gross, J. L. & Saint Fort, R. (1993) Investigation of side-chain SAR, formulation and injection site toleration of pyrazol(3,4-f)quinoline derivatives: a potent series of *in vivo* active immunostimulants, *Bioorganic & Med. Chem. Lett. 3*, 1379–1384.
132. Borghi, M. O., Fain, C., Barcellini, W., Del Papa, N., La Rosa, L., Nicoletti, F., Uslenghi, C. & Meroni, P. L. (1994) *Ex vivo* effect of pidotimod on peripheral blood mononuclear cell immune functions: study of an elderly population, *Int. J. Immunother. 10*, 35–39.
133. Migliorati, G., D'Adamio, L., Coppi, G., Nicoletti, I. & Riccardi, C. (1992) Pidotimod stimulates natural killer cell activity and inhibits thymocyte cell death, *Immunopharmacol. Immunotoxicol. 14*, 737–748.
134. Vacca, A., DiStefano, R., Serio, G. & Damacco, F. (1993) Stimulation of residual B-cell immunity in nude mice by pidotimod: an immuno-histological study, *Int. J. Immunother. 9*, 85–93.
135. Di Marco, R., Condorelli, F., Girardello, R., Uslenghi, C., Chisari, G. & Di Mauro, M. (1992) Increased rate of survival in *Streptococcus pneumoniae*-infected rats with the new immunomodulator pidotimod, *Scand. J. Infect. Dis. 24*, 821–823.
136. Capsoni, F., Minonzio, F., Ongari, A. M., Girardello, R., Di Bello, M. & Zanussi, C. (1991) *In vitro* and *ex vivo* enhancement of neutrophil functions by PGT/1 A, a new immunostimulating peptide, *J. Chemother. 3 (suppl. 3)*, 147–149.
137. Careddu, P., Alfano, S. & Zavattini, G. (1992) Pidotimod in the prophylaxis of recurrent acute tonsillitis in childhood, *Adv. Ther. 9*, 174–183.
138. Auteris, A., Pasqui, A. L., Gotti, G., Bruni, F., Saletti, M., Di Renzo, M., Bova, G., Borlini, G., Gori, S., Fanetti, G., Campoccia, G., Maggiore, D. & Girardello, R. (1993) The effect of a new biological response modifier (pidotimod) on surgery-associated immunodeficiency, *Int. J. Immunother. 9*, 95–102.
139. Rhodes, J., Chen, H., Hall, S. R., Beesley, J. E., Jenkins, D. C., Collins, P. & Zheng, B. (1995) Therapeutic potentiation of the immune system by costimulatory Schiff-base-forming drugs, *Nature 377*, 71–75.
140. High, W. B., Bugelski, P. J., Nichols, M. E., Swift, B. A., Solleveld, H. A. & Badger, A. M. (1994) Effect of a novel azaspirane on the arthritic lesions in the adjuvant Lewis rat. Attenuation of the inflammatory process and preservation of skeletal integrity, *J. Rheumatol. 21*, 476–483.
141. King, A. G. & Badger, A. M. (1991) Administration of an immunomodulatory azaspirane (SKF 105685) or recombinant human interleukin 1 stimulates myelopoiesis and enhances survival from lethal irradition in C 57 Bl/6 mice, *Exp. Haematol. 19*, 624–628.
142. Audibert, F. M. & Lise, L. D. (1993) Adjuvants: current status, clinical perspectives and future prospects, *Immunol. Today 14*, 281–284.
143. Gupta, R. K., Relyveld, E. H., Lindblad, E. B., Bizzini, B., Ben-Efraim, S. & Gupta, C. K. (1993) Adjuvants, a balance between toxicity and adjuvanticity, *Vaccine 11*, 293–306.
144. Johnson, A. G. (1994) Molecular adjuvants and immunomodulators: new approaches to immunization, *Clin. Microbiol. Rev. 7*, 277–289.
145. Gupta, K. & Siber, G. R. (1995) Adjuvants for human vaccines. Current status, problems and future prospects, *Vaccine 13*, 1263–1275.
146. Kensil, C. R., Wu, J. Y. & Soltsysik, S. (1995) Structural and immunological characterization of the vaccine adjuvant QS-21, in *Vaccine design, the subunit and adjuvant approach* (Powell, M. F. & Newman, M. J., eds) pp. 525–541, Plenum Publishing Corporation, New York.
147. Gjata, B., Hannoun, C., Boulouis, H. J., Neway, T. & Pilet, C. (1994) Activité adjuvante des glycopeptidolipides polaires de *Mycobacterium chelonae* (GPLp-Mc) sur l'immunogénicité et l'effet protecteur d'un vaccin grippal inactivé, *C.R. Acad. Sci. Paris 317*, 257–263.
148. Afonso, L. C. C., Scharton, T. M., Vieira, L. Q., Wysocka, M., Trinchieri, G. & Scott, P. (1994) The adjuvant effect of interleukin-12 in a vaccine against *Leishmania major*, *Science 263*, 235–237.
149. Duits, A. J., Van Puijenboek, A., Vermeulen, H., Hofhuis, F. M. A., Van de Winkel, G. J. & Capel, P. J. A. (1993) Immunoadjuvant activity of a liposomal IL-6 formulation, *Vaccine 11*, 777–781.
150. Murray, P. J., Aldovini, A. & Young, R. A. (1996) Manipulation and potentiation of antimycobacterial immunity using recombinant bacille Calmette-Guérin strains that secrete cytokines, *Proc. Natl Acad. Sci. USA 93*, 934–939.
151. Dempsey, P. W., Allison, M. E. D., Akkaraju, S., Goodnow, C. C. & Fearon, D. T. (1996) C3d of complement as a molecular adjuvant: bridging innate and acquired immunity, *Science 271*, 348–350.
152. Boon, T. (1995) Tumor antigens and perspectives for cancer immunotherapy, *The immunologist 3*, 262–263.
153. Williams, N. (1996) An immune boost to the war on cancer, *Science 272*, 28–30.
154. Kleinerman, E. S., Gano, J. B., Johnston, D. A., Bendamin, R. L. & Jaffe, N. (1995) Efficacy of liposomal muramyl-tripeptide in the treatment of relapsed osteosarcoma, *Am. J. Clin. Oncol. Cancer Clin. Trials 18*, 93–101.
155. Sigel, K. & Rosenbaum, D. M. (1996) Potential role of hemopoietic cytokines in neuronal survival, *Drug News Perspect. 9*, 142–148.
156. Thomas, L. (1980) *The medusa and the snail*, Bantam Books, Toronto, New York, p. 143.

Eur. J. Biochem. 242, 20–28 (1996)

Review

Locations of functional domains in the RecA protein

Overlap of domains and regulation of activities

Masayuki TAKAHASHI[1], Fabrice MARABOEUF[1] and Bengt NORDÉN[2]

[1] Groupe d'Etude Mutagénèse et Cancérogénèse, UMR 216 CNRS et Institut Curie, Orsay, France
[2] Department of Physical Chemistry, Chalmers University of Technology, Gothenburg, Sweden

(Received 2 August 1996) – EJB 96 1154/0

We review the locations of various functional domains of the RecA protein of *Escherichia coli*, including how they have been assigned, and discuss the potential regulatory roles of spatial overlap between different domains. RecA is a multifunctional and ubiquitous protein involved both in general genetic recombination and in DNA repair: it regulates the synthesis and activity of DNA repair enzymes (SOS induction) and catalyses homologous recombination and mutagenesis. For these activities RecA interacts with a nucleotide cofactor, single-stranded and double-stranded DNAs, the LexA repressor, UmuD protein, the UmuD$'_2$C complex as well as with RecA itself in forming the catalytically active nucleofilament. Attempts to locate the respective interaction sites have been advanced in order to understand the various functions of RecA. An intriguing question is how these numerous functional sites are contained within this rather small protein (38 kDa). To assess more clearly the roles of the respective sites and to what extent the sites may be interacting with each other, we review and compare the results obtained from various biological, biochemical and physico-chemical approaches. From a three-dimensional model it is concluded that all sites are concentrated to one part of the protein. As a consequence there are significant overlaps between the sites and it is speculated that corresponding interactions may play important roles in regulating RecA activities.

Keywords: RecA protein; DNA-binding site; functional domain; homologous recombination; DNA repair.

RecA is a multi-functional and ubiquitous enzyme that plays a crucial role for several different steps of DNA repair and which exists in various organisms (For reviews, Walker, 1984; Smith and Wang, 1989; Roca and Cox, 1990; Devoret, 1992). A homologous protein, Rad51, has been discovered in eucaryotes, including human, and been shown to be involved in the homologous recombination just as RecA is in *Escherichia coli* (Aboussekhra et al., 1992; Shinohara et al., 1993). In the case of *E. coli* RecA, genetic analysis has shown that the protein has also other functions for DNA repair and its regulation besides the catalysis of homologous recombination: RecA regulates the activity and the synthesis of DNA-repair proteins (SOS induction; George et al., 1974; Little and Mount, 1982) and helps the replisome to pass through the DNA lesion (mutational repair; Sommer et al., 1993). RecA also switches the repair mode from excision to recombinational repair and from this to mutational repair (Devoret, 1992; Sommer et al., 1993). The purified RecA can mimic these activities *in vitro* systems in the presence of ATP as a cofactor (Shibata et al., 1979; McEntee et al., 1979; Little and Mount, 1982; Burckhardt et al., 1988; Frank et al., 1993).

Correspondence to M. Takahashi, Institut Curie, Bat. 110, Centre Université Paris-Sud, F-91405 Orsay, France

Abbreviations. SANS, small-angle neutron scattering; LD, linear dichroism.

Note. This Review will be reprinted in *EJB Reviews 1996*, to appear in April 1997.

In vivo and *in vitro* analyzes suggest that the DNA repair proceeds in the following way in the cell (Devoret, 1992). As a first step, RecA binds cooperatively to a single-stranded part of DNA, created by DNA damage, and forms a helical polymer around the DNA. This nucleofilament interacts with the LexA repressor and inactivates it by stimulating its autocleavage. In this way, RecA increases the synthesis of more than 20 DNA-repair enzymes including RecA itself (SOS induction). In this process, the induction of UvrA protein, which is involved in the excision repair, occurs at a first stage probably because of a low affinity of the LexA repressor for the promoter of the *uvrA* gene (Peterson and Mount, 1987). The excision process stimulated in this way repairs most of the damages.

Remaining damage is then repaired by recombinational processes. By continuous induction of the SOS system the number of RecA molecules increases with time. The RecA-DNA nucleofilament becomes large enough to bind a second DNA molecule and to promote the strand-exchange reaction. At this stage the cleavage of LexA is probably reduced and SOS induction ceases. The recombinational repair becomes the main process in place of the excision repair.

Finally, mutational repair takes place. For this process, RecA acts in three steps: RecA accelerates the synthesis of UmuD protein by SOS induction (Elledge and Walker, 1983; Shinagawa et al., 1983), activates it by stimulating its autocleavage to the UmuD' form (Burckhardt et al., 1988), and finally interacts with the UmuD$'_2$C complex to place it at a DNA lesion and facilitate the replisome passing through the lesion (Frank et al., 1993; Sommer et al., 1993). This process, which could create a mutation, appears only at a late phase of DNA repair. This is probably because the affinity of LexA for the promoter of this gene is high and the induction occurs only lately (Peterson and Mount, 1987) and the cleavage of UmuD to the active UmuD′ form is very slow (Burckhardt et al., 1988). The interaction of the UmuD$'_2$C complex with RecA prevents the recombination reaction probably by disrupting the RecA-DNA filament. In this way, RecA switches the repair mode from recombinational to mutational repair (Devoret, 1992).

For these activities, RecA binds a nucleotide cofactor (ATP or dATP) and one ssDNA and one dsDNA (Takahashi et al., 1989; Kubista et al., 1990). RecA protein molecules form a nucleofilament by organizing themselves in a helical manner around the DNA in a head-to-tail arrangement (DiCapua et al., 1982; Egelman, 1993). The RecA nucleofilament interacts with the repressors and the UmuD protein to stimulate their autocleavage for the induction of the SOS system (Little and Mount, 1982) and for the mutagenesis (Burckhardt et al., 1988). This nucleofilament can also interact with the UmuD$'_2$C complex (Frank et al., 1993). Based on these functions and biochemical analyses one may conclude that RecA possesses one ATP-binding site, three DNA-binding sites, two subunit-subunit interaction parts, one or more repressor-binding site(s) and probably one UmuD$'_2$C-binding site.

An investigation of the locations of these sites in the three-dimensional structure of the protein should be useful for building a model of the catalytically active ATP-RecA-DNA complex and for understanding the molecular mechanisms of the various RecA actions and their regulation (Roca and Cox, 1990; Kowalczykowski et al., 1994; Takahashi and Nordén, 1994). The assignment of these locations would also address the pertinent question how these numerous sites may be organized in this rather small protein (38 kDa). Especially, knowledge about the location of the two DNA-binding sites could be helpful, as it may clarify how the two DNA chains interact with each other during the strand-exchange reaction.

The analyses have been advanced by use of various biological and physical approaches to RecA-DNA complexes. In addition, information from X-ray crystallography on the three-dimensional structure of RecA (Story et al., 1992) has been useful. No crystal of the active form of RecA (RecA-ATP complex) or any RecA-DNA complex has yet become available for X-ray analysis, so the structure of RecA in these contexts is not known at any detail. NMR cannot be used in the case of RecA, at least not on the entire protein, because RecA forms very large complexes which make the NMR signals too broad to be useful. Instead, structural analysis has been advanced by other approaches; among physical methods small-angle neutron scattering (SANS) and electron microscopy provide useful information at low resolution about the textural structure of RecA complexes and suggest the location of interaction sites. Biochemical approaches such as protease, chemical interference analysis and photocross-link analyses, as well as fluorescence measurements of tryptophan probes inserted in RecA by gene engineering, and activity measurements of RecA fragments, have advanced our knowledge as will be discussed below. Finally, molecular biology and genetics have contributed by analyzing the effects of site-directed mutagenesis on the isolated RecA activities in relation to the location of mutation, and by comparing sequences with those of other proteins having similar activity.

The determination of functional domains of RecA is, however, difficult because RecA is active in a form of a polymer and, therefore, one cannot generally isolate a single site without affecting the function of the entire protein. Thus, a mutation aiming for a certain activity could affect other activities as well; e.g., a mutant that suppresses polymerization will also affect DNA binding and LexA cleavage. By comparing and critically reviewing, from a global point of view, the results from the different approaches, recalling that each technique has its own limitations but may complement the others, a consensus about structure and function can be obtained. We here present the results and conclusions for each of the sites, and discuss the respective limitations of each technique as such know-how could be helpful in the corresponding studies of other proteins, especially Rad51. We note some overlap of the functional domains, the interactions of which could play a role in the regulation of RecA activities.

Subunit-subunit interaction parts

Structural information about RecA and the RecA-DNA filaments has been obtained mainly from electron microscopy studies (DiCapua et al., 1982; Williams and Spengler, 1986; Heuser and Griffith, 1989; Egelman, 1993) and small-angle neutron scattering (SANS) measurements (DiCapua et al., 1989, 1992; Ellouze et al., 1995). The monomer RecA subunits in the filament are organized in a helical manner with a periodicity of about 6 subunits/turn both in the presence and absence of DNA (DiCapua et al., 1982; Williams and Spengler, 1986). The filament has some structural and electric polarity (Egelman, 1993; Jonsson et al., 1993) which supports the suggestion that the filament is formed by a head-to-tail orientation of the RecA subunits. In the crystal of RecA alone (Story et al., 1992) the monomers are packed in a way which appears to be very similar to that in the RecA and RecA-DNA filaments in solution: a helical organization constituted by 6 subunits/turn. The orientation of Trp290, used as an internal chromophore reporter and studied by combined anisotropic light absorption and neutron-scattering techniques, indicates the same orientation of protein subunits in the RecA crystal and nucleofilament (Hagmar et al., 1992). Thus the subunit-subunit contacts in the RecA crystal probably also reflect the subunit-subunit interactions in the RecA and RecA-DNA filaments in solution. The size of the helical pitch is slightly smaller in solution than in the crystal in absence of cofactor (7.6 nm instead of 8.3 nm), but, SANS and electron microscopy both show that the helical pitch can vary, for example, upon the binding of nucleotide (DiCapua et al., 1989, 1992; Ellouze et al., 1995) apparently without any large modification of the contact parts (Morimatsu et al., 1996). The pitch size of RecA in the crystal is close to that observed for the RecA-ADP complex in solution (Ellouze et al., 1995). Probably a conformation of RecA corresponding to that of its complex with ADP is preferred in the crystal state.

In the crystal, the α-helices A and G, β-sheet 0 and turn 246–256 of one subunit are in contact with loop 95–101, β-sheet 3 and α-helices D and E of the adjacent subunit (Fig. 1; Story et al., 1992). Furthermore, deletion analysis has shown that the elimination of α-helix A prevents the subunit-subunit interaction *in vitro* (Mikawa et al., 1995). It was also observed *in vivo* that the addition of a mutated *recA* gene containing this deletion does not interfere with the activities of wild-type RecA (Horii et al., 1992) in contrast to the finding that mutated *recA* gene without the deletion frequently affects the activities, probably by forming a mixed filament (Lauder and Kowalczykowski, 1993). The latter result supports the notion that the same subunit-subunit contacts occur also in the active ATP-RecA-DNA complex. Since this α-helix A was proposed also as a DNA-binding site (Kawashima et al., 1984), Morimatsu and Horii examined the effect of mutation in this part of the protein and

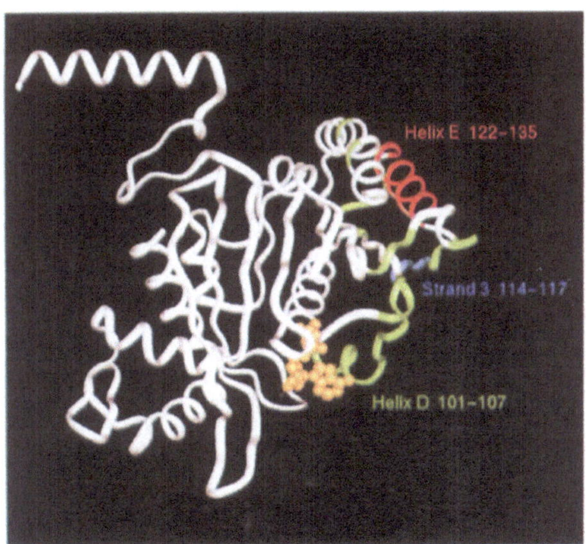

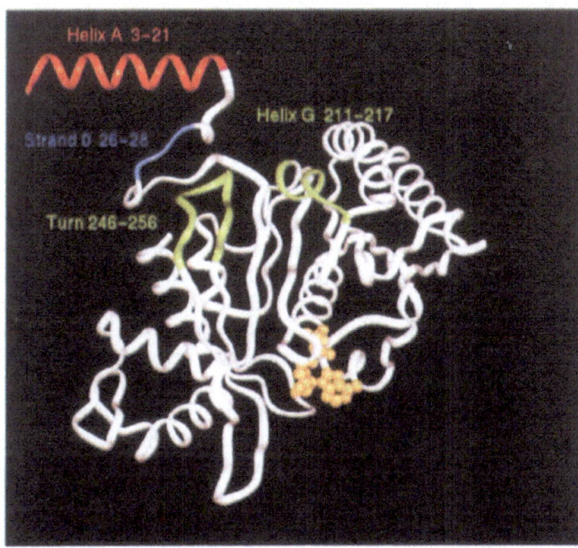

Fig. 1. Subunit-subunit contact parts of RecA [according to the crystal study of Story et al. (1992)]. The parts denoted in red (α-helix A), blue (β-strand 0) and green of in one subunit (A) are in contact with the parts, respectively, in red (α-helix E), blue (β-strand 3) and green of the neighbour subunit (B). The contacts occur between parts having the same colour. ADP is coloured orange.

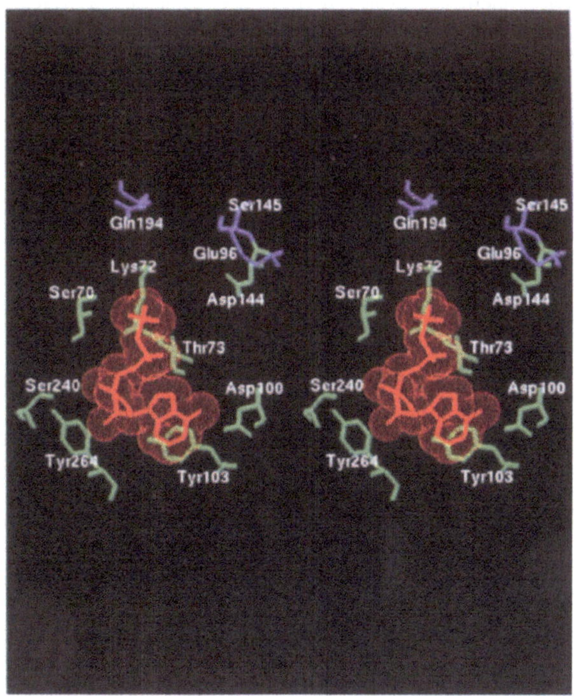

Fig. 2. Stereoscopic presentation of nucleotide cofactor binding site of RecA [ADP interaction site as assigned from the crystal study by Story and Steitz (1992)]. ADP coloured red and interacting residues green. The residues coloured blue were proposed by Story and Steitz (1992) to interact with the γ-phosphate of ATP.

observed that the mutation clearly affects the self-association but not significantly the intrinsic DNA binding (Morimatsu and Horii, 1995a; Takahashi, M., Morimatsu, K. and Horii, T., unpublished results). This α-helix is therefore very probably involved in the subunit-subunit interaction. Skiba and Knight (1994) confirmed by site-directed mutagenesis that the α-helix G is also important for the subunit-subunit interaction.

By contrast, little is known about the opposite side (β-sheet 3) of the subunit-subunit contact interface. Very large deletions from the C-terminal, including the β-sheet 3 part, apparently affect the polymerization while smaller deletions without this part do not affect the subunit-subunit interaction *in vivo* (Horii et al., 1992). However, such large deletions could also affect the overall structure. Bryant and colleagues (Nguyen et al., 1993) claim that the modification of His97 as well as Lys248, which are involved in the subunit-subunit contact according to the crystal structure, may affect the activity of RecA, but more systematic analysis seems required to establish this conclusion.

In general, further analysis of the contact parts will be required to understand how the size of the helical pitch is modified upon the cofactor binding. In this regard, Kuramitsu et al. (1984) observed that the accessibility to chemical reagent of Cys116, which is in the subunit-subunit interface, increased upon ATP binding. We also note that α-helix D (101–107) which is involved in the subunit-subunit interaction has a contact with the nucleotide cofactor at Tyr103 (Story and Steitz, 1992).

Nucleotide binding site

RecA requires the association of ATP for all activities (Shibata et al., 1979; Weinstock and McEntee, 1981; Little and Mount, 1982; Menge and Bryant, 1992) and gives rise to hydrolysis of ATP to ADP in the presence of DNA (Ogawa et al., 1979; McEntee et al., 1979), although the exact role of ATP hydrolysis in the activities of RecA is not yet quite clear (Menetski et al., 1990; Rosselli and Stasiak, 1991; Jain et al., 1994). RecA appears to bind ATP and ADP at the same site, biochemical analysis showing competitive binding between ATP and ADP (Cotterill et al., 1982).

A RecA crystal soaked in ADP solution has been found able to bind ADP without modifying the crystal packing (Story and Steitz, 1992). The structure of the ADP-RecA complex has thus been solved and the ADP-binding site determined (Story and Steitz, 1992; Fig. 2). The adenine base of ADP is π-stacked upon the Tyr103 of RecA and the phosphate groups are in contact with a loop (66–73) containing a Walker A (or P-loop,

GKT) motif and β-sheet 4 (140–144) containing a Walker B motif, motifs which are conserved in many nucleotide-binding proteins (Saraste et al., 1990). An ADP analog, azidoADP, has been found to cross-link to the region containing residues 110–117, which are rather close to the Tyr103 site, although any cross-link to the Tyr103 itself was not observed (Ogawa and Ogawa, 1986). The fluorescence of a tryptophan residue inserted at the 103 position was completely quenched upon ADP binding indicating a stacking interaction between this residue and ADP also in solution (Morimatsu et al., 1995). An ultraviolet absorption study of the ADP-wild-type RecA complex showed a large hypochromicity (Ellouze, C., Nordén, B. and Takahashi, M., unpublished results), supporting a stacking interaction, which is in accord with the crystal structure.

The ATP-RecA complex has not yet been crystallized and its structure is not known at any detail. As a first approach it would seem reasonable to assume that ATP binds to RecA in a similar way as ADP, although ADP inactivates RecA in contrast to ATP (Weinstock and McEntee, 1981; Menetski and Kowalczykowski, 1985). CD measurements suggest that the conformation of the two nucleotides is similar in their complex with RecA (Watanabe et al., 1994; Wittung et al., 1995). Extensive mutations performed in the P-loop, which is in the ADP-binding domain, are found to affect various RecA activities (Konola et al., 1994a, b), supporting that ATP may indeed bind to the same site as ADP. However, photocross-link experiments and spectroscopic measurements indicate certain differences in the binding modes between ATP and ADP. ATP analog, azidoATP, is cross-linked to Tyr264 (Knight and McEntee, 1985) while ADP analog, azidoADP to the 110–117 region (Ogawa and Ogawa, 1986). Although ATP has been found to be cross-linked to a region (116–170) rather close to that of azidoADP (Banks and Sedgwick, 1986), the difference between azidoATP and azidoADP may reflect a difference in the binding modes. Furthermore, the fluorescence of the tryptophan residue at position 103 is, not completely quenched by ATP binding, in contrast to that by ADP (Morimatsu et al., 1995). Furthermore, the ultraviolet absorption spectrum of the ATP-RecA complex does not exhibit any marked hypochromic effect (Ellouze, C., Nordén, B. and Takahashi, M., unpublished results) contradicting significant stacking interactions.

McEntee and colleagues have shown that ATP binds to the fragment of RecA that contains the Tyr264 residue (Knight and McEntee, 1986) and that the mutation at Tyr264 markedly affects the ATPase activity (Freitag and McEntee, 1991) indicating the importance of this residue. In the crystal structure of the RecA-ADP complex Tyr264 is close to ADP but not in direct contact (Fig. 2; Story and Steitz, 1992). The fluorescence of the tryptophan residue inserted at 264 is strongly quenched upon ATP binding (Morimatsu et al., 1995). The quenching is not, however, complete and of an extent similar to that observed by ADP. These results indicate that Tyr264 is in the vicinity of ATP but not in stacking contact with it. Story and Steitz (1992) proposed that the γ phosphate would be in interaction contact with the disordered L2 loop. However, the fluorescence of a tryptophan residue inserted at the middle of the L2 loop is unaffected by the binding of ATP (Maraboeuf et al., 1995).

More information about the ATP-RecA complex is clearly required to understand the role of ATP in the activation of RecA. Using a protease, Kobayashi and colleagues (1987) have shown that, upon cofactor binding, the region 150–180 (including the L1 loop) becomes more accessible. Since this part is far from the cofactor-binding site, the effect could reflect some conformational change of RecA. Furthermore, one of the two histidine residues of RecA becomes more accessible to chemical modification by diethylpyrocarbonate upon the binding of ATP analog,

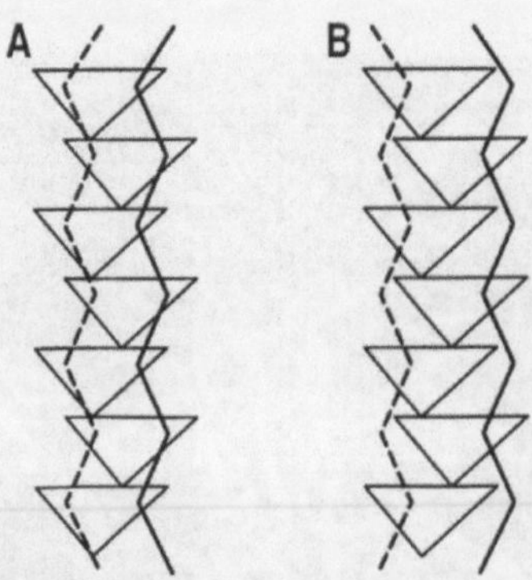

Fig. 3. Two models for binding two DNA strands by the RecA polymer filament. Model A, each subunit interacts with two DNA strands. Model B, each subunit interacts with only one of the two strands at a time.

adenosine 5′-O-3-thiotriphosphate (Takahashi and Nordén, 1993). The result is in accord with the protease analysis as one of the two histidine residues (163) is in the L1 loop. Bryant and colleague observed that the fluorescence of a tryptophan residue at this histidine position (163) is affected by ATP-binding (Stole and Bryant, 1994) and that the change by ATP binding is stronger than that by ADP.

DNA-binding sites

The RecA filament can bind two DNA molecules (one ssDNA and one dsDNA) and has probably three different DNA-binding sites (Kubista et al., 1990; Kurumizaka et al., 1994; Cox, 1995). Each site binds one strand of DNA covering three bases/monomer subunit of RecA. Since RecA forms a polymer on the DNA, this observation does not necessarily mean that every subunit has three DNA-binding sites. It is thus possible that several units are needed to constitute a site or that only one out of two or three subunits binds one DNA chain while another subunit binds the next DNA chain (Fig. 3). One must consider this point and also recall that two parts that may be distant from each other in the monomer of RecA could be close to each other in the polymer (Fig. 1): one part of a subunit may be close to another part of a neighbour subunit. Since RecA can bind one ssDNA without the need of cofactor (McEntee et al., 1981), a subsequent question is whether the binding in the presence and absence of ATP occurs at the same site or not.

No RecA-DNA complex has yet been crystallized. Phase contrast SANS measurements (DiCapua et al., 1989) and advanced image analysis of electron microscopy (Egelman and Yu, 1989) have shown that the primary DNA is in the interior of the RecA filament. Story et al. (1992) proposed from their crystal structure of the RecA filament without DNA that the L1 and L2 loops could be involved in the DNA binding because these parts were facing the central cavity of the RecA filament. However, one should also note that the structures of the L1 and L2 loops were not resolved in the X-ray study, probably because these loops are mobile or of varying conformation, and that the crystal structure of the pure RecA filament corresponds to the RecA-ADP complex, which cannot efficiently bind any DNA (Menetski and Kowalczykowski, 1985; Ellouze, C. and Takahashi, M., unpublished results). The distance between the extremities of the L1 loop is rather large, being of a similar size as the whole peptide (Fig. 4A). Possible positions of these residues in the protein are, therefore, rather limited. By contrast, the distance between the extremities of the L2 loop is much smaller than the

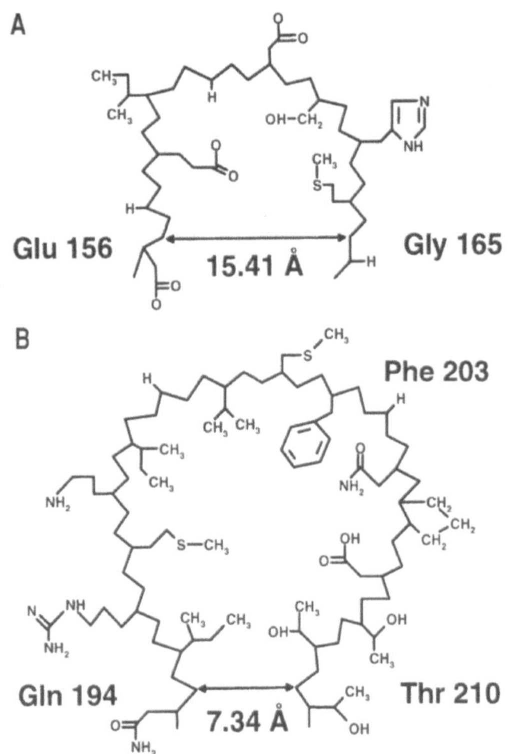

Fig. 4. Schematic presentation of L1 and L2 loops. The size of the L1 (A) and L2 (B) loops is indicated by their amino acid backbones (roughly shown as peptide bonds). The distance between their two extremities are obtained from the crystal structure of RecA (Story et al., 1992).

size of the peptide (Fig. 4B), and thus the position of the middle parts of L2 loop in the protein could not be estimated and might be very distant from its extremities.

Other propositions about the DNA-binding sites have been made by analogies from the comparison of the sequence with other ssDNA binding proteins. One was made simply by comparing sequences without inclusion of any structural information (Kawashima et al., 1984) while another was based on the structure of a DNA-binding domain of GP5 protein (DNA-binding wing; Prasad and Chiu, 1987; Morimatsu and Horii, 1995a). The first authors proposed α-helix A and the latter β-sheets 7 and 8. However, both parts are involved in the subunit-subunit interaction in the crystal structure of RecA filament (Story et al., 1992). Although a peptide that contains α-helix A can interact with DNA (Zlotnick and Brenner, 1989), the interaction is very weak (Gardner et al., 1995), and analysis by site-directed mutagenesis (Morimatsu and Horii, 1995a) and deletion analysis (Horii et al., 1992) indicate that this part is not a DNA-binding site but in fact the subunit-subunit contact part also in the nucleofilament. By contrast, regarding the β-sheets 7 and 8 part, which is also a subunit-subunit interface in the crystal, Morimatsu and Horii claim from photocross-link experiments (1995c) and site-directed mutagenesis analysis (1995a) that this part is really involved in the DNA binding.

The results of cross-link analysis, however depend upon the substrate. The laboratories of Horii (Morimatsu and Horii, 1995b, c) and Kowalczykowski (Rehrauer and Kowalczykowski, 1996) used oligo(dT) or dT-rich oligonucleotides and found that the regions around 65 and 178–183 are cross-linked to DNA (Table 1). The cross-linking parts are not so much affected by the presence of nucleotide. By contrast, the groups of Adzuma (Wang and Adzuma, 1996) and Camerini-Otero (Malkov and Camerini-Otero, 1995) used IdU-containing oligonucleotides and both found cross-liking with L1 and L2 loops. However, there is a difference between their results: while Malkov and Camerini-Otero found that L1 and L2 are both cross-linked to the first DNA, Wang and Adzuma found that L1 is cross-linked to the first DNA and L2 to the second DNA. However, as observed in a part of the nucleotide-binding site, the cross-link does not always occur at the residues in real contact but rather at their proximities.

The discrepancy between the results of Adzuma's and Camerini-Otero's groups could be explained by the closeness of the two sites. By using fluorescence of tryptophan residues inserted at these parts, Takahashi and colleagues have confirmed that the residues Tyr65, Tyr103, Tyr264 and Phe203 could be close to the primary DNA-binding site (Morimatsu et al., 1995; Maraboeuf et al., 1995; Morimatsu, K., Takahashi, M. and Horii, T., unpublished results). They observed that the fluorescence of the tryptophan residue inserted at position 203 is modified both by the first and the second binding DNA strand (Maraboeuf et al., 1995). The results suggest that the first and second binding sites are close to each other, although this demonstration is not in itself an unambiguous proof of closeness because the fluorescence change might be caused indirectly by some conformational change of the protein upon the DNA binding.

Another piece of evidence that supports the involvement of the L2 loop in the DNA binding is the finding that the isolated L2 loop peptide can interact on its own with the DNA (Gardner et al., 1995) and even promote homologous DNA pairing reaction (Voloshin et al., 1996). The latter result supports again the idea that this loop interacts both with the primary and secondary binding DNAs. In addition, mutations in this loop have been observed to affect the DNA binding, although the effect of muta-

Table 1. DNA cross-linked residues of RecA. n.d., not determined.

Research group	Residues cross-linked to			Oligonucleotide used
	−ATP	first DNA	second DNA	
Horii[a]	65, 103, 178–183, 199–216, 264	65, 103, 178–183, 199–216, 264	65, 103	oligo(dT), dT-rich
Kowalczykowski[b]	61–72, 178–183, 233–243	61–72, 178–183, 233–243	n.d.	oligo(dT)
Camerini-Otero[c]	n.d.	164, 203	n.d.	oligo(dIU)
Adzuma[d]	n.d.	164	202, 203	oligo(dIU)

[a] Morimatsu and Horii (1995b, c) and unpublished data.
[b] Rehrauer and Kowalczykowski (1996).
[c] Malkov and Camerini-Otero (1995).
[d] Wang and Adzuma (1996).

tion in the DNA binding is rather modest (Cazaux et al., 1994) and thus indicates that other parts may also be involved in the DNA binding.

By contrast, the isolated L1 loop peptide cannot bind DNA (Gardner et al., 1995) even when the peptide contains a Thr-Pro-Lys-Ala (150–153) sequence (Takahashi and Nordén, 1994) which is similar to the S(T) PKK histone DNA-binding motif (Suzuki, 1989). Furthermore, the L1 loop has exhibited large mutational flexibility (extensive study on 152–159 part) and only Glu154 was found to be a critical residue (Nastri and Knight, 1994). Muench and Bryant (1991) observed that modifications of the L1 loop at position 160 or 163 affect the strand-exchange activity without affecting the ssDNA-dependent ATPase activity. This could indicate that the L1 loop is involved in the second DNA binding. Chemical interference analysis on histidine residues also support that His163 in L1 loop could be close to DNA (Takahashi and Nordén, 1993), but the protection by dsDNA is more pronounced than that by ssDNA suggesting that the L1 loop is close to the secondary DNA-binding site. This is in agreement with the proposition of Story et al. (1992) but conflict with the result of Wang and Adzuma (1996). More work is clearly required to understand the potential role of this part of RecA in the DNA binding.

The modification of Tyr103 and Tyr264 affects the binding of DNA even in the absence of cofactor which further supports the suggestion that these sites are involved in the DNA binding (Eriksson et al., 1993; Morimatsu and Horii, 1995a). Tyr103 and Tyr264 are, as discussed previously, involved in the cofactor binding. There seems, thus, to be at least a partial overlap between the DNA and cofactor-binding sites. Another DNA-binding part around position 65 which was found to be cross-linked to DNA by Horii's and Kowalczykowski's groups (Morimatsu and Horii, 1995b; Rehrauer and Kowalczykowski, 1996) is also close to the P-loop of the nucleotide binding site.

The results presented so far have concerned the first and second DNA-binding sites. There is little data about the putative third DNA-binding site. It was originally proposed based on the finding that the binding of an additional (third) strand of DNA prevents the cleavage of LexA (Takahashi and Schnarr, 1989) and the observation that the third DNA bound to RecA is accessible to DNase (Kubista et al., 1990), the third site therefore being inferred to be outside or at the mouth of the groove of the RecA helix. This site may overlap with the LexA-binding site (see below). Shibata and colleagues proposed from the DNA-binding character of a chimera RecA that the part 58–110 is the third DNA-binding site (Kurumizaka et al., 1994). This part, however, includes the primary binding site, Tyr103. All three DNA-binding sites may be close to each other.

Parts of interaction with other proteins

An electron microscopy study of the LexA-RecA filament complex has shown that LexA is bound at the groove of the RecA filament (Yu and Egelman, 1993). Comparison with the crystal structure of RecA suggests that the 156–165 and 229–243 regions are the LexA interacting domain. Tessman and colleagues have selected a rather large number of SOS constitutive mutants and found that the mutations are located at the 25–39, 157–184 and 298–301 parts (Wang and Tessman, 1986). These results indicate that the 156–165 part is indeed the LexA-binding site. Note that this part is the L1 loop. Nastri and Knight (1994) observed that substitutions of Gly157 create constitutive coprotease mutants but decrease recombination activity. These results indicate that the L1 loop may be the second DNA-binding site and that it overlaps with the LexA-binding site. The 25–39 part is in the subunit-subunit interface. Some modification of subunit-subunit contact may affect the coprotease activity because LexA binds to the groove of RecA filament and has contacts with two RecA subunits. Knight has observed that modification in the 213–222 part, which is also the subunit-subunit interface, affects coprotease activity (personal communication). Concerning the 298–301 part, the reason for the mutation is not clear. It is not yet clear whether the interaction with UmuD occurs at the same site for its auto-cleavage.

Concerning the binding site for the UmuD$'_2$C complex, the group of Devoret isolated a large number of RecA mutants for which the recombination activity is not inhibited by UmuD′ (Sommer et al. 1995). The mutations are mainly located at the growing extremity of the RecA filament. The result explains the inhibition of recombination by UmuD$'_2$C: the complex interacts with this part of RecA and prevents the growth of the RecA filament and stops the recombination reaction. Some minor mutants are situated at the opposite side of the subunit-subunit interface. Probably the mutant RecA proteins more efficiently polymerize and compete with the binding of the UmuD$'_2$C complex. Here again we can see an overlap between two functional domains (subunit-subunit interaction site and UmuD$'_2$C-binding site). This overlap may play a role for the switch of DNA repair mode from a recombination mode to a mutational repair mode.

Conclusion

The X-ray crystallographic three-dimensional structure of the RecA homopolymer indicates which parts may be active in filament formation and nucleotide binding. This structure is an important basis also for the consideration of other functional parts of RecA. The knowledge has been further advanced by a combination of photocross-link analysis, site-directed mutagenesis and fluorescence measurement of tryptophan inserted at different parts of RecA. The locations of the functional domains and their overlaps are shown in Fig. 5.

The subunit-subunit interfaces and nucleotide-cofactor-binding sites appear to be relatively clear from the X-ray crystallographic analysis. The suggested positions are supported by the various biochemical and spectroscopic analyses although certain details still remain to be revealed, such as how the change of filament structure is affected by cofactor binding and what conformational differences exist between the RecA complexes with ATP and ADP. The DNA-binding sites have not yet been pinpointed because of lack of any RecA-DNA complex crystal, although some data about the parts involved begin to accumulate: the L1 and L2 loops are involved. All three DNA strands seem to interact with the same part of RecA. This means, if each subunit has contacts with all DNA strands (Model A of Fig. 3), that the three strands are extremely close to each other within the RecA filament. One must clarify whether each subunit has contacts with all three DNA strands (Model A) or only one or two strands (Model B).

An interesting question is the potential overlap of domains with different functions. Fig. 5 shows that all functional domains are localized in a small part of the RecA protein and thus close to each other in space. DNA binding and nucleotide-cofactor-binding share the Tyr103-Tyr264 and Tyr65 regions. One possibility is that the occupation of the nucleotide at this site affects, by direct interaction, the binding mode of DNA to RecA. Further, the subunit-subunit interfaces are also close to the DNA-binding sites (Tyr65, L1 loop) and the nucleotide-binding site (residues 103, 144 and 145). The binding of nucleotide cofactor can thus affect the subunit-subunit contact, whose change, in its turn, would affect the DNA binding. The third DNA binding could be close to the repressor binding site in the L1 loop. Prob-

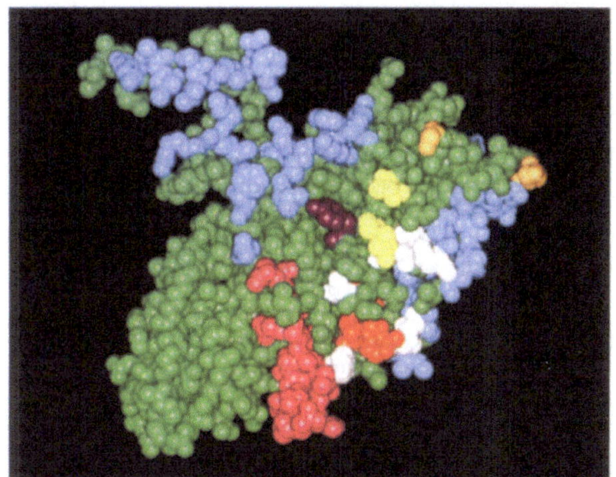

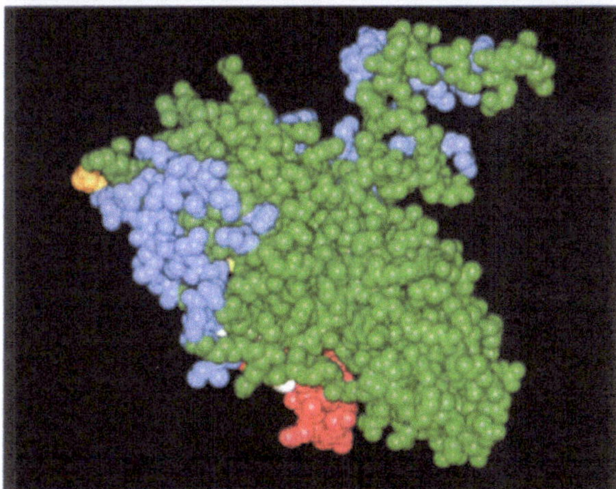

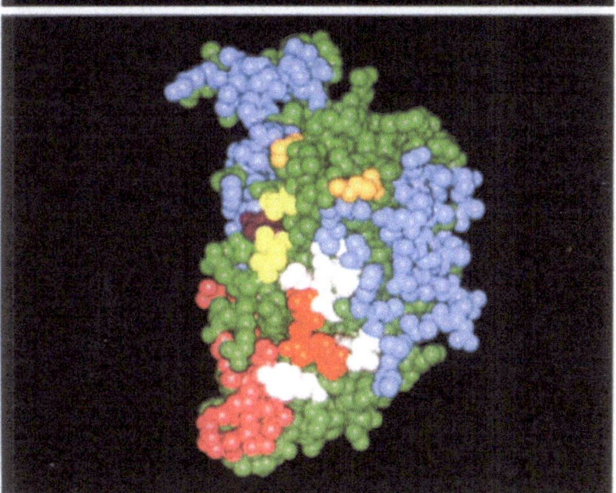

Fig. 5. Locations of functional domains of RecA. Subunit-subunit interaction interfaces (light blue), cofactor binding site (white), LexA binding parts (pink) are shown in the three-dimensional structure of RecA. Putative DNA-binding domains are coloured yellow (L2 loop), orange (L1 loop) and purple. Only the extremities of the L1 and L2 loops are highlighted as these loops are not resolved in the X-ray analysis. The L1 loop interacts also with LexA repressor. In (A) RecA is seen from the inside of the filament and in (B) from the outside. (C) presents the profile. Note the high concentration of functional domains on the inside (A) in contrast to the empty outside (B).

ably, in this way the binding of a third DNA strand from the second DNA (dsDNA) inhibits LexA cleavage. This stops the SOS induction and thus switches the repair mode from excision repair to recombinational repair. The UmuD$'_2$C-binding site is adjacent to the subunit-subunit interface. This could explain why the binding of UmuD$'_2$C prevents the cooperative binding of RecA to DNA. It inhibits the recombination reaction and switches the repair mode from recombinational repair to mutational repair.

In conclusion, we find overlaps between functional domains a very interesting issue and we propose that they could have important roles in regulating several of the activities of RecA.

We thank Drs K. Adzuma, A. Bailone, D. Camerini-Otero, M. Cox, T. Horii, K. Knight, S. Kowalczykowski, K. Morimatsu, T. Shibata and S. Sommer for communicating manuscripts and data prior to publication and we also thank Dr R. Devoret for valuable discussion. This work has been supported by *Institute Curie*, *Centre National de la Recherche Scientifique*, Swedish Research Council and Japan Society for the Promotion of Science.

REFERENCES

Aboussekhra, A., Chanet, R., Adjiri, A. & Fabre, F. (1992) Semidominant suppressors of Srs2 helicase mutations of *Saccharomyces cerevisiase* map in the Rad51 gene, whose sequence predicts a protein with similarities to procaryotic RecA proteins, *Mol. Cell. Biol. 12*, 3224–3234.

Banks, G. R. & Sedgwick, S. G. (1986) Direct ATP photolabelling of *Escherichia coli* RecA proteins: identification of regions required for ATP binding, *Biochemistry 25*, 5882–5889.

Burckhardt, S. E., Woodgate, R., Scheuermann, R. H. & Echols, H. (1988) UmuD mutagenesis protein of *Escherichia coli*: over-production, purification and cleavage by RecA, *Proc. Natl Acad. Sci. USA 85*, 1811–1815.

Cazaux, C., Larminat, F., Villani, G., Johnson, N. P., Schnarr, M. & Defais, M. (1994) Purification and biochemical characterization of *Escherichia coli* RecA proteins mutated in the putative DNA binding site, *J. Biol. Chem. 269*, 8246–8254.

Cotterill, S. M., Satterthwait, A. C. & Fersht, A. R. (1982) recA protein from *Escherichia coli*. A very rapid and simple purification procedure: binding of adenosine 5′-triphosphate and adenosine 5′-diphosphate by the homogeneous protein, *Biochemistry 21*, 4332–4337.

Cox, M. M. (1995) Alignment of 3 (but not 4) DNA strands within a RecA protein filament, *J. Biol. Chem. 270*, 26021–26024.

Devoret, R. (1992) *Les fonctions SOS ou comment les bactéries survivent aux lésions de leur ADN*, *Ann. Inst. Pasteur (Paris) 1*, 11–20.

DiCapua, E., Engel, A., Stasiak, A. & Koller, T. (1982) Characterization of complexes between RecA protein and duplex DNA by electron microscopy, *J. Mol. Biol. 157*, 87–103.

DiCapua, E., Schnarr, M. & Timmins, P. A. (1989) The location of DNA in complexes of recA protein with double-stranded DNA. A neutron scattering study, *Biochemistry 28*, 3287–3292.

DiCapua, E., Cuillel, M., Hewat, E., Schnarr, M., Timmins, P. A. & Ruigrok, R. W. H. (1992) Activation of RecA protein. The open helix model for LexA cleavage, *J. Mol. Biol. 226*, 707–719.

Egelman, E. H. (1993) What do X-ray crystallographic and electron microscopic structural studies of the RecA protein tell us about recombination? *Curr. Opin. Struct. Biol. 3*, 189–197.

Egelman, E. H. & Yu, X. (1989) The location of DNA in RecA-DNA helical filaments, *Science 245*, 404–407.

Elledge, S. J. & Walker, G. C. (1983) Protein required for ultraviolet light and chemical mutagenesis, *J. Mol. Biol. 164*, 175–192.

Ellouze, C., Takahashi, M., Wittung, P., Mortensen, K., Schnarr, M. & Nordén, B. (1995) Elongation of helical pitch of RecA filament upon ATP and ADP binding evidenced from small angle neutron scattering data, *Eur. J. Biochem. 233*, 579–583.

Eriksson, S., Nordén, B., Morimatsu, K., Horii, T. & Takahashi, M. (1993) Role of tyrosine residue 264 of RecA for the binding of cofactor and DNA, *J. Biol. Chem. 268*, 1811–1816.

Frank, E. G., Hauser, J., Levine, A. S. & Woodgate, R. (1993) Targeting of the UmuD, UmuD', and MucA' mutagenesis proteins to DNA by RecA protein, *Proc. Natl Acad. Sci. USA 90*, 8169–8173.

Freitag, N. E. & McEntee, K. (1991) Site-directed mutagenesis of the RecA protein of *Escherichia coli*, *J. Biol. Chem. 266*, 7058–7066.

Gardner, R. V., Voloshin, O. N. & Camerini-Otero, R. D. (1995) The identification of the single-stranded DNA binding domain of the *Escherichia coli* RecA protein, *Eur. J. Biochem. 233*, 419–425.

George, J., Devoret, R. & Radman, M. (1974) Indirect ultraviolet-reactivation of phage λ, *Proc. Natl Acad. Sci. USA 71*, 144–147.

Hagmar, P., Nordén, B., Baty, D., Chartier, M. & Takahashi, M. (1992) Structure of DNA-RecA complexes studied by residue differential linear dichroism and fluorescence spectroscopy for a genetically engineered RecA protein, *J. Mol. Biol. 226*, 1193–1205.

Heuser, J. & Griffith, J. (1989) Visualization of RecA protein and its complexes with DNA by quick-freeze/deep-etch electron microscopy, *J. Mol. Biol. 210*, 473–484.

Horii, T., Ozawa, N., Ogawa, T. & Ogawa, H. (1992) Inhibitory effects of N- and C-terminal truncated *Escherichia coli* recA gene products on functions of the wild-type *recA* gene, *J. Mol. Biol. 223*, 105–114.

Jain, S. K., Cox, M. M. & Inman, R. B. (1994) On the role of ATP hydrolysis in RecA protein-mediated DNA strand exchange, *J. Biol. Chem. 269*, 20653–20661.

Jonsson, M., Jacobsson, U., Takahashi, M. & Nordén, B. (1993) Orientation of large DNA during free solution electrophoresis studied by linear dichroism, *J. Chem. Soc. Faraday Trans. 89*, 2791–2798.

Kawashima, H., Horii, T., Ogawa, T. & Ogawa, H. (1984) Functional domains of *Escherichia coli* recA protein deduced from the mutational sites in the gene, *Mol. Gen. Genet. 193*, 288–292.

Knight, K. L. & McEntee, K. (1985) Tyrosine 264 in the recA protein from *Escherichia coli* is the site of modification by photoaffinity label 8-azidoadenosine 5'-triphosphate, *J. Biol. Chem. 260*, 10185–10191.

Knight, K. L. & McEntee, K. (1986) Nucleotide binding by a 24-residue peptide from the RecA protein of *Escherichia coli*, *Proc. Natl Acad. Sci. USA 83*, 9289–9293.

Kobayashi, N., Knight, K. & McEntee, K. (1987) Evidence for nucleotide-mediated changes in the domain structure of the RecA protein from *Escherichia coli*, *Biochemistry 26*, 6801–6810.

Konola, J. T., Logan, K. M. & Knight, K. L. (1994a) Functional characterization of residues in the P-loop motif of the RecA protein ATP binding site, *J. Mol. Biol. 237*, 20–34.

Konola, J. T., Nastri, H. G., Logan, K. M. & Knight, K. L. (1994b) Mutations at Pro67 in the RecA protein P-loop motif differentially modify coprotease function and separate coprotease from recombination activities, *J. Biol. Chem. 270*, 8411–8419.

Kowalczykowski, S. C., Dixon, D. A., Eggleston, A. K., Lauder, S. D. & Rehrauer, W. M. (1994) Biochemistry of homologous recombination in *Escherichia coli*, *Microbiol. Rev. 58*, 401–465.

Kubista, M., Takahashi, M. & Nordén, B. (1990) Stoichiometry, base orientation, and nuclease accessibility of RecA-DNA complexes seen by polarized light in flow oriented solution. Implication for mechanism of genetic recombination, *J. Biol. Chem. 265*, 18891–18897.

Kuramitsu, S., Hamaguchi, K., Tachibana, H., Horii, T., Ogawa, T. & Ogawa, H. (1984) Cysteinyl residues of *Escherichia coli* recA protein, *Biochemistry 23*, 2363–2367.

Kurumizaka, H., Ikawa, S., Ikeya, T., Ogawa, T. & Shibata, T. (1994) A chimeric RecA protein exhibits altered double-stranded DNA binding, *J. Biol. Chem. 269*, 3068–3075.

Lauder, S. D. & Kowalczykowski, S. C. (1993) Negative co-dominant inhibition of recA protein function, *J. Mol. Biol. 234*, 72–86.

Little, J. W. & Mount, D. W. (1982) The SOS regulatory system of *Escherichia coli*, *Cell 29*, 11–22.

Malkov, V. A. & Camerini-Otero, R. D. (1995) Photocross-links between single-stranded DNA and *Escherichia coli* RecA protein map to loops L1 and L2, *J. Biol. Chem. 270*, 30230–30233.

Maraboeuf, F., Voloshin, O., Camerini-Otero, D. & Takahashi, M. (1995) The central aromatic residue in loop L2 of RecA interacts with DNA: quenching of the fluorescence of tryptophan reporter inserted L2 upon binding to DNA, *J. Biol. Chem. 270*, 30927–30932.

McEntee, K., Weinstock, G. M. & Lehman, I. R. (1979) Initiation of general recombination catalyzed *in vitro* by the *recA* protein *of Escherichia coli*, *Proc. Natl Acad. Sci. USA 76*, 2615–2619.

McEntee, K., Weinstock, G. M. & Lehman, I. R. (1981) Binding of the RecA protein of *Escherichia coli* to single- and double-stranded DNA, *J. Biol. Chem. 256*, 8835–8844.

Menetski, J. P. & Kowalczykowski, S. C. (1985) Interaction of recA protein with single-stranded DNA. Quantitative aspect of binding affinity modulation by nucleotide cofactors, *J. Mol. Biol. 181*, 281–295.

Menetski, J. P., Bear, D. G. & Kowalczykowski, S. C. (1990) Stable DNA heteroduplex formation catalyzed by the *Escherichia coli* RecA protein in the absence of ATP hydrolysis, *Proc. Natl Acad. Sci. USA 87*, 21–25.

Menge, K. L. & Bryant, F. R. (1992) Effect of nueelotide cofactor structure on RecA protein-promoted DNA pairing, *Biochemistry 31*, 5151–5157.

Mikawa, T., Masui, R., Ogawa, T., Ogawa, H. & Kuramitsu, S. (1995) N-terminal 33 amino acid residues of *Escherichia coli* RecA protein contribute to its self-assembly, *J. Mol. Biol. 250*, 471–483.

Morimatsu, K. & Horii, T. (1995a) Analysis of the DNA binding site of *Escherichia coli* RecA protein, *Adv. Biophys. 31*, 23–48.

Morimatsu, K. & Horii, T. (1995b) The DNA binding site of the RecA protein: photochemical cross-linking of Tyr103 to single-stranded DNA, *Eur. J. Biochem. 228*, 772–778.

Morimatsu, K. & Horii, T. (1995c) DNA binding surface of the RecA protein: photochemical cross-linking of the first DNA site on RecA filament, *Eur. J. Biochem. 234*, 695–705.

Morimatsu, K., Horii, T. & Takahashi, M. (1995) Interaction of tyrosine residues 103 and 264 of RecA protein with nucleotide cofactors and DNA: fluorescence study of engineered proteins, *Eur. J. Biochem. 228*, 779–785.

Morimatsu, K., Maraboeuf, F., Hagmar, P., Nordén, B., Horii, T. & Takahashi, M. (1996) Role of Tyr103 and Tyr264 in the regulation of RecA-DNA interactions by nucleotide cofactor, *Eur. J. Biochem. 240*, 91–97.

Muench, K. A. & Bryant, F. R. (1991) Disruption of an ATP-dependent isomerization of the recA protein by mutation of histidine 163, *J. Biol. Chem. 266*, 844–850.

Nastri, H. G. & Knight, K. L. (1994) Identification of residues in the L1 region of the RecA protein which are important to recombination or coprotease activities, *J. Biol. Chem. 269*, 26311–26322.

Nguyen, T. T., Muench, K. A. & Bryant, F. R. (1993) Inactivation of the recA protein by mutation of histidine 97 or lysine 248 at the subunit interface, *J. Biol. Chem. 268*, 3107–3113.

Ogawa, T., Wabiko, H., Tsurimoto, T., Hori, T., Masukata, H. & Ogawa, H. (1979) Characteristics of purified RecA protein and the regulation of its synthesis *in vitro*, *Cold Spring Harbor Symp. Quant. Biol. 43*, 909–914.

Ogawa, T. & Ogawa, H. (1986) General recombination: functions and structure of RecA protein, *Adv. Biophys. 21*, 135–148.

Peterson, K. R. & Mount, D. W. (1987) Differential repression of SOS genes by unstable, LexA41 (Tsl-1) protein causes a 'split-phenotype' in *Escherichia coli* K-12, *J. Mol. Biol. 193*, 27–40.

Prasad, B. V. & Chiu, W. (1987) Sequence comparison of single-stranded DNA binding proteins and its structural implications, *J. Mol. Biol. 193*, 579–584.

Rehrauer, W. M. & Kowalczykowski, S. C. (1996) The DNA binding site(s) of the *Escherichia coli* RecA protein, *J. Biol. Chem. 271*, 11996–12002.

Roca, A. I. & Cox, M. M. (1990) The RecA protein: structure and function, *Critic. Rev. Biochem. Mol. Biol. 25*, 415–456.

Rosselli, W. & Stasiak, A. (1991) The ATPase activity of RecA is needed to push the DNA strand exchange through heterologous regions, *EMBO J. 10*, 4391–4396.

Saraste, M., Sibbald, P. R. & Wittghofer, A. (1990) The P-loop – a common motif in ATP- and GTP-binding proteins, *Trends Biochem. Sci. 45*, 430–434.

Shibata, T., Das Gupta, C., Cunningham, R. P. & Radding, C. M. (1979) Purified *Escherichia coli* RecA protein catalyzes homologous pairing of superhelical DNA and single-stranded fragments, *Proc. Natl Acad. Sci. USA 76*, 1638–1642.

Shinagawa, H., Kato, T., Ise, T., Makino, K. & Nakata, A. (1983) Cloning and characterization of the umuD operon responsible for inducible mutagenesis in *Escherichia coli*, *Gene* (*Amst.*) *23*, 167–174.

Shinohara, A., Ogawa, H., Matsuda, Y., Ushio, N., Ikeo, K. & Ogawa, T. (1993) Cloning of human, mouse and fission yeast genes homologous Rad51 and recA, *Nat. Genet. 4*, 239−243.

Skiba, M. C. & Knight, K. L. (1994) Functionally important residues at a subunit interface site in the RecA protein from *Escherichia coli*, *J. Biol. Chem. 269*, 3823−3828.

Smith, K. C. & Wang, T. V. (1989) *recA*-dependent DNA repair processes, *Bioessays 10*, 12−16.

Sommer, S., Bailone, A. & Devoret, R. (1993) The appearance of the UmuD'C protein complex in *Escherichia coli* switches repair from homologous recombination to SOS mutagenesis, *Mol. Microbiol. 10*, 963−971.

Stole, E. & Bryant, F. R. (1994) Introduction of a tryptophan reporter into loop1 of recA protein. Examination of the conformational states of the recA-ssDNA complex by fluorescence spectroscopy, *J. Biol. Chem. 269*, 7919−7925.

Story, R. M. & Steitz, T. A. (1992) Structure of the recA protein-ADP complex, *Nature 355*, 374−376.

Story, R. M., Weber, I. T. & Steitz, T. A. (1992) The structure of the *E. coli recA* protein monomer and polymer, *Nature 355*, 318−325.

Suzuki, M. (1989) SPKK, a new nucleic acid binding unit of protein found in histone, *EMBO J. 8*, 797−804.

Takahashi, M. & Schnarr, M. (1989) Investigation of RecA-polynucleotide interactions from the measurement of LexA repressor cleavage kinetics. Presence of different types of complex, *Eur. J. Biochem. 183*, 617−622.

Takahashi, M., Kubista, M. & Nordén, B. (1989) Binding stoichiometry and structure of RecA-DNA complexes studied by the linear dichroism and fluorescence, *J. Mol. Biol. 205*, 137−147.

Takahashi, M. & Nordén, B. (1993) Accessibility to modification reagent of histidine residues of RecA protein upon DNA and cofactor binding, *Eur. J. Biochem. 217*, 665−670.

Takahashi, M. & Nordén, B. (1994) Structure of RecA-DNA complex and mechanism of DNA strand exchange reaction in homologous recombination, *Adv. Biophys. 30*, 1−35.

Voloshin, O. N., Wang, L. & Camerini-Otero, R. D. (1996) Homologous DNA pairing promoted by a 20-amino acid peptide derived from RecA, *Science 272*, 868−872.

Walker, G. C. (1984) Mutagenesis and inducible responses to deoxyribonucleic acid damage in *Escherichia coli*, *Microbiol. Rev. 48*, 60−93.

Wang, W.-B. & Tessman, E. S. (1986) Location of functional regions of the *Escherichia coli* RecA protein by DNA sequence analysis of RecA protease-constitutive mutants, *J. Bacteriol. 168*, 901−910.

Wang, Y. & Adzuma, K. (1996) Differential proximity probing of two DNA binding sites in the *Escherichia coli* RecA protein using photo-cross-linking methods, *Biochemistry 35*, 3563−3571.

Watanabe, R., Matsui, R., Mikawa, T., Takamatsu, S., Kato, R. & Kuramitsu, S. (1994) Interaction of *Escherichia coli* RecA protein with ATP and its analogues, *J. Biochem. (Tokyo) 116*, 960−966.

Weinstock, G. M. & McEntee, K. (1981) RecA protein-dependent proteolysis of bacteriophage λ repressor, *J. Biol. Chem. 256*, 10883−10888.

Williams, R. C. & Spengler, S. J. (1986) Fibers of RecA protein and complexes of RecA protein and single-stranded ϕX174 DNA as visualized by negative-stain electron microscopy, *J. Mol. Biol. 187*, 109−118.

Wittung, P., Nordén, B. & Takahashi, M. (1995) Secondary structure of RecA in solution. The effect of cofactor, DNA and ionic conditions, *Eur. J. Biochem. 228*, 149−154.

Yu, X. & Egelman, E. H. (1993) The LexA repressor binds within the deep helical groove of the activated RecA filament, *J. Mol. Biol. 231*, 29−40.

Zlotnick, A. & Brenner, S. L. (1989) An alpha-helical peptide model for electrostatic interactions of proteins with RecA. The N-terminus of RecA, *J. Mol. Biol. 209*, 447−457.

Eur. J. Biochem. 242, 171–185 (1996)

Review

Regulation of phosphorylation pathways by p21 GTPases

The p21 Ras-related Rho subfamily and its role in phosphorylation signalling pathways

Louis LIM[1,2], Edward MANSER[2], Thomas LEUNG[2] and Christine HALL[1]

[1] Institute of Neurology, London, UK

[2] Glaxo-IMCB Group, Institute of Molecular and Cell Biology, National University of Singapore, Singapore

(Received 16 August/27 September 1996) – EJB 96 1224/0

The oncogenic Ras p21 GTPases regulate phosphorylation pathways that underlie a wealth of activities, including growth and differentiation, in organisms ranging from yeast to human. In metazoa, growth factors trigger conversion of Ras from an inactive GDP-bound form to an active GTP-bound form. This activation of Ras leads to activation of Raf. Raf is one of the initial kinases in the cytoplasmic mitogen-activated protein kinase (MAPK) cascade, involving extracellular-signal-regulated kinases (ERK), which culminates in nuclear transcription. The Ras-related subfamily of Rho p21s, including Rho, Rac and Cdc42 are similarly active in their GTP-bound forms. These p21s mediate growth-factor-induced morphological changes involving actin-based cellular structures. For example, in mammalian fibroblasts, Rho mediates the formation of cytoskeletal stress fibres induced by lysophosphatidic acid, while Rac mediates the formation of membrane ruffles induced by platelet-derived growth factor, and Cdc42 mediates the formation of peripheral filopodia by bradykinin. In some cases, factor-induced Rac activation results in Rho activation, and factor-induced Cdc42 activation leads to Rac activation, as determined by specific morphological changes. Although separate Cdc42/Rac and Rac/Rho hierarchies exist, these might not extend into a linear form (i.e. Cdc42→Rac→Rho) since Cdc42 and Rho activities may be competitive or even antagonistic. Thus Cdc42-mediated formation of filopodia is accompanied by loss of stress fibres (whose formation is mediated by Rho). Recently, mammalian kinases that bind to the GTP-bound forms of Rho p21s have been isolated. These kinases include the p21-activated serine/threonine kinase (PAK), which is stimulated by binding to Cdc42 and Rac, and the Rho-binding serine/threonine kinase (ROK), which is not as strongly stimulated by binding. These kinases act as effectors for their p21 partners since they can directly affect the reorganization of the relevant actin-containing structures. ROK promotes the formation of Rho-induced actin-containing stress fibres and focal-adhesion complexes, to which the ends of the stress fibres attach. PAK stimulates the disassembly of stress fibres, which has been shown to accompany formation of Cdc42-induced peripheral-actin-containing structures, including filopodia, which with Rac-induced membrane ruffles play a role in cell movement. PAK also fosters loss of focal-adhesion complexes. Thus, there is cooperation between different Rho p21s as well as antagonism, with their associated kinases having a role in the integration of the reorganization of the actin cytoskeleton. The similarity of PAK to the *Saccharomyces cerevisiae* kinase Ste20p, which initiates the yeast mating/pheromone MAPK cascade, led to experiments showing that Cdc42 regulates Ste20p in this MAPK pathway. This similarity has also led to the demonstration that mammalian Cdc42 and Rac can signal to the nucleus through MAPK pathways. However, c-Jun N-terminal kinase (JNK, stress-activated protein kinase) rather than ERK, is involved. PAK have been implicated in the JNK pathway, but their exact roles are uncertain. Thus members of the Rho subfamily, and kinases that bind to these p21s are intimately involved in immediate morphological processes as well as long-term transcriptional events.

Keywords: mitogen-activated kinase pathways; morphology; p21-activated kinase; Rho guanosine triphosphatase; Rho-binding kinase.

Correspondence to L. Lim, Institute of Neurology, 1 Wakefield St., London WC1N 1PJ, UK

Fax: +44 171 278 7045.

Abbreviations. ACK, activated Cdc42-binding kinase; Bcr, breakpoint cluster region gene product; ERK, extracellular-signal-regulated kinase; GEF, guanine-nucleotide-exchange factor; JNK, c-Jun N-terminal kinase; MAPK, mitogen-activated protein kinase; MBS, myosin-binding subunit of myosin phosphatase; MLC, myosin light chain; PAK, p21 (Cdc42/Rac)-activated kinase; PAM, peripheral actin microspikes; PDGF, platelet-derived growth factor; PH, pleckstrin homology; PKC, protein kinase C; PKN, protein kinase N; ROK, Rho-binding kinase; RTK, receptor tyrosine kinase; SH, Src-homology region; SRE, serum-response element; EGF, epidermal-growth factor; GAP, GTPase-activating protein; FAK, focal-adhesion kinase; PYK, proline-rich tyrosine kinase; MST, mammalian sterile-twenty-like.

This review deals with the regulation by Ras-related Rho p21s of phosphorylation pathways, including those that lead to changes in actin-based morphology.

The role of Ras p21s in mediating the effects of growth factors on metabolic pathways that are involved in various aspects of growth and differentiation became clearer with the realization that they performed similar roles in diverse organisms, ranging from unicellular yeast to *Drosophila* to human. In many cases, the critical biochemical event in Ras-signalling pathways is the activation of specific receptor tyrosine kinases (RTK), which through Ras initiate a sequential series of phosphorylation events, the kinase cascades (McCormick, 1994; Marshall, 1994, 1996). Ras signalling can also involve receptors linked to non-receptor kinases or heterotrimeric G-proteins.

All members of the Ras superfamily are GTP-binding proteins that possess intrinsic GTPase activity (Boguski and McCormick, 1993). These p21s act as molecular switches, being active in the GTP-bound form and inactive in the GDP-bound form to which they are converted by the GTPase activity. Different members of the family have different rates of GTPase activity, which can thus determine the duration and possibly the extent of the signal. The GTP/GDP cycle is modulated by interaction with guanine-nucleotide-exchange factors (GEF), which upregulate p21s by stimulating GTP loading, GDP-dissociation inhibitors, which inhibit exchange and maintain the GDP-bound p21 in a soluble complex, and GTPase-activating proteins (GAP), which downregulate p21s by stimulating the intrinsic GTPase activity.

Ligand binding to RTK results in the formation of GTP-Ras. Active Ras is responsible for the activation of Raf-1, a serine/threonine kinase. Phosphorylation by Raf of mitogen-activated protein kinase (MAPK) kinase leads to activation of MAPK. This dual specificity kinase phosphorylates Thr and Ser residues on the Ets family of transcriptional factors, which allows them to bind specific DNA motifs, sometimes in association with other factors, after translocation into the nucleus. The phosphorylation pathways in *Drosophila melanogaster* utilized in development include those of *sevenless* and *torso* for determination of photoreceptor cells and of terminal structures, respectively. In *Caenorhabditis elegans*, vulval development is the endpoint of the *let-23* pathway. Although these developmental events are more well defined than in mammalian systems, the identities of the causative agents/ligands responsible for the activation of these RTK, which trigger the Ras-MAPK pathway, are unknown. In mammalian cells, the ligands that trigger RTK activation include growth factors.

The manner in which RTK transmit their signal to Ras involves ligand-induced phosphorylation, resulting in the binding of Src-homology region (SH)2/SH3-containing adapter molecules (e.g. Grb) which in turn recruit the Cdc25-related GEF to the membrane, where it promotes the formation of the GTP-bound form of Ras. The binding of Raf-1 by Ras (which must be prenylated on its C-terminal tail) leads to a membrane disposition of Raf, fostering its phosphorylation (and activation) by a tyrosine kinase present in the membranes. Whether the kinase is Src remains to be established. [Factors other than kinases may also be required, and Raf-B appears to be directly activated by Ras itself (Yamamori et al., 1995)]. The specific region within Raf-1 necessary for Ras association includes a Ras-binding domain and the cysteine-rich domain just N-terminal of this. The cysteine-rich domain is similar to the C1 region of protein kinase C (PKC), counterparts of which are present in a variety of transduction molecules, including several Rho GAP. Whether the lipid-binding properties of this domain are involved in its activation is not known.

Ras-related subfamily of Rho GTPases

This subfamily is represented in all organisms that contain Ras. In mammals, the subfamily includes Rho A, B and C, Rac 1 and 2 and two isoforms of Cdc42. The isoforms exhibit 80–90% identity within each class. Lower organisms have their own set of specific isoforms within the three classes. For example, all three classes are represented in *C. elegans* and *Drosophila* while *Saccharomyces cerevisiae* appears not to contain Rac but more types of Rho. These p21s have been implicated in a wide variety of biological activities, including morphological changes (Hall, 1994), motility and mitosis (Takai et al., 1995), apoptosis (Jimenez et al., 1995), cell-cycle progression (Olson et al., 1995) and transformation (Prendergast et al., 1995; Symons, 1996). As with Ras, members of this family have been shown to be intimately involved in phosphorylation signalling pathways. The isolation and characterization of kinases associated with this subfamily has provided much of the impetus in implicating these GTPases in several metabolic events with which they were previously not associated.

There is a large family of sequence-related GAP for the Rho-family proteins, which have Bcr-homology domains (Diekmann et al., 1991), including Abr, α-chimaerin, β-chimaerin and p190 RhoGAP (Lamarche and Hall, 1994; Lim et al., 1995). Many of the GEF are related to the product of the oncogene *dbl* (Hart et al., 1991) e.g. dbs, ost, lbc, tiam-1 (Cerione and Zheng, 1996). These GEF are growth-regulatory molecules and have cell-transforming potential. The GAP and GEF are characterized by their multidomain structure, including SH2, SH3 and pleckstrin-homology (PH) domains, proline-rich sequences and cysteine-rich domains, in addition to one or more catalytic domain. Intriguingly, *n*-chimaerin can also act as an effector (Kozma et al., 1996).

The group of proteins that interact only with GTP-bound form of the p21s and therefore represent targets/effectors include kinases, the prototypes being the activated Cdc42-binding kinase (ACK, a tyrosine kinase; Manser et al., 1993) and the p21(Cdc42/Rac)-activated kinase (PAK, a serine/threonine protein kinase; Manser et al., 1994), which have related p21-binding domains, the Rho-binding kinase (ROK, a serine/threonine kinase; Leung et al., 1995) and protein kinase N (PKN, a PKC-related kinase with a separate Rho-binding domain; Amano et al., 1996; Watanabe et al., 1996). Non-kinases that interact with GTP-p21s include the Wiskott-Aldrich Syndrome protein, whose Cdc42-binding domain is related to those of ACK and PAK (Burbelo et al., 1995; Aspenstrom et al., 1996; Symons et al., 1996). Other targets of the Rho-family p21s include the lipid kinases, phosphatidyl inositol 3-kinase (PtdIns 3-kinase) and phosphatidyl inositol-4-phosphate 5-kinase (PtdIns*P* 5-kinase) (Zheng et al., 1994; Chong et al., 1994; Tolias et al., 1995).

The most extensively studied function of the Rho subfamily concerns its effects on mammalian morphology, and much experimental use has been made of p21 mutants that are dominant negative or dominant positive. These mutants have been generated on the basis of similar mutations of Ras.

p21 effector domains, dominant-positive mutants and dominant-negative mutants

The determination of the crystalline structures of inactive (GDP bound) and active (GTP bound) Ras has helped to establish the functional specification of its various regions, which had been ascribed previously through mutational analysis (Wittinghofer and Valencia, 1995). There is an effector domain consisting of residues 32–40, which in certain cases requires additional C-terminal residues for its effector activities. Within this region there are residues that are responsible for specifying individual responses. Some of the more interesting mutations outside this 'effector' domain affect the interaction of the GTPase with the guanine nucleotides whose binding determines the activity of Ras. The mutation of glycine to valine in the 12 position (G12→V) results in a decrease in the intrinsic GTPase activity which ensures that the bound GTP is hydrolyzed at a very much decreased rate, even in the presence of GAP, thus maintaining Ras in its active form for much longer. The Q61→L mutant behaves in a similar fashion. These are considered to be dominant-positive mutants, since they are 'locked' in the active form and presumed to be interacting continually with their effector molecules. The T17→N mutation results in a conformation in which binding of GDP, rather than GTP, is favoured, thus maintaining the inactive form of Ras. This form is considered to be the dominant-negative mutant, which, by interacting continually with associated regulatory proteins, such as GEF, results in a decreased concentration of the latter proteins and an attendant reduction of the activation of wild-type GDP-Ras.

Similar mutations in the Rho subfamily result in somewhat similar outcomes in the GTP-binding and GDP-binding activities. [V12]Rac, [V12]Cdc42 and the corresponding [V14]-Rho mutants are dominant-positive, as are [L61]Rac and [L61]Cdc42, while [N17]Rac, [N17]Cdc42 and [N19]Rho are dominant negative. Although the specificity of these mutants has been analyzed to some extent, it has not been as extensively studied as for the Ras mutants. Consequently, some conclusions regarding Rho p21 activities must remain tentative.

Actin cytoskeleton and cell morphology

In their morphological role, the Rho p21s affect the organization of actin-containing structures and a brief description of some of the various structures is presented for readers new to the field.

An actin network underlies and closely interconnects with the plasma membrane. Forces generating movement in the actin cytoskeleton arise from actin-filament polymerization, with associated ATP hydrolysis, and from forces generated by motor proteins, such as myosin. Other cytoskeletal components, including microtubules, motor proteins such as dynein, spindle-assembly components, as well as intermediate filaments, which function independently in processes affecting cell morphology, intercommunicate with the actin cytoskeleton.

Local actin polymerization is associated with the formation of membrane protrusions. These include the fine needle-like filopodia, which may have a sensory function at the leading edge of motile cells (Albrecht-Buehler, 1976) and in neuronal growth cones (Chien et al., 1993; Hynes and Lander, 1992). Lamellipodia are peripheral protrusions, which may form a web between two filopodia and generate ruffles. Actin filaments run in a single long bundle within filopodia, whereas lamellipodia have a criss-cross network.

Actin stress fibres are long bundles that end in focal adhesions, points of attachment of the plasma membranes to the substratum. These focal adhesions contain a complex of proteins including the tyrosine kinases Src and focal-adhesion kinase (FAK), and the actin-binding proteins vinculin, paxillin, α-actinin and talin, which are linked to integrins at the cytoplasmic face of the membrane. The function of stress fibres is poorly understood. They disappear when cells round up at mitosis, when an actin-containing contractile ring forms at the cleavage furrow during cytokinesis, for which Rho is required (Takai et al., 1995).

Rho, Rac, Cdc42 and morphological changes

Fibroblasts exposed to various growth factors respond with a variety of morphological changes, and it is presumed that these factors are present in serum. The effect of Rho on the actin cytoskeleton was first indicated by loss of stress fibres in cells injected with *Clostridium botulinum* C3 transferase which ADP-ribosylates Rho on Asn41 in the effector domain, which inactivates the protein (Chardin et al., 1989; Paterson et al., 1990). Subsequent studies have shown that Rho and its close relatives mediate factor-dependent morphological changes. These factors, acting through RTK [e.g. platelet-derived growth factor (PDGF), epidermal growth factor] or heterotrimeric-G-protein−coupled serpentine receptors (e.g. lysophosphatidic acid, bombesin, bradykinin), feed into signalling pathways that activate Rho-family proteins.

The direct involvement of individual p21s was demonstrated in Swiss 3T3 fibroblasts microinjected with recombinant Rho, Rac, Cdc42 and/or constitutively-activated and dominant-negative mutants. Stress-fibre and focal-adhesion formation is rapidly generated (1−2 min) by lysophosphatidic acid (Ridley and Hall, 1992), a powerful mitogen in fibroblasts whose action was inhibited by C3 transferase. This formation was mimicked by microinjection of [V14]Rho into Swiss 3T3 fibroblasts. The effects of lysophosphatidic acid, but not those of microinjected Rho, can be inhibited by tyrphostin, suggesting that there are tyrosine-kinase activities downstream (Ridley and Hall, 1994) and upstream (Nobes et al., 1995) of Rho in this pathway.

The formation of Rho-dependent stress fibres also occurred after the formation of lamellipodia or membrane ruffling, but as a later response (20−30 min) to PDGF, EGF, insulin or microinjected [V12]Rac1. These stress fibres do not accumulate to such high density as those elicited as an immediate response to lysophosphatidic acid. The effects of the growth factors were prevented by injection of inhibitory [V12, N17]Rac1 (Ridley et al., 1992).

These findings prompted the concept that the p21s were hierarchically related, with Rho being activated by certain factors, including PDGF, via Rac (Ridley et al., 1992) [Other factors, such as bombesin, can activate Rac and Rho independently in fibroblasts]. This signalling of Rac to Rho may involve Rac-mediated arachidonic-acid production and subsequent leukotriene-dependent activation of Rho (Peppelenbosch et al., 1995).

Ras, which also induces membrane ruffling (Bar-Sagi and Feramisco, 1986), acts upstream of Rac in fibroblasts (Ridley et al., 1992). Effector mutations of Ras indicate that induction of ruffling represents a separate signalling pathway from Ras activation of the MAPK cascade, but that inputs through both are required for Ras-induced DNA synthesis (Joneson et al., 1996). The involvement of Rho and Rac in Ras-dependent transformation (Khosravi-Far et al., 1995; Qiu et al., 1995a,b) has been recently reviewed in relation to morphological changes (Symons, 1996).

Cdc42 relationships with Rac and Rho

Upon microinjection of the family member Cdc42Hs, large-scale formation of peripheral-actin microspikes (PAM), which include filopodia, occurred in fibroblasts (Kozma et al., 1995). This was followed by lamellipodia formation and membrane ruffling. Although PAM formation was unaffected, ruffling did not occur when the inhibitor [N17]Rac1 was injected with Cdc42. Bradykinin treatment elicited similar morphological changes and was ineffective when cells were microinjected with the inhibitor [N17]Cdc42Hs before treatment. However prior injection of [N17]Rac1 did not affect Cdc42-induced PAM formation, al-

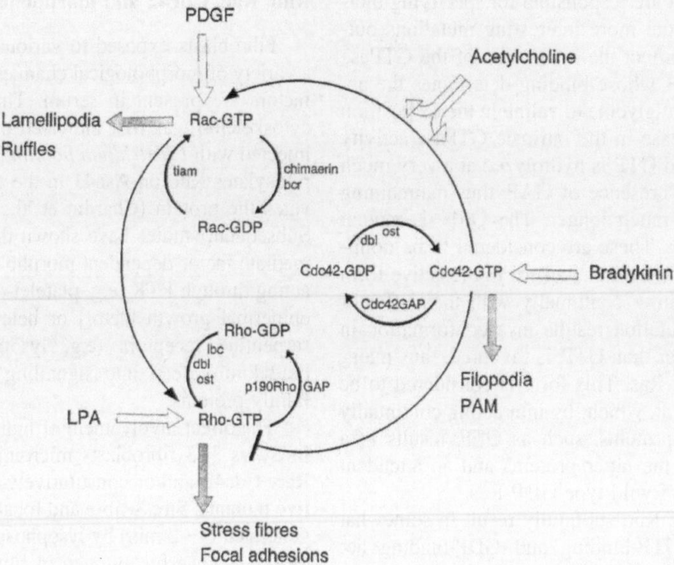

Fig. 1. Rho p21s and morphology. In fibroblasts, bradykinin stimulates Cdc42 pathways, leading to formation of PAM, including filopodia (Kozma et al., 1995). PAM formation is followed by stimulation of Rac pathways that lead to formation of ruffles and lamellipodia. PDGF stimulates Rac pathways leading to lamellipodia and ruffles, and subsequently to stimulation of Rho pathways leading to formation of stress fibres and focal adhesions (Ridley et al., 1992). The latter pathway is stimulated by lysophosphatidic acid (LPA) (Ridley and Hall, 1992). Cdc42-induced PAM formation is associated with loss of stress fibres, suggesting competition/antagonism between Cdc42 and Rho. Thus, although separate Cdc42-Rac and Rac-Rho heirarchies exist, these may not extend into a Cdc42-Rac-Rho hierarchy. In neuroblastoma cells, acetylcholine can induce Cdc42 and Rac pathways, either sequentially or independently (Kozma, R., Sarner, S., Ahmed, S. and Lim, L., unpublished results). The factors may act at the level of GEF, which catalyze formation of active GTP-bound forms of the p21. Specific examples of upregulating and downregulating GEF and GAP are shown on the basis of activities *in vitro*; their exact *in vivo* specificities are not established.

though ruffling/lamellipodia formation was affected. Thus bradykinin appears to act by activating Cdc42, which in turn activates Rac, implying a hierarchical relationship between these two p21s.

In these experiments, the formation of stress fibres did not follow ruffling, even after extended periods. On the contrary, microinjected Cdc42, in stimulating PAM formation, also promoted a loss of stress fibres. Cdc42 appears to inhibit Rho-mediated actions (Fig. 1). This inhibition by Cdc42 of Rho-mediated effects may be due to competition for common components in filopodia and stress fibres, such as actin or actin-binding proteins.

Some evidence for antagonism between Cdc42 and Rho can also be found in the experiments of Nobes and Hall (1995), where Cdc42-type effects are enhanced when Rho is inhibited. Small focal complexes, distinct from focal adhesions induced by Rho, are formed along filopodia and at the cell periphery on injection of [V12]Cdc42 with inhibitory [N17]Rac and C3 transferase. (Other small focal complexes are formed around lamellipodia upon microinjection of [V12]Rac). Microinjection of this combination led to formation of filopodia within a short period (5–10 min). In the absence of inhibitory [N17]Rac, [V12]Cdc42 and C3 transferase promoted formation of lamellipodia and filopodia also within 5–10 min. When the Rho inhibitor was omitted, the formation of filopodia by [V12]Cdc42 (with or without [N17]Rac) was delayed.

Kinases binding to Rho subfamily

Tyrosine kinases. The isolation and identification of the Rho-family-associated kinases occurred through characterization of proteins that bound to the GTP-bound forms of Rho p21s and represented potential downstream targets (Manser et al., 1992). The first to be described was the p120 ACK, which was isolated by expression screening with GTP-Cdc42 (Manser et al., 1993). This tyrosine kinase contained an SH3 domain, as well as a C-terminal segment with several proline-rich regions. Intriguingly, there is considerable sequence similarity between ACK and the insulin-induced gene-product 33 in the proline-rich C-terminal domain. The function of this kinase is not known, although *in vitro* it appears to inhibit the intrinsic GTPase activity of Cdc42 and to block the action of the GAP, Bcr (breakpoint cluster region gene product). ACK expression appears to be increased on lowering the temperature (Satoh et al., 1996).

A search for proteins similar to ACK uncovered proline-rich tyrosine kinase (PYK) 1 (an isoform of ACK) and subsequently PYK2, which is more related to FAK than ACK. PYK-2, which does not have the p21-binding motif, appears to be involved in regulation of ion channels and MAPK-mediated functions (Lev et al., 1995). The ACK-related kinase, CD38-negative kinase (Tnk1), is highly expressed in foetal tissues (Hoehn et al., 1996).

The PAK family of serine/threonine kinases. Three mammalian PAK isoforms have been described: the brain-enriched 68-kDa rat α-PAK (Manser et al., 1994), which corresponds to human hPAK1 (Brown et al., 1996); the 65-kDa rat β-PAK isoform (Manser et al., 1995), which is essentially identical to mPAK3 whose DNA was isolated from a mouse fibroblast library (Bagrodia et al., 1995b); and the ubiquitous 62-kDa γ-PAK (Teo et al., 1995), also referred to as hPAK65/PAK2 (Martin et al., 1995). γ-PAK has been described as a protease-activated kinase (Jakobi et al., 1996) and a H4/S6 kinase (Benner et al., 1995). All these proteins contain highly conserved kinase and p21-binding domains (Fig. 2). In addition, PAK from multicellular organisms share conserved N-terminal sequences, which contain polyproline motifs that might function to bind SH3-con-

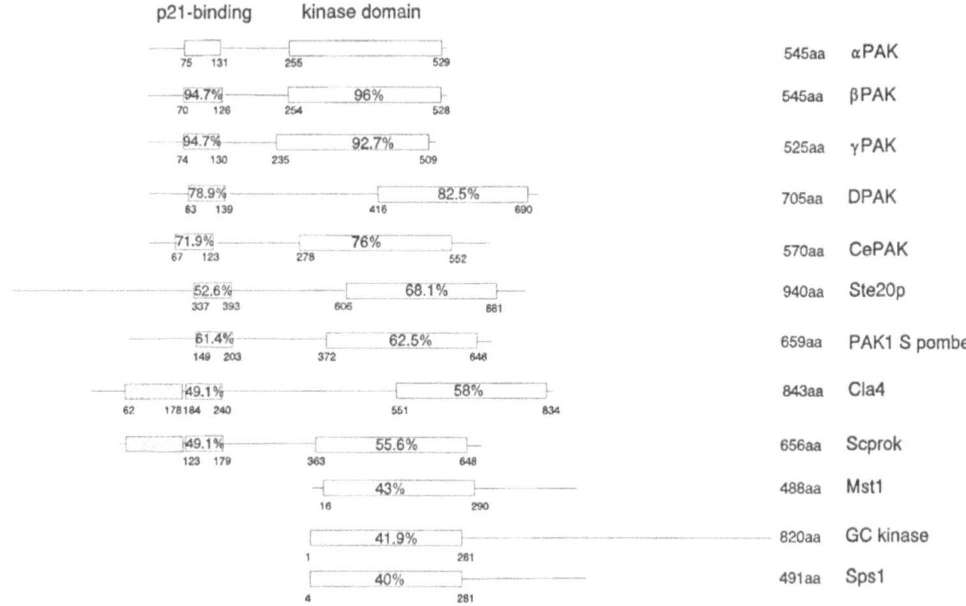

Fig. 2. Relationship of the PAK family of kinases. The sequence identities (%) of the p21-binding and kinase domains are given in relation to rat α-PAK. These were determined by means of the DNASTAR AALIGN programme. Numerals in lower case denote the positions of amino acid residues (aa). The stippled regions indicate PH domains of Cla4p and Scprok. The p21-binding domain is not present in the lowest three kinases. Accession numbers are as follows: rat α-PAK, U23443 (Manser et al., 1994); rat β-PAK, U33314 (Manser et al., 1995); rat γ-PAK, U35345 (Teo et al., 1995); *Drosophila* PAK (DPAK), U49446 (Harden et al., 1996); *C. elegans* PAK (CePAK), U63744 (Chen et al., 1996); *S. pombe* PAK1, U22371 (Ottilie et al., 1995); *S. cerevisiae Ste20p*, L04655 (Ramer and Davis, 1993); *S. cerevisiae* Scprok, X69322 (also referred to as ScPAK or Scprokin); *S. cerevisiae* Cla4, X82499 (Cvrckova et al., 1995); human Mst1, U18297 (Creasy and Chernoff, 1995); human germinal-centre (GC) kinase, U07349 (Katz et al., 1994); and *S. cerevisiae* Sps1, U13018 (Friesen et al., 1994).

taining proteins, such as phospholipase Cγ and Nck (Bagrodia et al., 1995b).

PAK-family kinases are present in many experimentally useful organisms: three forms of the kinase have been identified in *S. cerevisiae*, namely Ste20p, Cla4p and *S. cerevisiae* PAK. Cla4p and *S. cerevisiae* PAK are unique among PAK because of the presence of an N-terminal PH domain (Cvrckova et al., 1995). The approximately 100 kDa Ste20p can be activated *in vitro* by GTP-Cdc42 (Simon et al., 1995), and *Ste20⁻* strains can be complemented by mammalian PAK (Bagrodia et al., 1995a), indicating similar mechanisms of activation *in vivo*. A form of the kinase has been reported in *Schizosaccharomyces pombe* and referred to as PAK1 (Ottilie et al., 1995) or SHK1 (Marcus et al., 1995). As might be expected, the *C. elegans* PAK (Chen et al., 1996) and *Drosophila* PAK (Harden et al., 1996) are more similar to the mammalian kinases than the yeast enzymes. The kinase domains of PAK are also similar to the mammalian germinal-centre kinase (Katz et al., 1994) the mammalian sterile-twenty-like (MST)1 kinase (Creasy and Chernoff, 1995), and the putative *S. cerevisiae* sporulation-specific SPS1 kinase (Friesen et al., 1994). In these cases their mechanism of activation is unknown, since the absence of recognizable p21-binding domains suggests that activation is not mediated by Cdc42 or Rac GTPases. The relationships of the PAK family are shown in Figs 2 and 3.

The upregulation of the PAK-family kinases *in vitro* requires a single interaction, namely binding of GTP-Cdc42 or GTP-Rac, leading to autophosphorylation and subsequent activation of the kinase. Two lines of evidence suggest that this process involves detachment of an inhibitory domain from the N-terminus of the kinase. Firstly, activation of purified PAK can be achieved by limited proteolysis of the N-terminus followed by autophosphorylation (Benner et al., 1995), and secondly, N-terminally truncated Ste20p has been found to be activated *in vivo* (Ramer and Davis, 1993). Whether inhibition occurs through a pseudosubstrate-like sequence remains to be established. PAK autophosphorylation occurs on serine and threonine residues (Manser et al., 1994). Although it has been reported that hPAK65 is exclusively phosphorylated on serine (Martin et al., 1995), this is unlikely given that threonine phosphorylation occurs in the regulatory loop of the kinase domain (T422 in α-PAK; Manser, E., Chong, C., Leung, T. and Lim, L., unpublished results) at a resi-

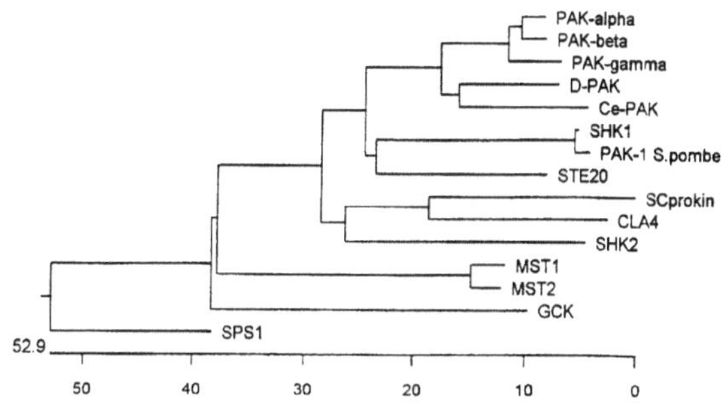

Fig. 3. Phylogeny of kinase domains of PAK. The brain specific/enriched α-PAK and β-PAK are the most recently evolved, followed by the ubiquitous γ-PAK. The metazoan kinases appear to be closer to the *S. pombe* Shk1 and Pak1 than to *S. cerevisiae* Ste20p. Sequence relationships were determined using the DNASTAR MEGALIGN programme. Accession numbers are given in the legend to Fig. 2 or are as follows: *S. pombe* SHK1, L41552 (Marcus et al., 1995); SHK2, U45981; and MST2, U26424.

Fig. 4. Structural and functional analysis of α-PAK. The boxes above the model of PAK denote mutations resulting in constitutively activated kinase. The first box refers to those of Bagrodia et al. (1995a) and the second to that of Brown et al. (1996), retaining the original numbering. The third box refers to the conserved threonine in the regulatory loop between motifs VII and VIII, which is mutated to glutamic acid; p, autophosphorylation sites; PR, proline-rich domains [PR1 and PR2 synergize with the p21-binding domain to target PAK to focal complexes (Manser, E., Chong, E., Leung, T. and Lim, L., unpublished results)].

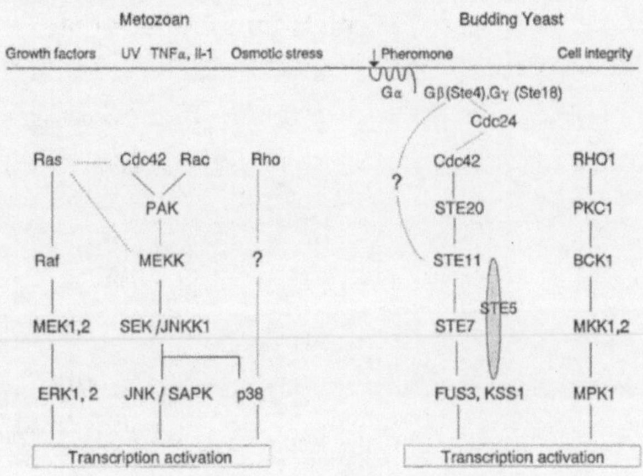

Fig. 5. Rho p21s and kinases in MAPK pathways. In metazoans, Ras activates the MAPK, ERK1 and ERK2 through Raf (Marshall, 1994). Cdc42 and Rac activate JNK and p38 MAPK pathways (Coso et al., 1995; Minden et al., 1995); PAK may be in this pathway (Bagrodia et al., 1995a; Zhang et al., 1995). Ras can also, albeit less effectively, activate JNK, either independently or through Cdc42/Rac (Minden et al., 1995). In the yeast pheromone pathway, Cdc42 mediates signalling from the receptor-linked $\beta\gamma$ G-proteins, Ste4p and Ste18p (Zhao et al., 1995; Simon et al., 1995) by activating Ste20p, which is upstream of the Ste11p (Leberer et al., 1992; Ramer and Davis, 1993). Cdc24p, the GEF for Cdc42, interacts with Ste4p *in vitro* (Zhao et al., 1995). Ste5p is considered to tether together the relevant kinases of this MAPK pathway (Eilon, 1995). Signalling from the $\beta\gamma$ G-proteins may bypass the Cdc42/Ste20p components (Akada et al., 1996). Yeast rho1p is upstream of pkc1 in the MAPK pathway involved in cell integrity (Nonaka et al., 1995). In mammalian cells, Rho activates a p38 MAPK pathway subserving the serum-response (transcription) factor (Hill et al., 1995) but the identities of the intermediatory kinases involved have not been established. The Rac/JNK pathways may also involve $\beta\gamma$ G-proteins (Coso et al., 1996). SAPK, stress-activated protein kinase; MEK, MAPK/ERK kinase; MEKK, MEK kinase; SEK, SAPK kinase; JNKK, JNK kinase; UV, ultraviolet; TNF, tumour-necrosis factor; IL-1, interleukin-1.

due that is completely conserved among the PAK family. In Ste20p, mutation of the corresponding threonine 777 to alanine inactivates the kinase (Wu et al., 1995). Phosphorylation within this regulatory loop (between the conserved motifs VII and VIII) is a feature in the activation of many kinases (Johnson et al., 1996); however, we have found that the N-terminal serine-autophosphorylation sites contribute to kinase activation (Manser, E., Chong, C., Leung, T. and Lim, L., unpublished results). Nevertheless, we can generate a constitutively active PAK mutant by substituting threonine 422 with glutamic acid (Fig. 4). In addition to kinase activation, autophosphorylation of the β-PAK isoform (but not α-PAK) leads to downregulation of p21 binding (Manser et al., 1995), perhaps allowing dissociation of the kinase from the signalling complex to act at a different site, and further rounds of kinase activation by the GTP-p21.

Roles for PAK-related kinases

Involvement of Ste20p in the yeast pheromone pathway. Generally, genetic studies on yeast have provided clues to the workings of mammalian systems. However, in the case of the mating signalling pathway in *S. cerevisiae*, it was studies on mammalian systems that led to the realization that Cdc42, hitherto regarded only as being important for cytoskeletal reorganization events in bud formation, was also important in the pheromone response. The similarity between mammalian PAK and yeast Ste20p, which were both shown to bind Cdc42 via similar motifs, prompted the suggestion the Cdc42 had a role in the yeast pathway (Manser et al., 1994). We and others have now shown *Cdc42* to be upstream of *Ste20* in this pathway (Zhao et al., 1995; Simon et al., 1995).

The mating-pheromone pathway in budding yeast is one of the most well-studied phosphorylating pathways. There are two mating types, which are characterized by their ability to respond to one or the other of the two peptide pheromones, *α* and **a** factors. Mating between the different types is engendered by individual yeast cells expressing and secreting one type of factor while possessing receptors only for the other factor.

These serpentine receptors are linked to heterotrimeric G ($\alpha\beta\gamma$) proteins and, unusually, the $\beta\gamma$ subunits released from the G-protein · receptor complex upon binding of the mating factors are the mediatory elements in the mating pathway. In other pathways, the GTP-binding α subunit is the mediator. The following description of the mating pathway is derived from studies on yeast mutants which either lack or have thermolabile gene products that are components of the pathway.

The membrane-bound $\beta\gamma$ complex activates Cdc42, probably by recruiting the GEF Cdc24 in a manner somewhat reminiscent of Cdc25/Ras activation. Two-hybrid studies have shown that the β subunit is capable of binding Cdc24, but whether this occurs *in vivo* has not been established (Zhao et al., 1995). The binding domain in Cdc24 is its N-terminal region, rather than the PH domain, which in other proteins can serve as the interacting domain. The activated GTP-Cdc42 in turn binds to and activates Ste20p. This kinase is the first in the kinase cascade which includes Ste11p, Ste7p and Fus3p. The latter MAPK homologue then promotes phosphorylation and activation of Ste12p, the transcription factor that binds to the pheromone-responsive elements within the yeast genome (reviewed by Herskowitz, 1995; Fig. 5). The stimulation of genes containing these elements results in expression of whole families of proteins involved in mediating the mating process.

The sterile-20 gene *STE20* was detected as a multicopy suppressor of a dominant-inhibitory Ste4p mutant in a screen designed to detect proteins that might interact directly with the β subunit (Leberer et al., 1992). The reason that the kinase was not detected in previous exhaustive searches for mating-deficient strains is probably due to its partial complementation by *CLA4*. Direct evidence for such an interaction between Ste20p and Ste4p is lacking, but the kinase interacts with the SH3-containing protein Bem1p (Leeuw et al., 1995). This protein participates with Cdc24p and Cdc42p to establish polarized cell growth and budding (Kao et al., 1996). Mutants of Bem1p exhibiting mating-specific defects are unable to bind Ste20p but interact normally with Ste5p and actin (Leeuw et al., 1995). Ste5p (Leberer et al., 1993) has been described as a 'scaffold' protein that physically links the three components of the MAPK cascade, Ste11p/Ste7p/Kss1p (Kss1p is a Fus3p homologue) (Choi et al., 1994), and has been reported to bind Ste4p (Leeuw et al., 1995). Our suggestion that the pheromone-responsive pathway is branched at the level of Ste4p/Cdc24p/Cdc42p (Zhao et al., 1995) has been confirmed by more recent epistatic analysis (Akada et al., 1996).

Mammalian PAK and activation of JNK/SAPK and p38 MAPK. The realization that PAK was activated by Rho GTPases, coupled with its similarity to Ste20p, a member of a well-defined MAPK pathway prompted several groups to investigate whether a similar pathway existed in mammalian systems. Minden et al. (1995) and Coso et al. (1995) showed that Cdc42 and Rac were involved in mediating extracellular signals not to extracellular-signal-regulated kinase but to c-Jun N-terminal kinase (JNK, stress-activated protein kinase) (Fig. 5). These experiments relied on transfection of tagged JNK1 and ERK2 with dominant-positive and dominant-negative Cdc42 and Rac mutants. Coso et al. (1995) showed that [L61]Cdc42 and [L61]Rac activated JNK1 to the same extent as the potent anisomycin (Cano and Mahadevan, 1995). Expression of *ost* and *dbl* proteins, which contain GEF domains for the Rho p21s, was also effective in activating JNK. Correspondingly, [N17]Cdc42 and [N17]Rac1 blocked JNK activation by EGF and tumour-necrosis factor α but not by anisomycin. The effect of *dbl* but not of [L61]Cdc42 was blocked by Rho GDP-dissociation inhibitor and p190 RhoGAP. Essentially similar results for a Cdc42 and Rac1/Rac2 pathway to JNK were obtained by Minden et al. (1995). For example, [V12]Rac1, [L61]Rac2 and [V12]Cdc42 activated JNK1 as effectively as EGF, with the EGF response being blocked by [N17]Rac1. Transfection of [V12]Cdc42 or the GEF domain of *dbl* also led to JNK activation, although this could also be blocked by [N17]Rac1. These and other data suggest that there is no hierarchical relation between Cdc42 and Rac1 with respect to JNK pathways, although the two pathways are clearly linked. Such a situation may reflect the dual activation of PAK by these p21s.

Minden et al. (1995) additionally showed the constitutively active [V12]Cdc42 and [V12]Rac1 to activate another MAPK, p38. Furthermore, [N17]Rac1 was able to block the activating effects of [L61]Ha-Ras on JNK, placing Rac as being possibly downstream of Ras in the Ras-JNK pathway.

Constitutively active PAK mutants behave as activators of JNK and p38 (Bagrodia et al., 1995a; Brown et al., 1996) and overexpression of wild-type PAK can upregulate the pathway (Frost et al., 1996). These results have been interpreted as showing that PAK lies downstream of Cdc42 and Rac, although co-transfection of dominant-inhibitory (T17→N) mutants has not been used in these experiments to test this. The germinal-centre kinase, which is related to PAK only in the kinase domain, also acts as a strong JNK activator (Pombo et al., 1995), perhaps independently of Cdc42 and Rac. Although activated versions of Cdc42, Rac or PAK do not upregulate the ERK1/ERK2 pathway, both Ste20p and *S. pombe* Pak1 (in their N-terminally truncated activated forms) are capable of activating ERK2 in *Xenopus* oocyte extracts (Polverino et al., 1995). In this cell-free system, all three kinases, PAK (N-truncated and active), Ste20p and Pak1, were found to activate JNK. The observation that [N17]Rac2 or an N-terminal PAK construct can inhibit Ras-mediated activation of ERK2 (Frost et al., 1996) indicates that Rac (and perhaps Cdc42) has a permissive role in the control of the ERK pathway, and suggests that signal transduction does not occur through a linear pathway. This situation is reminiscent of the role of Cdc42 as a regulator of Ste20p in the pheromone-responsive MAPK cascade, where activated [V12]Cdc42Sc is unable to directly activate the pathway (Zhao et al., 1995).

Rabbit PAKI (corresponding to γ-PAK) can be isolated as an endogenously active (58 kDa) or inactive (60 kDa) form; limited protease treatment of the latter generates a 37-kDa form containing the active kinase domain (Jacobi et al., 1996). Injection of the active PAK forms into one blastomere of a two-cell embryo of the frog *Lepidobatrachus laevis* arrested cleavage of the injected but not the uninjected cell (Rooney et al., 1996). Although present in oocytes and other stages in development, frog proteins of 58−60 kDa that were immunoreactive towards the PAKI antibody were virtually absent in the two-cell embryo, as was putative PAK activity. The proposal that one action of PAK is to maintain cells in a non-dividing state awaits characterization and analysis of the frog PAK. In line with this proposal, α-PAK and β-PAK are highly enriched in the brain at protein and mRNA levels, with a selective neuronal distribution (Manser et al., 1994, 1995).

Morphological roles of PAK. A number of lines of evidence suggest roles for PAK-family kinases in cell morphology. In haploid *S. cerevisiae* cells, Ste20p appears to have overlapping functions with Cla4p: *CLA4* mutants are characterized by cells with elongated and misshaped buds (Cvrckova et al., 1995) but loss of both kinases is lethal. Such double mutants exhibit defects in cytokinesis and the localization of the septin ring that marks the bud site. In its diploid state, Ste20p is required to initiate pseudohyphal growth under conditions of low nutrients, as are a number of other components of the MAPK cascade, including Ste11p and Ste7p, but not the 'scaffold' protein Ste5p (Roberts and Fink, 1994). The *Ras2* gene is upstream of *Cdc42* in this pathway (Mosch et al., 1996). The *S. pombe pak1* gene product is a 72-kDa protein that is able to complement the function of Ste20p in the mating pathway (Marcus et al., 1995). Overexpression of an inactive form of *pak1* in *S. pombe* gives rise to cells with abnormal shape and actin distribution (Ottilie et al., 1995).

The localization of *Drosophila* PAK indicates its role in the morphological process of dorsal closure during embryonic development. mRNA and protein levels of the kinase are enriched in leading-edge epithelial cells (Harden et al., 1996). The protein level is markedly higher in those cells positioned at segment borders. The leading edge exhibits elevated levels of polymerized actin, myosin and tyrosine-phosphorylated proteins, which appear to form structures resembling the focal complexes found in cultured mammalian cells. The expression of the inhibitory *Drosophila* [N17]Rac1, during embryonic development leads to defects in dorsal closure (Harden et al., 1995), perhaps by blocking signals from endogenous Rac to PAK. The induced expression of [N17]Rac during dorsal closure results in loss of PAK and tyrosine-phosphorylated proteins in these cells. The involvement of a JNK kinase homologue, *hemipterous*, in dorsal closure (Glise et al., 1995) further establishes that MAPK modules can function in signal-transduction pathways linked to regulation of morphology. Studies of yeast and mammalian PAK would indicate that *Drosophila* PAK lies upstream of the *Drosophila* kinase cascade, but genetic evidence for this has not been obtained. *C. elegans* PAK is also particularly enriched in the boundary of hypodermal cells undergoing embryonic elongation (Chen et al., 1996), as are Cdc42 and Rac1.

In mammalian epithelial cells, α-PAK has been found to be associated with Cdc42 and Rac1-dependent focal complexes, but

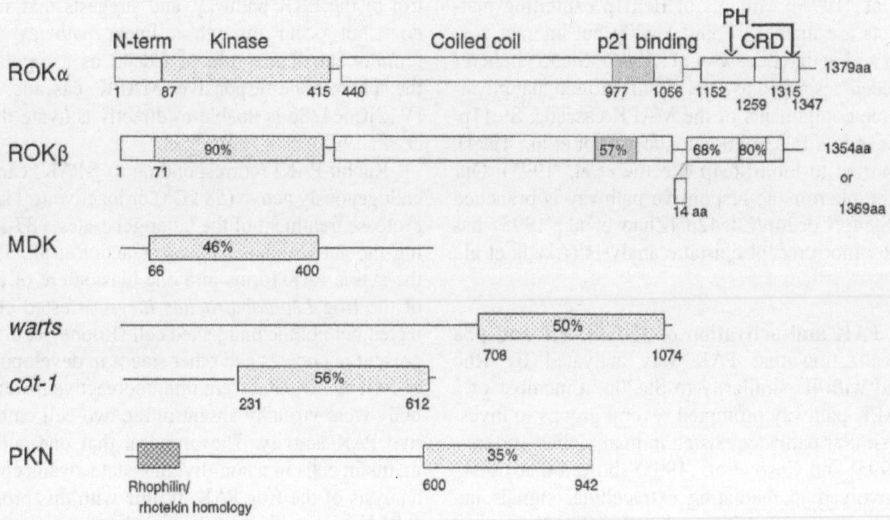

Fig. 6. Structure of Rho-binding kinases. Rat ROKα and ROKβ (Leung et al., 1995, 1996), products of different genes, contain kinase domains with high identity to the myotonic-dystrophy-kinase family, which includes *warts* and *cot-1* gene products. The kinase domain is followed by a coiled-coil structure, which contains the Rho-binding domain near the C-terminus of the coil. The Rho-binding domain of ROKα and ROKβ have identical stretches of 20 amino acids. The PH domain is split by a cysteine-rich domain (CRD) related to the cysteine-rich domains of the C1 region of PKCα and chimaerins. The bovine Rho kinase (Matsui et al., 1996; U36909) is a homologue of ROKα. Human p160^{ROCK} (Ishizawa et al., 1996; U43195) is a homologue of ROKβ, but lacks a 14-amino-acid exon just N-terminal of the PH domain. PKN has an unrelated Rho-binding domain (Amano et al., 1996; Watanabe et al., 1996), which is related to that of several non-kinases; its kinase domain is more related to that of PKCδ (52% identity). Accession numbers are as follows: rat ROKα, U38481 (Leung et al., 1995); rat ROKβ, U61266 (Leung et al., 1996); human myotonic-dystrophy kinase (MDK), L08835; *N. crassa cot-1*, M83093 (Yarden et al., 1992); *Drosophila warts*, L39837 (Justice et al., 1995); and human PKN, D26181 (Mukai and Ono, 1994). aa, amino acids.

not RhoA-dependent structures (Manser, E., Chen, X., Leung, T. and Lim, L., unpublished results). The association with these complexes is mediated by interaction of the kinase with GTP-Cdc42 or GTP-Rac and additional interactions within the first 60 residues of the kinase (N-terminal to the p21-binding domain). This region contains conserved motifs found in *Drosophila* PAK and *C. elegans* PAK; these include proline-rich sequences with consensus SH3-binding sites (Fig. 4). Since α-PAK is not tyrosine phosphorylated, focal-complex localization could be mediated via weaker SH3 interactions. What could be the function of PAK in these structures? Overexpression of constitutively activated PAK mutants has revealed that the kinase can cause disassembly of focal adhesions, but not the formation of filopodia or ruffles (Manser, E., Chen, X., Leung, T. and Lim, L., unpublished results). Additionally, there is a loss of stress fibres, as also occurs in cells expressing [V12]Cdc42 or [V12]Rac. These observations imply that PAK does not participate in the dynamic formation of focal complexes or actin polymerization downstream of Cdc42 or Rac, but rather the opposite: PAK promotes turnover of these structures thus allowing continual regeneration of these components.

Kinases binding to Rho

Several Rho-GTP–binding proteins have recently been isolated by several groups either by protein purification, peptide-sequence determination and subsequent cDNA isolation or by use of the yeast two-hybrid expression-cloning system for selection of interactive proteins. These include rat ROKα (Leung et al., 1995, 1996) and its bovine homologue (Matsui et al., 1996), the closely related ROKβ (Leung et al., 1996), which has an additional exon compared with its human homologue p160^{ROCK} (Ishizaki et al., 1996), and PKN (Mukai and Ono, 1994; Watanabe et al., 1996; Amano et al., 1996). Their structural relationships are indicated in Fig. 6. ROK are members of a kinase family including the myotonic dystrophy kinase (Brook et al., 1992), the *Drosophila* protein *warts*, which controls cell growth and morphology (Justice et al., 1995) and the *Neurospora* protein *cot-1* (Yarden et al., 1992), which is essential for hyphal elongation.

Rat ROKα, which may correspond to a previously detected 150–160-kDa RhoA-binding protein (Manser et al., 1994) and which prompted its isolation, was purified on the basis of its affinity for the GTP-form of RhoA (Leung et al., 1995). The corresponding cDNA encoded an N-terminal serine/threonine-kinase domain highly related to the kinase domain of myotonic dystrophy kinase. At the C-terminus, there is a cysteine-rich domain related to those of the C1 regions of PKCα and chimaerin, which splits a PH domain. The centre of the molecule comprises a long stretch of coiled-coil structure and contains the Rho-binding domain. The latter has no sequence similarity to the Cdc42/Rac-binding domains of the PAK group of kinases. Native ROK was autophosphorylated. Whereas Cdc42 stimulates PAK activity towards myelin basic protein up to 200-fold, the activity of native purified ROK is increased only slightly (2–5-fold) by RhoA. The bovine homologue of ROKα (Matsui et al., 1996) and the human homologue of ROKβ (Ishizaki et al., 1996) displayed essentially similar characteristics, except that bovine ROK increased phosphorylation of a fragment of the myosin-binding subunit (MBS) of myosin phosphatase greater than 15-fold. ROKα is ubiquitously expressed and is cytosolic in COS cells, but is translocated to membranes upon transfection of COS cells with RhoA or [V14]RhoA, where it is localized with actin microfilaments at the cell periphery and at the cleavage furrow in mitotic cells (Leung et al., 1995), consistent with a role as an effector of Rho (Takai et al., 1995). Ishizaki et al. (1996) also

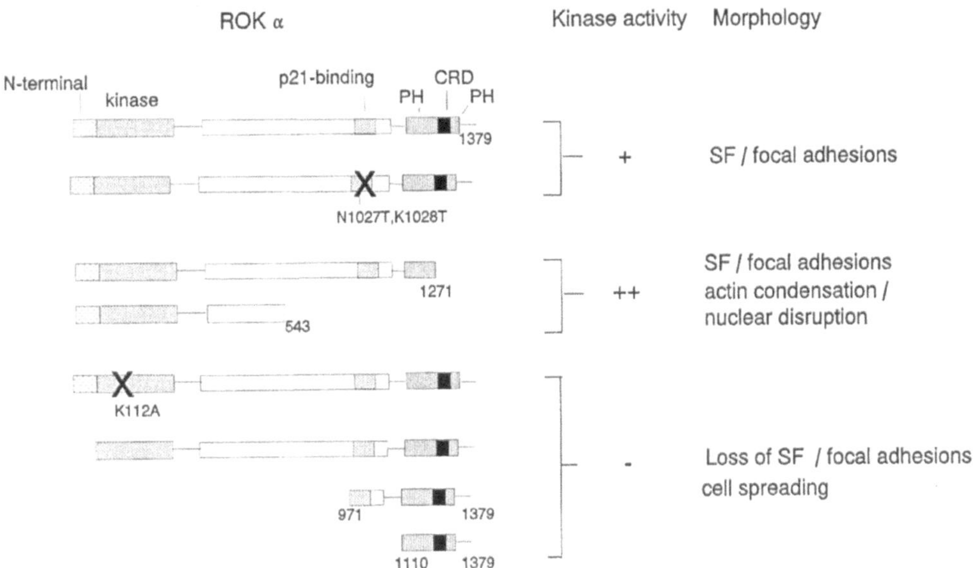

Fig. 7. Morphological effects of the kinase and other domains of ROKα. In HeLa cells transfected with ROKα DNA, wild-type ROKα promotes formation of stress fibres (SF) and focal adhesions (Leung et al., 1996). [T1027, T1028]ROKα, which has impaired Rho binding, can still cause this formation, perhaps as a result of overexpression in this system (see text). Deletion of ROKα C-terminal region ROKα-(1−543)-peptide results in hyperactive kinase with an accompanying increase in formation of SF and focal adhesions; this leads to extensive actin condensation and nuclear disruption. Conversely, deletion of the ROKα N-terminal region results in loss of kinase activity. All inactive mutants, including [A112]ROKα, promote loss of SF/focal adhesions, probably through acting as dominant-negative mutants that inhibit endogenous ROK. This loss can be promoted by the C-terminus containing the PH domain; inclusion of the binding domain enhances the inhibition. The C-terminal and N-terminal regions of ROKα appear to play positive and negative regulatory roles, respectively, in kinase and morphological activities.

reported that RhoA induced ROK translocation to peripheral membranes.

Morphological activities of ROKα and the roles of its kinase and other domains

The expression of transfected full-length ROKα in serum-starved HeLa cells (Leung et al., 1996) resulted in the formation of stress fibres and focal adhesions. In these epithelial cells, as in fibroblasts, RhoA promotes stress-fibre formation, which is inhibited by C3 transferase. ROKα was effective in promoting its effects, even in the presence of the Rho inhibitors [N19]Rho or C3 transferase, consistent with its role as a downstream effector of Rho.

The proper regulation of its kinase activity is important to the role of ROKα in stress-fibre and focal-adhesion formation. While mutants lacking kinase activity did not induce stress-fibre formation, hyperactive mutants induced more stress fibres than wild-type ROKα. The exaggerated formation of stress fibres was associated with central condensation of actin fibres and nuclear distortion. The N-terminal and C-terminal regions appear to have positive and negative, respectively, effects on kinase activities and thus their functional correlates. ROKα deleted of its N-terminal exhibited loss of kinase activity and the ability to promote stress fibres. Conversely, C-terminal deletion was associated with an increase in kinase activity and in ROKα's effects on stress fibres (Fig. 7).

In microinjection studies, the binding domain was not required since a ROKα mutant with impaired RhoA binding was capable of promoting the formation of stress fibres and focal-adhesion complexes. It appears that overexpression can generate binding-deficient ROKα in a physiologically active and appropriate form, provided other domains are intact. With wild-type ROKα, there was no substantial increase in kinase activity on coexpression of activated [V14]RhoA; this is consistent with previous findings that RhoA does not extensively stimulate native ROK activity *in vitro* (Leung et al., 1995). Nevertheless, exogenous [V14]RhoA promotes translocation of ROKα from the cytosol to peripheral membranes. Under conditions where the endogenous ROKα is expressed at normal levels, [V14]RhoA perhaps serves to direct the kinase to sites in the membrane where it is activated, very much like the Ras-Raf paradigm, and where it operates. Because measurements of enzymatic activity were made on cytosolic ROK immunoprecipitates, we cannot discount the presence of a highly activated membrane-associated ROK. Certainly, the kinase activity of ROK can be substantially increased, the cytosolic C-terminal-deleted ROKα mutant having eightfold more activity than wild-type ROKα.

In serum-fed cells, transfection of ROKα mutants lacking kinase activity, including kinase-dead [A112]ROKα, resulted in the disassembly of stress fibres and focal-adhesion complexes. These mutants may have interfered with the endogenous ROK responsible for formation of pre-existing stress fibres, perhaps acting as dominant-negative forms whose various domains bind and block the downstream targets of ROK or by inhibiting kinase activity through their C-terminus acting as a negative regulator. The C-terminal region with only the PH domain and cysteine-rich domain (which can potentially associate with membranes) was effective in promoting disassembly (Fig. 7). When the C-terminal fragment included the Rho-binding domain, the extent of disassembly was greater. The synergistic effect of the Rho-binding domain could be explained by its specifically directing the inhibitory fragment to the membrane sites containing activated RhoA responsible for binding endogenous ROKα.

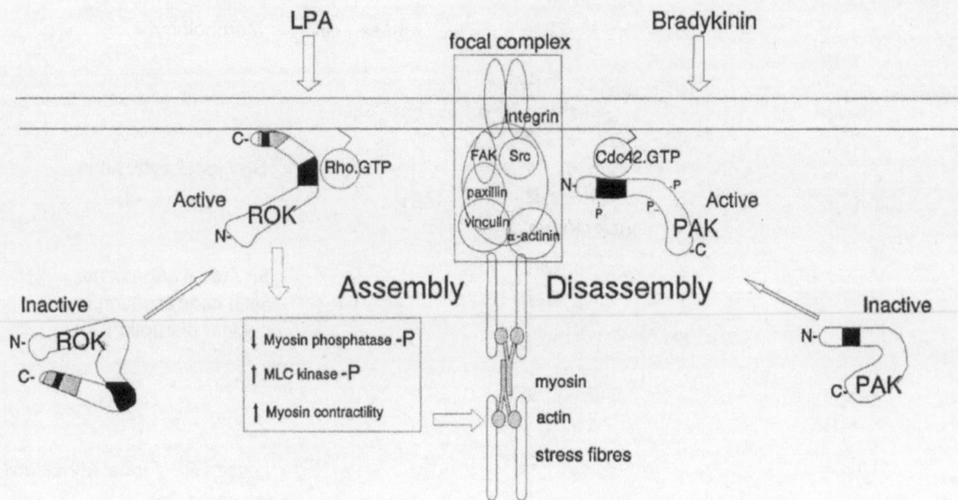

Fig. 8. A model for the roles of PAK and ROK in assembly and disassembly of focal adhesions/stress fibres and in Cdc42-Rho antagonism. Binding of Rho targets ROK to the membrane where it is activated by GTP-Rho. The C-terminal PH domain (stippled), including the cysteine-rich domain (bold) may facilitate localized membrane disposition. ROK phosphorylation of the MBS of the MLC phosphatase leads to phosphatase inhibition and increased phosphorylation status of MLC kinase. The resultant myosin contractability leads to assembly of stress fibres and to focal adhesions through myosin-actin interactions. The integrin-containing focal complex involved is shown to include several interacting proteins; however, their exact organization is not known. This model encompasses findings and models of Leung et al. (1995, 1996), Chrzanowska-Wodnicka and Burridge (1996) and Kimura et al. (1996) (see text for details). Whether ROK directly affects MLC is not known. PAK is activated on binding to GTP-Cdc42 (Manser et al., 1994). Autophosphorylated active PAK interacts with the focal-adhesion complex through its N-terminal region, which contains proline-rich regions and the p21-binding domain. PAK phosphorylation of a component(s) results in dissolution of the focal complexes and disassembly of stress fibres (Manser, E., Chen, X., Leung, T. and Lim, L., unpublished results). This disassembly perhaps releases structural components participating in filopodia formation and removes cytoarchitectural constraints necessary for the formation. Other effectors of Cdc42 required for this formation remain to be identified.

Since Rho directly affects the actin cytoskeleton, it would be interesting to determine whether the related kinase domains of myotonic-dystrophy kinase and proteins involved in morphogenetic events (e.g. *warts* and *cot-1* gene products) can directly affect cytoskeletal structures. Equally interesting is whether mammalian ROK can play a morphogenetic role, since overexpression of myotonic-dystrophy kinase has been reported to stimulate muscle-cell differentiation (Bush et al., 1996).

ROK and Rho, actin and myosin

There is evidence that Rho, through ROK, can effect changes in cytoskeletal structures containing actin through its binding and functional partner, myosin.

In smooth muscles, phosphorylation of the 20-kDa myosin light chain (MLC) by a Ca^{2+}/calmodulin-dependent kinase results in its interaction with actin and subsequent contraction. Guanosine 5′-[γ-thio]-triphosphate stimulation of Ca^{2+}-mediated phosphorylation of MLC was abolished by C3 transferase, which implies the involvement of Rho (Hirata et al., 1992). The increase in the phosphorylation status of MLC is associated with inhibition of the MLC phosphatase (Kitazawa et al., 1991; Noda et al., 1995). The phosphatase comprises three subunits, including 20-kDa and 110-kDa regulatory subunits.

The putative bovine p138 homologue of the 110-kDa myosin-binding subunit (MBS) specifically binds to RhoA in its GTP form (Kimura et al., 1996). This subunit was previously reported to be a preferred substrate of bovine ROKα *in vitro* (Matsui et al., 1996). Transfection experiments with COS-7 cells showed that [V14]RhoA promotes MBS association with membranes. *In vitro*, ROKα phosphorylated MBS, with an associated inhibition of MLC-phosphatase activity. RhoA enhanced both these effects. While Rho itself does not directly affect MLC-phosphatase activity, it has not been determined whether ROK directly phosphorylates MLC.

Some support for a role for Rho in MLC phosphorylation *in vivo* was provided by means of NIH3T3 cells transfected with inducible [V14]RhoA, whose expression resulted in slight increase in MBS phosphorylation. It was suggested that activation of Rho results in its interaction with ROK as well as MBS; activated ROK then phosphorylated MBS, which inhibited the MLC phosphatase with a resultant increase in net MLC phosphorylation (by other kinases). This leads to muscle contraction or actin-myosin interactions that underlie stress-fibre formation in non-muscle cells.

It has been demonstrated recently that myosin-driven contraction occurs prior to, and is probably the driving force that promotes, the formation of stress fibres and adhesion complexes in lysophosphatidic-acid-treated fibroblasts (Chrzanowska-Wodnicka and Burridge, 1996). Several inhibitors of actin-myosin interaction, including the MLC-kinase inhibitor KT5926, also blocked contraction, formation of stress fibres and focal adhesions, and tyrosine phosphorylation. Butanedione-2-monoxine, another inhibitor of actin-myosin interaction also blocked [V14]RhoA-induced formation of stress fibres. It was proposed that activation of Rho results in myosin being transformed from a relaxed to an elongated conformation, promoted by MLC phosphorylation. The ensuing alignment of actin and the bipolar myosin filaments results in aggregation of membrane integrins to which actin is attached at the cytoplasmic face (through binding proteins such as α-actinin or talin) (Fig. 8). It was also suggested that this formation of integrin-containing focal adhesions, is upstream of FAK phosphorylation, which accompanies the formation. FAK apparently is not involved in formation either of stress fibres or focal adhesions (Wilson et al., 1995; Ilic et al., 1995), and the tyrosine-kinase inhibitors that block this formation (Burridge et al., 1992; Ridley and Hall, 1994) and FAK phosphorylation probably act upstream of MLC phosphorylation. Whether this involves a tyrosine kinase that activates ROK at the peripheral membranes, to which it is targeted by Rho, merits investigation.

Use of the inhibitor KT5926 in neural-like PC12 cells led Jalink et al. (1994) to suggest that RhoA and its downstream effectors regulated lysophosphatidic-acid-induced actin-myosin contractability through alteration of the balance between MLC-kinase and MLC phosphatase activities. Reference was made to unpublished data of Eichholtz and Molenaar, which showed that the effects of lysophosphatidic acid were accompanied by increased phosphorylation of a protein whose characteristics on two-dimensional gel analysis resembled those MLC. It may turn out that MLC phosphorylation underlies Rho-mediated changes in actin-based morphology wherever these occur.

Tyrosine phosphorylation has been implicated in Rho-mediated signalling pathways that affect morphology. Because the role of the tyrosine kinases has not been established, and since no direct interaction of Rho with a tyrosine kinase has been reported, this aspect has not been discussed in any detail in this review for reasons of brevity and clarity. For similar reasons, serine/threonine kinases that are downstream of Rho, Rac and Cdc42 but have not been shown to interact directly with the p21s are not discussed.

Rho and PKN

PKN, isolated from a human hippocampus cDNA library, is an approximately 120-kDa protein with a kinase domain highly related to that of PKC, and contains a leucine zipper (Mukai and Ono, 1994). Two Rho-interactive proteins isolated by the yeast two-hybrid system, rhophilin and rhotekin, have N-terminal-sequence similarity ($\approx$40%) to a presumed regulatory region of PKN. This region in all three proteins contains a Rho-binding site, but is unrelated to the Cdc42/Rac-binding motif of the PAK group, and differs from the Rho-binding site of ROKα. The binding region can inhibit intrinsic and GAP-stimulated GTPase activity (Reid et al., 1996). PKN is slightly activated (2−3-fold) by expression with [V14]Rho in COS cells and by lysophosphatidic-acid treatment (Watanabe et al., 1996; Amano et al., 1996). Neurofilament subunits are targets of PKN; they bind to the N-terminal region of PKN and their head-rod domains are phosphorylated, suggesting that the action of PKN, downstream of Rho, can affect neurofilament function in neuronal cell types (Mukai et al., 1996). Unlike PKN, rhophilin and rhotekin are not kinases, and their function is unknown.

Yeast RHO and PKC

In *S. cerevisiae*, rho1p (a RhoA homologue) was shown to be upstream of pkc1 (a PKC homologue), which regulates a MAPK pathway associated with bud formation (Nonaka et al., 1995). Two-hybrid analysis provided evidence for a direct interaction between yeast pkc1 and dominant-positive [L68]rho1p (equivalent to [L61]RhoA). The binding domain in pkc1 appears to be a region containing the pseudosubstrate site and the cysteine-rich domain. However, no direct activation of pkc1 *in vitro* could be demonstrated, leading to the suggestion that the p21 may serve to direct the kinase to actin-containing peripheral sites. Recently, Kamada et al. (1996) showed that rho1p binding results in pkc1 stimulation in yeast extracts, but only when phosphatidylserine was present. Because no synergism with diacylglycerol (or phorbol ester) occurred, it was suggested that rho1 was able to replace the diacylglycerol.

Rho and lipid kinases

The relationship of Rho (and Rac) to phospholipids has attracted much recent attraction, with targets of the p21s being shown to include the lipid kinases. For example, Rho associates with PtdIns*P* 5-kinase in fibroblasts, with the GTP-bound form stimulating kinase activity (Ren et al., 1996; Chong et al., 1994). Rac also binds PtdIns*P* 5-kinase in GDP-bound and GTP-bound forms *in vivo* and *in vitro* (Tolias et al., 1995). It has been suggested that the Rho-stimulated increase in PtdIns*P* 5-kinase activity, which leads to elevated PtdInsP_2 levels, reveals cryptic binding sites in the focal-adhesion protein vinculin, which facilitates binding of talin and actin. This would enhance actin polymerization and focal-adhesion formation (Gilmore and Burridge, 1996).

PtdIns 3-kinase activity is stimulated by Rho in soluble preparations of platelets (Zhang et al., 1993). PtdIns 3-kinase is bound by GTP-Rac and by GTP-Cdc42, but not by Rho (Zheng et al., 1994; Tolias et al., 1995; Bokoch et al., 1996), and the associations occur in response to PDGF stimulation in fibroblasts. This stimulation of PtdIns-3 kinase activity appears to lead to Rac activation in the ruffling pathway (Hawkins et al., 1995). Membrane-bound phospholipase D, which hydrolyses phosphatidylcholine to phosphatidic acid, is activated by GTP-Rho (Malcolm et al., 1994; Siddiqi et al., 1995).

Rho and transcription

Rho (along with Cdc42 and Rac) was found to affect serum-response element (SRE)-mediated processes initiated by *c-fos* (Hill et al., 1995). The SRE DNA sequence is present in other growth-factor-regulated promoters, binds serum-response factor and forms a complex with ternary-complex factor, which binds to an adjacent genomic site. By means of transfected cells, SRE was shown to be activated by factors such as lysophosphatidic acid, serum and the heterotrimeric G-proteins. In the absence of these factors, [V12]Rac1, [V14]RhoA and [V12]Cdc42Hs activated SRE, through serum-response factor, in a ternary-complex-factor-independent manner. Experiments with C3 transferase and the ADP-ribosylation mutant [41I]RhoA showed that RhoA mediates the growth-factor response, and that the RhoA pathway was distinct from that mediated by Cdc42 and Rac1.

By analogy with PAK involvement in the Cdc42/Rac-mediated JNK pathway, it is possible that there are Rho-binding kinases downstream of Rho in the SRE pathway. It would be interesting to determine whether these are ROK and/or PKN, given the different Rho-binding and kinase domains of these enzymes.

p21 hierarchies, antagonism, morphology and kinases

Cdc42, Rac and Rho can respond to signals separately, with each having its own pathway. There is good evidence that in pathways leading to changes in morphology, Cdc42-Rac (Kozma et al., 1995) and Rac-Rho (Ridley et al., 1992) hierarchies exist. However, these do not appear to extend to a hierarchical cascade, since Cdc42-induced PAM are formed at the expense of stress fibres, which are RhoA induced (Kozma et al., 1995). This apparent antagonism between Cdc42 and Rho can also be seen in the morphological data of Chrzanowska-Wodnicka and Burridge (1996), which show that KT5926, while blocking lysophosphatidic-acid-induced RhoA−mediated stress-fibre formation, can stimulate PAM formation.

In NIE-115 neuroblastoma and PC12 cells, lysophosphatidic acid causes neurite growth, cone collapse and retraction. The lysophosphatidic-acid-induced effects are abolished on treatment with C3 transferase, which can itself promote neurite outgrowth (Jalink et al., 1994). In neuroblastoma cells with injected recombinant proteins, Cdc42Hs-induced formation of filopodia in growth cones and along neurites was blocked by RhoA (Kozma, R., Sarner, S., Ahmed, S. and Lim, L., unpublished results). C3 transferase promoted this formation but not when injected with inhibitory [N17]Cdc42Hs. Treatment with the neurotransmitter acetylcholine also resulted in filopodia formation, with Cdc42Hs being the mediator. Acetylcholine was found to counteract the neurite-retraction effects of lysophosphatidic acid. As in fibroblasts, Cdc42 and Rho have apparently competing morphological roles; Cdc42 inducing outgrowth and Rho retraction of neurites.

ROK participates in the assembly of stress fibres and focal adhesions. PAK may also be an important player in the reorganization of these actin-containing structures. A model of how ROK and PAK function in assembly and disassembly of stress fibres and focal adhesions is shown in Fig. 8. For filopodia formation to occur, it may be necessary because of cytomechanical reasons and perhaps because of shared utilization of structural protein components, that focal adhesions and stress fibres must disassemble prior to or during filopodia formation. Cdc42 activation of PAK facilitates the former event with other Cdc42 effectors accomplishing the latter (since PAK does not appear to promote filopodia formation). It is possible that ROK is concurrently inhibited, although it has not been established whether ROK is a substrate of PAK and vice versa.

Compartmentation of PAK must also play a part in its morphological activities, since, like Cdc42, Rac activates PAK and, unlike Cdc42, can apparently stimulate Rho-induced pathways. Rac and Cdc42 distributions appear to be different, although there is some overlap, as revealed by antibody staining and expression of tagged p21s, which may be compromised by artefactual effects of overexpression. However, these methods cannot distinguish between the native active GTP-bound forms, which target the kinases, from other forms.

Functional compartmentation of kinases common to several pathways could perhaps be achieved by tethering/scaffolding molecules similar to Ste5p, each of which is specific to a different kinase complex (Elion, 1995). Ste5p is expressed only in haploid yeast cells; it is essential for the Cdc42-Ste20p-MAPK pheromone pathway, but not for the Cdc42-Ste20p-MAPK morphological pathway that is activated by Ras in diploid cells (Mosch et al., 1996). The latter pathway includes the component kinases Ste11p, Ste7p and the transcription factor Ste12p, but not the MAPK Fus3p or Kss1p. It is possible that molecules such as the Cdc42-binding protein, Wiskott-Aldrich syndrome protein, and the Rac-binding protein POR1 (partner of Rac1) (Van Aelst et al., 1996) direct the p21s to the different complexes. The different GEF could bind and contribute to these complexes, facilitating either individual or hierarchical responses. For example, lbc, the Rho-GEF may only contribute to complexes serving Rho, while the Rho-GEF and Cdc42-GEF ost can contribute to different complexes serving Rho and Cdc42 pathways, and its preferential binding to one or the other complex could explain Rho-Cdc42 antagonism. Hierarchies could arise from a GEF being part of the signalling complex for one p21 and by its being activated by a kinase component of another p21 complex. Proteins with both GEF and GAP domains, such as Bcr and Abr, could upregulate or downregulate the pathway depending on the p21 complex to which they bind. Even though differential effects have been observed with GAP and GEF domains *in vitro*, the specificities of GAP and GEF activities *in vivo* of these various multidomain proteins in an intact form have to be established. Once this is achieved, it should provide information on the mechanisms that underlie p21 regulation of phosphorylation pathways. It is highly probable that a target for a kinase associated with one Rho p21 may be a GEF for another Rho p21.

Closing remarks

The Rho subfamily plays a major role in the organization of actin-based morphologies and in the activation of transcriptional pathways stimulated by growth factors. The effectors for these p21s include kinases such as PAK, which are activated on binding of p21s and others, such as ROK, which may be targeted to appropriate sites for their activation. The p21s may be responsible for directing the kinases to their specific sites of action. The kinases are involved in mediating the morphological activities of the p21s and evidence is increasing for a nuclear role. It has been reported recently that through its binding to the adaptor protein Nck, PAK may be linked to RTK (Galisteo et al., 1996). Clearly, the p21-associated kinases play an important part in integrating the various biochemical activities of the p21s. The kinases identified are conserved throughout evolution, and analysis of their roles in organisms such as yeast and *Drosophila* promises to provide a better understanding of their mechanism of action, including their regulation by the p21s.

The help of Clinton Monfries in compiling the bibliography is gratefully acknowledged. We also thank the Glaxo-Singapore Research Fund for its generous support.

REFERENCES

Albrecht-Buehler, G. (1976) Filopodia of spreading 3T3 cells. Do they have a substrate-exploring function? *J. Cell Biol. 69*, 275–286.

Akada, R., Kallal, L., Johnson, D. I. & Kurjan, J. (1996) Genetic relationships between the G protein $\beta\gamma$ complex, Ste5p, Ste20p and Cdc42p: investigation of effector roles in the yeast pheromone response pathway, *Genetics 143*, 103–117.

Amano, M., Mukai, H., Ono, Y., Chihara, K., Matsui, T., Hamajima, Y., Okawa, K., Iwamatsu, A. & Kaibuchi, K. (1996) Identification of a putative target for Rho as the serine-threonine kinase protein kinase N, *Science 271*, 648–650.

Aspenstrom, P., Lindberg, U. & Hall, A. (1996) Two GTPases, Cdc42 and Rac, bind directly to a protein implicated in the immunodeficiency disorder Wiskott-Aldrich syndrome, *Curr. Biol. 6*, 70–75.

Bagrodia, S., Derijard, B., Davis, R. J. & Cerione, R. A. (1995a) Cdc42 and PAK-mediated signaling leads to Jun kinase and p38 mitogen-activated protein kinase activation, *J. Biol. Chem. 270*, 27995–27998.

Bagrodia, S., Taylor, S. J., Creasy, C. L., Chernoff, J. & Cerione, R. A. (1995b) Identification of a mouse p21(cdc42/rac) activated kinase, *J. Biol. Chem. 270*, 22731–22737.

Bar-Sagi, D. & Feramisco, J. R. (1986) Induction of membrane ruffling and fluid-phase pinocytosis in quiescent fibroblasts by Ras proteins, *Science 233*, 1061–1068.

Benner, G. E., Dennis, P. B. & Masaracchia, R. A. (1995) Activation of an S6/H4 kinase (PAK-65) from human placenta by intramolecular and intermolecular autophosphorylation, *J. Biol. Chem. 270*, 21121–21128.

Boguski, M. S. & McCormick, F. (1993) Proteins regulating Ras and its relatives, *Nature 366*, 643–654.

Bokoch, G. M., Vlahos, C. J., Wang, Y., Knaus, U. G. & Traynorkaplan, A. E. (1996) Rac GTPase interacts specifically with phosphatidylinositol 3-kinase, *Biochem. J. 315*, 775–779.

Brook, J. D., McCurrach, M. E., Harley, H. G., Buckler, A. J., Church, D., Aburatani, H., Hunter, K., Stanton, V. P., Thirion, J. P., Hudson, T., Sohn, R., Zemelman, B., Snell, R. G., Rundle, S. A., Crow, S., Davies, J., Shelbourne, P., Buxton, J., Jones, C., Juvonen, V., Johnson, K., Harper, P. S., Shaw, D. J. & Housman, D. E. (1992) Molecular basis of myotonic dystrophy: expansion of a trinucleotide (CTG) repeat at the 3′ end of a transcript encoding a protein kinase family member, *Cell 68*, 799–808.

Brown, J. L., Stowers, L., Baer, M., Trejo, J., Coughlin, S. & Chant, J. (1996) Human Ste20 homolog hPAK1 links GTPases to the JNK MAP kinase pathway, *Curr. Biol. 6*, 598–605.

Burbelo, P. D., Drechsel, D. & Hall, A. (1995) A conserved binding motif defines numerous candidate target proteins for both Cdc42 and Rac GTPases, *J. Biol. Chem. 270*, 29071–29074.

Burridge, K., Turner, C. E. & Romer, L. H. (1992) Tyrosine phosphorylation of paxillin and pp125(FAK) accompanies cell adhesion to extracellular matrix – a role in cytoskeletal assembly, *J. Cell Biol. 119*, 893–903.

Bush, E. W., Taft, C. S., Meixell, G. E. & Perryman, M. B. (1996) Overexpression of myotonic-dystrophy kinase in BC_3H1 cells induces the skeletal-muscle phenotype, *J. Biol. Chem. 271*, 548–552.

Cano, E. & Mahadevan, L. C. (1995) Parallel signal processing among mammalian MAPKs, *Trends Biochem. Sci. 20*, 117–122.

Cerione, R. A. & Zheng, Y. (1996) The *dbl* family of oncogenes, *Curr. Opin. Cell Biol. 8*, 216–222.

Chardin, P., Boquet, P., Madaule, P., Popoff, M. R., Rubin, E. J. & Gill, D. M. (1989) The mammalian G-protein RhoC is ADP-ribosylated by *Clostridium botulinum* exoenzyme C3 and affects actin microfilaments in vero cells, *EMBO J. 8*, 1087–1092.

Chen, W., Chen, S., Yap, S. F. & Lim, L. (1996) The *Caenorhabditis elegans* p21–activated kinase (CePAK) colocalizes with CeRac1 and CDC42Ce at hypodermal cell boundaries during embryo elongation, *J. Biol. Chem. 271*, 26362–26368.

Chien, C. B., Rosenthal, D. E., Harris, W. A. & Holt, C. E. (1993) Navigational errors made by growth cones without filopodia in the embryonic *Xenopus* brain, *Neuron 11*, 237–251.

Chong, L. D., Traynor Kaplan, A., Bokoch, G. M. & Schwartz, M. A. (1994) The small GTP-binding protein Rho regulates a phosphatidylinositol 4-phosphate 5-kinase in mammalian cells, *Cell 79*, 507–513.

Choi, K. Y., Stratterberg, B., Lyons, D. M. & Elion, E. A. (1994) Ste5 tethers multiple protein kinases in the MAP kinase cascade required for mating in *S. cerevisiae*, *Cell 78*, 499–512.

Chrzanowska-Wodnicka, M. & Burridge, K. (1996) Rho-stimulated contractility drives the formation of stress fibers and focal adhesions, *J. Cell Biol. 133*, 1403–1415.

Coso, O. A., Chiariello, M., Yu, J. C., Teramoto, H., Crespo, P., Xu, N., Miki, T. & Gutkind, J. S. (1995) The small GTP-binding proteins Rac1 and Cdc42 regulate the activity of the JNK/SAPK signaling pathway, *Cell 81*, 1137–1146.

Coso, O. A., Teramoto, H., Simonds, W. F. & Gutkind, J. S. (1996) Signaling from G-protein-coupled receptors to c-jun kinase involves $\beta\gamma$-subunits of heterotrimeric G-proteins acting on a Ras and Rac1-dependent pathway, *J. Biol. Chem. 271*, 3963–3966.

Creasy, C. L. & Chernoff, J. (1995) Cloning and characterization of a human protein kinase with homology to Ste20, *J. Biol. Chem. 270*, 21695–21700.

Cvrckova, F., De Virgilio, C., Manser, E., Pringle, J. R. & Nasmyth, K. (1995) Ste20-like protein kinases are required for normal localization of cell growth and for cytokinesis in budding yeast, *Genes & Dev. 9*, 1817–1830.

Diekmann, D., Brill, S., Garrett, M. D., Totty, N., Hsuan, J., Monfries, C., Hall, C., Lim, L. & Hall, A. (1991) Bcr encodes a GTPase-activating protein for $p21^{rac}$, *Nature 351*, 400–402.

Elion, E. A. (1995) Ste5 – a meeting place for MAP kinases and their associates, *Trends Cell Biol. 5*, 322–327.

Friesen, H., Lunz, R., Doyle, S. & Segall, J. (1994) Mutation of the *SPS1*-encoded protein kinase of *Saccharomyces cerevisiae* leads to defects in transcription and morphology during spore formation, *Genes & Dev. 8*, 2162–2175.

Frost, J. A., Xu, S. C., Hutchison, M. R., Marcus, S. & Cobb, M. H. (1996) Actions of Rho-family small G-proteins and p21-activated protein kinases on mitogen activated protein kinase family members, *Mol. Cell Biol. 16*, 3707–3713.

Galisteo, M. L., Chernoff, J., Su, Y.-C., Skolnik, E. Y. & Schlessinger, J. (1996) The adaptor protein Nck links receptor tyrosine kinases with the serine threonine kinase PAK1, *J. Biol. Chem. 271*, 20997–21000.

Gilmore, A. P. & Burridge, K. (1996) Regulation of vinculin binding to talin and actin by phosphatidyl-inositol-4,5-bisphosphate, *Nature 381*, 531–535.

Glise, B., Bourbon, H. & Noselli, S. (1995) Hemipterous encodes a novel *Drosophila* MAP kinase kinase, required for epithelial cell sheet movement, *Cell 83*, 451–461.

Hall, A. (1994) Small GTP-binding proteins and the regulation of the actin cytoskeleton, *Annu. Rev. Cell Biol. 10*, 31–54.

Harden, N., Loh, H. Y., Chia, W. & Lim, L. (1995) A dominant inhibitory version of the small GTP-binding protein Rac disrupts cytoskeletal structures and inhibits developmental cell shape changes in *Drosophila*, *Development (Camb.) 121*, 903–914.

Harden, N., Lee, J., Loh, H. Y., Ong, Y. M., Tan, I., Leung, T., Manser, E. & Lim, L. (1996) A *Drosophila* homolog of the Rac-activated and Cdc42-activated serine/threonine kinase PAK is a potential focal adhesion and focal complex protein that colocalizes with dynamic actin structures, *Mol. Cell Biol. 16*, 1896–1908.

Hart, M. J., Eva, A., Evans, T., Aaronson, S. A. & Cerione, R. A. (1991) Catalysis of guanine nucleotide exchange on the CDC42Hs protein by the *dbl* oncogene product, *Nature 354*, 311–314.

Hawkins, P. T., Eguinoa, A., Qiu, R. G., Stokoe, D., Cooke, F. T., Walters, R., Wennstrom, S., Claesson Welsh, L., Evans, T., Symons, M. & Stephens, L. (1995) PDGF stimulates an increase in GTP-Rac via activation of phosphoinositide 3-kinase, *Curr. Biol. 5*, 393–403.

Herskowitz, I. (1995) MAP kinase pathways in yeast: for mating and more, *Cell 80*, 187–197.

Hill, C. S., Wynne, J. & Treisman, R. (1995) The Rho family GTPases RhoA, Rac1, and Cdc42Hs regulate transcriptional activation by SRF, *Cell 81*, 1159–1170.

Hirata, K., Kikuchi, A., Sasaki, T., Kuroda, S., Kaibuchi, K., Matsuura, Y., Seki, H., Saida, K. & Takai, Y. (1992) Involvement of Rho p21 in the GTP-enhanced calcium ion sensitivity of smooth muscle contraction, *J. Biol. Chem. 267*, 8719–8722.

Hoehn, G. T., Stokland, T., Amin, S., Ramirez, M., Hawkins, A., Griffin, C. A., Small, D. & Civin, C. I. (1996) Tnk1: a novel intracellular tyrosine kinase gene isolated from human umbilical cord blood $CD34^+/Lin^-CD38^-$ stem/progenitor cells, *Oncogene 12*, 903–913.

Hynes, R. O. & Lander, A. D. (1992) Contact and adhesive specificities in the associations, migrations, and targeting of cells and axons, *Cell 68*, 303–322.

Ilic, D., Furuta, Y., Kanazawa, S., Takeda, N., Sobue, K., Nakatsuji, N., Nomura, S., Fujimoto, J., Okada, M., Yamamoto, T. & Aizawa, S. (1995) Reduced cell motility and enhanced focal adhesion contact formation in cells from FAK-deficient mice, *Nature 377*, 539–544.

Ishizaki, T., Maekawa, M., Fujisawa, K., Okawa, K., Iwamatsu, A., Fujita, A., Watanabe, N., Saito, Y., Kakizuka, A., Morii, N. & Narumiya, S. (1996) The small GTP-binding protein Rho binds to and activates a 160 kDa ser/thr protein kinase homologous to myotonic dystrophy kinase, *EMBO J. 15*, 1885–1893.

Jakobi, R., Chen, C., Tuazon, P. T. & Traugh, J. A. (1996) Molecular cloning and sequencing of the cytostatic G protein-activated protein kinase PAK I, *J. Biol. Chem. 271*, 6206–6211.

Jalink, K., van Corven, E. J., Hengeveld, T., Morii, N., Narumiya, S. & Moolenaar, W. H. (1994) Inhibition of lysophosphatidate- and thrombin-induced neurite retraction and neuronal cell rounding by ADP ribosylation of the small GTP-binding protein Rho, *J. Cell Biol. 126*, 801–810.

Jimenez, B., Arends, M., Esteve, P., Perona, R., Sanchez, R., Ramon, C., Wyllie, A. & Lacal, J. C. (1995) Induction of apoptosis in NIH3T3 cells after serum deprivation by overexpression of Rho-p21, a GTPase protein of the Ras superfamily, *Oncogene 10*, 811–816.

Johnson, L. N., Noble, M. E. M. & Owen, D. J. (1996) Active and inactive protein kinases – structural basis for regulation, *Cell 85*, 149–158.

Joneson, T., White, M. A., Wigler, M. H. & Bar-Sagi, D. (1996) Stimulation of membrane ruffling and MAP kinase activation by distinct effectors of Ras, *Science 271*, 810–812.

Justice, R. W., Zilian, O., Woods, D. F., Noll, M. & Bryant, P. J. (1995) The *Drosophila* tumor suppressor gene warts encodes a homolog of human myotonic dystrophy kinase and is required for the control of cell shape and proliferation, *Genes & Dev. 9*, 534–546.

Kamada, Y., Qadota, H., Python, C. P., Anraku, Y., Ohya, Y. & Levin, D. E. (1996) Activation of yeast protein kinase C by Rho1 GTPase, *J. Biol. Chem. 271*, 9193–9196.

Kao, L. R., Peterson, J., Ji, R., Bender, L. & Bender, A. (1996) Interactions between the ankyrin repeat-containing protein Akr1p and the pheromone response pathway in *Saccharomyces cerevisiae*, *Mol. Cell. Biol. 16*, 168–178.

Katz, P., Whalen, G. & Kehrl, J. H. (1994) Differential expression of a novel protein kinase in human B lymphocytes – preferential localization in the germinal centre, *J. Biol. Chem. 269*, 16802–16809.

Khosravi-Far, R., Solski, P. A., Clark, G. J., Kinch, M. S. & Der, C. J. (1995) Activation of Rac1, RhoA, and mitogen activated protein kinases is required for Ras transformation, *Mol. Cell. Biol. 15*, 6443–6453.

Kimura, K., Ito, M., Amano, M., Chihara, K., Fukata, Y., Nakafuku, M., Yamamori, B., Feng, J. H., Nakano, T., Okawa, K., Iwamatsu, A. & Kaibuchi, K. (1996) Regulation of myosin phosphatase by Rho and Rho-associated kinase (Rho-kinase), *Science 273*, 245–248.

Kitazawa, T., Masuo, M. & Somlyo, A. P. (1991) G-protein-mediated inhibition of myosin light chain phosphatase in vascular smooth muscle, *Proc. Natl Acad. Sci. USA 88*, 9307–9310.

Kozma, R., Ahmed, S., Best, A. & Lim, L. (1995) The Ras-related protein Cdc42Hs and bradykinin promote formation of peripheral actin microspikes and filopodia in Swiss 3T3 fibroblasts, *Mol. Cell. Biol. 15*, 1942–1952.

Kozma, R., Ahmed, S., Best, A. & Lim, L. (1996) The GTPase activating protein n-chimaerin co-operates with Rac1 and Cdc42Hs to induce the formation of lamellipodia and filopodia, *Mol. Cell. Biol. 16*, 5069–5080.

Lamarche, N. & Hall, A. (1994) GAPs for Rho-related GTPases, *Trends Genet. 10*, 436–440.

Leberer, E., Dignard, D., Harcus, D., Thomas, D. Y. & Whiteway, M. (1992) The protein kinase homologue Ste20p is required to link the yeast pheromone response G-protein $\beta\gamma$-subunits to downstream signaling components, *EMBO J. 11*, 4815–4824.

Leberer, E., Dignard, D., Harcus, D., Hougan, L., Whiteway, M. & Thomas, D. Y. (1993) Cloning of *Saccharomyces cerevisiae* Ste5 as a suppressor of a Ste20 protein kinase mutant – structural and functional similarity of Ste5 to Far1, *Mol. & Gen. Genet. 241*, 241–254.

Leeuw, T., Fourestlieuvin, A., Wu, C. L., Chenevert, J., Clark, K., Whiteway, M., Thomas, D. Y. & Leberer, E. (1995) Pheromone response in yeast – association of Bem1p with proteins of the MAP kinase cascade and actin, *Science 270*, 1210–1213.

Leung, T., Manser, E., Tan, L. & Lim, L. (1995) A novel serine/threonine kinase binding the Ras-related RhoA GTPase which translocates the kinase to peripheral membranes, *J. Biol. Chem. 270*, 29051–29054.

Leung, T., Chen, X., Manser, E. & Lim, L. (1996) The p160 RhoA-binding kinase ROKα is a member of a kinase family and is involved in the reorganization of the cytoskeleton, *Mol. Cell. Biol. 16*, 5313–5327.

Lev, S., Moreno, H., Martinez, R., Canoll, P., Peles, E., Musacchio, J. M., Plowman, G. D., Rudy, B. & Schlessinger, J. (1995) Protein-tyrosine kinase PYK2 involved in Ca^{2+}-induced regulation of ion channel and MAP kinase functions, *Nature 376*, 737–745.

Lim, L., Hall, C., Leung, T. & Manser, E. (1995) The chimaerin and BCR families, in *Guidebook to the small GTPases* (Zerial, M. & Huber, L. A., eds) pp. 246–260, Oxford University Press, Oxford.

Malcolm, K. C., Ross, A. H., Qiu, R.-G., Symons, M. & Exton, J. H. (1994) Activation of rat liver phospholipase D by the small GTP-binding protein RhoA, *J. Biol. Chem. 269*, 25951–25954.

Manser, E., Leung, T., Monfries, C., Teo, M., Hall, C. & Lim, L. (1992) Diversity and versatility of GTPase activating proteins for the $p21^{rho}$ subfamily of Ras G proteins detected by a novel overlay assay, *J. Biol. Chem. 267*, 16025–16028.

Manser, E., Leung, T., Salihuddin, H., Tan, L. & Lim, L. (1993) A non-receptor tyrosine kinase that inhibits the GTPase activity of $p21^{Cdc42}$, *Nature 363*, 364–367.

Manser, E., Leung, T., Salihuddin, H., Zhao, Z. S. & Lim, L. (1994) A brain serine/threonine protein kinase activated by Cdc42 and Rac1, *Nature 367*, 40–46.

Manser, E., Chong, C., Zhao, Z. S., Leung, T., Michael, G., Hall, C. & Lim, L. (1995) Molecular cloning of a new member of the p21-Cdc42/Rac-activated kinase (PAK) family, *J. Biol. Chem. 270*, 25070–25078.

Marcus, S., Polverino, A., Chang, E., Robbins, D., Cobb, M. H. & Wigler, M. H. (1995) Shk1, a homolog of the *Saccharomyces cerevisiae* Ste20 and mammalian p65PAK protein kinases, is a component of a Ras/Cdc42 signaling module in the fission yeast *Schizosaccharomyces pombe*, *Proc. Natl Acad. Sci. USA 92*, 6180–6184.

Marshall, C. J. (1994) MAP kinase kinase kinase, MAP kinase kinase and MAP kinase, *Curr. Opin. Genet. Dev. 4*, 82–89.

Marshall, C. J. (1996) Ras effectors, *Curr. Opin. Cell Biol. 8*, 197–204.

Martin, G. A., Bollag, G., McCormick, F. & Abo, A. (1995) A novel serine kinase activated by Rac1/Cdc42Hs-dependent autophosphorylation is related to PAK65 and STE20, *EMBO J. 14*, 1970–1978.

Matsui, T., Amano, M., Yamamoto, T., Chihara, K., Nakafuku, M., Ito, M., Nakano, T., Okawa, K., Iwamatsu, A. & Kaibuchi, K. (1996) Rho-associated kinase, a novel serine threonine kinase, as a putative target for the small GTP-binding protein Rho, *EMBO J. 15*, 2208–2216.

McCormick, F. (1994) Activators and effectors of Ras p21 proteins, *Curr. Opin. Genet. Dev. 4*, 71–76.

Minden, A., Lin, A., Claret, F. X., Abo, A. & Karin, M. (1995) Selective activation of the JNK signaling cascade and c-Jun transcriptional activity by the small GTPases Rac and Cdc42Hs, *Cell 81*, 1147–1157.

Mosch, H., Roberts, R. L. & Fink, G. R. (1996) Ras2 signals via the Cdc42/Ste20/mitogen-activated protein kinase module to induce filamentous growth in *Saccharomyces cerevisiae*, *Proc. Natl Acad. Sci. USA 93*, 5352–5356.

Mukai, H., Toshimori, M., Shibata, H., Kitagawa, M., Shimakawa, M., Miyahara, M., Sunakawa, H. & Ono, Y. (1996) PKN associates and phosphorylates the head-rod domain of neurofilament protein, *J. Biol. Chem. 271*, 9816–9822.

Mukai, H. & Ono, Y. (1994) A novel protein kinase with leucine zipper-like sequences – its catalytic domain is highly homologous to that of protein kinase C, *Biochem. Biophys. Res. Commun. 199*, 897–904.

Nobes, C. D. & Hall, A. (1995) Rho, Rac, and Cdc42 GTPases regulate the assembly of multimolecular focal complexes associated with actin stress fibers, lamellipodia, and filopodia, *Cell 81*, 53–62.

Nobes, C. D., Hawkins, P., Stephens, L. & Hall, A. (1995) Activation of the small GTP-binding proteins rho and rac by growth factor receptors, *J. Cell Sci. 108*, 225–233

Noda, M., Yasuda Fukazawa, C., Moriishi, K., Kato, T., Okuda, T., Kurokawa, K. & Takuwa, Y. (1995) Involvement of *rho* in GTPγS-induced enhancement of phosphorylation of 20 kDa myosin light chain in vascular smooth muscle cells: inhibition of phosphatase activity, *FEBS Lett. 367*, 246–250.

Nonaka, H., Tanaka, K., Hirano, H., Fujiwara, T., Kohno, H., Umikawa, M., Mino, A. & Takai, Y. (1995) A downstream target of rho1 small GTP-binding protein is pkc1, a homolog of protein kinase c, which leads to activation of the MAP kinase cascade in *Saccharomyces cerevisiae*, *EMBO J. 14*, 5931–5938.

Olson, M. F., Ashworth, A. & Hall, A. (1995) An essential role for Rho, Rac, and Cdc42 GTPases in cell cycle progression through G1, *Science 269*, 1270–1272.

Ottilie, S., Miller, P. J., Johnson, D. I., Creasy, C. L., Sells, M. A., Bagrodia, S., Forsburg, S. L. & Chernoff, J. (1995) Fission yeast *PAK1*$^+$ encodes a protein-kinase that interacts with Cdc42p and is involved in the control of cell polarity and mating, *EMBO J. 14*, 5908–5919.

Paterson, H. F., Self, A. J., Garrett, M. D., Just, I., Aktories, K. & Hall, A. (1990) Microinjection of recombinant p21rho induces rapid changes in cell morphology, *J. Cell Biol. 111*, 1001–1007.

Peppelenbosch, M. P., Qiu, R. G., de Vries Smits, A. M., Tertoolen, L. G., de Laat, S. W., McCormick, F., Hall, A., Symons, M. H. & Bos, J. L. (1995) Rac mediates growth factor-induced arachidonic acid release, *Cell 81*, 849–856.

Polverino, A., Frost, J., Yang, P., Hutchison, M., Neiman, A. M., Cobb, M. H. & Marcus, S. (1995) Activation of mitogen activated protein kinase cascades by p21-activated protein kinases in cell-free extracts of *Xenopus* oocytes, *J. Biol. Chem. 270*, 26067–26070.

Pombo, C. M., Kehrl, J. H., Sanchez, I., Katz, P., Avruch, J., Zon, L. I., Woodgett, J. R., Force, T. & Kyriakis, J. M. (1995) Activation of the SAPK pathway by the human STE20 homologue germinal centre kinase, *Nature 377*, 750–754.

Prendergast, G. C., Khosravifar, R., Solski, P. A., Kurzawa, H., Lebowitz, P. F. & Der, C. J. (1995) Critical role of Rho in cell transformation by oncogenic Ras, *Oncogene 10*, 2289–2296.

Qiu, R. G., Chen, J., Kirn, D., McCormick, F. & Symons, M. (1995a) An essential role for Rac in Ras transformation, *Nature 374*, 457–459.

Qiu, R. G., Chen, J., McCormick, F. & Symons, M. (1995b) A role for Rho in Ras transformation, *Proc. Natl Acad. Sci. USA 92*, 11781–11785.

Ramer, S. W. & Davis, R. W. (1993) A dominant truncation allele identifies a gene, *Ste20*, that encodes a putative protein kinase necessary for mating in *Saccharomyces cerevisiae*, *Proc. Natl Acad. Sci. USA 90*, 452–456.

Reid, T., Furayashiki, T., Ishizaki, T., Watanabe, G., Watanabe, N., Fujisawa, K., Morii, N., Madaule, P. & Narumiya, S. (1996) Rhotekin, a new putative target for Rho-bearing homology to a serine/threonine kinase, PKN, and rhophilin in the Rho-binding domain, *J. Biol. Chem. 271*, 13556–13560.

Ren, X. D., Bokoch, G. M., Traynorkaplan, A., Jenkins, G. H., Anderson, R. A. & Schwartz, M. A. (1996) Physical association of the small GTPase Rho with a 68 kDa phosphatidylinositol 4-phosphate 5-kinase in Swiss 3T3 cells, *Mol. Biol. Cell 7*, 435–442.

Ridley, A. J., Paterson, H. F., Johnston, C. L., Diekmann, D. & Hall, A. (1992) The small GTP-binding protein Rac regulates growth factor-induced membrane ruffling, *Cell 70*, 401–410.

Ridley, A. J. & Hall, A. (1992) The small GTP-binding protein Rho regulates the assembly of focal adhesions and actin stress fibers in response to growth factors, *Cell 70*, 389–399.

Ridley, A. J. & Hall, A. (1994) Signal transduction pathways regulating Rho-mediated stress fibre formation: requirement for a tyrosine kinase, *EMBO J. 13*, 2600–2610.

Roberts, R. L. & Fink, G. R. (1994) Elements of a single MAP kinase cascade in *Saccharomyces cerevisiae* mediate two developmental programs in the same cell type – mating and invasive growth, *Genes & Dev. 8*, 2974–2985.

Rooney, R. D., Tuazon, P. T., Meek, W. E., Carroll, E. J., Hagen, J. J., Gump, E. L., Monnig, C. A., Lugo, T. & Traugh, J. A. (1996) Cleavage arrest of early frog embryos by the G-protein activated protein kinase PAKI, *J. Biol. Chem.* 271, 21498–21504.

Satoh, T., Kato, J., Nishida, K. & Kaziro, Y. (1996) Tyrosine phosphorylation of ACK in response to temperature shift-down, hyperosmotic shock, and epidermal growth factor stimulation, *FEBS Lett. 386*, 230–234.

Simon, M. N., De Virgilio, C., Souza, B., Pringle, J. R., Abo, A. & Reed, S. I. (1995) Role for the Rho-family GTPase Cdc42 in yeast mating-pheromone signal pathway, *Nature 376*, 702–705.

Siddiqi, A. R., Smith, J. L., Ross, A. H., Qiu, R.-G., Symons, M. & Exton, J. H. (1995) Regulation of phospholipase D in HL60 cells, *J. Biol. Chem. 269*, 8466–8473.

Symons, M. (1996) Rho-family GTPases – the cytoskeleton and beyond, *Trends Biochem. Sci. 21*, 178–181.

Symons, M., Derry, J. M. J., Karlak, B., Jiang, S., Lemahieu, V., McCormick, F., Francke, U. & Abo, A. (1996) Wiskott-Aldrich syndrome protein, a novel effector for the GTPase Cdc42Hs, is implicated in actin polymerization, *Cell 84*, 723–734.

Takai, Y., Sasaki, T., Tanaka, K. & Nakanishi, H. (1995) Rho as a regulator of the cytoskeleton, *Trends Biochem. Sci. 20*, 227–231.

Teo, M., Manser, E. & Lim, L. (1995) Identification and molecular cloning of a p21Cdc42/Rac1-activated serine/threonine kinase that is rapidly activated by thrombin in platelets, *J. Biol. Chem. 270*, 26690–26697.

Tolias, K. F., Cantley, L. C. & Carpenter, C. L. (1995) Rho family GTPases bind to phosphoinositide kinases, *J. Biol. Chem. 270*, 17656–17659.

Van Aelst, L., Joneson, T. & Bar-Sagi, D. (1996) Identification of a novel Rac1–interacting protein involved in membrane ruffling, *EMBO J. 15*, 3778–3786.

Watanabe, G., Saito, Y., Madaule, P., Ishizaki, T., Fujisawa, K., Morii, N., Mukai, H., Ono, Y., Kakizuka, A. & Narumiya, S. (1996) Protein kinase N (PKN) and PKN-related protein rhophilin as targets of small GTPase Rho, *Science 271*, 645–648.

Wilson, L., Carrier, M. J. & Kellie, S. (1995) $Pp125^{fak}$ tyrosine kinase-activity is not required for the assembly of F-actin stress fibers and focal adhesions in cultured mouse aortic smooth-muscle cells, *J. Cell Sci. 108*, 2381–2391.

Wittinghofer, A. & Valencia, A. (1995) Three-dimensional structure of Ras and Ras-related proteins, in *Guidebook to the small GTPases* (Zerial, M. & Huber, L. A., eds) pp. 20–29, Oxford University Press, Oxford.

Wu, C., Whiteway, M., Thomas, D. Y. & Leberer, E. (1995) Molecular characterization of Ste20p, a potential mitogen-activated protein or extracellular signal-regulated kinase kinase (MEK) kinase kinase from *Saccharomyces cerevisiae*, *J. Biol. Chem. 270*, 15984–15992.

Yamamori, B., Kuroda, S., Shimizu, K., Fukui, K., Ohtsuka, T. & Takai, Y. (1995) Purification of a Ras-dependent mitogen activated protein kinase kinase kinase from bovine brain cytosol and its identification as a complex of B-Raf and 14-3-3 proteins, *J. Biol. Chem. 270*, 11723–11726.

Yarden, O., Plamann, M., Ebbole, D. J. & Yanofsky, C. (1992) *Cot-1*, a gene required for hyphal elongation in *Neurospora crassa*, encodes a protein kinase, *EMBO J. 11*, 2159–2166.

Zhang, J., King, W. G., Dillon, S., Hall, A., Feig, L. & Rittenhouse, S. E. (1993) Activation of platelet phosphatidylinositide 3-kinase requires the small GTP-binding protein Rho, *J. Biol. Chem. 268*, 22251–22254.

Zhang, S., Han, J., Sells, M. A., Chernoff, J., Knaus, U. G., Ulevitch, R. J. & Bokoch, G. M. (1995) Rho family GTPases regulate p38 mitogen-activated protein kinase through the downstream mediator Pak1, *J. Biol. Chem. 270*, 23934–23936.

Zhao, Z. S., Leung, T., Manser, E. & Lim, L. (1995) Pheromone signalling in *Saccharomyces cerevisiae* requires the small GTP-binding protein Cdc42p and its activator *CDC24*, *Mol. Cell Biol. 15*, 5246–5257.

Zheng, Y., Bagrodia, S. & Cerione, R. A. (1994) Activation of phosphoinositide 3–kinase activity by Cdc42Hs binding to p85, *J. Biol. Chem. 269*, 18727–18730.

Eur. J. Biochem. *242*, 435–445 (1996)

Review

Pancreatic development and maturation of the islet B cell

Studies of pluripotent islet cultures

Ole Dragsbæk MADSEN[1], Jan JENSEN[1], Niels BLUME[1], Helle V. PETERSEN[1], Kaare LUND[1], Christina KARLSEN[1], Frank G. ANDERSEN[1], Per B. JENSEN[1], Lars-Inge LARSSON[2] and Palle SERUP[1]

[1] Hagedorn Research Institute, Gentofte, Denmark

[2] Department of Molecular Cell Biology, State Serum Institute, Copenhagen, Denmark

(Received 30 July 1996) – EJB 96 1135/0

Pancreas organogenesis is a highly regulated process, in which two anlage evaginate from the primitive gut. They later fuse, and, under the influence of the surrounding mesenchyme, the mature organ develops, being mainly composed of ductal, exocrine and endocrine compartments. Early buds are characterized by a branching morphogenesis of the ductal epithelium from which endocrine and exocrine precursor cells bud to eventually form the two other compartments. The three compartments are thought to be of common endodermal origin; in contrast to earlier hypotheses, which suggested that the endocrine compartment was of neuroectodermal origin. It is thus generally believed that the pancreatic endocrine-lineage possesses the ability to mature along a differentiation pathway that shares many characteristics with those of neuronal differentiation. During recent years, studies of insulin-gene regulation and, in particular, the tissue-specific transcriptional control of insulin-gene activity have provided information on pancreas development in general. The present review summarizes these findings, with a special focus on our own studies on pluripotent endocrine cultures of rat pancreas.

Keywords: islet; pancreas; insulin; B cell; insulin-promoter factor 1.

Introduction

Pancreatic structure and function. The mammalian pancreas consists of three main epithelial compartments: the ductular system; the exocrine acinar cells; and the endocrine islets of Langerhans. The delicately branched ductal system drains the digestive proenzymes produced by exocrine acinar tissue into the duodenum. Exocrine cells are arranged into acini, in which each cell is highly polarized so that the zymogen granules accumulate around the border of centroacinar cell, which constitutes the finest branch of the ductal tree. One characteristic of the duct cells is their production of bicarbonate, which is used for neutralizing the acidic input from the stomach. In many species, the main collecting duct drains into the duodenum together with the common bile duct. Proenzymes contained in the zymogen granules become activated by limited proteolysis after entering the duodenum.

The endocrine islets of Langerhans are scattered throughout the pancreas and have a highly developed blood supply, whereby up to a few arteries enter to the center of the islet, from which they branch into glomerular structures, where collecting venules eventually drain into the portal vein (Bonner-Weir and Orci, 1982; Samols et al., 1988). The islets of rats contain a few thousand cells and constitutes approximately 1–2% of the pancreatic mass. Four main cell types are found in the islets (A, B, D and PP cells), each producing a particular hormone (glucagon, insulin, somatostatin and pancreatic polypeptide, respectively). A primary function of insulin is the stimulation of glucose uptake in peripheral tissues after a meal. It is released by the B cells in response to elevated blood-glucose levels, a process that is further boosted by the gut hormones gastic inhibitory peptide (GIP) and glucagon-like peptide (GLP-1), which also are released upon food intake. In contrast, glucagon release by the A cell is triggered by low blood-glucose levels, and a major site of glucagon action is the liver, where glycogen breakdown serves to restore blood-glucose levels. Pancreatic somatostatin may primarily work in a paracrine fashion to modulate insulin and glucagon secretion. Analysis of the microvascular-flow pattern in the islet suggests that secretion from the acinar cells may be inhibited by islet somatostatin (Samols and Stagner, 1990). The biological function of pancreatic polypeptide is poorly understood. PP cells occur in different pancreatic endocrine tumors as well as in silent tumors (Larsson, 1978). No specific syndrome associated with PP hypersecretion has been defined.

Pancreas formation. The pancreas develops as two anlage that bud from the duodenal wall (reviewed by Slack, 1995). The dorsal anlage buds directly from the duodenal wall and precedes budding of the ventral anlage, which can be considered as an independent branch of the liver bud (Pictet and Rutter, 1972). Later in development, the ventral anlage reposition and fuse with

Correspondence to O. D. Madsen, Hagedorn Research Institute, Niels Steensensvej 6, DK-2820 Gentofte, Denmark

Abbreviations. PYY, polypeptide YY; GLUT-2, glucose transporter 2; IPF1, insulin-promoter factor 1; GLP-1; glucagon-like peptide 1; IEF1, insulin-enhancer factor 1; SV, simian virus; IAPP, islet amyloid polypeptide; RIP, rat insulin promoter.

Note. This Review will be reprinted in *EJB Reviews* 1996, to appear in April 1997.

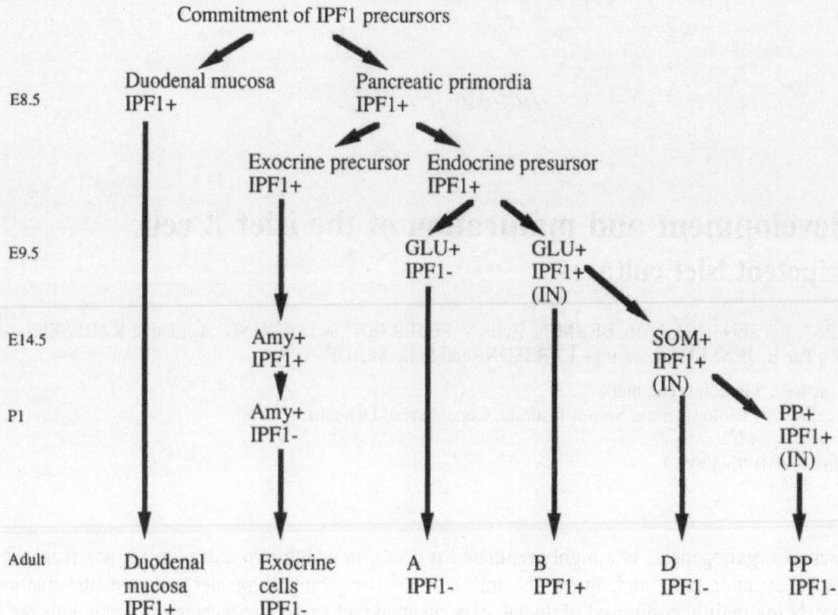

Fig. 1. The Teitelman model of islet development (Teitelman, 1990, 1991, 1993). Based on studies of the co-localizafion of islet hormones during mouse pancreas ontogeny (Alpert et al., 1988), Dr Teitelman proposed a cell-lineage model for the various pancreatic cell types. The model was recently refined by inclusion of expression data of IPF1 (Guz et al., 1995). The version shown was adapted from Guz et al. (1995). Contrasting data on the lack of IPF1/glucagon−double-positive cells during early development was recently reported (Ahlgren et al., 1996). Moreover, early duodenal glucagon and insulin cells form in IPF1-deficient mice (Offield et al., 1996; Ahlgren et al., 1996). GLU, glucagon; SOM, somatostatin; Amy, IAPP.

the dorsal part to form the mature pancreas. During this process, the dorsal duct/duodenal connection in most cases withers off and as a result of the intra-pancreatic fusion the original ventral duct now drains the excretory products of the entire pancreas via the common bile duct into the duodenum. Anatomical variations in the main-duct structure of mouse pancreas was recently examined (Watanabe et al., 1995).

Endodermal origin of the pancreas. It is now generally accepted that both the endocrine and exocrine pancreas are of endodermal origin (Pictet and Rutter, 1972; Pictet et al., 1976; Douarin, 1988). Based on the common features of the endocrine cells and neurons, Pearse formulated the APUD theory (amine-precursor uptake and decarboxylation), which claimed that the diffuse endocrine system of the gastro-intestinal tract was of ectodermal origin and that such cells would migrate from the neural crest to the endoderm during ontogeny (Pearse, 1982). This theory was never verified and it is now evident that endodermal cells can be induced to follow differentiation pathways that share many characteristics with those of neuro-ectodermal differentiation (Kanaka-Gantenbein et al., 1995a).

In the gut, transgenic experiments (chimeric mice) to investigate the cell-lineage relationship of the various cell types in the intestinal epithelium demonstrated that endocrine, columnar, goblet and Paneth cells arose from single endodermal-crypt stem cells. Thus, in chimeric mice the scattered endocrine cells always share the clonal marker of the pluripotent crypt cell, which is a strong indication that gastro-intestinal endocrine cells derive from the endoderm (Ponder et al., 1985; Gordon, 1989).

Similarly, the exocrine and endocrine pancreas are believed to be derived from common protodifferentiated progenitor cells of ductal origin (Fig. 1), which, under appropriate stimuli, may differentiate into acinar or islet cells (Pictet and Rutter, 1972). The major part of the ductal epithelium itself represents a third type of differentiated cells (Bouwens et al., 1994, 1995; Gittes et al., 1996). The phenotype of a putative pancreatic stem cell has not been defined although a remarkable regeneration potential exists in ductal tissue, even in the adult pancreas (Dudek et al., 1991) as observed after pancreatectomy (Bonner-Weir et al., 1993), duct ligation (Wang et al., 1995), or wrapping of the pancreas with cellophane (Rosenberg and Vinik, 1992). In such cases, endocrine-cell and exocrine-cell renewal have been reported to occur from the ductal tissue. Scattered hormone-producing endocrine cells are often found in the adult ductular system. Such cells still express ductal markers (e.g. DA39+) and are thus distinguishable from the islet phenotypes (Contreas et al., 1990). Not only restricted to pancreas determination, ductal epithelium harbours the capacity to also transdifferentiate into genuine hepatocytes, as observed in studies of rats maintained for a certain period on a copper-deficient diet (Rao et al., 1990).

Studies of transformed pancreatic cell lines (Jessop and Hay, 1980; Madsen et al., 1986; Philippe et al., 1987) and of a transgenic-mouse model for islet regeneration (Gu and Sarvetnick, 1993) have demonstrated the existence of mixed ductal/acinar/islet phenotypes (Rosewicz et al., 1992; Beck and Madsen, 1989; Gu et al., 1994; Jensen et al., 1996). Such observations strongly support the hypothesis of a developmental relationship between the three types of pancreatic tissue (Fig. 1).

Studies in chimeric mice with a B-cell specific marker (human proinsulin C-peptide immunoreactivity) revealed that individual islets retain chimerism, which directly demonstrates that islet formation is complex and not just the result of a clonal expansion from a single precursor cell (Deltour et al., 1991). This is consistent with morphological observations that suggest that larger endocrine structures in the fetal pancreas segregate to form mature islets (Møller et al., 1992).

Early studies implicated an important interaction between the mesenchyme and endoderm in pancreatic development, and

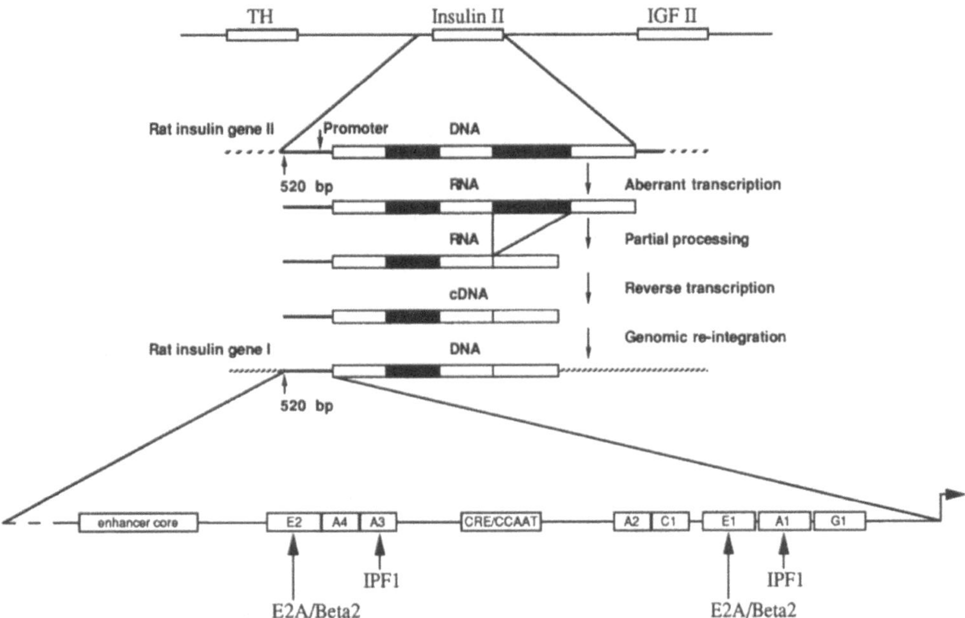

Fig. 2. Insulin-gene duplication in rat and mouse. The insulin-I gene in rat and mouse was duplicated from the ancestral insulin-II gene. this duplication was probably an RNA-mediated single event taking place in a common ancestor of mice and rats prior to their segregation as independent species (Soares et al., 1985). Like the single human insulin gene, the rodent insulin-II gene is located between the genes encoding tyrosine hydroxylase (TH) and insulin-like growth factor (IGF) II (Bell et al., 1985; O'Malley and Rotwein, 1988). The nomenclature of the elements within the promoter/enhancer region of the insulin-I gene follows the proposal of German et al. (1995).

the presence of mesenchymal factor(s) necessary for both B-cell and acinar-cell formation was predicted (Pictet and Rutter, 1972; Pictet et al., 1976). These factors have remained elusive, although roles for nerve-growth factor (Kanaka-Gantenbein et al., 1995a) and hepatocyte-growth factor (Otonkoski et al., 1996) have been implicated in islet development. Recent studies, however, involving recombination of microdissected embryonic pancreatic epithelium and mesenchyme suggest that endocrine development can occur in the absence of mesenchyme, which is in contrast to exocrine development (Gittes et al., 1996).

Islet development

Maturation of tissue-specific hormone-gene regulation. The hallmark of cell differentiation is the acquired efficiency by which mature cells can transcribe a particular subset of the gene pool common to all cells of the organism. The mammalian pancreatic islet B-cell is an extreme example, in which the insulin gene becomes activated to such an extent that insulin mRNA constitutes more than 10% of the total mRNA while being completely silent in other cell types. Similarly, glucagon, somatostatin and pancreatic-polypeptide genes are specifically activated in islet A, D and PP cells, respectively. However, the expression of these three hormone genes is not as restricted to the islets as that of insulin. The proglucagon gene is efficiently transcribed in the GLP-1–producing L-cell of intestinal tissue as well as in the brain stem (Lee et al., 1990). Somatostatin is produced in scattered D-cells of the stomach and gastrointestinal tract, in thyroid C-cells, and in many neurons of the central nervous system including sites in the hypothalamus, the cerebral cortex, the limbic system and the spinal cord (Patel, 1992). Although the different islet cells can be distinguished from each other based on their hormone production, they still share the expression of many neuro-endocrine–specific genes that distinguish the islet from the surrounding exocrine tissue.

Since the discovery that transcriptional control is in part mediated by *cis*-acting DNA elements located in the vicinity of transcriptional start sites (enhancer/promoter elements) of particular genes, much focus has been concentrated on the insulin gene (Walker et al., 1983). The insulin promoter proved to be a very useful tool in studies of several different aspects of islet cell as well as of pancreatic development. Mechanisms of the transcriptional regulation of insulin and glucagon gene transcription have been reviewed recently (Philippe, 1991, 1994; Docherty and Clark, 1994; Stein, 1993; German et al., 1995).

The insulin promoter in transgenic studies. The very elegant studies by Hanahan (1985) demonstrated that a sequence, comprising about 660 bp upstream of the transcriptional start site of the rat insulin-II gene, was able to guide B-cell–specific expression of the simian virus (SV) 40 T antigen of a fusion gene construct [rat insulin promoter (RIP)–T-antigen] in transgenic mice resulting in heritable formation of insulinomas. The equivalent upstream region for the rat insulin-I gene was previously shown, by transient-transfection experiments *in vitro*, to contain functional and tissue-specific enhancer/promoter activity (Walker et al., 1983; Edlund et al., 1985). Hanahan thus confirmed that a similar short sequence upstream of the rat insulin-II gene was sufficient to ensure a B-cell–specific expression pattern *in vivo*.

The transgenic-mouse experiment performed by Hanahan was essentially a replication of an evolutionary event involving the insulin-gene duplication that occurred more than 25 million years ago in an ancestral rodent from which both mice, and rats evolved (Fig. 2). In this common ancestor, the insulin-I gene was probably formed from an aberrantly initiated and partially processed transcript of the ancestral insulin-II gene, which contained approximately 520 bp of the upstream promoter and lacked one of two introns. This fragment was retropositioned into the genome to yield a second functional insulin gene

(Soares et al., 1985). The potential evolutionary advantage of maintaining two functional and very similar insulin genes is not clear, but both genes are expressed in a very similar fashion in normal rat and mouse islets (Blume et al., 1992), while the ancestral insulin-II gene selectively produces transcripts in certain areas of the brain (Deltour et al., 1993).

It was later demonstrated that integration of a genomic fragment containing the human insulin gene in transgenic mice (Selden et al., 1986; Bucchini et al., 1986, 1989) or in pluripotent rat islet tumor cells (Madsen et al., 1988) resulted in a B-cell–specific activation of the human insulin gene. Such results implicated functional conservation among the mouse, rat and human insulin promoters, a feature which is also demonstrated by the structural conservation of the 5′ flanking region of various insulin genes (Steiner et al., 1985; German et al., 1995). Promoter-deletion studies to further characterize the human insulin promoter in transgenic mice (Fromont-Racine et al., 1990) largely confirmed previous data obtained from *in vitro* analyses. Transgenic analysis of the expression of novel or artificial B-cell–specific markers (such as the SV40 T antigen) driven by the insulin promoter led to more detailed studies of the early process of islet ontogeny.

The Teitelman model of islet development. Although RIP–T-antigen transgenic mice selectively developed B-cell hyperplasia due to B-cell–specific expression of the T antigen (Hanahan, 1985), it was discovered that T-antigen and insulin expression was not strictly coupled during ontogeny and that T antigen could be detected in cells expressing other islet hormones. These observations led to the important study of co-expression of different islet hormones during ontogeny in Dr Teitelman's laboratory and has given rise to a cell-lineage model of development based on the sequential appearance of hormones first in mixed phenotypes and later in mature single-hormone-expressing phenotypes (Alpert et al., 1988; Guz et al., 1995; Fig. 1). Several additional studies have confirmed and substantiated the presence of mixed phenotypes during development in rat (Hashimoto et al., 1988) and man (DeKrijger et al., 1992; Larsson and Hougaard, 1994). Recent contradicting studies on the expression of polypeptide YY (PYY) (Upchurch et al., 1994), neuropeptide Y (Teitelman et al., 1993) and pancreatic polypeptide (Herrera et al., 1991) during development are likely explained by cross-reacting antisera against these related hormones (Jackerott et al., 1996). Thus, PYY is one of the earliest markers expressed during islet development and is expressed with each of the principal hormones when they are first detected (Upchurch et al., 1994; Jackerott et al., 1996). Furthermore, it is known that a wide array of additional polypeptide hormones are produced by the immature or developing islet cells. In addition to PYY and neuropeptide Y, other factors include thyroliberin (Ebiou et al., 1992), calcitonin-gene-related peptide (CGRP) β (Bretherton-Watt et al., 1992), and gastrin (Larsson et al., 1976). The expression levels of several of these factors seem to decline after birth, and their functions during early islet development are not known. However, the fact that these polypeptide products often are found expressed in islet-cell tumors (Larsson et al., 1975; Bordi and Bussolati, 1974; Larsson, 1978) may suggest that the transformed cells may be reminiscent of proliferating endocrine progenitor cells.

Although there are several discrepancies in the literature, it may be concluded that the sequential order of appearance of the four islet hormones is glucagon, insulin, somatostatin and pancreatic polypeptide, that these hormones tend to be expressed with other hormones early in development, and that mature islet cells have little or no co-expression of the four classical hormones. It can thus be anticipated that a first step in differentiation involves the activation of key regulatory factors that may operate on many endocrine specific genes. Further maturation may be characterized by selective inactivation of inappropriate hormone gene products thus resulting in the more restricted expression pattern of the mature islet phenotypes. However, studies of co-localization of hormones do not provide a formal proof of a cell-lineage relationship, and the existence of alternative pathways cannot be excluded. Studies, by means of bromodeoxyuridine incorporation (Jackerott et al., 1996) and [^{3}H]thymidine incorporation (Jackerott, M. and Larsson, L. I., unpublished results) show that developing cells storing insulin, glucagon and PYY have a high degree of DNA synthesis, whereas cells storing only insulin, or insulin and PYY, show almost negligible activity.

Endocrine cells of the intestinal epithelium mature by passage passing through stages in which co-expression of distinct combinations of hormones occur (Larsson and Jørgensen, 1978; Roth et al., 1990, 1992; Aiken et al., 1994; Larsson et al., 1995). Again PYY is an early marker for most if not for all distinct endocrine cell types of the colon (Upchurch et al., 1996).

Glucose transporter 2 (GLUT-2). In the adult pancreas, expression of glucose transporter 2 is restricted to the islet and, moreover, confined to the islet B-cell of rat and mouse (Heimberg et al., 1995), but presumably absent from human pancreas (Devos et al., 1995). During ontogeny, a large mass of apparently undifferentiated (hormone-negative) epithelium express GLUT-2 immunoreactivity (Pang et al., 1994). This cell population may give rise to the major burst of B-cells (GLUT-2 positive) and of acinar cells (GLUT-2 negative) around day 15–18 of the developing rat pancreas, also known as the secondary-transition phase of pancreatic development (Pictet and Rutter, 1972). It has thus been proposed that an alternative pathway for B-cell formation may exist during pancreas ontogeny (Pang et al., 1994), whereby hormone-negative GLUT-2–positive cells could be immediate precursors for a larger fraction of B-cells. These GLUT-2 positive cells probably also express insulin-promoter-factor 1 (IPF1) (Guz et al., 1995), as well as the nerve-growth-factor receptor, Trk-A (Kanaka-Gantenbein et al., 1995a, 1995b), and the hepatocyte-growth factor, c-*met* (Otonkoski et al., 1996), all of which become restricted within the mature islet to the B-cell compartment.

The transcription factor IPF1 and pancreas formation. IPF1 was identified as a B-cell–specific transcription factor (Ohlsson et al., 1991) based on studies involving *in vitro*-cultured mouse B-cell and A-cell tumors, β-TC and α-TC [derived from transgenic RIP–T-antigen (Efrat et al., 1988a) or glucagon-promoter–antigen (Efrat et al., 1988b) mice that develop insulinomas and glucagonomas, respectively]. The factor was cloned from mouse β-TC cDNA and claimed to be strictly B-cell specific (Ohlsson et al., 1993). It was expressed very early in development at the site of pancreatic-bud formation (Jonsson et al., 1994). The factor shared strong similarity to X1Hbox8 from *Xenopus laevis* (Wright et al., 1988), which was recently demonstrated to be involved in very early autonomous endodermal determination in *Xenopus* embryogenesis (Gamer and Wright, 1995). Almost simultaneously with the identification of IPF1, the rat cDNA was cloned by two groups and reported to encode a somatostatin-*trans*activating factor (STF-1) (Leonard et al., 1993), or islet duodenal homeodomain protein (IDX-1) (Miller et al., 1994). This protein was found to be expressed in the duodenal region in adult rats (Leonard et al., 1993; Miller et al., 1994). In the present review, we will use the name IPF1; it should however be mentioned that PDX-1 (pancreatic duodenal homeodomain protein) was recently proposed as a unifying term by the groups involved in cloning the *Xenopus* and mouse forms (Offield et al., 1996; Ahlgren et al., 1996). In summary, IPF1 works as an insulin-transcription factor Ohlsson et al., 1993; Petersen et al., 1994; Peers et al., 1994; Serup et al., 1995) as well as a somatostatin-transcription factor (Leonard et al., 1993; Miller et al., 1994) as judged by *in vitro* studies; it becomes restricted within the islet during development to B-cells and a few D-cells (Fig. 1; Guz et al., 1995; Serup et al., 1995); while wide-spread duodenal-epithelial expression is maintained from fetus to adult.

Homologous-recombination experiments in transgenic mice demonstrated that IPF1 was required for pancreas formation (Jonsson et al., 1994; Offield et al., 1996). In addition to abnormalities in the duodenum and pyloric sphincter, IPF1-null-mutant transgenic mice failed to develop any pancreas. Both studies, however, found that a short ductular structure was formed in place of the normal dorsal outgrowth and that this structure contained scattered glucagon cells (Offield et al., 1996) and few insulin-immunoreactive cells (Ahlgren et al., 1996). The structure never matures further, which indicates that morphogenesis in some way is controlled by the IPF1 protein. The pancreatic mesenchyme of null-mutant mice developed normally and would moreover stimulate *in vitro* differentiation of dorsal pancreatic epithelium from day-10 wild-type embryos. In contrast, null-mutant pancreatic epithelium could not be stimulated by normal mesenchyme (Ahlgren et al., 1996). Such data convincingly demonstrate a crucial role for this homeodomain protein in primitive foregut development/determination through evolution from *Xenopus* to mammals (Gamer and Wright, 1995; Offield et al., 1996; Jonsson et al., 1994; Ahlgren et al., 1996), but also show that some endocrine differentiation may be possible in its absence (Offield et al., 1996; Ahlgren et al., 1996).

Summarizing the involvement of IPF1 in pancreas development, it may be hypothesized that, in addition to a well-documented role in specifying pancreas formation, IPF1 may in the early pancreas also serve as an important regulator/activator of several endocrine-specific traits and could potentially be involved in specifying the multihormonal transient phenotypes (Petersen et al., 1994; Guz et al., 1995; Fig. 1). Data from our laboratory and others suggest that IPF1 is required for differentiated B-cell function such as involvement in glucose-mediated regulation of insulin-gene transcription in normal islet cultures (Melloul et al., 1993; Petersen et al., 1994; MacFarlane et al., 1994). Such observations correlate with a gradual restriction of the expression pattern of IPF1 during pancreatic development, where the mature islet B-cell is by far the major site of expression (Guz et al., 1995; Serup et al., 1995).

Studies of pluripotent transformed rat islet cells in relation to normal islet maturation.

Islet cell cultures. As a consequence of the cumbersome islet-isolation procedure to obtain material for *in vitro* studies of islet function, much effort has been put into the generation and establishment of continuous cultures of islet cells. Cultivation of untransformed freshly isolated islet cells is feasible, but the proliferative potential is limited (Nielsen et al., 1989; Andersson and Hellerström, 1972). Transformed cultures have been achieved by SV40 infection of hamster islet cells (HIT cells) (Santerre et al., 1981). A radiation-induced insulinoma was produced in NEDH rats (Chick et al., 1977), from which continuous cultures were established (RINm and RINr cells) (Gazdar et al., 1980). A BK-virus-induced Syrian hamster insulinoma also gave rise to established beta-cell cultures (In-111-R cells) (Uchida et al., 1979; Okada and Ando, 1980). Common to all culture systems was the lack of stability with which differentiated phenotypes could be maintained for prolonged periods. Although the tumors were scored initially as insulinoma phenotypes, it was later possible to derive variants with an apparent change in phenotype (Oie et al., 1983; Takaki et al., 1986; Madsen et al., 1986; Takaki, 1989).

The establishment of the transgenic lines with a transforming potential linked to a tissue-specific promoter (insulin or glucagon) would be thought to result in the formation of a homogeneous-tumor phenotype. With respect to RIP−T-antigen and glucagon−T-antigen this was found to occur, exemplified by the heritable insulinomas and pancreatic glucagonomas, respectively (Hanahan, 1985; Efrat et al., 1988b). By extending the glucagon promoter further upstream of the transcriptional start site, it was demonstrated that tumors preferentially formed in the intestines, presumably representing transformed L-cells, which is the endocrine source of proglucagon-derived GLP-1 (Brubaker et al., 1992). Even from cultures derived from such mice where the transforming potential is linked to a cell-type-specific promoter, it has been very difficult to retain homogeneous cultures of a defined phenotype. Most often, it is possible to detect subpopulations of other hormone-producing cell types in such clonal cultures (Madsen et al., 1989, 1991; Drucker et al., 1994; Upchurch et al., 1996). As mentioned above, it may be speculated that established gastro-intestinal endocrine cell cultures with enhanced proliferation potential share characteristics with the normal multihormonal counterpart found transiently during normal differentiation.

MSL cells are heterogeneous rat-islet-tumor cell cultures derived from a liver metastasis. A major focus of our laboratory has been to characterize islet-tumor cultures, which were established from a liver metastasis of the previously mentioned X-ray−induced transplantable insulinoma in the rat (Chick et al., 1977). A proliferating insulinoma was originally recovered from a single rat surviving an otherwise lethal X-radiation through parabiosis with a syngenic NEDH rat. Several groups have serially passaged this tumor either in NEDH rats or in nude mice with the aim to establish continuous insulin-producing cultures. In this way, RINr (from passage in rat) and RINm (from passage in nude mouse) cell lines were established (Gazdar et al., 1980). In Dr Steiner's NEDH-rat-tumor colony at The University of Chicago, we observed, in a single rat, metastatic spread to the liver (MSL) and found that this tumor tissue could be established in continuous cultures more readily than the subcutaneously passaged parental insulinoma (Madsen et al., 1986). Although these cultures were very similar in appearance to normal islet monolayer cultures, they either produced very little or no insulin. This was in contrast to several of the RIN cultures (Madsen et al., 1986). Surprisingly, MSL cells, in particular the MSL-G2 culture, were highly heterogeneous with respect to hormone production and a large variety of pancreatic and gut polypeptide hormones could be detected in discrete subpopulations of monolayer cells. The predominating hormones found in MSL-G2 were glucagon, islet amyloid polypeptide (IAPP) and cholecystokinin, while insulin-expressing cells were detectable but very sparse. We proposed that cultures such as MSL-G2 could represent a pluripotent islet phenotype related to the normal transient multihormonal phenotype mentioned above (Figs 1 and 3; Madsen et al., 1986).

Insulin-gene activation as a measure of B-cell differentiation. Studies involving *in vivo* passage of MSL-G2 cells and related cultures in the syngenic NEDH rat further demonstrated a differentiation potential of MSL-cells. First, we observed that MSL-G2 would primarily form insulinomas when serially transplanted subcutaneously. Prolonged serial *in vivo* passage resulted in a stable insulinoma phenotype, MSL-G2-IN, characterized as a small tumor associated with profound hyperinsulinemia and lethal hypoglycemia (Madsen et al., 1988, 1989). To investigate whether these effects occurred during the genuine processes of differentiation or during preferential proliferation of the low frequency of insulin-producing cells present, we performed stable transfection studies to introduce informative clonal markers into subclones of MSL-G2 cells. Introduction of a genomic fragment containing the human insulin gene into subclones of MSL-G2 by transfection and selection for neomycin resistance allowed subsequent studies of the transcriptional activation of the human insulin gene during the *in vitro*/*in vivo* transition. Two such clones, NHI-6F and NHI-5B, were tested in detail, and both were found to have retained the heterogeneity profile of the mother clone with a dominant expression of glucagon *in vitro*. From both of these clones, it was possible to derive insulinomas by serial *in vivo* passage, and in both cases transcriptional activation of the xenogenic human insulin gene contributed substantially to the highly elevated circulating levels of insulin in these tumor-bearing animals (Madsen et al., 1988, 1989).

Common clonal origin of insulinomas and glucagonomas supports the existence of a cell lineage relationship between A and B cells. Following inoculation of MSL cells in NEDH rats, we occasionally observed severe weight loss and anorexia in combination with hyperinsulinemia-induced hypoglycemia in the host rat. We hypothesized that this might reflect the dominance of cell phenotype other than B cells. After dissection and transplantation of such tissue, we were able to segregate anorexia from hypoglycemia and eventually established stable transplantable anorectic tumor lines (e.g. MSL-G-AN) which could be cryopreserved and grown in large amounts *in vivo* for further analyses. Anorectic tumors were characterized as glucagonomas (Madsen et al., 1993b), where proglucagon mRNA was the overall dominating islet-hormone transcript that could be detected (Madsen et al., 1991, 1993a, 1993b, 1994). To investigate whether insulinomas and glucagonomas could be derived from common clonal origin, we used subclones NHI-5B and NHI-6F, which carry human insulin and neomycin genes as unique clonal markers. In both subclones, we were able to successfully establish stable anorectic lines NHI-5B-AN (Madsen et al., 1993b) and NHI-6F-AN (Madsen, O. D., unpublished results). The anorectic-tumor phenotype had highly active glucagon genes, and the anorectic rats were hyperglucagonemic and hypoinsulinemic (Blume et al., 1995; Madsen et al., 1993b). We have thus shown that A-cell and B-cell tumors can be derived from common clonal origin (Fig. 3), thus supporting a cell-lineage relationship between the two cell types (Fig. 1).

Tumor-phenotype homogeneity *in vivo* suggests processes of islet maturation. The homogeneous segregation of active insulin and glucagon genes with the insulinoma and glucagonoma phenotypes (Madsen et al., 1988, 1991, 1993b), respectively, could suggest that these tumors represent a later stage in islet differentiation towards mature B and A cells. To verify this hypothesis, the expression profile of a larger number of genes were tested, including several of those that are known to segregate between normal A and B cells. First, we observed that IAPP mRNA expression was confined to the insulinoma phenotype and not detectable in the gluconoma, although precursor cultures such as MSL-G2 efficiently expressed IAPP and glucagon (Madsen et al., 1991). Furthermore, the neuroendocrine protein chromogranin B, previously shown to be a marker for the islet A cells, was isolated in large quantities from glucagonoma extracts (Nielsen et al., 1991). A more recent study, based on semiquantitation of a large number of mRNAs by reverse transcription/

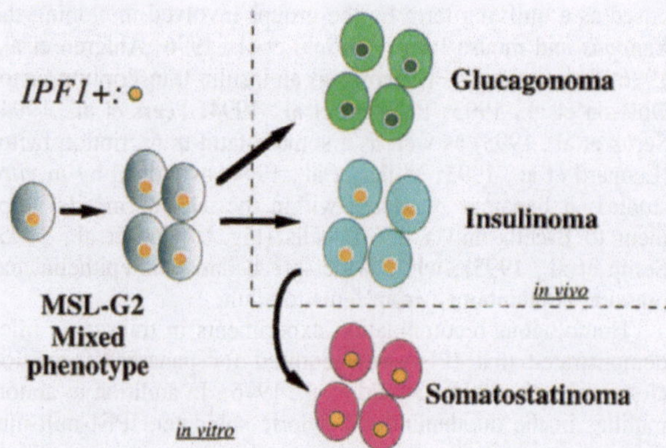

Fig. 3. MSL cells and the tumor phenotypes. MSL cells were derived from a liver metastasis (Madsen et al., 1986) of an X-ray−induced transplantable rat insulinoma (Chick et al., 1977). Clonal cultures MSL cells (MSL-G2) remained multihormonal and heterogeneous when kept in *in vitro* cultures (Madsen et al., 1986) and may thus resemble the transient multihormonal phenotype present during normal pancreas development (Alpert et al., 1988). We demonstrated that transplantable, homogeneous insulinomas (Madsen et al., 1988), glucagonomas (Madsen et al., 1993b) and a somatostatinoma culture (Serup et al., 1992) could be derived from a common clonal origin. The red nuclei indicate IPF1 expression. This transcription factor can be by localized by immunocytochemistry to the nucleus (Fig. 4). The IPF1 gene is turned off in the glucagonoma, as in the normal A cell, while expressed in the insulinoma the somatostatinoma (Serup et al., 1995, 1996; Jensen et al., 1996). The somatostatinoma culture, MSL-G2-Tu6, was derived from hypoglycemic insulinoma tumor tissue (Serup et al., 1992) and, although it has a fairly homogeneous expression of somatostatin, insulin is still expressed in a subpopulation of these cells. The rat homologue of mouse IPF1 (Ohlsson et al., 1993), was cloned as the somatostatin *trans*acting factor-1 from this culture (Leonard et al., 1993).

PCR, demonstrated that insulinomas and glucagonomas share substantial resemblance to mature B and A cells (Jensen et al., 1996). For example, the insulinoma tumor selectively expresses the GLP-1−receptor and GLUT-2 mRNAs. Most importantly, however, this study demonstrated that the homeodomain factor IPF-1 segregated with the insulinoma, while being completely switched off in the glucagonoma.

A recent comparative transplantation study involving the MSL-G2-IN insulinoma and the MSL-G-AN glucagonoma revealed a strikingly selective inhibitory effect on the corresponding normal islet-cell phenotype (Blume et al., 1995). In insulinoma-bearing rats, endogenous islet insulin-gene transcription was markedly reduced, insulin stores and IPF1 immunoreactivity disappeared and islet B-cell volume was selectively reduced due to extensive B-cell-specific apoptosis. In glucagonoma-bearing rats, endogenous glucagon-gene transcription was turned off, and A-cell volume was selectively reduced by cytoplasmic retraction. Apoptosis was confined in these animals to the acinar cells, possibly linked to anorexia. Islet B-cell content of insulin-I and insulin-II mRNA appeared normal in the glucagonoma-bearing rats, despite hypoinsulinemia (Blume et al., 1995). In summary, these selective inhibitory effects of our established tumor lines demonstrate their degree of homogeneity, which is taken to represent a more mature stage towards differentiated A and B cells (Fig. 3).

IPF1 expression in pluripotent MSL cells and MSL-derived tumors. When immunocytochemically testing for IPF1 expression in the heterogeneous precursor cultures MSL-G2 and NHI-6F, from which transplantable insulinomas and glucagono-

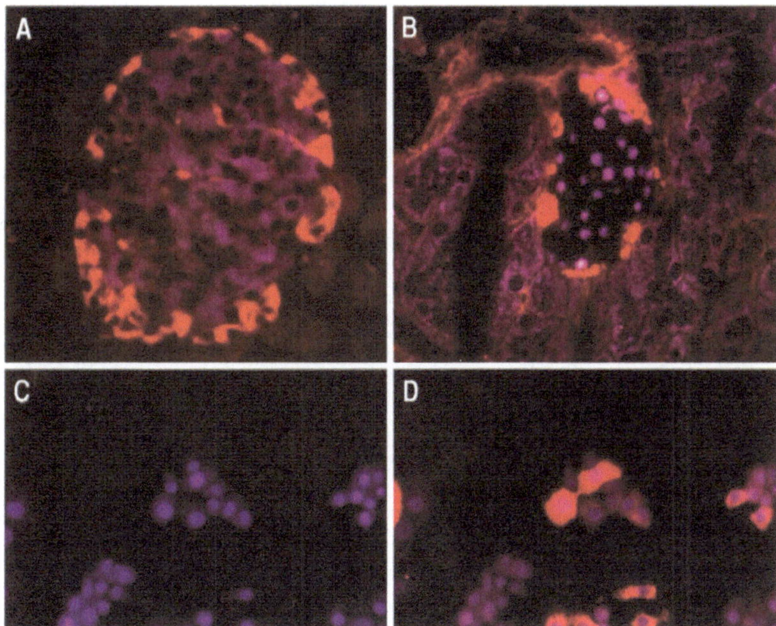

Fig. 4. Immunochemical detection of IPF1 expression. (A) The rat islet of Langerhans is of the mantle-type, where the insulin-producing B cells are located centrally (shown in green; section of pancreas) while most of the A, D and PP cells are found in the periphery. Glucagon-producing A cells are shown in red. (B) Section of rat pancreas stained for IPF1 (bright green) and glucagon (red). Only nuclei of centrally located islet cells (B cells) are stained by the IPF1 antibody. None of the glucagon-producing islet A cells (shown in red) have nuclear staining for IPF1. Double staining for somatostatin revealed heterogeneity where a subpopulation of D cells would express IPF1 immunoreactivity (Serup et al., 1995). (C) and (D) Monolayer cultures of MSL-G2 derived culture NHI-6F-Tu28 (Madsen et al., 1988; Serup et al., 1995; Blume et al., 1992). This culture is highly heterogeneous with respect to hormone production (Blume et al., 1992), and the insulin-producing subpopulation of the monolayer in (C) is shown in (D) (red). The culture has a homogeneous expression of IPF1 in all cells of the monolayer, irrespective of the hormone produced (bright green nuclei in C) (Serup et al., 1995).

mas have been derived, we found that almost 100% of cells in such cultures displayed nuclear staining for IPF1 (Fig. 4). IPF1 was thus expressed irrespective of the hormone combination found in a particular cell. When testing the insulinoma MSL-G2-IN and the glucagonoma MSL-G-AN, it became evident that IPF1 expression selectively followed insulin-gene expression and that the IPF1 gene thus was completely inactive in the glucagonoma (Jensen et al., 1996). These data correlate well with the lack of IPF1 expression in the vast majority of adult islet A cells while all B cells are positive (Fig. 4; Serup et al., 1995). From these studies, it could thus be anticipated that A-cell maturation is characterized in part by IPF1-gene inactivation of the transient multihormonal phenotype in normal islet development (Fig. 1; Guz et al., 1995). This finding is, however, in contrast with that reported in IPF1-null-mutant mice (Offield et al., 1996; Ahlgren et al., 1996); these apancreatic mice develop a rudimentary dorsal bud in which glucagon-positive cells and transient insulin-positive cells are seen (Ahlgren et al., 1996). Thus, the very early cells that produce glucagon and/or insulin mRNA transcripts (Gittes and Rutter, 1992) and immunoreactive protein in the ductal epithelium prior to bud formation can be activated without the involvement of active IPF1 genes (Ahlgren et al., 1996; Offield et al., 1996). Whether this glucagon cell resembles the pancreatic A cell (which never forms) or the intestinal L cell, which remains devoid of IPF1 expression, is unknown.

IPF1 is actively involved in specifying part of the B-cell phenotype. From previous experiments by us (Petersen et al., 1994; Serup et al., 1995) and others (Ohlsson et al., 1991, 1993; Peers et al., 1994), IPF-1 is probably actively engaged in transcriptional regulation of the insulin gene. Moreover, activation of the insulin gene by IPF1 appears to be dependent on cooperativity with insulin-enhancer factor-1 (IEF-1; Peers et al., 1994; Serup et al., 1995). IEF-1 is a heterodimer composed of the helix-loop-helix proteins Beta2 (the hamster homologue of mouse NeuroD), which is expressed is neurons (Lee et al., 1995) and in pancreatic endocrine cells (Naya et al., 1995), and the ubiquitous products of the E2A gene, E47 and E12 (Aronheim et al., 1991; German et al., 1991; Park and Walker, 1992; Robinson et al., 1994; Naya et al., 1995). Thus, two proteins, which are members of a family of transcription factors that are essential for the differentiation of a large number of cell types, converge on the insulin promoter in the islet B cell. However, even though IEF-1 is capable of conferring activity to a transfected insulin promoter, it is not sufficient to activate endogenous insulin-gene expression (Robinson et al., 1994). To test whether introduction of IPF1 into an insulin-negative IEF-1–containing cell would activate insulin production, we took advantage of our transplantable glucagonoma which had been reintroduced to *in vitro* culture. By generating stable transfected subclones of these cultures by means of IPFI-expression plasmids, we investigated whether this could induce insulin-gene expression (Fig. 5). We found that insulin and IAPP genes became activated in such clones with a reconstituted IPF1 expression. We could thus conclude that IPF1 supplies a crucial activity for the transcriptional activation of the B-cell–specific genes insulin and IAPP (Serup et al., 1996). We furthermore tested the expression profile of other genes known to be expressed in normal islet B cells. We found that GLUT-2 and the GLP-1 receptor were not induced, which is in contrast to expression in the transplantable insulinoma, while the glucokinase gene was already active in the untransfected glucagonoma cul-

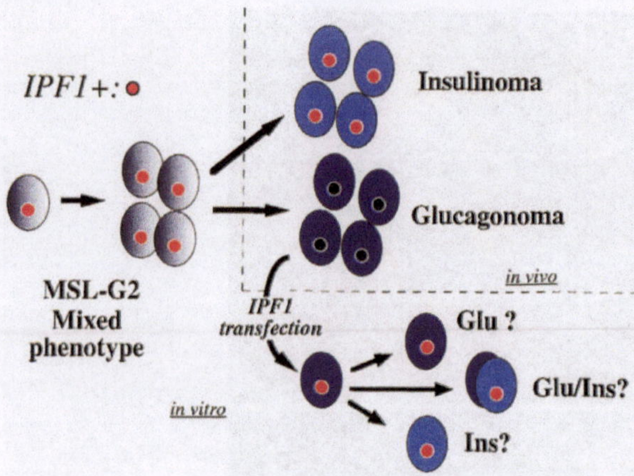

Fig. 5. Induction of insulin-gene and IAPP-gene expression. Since glucagonomas and insulinomas of common origin differ markedly in their IPF1 expression (Jensen et al., 1996) (Fig. 3), we wanted to determine whether IPF1 could have a direct role in controlling hormon-gene activity. By introducing IPF1 expression into subcultures of IPF1-negative glucagonomas, we could investigate whether such cultures would remain unchanged as a glucagonoma, undergo a phenotypic shift and become an insulinoma or represent a mixed insulinoma/glucagonoma phenotype. The isolation and characterization of a series of independent cultures stably transfected with an IPF1-expression vector showed that the insulin genes and IAPP genes became fully activated, but did not result in a complete phenotypic shift towards the B cell, since the glucagonoma phenotype was maintained when these clones were passaged *in vivo*, which resulted in anorexia as well as hypoglycemia (Serup et al., 1996).

tures and the expression level was unchanged after IPF1 transfection (Serup et al., 1996). We conclude that IPF1 is involved in specifying part of the islet-B-cell phenotype, including the ability to activate insulin and IAPP genes, and that other unknown factors, in concert with IPF1, are responsible for inducing the unique nature of the pancreatic B cell.

Our ongoing tumor-phenotype mRNA-profiling studies have identified the Nk-homeodomain factor, Nkx 6.1, as a candidate transcription factor to serve the above purpose. This factor was originally cloned from hamster insulinoma cDNA by Rudnick et al. (1994). We found that this gene was selectively induced with insulin when insulinomas were formed from the MSL-G2 culture (Fig. 3) and thus not active in the glucagonomas. As predicted from these tumor analyses, the islet expression of Nkx 6.1 was highly B-cell specific. Moreover, Nkx 6.1 was expressed in high amounts only in islets, barely detectable in the stomach, and absent from 20 other tissues tested (Jensen et al., 1996). Target genes for Nkx 6.1 are yet to be defined. Nkx 6.1 is not necessary for the transcriptional activation of insulin and IAPP genes in glucagonoma cells, since it was not activated with insulin by IPF1 transfection (Serup et al., 1996).

Concluding remarks

IPF1 appears to have pleiotropic effects during pancreatic development. As IPF1 was identified by the use of degenerate PCR in a search for homeodomain proteins expressed in the islets, the effect of the IPF1−knock-out mouse is surprising. How could one expect that a tissue-restricted transcription factor could abrogate development of all cell types of the organ? It turned out that IPF1 is not as tissue restricted as first anticipated, and immunocytochemical analyses has shown IPF1 expression in not only most endodermal cells of the pancreatic anlage, but also in certain cells of the duodenum. Therefore, it must be concluded that IPF1 has multiple roles, in addition to its effect on insulin-gene and IAPP-gene activation. Future experiments concerning this factor are necessary to outline these functions. In this regard, specification experiments, in which IPF1 is expressed under various tissue-specific promoters, e.g. hepatic or intestinal, may provide further insight into how the specification of the pancreatic cells is influenced by this factor, as it may easily be envisioned that pancreatic characteristics could be induced elsewhere along the primitive gut. Secondly, functional characterisation of IPF1 *in vivo* by means of mutated forms is possible by use of the functional assay of insulin-gene or IAPP-gene induction observed in our glucagonoma cells. The contribution of Nkx 6.1 in the specification of the B-cell phenotype remains to be identified. A next step of interest will be the identification of the growth factors and signalling pathways that control the transcriptional activation of key regulators such as IPF1 and Nkx 6.1.

We thank Dr Jens H. Nielsen for critically reading of this manuscript. We also thank Erna Petersen for skilful technical assistance in immunocytochemisty. This work was made possible by financial support from The Danish National Research Foundation, Juvenile Diabetes Foundation International, USA, the Danish Diabetes Association and the Danish Cancer Society, Denmark.

REFERENCES

Ahlgren, U., Jonsson, J. & Edlund, H. (1996) The morphogenesis of the pancreatic mesenchyme is uncoupled from that of the pancreatic epithelium in IPF1/PDX1-deficient mice, *Development (Camb.) 122*, 1409−1416.

Aiken, K. D., Kisslinger, J. A. & Roth, K. A. (1994) Immunohistochemical studies indicate multiple enteroendocrine cell differentiation pathways in the mouse proximal small intestine, *Dev. Dyn. 201*, 63−70.

Alpert, S., Hanahan, D. & Teitelman, G. (1988) Hybrid insulin genes reveal a developmental lineage for pancreatic endocrine cells and imply a relationship with neurons, *Cell 53*, 295−308.

Andersson, A. & Hellerström, C. (1972) Metabolic characteristics of isolated pancreatic islets in tissue culture, *Diabetes 21 (Suppl. 2)*, 546−554.

Aronheim, A., Ohlsson, H., Park, C. W., Edlund, T. & Walker, M. D. (1991) Distribution and characterisation of helix-loop-helix enhancer binding proteins from pancreatic beta-cells and lymphocytes, *Nucleic Acids Res. 19*, 3893−3899.

Beck, T. C. & Madsen, O. D. (1989) Monoclonal antibodies as probes to the differentiated exocrine pancreas react to monoclonal islet tumor tissue, *Exp. Clin. Endocrinol. 93*, 255−260.

Bell, G. I., Gerhard, D. S., Fong, N. M., Sanchez-Pescador, R. & Rall, L. B. (1985) Isolation of the human insulin-like growth factor genes: insulin-like growth factor II and insulin genes are contiguous, *Proc. Natl Acad. Sci. USA 82*, 6450−6454.

Blume, N., Petersen, J. S., Andersen, L. C., Kofod, H., Dyrberg, T., Michelsen, B. K., Serup, P. & Madsen, O. D. (1992) Immature transformed islet β-cells differentially express C-peptides derived from genes coding for insulin I and II as well as a transfected human insulin gene, *Mol. Endocrinol. 6*, 299−307.

Blume, N., Skouv, J., Larsson, L.-I., Holst, J. J. & Madsen, O. D. (1995) Potent inhibitory effects of transplantable rat glucagonomas and insulinomas on the respective endogenous islet cells are associated with pancreatic apoptosis, *J. Clin. Invest. 96*, 2227−2235.

Bonner-Weir, S. & Orci, L. (1982) New perspectives on the microvasculature of the islet of Langerhans in the rat, *Diabetes 31*, 883−889.

Bonner-Weir, S., Baxter, L. A., Schuppin, G. T. & Smith, F. E. (1993) A 2nd for regeneration of adult exocrine and endocrine pancreas − a possible recapitulation of embryonic-development, *Diabetes 42*, 1715−1720.

Bordi, C. & Bussolati, G. (1974) Immunofluorescence, histochemical and ultrastructural studies for the detection of multiple endocrine polypeptide tumours of the pancreas, *Virchows Arch. Cell Pathol. 17*, 13–27.

Bouwens, L., Wang, R. N., Deblay, E., Pipeleers, D. G. & Klöppel, G. (1994) Cytokeratins as markers of ductal cell-differentiation and islet neogenesis in the neonatal rat pancreas, *Diabetes 43*, 1279–1283.

Bouwens, L., Braet, F. & Heimberg, H. (1995) Identification of rat pancreatic duct cells by their expression of cytokeratin-7, cytokeratin-19, and cytokeratin-20 *in-vivo* and after isolation and culture, *J. Histochem. Cytochem. 43*, 245–253.

Bretherton-Watt, D., Ghatei, M. A., Jamal, H., Gilbey, S. G., Jones, P. M. & Bloom, S. R. (1992) The physiology of calcitonin gene-related peptide in the islet compared with that of sielt amyloid polypeptide (amylin), *Ann. N.Y. Acad. Sci. 657*, 299–312.

Brubaker, P. L., Lee, Y. C. & Drucker, D. J. (1992) Alterations in proglucagon processing and inhibition of proglucagon gene expression in transgenic mice which contain a chimeric proglucagon-SV40 T antigen gene, *J. Biol. Chem. 267*, 20728–20733.

Bucchini, D., Ripoche, M.-A., Stinnakre, M.-G., Desbois, P., Lores, P., Monthioux, E., Absil, J., Lepesant, J.-A., Pictet, R. & Jami, J. (1986) Pancreatic expression of human insulin gene in transgenic mice, *Proc. Natl Acad. Sci. USA 83*, 2511–2515.

Bucchini, D., Madsen, O., Desbois, P., Pictet, R. & Jami, J. (1989) B islet cells of pancreas are the site of expression of the human insulin gene in transgenic mice, *Exp. Cell Res. 180*, 467–474.

Chick, W. L., Warren, S., Chute, R. N., Like, A. A., Lauris, V. & Kitchen, K. C. (1977) A transplantable insulinoma in the rat, *Proc. Natl Acad. Sci. USA 74*, 628–632.

Contreas, G., Jørgensen, J. & Madsen, O. D. (1990) Novel islet, duct and acinar cell markers defined by monoclonal autoantibodies from prediabetic BB-rats, *Pancreas 5*, 540–547.

DeKrijger, R. R., Aanstoot, H. J., Kranenburg, G., Reinhard, M., Visser, W. J. & Bruining, G. J. (1992) The midgestational human fetal pancreas contains cells coexpressing islet hormones, *Dev. Biol. 153*, 368–375.

Deltour, L., Leduque, P., Paldi, A., Ripoche, M. A., Dubois, P. & Jami, J. (1991) Polyclonal origin of pancreatic islets in aggregation mouse chimaeras, *Development (Camb.) 112*, 1115–1121.

Deltour, L., Leduque, P., Blume, N., Madsen, O., Dubois, P., Jami, J. & Bucchini, D. (1993) Differential expression of the two nonallelic proinsulin genes in the developing mouse embryo, *Proc. Natl Acad. Sci. USA 90*, 527–531.

Devos, A., Heimberg, H., Quartier, E., Huypens, P., Bouwens, L., Pipeleers, D. & Schuit, F. (1995) Human and rat beta-cells differ in glucose-transporter but not in glucokinase gene-expression, *J. Clin. Invest. 96*, 2489–2495.

Docherty, K. & Clark, A. R. (1994) Nutrient regulation of insulin gene expression, *FASEB J. 8*, 20–27.

Douarin, N. M. L. (1988) On the origin of pancreatic endocrine cells, *Cell 153*, 169–171.

Drucker, D. J., Jin, T., Asa, S. L., Young, T. A. & Brubaker, P. L. (1994) Activation of proglucagon gene transcription by protein kinase-A in a novel mouse enteroendocrine cell line, *Mol. Endocrinol. 8*, 1646–1655.

Dudek, R. W., Lawrence, I. E., Hill, R. S. & Johnson, R. C. (1991) Induction of islet cytodifferentiation by fetal mesenchyme in adult pancreatic ductal epithelium, *Diabetes 40*, 1041–1048.

Ebiou, J. C., Bulant, M., Nicolas, P. & Aratan-Spire, S. (1992) Pattern of thyrotropinreleasing hormone secretion from the adult and neonatal rat pancreas: comparison with insulin secretion, *Endocrinology 130*, 1371–1379.

Edlund, T., Walker, M. D., Barr, P. J. & Rutter, W. J. (1985) Cell-specific expression of the rat I insulin gene: evidence for role of two distinct 5′ flanking elements, *Science 230*, 912–916.

Efrat, S., Linde, S., Kofod, H., Spector, D., Delannoy, M., Grant, S., Baekkeskov, S. & Hanahan, D. (1988a) Beta-cell lines derived from transgenic mice expressing a hybrid insulin-oncogene, *Proc. Natl Acad. Sci. USA 85*, 9037–9041.

Efrat, S., Teitelman, G., Anwar, M., Ruggiero, D. & Hanahan, D. (1988b) Glucagon gene regulatory region directs oncoprotein expression to neurons and pancreatic α-cells, *Neuron 1*, 605–613.

Fromont-Racine, M., Bucchini, D., Madsen, O., Desbois, P., Linde, S., Nielsen, J. H., Saulnier, C., Ripoche, M.-A., Jami, J. & Pectet, R. (1990) Effect of 5′-flanking sequence deletions on expression of the human insulin gene in transgenic mice, *Mol. Endocrinol. 4*, 669–677.

Gamer, L. W. & Wright, C. V. E. (1995) Autonomous endodermal determination in *Xenopus*: regulation of expression of the pancreatic gene *XlHbox 8*, *Dev. Biol. 171*, 240–251.

Gazdar, A. F., Chick, W. L., Oie, H. K., Sims, H. L., King, D. L., Weir, G. C. & Lauris, V. (1980) Continous, clonal, insulin- and somatostatin secreting cell lines established from a transplantable rat islet cell tumor, *Proc. Natl Acad. Sci. USA 77*, 3519–3523.

German, M. S., Blanar, M. A., Nelson, C., Moss, L. G. & Rutter, W. J. (1991) Two related helix-loop-helix proteins participate in separate cell-specific complexes that bind the insulin enhancer, *Mol. Endocrinol. 5*, 292–299.

German, M., Ashcroft, S., Docherty, K., Edlund, H., Edlund, T., Goodison, S., Imura, H., Kennedy, G., Madsen, O., Melloul, D., Moss, L., Olson, K., Permutt, A., Philippe, J., Robertson, R. P., Rutter, W. J., Serup, P., Stein, R., Steiner, D., Tsai, M.-J. & Walker, M. (1995) The insulin gene promoter: a simplified nomenclature, *Diabetes 44*, 1002–1004.

Gittes, G. K. & Rutter, W. J. (1992) Onset of cell-specific gene expression in the developing mouse pancreas, *Proc. Natl Acad. Sci. USA 89*, 1128–1132.

Gittes, G. K., Galante, P. E., Hanahan, D., Rutter, W. J. & Debas, H. T. (1996) Lineage-specific morphogenesis in the developing pancreas: role of mesenchymal factors, *Development (Camb.) 122*, 439–447.

Gordon, J. I. (1989) Intestinal epithelial differentiation: new insights from chimeric and transgenic mice, *J. Cell Biol. 108*, 1187–1194.

Gu, D. & Sarvetnick, N. (1993) Epithelial cell proliferation and islet neogenesis in IFNγ transgenic mice, *Development (Camb.) 118*, 33–46.

Gu, D., Lee, M. S., Krahl, T. & Sarvetnick, N. (1994) Transitional cells in the regenerating pancreas, *Development (Camb.) 120*, 1873–1881.

Guz, Y., Montminy, M. R., Stein, R., Leonard, J., Gamer, L. W., Wright, C. V. E. & Teitelman, G. (1995) Expression of Stf-1, a putative insulin gene transcription factor, in b-cells of pancreas, duodenal epithelium and pancreatic exocrine and endocrine progenitors during ontogeny, *Development (Camb.) 121*, 11–18.

Hanahan, D. (1985) Heritable formation of pancreatic β-cell tumors in transgenic mice expression recombinant insulin/simial virus 40 oncogenes, *Nature 315*, 115–122.

Hashimoto, T., Kawano, H., Daikoku, S., Shima, K., Taniguchi, H. & Baba, S. (1988) Transient coappearance of glucagon and insulin in the progenitor cells of the rat pancreatic islets, *Anat. Embryol. 178*, 489–497.

Heimberg, H., Devos, A., Pipeleers, D., Thorens, B. & Schuit, F. (1995) Differences in glucose-transporter gene-expression between rat pancreatic alpha-cells and beta-cells are correlated to differences in glucose-transport but not in glucose-utilization, *J. Biol. Chem. 270*, 8971–8975.

Herrera, P.-L., Huarte, J., Sanvito, F., Meda, P., Orci, L. & Vassali, J.-D. (1991) Embryogenesis of the murine endocrine pancreas; early expression of the pancreatic polypeptide gene, *Develoment (Camb.) 113*, 1257–1265.

Jackerott, M., Øster, A. & Larsson, L.-I. (1996) PYY in developing murine islet cells: Comparisons to development of islet hormones NPY, and BrdU incorporation, *J. Histochem. Cytochem. 44*, 809–817.

Jensen, J., Serup, P., Karlsen, C., Funder, T. F. & Madsen, O. D. (1996) mRNA profiling of rat islet tumors reveals Nkx 6.1 as a β-cell specific homeodomain transcription factor, *J. Biol. Chem. 271*, 18749–18758.

Jessop, N. W. & Hay, R. J. (1980) Characteristics of two rat pancreatic exocrine cell lines derived from transplantable tumors, *In Vitro (Rockville) 16*, 212–216.

Jonsson, J., Carlsson, L., Edlund, T. & Edlund, H. (1994) Insulin-promoter-factor 1 is required for pancreas development in mice, *Nature 371*, 606–609.

Kanaka-Gantenbein, C., Dicou, E., Czernichow, P. & Scharfmann, R. (1995a) Presence of nerve growth factor and its receptors in an *in vitro* model of islet cell development: implication in normal islet morphogenesis, *Endocrinology 136*, 3154–3162.

Kanaka-Gantenbein, C., Tazi, A., Czernichow, P. & Scharfmann, R. (1995b) *In vivo* presence of the high-affinity NGF receptor TRK-A in the rat pancreas: differential localization during pancreatic development, *Endocrinology 136*, 761–769.

Larsson, L.-I., Grimelius, L., Håkanson, R., Rehfeld, J. F., Stadil, F., Holst, J., Angervall, L. & Sundler, F. (1975) Mixed endocrine pancreatic tumors producing several peptide hormones, *Am. J. Pathol. 79*, 271–284.

Larsson, L.-I., Rehfeld, J. F., Sundler, F. & Hdkanson, R. (1976) Pancreatic gastrin in neonatal rats, *Nature 262*, 609–610.

Larsson, L.-I. (1978) Endocrine pancreatic tumors, *Hum. Pathol. 9*, 401–416.

Larsson, L.-I. & Jorgensen, L. M. (1978) Ultrastructural and cytochemical studies on the cytodifferentiation of duodenal endocrine cells, *Cell & Tissue Res. 194*, 79–102.

Larsson, L.-I. & Hougaard, D. M. (1994) Coexpression of islet hormones and messenger RNA's in the human foetal pancreas, *Endocrine 2*, 759–765.

Larsson, L.-I., Tingstedt, J.-E., Madsen, O. D., Serup, P. & Hougaard, D. (1995) The LIM-homeodomain protein Isl-1 segregates with somatostatin but not with gastrin expression during differentiation of somatostatin/gastrin precursor cells, *Endocrine 3*, 519–524.

Lee, J. E., Hollenberg, S. M., Snider, L., Turner, D. L., Lipnick, N. & Weintraub, H. (1995) Conversion of *Xenopus* ectoderm into neurons by NeuroD, a basic helix-loop-helix protein, *Science 268*, 836–844.

Lee, Y. C., Brubaker, P. L. & Drucker, D. J. (1990) Developmental and tissue-specific regulation of proglucagon gene expression, *Endocrinology 127*, 2217–2222.

Leonard, J., Peers, B., Johnson, T., Fefferi, K., Lee, S. & Montminy, M. R. (1993) Characterization of somatostatin transactivating factor-1, a novel homeobox factor that stimulates somatostatin expression in pancreatic islet cells, *Mol. Endocrinol. 7*, 1275–1283.

MacFarlane, W. M., Read, M. L., Gilligan, M., Bujalska, I. & Docherty, K. (1994) Glucose modulates the binding activity of the β-cell transcription factor IUFL in a phosphorylation-dependent manner, *Biochem. J. 303*, 625–631.

Madsen, O. D., Larsson, L.-I., Rehfeld, J. F., Schwartz, T., Lemmark, Å., Labrecque, A. & Steiner, D. F. (1986) Cloned cell lines from a transplantable insulinoma are heterogeneous and express cholecystokinin in addition to islet hormones, *J. Cell Biol. 103*, 2025–2034.

Madsen, O. D., Andersen, L. C., Michelsen, B., Owerbach, D., Larsson, L.-I., Lemmark, A. & Steiner, D. F. (1988) Tissue-specific expression of transfected human insulin genes in pluripotent clonal rat insulinoma lines induced during passage *in vivo*, *Proc. Natl Acad. Sci. USA 85*, 6652–6656.

Madsen, O. D., Andersen, L. C., Serup, P. & Michelsen, B. (1989) Control of insulin gene expression in pluripotent rat islet tumor cells, in *Genes and gene products in the development of diabetes mellitus* (Nerup, J., Mandrup-Poulsen, T. & Hökfelt, B., eds) pp. 141–153, Elsevier Science Publishers, Amsterdam.

Madsen, O. D., Nielsen, J. H., Michelsen, B., Westermark, P., Betsholtz, C., Nishi, M. & Steiner, D. F. (1991) Islet amyloid polypeptide and insulin expression are controlled differently in primary and transformed islet cells, *Mol. Endocrinol. 5*, 143–148.

Madsen, O. D., Karlsen, C., Nielsen, E. & Holst, J. J. (1993a) Proglucagon processing in transplantable glucagonomas associated with severe anorexia, *Digestion 54*, 361–363.

Madsen, O. D., Karlsen, C., Nielsen, E., Lund, K., Kofod, H., Welinder, B., Rehfeld, J. F., Larsson, L.-I., Steiner, D. F., Holst, J. J. & Michelsen, B. K. (1993b) The dissociation of tumor-induced weight loss from hypoglycemia in a transplantable pluripotent rat islet tumor results in the segregation of stable α- and β-cell tumor phenotypes, *Endocrinology 133*, 2022–2030.

Madsen, O., Karlsen, C., Blume, N., Jensen, H., Larsson, L.-I. & Holst, J. (1994) Transplantable glucagonomas derived from pluripotent rat islet tumor tissue cause severe anorexia and adipsia, *Scand. J. Clin. Lab. Invest. 55* (*Suppl. 220*), 27–36.

Melloul, D., Ben-Neriah, Y. & Cerasi, E. (1993) Glucose modulates the binding of an islet-specific factor to a conserved sequence within the rat I and the human insulin gene promoters, *Proc. Natl Acad. Sci. USA 90*, 3865–3869.

Miller, C. P., McGehee, R. E. & Habener, J. F. (1994) IDX-1: a new homeodomain transcription factor expressed in rat pancreatic islets and duodenum that transactivates the somatostatin gene, *EMBO J. 13*, 1145–1156.

Møller, C. J., Christgau, S., Williamson, M. R., Madsen, O. D., Zhan-Po, N., Bock, E. & Baekkeskov, S. (1992) Differential expression of neural cell adhesion molecule and cadherins in pancreatic islets, glucagonomas, and insulinomas, *Mol. Endocrinol. 6*, 1332–1342.

Naya, F. J., Stellrecht, C. M. & Tsai, M. J. (1995) Tissue-specific regulation of the insulin gene by a novel basic helix-loop-helix transcription factor, *Genes & Dev. 9*, 1009–1019.

Nielsen, E., Welinder, B. S. & Madsen, O. D. (1991) Chromoganin-B, a putative precursor of eight novel rat glucagonoma peptides through processing at mono-, di-, or tribasic residues, *Endocrinology 129*, 3147–3156.

Nielsen, J. H., Linde, S., Welinder, B. S., Billestrup, N. & Madsen, O. D. (1989) Growth hormone is a growth factor for the differentiated pancreatic β-cell, *Mol. Endocrinol. 13*, 165–173.

O'Malley, K. L. & Rotwein, P. (1988) Human tyrosine hydroxylase and insulin genes are contiguous on chromosome 11, *Nucleic Acids Res. 16*, 4437–4446.

Offield, M. F., Jetton, T. L., Labosky, P. A., Stein, R., Magnuson, M. A., Hogan, B. L. M. & Wright, C. V. E. (1996) Pdx-1 is required for pancreatic outgrowth and differentiation of the rostral duodenum, *Development* (*Camb.*) *112*, 983–995.

Ohlsson, H., Thor, S. & Edlund, T. (1991) Novel insulin promoter- and enhancer-binding proteins that discriminate between pancreatic α- and β-cells, *Mol. Endocrinol. 5*, 897–904.

Ohlsson, H., Karlsson, K. & Edlund, T. (1993) IPF1, a homeodomain-containing transactivator of the insulin gene, *EMBO J. 12*, 4251–4259.

Oie, H. K., Gazdar, A. F., Minna, J. D., Weir, G. & Baylin, S. B. (1983) Clonal analysis of insulin and somatostatin secretion and L-dopa decarboxylase expression by a rat islet cell tumor, *Endocrinology 112*, 1070–1075.

Okada, Y. & Ando, T. (1980) Properties of hamster insulinoma cell (In-111-R), *J. Physiol. Soc. Jpn 42*, 358.

Otonkoski, T., Cirulli, V., Beattie, G. M., Mally, M. I., Soto, G., Rubin, J. S. & Hayek, A. (1996) A role for hepatocyte growth factor/scatter factor in fetal mesenchyme-induced pancreatic β-cell growth, *Development* (*Camb.*) *137*, 3131–3139.

Pang, K., Mukonoweshuro, C. & Wong, G. G. (1994) Beta cells arise from glucose transporter type 2 (Glut2)-expressing epithelial cells of the developing rat pancreas, *Proc. Natl Acad. Sci. USA 91*, 9559–9563.

Park, C. W. & Walker, M. D. (1992) Subunit structure of cell-specific E-box binding proteins analysed by quantitation of electrophoretic mobility shift, *J. Biol. Chem. 267*, 15642–15649.

Patel, Y. C. (1992) General aspects of the biology and function of somatostatin, in *Somatostatin* (Weil, C., Muller, E. E. & Thomer, M. O., eds) pp. 1–16, Springer-Verlag, Berlin.

Pearse, A. G. E. (1982) Islet cell precursors are neurons, *Nature 295*, 96–97.

Peers, B., Leonard, J., Sharrna, S., Teitelman, G. & Montminy, M. R. (1994) Insulin expression in pancreatic islet cells relies in cooperative interactions between the helix-loop-helix factor E47 and the homeobox factor STF-1, *Mol Endocrinol. 8*, 1798–1806.

Petersen, H. V., Serup, P., Leonard, J., Michelsen, B. K. & Madsen, O. D. (1994) Transcriptional regulation of the human insulin gene is dependent on the homeodomain protein STF1/IPF1 acting through the CT boxes, *Proc. Natl Acad. Sci. USA 91*, 10465–10469.

Philippe, J., Chick, W. L. & Habener, J. F. (1987) Multipotential phenotypic expression of genes encoding peptide hormones in clonal rat insulinoma cell lines, *J. Clin. Invest. 79*, 351–358.

Philippe, J. (1991) Structure and pancreatic expression of the insulin and glucagon genes, *Endocr. Rev. 12*, 252–271.

Philippe, J. (1994) Pancreatic expression of the insulin and glucagon genes: update 1994, *Endocr. Rev. 2*, 21–27.

Pictet, R. L. & Runer, W. J. (1972) Development of the embryonic endocrine pancreas, *Handb. Physiol. 1*, 25–66.

Pictet, L. P., Rall, L. B., Phelps, P. & Rutter, W. J. (1976) The neural crest and the origin of the insulin-producing and other gastrointestinal hormone-producing cells, *Science 191*, 191–192.

Ponder, B. A. J., Schmidt, G. H., Wilkinson, M. M., Wood, M. J., Monk, M. & Reid, A. (1985) Derivation of mouse intestinal crypts from single progenitor cells, *Nature 313*, 689–691.

Rao, M. S., Yeldandi, A. V. & Reddy, J. K. (1990) Stem cell potential of ductular and periductular cells in the adult rat pancreas, *Cell Differ. Dev. 29*, 155–163.

Robinson, G. L. W. G., Peshavaria, M., Henderson, E., Shieh, S.-Y., Tsai, M. J., Teitelman, G. & Stein, R. (1994) Expression of the *trans*-active factors that stimulate insulin control element-mediated activity appear to precede insulin gene transcription, *J. Biol. Chem. 269*, 2452–2460.

Rosenberg, L. & Vinik, A. I. (1992) Trophic stimulation of the ductular-islet cell axis: a new approach to the treatment of diabetes, *Adv. Exp. Med. Biol. 321*, 95–104.

Rosewicz, S., Vogt, D., Harth, N., Grund, C., Franke, W. W., Ruppert, S., Schweitzer, R., Riecken, E.-O. & Wiedenmann, B. (1992) An amphicrine pancreatic cell line: AR42J cells combine exocrine and neuroendocrine properties, *Eur. J. Cell Biol. 59*, 80–91.

Roth, K. A., Hertz, J. M. & Gordon, J. I. (1990) Mapping enteroendocrine cell populations in transgenic mice reveals an unexpected degree of complexity in cellular differentiation within the gastrointestinal tract, *J. Cell Biol. 110*, 1791–1801.

Roth, K. A., Kim, S. & Gordon, J. I. (1992) Immununocytochemical studies suggest two pathways for enteroendocrine cell differentiation in the colon, *Am. J. Physiol. 263*, G174–G180.

Rudnick, A., Ling, T. Y., Odagiri, H., Rutter, W. J. & German, M. S. (1994) Pancreatic beta cells express a diverse set of homeobox genes, *Proc Natl Acad. Sci. USA 91*, 12203–12207.

Samols, E., Stagner, J. I., Ewart, R. B. L. & Marks, V. (1988) The order of islet microvascular cellular perfusion is B to A to D in the perfused rat pancreas, *J. Clin. Invest. 82*, 350–353.

Samols, E. & Stagner, J. I. (1990) Islet somatostatin – microvascular, paracrine and pulsatile regulation, *Metabolism 39* (*Suppl. 2*), 55–60.

Santerre, R. F., Cook, R. A., Crisel, R. M. D., Sharp, J. D., Schmidt, R. J., Williams, D. C. & Wilson, C. P. (1981) Insulin synthesis in a clonal cell line of simian virus 40-transformed hamster pancreatic beta cells, *Proc. Natl Acad. Sci. USA 78*, 4339–4343.

Selden, R. F., Skoskiewicz, M. J., Howie, K. B., Russel, P. S. & Goodman, H. M. (1986) Regulation of human insulin gene expression in transgenic mice, *Nature 321*, 525–528.

Serup, P., Andersen, F. G., Petersen, E. E. & Madsen, O. D. (1992) Generation of a somatostatinoma cell line derived from a pluripotent transformed islet culture, *Diabetologia 35* (*Suppl. 1*), A118.

Serup, P., Petersen, H. V., Petersen, E. E., Edlund, H., Leonard, J., Petersen, J. S., Larsson, L.-I. & Madsen, O. D. (1995) The homeodomain protein IPF1/STF1 is expressed in a subset of islet cells and promotes rat insulin 1 gene expression dependent on an intact E1 helix-loop-helix factor binding site, *Biochem. J. 310*, 997–1003.

Serup, P., Jensen, J., Andersen, F. G., Jørgensen, M. C., Blume, N., Holst, J. J. & Madsen, O. D. (1996) Induction of insulin and IAPP production in pancreatic islet glucagonoma cells by insulin promoter factor 1, *Proc. Natl Acad. Sci. USA 93*, 9015–9020.

Slack, J. M. W. (1995) Developmental biology of the pancreas, *Development (Camb.) 121*, 1569–1580.

Soares, M. B., Schon, E., Henderson, A., Karathanasis, S. K., Cate, R., Zeitlin, S., Chirgwin, J. & Efstratiadis, A. (1985) RNA-mediated gene duplication: the rat preproinsulin gene is a functional retroposon, *Mol. Cell. Biol. 5*, 2090–2103.

Stein, R. (1993) Regulation of insulin gene transcription, *Trends Endocrinol. Metab. 4*, 96–101.

Steiner, D. F., Chan, S. J., Welsh, J. M. & Kwok, S. C. M. (1985) Structure and evolution of the insulin gene, *Annu. Rev. Genet. 19*, 463–484.

Takaki, R., Ono, J., Nakamura, M., Tokogawa, Y., Kumae, S., Hiraoko, T., Yamaguchi, K., Hamaguchi, K. & Uchida, S. (1986) Isolation of glucagon-secreting cell lines by cloning insulinoma cells, *In Vitro (Rockville) 22*, 120–126.

Takaki, R. (1989) Culture of pancreatic islet cells and islet producing cell lines: 'morphological and functional integrity in culture', *In Vitro (Rockville) 25*, 763–769.

Teitelman, G. (1990) Phenotypic plasticity of pancreatic islet cells, in *Frontiers in diabetes research. Lessons from animal diabetes III* (Shafrir, E., ed.) pp. 520–522, Smith-Gordon, London.

Teitelman, G. (1991) Cellular and molecular analysis of pancreatic islet cell lineage and differentiation, in *Recent progress in hormone research* (Bardin, C. W., ed.) pp. 259–294, Academic Press, Inc., San Diego.

Teitelman, G. (1993) On the origin of pancreatic endocrine cells, proliferation and neoplastic transformation, *Tumor Biol. 14*, 167–173.

Teitelman, G., Alpert, S., Polak, J. M., Martinez, A. & Hanahan, D. (1993) Precursor cells of mouse endocrine pancreas coexpress insulin, glucagon and the neuronal proteins tyrosine hydroxylase and neuropeptide Y, but not pancreatic polypeptide, *Development (Camb.) 118*, 1031–1039.

Uchida, S., Watanabe, S., Aizawa, T., Furuno, A. & Muto, T. (1979) Polyoncogenicity and insulinoma-induced ability of BK virus, a human papovavirus, in Syrian golden hamsters, *J. Natl Cancer Inst. 63*, 119–126.

Upchurch, B. H., Aponte, G. W. & Leiter, A. B. (1994) Expression of peptide YY in all four islet cell types in the developing mouse pancreas suggests a common peptide YY-producing progenitor, *Development (Camb.) 120*, 245–252.

Upchurch, B. H., Fung, B. P., Rindi, G., Ronco, A. & Leiter, A. B. (1996) Peptide YY expression is an early event in colonic endocrine cell differentiation: evidence from normal and transgenic mice, *Development (Camb.) 22*, 1157–1163.

Walker, M. D., Edlund, T., Boulet, A. M. & Rutter, W. J. (1983) Cell-specific expression controlled by the 5′ flanking regions of the insulin and chymotrypsin genes, *Nature 306*, 557–561.

Wang, R. N., Klöppel, G. & Bouwens, L. (1995) Duct-cell to islet-cell differentiation and islet growth in the pancreas of duct-ligated adult rats, *Diabetologia 38*, 1404–1411.

Watanabe, S., Abe, K., Anbo, Y. & Katoh, H. (1995) Changes in the mouse exocrine pancreas after pancreatic duct ligation: a qualitative and quantitative histological study, *Arch. Histol. Cytol. 58*, 365–374.

Wright, C. V. E., Schnegelsberg, P. & De Robertis, E. M. (1988) XlHbox 8: a novel *Xenopus* homeoprotein restricted to a narrow band of endoderm, *Development (Camb.) 104*, 787–794.

Author index for EJB Reviews 1989–1996

Subject index